R

E5000 Singularity

FUNDAMENTALS OF AIRCRAFT STRUCTURAL ANALYSIS

FUNDAMENTALS OF AIRCRAFT STRUCTURAL ANALYSIS

Howard D. Curtis

Embry-Riddle Aeronautical University
Daytona Beach, Florida

Boston, Massachusetts   Burr Ridge, Illinois   Dubuque, Iowa
Madison, Wisconsin   New York, New York   San Francisco, California   St. Louis, Missouri

WCB/McGraw-Hill

A Division of The **McGraw·Hill** *Companies*

© Richard D. Irwin, a Times Mirror Higher Education Group, Inc. company, 1997

Irwin Book Team

Publisher: *Tom Casson*
Senior sponsoring editor: *Elizabeth A. Jones*
Senior developmental editor: *Kelley Butcher*
Project supervisor: *Beth Cigler*
Production supervisor: *Laurie Sander*
Designer: *Bethany Stubbe*
Coordinator, Graphics and Desktop Services: *Keri Johnson*
Director, Prepress Purchasing: *Kimberly Meriwether David*
Compositor: *Interactive Composition Corporation*
Typeface: *10/12 Times Roman*
Printer: *Times Mirror Higher Education Group, Inc., Print Group*

Library of Congress Cataloging-in-Publication Data

Curtis, Howard D.
    Fundamentals of aircraft structural analysis / Howard D. Curtis
        p.   cm.
    Includes index.

    ISBN-13: 978-0-256-19260-5
    ISBN-10: 0-256-19260-X

    1. Airframes, 2. Structural analysis (Engineering)   I. Title
  TL671.6.C87 1997
629.134'31—dc20 95–47845
Printed in the United States of America
    5 6 7 8 9 10 DOC/DOC 0 9 8 7 6

To my parents, Rondo and Geraldine, and my wife, Connie Dee.

PREFACE

This book grew out of lecture material developed by my colleagues and me over a ten-year period of teaching a three-course sequence in aircraft structures to undergraduate aerospace engineering students at Embry-Riddle Aeronautical University's Daytona Beach, Florida, campus. The first two courses are, essentially, statically determinate and statically indeterminate structures, respectively. The third course is an introduction to matrix methods and finite element concepts.

This book is not a design manual. It is a text on the elements of aircraft structural analysis written for the undergraduate student who has little or no previous exposure to the subject.

I have included the theory required for understanding linear, static structural behavior and the classical methods of analysis. It is my experience that students learn the fine points of theory by seeing it applied to practical problems. Therefore, numerous worked-out examples are found throughout the book, and they are a reflection of my "teach by example" classroom methodology.

Modern computational structures technology places powerful tools in the hands of today's structural designers and analysts. Some may argue that methods and approximations designed for hand computations have been rendered obsolete and have no place in a modern curriculum. My contention is that we must imbue the undergraduate engineering student with the basics. Teaching simplified procedures and idealizations is appropriate if it fosters understanding and provides a springboard to advanced concepts. I encourage the creative use of commercial computer packages in parallel with the topics in this text. On the other hand, I discourage students' from becoming dependent solely on canned software to solve problems they should be able to do with (at most) the aid of a hand-held calculator. The fact remains that students need to master neither statics nor algebra in order to mouse-click their way through sophisticated structural analyses on menu-driven desktop computers.

The book comprises 12 chapters, and a brief discussion of each of them follows.

Chapter 1 outlines the major developments in aircraft structures since 1903 and discusses the nature and composition of modern aircraft structural design. It also introduces the major elements used to model structures for preliminary design calculations.

Chapter 2 provides a review and a reinforcement of statics concepts, while introducing aspects of structures peculiar to aircraft. Statically determinate versions of structures typically encountered in aircraft design are discussed, including trusses, simple beams, stiffened shear webs, torque boxes and frames. Chapter 2 also covers the concept of shear flow along with the calculation of shear flow distribution in stiffened web structures of both closed and open sections.

Chapter 3 is a fairly complete reference on those topics in the theory of elasticity that are pertinent to the remainder of the text. The concepts of stress, deformation and strain in three dimensions are developed from first principles. Chapter 3 also discusses material anisotropy, as well as stress concentration and fatigue.

Chapter 4 explores beam theory, starting with the Bernoulli-Euler hypothesis. The purpose is to develop the methodology for computing stresses in statically determinate thin-walled box beams and beams simplified or idealized into an assembly of shear panels and stiffeners. The chapter also covers the notion of shear center as well as tapered idealized beams.

Chapter 5 addresses the subject of load transfer from ribs, frames, and stiffeners into attached shear webs. It also treats the influence on load paths of cutouts in box beams and stiffened web structures.

Chapter 6 delves into the subject of virtual work. Because of the central role virtual work concepts play throughout structural analysis, they are developed in careful detail. The rest of the text relies on the principles of virtual work and complementary virtual work. Since the unit load method, unit displacement method, Castigliano's theorems, and the minimum potential energy theorems are all special cases of virtual and complementary virtual work methods, they are discussed, but not emphasized.

Chapter 7 implements the principle of complementary virtual work in the form of the classical force method of computing loads and deflections in trusses, beams and frames, and rings. Chapter 8 continues with the force method, applying it to idealized thin-walled structures, with emphasis on box beams. In both of these chapters I resist the temptation to introduce matrix methods.

Matrix notation is finally adopted in Chapter 9, which is devoted to basic matrix operations and the displacement method of structural analysis. General formulas for the stiffness equations of elastic structures are obtained by means of the principle of virtual work. The procedure for computing element stiffness matrices, assembling them into global arrays, and applying constraints is explained.

Chapter 10 applies the displacement method to trusses, beams, and frames. The method is extended into the finite element domain in Chapter 11, which treats the shear panel as a hybrid element, and covers the classical constant strain triangle in depth. There is also a brief introduction to substructuring and static condensation.

The book concludes with Chapter 12, which covers buckling, with the focus on columns and stiffened panels. Tension and semitension beams are treated, as is static wing divergence.

I am grateful to several Embry-Riddle colleagues for their helpful criticism and assistance during the writing of this book. Charles N. Eastlake and James G. Ladesic provided historical details for Chapter 1, guidance for the fatigue and stress concentration content of Chapter 3, and lots of encouragement along the way. Frank J. Radosta, who was a dependable sounding board, composed a number of the homework exercises for Chapter 2, and his notes were the basis of a chapter of manuscript on aerodynamic loads. Habib Eslami was a big supporter of this project, and his notes on structural stability proved valuable in the preparation of Chapter 12. Glenn P. Greiner taught from portions of the evolving manuscript and gave me useful feedback. Wayne V. Nack, now with General Motors, was the source of many ideas for the book, especially for the material in Chapters 9, 10 and 11.

In addition, the manuscript was reviewed at several stages by the following people, to whom I am most obliged: Fuh-Gwo Yuan, North Carolina State University; Harold F. Zimbelman, University of Colorado-Boulder; Grzegorz Kawiecki, University of Tennessee-Knoxville; Earl A. Thornton, University of Virginia; William Bickford, Arizona State University; Stephen Batill, University of Notre Dame; Hayrani Oz, The Ohio State University; Alfred G. Stitz, University of Oklahoma; Edward C. Smith, The Pennsylvania State University; and James P. Thomas, University of Notre Dame. The comments and suggestions of this panel of reviewers resulted in significant restructuring and improvement of the manuscript.

I hope *Fundamentals of Aircraft Structural Analysis* proves useful to both students and teachers of the subject. I welcome any comments and suggestions.

A detailed solutions manual is available from the publisher to those who adopt the book.

Howard D. Curtis

Embry-Riddle Aeronautical University
Daytona Beach, Florida

CONTENTS

1

Historical Perspective

Engineering is the coupling of technology and the marketplace, the transition of discovery into cost-effective products and services.— F. W. Garry

---

1.1   INTRODUCTION

The airplane, that most fascinating and complex of man-made systems, epitomizes engineering at its finest. We focus in this text on the airframe, the "skeleton" of ribs and spars and other assorted structural members hidden beneath the shiny, fragile skin of the modern airplane. The airframe must be strong, rigid, and durable, yet as light in weight as safety will allow. It must also fit within the streamlined shape defined by aerodynamics and the mission the airplane was built to serve.

To understand the structures of today's aircraft, it is helpful to look back over the evolution of the airframe since the beginning of the twentieth century*.**. It is sufficient to restrict our attention to the development of fixed-wing aircraft. We

---

*Tom Rhodes, "History of the Airframe," Parts 1 through 4, *Aerospace Engineering*, May 1989 (pp. 11–15); June 1989 (pp. 11–17); September 1989 (pp. 27–32); and October 1989 (pp. 16-21).

**Proceedings, *Evolutions of Aircraft/Aerospace Structures and Materials Symposium*, Dayton-Cincinnati Section, American Institute of Aeronautics and Astronautics, Air Force Museum, Dayton, Ohio, April 24–25, 1985.

nevertheless acknowledge the significant structural dynamics problems that have attended the evolution of rotary-wing aircraft, which began in 1907.[1]

1.2 SURVEY OF DEVELOPMENTS IN AIRCRAFT STRUCTURES SINCE 1903

On December 17, 1903, a manned, self-propelled, heavier-than-air vehicle lifted off the earth's surface, traveled 120 feet through the air in controlled flight, and landed intact at a point no lower in elevation than where it started. Orville Wright, by flip of the coin, was the 32-year-old pilot on that brief but historic journey, while his older brother Wilbur, 36, looked on (Figure 1.2.1). Taking turns, the Wright brothers, whose trade was manufacturing bicycles in Dayton, Ohio, made three more, successively longer, flights that day on the sand dunes near Kitty Hawk, North Carolina. Wilbur flew the final one, which lasted 59 seconds and covered a distance of 852 feet. The modern aviation era sprang from the events of that day.

The Wright brothers' success was the culmination of four years of their own systematic and persistent study, research, test, and development, and was due in no small part to their synergistic inventiveness, common sense, and mechanical skill. They based the design of their series of gliders leading up to the 1903 *Flyer* on that of fellow countryman, friend, and confidant, Octave Chanute. Chanute, nearly 40 years older than the Wrights, took an interest in aeronautics relatively late in life and became an expert on current and historical developments in heavier-than-air flight worldwide. Having previously earned a reputation as a brilliant civil engineer, Chanute incorporated bridge truss design concepts into his successful, straight-wing, biplane hang glider of 1896 (Figure 1.2.2). He abandoned the bird or bat wing appearance of other successful contemporary manned gliders, such as Otto Lilienthal's in Germany and Percy Pilcher's in Scotland. The Wright brothers chose Chanute's trussed biplane configuration not only because of its inherent strength, but also to obtain the largest amount of lifting surface. The configuration also allowed them

Figure 1.2.1   The first successful takeoff of the Wrights' *Flyer*, with Orville at the controls and Wilbur having just released his running hold on the right wingtip of the craft as it gathered speed into the wind, perched atop a small trolley rolling on a wooden rail.

[1] In September 1907, Frenchman Louis Breguet's ungainly four-rotor *Gyroplane No. 1* became the first manned rotary-wing craft to lift itself off the ground in *tethered* flight, to a height of two feet for one minute. Two months later, fellow countryman Paul Cornu made the world's first manned *free* flight (height of one foot for 20 seconds) in his twin-rotor helicopter.

to warp the outboard portion of the wings by varying the tension on the diagonal bracing wires, thereby providing a measure of controllability. This was probably the element of their research that differed most significantly from that of their contemporaries. Other designs would be capable of flight, but the Wrights' had better control.

Two common truss arrangements are illustrated in Figure 1.2.3. The Pratt truss and its variants are among the best strength-to-weight structures, an important attribute for aircraft design, where minimizing structural weight is a prime objective. For the uniform loadings shown in the figure, the methods described in Chapter 2 can be used to verify that each truss member is in tension, compression, or load-free as shown. To make the truss as light as possible, the members in tension may be made of thin wire or cable, which is too flexible to support compression.

Figure 1.2.2   Successful flight test in August 1896 of the glider designed jointly by Octave Chanute and his associate Augustus Herring.

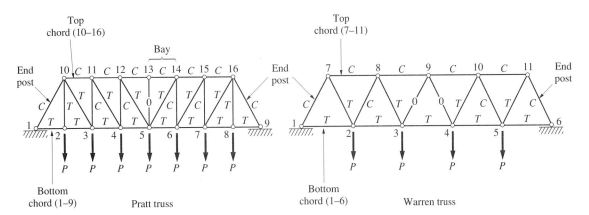

Figure 1.2.3   Examples of trusses, with some nomenclature: T = tension, C = compression, and zero means the member is load-free.

"Chord" in truss terminology has a different meaning than in aerodynamics, in which "chord" is the straight line connecting the leading and trailing edges of an airfoil.

The compression members must be thick and sturdy enough that they will not buckle (the subject of Chapter 12). In the Pratt truss of Figure 1.2.3, all of the diagonal members except the end posts are in tension, whereas in the Warren truss, they may be in either tension or compression. If we reverse the direction of the applied loads so that they act up instead of down, all of the member loads will likewise be reversed: tension becomes compression and vice versa. (The grounded supports at each end of the structure restrain it vertically, in both the up and down directions.) For the trusses to remain rigid upon such a load reversal, all of the members must have a cross section thick enough to sustain both compression and tension. To keep the Pratt truss rigid under load reversal, while holding its weight to a minimum, we simply add another diagonal wire brace to each bay, as shown in Figure 1.2.4.

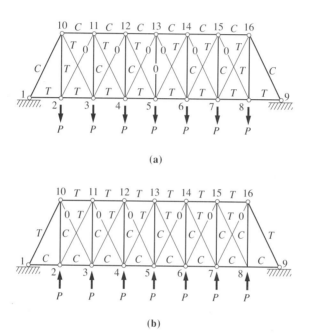

(a)

(b)

Figure 1.2.4   Pratt truss with each bay braced with two diagonal members, both of which are active in tension only. (a) Applied load directed down. (b) Applied load directed up.

A simplified sketch of the Wright 1903 *Flyer*, focusing on just its wing structure, is shown in Figure 1.2.5. The upper and lower wings were essentially identical, with a span of just over 40 feet and a chord of 78 inches. The wing's airfoil shape and aspect ratio were determined from the data collected by the Wright brothers in their homemade wind tunnel at their bicycle shop at Dayton. Each wing had a continuous front and rear spar made of spruce, as were the wing struts. The pilot lay prone on the lower wing in bay *E* with his hip in a cradle that was used to correct turning tendencies by being shifted side to side to warp the wings. The Wrights designed and built their own four-cylinder, water-cooled, 180-pound, 12-horsepower gasoline engine, which was mounted on the lower wing in bay *D* to the right of the pilot. Bicycle chains transferred power from the engine via sprockets to two counterrotating 8-foot-diameter pusher propellers, providing the *Flyer* with a cruising speed of 30 miles per hour. The two steel propeller shafts were supported between the wings by a separate, welded, tubular-steel truss attached to the rear wing spars. The Wright brothers designed and hand-carved the laminated spruce propellers.

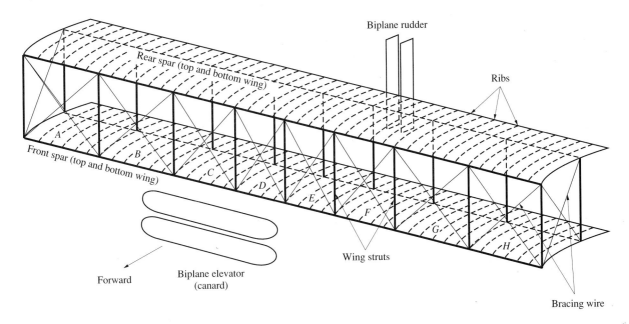

Figure 1.2.5   Simplified sketch of some of the structural features of the 1903 Wright *Flyer* wing, showing the relative position of the canard (forward) and rudder (aft).

Like the shape of the *Flyer*'s wings, that of the propellers was derived from Orville and Wilbur's own aerodynamic data and calculations.

There were some 40 ash wood ribs per wing, each one having been steamed and then formed into the thin airfoil shape designed by the Wrights (Figure 1.2.6). The ribs were attached to the two spruce wood spars. The front of the rib, which was flat, was attached to the rear face of the front spar with nails and glue and to the rear spar with tire cord. Untreated commercial cotton muslin fabric was used to cover the wing structure. The muslin, which was cut and sewn on the bias[2] into a bag that fit the shape of the wing, was attached to the rib caps with carpet tacks. The fabric carried not only the flight loads, but also the drag and antidrag[3] loads on the wing structure.

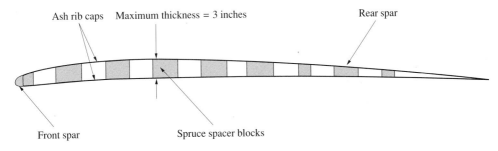

Figure 1.2.6   1903 Wright *Flyer* wing rib.

[2]The threads were at 45 degrees to the spars.
[3]These are inertia loads tending to throw the wing forward during rapid deceleration, as during landing.

In Figure 1.2.5, we can see the Pratt truss formed by the front spars of the upper and lower wing, the vertical compression struts, and the bracing wire. For structural rigidity, an identical vertical truss arrangement could be used between the rear spars and struts, and every fore–aft pair of struts could be braced like the outboard ones shown in the figure. However, to control the airplane in yaw and roll, the Wrights used a rudder mounted at the rear of their airplane, together with *wing warping*, which required the entire wingtip to rotate its trailing edge up or down. Wilbur had observed that when large soaring birds were partly upset by a wind gust, they regained lateral stability by twisting their wingtips slightly. The Wrights incorporated this control technique into their *Flyer* by means of a system of wires and pulleys (attached to the hip cradle) and by only partial braces on the outboard portions of the wing, to increase its twisting flexibility near the wingtips.

Following their success at Kitty Hawk, the Wright brothers returned to Ohio and began a series of refinements to their *Flyer* I design, to improve performance, stability, and control. They considered their 35-mile-per-hour 1905 *Flyer* III to be the first fully practical powered airplane. It was identical to the *Flyer* I in most ways. However, a more powerful engine, increased length, and larger control surface areas made it, unlike *Flyer* I, a completely controllable aircraft, which easily banked, turned, and performed figure-eight maneuvers. In their 1911 Wright Model B Military *Flyer,* they had eliminated the canard and placed the horizontal stabilizer at the rear, substituted hinged ailerons for wing warping, and added wheels for takeoff and landing. Furthermore, the pilot—plus a passenger—sat upright on the leading edge of the wing, instead of lying prone.

By this time, others throughout America and Europe had embarked upon the business of building successful heavier-than-air flying machines. Although many variations began to appear, the biplane/Pratt truss became by and large the world standard design. Spruce was adopted as the standard structural material, and the laminated spruce propeller became commonplace. Premium spruce wood was in good supply in the early years, readily available, easy to work with and repair, and pound for pound stronger than aluminum, which had not yet become a practical alternative. Unbleached cotton muslin was universally employed for aircraft covering. Fabric stretched over a wooden frame very nicely supported aerodynamic loads at the relatively low speeds flown by the early airplanes, and it was easily repaired if torn.

Most airplanes in the decade after 1903 resembled the Wright *Flyers*: they were trussed biplanes, many had canards, and the pilot, engine, fuel, payload, and practically everything else sat out in the open on the bottom wing. A new structural concept from Europe was the fuselage, invented and perfected in France, where there was a trend as well towards monoplane design. Figure 1.2.7 shows one of the best examples of the first use of the fuselage and a single wing for powered flight: the Blériot Model XI monoplane, designed by Louis Blériot and Raymond Saulnier. It made its first flight in January 1909 and had a top speed of 47 miles per hour, with a 25-horsepower, three-cylinder, radial, air-cooled engine. The Pratt truss design of the wooden fuselage, which was fabric-covered only part way along each side, is evident in Figure 1.2.8.

Like the Wright's *Flyers*, Blériot's fabric-covered wing was thin, with highly cambered upper and lower surfaces. The wing's two wooden spars were even thinner than the airfoil, since the rib caps passed over the top and bottom of the spars. The spars were far too thin to support lifting and landing loads by themselves, and they required the external wire bracing shown in the simplified sketch of Figure 1.2.9. During flight, lifting loads pushed the wing spar upwards, thereby unloading the top wires and inducing tension in the lower *lifting wires*, which were connected to the lower post at *A*. During landing, the inertia of the wing forced the spar downwards, unloading the lower wires while inducing tension in the top *landing wires*. The bracing-wire tension component perpendicular to the spar took out some of the shear loading in the spar, just as do the vertical struts of a biplane. Unfortunately, that same tensile force in the wire also had a component directed *along* the spar *towards* the fuselage. These bracing-wire forces, tugging inward, induced compression in the spar. That was a matter of concern at the time, because wood is weakest in compression; furthermore, like the bowing out of a meter stick pushed inward at both ends, increased compressive loading of the thin wooden spar could have led to its buckling and failing.

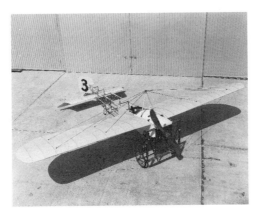

Figure 1.2.7   The Blériot Model XI monoplane, similar to the one in which Louis Blériot himself became the first person to cross the English channel in an airplane.

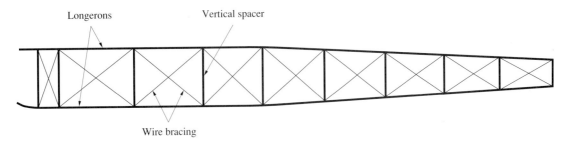

Figure 1.2.8   Pratt truss side panel of the Blériot XI wooden fuselage.

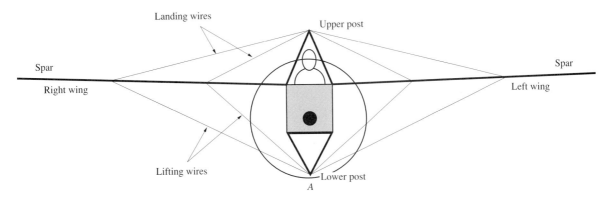

Figure 1.2.9   A simplified sketch of the Blériot monoplane, showing the wire bracing that joined points on one of the wing spars to posts above and below the fuselage. Both spars were braced in this fashion, but the lower post for the front spar was positioned ahead of that spar, to brace the wing against drag loading.

Blériot's Model XI wing was initially equipped with separate, pivoting ailerons, but the monoplane had poor lateral stability. On August 8, 1908, at a racetrack near Le Mans, France, Blériot witnessed Wilbur Wright repeatedly execute what were then startling, controlled, banked turns in the Wright *Flyer* "A" during one of many promotional demonstration flights. Blériot was so impressed that he copied the Wrights' wing warping system for the flexible wing of his Model XI monoplane, thereby endowing the aircraft with true lateral control. So confident was Louis Blériot in his flying machine's improved handling qualities that he set out in the rainy early dawn of July 25, 1909, to complete the first powered airplane crossing of the English Channel. The 24-mile flight from Calais to Dover required 36 minutes. The Model XI later went into production, and all five variants of it employed wing warping. By the outbreak of World War I, so many Model XIs had been manufactured—at the Blériot factory and under license worldwide—that it had become one of the world's most well-known airplanes. Higher-powered versions equipped with famous Gnome rotary[4] engines saw military service.

Other French monoplanes, of the same trussed fuselage, thin-airfoil, cable-braced wing design as Blériot's, included: the R.E.P. (Robert Esnault-Pelterie) Model N (1909); the Morane-Saulnier Models H (1913), L (1913), and N (1914); the Deperdussin TT (1914); and the Nieuport 6M (1914). Germany had the Fokker E-III (1915),[5] but went on to produce more advanced *internally* braced monoplanes before the war's end. Great Britain nearly shied away altogether from monoplane design in concern over the strength and durability of the single-wing configuration. During the war, England produced only one monoplane, the wire-braced Bristol M.1C (1917), which resembled the Moraine-Saulnier N. Nevertheless, the "stick and wire," fabric-covered biplane, with its strong, rigid, externally braced, aileron-equipped wings, and separate trussed fuselage, dominated the skies in World War I.

By the end of the war in 1918, certain aircraft structural design features and tendencies had become standard. To improve lift, wings were made thicker with less chordwise curvature of the lower surface (undercamber). This allowed a deeper spar, which improved the strength and rigidity of the wing (as can easily be demonstrated in the manner of Figure 1.2.10 with a ruler or meter stick). A wing that requires no external bracing of any kind is called a *cantilever* wing. The first cantilever wing fighter of World War I to go into production was the 115-mile-per-hour Fokker Dr-I (Figure 1.2.11), the famous triplane[6] in which the German ace, Baron Manfred von Richthofen, scored 21 of his 80 victories.

The airplane was designed by Rheinhold Platz, who joined the Fokker Company in 1912 as a welder, having spent years perfecting the new technique of acetylene welding. The 1917 Fokker Dr-I had no drag-producing wing bracing wire and just a single thin, streamlined, nonstructural, vertical, spruce strut joining each wingtip for

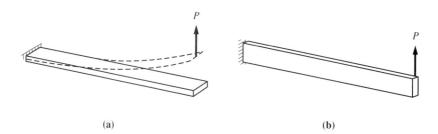

(a)   (b)

Figure 1.2.10   (a) With the load applied parallel to the thin dimension, the beam is flexible and weak. (b) With the load applied parallel to its wide dimension, the beam is relatively rigid and strong.

---

[4]The cylinders, crankcase, and propeller rotated as a unit to ensure adequate cooling; the crankshaft was stationary.

[5]The "E" stands for *Eindecker* (monoplane).

[6]The German air command had launched a massive triplane development effort to counter the threat of the fast and maneuverable British Sopwith Triplane, which, after only six months' service in 1917, was replaced by the more conventional Sopwith Camel.

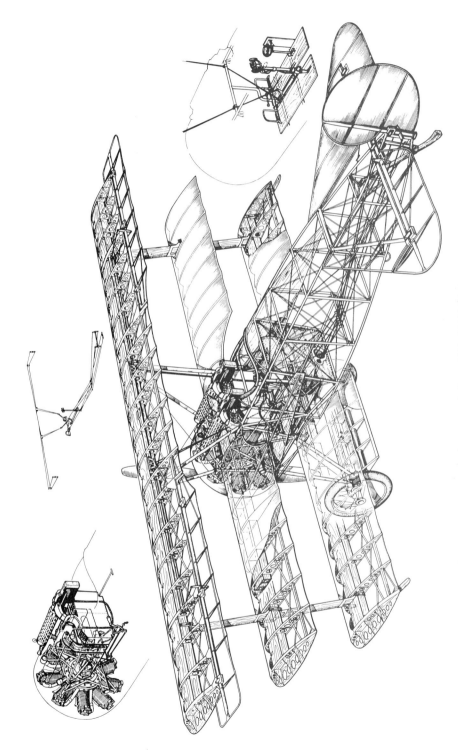

Figure 1.2.11 Germany's Fokker Dr-1, the first major cantilever wing fighter of World War I.

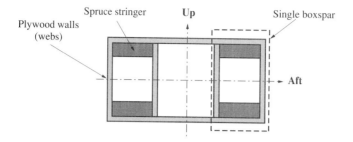

Figure 1.2.12   Cross section of the Fokker Dr-I spar ("spar box").

It is a composite of two boxspars. As such, it offers excellent resistance to shear, bending, and torsion.

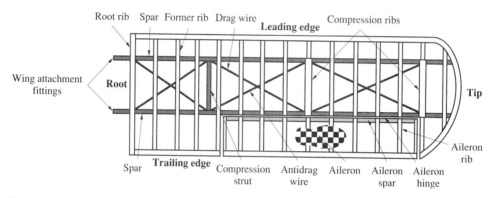

Figure 1.2.13   Structure typical of wooden, fabric-covered wings.

Note the rigid Pratt truss formed by the wing spars, the compression ribs/drag struts, and the diagonal bracing wire. Compression ribs are "beefed up" former ribs.

controlling wing flutter in some flight regimes. The wings owed their strength to Platz's design of its single spar, a cross section of which is sketched in Figure 1.2.12.

It is not surprising, given Platz's mastery of the art of welding, that in the Dr-I, the wooden truss members of the traditional box-shaped, fabric-covered fuselage of the time were replaced by steel tubing, welded together at the joints. However, truss bays were still cross braced with steel wire, as in wood construction (Figure 1.2.8). The innovative features of the Dr-I—thick, high-lift wings, box spars, no external wire bracing, and wire-braced welded steel fuselage—were incorporated by Platz into his design of the even more successful and effective Fokker D-VII biplane, which entered service in 1918. In these two outstanding Fokker fighters, we see the trend toward clean, unbraced wing design and the use of metal instead of wood for primary structure.

During the war, the evolution in wing structural design occurred not only in the spars, but also in the ribs and other details. As the speed of airplanes rapidly increased to over 100 miles per hour, drag loads on the wing became too large to be resisted by the fabric wing covering, as in the 30-mile-per-hour Wright *Flyer*. To make the fabric-covered wing rigid in its plane once again requires a truss structure. A typical arrangement is illustrated in the plan view of Figure 1.2.13. A typical rib layout is shown in Figure 1.2.14.

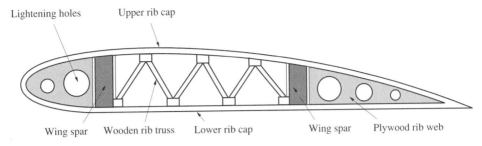

Figure 1.2.14   Typical wooden rib construction for fabric-covered wings.

Depending on the design, the space between the caps might have been all or part lattice or web. The lightening holes reduced the net weight of the many ribs and served as routing holes for cables, etc.

It must be pointed out that the thick, cantilever wing and the use of metal in airplane construction preceded the 1917–18 Fokker airplanes. In 1910, Dr. Hugo Junkers, a German thermodynamics professor and hot water heater manufacturer turned aviation enthusiast, patented the design of an internally braced flying wing. It was not built, but Junkers applied the same concepts to the design of a thick, completely internally braced cantilever wing for his 1915 J-1 monoplane. The ponderous, 105-mile-per-hour, single-seat Junkers J-1 was the first airplane to be made entirely of metal. The steel tubing structure was covered with thin sheet iron, from which arose the J-1's nickname, the "Tin Donkey." The unbraced wing was mounted midfuselage and tapered from a very thick root to a thin, highly undercambered tip. In spite of German military skepticism—and rejection—of his revolutionary design, Dr. Junkers was convinced that airplanes were destined to be made out of metal, and he pressed on with improved designs. Instead of sheet iron, he turned to the new, German-invented, strong and lightweight copper alloy of aluminum known as duraluminum to cover his 1917 all-metal J-4 sesquiplane.[7] The aluminum covering was corrugated for added strength, with the corrugations running fore–aft, in the streamwise direction, to minimize drag. The J-4 was picked up by the German military as an essentially armored ground attack weapon, which the Allies found extremely difficult to shoot down. By the end of the war in November 1918, the Junkers J-9 and J-10 all-metal, corrugated-aluminum skin, fully cantilevered wing, monoplane warships were in service. Both of these airplanes possessed the unique structural feature that the single wing was attached to the bottom of the fuselage rather than at the top or the middle, as in all previous monoplanes. Placing the energy-absorbing wing structure below the pilot appeared to provide a measure of safety in the event of a crash landing.

The accelerated pace of development in aircraft structural technology during 1914–1918 slackened considerably in the years thereafter without the lamentable stimulus of war. The wartime construction, by all participants, of over 150,000 mostly wood-and-fabric airplanes had significantly depleted the supply of prime wood. Toward the end of the war, the scarcity of spruce led designers to abandon solid wood in favor of laminated spruce for struts and spars, or to use built-up box spar construction similar to that illustrated in Figure 1.2.12. Unable to depend on wood as a strategic material, Great Britain's Air Ministry dictated that all future British airplane designs must be in metal. (Nevertheless, some famous World War II British fighters, like the De Havilland *Mosquito*, were made of plywood.)

With few exceptions, the changeover to all-steel welded or bolted aircraft structures through the late twenties was barely evident in the outward appearance of airplanes, since the wing and fuselage covering remained the same nonstructural muslin and/or plywood used in all-wooden construction. Furthermore, the biplane, with its drag-producing struts and bracing wire, continued to be the conservative design of choice for aircraft of all sizes.

---

[7]*Sesqui* means "one and a half." A sesquiplane is a biplane in which one wing (the lower one) has an area of less than half the other. Most of the series of French Nieuport WWI fighters, including the famous model 17, were sesquiplanes.

Meanwhile, in 1919, Hugo Junkers parlayed his wartime experience into the production of the single-engine F-13 civilian airliner, an all-metal, corrugated-aluminum skin, cantilevered low-wing monoplane, which carried a crew of two plus four passengers at 90 miles per hour. Three hundred fifty Junkers F-13s were built by 1932, when it was succeeded by the 15-passenger, 150-mile-per-hour, extremely successful Junkers JU-52 airliner, which eventually became a ubiquitous World War II workhorse transport for Hitler's Luftwaffe. The JU-52[8] was a low-wing monoplane bearing Junkers' boxy, corrugated-aluminum hallmark appearance. It had three cowled radial engines housed in nacelles, one on each wing and the third in the nose. In the United States, the Ford Motor Company had, since 1926, been manufacturing its own successful, equally boxy, all-metal, thick-wing, corrugated skin, 12-passenger airliner. The 110-mile-per-hour Ford *Trimotor* ("Tin Goose") was a high-winged monoplane, which, like the JU-52, had three (albeit uncowled) radial engines, one of them in the nose and the other two slung on struts below the massive wing. Although the corrugated aluminum of the Ford and Junkers airplanes was durable, the skin carried very little of the structural load, while producing a considerable drag. In terms of structural efficiency, the designs were dead ends. The next major breakthrough in aircraft structures required a temporary return to all-wood construction.

In 1916, 20-year-old John K. "Jack" Northrop, a draftsman, mechanic, and airplane enthusiast, joined the Loughead[9] Aircraft Manufacturing Company in Santa Barbara, California, to work on the production of a twin-engine, 10-passenger, biplane flying boat. Northrop began by designing the hull and went on to design the 75-foot-span upper wing. Named the F-1, the seaplane—the world's largest at the time—flew in March 1918. However, the Navy had just ordered the Curtiss HS-2L seaplane, so it contracted Loughead to build two modified versions of that aircraft. After the war, the Loughead brothers, Alan and Malcolm, with Jack Northrop still aboard, set out to produce the S-1, a small, lightweight, single-seat civilian biplane for the masses. Perhaps influenced by Northrop's experience with the boat-like seaplane hull design, the S-1 had a "monocoque" fuselage, its most significant feature.

The French word *monocoque* means "single-shell" (like an egg shell). A monocoque structure has a thin curved wall that is capable of supporting applied loads without the need for internal skeletal support. If the shell is attached to frames and stiffeners that share the load, then the structure is called "semimonocoque." In either case, a major load path through the structure is the thin, "stressed skin." The S-1 was not the first example of monocoque construction in an airplane. There were pre–World War I examples, one of the most notable being the French Deperdussin *Racer*, which in 1912 was the first airplane officially to exceed 100 miles per hour—and it went on to fly nearly 130 miles per hour. This fast, streamlined, monoplane had a molded plywood fuselage shell that was layed up in two lengthwise halves, which were subsequently joined together. The procedure was lengthy and tedious, requiring many manhours to align the crosshatched tulipwood strips properly. Like the Deperdussin *Racer*, the manufacture of the Loughead S-1 fuselage was done in two halves, but the process of aligning and gluing the three layers of spruce strips forming the laminated half-shell was dramatically reduced—to about 20 minutes. The S-1 flew in 1920, but did not sell, forcing the Loughead Company out of business. Jack Northrop moved down to the Douglas Aircraft Company in Santa Monica, California.

In 1926, Northrop left Douglas to join Alan Loughead in another new aircraft venture. The Lockheed Aircraft Company set up shop in Hollywood (moving, eventually, to Burbank), and work began on production of a Northrop-designed, seven-place, single-engine, high-wing monoplane, with a 1/4-inch-thick molded wooden monocoque fuselage (like the S-1), a fully cantilevered wing, and a completely enclosed cabin. The *Vega*, so named by Jack Northrop, made its first flight on July 4, 1928, and was at once a successful commercial aircraft (Figure 1.2.15). With a cruising speed of 155 miles per hour, the rugged, streamlined *Vega* went on to set a variety of speed and distance records. In 1929, Northrop left Lockheed to start his own company and pursue his own ideas on stressed-skin construction, while Lockheed continued to produce airplanes using the same airframe as the *Vega*. The 1931 Lockheed *Orion*[10] was

---

[8]JU-52/3m to be precise. The first JU-52 model, which appeared in 1930, had a single engine, like the F-13.

[9]Pronounced "Lock-heed, " not "Log-head."

[10]Orion is one of the brightest constellations in the northern sky.

Figure 1.2.15 Jack Northrop named the Lockheed 1 the *Vega*, after the brightest star in the constellation Lyra.

essentially a *Vega* with the wing relocated to the bottom of the fuselage. This made it possible to use shorter, lighter landing gear that could be folded up into the thick wing root. The *Orion* was the first commercial airliner equipped with retractable landing gear, and that was a major reason why the airplane cruised 50 miles per hour faster than the *Vega* with the same 550-horsepower radial engine. Indeed, its 200-plus miles per hour speed was faster than most of the military fighters of the day. Streamlined monocoque construction, a single cantilevered wing mounted low on the fuselage, and retractable landing gear were standard-setting features for the next generation of aircraft.

One of the first things Jack Northrop did after leaving Lockheed was to build an airplane resembling a flying wing, a portent of things to come from Northrop. The 1929 experimental craft consisted of a 30-foot-span wing continuously increasing in both depth and chord from the thin wingtips to the thick center section in which two open cockpits were embedded side-by-side, close to the centerline. A small fuselage-like nose protruding forward from the center of the wing contained the engine. Protruding aft of the wing were the center pusher propeller shaft and, outboard of the center wing section, twin slender booms supported the tail surfaces. Northrop was far from ready to eliminate the tail, which later disappeared from the flying wing designs, leading to the Northrop Corporation's graceful XB-35 and YB-49 of the late 1940s and eventually to its 600-mile-per-hour, all-composite B-2 stealth bomber, which first flew in 1989. The 1929 experimental airplane had a novel planform for that era, but the fact that it had an all-aluminum semimonocoque structure was much more significant. Northrop was clearly testing his theories of how to make efficient, lightweight, metal structures. For example, the wing had no conventional spars. Instead, there were six spanwise vertical webs about 15 inches apart, with a bend at each end to form flanges, by means of which the web was riveted to the upper and lower wing skin (Figure 1.2.16).

Although Northrop's fledgling flying wing was an experimental project, his new company soon turned out the first production airplane to combine an all-metal (aluminum), smooth-skin, flush-riveted, semimonocoque fuselage structure with an all-metal, stressed-skin, multicellular wing. It was the 1930 Northrop *Alpha*, a streamlined, six-passenger, faired low-wing monoplane with a NACA[11]-designed, drag-reducing cowling surrounding its single radial engine. This airplane signified the end of the use of metal to simply replace what was made of

---

[11]National Advisory Committee for Aeronautics, chartered by Congress on March 3, 1915, to "supervise and direct the scientific study of the problems of flight, with a view to their practical solution." On October 1, 1958, it became NASA, the National Aeronautics and Space Administration.

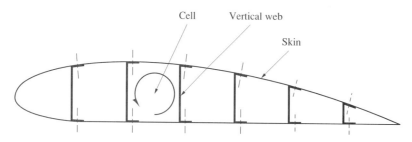

Figure 1.2.16   Multicell, all-metal wing box construction pioneered by Northrop in his 1929 experimental "flying wing."

wood in previous airplane designs. Single-wing, all-metal, semimonocoque, stressed-skin design became the norm from the early 1930s onward. This trend was accompanied and facilitated by the increased availability of strong aluminum alloys at lower cost and by advances in powerplant and propeller design.

In 1932, the Glenn L. Martin Company delivered its first B-10 bomber to the United States Army. The twin-engine, 200-mile-per-hour B-10 had an all-metal semimonocoque structure, and it pointed the way to a new generation of monoplane bombers with clean lines and retractable undercarriage. It superseded Boeing's semimonocoque, twin-engine B-9 "Flying Pencil," whereupon Boeing began development of a commercial version, and that led to its Model 247.

The Boeing 247 first flew in February 1933. It was a clean, all-metal, semimonocoque, low-wing monoplane, twin-engine, 10-passenger airliner with retractable landing gear. It is generally agreed that the 190-mile-per-hour Model 247 was the first modern transport airplane. Much faster than the Ford *Trimotor*, it could fly across the continental United States in less than a day. However, its slim fuselage required most passengers to move about in a crouch and to step over the rear wing spar, which ran right through the cabin. Nevertheless, Boeing could not manufacture its 247 fast enough to satisfy the airlines' orders for this remarkable airplane. Douglas Aircraft responded to the need when a design team, taking proper advantage of what was known about the unfinished Boeing 247, quickly came up with the prototype of the famous series of all-metal, low-wing, twin-engine, retractable-gear Douglas commercial ("DC") airliners. The maiden flight of the 12-passenger DC-1 in 1933 came only five months after that of the Boeing 247. The DC-1 had a flat-sided fuselage like the 247, but it was more spacious (no crouching required), and the wing spars passed beneath the cabin floor, as in all low-wing passenger planes since. The DC-1 was followed in 1934 by the stretched, slab-sided 14-passenger DC-2 and in 1935 by the 21-passenger DC-3, (Figure 1.2.17), with its pleasing oval-shaped fuselage cross section.

The DC-3's performance, payload, and ruggedness finally established the possibility of an airline's actually making a profit, and the DC-3 became the world's most widely used air transport by 1940. It was the first airliner to pay its own way, carrying passengers and cargo without government subsidies. Nearly 11,000 were built by the end of World War II, most of them in the military C-47 designation. Many features combined to make the DC-3 a landmark success: retractable gear; the NACA cowling (Boeing used its own cowl design on their model 247); wing flaps (lacking on the 247); efficient, supercharged engines; and controllable-pitch propellers. Certainly the Northrop-influenced airframe was one of the most inspired ever built. In the wing, there were not two but three spars, adding durability to the wing and pointing the way to "fail-safe" airframes with long fatigue life. All of the essential features of aluminum semimonocoque construction were embodied in the DC-3: formed aluminum skin riveted to frames, longerons, ribs, spars, stringers, and formers. Although subsequent airliners have gotten along on two wing spars, modern all-metal airframes differ from that of the DC-3 only in details, not in essence.

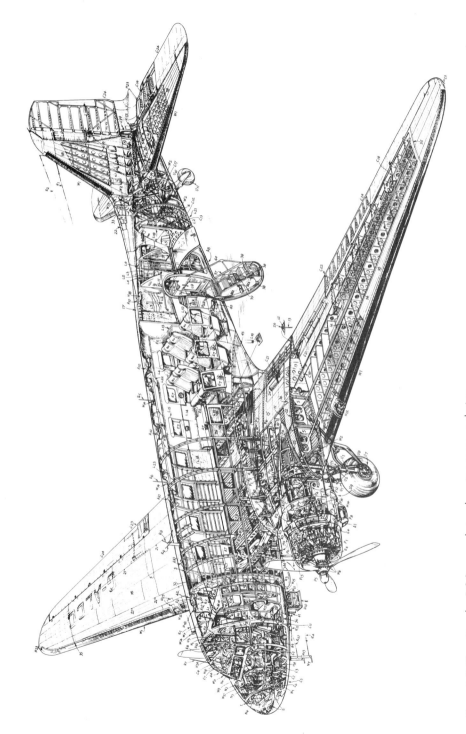

Figure 1.2.17   Douglas DC-3, an archetypal semimonocoque design.

During the 1930s, the application of all-metal, semimonocoque, monoplane design extended as well to military fighters. Although the biplane configuration persisted in several international designs through the beginning of World War II, the inevitable assertion of the monoplane fighter began in the United States in March 1932 with the debut of Boeing's 200-mile-per-hour P-26 "Peashooter," an all-metal, flush-riveted, open-cockpit, fixed-undercarriage, low-wing monoplane powered by a radial engine enclosed by the same cowling that Boeing used on its 247 airliner. The wing was not a complete cantilever, since it was externally braced by anachronistic lifting and landing wires. A true cantilever low-wing monoplane fighter followed three months later when the French Dewoitine D-500 made its first flight in June 1932. Like the P-26, the D-500 had fixed gear and an open cockpit and about the same performance, although it was powered by a V-12 water-cooled engine. The Soviet Union's 270-mile-per-hour Polikarpov I-16 appeared late in 1933 as the first cantilever monoplane fighter with retractable gear. Although it had metal spars and an aluminum leading-edge wing covering, its wing was primarily fabric-covered and was attached to a wooden monocoque fuselage. The first all-metal, flush-riveted, semi-monocoque, low-wing monoplane fighters with retractable undercarriage and enclosed cockpits appeared in 1935: the United States Curtiss Hawk 75A/P-36 (April), the German Messerschmitt Bf-109 (May), and the United States Seversky P-35 (August).

With few exceptions, the World War II fighter and bomber airframes were semimonocoque structures made mostly of aluminum alloys and steel. One of the major successful exceptions was Great Britain's De Havilland *Mosquito,* a 360-mile-per-hour, twin-engine, midwing monoplane fighter/bomber that first flew in November 1940. This light, agile airplane had an all-wood, stressed-skin structure, with a monocoque fuselage molded in two halves like the Lockheed *Vega.* But whereas the *Vega*'s 1/4-inch-thick fuselage skin consisted of three crossed plies of spruce strips glued together under pressure, the *Mosquito*'s much stronger fuselage shell had a balsa core sandwiched between two spruce veneers, as illustrated in Figure 1.2.18(a). This pioneering use of high-strength-to-weight composite sandwich construction had been employed previously in the design of the elegant 1937 all-wood De Havilland *Albatross* commercial airliner (Britain's answer to the DC-3), which was short-lived because of the exigencies of war. The all-wood wing of the *Mosquito* was strengthened in a manner reminiscent of that employed in the all-metal DC-3 and the Boeing B-17 *Flying Fortress,* whose upper wing panels were reinforced against buckling by thin, corrugated-aluminum plates attached to the inner surface. Figure 1.2.18(b) illustrates the *Mosquito* wing design's use of a number of spanwise solid-spruce stringers sandwiched between spruce plywood spacers. This arrangement, together with the spruce and plywood box spars, endowed the wing with excellent strength in bending and torsion. Almost 7,800 *Mosquitos* were produced before

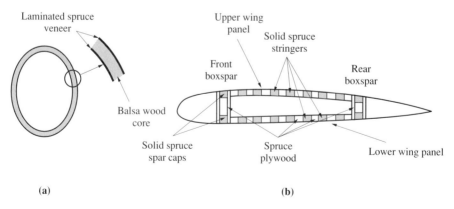

(a)   **(b)**

Figure 1.2.18   (a) A cross section of the De Havilland *Mosquito*'s tapered, wooden monocoque fuselage. (b) *Mosquito* wing cross section.

production ceased in 1950. The *Mosquito*, in spite of Germany's brief attempt to replicate it in the form of the mostly wood Focke-Wulf Ta-154 "Teutonik Moskito," was the last all-wood production airplane. It is worthy of note, however, that the Bellanca *Viking* high-performance general aviation airplane still uses an all-wood wing.

One cannot leave the subject of wooden airframes without mentioning the famous Hughes-Kaiser HK-1 *Hercules* flying boat, popularly known as the "Spruce Goose." Conceived during the early years of World War II as a solution to the problem German U-boats presented to the supply of Great Britain by sea, the *Spruce Goose* was built almost completely of birch plywood, a nonstrategic material for the United States. The behemoth, weighing 400,000 pounds, was propelled by eight 3000-horsepower radial engines, and it had the greatest wingspan—320 feet—of any airplane ever built. The prototype was not finished before the end of the war, whereupon government funding ceased. Nevertheless, the wealthy Howard Hughes pursued completion of the enormous craft. The airplane flew only once, on November 2, 1947. With Hughes at the controls, it covered about one mile at a speed of 80 miles per hour and an altitude of about 70 feet over Long Beach Harbor. One of the great oddities in aviation history, the *Spruce Goose* never flew again, eventually becoming a museum object.

The period from 1939 to 1947 might be called the turbojet revolution in aviation. With few exceptions,[12] regardless of outward appearance, the internal structure of airplanes introduced during these years differed little in essence from that of the DC-3. Of significance to future airframe design innovations was the introduction of the turbojet engine, developed independently during the late 1930s by Frank Whittle in England and Hans J. P. von Ohain in Germany. On August 27, 1939, the secret, experimental Heinkel He-178 flew successfully, becoming the world's first jet airplane, powered by von Ohain's 840-pound-thrust engine. Almost two years later, on May 15, 1941, Whittle's 860-pound-thrust centrifugal-flow turbojet powered the Gloster E-28/39 on a 17-minute flight that launched Great Britain into the jet age.[13] The world's first operational jet aircraft was the Messerschmitt Me-262 (Figure 1.2.19), which entered service with the Luftwaffe in June 1944. One month later, the Gloster *Meteor* began operations with the British Royal Air Force. Both airplanes had two wing-mounted turbojets and were capable of sustained speeds in excess of 500 miles per hour, faster than the speediest propeller-driven fighter. The

Figure 1.2.19   The Me-262 of World War II was the first production jet aircraft and the first jet fighter with swept-back wings.

---

[12]The wings and fuselage of Great Britain's Vickers *Wellington* twin-engine bomber had a unique, highly-redundant, basket-like, geodetic truss structure, making the fabric-covered airplane capable of surviving major damage. The *Wellington* first flew in 1936, and 11,400 were eventually built. Sir Barnes Wallis, designer of the *Wellington*, had first devised the geodetic form of structure for the R-100 airship (1929) and had applied it as well to the design of the Vickers *Wellesley* monoplane fighter (1935).

[13]Successful tests of the German and British turbojet engines both occured in 1937: Whittle's in April and von Ohain's six months later, in September. In England, the government did not begin fully backing Whittle's development efforts until 1940; the more purposeful German program proceeded at once to outpace the British.

Me-262's axial-flow engines were mounted in nacelles below its swept wing, while the centrifugal-flow engines of the *Meteor* were housed in its thick straight wing, just as reciprocating engines were. The Me-262 was the first production fighter with a swept-back wing, the purpose of which was to shift aftward the center of mass of the heavier-than-anticipated Jumo 004 turbojets. Coincidentally, the sweep-back yielded a reduction in drag at high speed, for reasons that were not widely understood at the beginning of the 1940s. In an intentional effort to reduce drag, the wings of the German jet were designed relatively thin (a portent of things to come in jet aircraft design). Although the main landing gear struts had room to retract into the wing root, the wheels themselves would not fit within the thin wing.[14] So the bottom of the Me-262 fuselage was widened to provide room for the wheels. This gave the jet its distinctive shark-like cross section: wider at the bottom than at the top. The third wheel of the very first models of the Me-262 was in the "conventional" location: under the tail. Later, it was moved to the nose where there was plenty of room to retract it into the forward fuselage. The Gloster *Meteor* had tricycle landing gear from the beginning.

The United States' first jet aircraft was Bell Aircraft's XP-59A *Airacomet*, which first flew in October 1942, powered by two General Electric jet engines, which were licensed copies of the Whittle turbojet. Capable of little more than 400 miles per hour, less than 100 variants of the P-59 were built before it went out of production in 1944 in favor of the superior, 560-mile-per-hour Lockheed P-80 *Shooting Star*. The P-80 was the first operational United States jet fighter. Production deliveries began in early 1945, and two YP-80As saw combat in Italy less than a month before the war in Europe ended in May of that year.

The Soviet Union was quick to parlay captured German turbojet engines, documents, and personnel into its first operational jet fighters, the Yakovlev Yak-15 and the Mikoyan-Gurevich MiG-9, prototypes of both of which flew on April 24, 1946. Like the P-80, these airplanes had straight wings. In fact, in the immediate postwar rush to convert from prop-driven to jet-powered fighters, conventional design prevailed. The outward appearance of airplanes was modified principally to take advantage of the more compact cross section of the jet engine and to utilize the space in the nose formally occupied by the reciprocating engine.

No radical structural design innovations were employed in the jet airplanes listed in Table 1.2.1. At first, there was a trend towards "burying" turbojet engines in the fuselage or in the wings, as typified in British designs not only by the Gloster *Meteor*, the English Electric *Canberra* light bomber, and the De Havilland *Comet* of the late 1940s, but also in the British "V-class" of heavy bombers: the Vickers *Valiant* (first flight, May 1951); the Avro *Vulcan* (August 1952), the world's first delta-wing bomber; and the Handley Page *Victor* (December 1952). Burying the engines is motivated primarily by the quest for aerodynamic performance, to the detriment of maintainability and serviceability. In the wings, a heavier structure is required to transfer the loads around the space occupied by the engines. Eventually, jet transports were designed with the engines attached to the exterior of the fuselage in pods or under the wings in nacelles for clear and easy access.

In a paper presented in October 1935 at a conference on high-speed aerodynamics at Rome, German engineer Adolf Busemann first proposed the swept wing as a means of reducing drag. Willy Messerschmitt took Busemann's theory seriously, and in 1944, under the leadership of his senior designer Woldemar Voigt, he began developing a swept-wing research plane, but the war ended before it flew. Meanwhile, in January 1945 at NACA's Langley, Virginia, laboratory, Robert T. Jones, who was unaware of Busemann's and Voigt's work, came up with his theory of "subsonic sweep" while working on the design of a guided missile. Several months of supersonic wind tunnel tests confirmed Jones's intuitive airflow theories. As WWII drew to a close, Jones recommended to his NACA superiors that swept-back wings be employed henceforth on all high-speed airplanes. After the war, Adolf Busemann together with copious wind tunnel data on swept-wing models, was brought to the United States to join the NACA staff at Langley. Busemann's work immediately influenced the design of Boeing's six-engine B-47 *Stratojet*, which debuted in 1947 as the world's first swept-wing jet bomber, as well as

---

| [14] Large tires were required so that the Me-262 could operate off rough, grass airfields.

Table 1.2.1   Post-World-War II operational jet aircraft of conventional (straight-wing) design.

Country	Airplane	Category	First Flight
USA	McDonnell FH-1 *Phantom*	Fighter	1945
USA	Republic F-84 *Thunderjet*	Fighter	1946
USA	McDonnell F2H *Banshee*	Fighter	1947
USA	North American B-45 *Tornado*	Bomber	1947
USA	Douglas F3D *Skynight*	Fighter	1948
USA	Northrop F-89 *Scorpion*	Fighter	1948
USA	Lockheed F-94 *Starfire*	Fighter	1949
Great Britain	Hawker *Sea Hawk*	Fighter	1947
Great Britain	English Electric *Canberra*	Bomber	1949
France	Dassault MD-450 Ouragan	Fighter	1949
Soviet Union	Tupolev Tu-14*	Bomber	1947
Soviet Union	Yakovlev Yak-23	Fighter	1947
Soviet Union	Ilyushin Il-2**	Bomber	1948

*Horizontal tail was swept.

**Horizontal and vertical tailplanes were swept.

the design of North American's F-86 *Sabre* jet fighter.[15] The wings of both the F-86 and B-47 had a 35 degree sweepback. The Soviets also had access to captured German data, which led to the development of the Mikoyan-Gurevich MiG-15, which was test flown in December 1947, just three months after the F-86's first flight.[16] The F-86 and MiG-15 inaugurated the second generation of jet fighters. Sweden's swept-wing SAAB-29 *Tunnan* prototype flew in 1948. France and Britain's production swept-wing entries followed in 1951 with the first flights of the Dassault *Mystère* and Hawker *Hunter*, respectively.

After the war and into the 1950s, NACA, the United States Air Force, and the United States Navy initiated research aimed at exploring the transonic flight regime and gathering data for the design of airplanes to fly faster than the speed of sound (Mach 1). A series of technology investigators and demonstrators were built. The bullet-shaped, rocket-powered Bell X-1 was the first supersonic airplane.[17] Advanced Bell X-1s exceeded Mach 2. The swept-wing, stainless steel Bell X-2 flew past Mach 3. The dartlike turbojet-powered Douglas X-3 *Stiletto* was a platform for research on inertial coupling, and it typified the generalized design configuration of early supersonic jet fighters. The twin-jet Northrop X-4 explored the characteristics of the swept-wing, semi-tailless configuration, like that of German designer Alexander M. Lippisch's 1941 rocket-powered Me-163 *Komet* interceptor and Britain's post-war De Havilland DH-108. The jet-powered Bell X-5, inspired by a Messerschmitt P-1011 variable geometry research aircraft captured by the United States at the end of the war, validated the concept of variable wing sweep (20 degrees to 60 degrees on the X-5). Although not part of the X-series, the Convair XF-92A, the Navy-sponsored Douglas D-558-1 and -2,[18] and the Northrop YB-49A

[15]North American completed the detail design of the F-86 with an unswept wing, just in case the swept wing did not work out. Not everyone was convinced at the time that the swept wing was a good idea.

[16]In April 1948, the F-86 became the first jet to exceed the speed of sound in a dive.

[17]USAF Maj. Charles E. "Chuck" Yeager piloted the X-1 to Mach 1.06 in October 1947.

[18]A rocket-powered Douglas D-558-2 *Skyrocket*, piloted by Scott Crossfield, was the first airplane to fly past Mach 2, in November 1953.

flying wing provided transonic design information on delta-wing, straight-wing, swept-wing and all-wing planforms. Having furnished convincing proof of the advantages of such items as the swept wing, reduced thickness-to-chord ratio, and the all-moving tail, this research and development fostered the supersonic breakthrough and led to airframe designs capable of taking full advantage of the performance available from the gas turbine engine.

One of the research efforts that had a significant influence on the supersonic breakthrough was reported by Richard T. Whitcomb in the 1952 NACA Research Memorandum L52H08, *A Study of the Zero Lift Drag Characteristics of Wing-Body Combinations Near the Speed of Sound*. In it, Whitcomb reported the results of wind tunnel tests, which supported his proposed *area rule* to reduce wave drag at transonic speeds. According to the area rule, in the region of the airplane where the wing joins the fuselage, the fuselage cross section should be reduced to compensate for the wing cross section. The total area (wing plus body) presented to the oncoming airstream should remain essentially constant, giving rise to a "pinched waist" or "Coca-Cola bottle" shaped fuselage. Convair was quick to redesign the straight-waisted fuselage of its disappointing F-102 *Delta Dagger*, which first flew in October 1953 with below-predicted performance. The indented-fuselage F-102A flew in December 1954, showing a 20 percent speed gain and easily surpassing Mach 1.

By the mid-1950s production jet aircraft, beginning with the North American F-100 *Super Sabre*, were capable of sustained, level flight at speeds in excess of Mach 1. In 1956, the Lockheed F-104 *Starfighter* (Figure 1.2.20) became the first to do so at speeds beyond Mach 2. The design solution for the thin wings (e.g., 3.4 percent thickness-to-chord ratio for the *Starfighter*) and high wing loadings of the second generation of jet fighters was often multispar/multicell wing structures, as, for example, in the Vought F-8 *Crusader* and the Republic F-105 *Thunderchief*, and, more recently, the Dassault-Breguet *Mirage* F1, the McDonnell-Douglas F-15 *Eagle* and F/A-18 *Hornet*, and the General Dynamics[19] F-16 *Fighting Falcon*. Wing structure was often machined from a single plate of raw material in the form of integrated skin/stringers, as in the North American A-5 *Vigilante*.

The years following World War II saw a rapid growth in commercial aviation. Hundreds of paved military airfields around the world became available for land-based commercial traffic, ushering out the prewar era of graceful transoceanic flying boats. High-performance propeller-driven military transports like the venerable Douglas C-47 *Dakota* and the Douglas C-54 *Skymaster* were converted to civilian use as the DC-3 and DC-4. Pressurized aircraft, such as the Lockheed *Constellation*, the Douglas DC-6, and the double-decked Boeing *Stratocruiser* (developed from the B-29 *Superfortress*) entered service by 1950. These airplanes could cruise up to 330 miles per hour at 20,000 feet while maintaining a cabin pressure equivalent to an elevation of 8000 feet or less. This required the fuselage to contain a pressure differential of four to five pounds per square inch.

As a pressure vessel, the circular fuselage cross section typical of pressurized aircraft is the most structurally efficient shape, because the pressure causes mainly tensile stresses in the thin skin. The pressure stress is superimposed on the stresses due to loads coming into the semimonocoque fuselage from the wings, tail, equipment, and payload. The complex stress field is exacerbated by cutouts such as windows and doors, which are reinforced by surrounding framework and thicker skin (doublers). The corners of fuselage doors and windows are rounded to reduce stress concentrations. In fact, one can usually infer from the shape of the windows whether an airplane is pressurized (round or rounded windows and openings) or not (rectangular windows and doors).

In the United States, the first use of jet propulsion was confined to military aircraft; in Great Britain, plans for a jet-powered commercial transport were underway before the end of World War II. On July 27, 1949, the De Havilland *Comet* 1 first took to the air. This revolutionary airliner had four 4500-pound-thrust jet engines buried in the root of its 20 degree swept wings and, with a crew of four, it flew 36 passengers seated four abreast, at a speed of 490 miles per hour. The *Comet* cruised high above the weather at over 35,000 feet, where

---

[19]General Dynamics Fighter Division was acquired in 1992 by Lockheed, which merged with Martin-Marietta in 1995 to become the Lockheed Martin Corp.

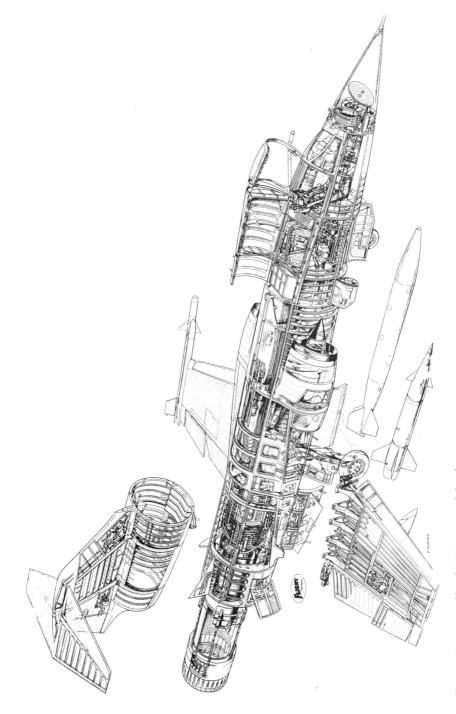

Figure 1.2.20   Lockheed F-104 structural details.

Note the multispar construction of the ultrathin wing.

Figure 1.2.21 Crack initiated by fatigue failure due to cycling the *Comet* 1 fuselage through repeated internal pressurization for the equivalent of 9000 flight hours.

the atmospheric pressure is one-fifth that at sea level. The first scheduled jet passenger service commenced in May 1952 with the flight of a British Overseas Airways Corporation *Comet* 1 from London to Johannesburg, including several refueling stops along the way. The promising career of the *Comet* 1 ended in tragedy in 1954 when two of them—the first in January, the second in April—mysteriously disintegrated while climbing to cruise altitude after departing Rome's Ciampino Airport. The fleet of *Comet*s was permanently grounded and an intensive investigation was launched into the cause of the twin disasters. At the Royal Aircraft Establishment in Farnborough, England, one of the grounded *Comet*s was immersed in a specially constructed water tank and subjected to around-the-clock cyclic internal pressurization of its fuselage, as well as flexing of its wings. After the equivalent of 9000 flight hours (about 3000 flights), the fuselage ruptured (Figure 1.2.21). A crack, which formed at the corner of a square window, propagated for eight feet through the 0.028-inch-thick skin, passing through window and fuselage frames until stopped by more massive bulkheads. A similar crack was subsequently found in a recovered portion of the cabin roof of the first downed *Comet*, the remains of which had been fished from the Mediterranean Sea and pieced back together in a hanger at Farnborough.

The cause of the *Comet*s' demise was low-cycle metal fatigue. On each and every flight, the fuselage was pressurized and depressurized as the airplane climbed to and descended from its lofty cruising altitudes. This initiated the formation and propagation of tiny microcracks in regions of the thin skin where stresses were concentrated, as at the corner of a square window. Over time, the tiny cracks coalesced until the crack reached a critical finite length, whereupon it ripped through the bulk of the skin, accompanied by an explosive decompression much like the bursting of a balloon. Although pressurized military and civilian aircraft were common at the time of the *Comet* failures in 1954, conventional airliners had neither the service ceiling nor the climb rate of the *Comet,* and military bombers and transports endured far fewer pressurization cycles per flight hour (and had fewer windows).

A British jet airliner would not carry passengers again for more than four years after the government forbade the *Comet*s to fly. Meanwhile, the Soviet Union introduced its first passenger jet to a surprised West when the prototype Tupolev Tu-104 arrived in London on a diplomatic mission in March 1956. This twin-jet, 560-mile-per-hour derivative of the Tu-16 *Badger* bomber carried 50 passengers and a crew of five and had about the same range as the *Comet* 1 (1,700 miles). The Tu-104 was the world's only operational commercial jet airliner until October 1958, when the much-improved, long-range, 67-passenger De Havilland *Comet* 4 entered regular service.

In 1950, Boeing commenced design of the prototype of a jet-powered transport to serve in one version as a civilian jet airliner (the 707) and in the other as a military tanker (KC-135) to replace its propeller-driven KC-97 (the civilian version of which was the *Stratocruiser*). Like Boeing's six-engine B-47 and eight-engine B-52 bombers, the 707 was to have its four turbojets slung beneath the wing. Boeing chose a skin thickness over four and a half times that of the *Comet* 1, and the windows were rounded to reduce stress concentrations. Fail-safe design features included a multiple load path structure. In the wings, multiple splices were employed, with splice members heavy enough to arrest dynamic crack propagation, thereby limiting the length of cracks that could occur. In the fuselage, titanium tearstraps were riveted to the skin under the circular frames to block crack propagation. Throughout the structure, materials were selected that had high fatigue life, slow crack propagation rates, and high tear resistance (fracture toughness). The fuselage, designed to provide six-abreast seating for ultimately up to 179 passengers, was hydro-fatigue tested for 50,000 cycles without failure. In October 1958, three weeks after the *Comet* 4's first operational flight, the 570-mile-per-hour Boeing 707 entered service as the United States' first commercial jetliner and the world's largest airliner at the time. The four-jet Douglas DC-8 and Convair 880 airliners followed the 707 into commercial service in September 1959 and May 1960, respectively.

To design against fatigue, two structural concepts evolved. A "fail-safe" structure is one like the 707 in which primary load-carrying members are duplicated, so that if one should fail, the load is redistributed to the remaining structure in such a way that a disastrous chain reaction will not occur. A "safe-life" structure is designed to last a given period of time before a major failure could occur. It is assumed that cracking will develop in regions deemed particularly susceptible to fatigue, and those parts of the structure are designed such that the cracks will not develop into fatal flaws before they are detected during regular inspections.

The effort to combat fatigue also led to the development of better materials for aircraft construction. The copper-rich aluminum alloys[20] tend to be resistant to crack propagation, but are of relatively low strength, whereas the high-strength zinc-rich alloys may not be as resistant to fatigue. For example, on the Boeing 707, 2024 aluminum was used for the wing lower surface, the wing spars, the fuselage skin, and the empennage skin and spars. On the other hand, 7178 aluminum, having superior yield strength in compression, was employed in the upper wing surfaces and other regions of low tensile stress.

Welding (except in aluminum), bonding, chemical etching, and machined integrally stiffened skin panels are aircraft manufacturing techniques that have been developed to reduce weight and cost, as well as sources of fatigue failure. Welded aluminum has poor fatigue life (the welds crack) and is rarely used. All of these methods eliminate rivet holes, at which stress concentrations occur.

The United States' first supersonic bomber was the Convair/General Dynamics B-58 *Hustler*, which first flew in November 1956 and entered service in 1959. The four-engine, delta-wing B-58 was capable of sustained flight near Mach 2, at which temperatures as high as 250 ° F develop on parts of the skin due to aerodynamic heating. At other regions, the skin temperature was that of the atmosphere at cruise altitude, about –70 ° F. The airplane was designed to accommodate these temperature extremes. To provide the necessary strength and rigidity and prevent skin buckling while keeping weight to a minimum, extensive use was made of aluminum honeycomb sandwich construction, adhesive bonding, and the thinnest possible metal gauges (as thin as 0.015 in.). The essential features of a honeycomb sandwich panel, whether curved or flat, are illustrated in Figures 1.2.22 and 1.2.23. Thin ribbons of material are bonded together to form the hexagonal honeycomb pattern of the core. Facing skins are bonded to each side of the core, preventing the skins from wrinkling or buckling under in-plane compressive load. The cost of manufacturing the extremely efficient, all-aluminum honeycomb skin employed throughout the B-58 "bonded bomber" was very high. Indeed, the airframe was said to be literally "worth its weight in gold." The fragile facings were susceptible to accidental damage by maintenance activities and by corrosion. Field repair of the thin sheets

---

[20] In the United States, the major aluminum producers use a four-digit numbering system for wrought (i.e., rolled, extruded, drawn, or forged) aluminum alloys. Most commonly used in aircraft are the 2XXX series, in which copper is the major alloying element, and the 7XXX series, in which it is mostly zinc along with some copper. Both series contain magnesium as well, and some 7XXX alloys also have small amounts of chromium.

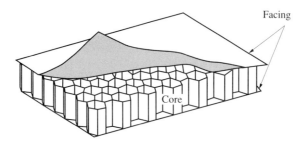

Figure 1.2.22 Honeycomb sandwich construction.

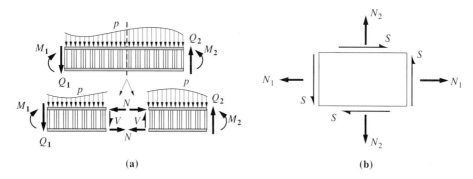

(a) (b)

Figure 1.2.23 (a) In a honeycomb sandwich skin, the core sustains transverse shear load *V* and the facings carry the tensile and compressive bending load *N*. (b) The facings also resist the shear *S* and normal *N* loads applied in the plane of the skin.

was time consuming and specialized in nature, requiring adhesive bonding under pressure at elevated temperature. Routine maintenance was slowed considerably because load-bearing access panels had to be attached with close-tolerance structural screws. The many problems associated with field maintenance of the honeycomb structure contributed to the early retirement of this advanced aircraft in 1969.

In September 1964, the first of two[21] North American Aviation XB-70A *Valkyrie* research aircraft made its first flight. This imposing, six-engine, delta-wing bomber was designed for Mach 3 cruising at 80,000 feet. At that speed, aerodynamic heating causes the temperature of portions of an airframe to reach 640 °F, and the entire airframe was designed to "soak" at around 530 ° F, far above the service temperature of any aluminum alloy. Therefore, for the airplane's forward fuselage, North American turned to new titanium alloys for the skin and frames and high-strength H-11 steel for the longitudinal stiffeners. For the intermediate fuselage and wing structure, brazed stainless steel honeycomb sandwich panels—which comprised 68 percent of the total airframe—formed the skin, engine inlet duct walls, and the frames and bulkheads. Tubular H-11 steel truss members supported the panels over the roof of the engine air-intake duct. The wing spars were titanium shear panels with a sine-wave web. In the aft fuselage, H-11 steel was used for the frames, longerons, and fittings, and the skin was made of titanium alloy. Most of the joints were welded, and the single weld joining each wing to the fuselage was over 80 feet long.

---

[21]Actually, three aircraft were built. The first one became a static test airframe because leakage of the integral fuel tanks could not be stopped.

Design and development of the XB-70 structure was supported by one of the first extensive digital computer implementations of matrix structural analysis. All major components of the airframe were investigated using the force method of linear elasticity. The three basic structural elements used to model the XB-70 structure were the rod, shear panel, and built-up beam. The procedure was written in assembly language and ran on an IBM 7094 with 192 kilobytes of core memory, 22.8 megabytes on hard disk storage, plus magnetic tape storage.

The first application of major finite element analysis to large structures was done in support of the design and development of the Boeing 747 jumbo jetliner, which began in the mid-1960s. Major portions of complex 747 substructures were analyzed by means of the Boeing finite element computer programs to obtain correct strength margins. The improved analysis capabilities, with their test-verified results, spurred the development of many new design concepts. Wing-body intersections and aeroelastic tailoring of the 747's nacelle structure to reduce flutter are examples of complex analyses that often established the state of the art. The 747, which first flew in February 1969, was largely of conventional construction and materials, yet it introduced the first major use of composite materials, in the form of fiberglass honeycomb skin on the rudder, elevators, and ailerons.

In the 747, as in other contemporary commercial and military aircraft, conventional aluminum-honeycomb sandwich panels were employed in the leading and trailing edges of the vertical and horizontal stabilizer and for the flaps and spoilers. By the mid-1960s the use of honeycomb sandwich panels or full-depth honeycomb construction had become common in secondary, thin-section substructures, because of its high strength-to-weight ratio, rigidity, and resistance to buckling and sonic fatigue. Figure 1.2.24 illustrates a typical honeycomb sandwich construction scenario.

Although the use of plastic composite materials in aircraft structures began in the mid-1960s, composites in general have been in use for quite some time. Note that composites consist of two or more different materials, and they exhibit properties that the individual constituents do not possess. As a matter of fact, wood is a composite of cellulose fibers and a matrix of lignin, a natural complex polymer that binds to the cellulose fibers to harden and strengthen the cell walls. We have already noted the 1940 De Havilland *Mosquito*'s fuselage was one of the first applications of high strength-to-weight ratio composite construction—spruce facings on a balsa wood core (see Figure 1.2.18). Metal-honeycomb sandwich panels might be classified as composites in which both the honeycomb and facings are made from thin gages of the same material (e.g., aluminum in the B-58 and stainless steel in the XB-70).

Compared to thick specimens, thin hair-like fibers of a material are much stronger, because the minute fiber cross sections contain far fewer microscopic flaws. Pioneering work on the strength of thin fibers of glass was carried out in the 1920s by A. A. Griffith and associates at the Royal Aircraft Establishment in England. The tensile strength of a glass fiber is nearly 200,000 pounds per square inch, which is comparable to that of the highest-strength steel alloy. However, as Griffith showed, the slightest amount of surface abrasion lowers this strength significantly. Fiberglass originated in the 1930s, after organic coatings were developed to protect the fiber surfaces, so that glass

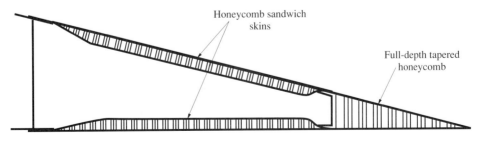

Figure 1.2.24   Typical use of honeycomb construction in thin sections, such as the trailing edge of an aileron, flap, elevator, or rudder.

could be spun and woven into thread, felt, and cloth, just like natural textiles, with little strength degradation. The fibrous glass material can be impregnated with liquid polymer resin and molded into a desired shape, after which the resin is cured (hardened) by the addition of a catalyst and often the application of heat. The resin matrix glues the fibers together and gives the composite structure its shape, as well as compression and shear strength. On the other hand, it contributes very little in the way of tensile strength or stiffness, which is the job of the fibers. The matrix typically constitutes 30 to 40 percent of the weight of a composite material.

One of the first aircraft applications of fiberglass/polyester composite appeared during World War II in the form of strong, durable radomes. The glass, of course, offered radar penetration. After the war, fiberglass went on to become commonly used in the manufacture of boat hulls, car bodies, swimming pools, bathroom fixtures, and numerous other molded, shell-like structures. Although fiberglass is relatively inexpensive, its modulus of elasticity is less than that of aluminum, making it too flexible for use in the primary structure of high-performance airplanes and space vehicles. It is also process sensitive. Environmental conditions, labor skill, and material variations can result in a large uncertainty in the net properties of the final product.

The pursuit of stiffer, stronger composite materials for aerospace applications led in 1958 to a process for producing boron fiber. Boron fiber is as light as aluminum and 20 percent stiffer than steel, and its tensile strength is twice that of glass fiber. It is also very expensive to produce. A 24-ply boron composite doubler was bonded to the surface of the steel wing pivot fittings on the General Dynamics F-111 variable-geometry fighter[22] to provide reinforcement in a critical stress location. This early application of plastic composites to production aircraft was accompanied on the same airplane by a carbon-fiber composite fairing to provide inspection access to the boron doubler.

Carbon fibers first appeared as filaments in electric lamp bulbs in the 1870s. Thomas Edison made his lamp filaments by carbonizing fibers of bamboo, but by the 1920s, incandescent lamp filaments had come to be made of metal. Work on carbon fibers was not revived until the early 1960s, when a process was developed for producing carbon fibers (actually, chains of ring-like graphite molecules) from filaments of the synthetic polymer polyacrylonitrile (PAN). Later, it was found to be cheaper to use pitch (tar) fibers as the raw material. Using epoxy resin as the matrix, carbon-fiber-reinforced plastics (CFRP) have become the most commonly used composite materials in aerospace vehicles. The carbon fibers are as stiff and strong as steel, but lighter in weight than aluminum. Epoxy resins are superior to polyesters in their stiffness and adhesiveness and in the fact that they shrink very little during curing (hardening). One of the first applications of graphite/epoxy composites to primary structures was in the McDonnell-Douglas AV-8B *Harrier* vertical/short-takeoff attack aircraft, which first flew in November 1978. The entire wing, horizontal tail, and forward fuselage of the AV-8B, which comprise 26 percent of the structural weight of the airplane, are of CFRP.

The *aramids* are modern grades of polyamides, synthetic polymers that have been around since the 1930s and are more popularly known collectively as nylon. *Kevlar* and *Nomex* (trademarks of the Du Pont Corporation) are aramids produced in the form of fibers. Kevlar fiber, which has been used for structural applications since the early 1970s, is lighter and cheaper than carbon, and it is tougher, with a better impact resistance. However, its relatively low compressive strength has limited its use in primary aircraft structures. Kevlar is also sensitive to ultraviolet light and degrades rapidly under continuous exposure. The aramids are known for their resistance to moderately high temperatures, and nonwoven Nomex felt is used for the strain isolation pads that lie between the aluminum skin and thermal protection tiles on the Space Shuttle Orbiters. Once ignited, Kevlar burns fairly well, so it cannot be used as a material for airliner interiors.

The 1968 Windecker *Eagle* was the first commercial production aircraft with an all-composite airframe. On January 1, 1981, the prototype Lear Fan 2100 business aircraft made its first flight. Designed by William P. Lear (designer of the classic 1964 twin-engine executive *Learjet*), this promising 77-percent CFRP, eight-passenger, twin-turboshaft, single-pusher propeller airplane did not complete FAA certification due to structural failures, and

---

[22] First flight: December 1964.

so did not enter production. The Raytheon Beechcraft *Starship 2000*, which first flew in February 1986, was the second all-composite commercial airplane to go into production (Figure 1.2.25). The twin-engine *Starship* cruises at 390 miles per hour, with a crew of two and eight passengers in a cabin not much smaller than that of the DC-1. It is an innovative design, characterized by a variable-sweep canard (forward-mounted elevator) and an aft-mounted 24 degree swept wing with a vertical stabilizer/rudder at each wingtip. The pusher turboshaft engines are installed on the wing inboard trailing edges. Seventy-two percent of the airframe weight is made up of composite materials. The structure consists mainly of large pieces bonded together with epoxy adhesives and a minimum of fasteners. There are some fiberglass/epoxy and Kevlar/epoxy details in the aircraft, but the majority of the structure is solid graphite/epoxy laminate or sandwich panels with a Nomex honeycomb core and graphite/epoxy face sheets. Most of the composite parts were laid up by hand and cured in an autoclave. Conventional metal alloys are used in the landing gear, engine mounts, pivot pins, wing-to-winglet fittings, and control-surface hinge brackets.[23]

Figure 1.2.25 The Raytheon/Beech *Starship 2000* had an all-composite airframe.

The strength-to-weight and stiffness-to weight ratios of fiber-reinforced plastics exceed those of the conventional metals of aircraft construction, which translates into increased payload and/or range for airplanes of composite design. Another feature of advanced composites is *aeroelastic tailoring*. Using the directional properties inherent in fiber-reinforced composites, one can design a structure to be stiffer in one mode of deformation than another. This is demonstrated in the Grumman[24] X-29A forward-swept wing research aircraft (Figure 1.2.26), which first flew in December 1984. This unique supersonic airplane has a close-coupled, variable-incidence canard just ahead of the very thin (5 to 6.7 percent) supercritical wing, which is swept forward 35 degrees. Forward-swept wings are not new. August 16, 1944, saw the maiden flight of the German Ju-287 swept-wing turbojet-powered bomber. Its wing was swept forward 18 degrees so that the center section would not penetrate the bomb bay. The wing of the 1964 West

---

[23] Unfortunately, the *Starship* was unable to compete financially with the conventional aluminum Beech *King Air* (which has the same number of seats), and *Starship* production ceased in 1995. This may be offered as some evidence that composites are not yet cost competitive. The time gap between the Windecker *Eagle* and the *Starship* is indicative of the technical uncertainty about composites.

[24] Grumman subsequently merged with Northrop to form the Northrop Grumman Corp.

Figure 1.2.26   The forward-swept wing of the X-29A technology demonstrator was aeroelastically tailored by means of graphite/epoxy wing covers.

German HFB 320 *Hansa* business jet also has an 18 degree forward sweep, in order to avoid penetration of the pressurized cabin by the carry-through structure. In 1947, during its post-war exploration of swept-wing technology, the Soviet Union flew the Tsybin LL-2 air-launched research glider, which had a 30 degree forward sweep configuration.

Whereas the stall of an aft-swept wing begins at the tip (where the ailerons are located), forward-swept wing stall begins inboard. Thus, forward-swept wing aircraft have better low-speed handling characteristics. However, at high speed, bending of the forward-swept wing induces an increase in the angle of attack, which in turn increases the loads in the wing. To counteract this *structural divergence* with a conventional structure would require making the wing thicker, which of course would add weight and drag. On the other hand, the angle of attack of aft-swept wings *decreases* at high speeds, so they do not have the divergence problem. Composite materials can be used to endow the wing surfaces with directional stiffness that yields a coupling between wing bending and wing twisting, which reduces or eliminates divergence. On the X-29A, the wing covers are made of 12-inch-wide 0.00525-inch-thick graphite/epoxy tape in a $0°, 90°, \pm 45°$ layup of plies to produce an orthotropic laminate. The 45 degree plies (−45 degrees in the left wing, +45 degrees in the right one) are oriented 9 degrees forward of the 40 percent chord axis, becoming the load-carrying plies and providing the bending-torsion coupling that prevents divergence. Mechanical fasteners join the composite wing covers to the substructure. The rest of the X-29A airframe, including the wing ribs and spars, is of conventional light metal alloy construction, with full-depth aluminum-honeycomb sandwich used for the rudder and the strake flap.

The Boeing 757 and 767 commercial airliners, which entered service in 1983 and 1982, respectively, have advanced composites (graphite or graphite/kevlar hybrid) in 30 percent of their wetted area. Six percent of the weight of the McDonnell-Douglas MD-11 wide-body aircraft is composites. Fifteen percent of the weight of the Boeing 777, which entered service in 1995, is carbon fiber composites. They are not only employed for control surfaces and other secondary structure, but they also form the primary structure in the horizontal and vertical stabilizers, as in the Airbus A330/340.

The use of advanced composite materials in airframes is expected to increase and to include more of the primary structure of both civil and military fixed and rotary wing aircraft,[25] since one sure way of improving aircraft per-

---

[25] Plastic composites are susceptible to impact damage and are more costly and time-consuming to repair than conventional aluminum skin, if the damage is detected. Dropping a tool on a graphite wing panel may yield no visible scar, yet may severely damage the internal plies or honeycomb. Repairability is a matter of serious concern to the airlines.

formance is to reduce structural weight. In the 1950s, weight savings were achieved by introducing high-strength aluminum alloys (the 2000 and 7000 series), titanium for high-temperature applications, and ultrahigh-strength steel. The B-58 *Hustler* represented structural efficiency carried to the extreme (only 16.5 percent of its takeoff gross weight was structure), the price of which could not be borne in the civil marketplace. In the 1960s, boron/epoxy and graphite/epoxy composites showed a potential weight savings of 20 percent over aluminum. Current materials now finding their way into aircraft structures are aluminum/lithium alloys (offering high strength, lower density, and greater stiffness than conventional aluminum alloys), aramid/aluminum laminates (ARALL), metal matrix composites, and improved organic matrix composites, including stitched graphite and thermoplastics.

1.3   NATURE AND COMPOSITION OF MODERN AIRCRAFT STRUCTURES

Figures 1.3.1 and 1.3.2 reveal the internal complexity of today's large commercial airliners. The major parts of a conventional fixed-wing aircraft are the wing, fuselage, empennage (tail), engine mounts, and undercarriage (landing gear). Most of the structure is covered by a thin skin, which gives the airplane its streamlined aerodynamic shape while helping to distribute a major portion of the internal loads. In flight, the wing supports the weight of the airplane (including that of the wing itself) plus maneuvering loads and gust loads. On the ground, landing loads and the weight of the airplane at rest are borne entirely by the undercarriage, the main wheels of which may be located in the wing, which thereby repeats its airborne load-bearing role. The horizontal tailplane with its moving elevator and the vertical tailplane (fin) with its pivoting rudder provide longitudinal and directional stability and control, in combination with the ailerons on the outboard trailing edges of the main wing. Since the mid-1950s, supersonic military airplanes have been equipped with a one-piece, all-moving horizontal tailplane with no elevator. All-moving vertical fins have also been employed (e.g., in the North American A-5 *Vigilante*, the Mach 3-*plus* Lockheed SR-71 *Blackbird*,[26] and the Lockheed F-117A *Night Hawk*[27] stealth fighter). In addition to ailerons, the wing may have movable trailing-edge flaps and leading-edge slats to increase lift (and therefore drag) at low speeds, as well as spoilers to "kill" lift (after touchdown, for example). Delta-wing aircraft usually have no horizontal tail,[28] and the elevons on the trailing edge of the delta planform serve the dual role of aileron and elevator, as well as flap. Engines may be mounted on the wings, on the fuselage, or both. Propellers are usually placed in the "tractor" position, ahead of the piston or turbine power unit. Notable exceptions include the giant Consolidated-Vultee B-36 *Peacemaker*[29] and the more recent Beech/Raytheon *Starship* previously discussed. Jet engines (efficient *turbofans* nowadays) are either housed within the aft fuselage (most military fighters) or mounted on the aft fuselage (commercial and general aviation jets). With few exceptions, wing-mounted jet engines on modern aircraft are slung beneath the wing in streamlined nacelles.

The major structural elements of a conventional semimonocoque fuselage are the transverse frames and the longitudinal stringers. The skin is attached (usually riveted) to both, and frames and stiffeners may themselves be fastened together where they intersect. *Bulkheads* are major transverse members that are more massive than a typical frame. *Pressure* bulkheads fill the entire fuselage cross section. Primary longitudinal load-bearing members are known as *longerons*. Large low-wing aircraft also have a *keelson*, a massive longeron that runs along the bottom of

---

[26] First flight: December 1964.

[27] First flight: June 1981.

[28] The delta-winged MiG-21 (first flight, June 1956) had a horizontal tail.

[29] First flight: August 1946. In 1949, the "D" model of the B-36 appeared with a pair of 5,200-lb-thrust turbojets added beneath each outer wing section. These jets, together with the six 3,800-horsepower piston pusher propeller engines, gave the 10-engine bomber a top speed of 439 miles per hour and a bomb load of 84,000 lbs (over four times that of a B-29 *Superfortress* and twice that of a B-2). Consolidated-Vultee became the Convair Division of General Dynamics in 1954. The last B-36 was retired from the USAF Strategic Air Command in February 1959.

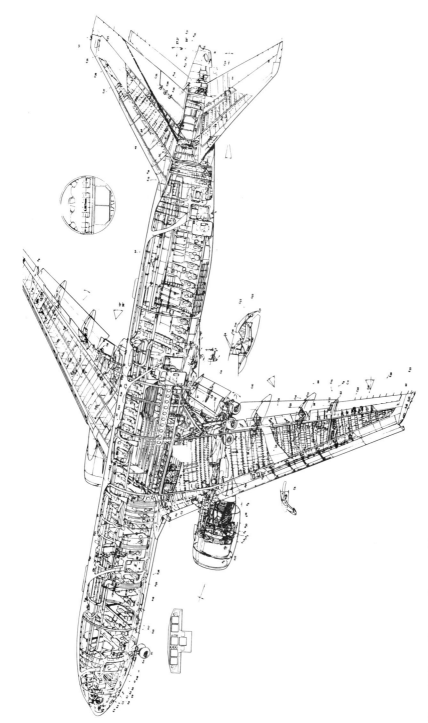

Figure 1.3.1 Cutaway drawing of the Boeing 767.

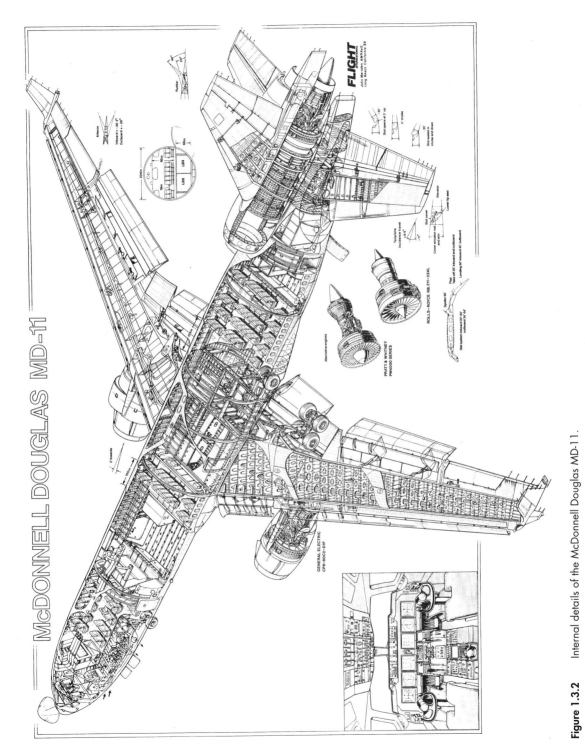

Figure 1.3.2 Internal details of the McDonnell Douglas MD-11.

the fuselage. Its purpose is to transmit bending loads across the central gap created by the landing gear wells and the wing carry-through structure. Bonded doublers or frames surround cutouts (e.g., windows and doors) to pick up and transmit the loads that would otherwise have been carried across the opening by the absent semimonocoque structure. Where the fuselage of a large transport contains mostly empty space, that of a typical modern military jet is crammed with engines, fuel cells, avionics, weaponry, and other gear. All of this must be accessed through numerous removable panels and access doors—so many, in fact, that a modern fighter fuselage cannot be considered as semimonocoque. Sizeable longerons and frames must be relied upon in stealthy, blended wing-body designs.

Ribs and spars are the major internal structural components of a wing. Like fuselage frames, the ribs define the shape of the attached envelope of skin, which is anchored as well to many spanwise stringers (stiffeners) on the upper and lower surfaces. Wing spars play much the same role as longerons in the fuselage, transmitting most of the bending load arising from the wing's support of the aircraft's weight. The vertical webs of the spars, in combination with the wing cover skins, act as torque boxes to resist the twist that also accompanies lift. The stiffeners pick up some of the bending load in both the wing and the fuselage, but their primary purpose is to aid the ribs and frames in stiffening the skin against buckling. Spars, ribs, and stringers also comprise the substructure of horizontal and vertical stabilizers.

The landing gear rank as one of the substructures most crucial to aircraft performance during the periods between touchdown and takeoff. During the early days of flight, "conventional gear" consisted of two main wheels and a small skid or wheel at the tail. (The Wrights were unique in using no wheels at all!) Suspended beneath the fuselage during flight, the landing gear produced drag, but it was small in comparison with that of the maze of wing struts and bracing wire common to the boxy, slow-flying biplanes. At the speeds made possible with the advent of more powerful engines and streamlined design by the 1930s, the drag of fixed landing gear became significant, and a retractable undercarriage was dictated in the name of aerodynamic efficiency. Today, fixed gear is found on only a few propeller-driven light utility transports and on low-speed general aviation airplanes. Tricycle undercarriage, in which the third wheel is located at the nose rather than under the tail—so that the airplane sits level—first appeared on production aircraft near the beginning of World War II. Some pre-1940 examples of tricycle gear aircraft (with their maiden flights) were the single-engine Bell P-39 *Airacobra* fighter (April 1938), the four-engine Douglas DC-4E transport (June 1938), the twin-engine Douglas A-20 *Havoc* medium bomber (October 1938), the twin-engine Lockheed P-38 *Lightning* fighter (January 1939), and the four-engine Consolidated B-24 *Liberator* heavy bomber (December 1939).

The appearance of jet propulsion signaled the end of "tail-dragging" fighters.[30] It also heralded some landing gear placement innovations, as in the United States' Boeing B-47 *Stratojet* (1947) and B-52 *Stratofortress* (1952), and the Lockheed U-2 (1955); the Soviet Union's Myasishchev Mya-4A *Bison* (1953) and Yakovlev Yak-28P *Firebar*[31] (1961); France's Sud-Ouest SO-4050 *Vautour* II-B (1956); and Britain's Hawker Siddeley V/STOL *Harrier* (1966). The retractable main landing gear on all of these airplanes was located in tandem (like a bicycle) in the fuselage, while small outrigger wheels were outboard on the wings for lateral stability and were retracted into the wing or an engine pod (except for the U-2, which drops its outriggers on takeoff).

Retractable landing gear are among the most complex of the mechanical systems on a modern airplane. The landing gear must be able to absorb several times the entire gross weight of the airplane—without undue shock to the airframe or its contents—while comprising less than 10 percent of the structural weight. Furthermore, it must quickly fold up into a compact space in the wing or fuselage for storage during flight and must as quickly and reliably deploy for landing. Over the years, thousands of landing gear systems have been designed to serve the multitude of purposes of many different aircraft. Figure 1.3.3 illustrates some of the features of a typical retractable undercarriage. The pur-

---

[30]Actually, the first versions of the German Me-262 were tail-draggers, with no propwash to help lift the horizontal tail off the ground during takeoff. On short, rough airfields, pilots of these early models had to resort to the tricky business of tapping the main landing gear brake pedals halfway down the runway to rotate the airplane out of the nose-up, high-drag attitude and gain takeoff speed.

[31]"Bison" and "Firebar" were NATO code names.

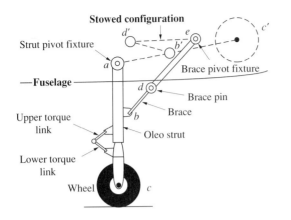

Figure 1.3.3 Typical components of a simple retractable landing gear.

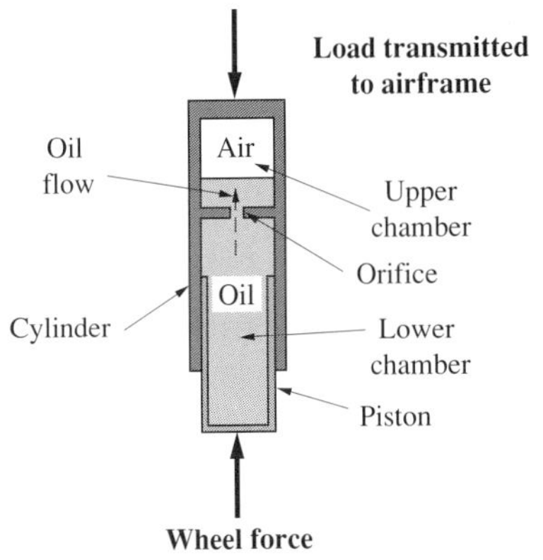

Figure 1.3.4 Basic features of an oleo strut (during touchdown)

Wheels of the main gear are equipped with disk brakes.

pose of the oleopneumatic shock absorber, or *oleo strut*, is to ease the loads on the airframe during landings and to cushion impact. Figure 1.3.4 shows the interior of an oleo strut, reduced to its bare essentials. During touchdown, an airplane settles onto its landing gear with an initially downward velocity component. To arrest the sink rate quickly, the runway must briefly exert a net upward force, through the wheels, that is greater than the weight of the airplane. (When the sink rate ceases, the airplane is in equilibrium, and the net upward force from the ground precisely *equals* the airplane's weight.) The vertical ground load pushes the piston into the cylinder of the oleo strut, which forces oil through the small orifice into the upper chamber, thereby further compressing the already high-pressure air above the oil. The metering effect of the small orifice causes the load in the strut to remain roughly constant throughout the compression stroke, after which the compressed air pushes the oil back through the orifice into the lower chamber, dropping the pressure in the cylinder (and hence the transmitted force) as it expands. The net effect is that much of the kinetic energy associated with the vertical component of velocity is rapidly dissipated in the shock struts. The energy is not stored and released in the form of repeated bounces, as is true of a coil spring. Design details prevent the compressed air in the upper chamber from forcing the piston out of the cylinder in the absence of ground loads.

High-strength steel alloys are used for the primary landing gear structure (e.g., the strut), while lighter aluminum alloys may be employed where the loads are relatively low (e.g., braces). Corrosion protection—a matter of concern throughout an airframe—is clearly required for the undercarriage. The landing gear attachment points (spars and bulkhead ribs on the wing, frames and beams in the fuselage) must be "hard," that is, designed to absorb the concentrated load and distribute it efficiently into the less massive surrounding structure.

The same is true where engines are attached to the airframe. These structures must resist the weight and moment of the powerplant, as well as the thrust, torsion, and vibration it produces. As with landing gear, there is a huge variety of engine mounts. The design clearly depends on the kind of engine it supports (piston, turboprop, turbojet) and the position where the engine is mounted; nearly all piston engines have four mounts. The layout typical of piston engine mounts is shown in Figure 1.3.5. Tubular-steel truss structures are common in this case. Rubber bushings or pads are required to minimize the transmission of engine vibration to the fuselage.

Jet engines are invariably attached to the airframe at just three points, by means of a support system that is statically determinate. This ensures that airframe deflections cannot transfer to the engine and that thermal expansion of the engine can occur freely. It also means the engine can be removed and installed quickly. Vibration is of little concern. Engines mounted inside the fuselage attach directly to the fuselage framework via

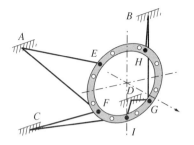

Figure 1.3.5 Essential features of a welded-tube truss piston-engine mount.

It is bolted to the firewall at *A, B, C,* and *D.* The rear of the engine attaches at *E, F, G,* and *H.* Old radial engines were bolted to the thick metal ring.

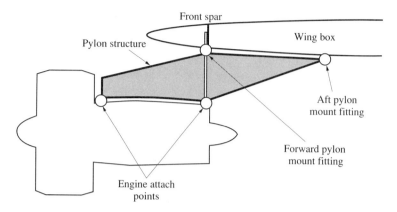

Figure 1.3.6 Turbofan engine wing-pylon mount configuration.

links and trunnions. Those mounted below the wings or on the aft fuselage are housed in protective streamlined nacelles and are fastened to pylons that in turn are joined to the wing or fuselage structure. The pylons are built-up, stiffened-web structures, or truss structures, or perhaps a combination of the two (see Figure 1.3.6).

1.4 IDEALIZATION OF AIRCRAFT STRUCTURES FOR PRELIMINARY ANALYSES

In order to calculate the strength or stiffness of a structure, we must make simplifying assumptions about its geometry and material properties, as well as about the way it is loaded and supported, so that established methods of engineering mechanics can be used. Rational simplifying assumptions yield a mathematical model that accurately reflects the essential features of the real structure. In engineering practice, we strive to employ mathematical models that are already widely accepted, so that the formulas or handbook data we choose can be used with confidence. Obviously, results based on erroneous or oversimplifying assumptions may be useless. On the other hand, calculations may become unwieldy if we try to include too many details in a model. It is important to endow the model with only the most essential features of the actual design.

In this text, we will focus on elastic structures that can be treated as trusses, beams, torsion bars, frames, or built-up assemblies of rods (stiffeners) and shear webs. The use of plate and shell theory is beyond our scope, except as it applies to simple pressure vessels and a few formulas for buckling. What we retain will be more than adequate for illustrating and implementing the basic methods of statically determinate and indeterminate analyses as they apply to typical aircraft structures in the preliminary design phase.

In preliminary design, we define the primary load paths while enforcing strength, stiffness, and weight constraints. This requires an understanding of how solids behave under load. The design effort also requires human judgement and insight, which become more crucial as considerations proceed to cost, manufacturability, safety, reliability, and maintainability. Eventually, computerized structural analysis methods—*computational structures technology* (CST)—may be called upon to verify and optimize complex, detailed design solutions. Analysis is done in support of, not in place of, the creative process of design: Design is the art of engineering.

EXERCISES

Write a paper on one of the individuals in the following list, or any other major contributer to the evolution of aircraft design. Give the date and place of birth, parents' names, and siblings. Describe the person's youth, early schooling, and interests. Trace the events that led to the individual's becoming involved with aviation. Trace the person through adolescence and into early adulthood. Indicate early jobs held, and the nature and place of any advanced schooling. Describe the contributions for which the individual is noted, and compare them with similar accomplishments of others, if any. Trace the person's adult life, immigrations, emigrations, and politics; describe his/her marriage, list the children. List failures, as well as successes.

Oleg Antonov

Walter Beech

Lawrence Dale Bell

Ronald Bishop

Louis Blériot

William Edward Boeing

George de Bothezat

Louis Charles Breguet

Ludwig Bölkow

George Cayley

Clyde V. Cessna

Octave Chanute

Juan de la Cierva

Paul Cornu

Glenn Curtiss

Marcel Dassault

Geoffrey De Havilland

Claudius Dornier

Donald W. Douglas

Anton Flettner

Henrich Focke

Anthony H. G. Fokker

Mikhail Gurevich

Ed Heinemann

Ernst Heinkel

Harry Hillaker

Stanley Hiller

Jiro Horikoshi

V. I. Ilyushin

Clarence L. "Kelly" Johnson

Robert T. Jones

Hugo Junkers

Charles Kaman

"Dutch" Kindelberger

Samuel P. Langley

William P. Lear

Otto Lilienthal

Alexander M. Lippisch

Paul MacCready

Glenn L. Martin

James S. McDonnell, Jr.

Willy Messerschmitt

Artem Mikoyan

Mikhail Mil

Reginald Mitchell

Sanford Moss

John K. Northrop

Hans von Ohain

Frank Piasecki

William Piper

Arthur Raymond

Alliot Verdon Roe

Burt Rutan

Igor Sikorsky

Andrei Nikolaevich Tupolev

Woldemar Voight

Chance B. Vought

Barnes Wallis

Helmuth Walter

Fred Weick

Richard T. Whitcomb

Frank Whittle

Robert J. Woods

Orville Wright

Wilbur Wright

Alexander Sergeivitch Yakovlev

Arthur Young

2

Statically Determinate Structures

CHAPTER OUTLINE

2.1   INTRODUCTION

The purpose of this chapter is to review and reinforce the principle of static equilibrium within the context of some basic types of aircraft structures. A structure may be defined as an assemblage of materials that is intended to sustain loads.[1] How well the intention is realized depends on the quality of the design, and that depends (among other things) on how well the shape of the structure and the properties of the selected materials accommodate the predicted internal loads. Therefore, it is important for a structural designer, in spite of—and aided by— digital computers, to develop a keen insight for predicting and visualizing load paths throughout a structure. The ability to do so largely depends on how well one has mastered the skills of sketching accurate free-body diagrams and properly applying the equilibrium equations to them, which will be one of our primary concerns here. Those prepared to move on to loftier topics can skip over or skim this chapter, as its contents are based almost exclusively on statics.

[1] J.E. Gordon, *The Science of Structures and Materials,* Scientific American Library, 1988.

A structure that is acted on by a set of external loads is in equilibrium if the vector sum of all of those forces ($\sum \mathbf{F}$) is zero and the vector sum of their moments about any point P ($\sum \mathbf{M}_P$) is also zero. Thus, for equilibrium we must have

$$\sum \mathbf{F} = 0 \quad \text{and} \quad \sum \mathbf{M}_P = 0 \qquad \text{[2.1.1]}$$

If, on the other hand, there is a net imbalance of forces and moments on the structure, we know from dynamics that the laws of motion require that

$$\sum \mathbf{F} = \dot{\mathbf{L}}_{c.m.} \quad \text{and} \quad \sum \mathbf{M}_{c.m.} = \dot{\mathbf{H}}_{c.m.} \qquad \text{[2.1.2]}$$

where $\dot{\mathbf{L}}_{c.m.}$ is the time rate of change of *linear* momentum of the structure's center of mass (*c.m.*) and $\dot{\mathbf{H}}_{c.m.}$ is the time rate of change of the structure's *angular* momentum about its center of mass (not just any arbitrary point!). The methods developed for analyzing problems of static equilibrium can be applied in dynamics situations by rewriting Equation 2.1.2 in the form

$$\sum \mathbf{F} + \left(-\dot{\mathbf{L}}_{c.m.}\right) = 0 \quad \text{and} \quad \sum \mathbf{M}_{c.m.} + \left(-\dot{\mathbf{H}}_{c.m.}\right) = 0 \qquad \text{[2.1.3]}$$

which looks like Equation 2.1.1 with the fictitious *inertial force* $-\dot{\mathbf{L}}_{c.m.}$ and *inertial couple* $-\dot{\mathbf{H}}_{c.m.}$ applied at the center of mass. Treating an accelerating object in this way—as though it were instantaneously in a state of "dynamic equilibrium"—is referred to as *D'Alembert's principle*. This approach is useful in the analysis of structural loads induced by aircraft maneuvers, landings, or gusts, as well as in the study of structural vibrations.

Structural vibration involves the complex interplay between externally applied loads, inertial loads, the elastic restoring forces that arise because of deformation, and damping due to dissipative mechanisms (e.g., friction and material hysteresis). A flexible structure prefers, or tends, to vibrate at one or more of its *natural* frequencies. If an oscillatory external load is applied with a frequency near one of these natural frequencies, *resonance* occurs, possibly accompanied by large-amplitude vibrations and internal stresses. The stresses and deflections arising from a load applied statically (i.e., very slowly) can be magnified significantly, if not catastrophically, if the same load is applied rapidly or cyclically. If the frequency of the applied load is less than about one-third of the lowest natural frequency of the structure, the inertial effects are negligible. In such cases, the driving forces may be treated as static loads.

Our focus in this chapter will be on *statically determinate structures* of the following types, which are typical of those found in aircraft:

- Pinned and rigid-jointed frames.
- Stiffened shear webs.
- Thin-walled beams and torque tubes.

Consideration of the last two classes of structures leads to the concept of *shear flow*.

In this chapter, the structures we discuss may be treated as rigid bodies incapable of deformation. This idealization means that material properties will not be required for computing the forces required for equilibrium. However, either by nature or by design, most real structures are statically *indeterminate*. For example, safety and durability require that aircraft structures contain extra members, so that if one of them should fail, the redundant element would provide an alternate load path to maintain structural integrity. To calculate the loads throughout a statically indeterminate structure, we must account for the fact that materials do deform under load. Coupling this *kinematics* of structures with the requirements of equilibrium underlies much of the remainder of this text. For the present, however, confining ourselves to statically determinate structures will allow us to review and become familiar with the essential, distinguishing features of the various types of structures listed, while keeping mathematical complexity to a minimum.

2.2 PLANE TRUSSES

A *truss*, also called a *pin-jointed frame*, is an idealized skeletal or "stick-like" structure composed of slender rods joined together by smooth pins at the joints, also called *nodes*. The joints of a truss may be pinned, as in Figure 2.2.1(a), or more likely welded or riveted together in the manner shown in Figure 2.2.1(b). If the rods of the truss are slender (very long compared to a characteristic cross-sectional dimension d), and if the centerlines (axes of centroids) of the joined members intersect at a common point A, then the pin-joint idealization (*ball joint* in three dimensions) is satisfactory. The pins must be located at the points of concurrency of the rod centerlines.

Ordinarily, external loads are applied only to the joints of a truss. Since, by definition, none of the smooth pins can apply a "couple" to the rods connected to it, the rods are generally two-force members. As such, each rod can only transmit a load that is parallel to its length and passes through the pin at each end. This means that a truss is a network of tensile and compressive forces, each having a known direction. Tensile forces pull on the joints and compressive forces push on the joints.

The simplest plane truss consists of three rods pinned together to form a rigid triangle, as shown in Figure 2.2.2(a). If j is the number of joints and m is the number of members, we see that for the triangular truss,

$$2j = m + 3 \qquad\qquad \text{[2.2.1]}$$

If we add two members and one joint to the triangle, as in Figure 2.2.2(b), this equality is maintained. We now have a rigid truss consisting of two triangles with a common side. Continuing in this fashion, we can generate plane trusses of arbitrary shape and size, as in Figure 2.2.2(c). The rigid triangular "building blocks" are evident in the larger truss, and the equality given by Equation 2.2.1 is maintained.

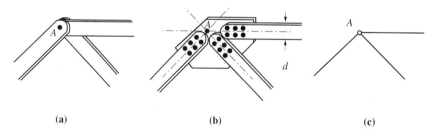

(a)   (b)   (c)

Figure 2.2.1   (a) Plane frame members held together by a single pin. (b) A riveted or bolted connection. (c) The idealized pin joint.

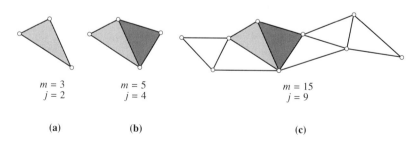

$m = 3$ $m = 5$ $m = 15$
$j = 2$ $j = 4$ $j = 9$

(a)   (b)   (c)

Figure 2.2.2   Stable trusses composed of triangular "building blocks."

Before we can apply external forces to a truss, it must somehow be supported. The supports must be capable of reacting to the applied loads in such a way that the truss will be in equilibrium. The truss must not be able to move as a rigid body. We can treat the entire truss as a free-body structure and use equilibrium equations to calculate the reactions at the supports. If the number of equilibrium equations equals the number of unknown reactions, the truss is *externally statically determinate*. In two dimensions, we can write precisely three independent equilibrium equations for a rigid body. Therefore, if the number of unknown reactions exceeds three, the truss is *externally statically indeterminate*.

A truss is *minimally stable* if it has the minimum number of rods required to support external loads and remain rigid. A truss composed of triangular subtrusses is minimally stable. If just one of the rods is removed, the truss will lose its rigidity and will become a mechanism. Rotation about one or more of the pins will occur, and the truss will collapse. A minimally stable plane truss is *internally statically determinate*. That means we can calculate the forces in all of the rods if we are given the external loads at the joints. Figure 2.2.3 shows three examples of minimally stable trusses. Observe that Equation 2.2.1 is satisfied in all three cases, even though the center truss is not built exclusively of triangles. These trusses are both internally and externally statically determinate. If we replace the roller supports with pin supports, a fourth unknown reaction will be added to each truss and they will become externally statically indeterminate.

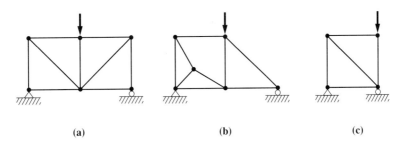

Figure 2.2.3   Examples of statically determinate, minimally stable trusses.

In Figure 2.2.4, there are three examples of unstable trusses. They are externally statically determinate, but they are not rigid. They will collapse if the loads are actually applied. Notice that Equation 2.2.1 is not satisfied: for each truss, $2j > m + 3$. These trusses can be made minimally stable by adding just one rod to each.

Figure 2.2.5 shows examples of internally statically determinate trusses ($2j = m + 3$) that are inadequately supported. The supports on the left truss cannot prevent rigid-body horizontal translation. One of the rollers should be replaced by a pin. The truss in part (b) of the figure seems to have the right number of supports.

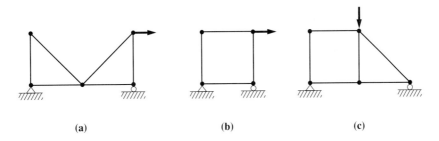

Figure 2.2.4   Examples of unstable trusses.

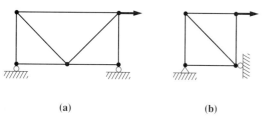

(a)   **(b)**

Figure 2.2.5   Examples of supports inadequate to restrain
rigid-body translation (a) and rotation (b).

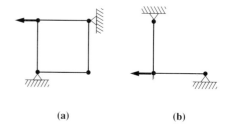

(a)   **(b)**

Figure 2.2.6   Unstable trusses made rigid by properly located additional supports.

However, the roller at the wall cannot exert the force required to balance out the moment of the applied force about the pin. The roller should be on the floor to prevent rigid-body rotation around the pin.

At the outset of a truss analysis, we know the applied loads. The problem is to compute the forces in the rods and the reactions at the supports. We can resolve each of the reactive loads into orthogonal components. Let r be the total number of reactions, and let j be the total number of joints. Pick any joint and isolate it as a free body. In two dimensions, we can write two equations of equilibrium for the joint. Doing this for every joint on the truss, we come up with $2j$ equations of equilibrium. If

$$2j = m + r \qquad\qquad\qquad \text{[2.2.2]}$$

and the supports restrain rigid-body motion, then the problem is statically determinate. For space trusses, the analogous expression is given by Equation 2.3.1.

Figure 2.2.6 shows how stable, statically determinate truss structures can be created from unstable rod assemblies by adding a minimum number of properly located supports instead of adding members. In both cases, Equation 2.2.2 is satisfied.

If the cross-sectional area of a truss member is A, then the axial load N applied by the smooth pins at each end produces a uniform normal stress (load intensity)[2]

$$\sigma = \frac{N}{A} \qquad\qquad\qquad \text{[2.2.3]}$$

on cross sections throughout the bulk of the rod (see Figure 2.2.7). To avoid mechanical failure (damage) of the rod, the value of the normal stress σ must remain within limits dictated by the strength of the material from which the rod is made. Furthermore, truss members that are in compression act like columns and may buckle, which is another form of failure to be avoided. In Chapter 12, we show that a slender pin-supported rod of length L buckles at a critical load N_{cr} given by the classic *Euler column formula*

$$N_{cr} = \frac{\pi^2 E I}{L^2} \qquad\qquad\qquad \text{[2.2.4]}$$

where E is Young's modulus, the property that indicates a material's resistance to deformation (see Chapter 3), and I is the area *moment of inertia*, a geometric property of the cross section. Table 2.2.1 lists several examples of this section property, which is discussed in detail in Chapter 4 on beams. Equation 2.2.4 is in accordance with what we know from experience, namely, that long columns of a given cross section buckle more easily than shorter ones.

| [2]Although the reader should have previously encountered the concepts of stress and strength, they are reviewed in detail in Chapter 3.

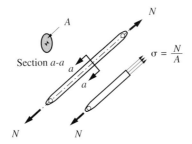

Figure 2.2.7   Uniform, uniaxial stress in a two-force rod element.

Table 2.2.1   Moments of inertia for some simple sections.

Cross section	![circle]	![square]	![circular tube] $t^2 \ll d^2$	![square tube] $t^2 \ll d^2$
Area	$\dfrac{\pi d^2}{4}$	d^2	$\pi t d$	$4td$
Moment of inertia	$\dfrac{\pi d^4}{64}$	$\dfrac{d^4}{12}$	$\dfrac{\pi t d^3}{8}$	$\dfrac{2td^3}{3}$

Figure 2.2.8(a) shows a truss for which $j = 7$, $m = 11$, and $r = 3$, so that, according to Equation 2.2.2, the structure is statically determinate. To calculate the axial force in every member of the truss, we can employ the joint method. This procedure requires isolating each joint as a point in equilibrium under the action of the internal loads applied by the attached rods and the externally applied loads, possibly including those due to supports. Therefore, at each node, we obtain $\sum F_x = 0$ and $\sum F_y = 0$. The seven pairs of nodal equilibrium equations, as a system of 14 equations, will yield values for the 11 unknown member loads and the 3 unknown reactions.

We can write all 14 equilibrium equations and then solve them with the aid of a linear equation solver on a personal computer or hand-held calculator. This might be called the *brute force* method. Alternatively, we can begin by drawing a free-body diagram of the whole structure, as shown in Figure 2.2.8(b). Using that sketch, we write the three equilibrium equations of the truss, $\sum F_x = 0$, $\sum F_y = 0$, and $\sum M_P = 0$, where P is any point in the plane of the truss. These three equations yield the unknown reactions X_4, Y_4, and Y_6. Note that these three truss equilibrium equations are not independent of the 14 joint equilibrium equations. In fact, we can use any 11 of the joint equilibrium equations to find the 11 member loads in terms of the reactions. The remaining three equations, involving just the reactions X_4, Y_4, and Y_6, will be equivalent to (i.e., linear combinations of) the three overall truss equilibrium equations.

Thus, as an alternative to the brute force approach, we first obtain the reactions from Figure 2.2.8(b):

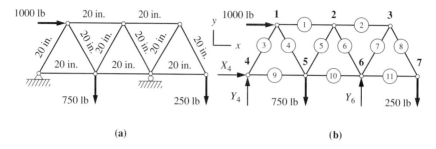

Figure 2.2.8 (a) Truss with loads and dimensions. (b) Truss as a free body, showing support reactions, and the chosen joint and member numbering scheme.

$$\sum F_x = 0: \qquad\qquad X_4 + 1000 = 0 \quad X_4 = -1000 \text{ lb}$$

$$\sum M_4 = 0: \quad 40Y_6 - 20 \times 750 - 60 \times 250 \;-\; (20\sin 60) \times 1000 = 0 \quad Y_6 = 1183 \text{ lb}$$

$$\sum F_y = 0: \qquad\qquad Y_4 - 750 + 1183 - 250 = 0 \quad Y_4 = -183.0 \text{ lb}$$

We then proceed from node to node, using the remaining 11 equilibrium equations to find the values of the member loads. If two unknown forces at most act at a joint, their values may be found by solving the two equilibrium equations at that point. By choosing the nodes judiciously, we can obtain all of the member loads "on the fly" as we work our way around the truss. In this case, joints 4 and 7 are candidate starting points for the solution process. Setting joint 4 in equilibrium yields the member forces N_3 and N_9:

$$\sum F_y = 0: \qquad\qquad 0.8660N_3 - 183.0 = 0 \quad N_3 = 211.3 \text{ lb}$$

$$\sum F_x = 0: \quad 0.5000(211.3) + N_9 - 1000 = 0 \quad N_9 = 894.2 \text{ lb}$$

We then proceed to node 1, where the two member forces N_1 and N_4 are the unknowns:

$$\sum F_y = 0: \qquad\qquad -0.8660(211.3) - 0.8660N_4 = 0 \quad N_4 = -211.3 \text{ lb}$$

$$\sum F_x = 0: \quad 1000 + N_1 - 0.5000(211.3) + 0.5000(-211.3) = 0 \quad N_1 = -788.8 \text{ lb}$$

In similar fashion, we obtain N_5 and N_{10} at node 5, N_2 and N_6 at node 2, and N_7 and N_{11} at node 6. At that point, we will have used up all but one of the independent equilibrium equations. We may employ it at either joint 3 or joint 7 to solve for the remaining member force, N_8.

The results of this joint method of analysis are summarized in Figure 2.2.9. Since we assumed at the outset that all of the member loads were tensile (acting away from the joints), a minus sign means the member is in compression. Although we assumed that the structure was rigid, truss members do deform when loaded; therefore, those in compression are in jeopardy of buckling.

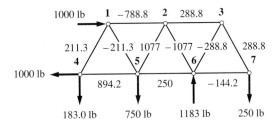

Figure 2.2.9   Solution of the truss problem in Figure 2.2.7.

Member loads are given in pounds (+ = tension, − = compression).

Table 2.2.2   Properties of round steel tubing.

Diameter, d (in.)	Wall Thickness (Gage), t (in.)	Area, A (in.^{2})	Moment of Inertia, I (in.^{4})	t/d	Weight (lb/in.)
0.250	0.022	0.01576	0.000103	0.0879	0.0045
	0.028	0.01953	0.000122	0.112	0.0055
0.375	0.028	0.03063	0.000462	0.0747	0.0086
	0.035	0.03739	0.000546	0.0933	0.0106
	0.049	0.06018	0.000682	0.131	0.0143
0.500	0.028	0.04152	0.001160	0.0560	0.0117
	0.035	0.05113	0.001390	0.0700	0.0145
	0.049	0.06943	0.001724	0.0980	0.0196
0.625	0.028	0.05252	0.002345	0.0448	0.0149
	0.035	0.06487	0.002833	0.0560	0.0184
	0.049	0.08867	0.003704	0.0783	0.0251

Example 2.2.1   All members of the truss in Figure 2.2.8 are to be fabricated from the same stock of thin-walled, round, steel tubing, the section properties of which are listed in Table 2.2.2. Select the lightest weight tubing for which the axial stress in any rod of the truss does not exceed 25,000 psi in tension or compression and the critical buckling load is not exceeded. For steel, $E = 30 \times 10^6$ psi.

Let us assume that the buckling criterion controls. Since all members of this truss have the same length, we examine Figure 2.2.9 for the largest compressive load, which is 1077 lb in rod 6. This force also happens to be the largest tensile load in the truss (rod 5). Therefore, sizing just member 6 will do for the entire structure. (Reality is rarely this simple!) From Equation 2.2.4, we have for rod 6,

$$1077 = \frac{\pi^2 (30 \times 10^6) I}{20^2}$$

so that, for the tube to be on the verge of buckling,

$$I = 0.001455 \text{ in.}^4$$

The area moment of inertia we select can be no smaller than this value. According to Table 2.2.2, the lightest weight tubing for which $I > 0.001455$ in.^{4} is the 0.625 in. diameter, 0.028 gage stock (0.0149 lb/in.). The maximum tensile and compressive stress in the truss for this choice of material is

$$|\sigma_{max}| = \frac{1077 \text{ lb}}{0.05252 \text{ in.}^2} = 20,500 \text{ psi}$$

This is well within the 25,000 psi limits.

An alternative to the joint method of plane truss analysis is the *method of sections*. Because this method avoids having to work through the entire structure, it is the method of choice whenever we need to find the force in just a few members of a statically determinate truss. After identifying the member of interest, we section the truss into two free bodies so as to expose the force in that member. We then write the equations of equilibrium for the free body on either side of the section and solve them for the unknown force. For example, to find the member forces in the center bay of the cantilever truss in Figure 2.2.10(a), make an imaginary vertical cut through that bay and isolate the portion of the truss to the left of the cut so as to form the free-body diagram in Figure 2.2.10(b). The three equations of equilibrium for this free body yield the three unknown member forces:

$$\sum F_y = 0 \quad \Rightarrow \quad N_{5,8} = \sqrt{2}P$$
$$\sum M_5 = 0 \quad \Rightarrow \quad N_{6,8} = 2P$$
$$\sum F_x = 0 \quad \Rightarrow \quad N_{5,7} = -3P$$

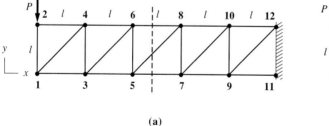
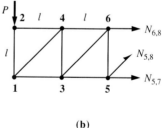

(a)   (b)

Figure 2.2.10   (a) Cantilevered truss with a transverse section through the center bay. (b) Free-body diagram to the left of the cut, revealing the member forces in that bay.

2.3 SPACE TRUSSES

Just as for a plane truss, a space truss (three dimensional) must be supported in such a way that rigid-body translation and rotation are prohibited. In three dimensions, an unconstrained body can translate in three orthogonal directions and can rotate about each of those three axes. In other words, there are six rigid-body degrees of freedom in three-dimensional space.

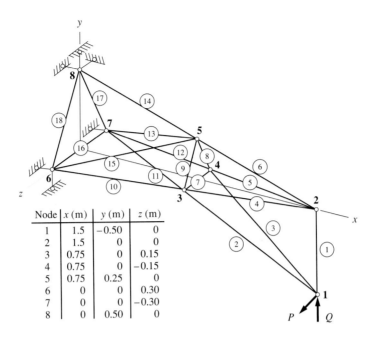

Node	x (m)	y (m)	z (m)
1	1.5	−0.50	0
2	1.5	0	0
3	0.75	0	0.15
4	0.75	0	−0.15
5	0.75	0.25	0
6	0	0	0.30
7	0	0	−0.30
8	0	0.50	0

Figure 2.3.1   A space truss that is both internally and externally statically determinate.

The nodal coordinates are given in the table.

Figure 2.3.1 shows a cantilevered space truss. The supports at the wall are represented by short links, which are rigid two-force members. The node to which a link is attached cannot move in the direction of the link. The three links at node 8 of the truss represent a ball joint (the three-dimensional version of the pin), which constrains the point in all three directions. The two links at node 6 represent a collar constrained to slide on the z-axis, so that only x and y displacements are prevented. The single link at node 7 represents a support that rolls on the y-z plane, preventing displacement in the x-direction. The supports at node 8 prevent rigid-body translation. The vertical support at node 6 prevents rotation around the x-axis, and the horizontal support restrains rotation around the line connecting nodes 7 and 8. Finally, the link at node 7 prevents rotation about the line connecting nodes 6 and 8. The truss is externally statically determinate because the six forces exerted by the links on the truss can be found in terms of the applied loads P and Q by means of the six equations of equilibrium ($\sum \mathbf{F} = 0$, $\sum \mathbf{M} = 0$). Furthermore, the truss is internally statically determinate since the number of rods (18) plus the number of reactions (6) equals 24, which is the total number of joint equilibrium equations (8 nodes, 3 equations per node). The truss is clearly minimally stable, and we can easily see the triangular subtrusses of which the structure is built. Observe that in three dimensions, Equation 2.2.2 is replaced by

$$3j = m + r \tag{2.3.1}$$

since there are three equations of equilibrium per joint instead of just the two available in plane truss analysis.

After concluding that a space truss is statically determinate, we can solve for the member forces, using either the method of joints or the method of sections. Because of the additional dimension, there is more bookkeeping required than in plane truss analysis. Vector notation is very useful for three dimensions, as illustrated in the following example.

Example 2.3.1   Using the method of joints, calculate all of the member loads of the truss in Figure 2.3.1 in terms of the loads P and Q applied as shown.

We could begin by writing the equilibrium equations for the truss as a whole and then solving for the reactions before proceeding to the member loads. Alternatively, we start with node 1 at the unsupported end of the structure and write the equilibrium equations at each of the eight nodes, aided by Figure 2.3.2. The geometry of the truss is such that we can find three of the unknowns from each nodal equilibrium equation, if we select the nodes in the right order. (On the other hand, we could select the eight nodes in no particular sequence, write the 24 equilibrium equations, and solve them simultaneously with the aid of a computer.)

Assume, as before, that all of the unknown forces in the rods are tensile and therefore pull on (i.e., act away from) the nodes to which they are attached. The force exerted by a rod on a node acts in the direction of the rod. Therefore, to express the force in vector form, we must calculate the direction cosines l, m, and n of the rod. These are the components of a unit vector $\mathbf{e}$ pointing from one end of the rod toward the other end: $\mathbf{e} = l\mathbf{i} + m\mathbf{j} + n\mathbf{k}$, where $\mathbf{i}$, $\mathbf{j}$, and $\mathbf{k}$ are the unit vectors along the x, y, and z axes, respectively. Before calculating l, m, and n, we must decide which way to draw the unit vector on the rod. Either end can be named the beginning (or "near") end of the rod. The other end becomes the "far" end. The unit vector $\mathbf{e}$ points toward the far end.

The first three columns of Table 2.3.1 show the nodes of the truss to which the near and far ends of each rod are connected. The next three columns give the components of the vector drawn from the near end to the far end of each rod, from which the length of the rod can be found (column 7). Dividing the x, y, and z components of the vector by its length yields its direction cosines, the components of $\mathbf{e}$, which are listed in the last three columns. With this information and vector notation as a bookkeeping aid, we can conveniently write the equilibrium equations at each node of the truss.

Node 1

If $\mathbf{e}_i$ represents the unit vector along rod i (from the near node to the far node), then according to Figure 2.3.2, the equilibrium of node 1 requires that

$$N_1\mathbf{e}_1 + N_2\mathbf{e}_2 + N_3\mathbf{e}_3 + P\mathbf{k} + Q\mathbf{j} = 0 \qquad \text{[a]}$$

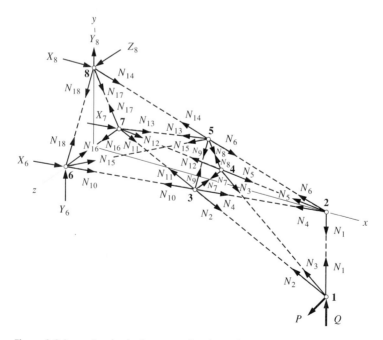

Figure 2.3.2   Free-body diagrams of nodes 1 through 5 of the truss in Figure 2.3.1.

Table 2.3.1   Nodal connectivity of the rods in the truss shown in Figure 2.3.1.

Rod	Near Node	Far Node	Δx (m)	Δy (m)	Δz (m)	L (m)	l	m	n
1	1	2	0	0.5	0	0.5	0	1.000	0
2	1	3	−0.75	0.5	0.15	0.9138	−0.8208	0.5472	0.1642
3	1	4	−0.75	0.5	−0.15	0.9138	−0.8208	0.5472	−0.1642
4	2	3	−0.75	0	0.15	0.7648	−0.9806	0	0.1961
5	2	4	−0.75	0	−0.15	0.7648	−0.9806	0	−0.1961
6	2	5	−0.75	0.25	0	0.7906	−0.9487	0.3162	0
7	3	4	0	0	−0.30	0.3	0	0	−1
8	4	5	0	0.25	0.15	0.2916	0	0.8575	0.5145
9	3	5	0	0.25	−0.15	0.2916	0	0.8575	−0.5145
10	3	6	−0.75	0	0.15	0.7648	−0.9806	0	0.1961
11	3	7	−0.75	0	−0.45	0.8746	−0.8575	0	−0.5145
12	4	7	−0.75	0	−0.15	0.7648	−0.9806	0	−0.1961
13	5	7	−0.75	−0.25	−0.30	0.8456	−0.8870	−0.2957	−0.3548
14	5	8	−0.75	0.25	0	0.7906	−0.9487	0.3162	0
15	5	6	−0.75	−0.25	0.30	0.8456	−0.8870	−0.2957	0.3548
16	6	7	0	0	−0.60	0.6	0	0	−1
17	7	8	0	0.50	0.30	0.5831	0	0.8575	0.5145
18	6	8	0	0.50	−0.30	0.5831	0	0.8575	−0.5145

Since the components of $\mathbf{e}_i$ are the direction cosines of rod i, we can go to Table 2.3.1 to find

$$\mathbf{e}_1 = \mathbf{j} \qquad \mathbf{e}_2 = -0.8208\mathbf{i} + 0.5472\mathbf{j} + 0.1642\mathbf{k} \qquad \mathbf{e}_3 = -0.8208\mathbf{i} + 0.5472\mathbf{j} - 0.1642\mathbf{k}$$

Substituting these unit vectors into Equation (a) and collecting terms, we obtain

$$(-0.8208N_2 - 0.8208N_3)\mathbf{i} + (N_1 + 0.5472N_2 + 0.5472N_3 + Q)\mathbf{j} + (0.1642N_2 - 0.1642N_3 + P)\mathbf{k} = 0$$

The three scalar equilibrium equations at node 1 are obtained by setting the x, y, and z components of this vector equation equal to zero:

$$-0.8208N_2 - 0.8208N_3 = 0$$
$$N_1 + 0.5472N_2 + 0.5472N_3 = -Q$$
$$0.1642N_2 - 0.1642N_3 = -P$$

The solution of this system of equations is

$$N_1 = -Q \qquad N_2 = -3.046P \qquad N_3 = 3.046P$$

Node 2

We use the free-body diagram in Figure 2.3.2 to write

$$N_1(-\mathbf{e}_1) + N_4\mathbf{e}_4 + N_5\mathbf{e}_5 + N_6\mathbf{e}_6 = 0$$

Since N_1 acts away from node 2, in the direction opposite to that assumed for $\mathbf{e}_1$, this explains the minus sign in the first term. As before, we use the information in the table to express the unit vectors in terms of their $\mathbf{i}$, $\mathbf{j}$, and $\mathbf{k}$ components.

Substituting those expressions into this equation, we get three scalar equations, as follows:

$$-0.9806N_4 - 0.9806N_5 - 0.9487N_6 = 0$$
$$N_1 \qquad\qquad + 0.3162N_5 \qquad\qquad = 0$$
$$0.1961N_4 - 0.1961N_5 \qquad\qquad = 0$$

We already found that $N_1 = -Q$. Therefore, from this set of equilibrium equations, we readily obtain

$$N_4 = 1.530Q \quad N_5 = 1.530Q \quad N_6 = -3.162Q$$

Node 4

We go to this node next since only three unknown forces, N_7, N_8, and N_{12}, act on it, whereas there are four unknowns at node 3 and five at node 5. The equilibrium equation for node 4 is

$$N_3(-\mathbf{e}_3) + N_5(-\mathbf{e}_5) + N_7(-\mathbf{e}_7) + N_8\mathbf{e}_8 + N_{12}\mathbf{e}_{12} = 0$$

Substituting $N_3 = 3.046P$ and $N_5 = 1.530Q$, as just determined, and using the data in Table 2.3.1, we get, after simplification,

$$-0.9806N_{12} = -2.5P - 1.5Q$$
$$0.8575N_8 \qquad\qquad = 1.667P$$
$$N_7 + 0.5145N_8 - 0.1961N_{12} = -0.5P - 0.3Q$$

These immediately yield N_7, N_8, and N_{12} (see Table 2.3.2).

Node 3

Referring to Figure 2.3.2, we write

$$N_2(-\mathbf{e}_2) + N_4(-\mathbf{e}_4) + N_7\mathbf{e}_7 + N_9\mathbf{e}_9 + N_{10}\mathbf{e}_{10} + N_{11}\mathbf{e}_{11} = 0$$

Writing out the unit vectors using Table 2.3.1 and using the values of N_2, N_4, and N_7 just obtained yields three equations in the three unknowns N_9, N_{10}, and N_{11}.

Continuing in this fashion, we proceed next to node 5 to find N_{13}, N_{14}, and N_{15}, and then to node 7, which yields N_{16}, N_{17}, and the reaction $X_7 = -2.5P - 1.5Q$. At node 6, we then obtain N_{18} and the reactions $X_6 = 2.5P - 1.5Q$ and $Y_6 = -3.333P$. Finally, at node 8, we solve for the last of the 24 original unknowns, namely the reaction components:

$$X_8 = 3Q \quad Y_8 = 3.333P - Q \quad Z_8 = -P$$

The internal loads throughout the structure are summarized in Table 2.3.2.

Table 2.3.2   Axial loads in the members of the space truss in Figure 2.3.1.

Member	Axial Force
1	Q
2	$-3.046P$
3	$3.046P$
4	$1.530Q$
5	$1.530Q$
6	$-3.162Q$
7	$-P$
8	$1.944P$

Table 2.3.2   Concluded

Member	Axial Force
9	21.944P
10	−5.099P+1.530P
11	2.916P
12	2.540P+1.530Q
13	−2.819P
14	−3.162Q
15	2.819P
16	−1.500P−0.3000Q
17	0.9718P
18	2.916P

Figure 2.3.3(a) shows another minimally stable, statically determinate space truss. It is supported against rigid-body translations and rotations. To find the forces in the members in terms of the applied 1000 lb load, we must write the equilibrium equations at each node. This will give us 24 equations for the 18 member loads and the 6 reactions. Unlike the previous example, on this truss, there is no joint with fewer than four rods framing into it. Therefore, the 24 equations do not uncouple into smaller, independent groups of equations that will yield the unknowns as we proceed from node to node around the structure. Instead they must be solved simultaneously to find the forces shown in Figure 2.3.3(b).

The procedure for solving determinate space trusses is straightforward, but extremely lengthy and tedious. Yet, the joint method is easy to program for the most modest personal computer. However, such a program, by itself, would have limited application in "real world" situations, where trusses are likely to be statically indeterminate. In fact, it is difficult to design space trusses that are not statically indeterminate. For example, suppose we simply add

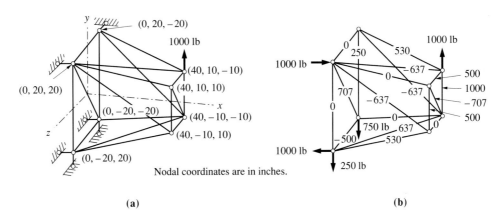

Nodal coordinates are in inches.

(a)   (b)

Figure 2.3.3   (a) Statically determinate space truss. (b) The reactions and member loads (in pounds).

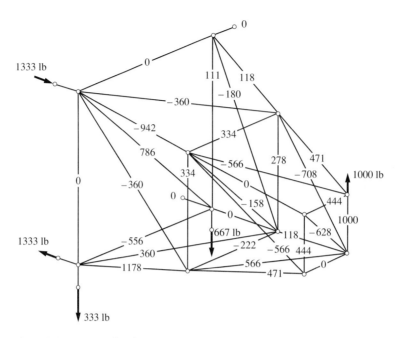

Figure 2.3.4 Member forces (in pounds) and the reactions of a two-bay space truss similar to the one in Figure 2.3.3.

All rods have the same cross sectional area and the same material properties.

another identical, tapered, minimally stable bay to the statically determinate truss in Figure 2.3.3(a) and support it in the same way, so that the number of reactions remains the same. The result is shown in Figure 2.3.4. The new bay adds 13 members to those already present. It also adds four nodes, which means we have only 12 additional joint equilibrium equations. For this expanded truss, we are one equation short of matching the total number of unknown member forces and reactions. If we add a third bay, we fall two equations behind, and so on. The degree of static indeterminacy increases with each additional bay.

The means of dealing with statically indeterminate trusses are explored in Chapters 7 and 10. The member forces and the reactions shown in Figure 2.3.4 are obtained by a procedure that requires dealing with the deformation of the truss. This procedure requires specifying the cross-sectional areas of the rods and their material properties. It is interesting to compare the loads in Figures 2.3.3(b) and 2.3.4 and to observe the redistribution apparent in the original, smaller bay with the larger second bay added on. The differences among the member loads may be attributed to the flexibility of the structure. Changing the cross-sectional dimensions of the rods or their individual material properties will further alter the load distribution. In contrast, statically determinate structures are insensitive to such modifications.

2.4 SIMPLE BEAMS

A simple beam is a slender, homogeneous bar that bends without twisting when acted upon by loads applied perpendicular to its axis and in a single plane containing the axis. Consider the cantilevered beam shown in Figure 2.4.1. Pass a cutting plane through the beam at a distance x from the free end, where the point load P is applied, and isolate the left side of the cut as a free body. For equilibrium, a downward force equal to P must

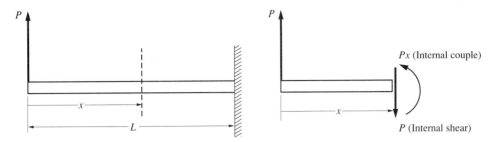

Figure 2.4.1   Cantilever beam and a free-body diagram showing the shear and moment at an arbitrary section.

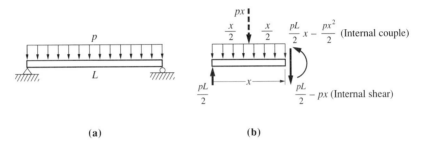

(a)   **(b)**

Figure 2.4.2   (a) Simply supported beam. (b) Internal shear and bending moment as a function of distance from the left end.

act on the cut. This is a *shearing force*, because it acts parallel to the surface of the cut. The pair of equal and oppositely directed forces, one at the free end and one at the section, form a couple whose moment is *Px*. In order to balance out the effect of this couple, a so-called *bending moment* (a localized couple) must develop at the section. The shear force on a section of this beam is the same, regardless of where we make the cut. The bending moment, on the other hand, increases linearly across the beam, to a maximum value at the support.

The internal shear force and bending moment that develop at a beam cross section are the result of internal stress distributions (see Chapter 4). These stresses are directly proportional to the shear and moment magnitudes. Knowing the stress distribution and the location where the largest stresses develop is crucial to assessing the structural integrity of a beam. Therefore, an understanding of internal shear and moment distribution in beams is essential.

Figure 2.4.2 shows a *simply supported* beam of length *L* carrying an external, uniformly distributed load per unit length *p*. The resultant of this load is *pL*, and it is distributed equally to the two supports. An equilibrium analysis of the free body in Figure 2.4.2(b), at an intermediate section, reveals a somewhat more complicated internal shear and moment distribution than that of the cantilever beam previously discussed. The shear force is zero at the midpoint of the beam, and it increases linearly to its maximum value at each end. The shear on a right-facing section, such as the right end of the free body in Figure 2.4.2(b), is directed downwards in the left half of the beam ($x < L/2$) and upwards in the right half. The bending moment varies parabolically across the span. It is zero at each end and reaches its maximum value at the center.

Since the internal shear on a section can act either up or down, and the bending moment can be either clockwise or counterclockwise, we need to establish a sign convention for these quantities. The choice is arbitrary. We shall use the system illustrated in Figure 2.4.3, because it is easy to extend to beams in three dimensions, as will be done in Chapter 4. The term *V* stands for the internal shear on a section and *M* is the bending moment. The

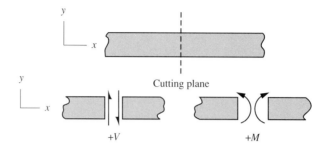

Figure 2.4.3   Sign convention for positive internal shear and bending moment.

shear is positive if it is directed upwards, in the positive y-direction, on the section oriented in the positive x-direction (right-facing surface). Similarly, the bending moment is positive if it is directed in (i.e., exerts a twist about) the positive z-direction on the section oriented in the positive x-direction. Using this sign convention, we can plot the internal shear and bending moment distributions across a beam, as shown in Figure 2.4.4.

In Figure 2.4.1 we have $V = -P$ and $M = Px$. In Figure 2.4.2, $V = -(pL/2 - px)$ and $M = pLx/2 - px^2/2$. In both cases, we see that $dM/dx = -V$ and $dV/dx = p$, where p is the distributed load on the beam (zero in Figure 2.4.1). This is not a coincidence. These are extremely useful, general relationships that can be derived from basic statics. Figure 2.4.5 shows a differential "slice" of a beam lying between two transverse sections that are a differential distance dx apart anywhere along the span. We will henceforth assume that a distributed load is positive if it is directed upwards. The portion of the distributed load acting on the differential segment is shown in the figure.

Summing the forces in the y-direction, we get

$$-V + (V + dV) + p\,dx = 0$$

$$dV = -p\,dx$$

$$\boxed{\frac{dV}{dx} = -p}$$

[2.4.1]

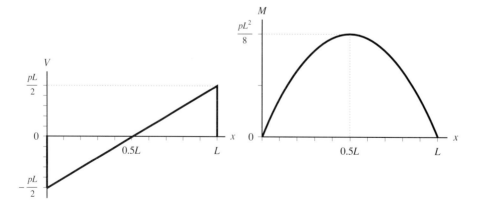

Figure 2.4.4   Shear (left) and moment diagrams for the beam in Figure 2.4.2.

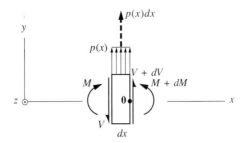

Figure 2.4.5   Free-body diagram of a differential beam segment, showing the internal shears and bending moments and a differential portion of the externally applied distributed load.

Clearly, the slope of the shear curve at a point is *minus* the value of the load curve at that point. If the shear at section 1 is known, then according to Equation 2.4.1, the shear at section 2 is

$$V_2 = V_1 - \int_{x_1}^{x_2} p(x)\,dx \qquad\qquad \textbf{[2.4.2]}$$

Summing the moments about **0** in Figure 2.4.5 yields

$$-M + V\,dx + (M + dM) - (p\,dx)\left(\frac{dx}{2}\right) = 0$$

$$V\,dx + dM - \frac{p\,dx^2}{2} = 0$$

Since second-order differentials are negligible compared to first-order ones, the last term may be dropped, yielding

$$\boxed{\frac{dM}{dx} = -V} \qquad\qquad \textbf{[2.4.3]}$$

The slope of the bending moment curve at a point equals minus the value of the shear curve at that point. Therefore, if the bending moment is known at section 1 of the beam, then at section 2 we have

$$M_2 = M_1 - \int_{x_1}^{x_2} V(x)\,dx \qquad\qquad \textbf{[2.4.4]}$$

Example 2.4.1   A simply-supported beam is loaded as shown at the top of Figure 2.4.6. Construct the internal shear and bending moment curves.

We first compute the reactions at the supports (points 1 and 5). Statics readily yields the values shown in Figure 2.4.6(a).

Starting at the left end of the beam where the shear is -8.111 kN, we see that the area under the uniformly distributed load between points 1 and 2 is $(-40\ \text{kN/m}) \times (0.75\ \text{m}) = -30$ kN. According to Equation 2.4.2, the change in the shear over that span is $+30$ kN, from -8.111 kN to $+21.89$ kN. The external upward load of 45 kN at point 2 is counteracted internally by a downward-directed shear, so that the shear diagram at point 2 jumps by -45 kN, from $+21.89$kN to -23.11 kN. The downward distributed load continues at a constant -45 kN/m beyond point 2 and up to point 3, which means that the shear

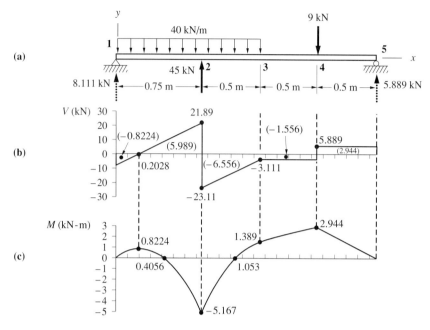

Figure 2.4.6 Shear (b) and moment (c) diagrams for a beam loaded as shown at the top of the figure (a).

Numbers in parentheses in (b) are the areas in kN-m under the shear diagram (positive above and negative below the x-axis).

diagram increases at the uniform rate of +45 kN/m over the 0.5 m span, becoming −23.11 kN − (−45 kN/m) ×(0.5 m) = −3.111 kN at point 3. The distributed load is zero beyond point 3, so the shear remains constant at −3.111 kN up to point 4, where the external, downward-directed 9 kN load is applied. That load is counteracted internally by an upward (positive) shear on the section, so that the shear diagram jumps in the positive direction at point 4 to a value of −3.111 kN + 9 kN = 5.889 kN. The internal shear is constant thereafter, as shown in part (b) of the figure.

Since the distributed load is either zero or constant over the entire 2.25 m span of the beam, the shear diagram turns out to be a piecewise-linear curve. Therefore, the area under each constant-slope portion of the shear curve is easy to determine by inspection. These areas are shown in parentheses in part (b) of Figure 2.4.6. Using these numerical values in Equation 2.4.4 helps us develop the bending moment curve.

The bending moment is zero at the left end (point 1). Between points 1 and 2, the shear increases linearly. Therefore, according to Equation 2.4.3, the slope of the bending moment curve decreases linearly and becomes zero where $V = 0$, which occurs at $x = 0.2028$ m. Thus, from $x = 0$ to $x = 0.2028$ m, the moment increases quadratically by an amount equal to minus the area under the shear diagram, which is −0.8224 kN-m. That is, according to Equation 2.4.4, the bending moment at $x = 0.2028$ m is $0 - (-0.8224$ kN-m$) = 0.8224$ kN-m. Between this point and point 2, the area under the linearly increasing shear diagram is +5.989 kN-m. This means then at point 2, the moment curve has decreased parabolically to $(0.8224$ kN-m$) - (5.989$ kN-m$) = -5.167$ kN-m, as shown in part (c) of Figure 2.4.6. At point 2, the slope of the moment diagram changes abruptly from −21.89 kN to +23.11 kN. From point 2 to point 3, we see from the shear diagram that the bending moment changes quadratically from −5.167 kN-m to $-5.167 - (-6.556) = 1.389$ kN-m. Since the area under the zero-slope shear diagram from point 3 to point 4 is −1.556 kN-m, the bending moment increases linearly to $(1.389$ kN-m$) - (-1.556$ kN-m$) = +2.944$ kN-m at point 4, where the slope suddenly changes from +3.111 kN to −5.889 kN due to the jump in the shear diagram. Beyond point 4, the bending moment decreases linearly at the constant rate of −5.889 kN for 0.5 m, reaching a value of zero at the end of the beam (point 5).

The "eyeball integration" procedure illustrated in the previous example relies on the distributed load being piecewise constant. If the distributed load has a more complex variation between the concentrated loads, then the integral in Equation 2.4.2 must be solved to obtain the variation of the shear, which in turn must be integrated according to Equation 2.4.4 to obtain the moment distribution. One must be careful to account for jumps in the shear diagram at concentrated loads and jumps in the moment curve at concentrated couples. It may be necessary to apply a numerical integration technique, such as the trapezoid rule or Simpson's rule, in order to find the shear and moment diagrams. This is illustrated by the next example, in which the load intensity on a beam is given at finite intervals across the span, rather than as a mathematical function of position.

Example 2.4.2   According to experimental data, the upward-directed load per unit length p at 17 points along a 240-inch cantilever beam is as shown in Figure 2.4.7(a). Find the internal shear and bending moment distribution for the beam.

We connect the data points in Figure 2.4.7(a) with straight lines, forming a series of trapezoids comprising the area under the approximated load diagram. The area of each trapezoid, $p_{\text{average}}\Delta x$, is noted in parentheses on the figure. According to

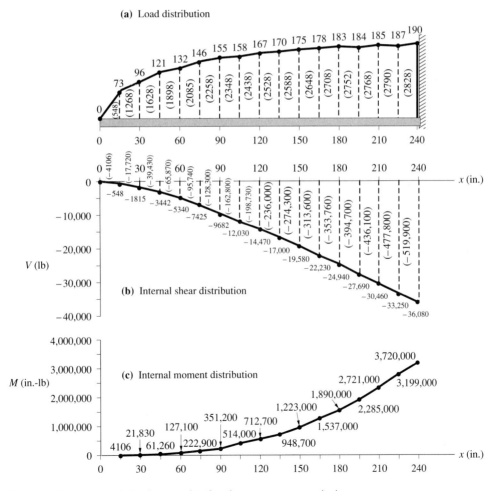

Figure 2.4.7   (a) Load, (b) shear, and (c) bending moment across the beam.

Equation 2.4.2, the change in the internal shear over each segment is the negative of the trapezoidal area. We use that information to obtain the values of the shear diagram at the same 17 points at which the load was defined. Joining these points with straight lines yields the shear diagram shown in Figure 2.4.7(b), on which the area of each trapezoidal segment is indicated. This area ($V_{\text{average}}\Delta x$) is used in a similar fashion to construct the bending moment diagram, Figure 2.4.7(c), by means of Equation 2.4.4.

2.5   STIFFENED SHEAR WEBS

The structure shown in Figure 2.5.1 consists of a thin rectangular sheet of material to which a rod is bonded along each edge. If the sheet were eliminated and the rods were pinned together, this would be an unstable truss. It could be made a statically determinate and stable truss by adding a diagonal member. Alternatively, adding just the sheet, or *panel*, also makes it a statically determinate structure, if we assume that the panel carries only shear load. If we make a transverse cut at an intermediate station, then we see a free-body diagram similar to that in Figure 2.4.1. In this case, the shear is carried entirely by the panel, and the moment at the section is due exclusively to the couple formed by the axial loads in the rods. The structure acts like a beam. It is an *idealized* beam because of the assumption that the panel does not participate in producing the internal bending moment at the section. (We will investigate the consequences of relaxing this assumption in Chapter 4.) The top and bottom rods are *flanges*, and the panel is the *web*. The vertical rods are referred to as *stiffeners*. The panel is called a *shear panel* because it has only shear forces on its edges.

Figure 2.5.2 shows a parallelogram-shaped panel with included acute angle θ. The edges of the panel are subjected to pure shear loads S_1, S_2, S_3, and S_4. If we subdivide a shear load S into infinitesimal segments and then divide each differential shear force dS by the differential length ds on which it acts, we get a finite load intensity, a shear force per unit length called the *shear flow*, q. On an edge of length l, the shear force can be expressed in terms of the shear flow, as follows

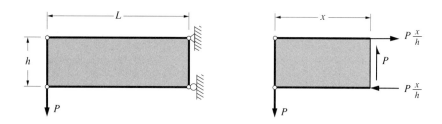

Figure 2.5.1   Stiffened shear web acting as an idealized beam.

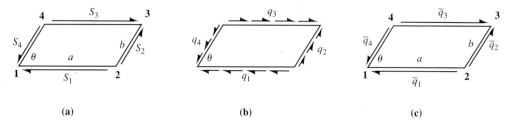

(a)   (b)   (c)

Figure 2.5.2   Parallelogram shear panel. (a) Shear forces on the edges. (b) Shear flows. (c) Average shear flows.

$$S = \int_0^l q\, ds \qquad\qquad \text{[2.5.1]}$$

where s is the distance from the beginning point of the side. The average shear flow, $\bar{q}$, is the resultant shear force S divided by the total length of the edge on which it acts, or

$$\bar{q}_1 = \frac{S_1}{l_1}, \qquad \bar{q}_2 = \frac{S_2}{l_2}, \qquad \text{etc.}$$

The subscripts refer to the four edges, numbered in counterclockwise order around the panel.

Summing moments about point 1 in Figure 2.5.2(a) and expressing the shear forces in terms of the average shear flows, we get

$$(S_2 \sin\theta)a = S_3(b\sin\theta) \qquad \Rightarrow \qquad (\bar{q}_2 b)a = (\bar{q}_3 a)b \qquad \Rightarrow \qquad \bar{q}_2 = \bar{q}_3$$

Summing moments about any two of the remaining three corners quickly leads to the conclusion that the average shear flow is constant around the panel; that is,

$$\bar{q}_1 = \bar{q}_2 = \bar{q}_3 = \bar{q}_4 = \bar{q}$$

This conclusion holds for a differential parallelogram surrounding *any* point in a parallelogram shear panel, as shown in Figure 2.5.3. The shear flow q is constant around such an element. If we extend the differential parallelogram in any direction and parallel to the sides of the panel, we see that the shear flow at the point of intersection with the edge of the panel must be q as well. This means that the shear flow throughout and around the sides of a parallelogram panel (and, hence, a rectangular one) is constant, which is why these are called *constant shear flow panels*.

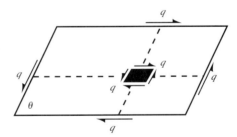

Figure 2.5.3   Shear flow on a differential parallelogram surrounding any point in a parallelogram shear panel.

Example 2.5.1   Find the shear flow in the web of the structure shown in Figure 2.5.4 and the flange loads at a section 75 cm from the left end.

Pass a vertical cutting plane through the structure at the 0.75 m station, as shown in Figure 2.5.4(b). Setting the moment about point A equal to zero yields the upper flange tensile load of 7.794 kN. Summing the forces in the horizontal direction shows that the lower flange compression load is 7.794 kN. Finally, setting the net vertical force equal to zero gives the 4.5 kN shear load in the panel. Dividing this force by the depth of the web, we find the shear flow to be 9 kN/m. This shear flow is uniform throughout the web. Of course, the flange loads vary with distance from the applied 4.5 kN load, reaching a maximum value (10.39 kN) at the wall. Note that since the flanges are parallel, they take out none of the vertical shear. In tapered panels, some of the shear load is supported by the flanges.

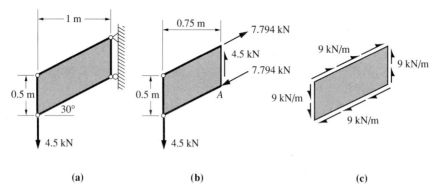

Figure 2.5.4   (a) Cantilevered parallelogram stiffened shear panel. (b) Shear and axial loads 0.75 m from free end. (c) Constant shear flows around the panel.

Figure 2.5.5(a) shows a tapered, trapezoid-shaped panel with its edges in pure shear. The parallel bases are separated by a distance L. The nonparallel sides are extended to intersect at the *vertex P*. The shaded element $aa'b'b$ is formed by two lines parallel to the bases of the trapezoid and two lines through P at a differential angle $d\theta$ apart. Following Garvey,[3] we assume that the edges of any such element in the panel are also in pure shear, as shown. Also, p and p' are the perpendicular distances to the parallel bases of the differential trapezoid. Treating the element as a free body and setting the moment about P equal to zero, we get

$$p'\left(q'\overline{a'b'}\right) = p\left(q\overline{ab}\right)$$ [2.5.2]

By similar triangles, the differential lengths $\overline{a'b'}$ and $\overline{ab}$ are related by

$$\frac{\overline{a'b'}}{p'} = \frac{\overline{ab}}{p}$$

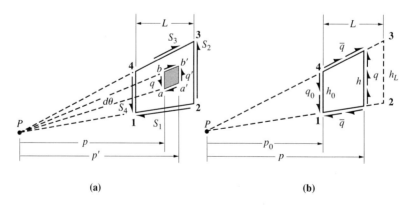

Figure 2.5.5   (a) Trapezoidal shear panel. (b) Internal shear flow q related to the base shear flow q_0.

| [3]S. J. Garvey, "The Quadrilateral Shear Panel," *Aircraft Engineering*, May 1951, pp. 134–135.

Substituting this into Equation 2.5.2 yields

$$p'^2 q' = p^2 q$$

This means that the shear flow in a trapezoidal shear panel is inversely proportional to the square of the distance from the trapezoid vertex P, or

$$q = \frac{\beta}{p^2} \qquad\qquad \text{[2.5.3]}$$

where β is a constant that depends on the boundary conditions. The shear flow is uniform along the panel's parallel edges, since p is constant on each of them.

To calculate the average shear flow along an inclined edge of length l such as that illustrated in Figure 2.5.6, we first substitute Equation 2.5.3 into Equation 2.5.1, as follows:

$$S = \int_0^l q\,ds = \int_{p_0}^{p_L} q\frac{dp}{\sin\alpha} = \int_{p_0}^{p_L} \frac{\beta}{p^2}\frac{dp}{\sin\alpha}$$

Carrying out the integration and substituting the limits yields

$$S = \frac{p_L - p_0}{\sin\alpha}\frac{\beta}{p_L p_0}$$

Since $(p_L - p_0)/\sin\alpha$ is the length l of the edge and S/l is the average shear flow $\bar{q}$, this expression yields

$$\bar{q} = \frac{\beta}{p_0 p_L}$$

Solving for β and using the fact that, by similar triangles, $p/h = p_0/h_0 = p_L/h_L$, we may write Equation 2.5.3 as

$$\boxed{q = \bar{q}\frac{h_0 h_L}{h^2}} \qquad\qquad \text{[2.5.4]}$$

This is a very useful formula for calculating the shear flow in a tapered shear panel. Bear in mind that it requires equilibrium under the assumption that the edges of the panel are in pure shear.

Let $\bar{q}(h)$ be the average shear flow up to the point where the panel width is h. Then,

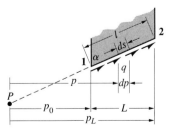

Figure 2.5.6   Shear flow along a tapered edge.

$$\bar{q}(h) = \frac{\beta}{p_0 p} = \frac{\bar{q} p_0 p_L}{p_0 p} = \bar{q}\frac{p_L}{p}$$

Since $p_L/p = h_L/h$, it follows that

$$\bar{q}(h) = \bar{q}\frac{h_L}{h} \qquad\qquad \text{[2.5.5]} \ A$$

From this, we see that the average shear flow varies inversely as the width of the panel, and it is constant if the panel is a parallelogram or rectangle (i.e., if $h = \text{constant}$).

$$\boxed{\bar{q}'(h) = \bar{q}\frac{h_o}{h}} \qquad\qquad [2.5.5] \ B$$

Example 2.5.2 Find the shear flow in the web of the tapered beam shown in Figure 2.5.7. Also, calculate the average shear flow on each edge of the panel.

The 20 kN load, applied to the vertical stiffener, is transferred as a 1000 kN/m shear flow along the left edge of the tapered shear web, as illustrated in Figure 2.5.8. Therefore,

$$q(0) = 1000\ \frac{\text{kN}}{\text{m}}$$

From Equation 2.5.4, we have

$$q_0 = \bar{q}\frac{h_L}{h_0}$$

Using this, we find the average shear flow to be

$$\bar{q} = q_0\frac{h_0}{h_L} = 1000\frac{\text{kN}}{\text{m}} \times \frac{0.2\,\text{mm}}{0.7\,\text{mm}} = 285.7\frac{\text{kN}}{\text{m}}$$

Given the geometry of the structure, it is easy to see that the width h of the panel as a function of x, both in meters, is $h(x) = \frac{1}{3}(x + 0.060)$. Thus, the shear flow in kN/m on any vertical section of the web, according to Equation 2.5.4, is

$$q(x) = \bar{q}\frac{h_0 h_L}{h^2(x)} = 285.7\frac{(0.020)(0.070)}{\frac{1}{9}(x + 0.060)^2} = \frac{3.60}{(x + 0.060)^2}\ \frac{\text{kN}}{\text{m}} \qquad \text{[a]}$$

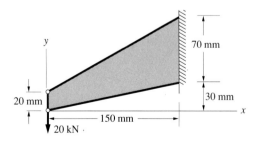

Figure 2.5.7 Idealized tapered, cantilevered beam.

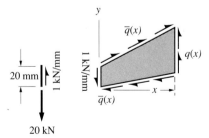

Figure 2.5.8 Free-body diagram of a portion of the web in Figure 2.5.7.

Finally, Equation 2.5.5 is used to obtain the average shear flow on the tapered edges of the panel up to the transverse section at x:

$$\bar{q}(x) = \bar{q}\,\frac{h(0.150)}{h(x)} = 285.7\frac{0.070}{\frac{1}{3}(x+0.060)} = \frac{60.0}{x+0.06}\,\frac{\text{kN}}{\text{m}} \qquad \textbf{[b]}$$

These two shear flow expressions—Equations (a) and (b)—are plotted in Figure 2.5.9.

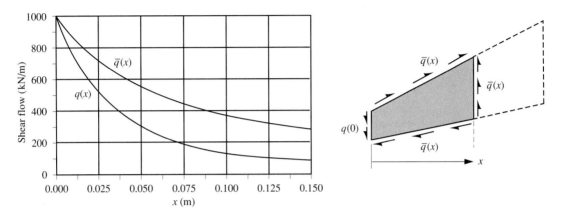

Figure 2.5.9   Shear flow on the edges of the shear panel in Figure 2.5.7

Shear flow is transferred to the edge of a panel by an adjoining rod. Direct loads are applied to the ends of the rod, and a distributed axial load per unit length develops to equilibrate (balance out) these loads, as shown in Figure 2.5.10.

From the figure, we see that

$$N_2 - N_1 + \int_0^L q(s)\,ds = 0 \qquad \textbf{[2.5.6a]}$$

where s is the distance measured from the left end of the rod. More practically, in terms of the average shear flow, we have

$$N_2 - N_1 + \bar{q}\,L = 0 \qquad \textbf{[2.5.6b]}$$

Consider a planar, stiffened web structure like that shown in Figure 2.5.11. The small circles represent the nodes, i.e., the points at which the stiffeners (rods) are joined together. The unknowns of this problem are the constant shear flows in the rectangular shear panels, the axial loads at each end of the rods, and the reactions at the supports. There

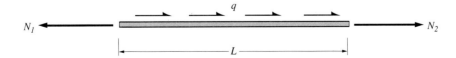

Figure 2.5.10   Rod in equilibrium under the direct loads applied at each end and the shear flow distributed along its length.

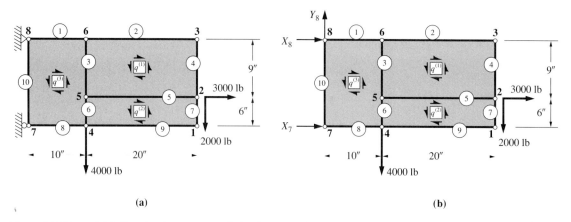

Figure 2.5.11   (a) Stiffened web structure, with the chosen node, rod, and panel numbering. (b) Free-body diagram, showing the applied loads and the reactions.

are two equilibrium equations per node, and one equilibrium equation (Equation 2.5.6) per rod. A problem such as this is statically determinate if the number of equilibrium equations equals the number of unknowns; that is, if

$$n_{rods} + 2n_{nodes} = 2n_{rods} + n_{panels} + n_{reactions}$$

or

$$n_{rods} + n_{panels} + n_{reactions} = 2n_{nodes} \qquad \text{[2.5.7]}$$

In Figure 2.5.11, there are 10 rods, 3 panels, and 3 reactions. Since there are 8 nodes, Equation 2.5.7 is satisfied and the problem is statically determinate. Figure 2.5.12 is an exploded view of the same structure, showing the nodes, rods, and shear panels as free bodies, revealing the unknown loads within the structure and the reactions at the supports.

Example 2.5.3   For the structure shown in Figure 2.5.11, calculate the shear flows in each of the three panels and the maximum load in the stiffeners.

We could proceed by isolating each of the 10 rods and 8 nodes as free bodies, as in Figure 2.5.12, then writing and solving the $10 + (2 \times 8) = 26$ equations of equilibrium for the 20 rod end loads, 3 panel shear flows, and 3 reactions. However, it is far more expedient to use the method of sections to arrive at the quantities needed.

We first find the three support reactions by considering the overall equilibrium of the structure, using the free-body diagram shown in Figure 2.5.11(b):

$$X_7 = 4867 \, lb$$

$$X_8 = -7867 \, lb$$

$$Y_8 = 6000 \, lb$$

Passing a vertical cutting plane through panel 3 yields the free-body diagram shown in Figure 2.5.13. Setting the net vertical force equal to zero yields $-15q^{(3)} - 4000 - 2000 = 0$, or $q^{(3)} = -400 \, lb/in$. The minus sign again means that the direction of the shear flow $q^{(3)}$ is opposite to that shown in Figure 2.5.13.

Equilibrium of the topmost stiffener (Figure 2.5.14) requires that $-7867 + (400 \times 10) - 20q^{(1)} = 0$, or $q^{(1)} = -193.4 \, lb/in$. The direction of the shear flow is therefore opposite to that originally assumed.

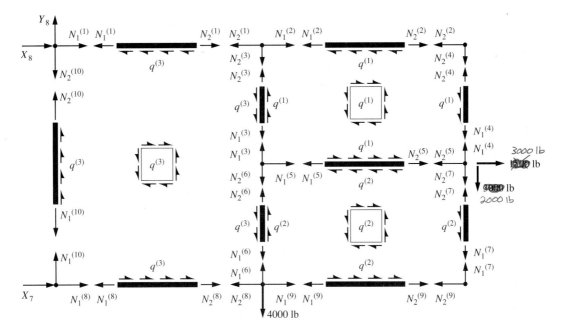

Figure 2.5.12   The unknown member forces and reactions in the stiffened web structure of Figure 2.5.11.

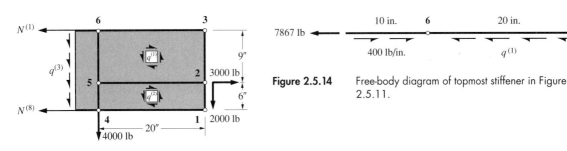

Figure 2.5.13   Free-body diagram resulting from a vertical section through panel 3.

Figure 2.5.14   Free-body diagram of topmost stiffener in Figure 2.5.11.

Finally, for equilibrium of the rightmost vertical stiffener (Figure 2.5.15), we have $9 \times 193.3 - 2000 - 6q^{(2)} = 0$, or $q^{(2)} = -43.33$ lb/in. Once again, the sign of the shear flow indicates that the assumed direction in Figure 2.5.13 was incorrect. The solution thus far is summarized in Figure 2.5.16.

Since the shear flow in a rectangular panel is constant, the shear flow along each rod of this structure is uniform. This means that the axial force in each rod varies linearly as we move from one end to the other (cf. Figure 2.5.10). Since we have now calculated all of the shear flows for this problem, all we have to do is plot axial load as a function of position along each rod, and then observe where the maximum occurs. This is done in Figure 2.5.17. The shaded figures show the internal axial load distributions in the rods.

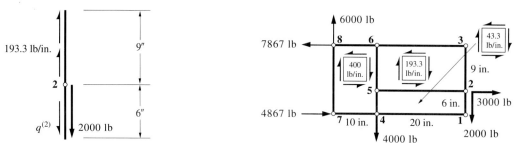

Figure 2.5.15 Free-body diagram of the rightmost vertical stiffener in Figure 2.5.11.

Figure 2.5.16 Constant shear flows in the panels and the reactions at the supports in Figure 2.5.11.

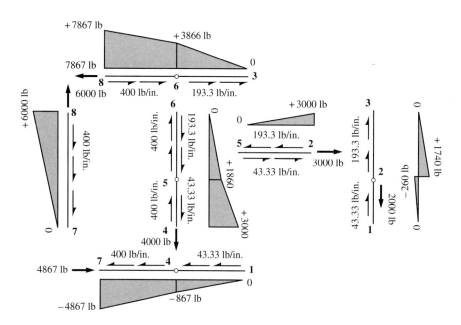

Figure 2.5.17 Axial load distribution in the stiffeners is Figure 2.5.11; negative indicates compression.

Let us next consider a thin cylindrical sheet that is in pure shear; that is, the edges of the differential rectangular element in Figure 2.5.18 formed by adjacent lines parallel to the cylinder axis and by adjacent sections through that axis have only shear components of force acting on them. Let q and q' be the shear force per unit length on the axially-oriented edges of the element separated by the distance ds as measured along the curved boundary of the cross section. As can be seen from Figure 2.5.18, element equilibrium in the axial direction requires that

$$\sum F_x = 0: \quad \Rightarrow \quad -q\,dx + q'\,dx = 0 \quad \Rightarrow \quad q' = q \qquad \text{[2.5.8]}$$

Thus, *the shear force per unit length (shear flow) is constant around the cross section* (which need not be circular).

This is true as well for a conical surface, which is described by a straight line (the *generator* or *ruling line*) pivoting about a fixed point P (the *vertex*) while its other end moves along an arbitrarily shaped curve (the *directrix*). (If the

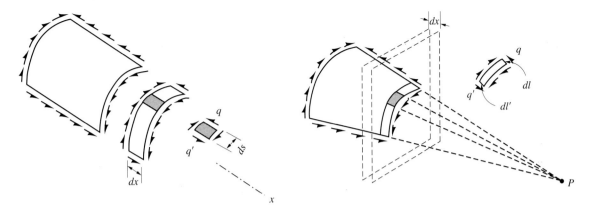

Figure 2.5.18   Cylindrical sheet in pure shear.

Figure 2.5.19   Conical surface, with vertex P, in pure shear.

vertex is infinitely removed from the directrix, we obtain a cylinder.) Let a transverse section be formed by a plane per-pendicular to the x-axis, which passes through the vertex P, and assume that such transverse sections are acted on only by shear, as shown in Figure 2.5.19. That figure also shows a small element of the curved wall formed by two adjacent transverse sections and two generators of the cone. For equilibrium, the net force in the x-direction must be zero,

$$(q'd\mathbf{l}') \cdot \mathbf{i} - (qd\mathbf{l}) \cdot \mathbf{i} = 0$$

where $d\mathbf{l}$ is the line segment directed towards the vertex along a generator. Since $d\mathbf{l} \cdot \mathbf{i} = d\mathbf{l}' \cdot \mathbf{i} = dx$, this equilibri-um equation is identical to Equation 2.5.8, so that $q' = q$; that is, the shear flow is constant around the cross section.

The fact that the shear flow is constant along a curve has several useful consequences. Consider the curve with constant shear flow q joining points B and C in Figure 2.5.20. At any point along the web, the differential shear force is $qd\mathbf{s}$, where $d\mathbf{s} = dz\mathbf{k} + dy\mathbf{j}$ is the vector joining two neighboring points on the curve. The resultant of the constant shear flow distribution is

$$\mathbf{R} = \int qd\mathbf{s} = q \int d\mathbf{s} = q\Delta\mathbf{s} = q(\Delta z\mathbf{k} + \Delta y\mathbf{j})$$

That is, the y and z components of the resultant force $\mathbf{R}$ are

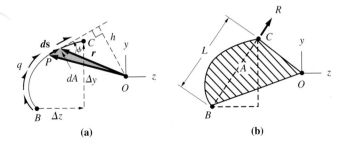

(a)   **(b)**

Figure 2.5.20   (a) Uniform shear flow q on a curved web. (b) Area enclosed by the web and the lines joining each end of the web to point O.

$$R_y = q\Delta y \qquad R_z = q\Delta z \qquad\qquad \text{[2.5.9]}$$

Since Δy and Δz are the projected lengths of the curve BC onto the y-axis and z-axis, respectively, we see that the force resultant of a shear flow distributed uniformly along a curve is the product of the shear flow times the projected length of the line joining the endpoints of the curve, and it lies in the direction of that line. The magnitude R of the force $\mathbf{R}$ is

$$R = \sqrt{R_y^2 + R_z^2} = \sqrt{q^2\Delta y^2 + q^2\Delta z^2} = q\sqrt{\Delta y^2 + \Delta z^2} = qL \qquad\qquad \text{[2.5.10]}$$

where L is the length of the line joining B to C.

Consider next the moment of the shear flow on the web about the arbitrary point O in Figure 2.5.20. The moment $d\mathbf{T}$ of the shear flow q acting on element ds at point P of the curve BC is

$$d\mathbf{T} = \mathbf{r} \times q d\mathbf{s}$$

However, according to the definition of the cross product,

$$\mathbf{r} \times d\mathbf{s} = (|\mathbf{r}||d\mathbf{s}|\sin\phi)\mathbf{i}$$

where $|\mathbf{r}| = r$ is the distance from O to the point on the curve; $|d\mathbf{s}| = ds$ is the differential arc length along the curve; and ϕ is the angle between $\mathbf{r}$ and the tangent to the curve at P, as shown in Figure 2.5.20(a). Thus,

$$d\mathbf{T} = q ds (r\sin\phi)\mathbf{i}$$

From the same figure, we see that $r\sin\phi$ equals h, the perpendicular distance from O to the tangent at P. Thus,

$$d\mathbf{T} = q(hds)\mathbf{i}$$

However, $hds/2$ is the area dA of the shaded triangle formed by joining O to each end of ds. Therefore, the differential moment of the shear flow about an x-axis through O is

$$dT = 2qdA$$

Since q is constant, the total moment about O of the shear flow on BC is given by

$$T = 2Aq \qquad\qquad \text{[2.5.11]}$$

where A is the area of the shaded region in Figure 2.5.20(b), which is formed by the curve BC and the lines joining each end of the web to point O.

We know that the line of action of the shear flow resultant $\mathbf{R}$ is parallel to BC. The perpendicular distance e from point O to $\mathbf{R}$ is found by setting the moment of $\mathbf{R}$ about O equal to that of the shear flow on the section. Thus, from Equations 2.5.10 and 2.5.11, $(qL)e = 2Aq$, or (cf. Figure 2.5.21)

$$e = \frac{2A}{L} \qquad\qquad \text{[2.5.12]}$$

2.6 IDEALIZED BEAMS: TORSIONAL AND SHEAR LOADING

Equation 2.5.11 is valid not only for constant shear flow around open sections, but for closed sections as well, in which case A is the area inside the closed web. To see this, pass a straight line a-a' of length l through the closed curve, thereby dividing it into two open sections with areas A_1 and A_2, as shown in Figure 2.6.1. The moment of

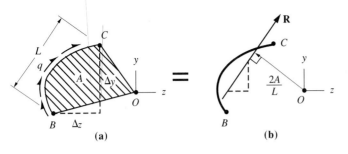

(a)

(b)

Figure 2.5.21   The resultant **R** of the constant shear flow in (a) has a line of action located as shown in (b).

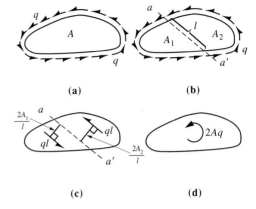

Figure 2.6.1   (a) Constant shear flow on a thin-walled closed section. (b) Closed section viewed as two open sections. (c) Shear flow resultants on each section. (d) Pure couple resultant for the closed section.

the shear flow about any point on line a-a' is $2A_1q + 2A_2q = 2Aq$. The resultant of the shear flow on the section to the left of a-a' is a force of magnitude ql parallel to a-a'. The shear flow on the right has a force resultant with the same magnitude but opposite direction. We conclude that *constant shear flow q on a closed section is equivalent to a pure couple of magnitude $2Aq$.*

It would appear that an idealized beam of open section composed of a single, constant shear flow web like that in Figure 2.5.20 cannot transmit pure torsion. Such a loading has a zero force resultant ($\mathbf{R} = 0$), and Equation 2.5.10 dictates that the shear flow q would therefore be zero. What actually happens is that a nonuniform shear stress field develops in the web, to produce the requisite torsional moment. Using just the analytical tools we have developed so far in this book, we can gain some insight into how this stress field develops to enable an open section to resist torsion.

Consider the closed, semicircular, thin-walled section of uniform wall thickness shown in Figure 2.6.2 (a). By *thin-walled* we mean that, in general, the depth of the section is much larger than the wall thickness t, which in this particular case means $2r \gg t$. The area enclosed by the walls is one-half that of a circle, $\pi r^2/2$, so that

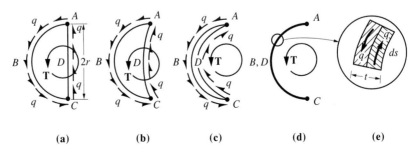

(a)   **(b)**   **(c)**   **(d)**   **(e)**

Figure 2.6.2   The same torque **T** is applied to a closed thin-walled section whose enclosed area approaches zero, moving from (a) to (d).

according to Equation 2.5.11, the constant shear flow around the cell is $T/\pi r^2$. Holding the torque constant, if we reduce the cell area, then the shear flow must increase. Therefore, as we move from (a) to (c) in Figure 2.6.2, the shear flow becomes progressively larger as wall ADC approaches wall ABC.

If we continue to reduce the area, holding the torque fixed, we approach the situation depicted in parts (d) and (e) of the figure, where the two walls are nearly in contact, separated by a gap of infinitesimal width and forming a composite wall. Let the thickness of the neighboring walls be $t/2$, so that the composite wall thickness is t. We then have what appears to be an open section. The equal and opposite shear flows in the composite wall are separated by a distance of $t/2$, and they produce a net shear flow of zero. However, at each point, they form a couple that yields a differential torque $dT = (qds) \times (t/2)$, where ds is an element of length along the wall. Integrating over the length of the wall, we get $T = \pi r t q/2$, which means that the shear flow in each part of the composite wall is $2T/\pi r t$.

The shear *stress* τ is found by dividing the shear flow by the wall thickness $t/2$, so that $\tau = 4T/\pi r t^2$. The ratio of the shear flow in the open semicircular section to that in its closed counterpart is $2r/t$. For example, if $r = 10$ in. and $t = 0.1$ in., our analysis suggests that the shear flow in the open section will be two orders of magnitude greater.

The described approach suffers from oversimplification, but it yields the right order of magnitude for the shear stress in the open web. An exact approach,[4] using the theory of elasticity, reveals that the shear stress varies linearly across the wall, except in small regions close to each end (cf. Figure 2.6.3). The shear stress is zero at the middle of the wall, and its maximum value, at the outer surfaces, is $3T/bt^2$, where b is the length of the wall. For a semicircular section, $b = \pi r$, so that $\tau_{max} = 3T/\pi r t^2$. This is 25 percent lower than our prediction.

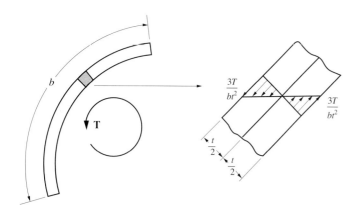

Figure 2.6.3 Shear stress due to torsion in a thin-walled open section.

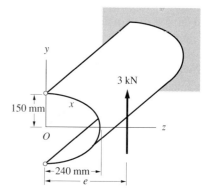

Figure 2.6.4 Idealized cantilever beam with a semielliptical web.

A 500 lb shear load is applied at the free end.

Example 2.6.1 Figure 2.6.4 shows an idealized beam composed of two flanges and a curved, thin web that has a semi-elliptical shape. A 3 kN vertical shear load is applied to the free end. Calculate the shear flow and find the horizontal location where the shear force must be applied to produce no torsion.

The shear force on any section between the free end and the wall is the same. Figure 2.6.5 shows such a section with the constant shear flow q whose resultant is a 3 kN upward force at a perpendicular distance e from O. The projected length of the web on the vertical y-axis is 0.3 m. Therefore, from Equation 2.5.9, $0.3q = 3$, or $q = 10$ kN/m. The area of an ellipse is

| [4]C. T. Wang, *Applied Elasticity*, Chapter 5, McGraw-Hill, 1953.

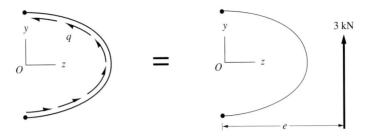

Figure 2.6.5   The shear flow on the left is statically equivalent to the vertical load located a distance e from O.

πab, where a and b are the lengths of the semimajor and semiminor axes. For the case at hand, $A = \pi ab/2 = (\pi \times 0.15 \times 0.24)/2 = 0.05655\,\text{m}^2$. Substituting this area into Equation 2.5.11, we get $3e = 2 \times 0.05655 \times 10$, so that $e = 0.3770\,\text{m}$. Thus, the 3 kN load must be applied 137 mm to the right of the web. This value of e is independent of the value of the shear load.

Example 2.6.2   Calculate the shear flow in the walls of the closed section subjected to pure torsion, shown in Figure 2.6.6.

The enclosed area is a composite of the triangle on the left, the rectangle in the middle, and the semicircle on the right. Thus, $A = (\frac{1}{2} \times 18 \times 12) + (29 \times 18) + (\pi \times 9^2) = 884.5\,\text{in}^2$. Since the torque on the section is 50,000 in.-lb, Equation 2.5.11 yields $q = 50,000/(2 \times 884.5) = 28.26\,\text{lb/in}$. Shear *stress* is the shear flow divided by the wall thickness. If the walls of this section have a uniform thickness of, say, 0.025 in., then the shear stress is $28.26/0.025 = 1130\,\text{psi}$.

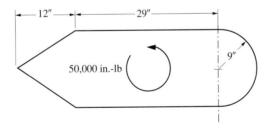

Figure 2.6.6   Closed section under pure torsion.

A rigid, minimally stable, three-dimensional, idealized beam is one that can transmit arbitrary shear and torsional loading. If such a beam is to be statically determinate as well, it must then have precisely three webs. The required number of flanges is therefore three for a closed section and four for an open section. The following example illustrates the solution procedure for such beams. Keep in mind that we are ignoring structural deformation; therefore, classical beam theory cannot be used (although it would yield the same results).

Example 2.6.3   Figure 2.6.7 shows a 50-inch span of a tapered box beam. At the left end, where the indicated loads are applied, there is a rigid rib at which the flange loads are zero. Other ribs (not shown) of varying size are spaced along the beam to maintain the form of the cross section. Calculate the shear flows and flange loads at the 50-inch station, which lies between two ribs.

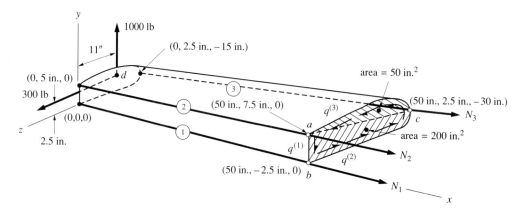

Figure 2.6.7   Free-body diagram of a tapered box beam, showing the three flange loads and three shear flows at the 50-inch station.

For convenience, we will use vector notation to handle the three-dimensional bookkeeping in this problem. The coordinates of the endpoints of the three stringers are given, so we can find the unit vectors $\mathbf{e}_1$, $\mathbf{e}_2$, and $\mathbf{e}_3$ along each stringer. First, calculate the lengths of the stringers, as follows:

$$L_1 = \sqrt{(50-0)^2 + (-2.5-0)^2 + (0-0)^2} = 50.06 \text{ in.}$$

$$L_2 = \sqrt{(50-0)^2 + (7.5-5)^2 + (0-0)^2} = 50.06 \text{ in.}$$

$$L_3 = \sqrt{(50-0)^2 + (2.5-2.5)^2 + [-30-(-15)]^2} = 52.20 \text{ in.}$$

Then,

$$\mathbf{e}_1 = \frac{1}{50.06}[(50-0)\mathbf{i} + (-2.5-0)\mathbf{j} + (0-0)\mathbf{k}] = 0.9988\mathbf{i} - 0.04994\mathbf{j}$$

Similarly, for the other two stringers, we get

$$\mathbf{e}_2 = 0.9988\mathbf{i} + 0.04994\mathbf{j}$$

$$\mathbf{e}_3 = 0.9578\mathbf{i} - 0.2874\mathbf{k}$$

Next, set the net force on the free body equal to zero, as follows:

$$N_1\mathbf{e}_1 + N_2\mathbf{e}_2 + N_3\mathbf{e}_3 - \left(10q^{(1)}\right)\mathbf{j} + q^{(2)}\mathbf{r}_{bc} + q^{(3)}\mathbf{r}_{ca} + (1000\mathbf{j} + 300\mathbf{k}) = 0$$

where $\mathbf{r}_{bc}$ and $\mathbf{r}_{ca}$ are the position vectors drawn from point b to c and from point c to a, respectively. That is,

$$\mathbf{r}_{bc} = 5\mathbf{j} - 30\mathbf{k} \qquad \mathbf{r}_{ca} = 5\mathbf{j} + 30\mathbf{k}$$

Therefore, we can write the equilibrium equation as follows:

$$(0.9988N_1 + 0.9988N_2 + 0.9578N_3)\,\mathbf{i}$$
$$+ \left(-0.04994N_1 + 0.04994N_2 - 10q^{(1)} + 5q^{(2)} + 5q^{(3)} + 1000\right)\mathbf{j}$$
$$+ \left(-0.2872N_3 - 30q^{(2)} + 30q^{(3)} + 300\right)\mathbf{k} = 0$$

The net moment about an arbitrary point must also vanish. Choosing point a, we have

$$\mathbf{r}_{ab} \times (N_1\mathbf{e}_1) + \mathbf{r}_{ac} \times (N_3\mathbf{e}_3) + \left[(2)(200)q^{(2)}\right]\mathbf{i} + \left[(2)(50)q^{(3)}\right]\mathbf{i} + \mathbf{r}_{ad} \times (1000\mathbf{j} + 300\mathbf{k}) = 0$$

Substituting terms and using the familiar determinant expansion method[5] to compute the cross products, we obtain

$$\begin{vmatrix} \mathbf{i} & \mathbf{j} & \mathbf{k} \\ 0 & -10 & 0 \\ 0.9988N_1 & -0.04994N_1 & 0 \end{vmatrix} + \begin{vmatrix} \mathbf{i} & \mathbf{j} & \mathbf{k} \\ 0 & -5 & -30 \\ 0.9578N_3 & 0 & -0.2874N_3 \end{vmatrix} + 400q^{(2)}\mathbf{i} + 100q^{(3)}\mathbf{i} + \begin{vmatrix} \mathbf{i} & \mathbf{j} & \mathbf{k} \\ -50 & -5 & -11 \\ 0 & 1000 & 300 \end{vmatrix} = 0$$

Expanding the determinants and collecting terms yields the following form of the moment equilibrium equation:

$$(1.437N_3 + 400q^{(2)} + 100q^{(3)} + 9500)\mathbf{i} + (-28.74N_3 + 15,000)\mathbf{j} + (9.988N_1 + 4.789N_3 - 50,000)\mathbf{k} = 0$$

Finally, we extract the scalar components of the force and moment equilibrium equations to obtain the following six equations:

$$
\begin{aligned}
0.9988N_1 &+ 0.9988N_2 + 0.9578N_3 & &= 0 \\
-0.04994N_1 &+ 0.04994N_2 & -10q^{(1)} + 5q^{(2)} + 5q^{(3)} &= -1000 \\
& -0.2874N_3 & -30q^{(2)} + 30q^{(3)} &= -300 \\
& +1.437N_3 & +400q^{(2)} + 100q^{(3)} &= -9500 \\
& -28.74N_3 & &= -15,000 \\
9.988N_1 & +4.789N_3 & &= 50,000
\end{aligned}
$$

The solution of this system is therefore

$$N_1 = 4756 \text{ lb} \qquad N_2 = -5257 \text{ lb} \qquad N_3 = 522.0 \text{ lb}$$
$$q^{(1)} = 28.0 \text{ lb/in.} \qquad q^{(2)} = 19.5 \text{ lb/in.} \qquad q^{(3)} = -24.5 \text{ lb/in.}$$

These quantities with their proper orientation are sketched in Figure 2.6.8.

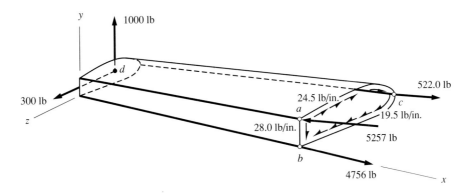

Figure 2.6.8   Solution of the problem illustrated in Figure 2.6.7.

---

[5]Determinants and other matrix operations are reviewed in Chapter 9.

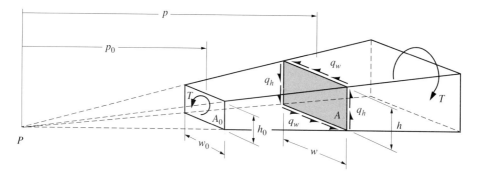

Figure 2.6.9   Torque box in which all four corners intersect in a common point P.

The equations for the shear flow in trapezoidal panels, derived in section 2.5, can be applied to the analysis of idealized single-cell tapered *torque boxes*. These structures are like that of Example 2.6.3, but are subjected to pure torsion at each end. To treat them as statically determinate, we assume that there are no ribs between the two ends of the box that take out any of the torsional loading. We then require that the panels and stringers be individually in equilibrium.

The torque box in Figure 2.6.9 is of rectangular cross section, and all four sides taper in such a way that the corners, extended, intersect at a common point P (recall Figure 2.5.19). Then, for any section of the box, the height h and width w of the cross section are related to those (h_0 and w_0) at the smallest end of the box by

$$\frac{h}{h_0} = \frac{w}{w_0} \qquad [2.6.1]$$

Since the axial load at each end of the stringers is zero, equilibrium requires that the net average shear flow exerted along the length of each stringer must be zero. Therefore, the average shear flow $\bar{q}_h$ in the side panels equals that of the top and bottom panels, $\bar{q}_w$. The average shear flows in the panels are related to those at the small end of the structure by the shear panel equilibrium relation, Equation 2.5.4:

$$\bar{q}_h = q_{h0}\left(\frac{h_0}{h_L}\right) \qquad \bar{q}_w = q_{w0}\left(\frac{w_0}{w_L}\right)$$

These two equations, together with Equation 2.6.1 and the fact that $\bar{q}_h = \bar{q}_w$, imply that $q_{h0} = q_{w0}$. It then follows, again from Equation 2.5.4, that $q_h(x) = q_w(x)$, that is, that the shear flow around any section of this torque box is constant. Since Equation 2.5.5 implies that $\bar{q}_h(x) = \bar{q}_w(x)$, the net average shear flow over any portion of a stringer is zero, which means that the flange loads are zero everywhere between the two ends of the torque box. Since the shear flow $q(x)$ is constant around every section, we can calculate it most easily using Equation 2.5.11, $q = T/2A$.

In general, the webs of a torque box do not all taper in the same ratio, which means that the shear flows at a section will *not* be the same in all of the webs and $q = T/2A$ cannot be used to find the shear flow. The difference in the shear flows between two adjacent webs gives rise to axial loads in the flange members to which they are attached. Figure 2.6.10 illustrates such a case. The box carries pure torsion, and the flange loads are zero at each end, where rigid ribs are placed to maintain the form of the cross section. As we move through a distance L from one end of the box to the other, the flange loads increase in magnitude and then return to zero. At intermediate sections, the in-plane components of the flange loads act together with the panel shear flows to carry the torsional load.

As pointed out previously, we assume that interior ribs, rigid in their own plane, maintain the shape of the cross section and exert no intermediate loads on the webs and stringers. Given the torque T, we can find the four

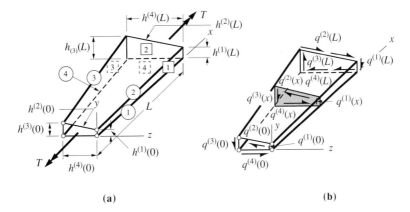

Figure 2.6.10   (a) Idealized torque box with pure torsion applied to each end. Webs are referenced by numbers enclosed by squares. Flange members are referenced by numbers enclosed by circles. (b) The corresponding shear flows at each end and on an intermediate section.

panel shear flows and the four stringer loads by requiring that each of the shear panels and each of the stringers be in equilibrium. Since the axial load in each stringer is zero at $x = 0$ and $x = L$, and since the force exerted by a shear panel on a stringer is the average shear flow in the panel times the length of the stringer, it follows that the average shear flow in adjacent webs must be equal. This means that the average shear flow is the same in every panel of the structure. Let $\bar{q}$ denote this common average shear flow. In terms of the average shear flow, the shear flow in panel i, according to Equation 2.5.4, is

$$q^{(i)}(x) = \bar{q}^{(i)} \frac{h^{(i)}(0)\, h^{(i)}(L)}{h^{(i)}(x)^2}$$ [2.6.2]

At station $x = 0$, where the flange loads are zero, the resultant of the shear flow distribution must be a pure torque T around the x-axis. Thus, summing moments about, say, flange 4 in the plane of the section, we get

$$2 \times \left[\tfrac{1}{2} h^{(1)}(0)\, h^{(4)}(0)\right] \times q^{(1)}(0) + 2 \times \left[\tfrac{1}{2} h^{(3)}(0)\, h^{(4)}(0)\right] \times q^{(2)}(0) = T$$

Substituting Equation 2.6.2, this becomes

$$h^{(1)}(0)\, h^{(4)}(0) \left(\bar{q}\frac{h^{(1)}(L)}{h^{(1)}(0)}\right) + h^{(3)}(0)\, h^{(4)}(0) \left(\bar{q}\frac{h^{(2)}(L)}{h^{(2)}(0)}\right) = T$$

Solving this equation yields the average shear flow $\bar{q}$ of the torque box in terms of the applied torque T:

$$\bar{q} = \frac{T}{2\bar{A}}$$ [2.6.3]

where,

$$\bar{A} = \left[\frac{h^{(1)}(L)\, h^{(2)}(0) + h^{(2)}(L)\, h^{(3)}(0)}{2}\right] \frac{h^{(4)}(0)}{h^{(2)}(0)}$$ [2.6.4]

To establish an analogy between Equation 2.6.3 and Equation 2.5.11, we may view $\bar{A}$ as the area of the fictitious intermediate cross section around which the shear flow has the constant value $\bar{q}$. Of course, the shear flow is generally *not* constant around any section of a tapered torque box. In fact, we use $\bar{q}$ together with Equation 2.6.2 to find the individual shear flows $q^{(1)}(x)$, $q^{(2)}(x)$, $q^{(3)}(x)$, and $q^{(4)}(x)$ in the shear panels of Figure 2.6.10.

The flange load at a section equals the net average shear flow up to that particular point of the stringer, as will be illustrated in the following example.

Example 2.6.4   The idealized, stiffened web torque box structure in Figure 2.6.11(a) is tapered in span, depth, and chord. Given that it transmits a pure torque of 42,000 in.-lb, calculate the shear flows and flange loads at 20-in. intervals.

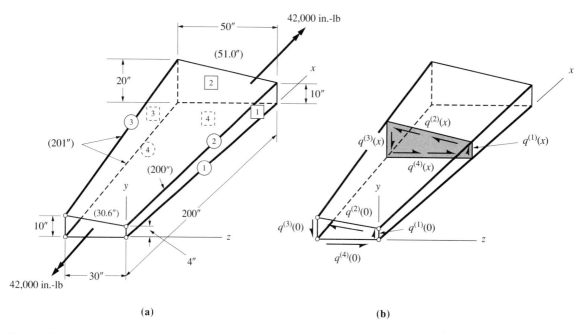

(a)   (b)

Figure 2.6.11   (a) Torque box with pure torsion applied to each end. (b) The shear flows on intermediate sections.

Numbers in parentheses are the lengths, in inches, of the inclined edges of the box.

This structure is similar to the one in Figure 2.6.10, so we can use Equation 2.6.4 to calculate $\bar{A}$ for the structure in Figure 2.6.11, as follows:

$$\bar{A} = \left[\frac{h^{(1)}(L)\,h^{(2)}(0) + h^{(2)}(L)\,h^{(3)}(0)}{2}\right]\frac{h^{(4)}(0)}{h^{(2)}(0)} = \left(\frac{10 \times 30.6 + 51 \times 10}{2}\right)\frac{30}{30.6} = 400 \text{ in.}^2$$

Therefore, according to Equation 2.6.3, the average shear flow for the torque box is

$$\bar{q} = \frac{42,000}{2 \times 400} = 52.5\frac{\text{lb}}{\text{in.}} \tag{a}$$

The linear variation of the *i*th panel's width with *x* is $h^{(i)}(x) = h^{(i)}(0) + \left[h^{(i)}(L) - h^{(i)}(0)\right]x/L$. Thus, for each of the panels, we have

$$h^{(1)}(x) = 4 + (10 - 4) \times \frac{x}{200} = 4 + 0.03x \qquad \text{[b]}$$

$$h^{(2)}(x) = 30.6 + (51 - 30.6) \times \frac{x}{200} = 30.6 + 0.102x$$

$$h^{(3)}(x) = 10 + (20 - 10) \times \frac{x}{200} = 10 + 0.05x$$

$$h^{(4)}(x) = 30 + (50 - 30) \times \frac{x}{200} = 30 + 0.1x$$

Substitute Equations (a) and (b) into Equation 2.6.2 to find the shear flows (lb/in.) within each of the panels.

$$q^{(1)}(x) = \bar{q}\frac{h^{(1)}(0)h^{(1)}(L)}{h^{(1)}(x)^2} = 52.5\frac{4 \times 10}{(4 + 0.03x)^2} = \frac{2100}{(4 + 0.03x)^2}$$

$$q^{(2)}(x) = \bar{q}\frac{h^{(2)}(0)h^{(2)}(L)}{h^{(2)}(x)^2} = 52.5\frac{30.6 \times 51}{(30.6 + 0.102x)^2} = \frac{81,930}{(30.6 + 0.102x)^2}$$

$$q^{(3)}(x) = \bar{q}\frac{h^{(3)}(0)h^{(3)}(L)}{h^{(3)}(x)^2} = 52.5\frac{10 \times 20}{(10 + 0.05x)^2} = \frac{10,500}{(10 + 0.05x)^2}$$

$$q^{(4)}(x) = \bar{q}\frac{h^{(4)}(0)h^{(4)}(L)}{h^{(4)}(x)^2} = 52.5\frac{30 \times 50}{(30 + 0.1x)^2} = \frac{78,750}{(30 + 0.1x)^2}$$

The average shear flow between the section at $x = 0$ and an intermediate station is found in a similar way, using Equation 2.5.5, so that for each of the four panels, we have,

$$\bar{q}^{(1)}(x) = \bar{q}\frac{h^{(1)}(L)}{h^{(1)}(x)} = 52.5\frac{10}{4 + 0.03x} = \frac{525}{4 + 0.03x}$$

$$\bar{q}^{(2)}(x) = \bar{q}\frac{h^{(2)}(L)}{h^{(2)}(x)} = 52.5\frac{51}{30.6 + 0.102x} = \frac{2678}{30.6 + 0.102x}$$

$$\bar{q}^{(3)}(x) = \bar{q}\frac{h^{(3)}(L)}{h^{(3)}(x)} = 52.5\frac{20}{10 + 0.05x} = \frac{1050}{10 + 0.05x}$$

$$\bar{q}^{(4)}(x) = \bar{q}\frac{h^{(4)}(L)}{h^{(4)}(x)} = 52.5\frac{50}{30 + 0.1x} = \frac{2625}{30 + 0.1x}$$

The force exerted on a portion of a stringer by an adjacent panel is the average shear flow in the panel times the length of the stringer over which it acts. Thus, to calculate the flange load at an intermediate section, we balance the net shear force from the attached panels up to that point against the axial stringer load at that point. For stringer 1, the picture is as shown in Figure 2.6.12, from which we see that $N_1 = [\bar{q}^{(1)} - \bar{q}^{(4)}]\,s$, where s is the distance measured along the stringer from the point where the flange load is zero (the near section of the torque box). Since this stringer lies on the x axis, $s = x$. In fact, for this torque box, all of the stringers are so slightly inclined to the x axis that for each of them, we may write $s = x$ with sufficient accuracy. Free-body diagrams like that of Figure 2.6.12 can be drawn for the other stringers to obtain the pertinent flange load relations. In summary, we find

$$N_1(x) = \left[\bar{q}^{(1)}(x) - \bar{q}^{(4)}(x)\right]x$$

$$N_2(x) = \left[\bar{q}^{(2)}(x) - \bar{q}^{(1)}(x)\right]x$$

$$N_3(x) = \left[\bar{q}^{(3)}(x) - \bar{q}^{(2)}(x)\right]x$$

$$N_4(x) = \left[\bar{q}^{(4)}(x) - \bar{q}^{(3)}(x)\right]x$$

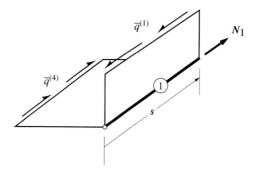

Figure 2.6.12   Relationship between flange load and average shear flows in adjacent webs (stringer 1 illustrated).

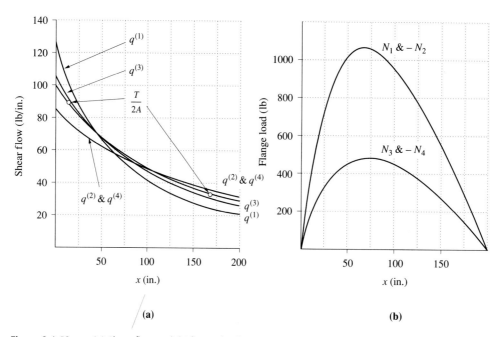

(a)   **(b)**

Figure 2.6.13   (a) Shear flow and (b) flange load variations with span for the torque box in Figure 2.6.11.

We now have all of the equations necessary to evaluate the shear flows and flange loads at any section of the structure. Before doing so, however, let us calculate, for comparison purposes, the shear flow at each section, using Equation 2.5.11, $q = T/2A$. This expression is valid only if the shear flow is constant around the cross section of area A. That is clearly not the case for the problem at hand, but it may be useful as an estimate of the average value for the shear flows around the section. Since the area of an intermediate section is that of a trapezoid, we have $A = h^{(4)}(x) \times \frac{1}{2}[h^{(1)}(x) + h^{(3)}(x)]$. Therefore, for this "constant" shear flow $q(x)$, we get

$$q(x) = \frac{T}{2A} = \frac{42,000}{2\left\{h^{(4)}(x) \times \frac{1}{2}\left[h^{(1)}(x) + h^{(3)}(x)\right]\right\}} = \frac{42,000}{(30 + 0.1x)(14 + 0.08x)}$$

The shear flows and flange loads are plotted in Figure 2.6.13. Notice that the flange loads reach their peak values at about one-third of the span from the smaller end. Note as well that $T/2A$ is, indeed, very nearly the average of the four shear flows at a given section.

In Chapter 8, after we become familiar with the means to compute the displacements accompanying torsion, we will see how torsion of a thin-walled tube generally produces not only rotation or *twist* around the torsion axis, as expected, but also *warping* of the cross sections. Warping is the deformation of an initially plane section into a curved surface, due to displacements normal to the section, in the axial direction. Restraint of warping—by supports, say—has a profound effect on the shear flow distribution, and it increases the twisting resistance (torsional stiffness).

2.7 FRAMES

Frames, like trusses, are skeletal structures composed of slender members. However, unlike trusses, the members of a frame transmit shear and bending, as well as axial loads. Therefore, the individual frame elements behave like a beam with a superimposed axial load. The joints of a frame are usually considered to be *rigid*, although pin joints may be found in the structure. Moments can be transferred from one member to another across a rigid joint, but not across a pin. Some examples of plane frames are shown in Figure 2.7.1.

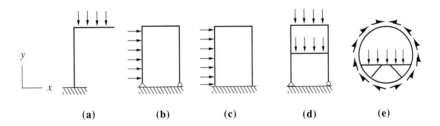

| (a) | (b) | (c) | (d) | (e) |

Figure 2.7.1   Examples of rigid-jointed plane frames.

In frame analysis, we seek the distribution of axial, transverse, and bending loads throughout each member of the structure. Statics is used to determine the forces within a given frame element, once the loads at either end of that element are known. To identify and calculate these loads may require drawing free-body diagrams of each frame member. The action–reaction principle is used to transfer loads across an unloaded node involving just two members. If an external load is applied to a node, or if more than three members frame into it, then the node must also be isolated and analyzed as a free body.

In a plane frame, there are two components of force and a couple at each end of a member, for a total of possibly six unknowns per element. There are three equations of equilibrium for each element of the frame and three for each node to be analyzed. Again, the structure is statically determinate only if the total number of equilibrium equations equals the total number of unknown forces and couples. In a three-dimensional frame, there are three components of force and three components of couple at each end of an element, yielding as many as 12 unknowns per member. On the other hand, there are only six equations of equilibrium for each element and each node.

The frames in Figure 2.7.1(a) and (b) are statically determinate. The three reactions at the supports are found in each case by means of the three equations of equilibrium for the frame: $\sum F_x = 0$, $\sum F_y = 0$, and $\sum M_z = 0$.

The frame in Figure 2.7.1(c) has three additional support reactions, for a total of six unknowns, three more than the number of equilibrium equations. Hence, that frame is statically indeterminate.

The structure in part (d) of the figure is also statically indeterminate. Although we can find the three external reactions using statics, the presence of an extra horizontal member prevents our finding the internal loads in the upper part of the frame. Figure 2.7.2 shows free-body diagrams of the individual members of that frame, as well as free-body diagrams of the two midside nodes where three members join together. Using the symbol Q for both forces and couples (generalized forces), it can be seen from the figure that there are 27 generalized forces comprising the list of unknowns for this problem. There are $6 \times 3 = 18$ equilibrium equations for the six beams and $2 \times 3 = 6$ for the two midside nodes, for a total of 24 equations. We are therefore three equations short of being able to analyze this frame by means of statics alone.

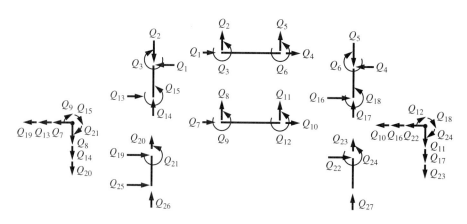

Figure 2.7.2   Free-body diagrams of elements of the frame in Figure 2.7.1(d).

The fuselage frame in Figure 2.7.1(e) is another example of a statically indeterminate frame. The means of dealing with such problems are presented in Chapters 7 and 10.

The following examples illustrate the use of statics for frame analysis, in which one of the principal objectives is often that of finding the shear and bending moment distributions. Since circular frames are common in aircraft structures, we will give them special attention.

Example 2.7.1   Find the location and value of the maximum bending moment in the semicircular frame shown in Figure 2.7.3.

From the free-body diagram in part (b) of the figure, we readily obtain the reactions at the supports: $X_A = 0.899P$ $Y_A = Y_B = 0.25P$.

Part (c) of Figure 2.7.3 shows the internal forces on a section at an angle of $\theta > 30°$ away from the support at B. Setting the net moment about a point on that section equal to zero, we obtain $M + 0.25P \times (r - r\cos\theta) = 0$ or

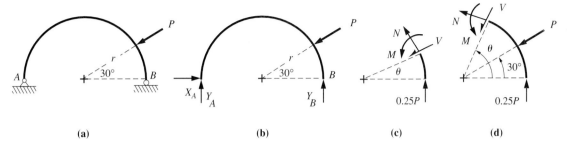

Figure 2.7.3   (a) Semicircular frame with an intermediate transverse point load. (b) Free-body diagram of the complete frame. (c) Free-body diagram of a section at $\theta < 30°$. (d) Free-body diagram for a section at $\theta > 30°$.

$$M = -0.25P \times r(1 - \cos\theta) \quad (\theta < 30°) \tag{a}$$

The free-body diagram in part (d) of the figure is for a section located at $\theta > 30°$. Summing the moments about such a point yields

$$M + 0.25P \times (r - r\cos\theta) - P \times r\sin(\theta - 30) = 0$$

or

$$M = -P \cdot r(0.25 + 0.25\cos\theta - 0.866\sin\theta) \quad (\theta > 30°) \tag{b}$$

The bending moment is zero at $\theta = 0$, and it increases to $0.0335Pr$ at $30°$, where the transverse point load is applied. At a point between $30°$ and $180°$, the bending moment achieves an extreme value before returning to zero at A. To find where that point is, we must differentiate expression (b) for M, valid for $\theta > 30°$, as follows:

$$\frac{dM}{d\theta} = P \cdot r(-0.25\sin\theta - 0.866\cos\theta)$$

Setting this derivative equal to zero and solving for θ, we get

$$\frac{\sin\theta}{\cos\theta} = \frac{-0.866}{0.25}$$

That is, $\tan\theta = -3.46$, which means that $\theta = 106.1°$. Substituting this value of θ into Equation (b) yields

$$M_{\max} = -0.651Pr$$

The bending moment distribution is sketched in Figure 2.7.4.

The skin of an airplane exerts a shear flow on the fuselage frames to which it is attached. The following example shows how to incorporate shear flow into a typical, statically determinate frame problem.

Example 2.7.2   Find the location and value of the maximum bending moment in the frame in Figure 2.7.5. The semicircular portion of the structure is acted on by a uniform shear flow, while point loads are applied at the endpoints A and A' of the horizontal elements. The given shear flow and point loads form a self-equilibrating set.

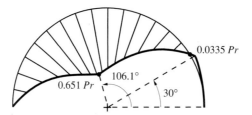

Figure 2.7.4 Bending moment as a function of position around the semicircular frame of Figure 2.7.3.

Compression occurs on the side of the frame on which the moment diagram is plotted.

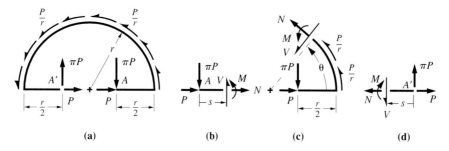

Figure 2.7.5 (a) Semicircular frame with self-equilibrating applied load system. (b), (c), & (d) Free-body diagrams of each element of the frame, revealing the internal forces on transverse sections.

Let us work our way around the frame counterclockwise from point A to point A', starting at point A. Assume that a positive bending moment produces compression on the inner surface of the frame. From Figure 2.7.5(b) we find $M = -\pi P s$ in the initial horizontal segment. Entering the circular portion, we use the free-body diagram in (c). In order to obtain the moment of the shear flow about the section, we need to calculate the area of the *segment* of the circle over which it acts. A segment of a circle is the shaded region shown in Figure 2.7.6. Its area is that of the *sector abcd* ($\frac{1}{2}r^2\theta$) minus that of the isosceles triangle *abc* ($\frac{1}{2}r^2 \sin\theta$). Therefore, setting the net moment about the section in Figure 2.7.5(c) equal to zero, assuming counterclockwise moments are positive, yields

$$M + 2\left[\frac{r^2}{2}(\theta - \sin\theta)\right]\left(\frac{P}{r}\right) + \pi P\left(r\cos\theta - \frac{r}{2}\right) + P \cdot r \sin\theta = 0$$

Upon simplification, this becomes

$$M = P \cdot r\left[\pi\left(\tfrac{1}{2} - \cos\theta\right) - \theta\right]$$ [a]

Clearly, $M = 0$ at the top of the frame, where $\theta = \pi/2$. To find the extreme values of M in the semicircular portion, we calculate its derivative, as follows:

$$\frac{dM}{d\theta} = P \cdot r(\pi \sin\theta - 1),$$

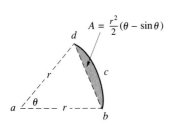

$$A = \frac{r^2}{2}(\theta - \sin\theta)$$

Figure 2.7.6   Area of the segment of a circle spanned by the angle θ.

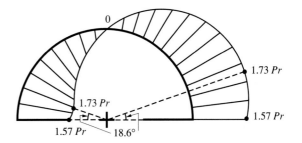

Figure 2.7.7   Bending moment distribution in the semicircular frame member.

Compression occurs on the side of the frame on which the bending moment diagram is plotted.

Setting $dM/d\theta$ equal to zero and solving for θ yields $\theta = \sin^{-1}(1/\pi)$. Thus, $\theta = 18.56°$ and $161.4°$. We substitute these two angles into Equation (a) to obtain, respectively, $M = -1.731Pr$ and $M = 1.731Pr$. The bending moment distribution throughout the frame is plotted in Figure 2.7.7.

When analyzing frames in three dimensions, we can take advantage of vector notation—as we do with trusses—to make our job easier, as illustrated in the following example.

Example 2.7.3   Figure 2.7.8(a) shows a frame built into a wall at point W, with a 20 kN load **P** applied at point D. Calculate the magnitudes of the shear force, bending moment, and torsional moment acting on a transverse section through the frame at point O, located at some distance from the built-in support.

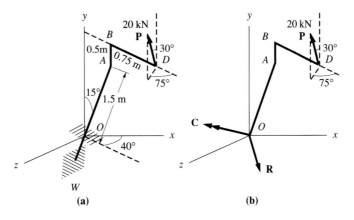

Figure 2.7.8   (a) A frame loaded at point C and built into the wall at W. (b) Free-body diagram of the frame to one side of the transverse section through O.

A free-body diagram of the portion of the frame on one side of the cut (Figure 2.7.8(b)) reveals the internal force **R** and the couple **C**, which act on the frame at point O. The shear force V is the component of **R** that lies in the plane of the transverse section (perpendicular to the frame axis). Likewise, the bending moment M is the component of **C** that is perpendicular to the frame axis. The torsional moment T is the component of **C** parallel to the axis.

We find **R** by setting the net force on the free body equal to zero. Thus,

$$\mathbf{R} + \mathbf{P} = 0$$

where

$$\mathbf{P} = 20[\sin 30 \cos 75(\cos 40\mathbf{i} + \sin 40\mathbf{k}) + \cos 30\mathbf{j} + \sin 30 \sin 75(-\sin 40\mathbf{i} + \cos 40\mathbf{k})]$$

that is,

$$\mathbf{P} = -4.226\mathbf{i} + 17.32\mathbf{j} + 9.063\mathbf{k} \quad (\text{kN})$$

Therefore,

$$\mathbf{R} = 4.226\mathbf{i} - 17.32\mathbf{j} - 9.063\mathbf{k} \quad (\text{kN})$$

The couple moment **C** is found by setting the net moment about point O equal to zero, or

$$\mathbf{C} + \mathbf{r}_{OD} \times \mathbf{P} = 0$$

where $\mathbf{r}_{OD}$, the position vector drawn from O to D, is given by

$$\mathbf{r}_{OD} = 1.5[\sin 15 \,(\cos 40\mathbf{i} + \sin 40\mathbf{k}) + \cos 15\,j] + 0.5\mathbf{j} + 0.75(\cos 40\mathbf{i} + \sin 40\mathbf{k})$$

That is,

$$\mathbf{r}_{OD} = 0.8719\mathbf{i} + 1.949\mathbf{j} + 0.7316\mathbf{k} \quad (\text{m})$$

Hence,

$$\mathbf{C} = -\mathbf{r}_{OD} \times \mathbf{P} = - \begin{vmatrix} \mathbf{i} & \mathbf{j} & \mathbf{k} \\ 0.8719 & 1.949 & 0.7316 \\ -4.226 & 17.32 & 9.063 \end{vmatrix} = -4.992\mathbf{i} + 10.99\mathbf{j} - 23.34\mathbf{k} \quad (\text{kN-m})$$

To find the shear component of **R**, we first obtain its axial component by projecting **R** onto the frame axis using the dot product operation. The unit vector $\mathbf{e}_{\|}$ pointing from A to O along the frame axis is

$$\mathbf{e}_{\|} = -[\sin 15 \,(\cos 40\mathbf{i} + \sin 40\mathbf{k}) + \cos 15\mathbf{j}] = -0.1983\mathbf{i} - 0.9659\mathbf{j} - 0.1664\mathbf{k}$$

Letting N represent the axial force at O, we therefore have,

$$N = \mathbf{R} \cdot \mathbf{e}_{\|} = 17.40 \quad (\text{kN})$$

N is positive, which means it is a tensile force directed away from the section. Since the shear force V and the axial force N are the two legs of the right triangle whose hypotenuse is **R**, we obtain the magnitude of the shear force V from the Pythagorean formula,

$$V = \sqrt{\mathbf{R} \cdot \mathbf{R} - N^2} = \sqrt{400 - 17.40^2} = 9.861 \quad (\text{kN})$$

We proceed in the same way to find the bending moment M. First, project **C** onto the axis of the member to obtain the torsional component T:

$$T = \mathbf{C} \cdot \mathbf{e}_{\|} = -5.742 \quad (\text{kN-m})$$

The minus sign means that the torque on the section is clockwise as viewed from the support W. The bending moment, by definition, is perpendicular to the twisting moment. Therefore, it follows that the magnitude of the bending moment is

$$M = \sqrt{\mathbf{C} \cdot \mathbf{C} - T^2} = \sqrt{690.4 - (-5.742)^2} = 26.90 \text{ (kN-m)}$$

Our final statically determinate frame example deals with shear and moment calculations in a three-dimensional context. The problem is simplified by the fact that the elements are at right angles to one another. This makes vector notation unnecessary, and we can deal directly with the components of forces and couples along cartesian coordinate axes aligned with the members of the frame. Observe that what is a bending moment to one element may be a twisting moment to its neighbor.

Example 2.7.4   Find the location and magnitude of the maximum bending moment in the frame of Figure 2.7.9. The support at 1 is capable of exerting reactive forces in the y- and z-directions and couples about those axes. The support at 2 can exert forces only in the x- and y-directions.

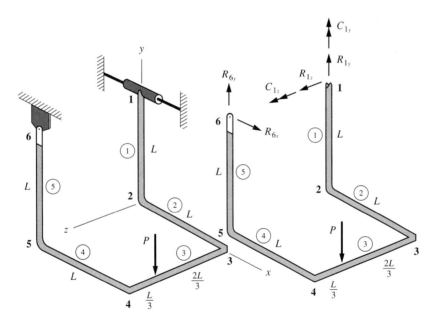

Figure 2.7.9   Statically determinate frame and its free-body diagram.

Using the free-body diagram in the figure, we find the reactions at the supports.

$$\sum F_x = 0: \qquad R_{6x} + 0 = 0 \qquad \Rightarrow \qquad R_{6x} = 0$$

$$\sum M_y = 0: \qquad C_{1y} + R_{6x}L = 0 \qquad \Rightarrow \qquad C_{1y} = 0$$

$$\sum M_x = 0: \qquad -R_{6y}L + \tfrac{2}{3}PL = 0 \qquad \Rightarrow \qquad R_{6y} = \tfrac{2}{3}P$$

$$\sum F_y = 0: \qquad R_{1y} + R_{6y} - P = 0 \qquad \Rightarrow \qquad R_{1y} = \tfrac{1}{3}P$$

$$\sum F_z = 0: \qquad R_{1z} + 0 = 0 \qquad \Rightarrow \qquad R_{1z} = 0$$

$$\sum M_z = 0: \qquad C_{1z} - PL = 0 \qquad \Rightarrow \qquad C_{1z} = PL$$

To determine the internal forces throughout the frame, we can start at point 6 and move from member to member, calculating the forces and moments at each section using the free-body diagrams drawn in Figure 2.7.10.

From Figure 2.7.10(a), it is evident that member 5 is subjected only to a uniform, tensile, axial load $N = \tfrac{2}{3}P$.

Figure 2.7.10(b) shows that member 4 is in a state of uniform shear, $V = \tfrac{2}{3}P$, with a linearly varying bending moment $M = \tfrac{2}{3}Ps$, where s is measured from the beginning of the member (point 5).

Figure 2.7.10(c) takes us into the first part of member 3, before we encounter the applied load P. Equilibrium of that free body reveals a constant downward shear force $V = \tfrac{2}{3}P$, a uniform torque $T = \tfrac{2}{3}PL$ in the direction shown, and a linearly varying bending moment $M = \tfrac{2}{3}Ps$, where s is measured from point 4.

Figure 2.7.10(d) shows the loads in member 3 at sections beyond the applied load P. Shifting the $s = 0$ point to P, we find from statics that

$$V = \tfrac{1}{3}P, \quad M = \tfrac{1}{3}P\left(\tfrac{2}{3}L - s\right), \quad \text{and} \quad T = \tfrac{2}{3}PL$$

all in the directions indicated.

From part (e) of the figure, we find that the shear in member 2 is $V = \tfrac{1}{3}P$ and the bending moment is $M = \tfrac{1}{3}P(2L + s)$, where s starts at point 3.

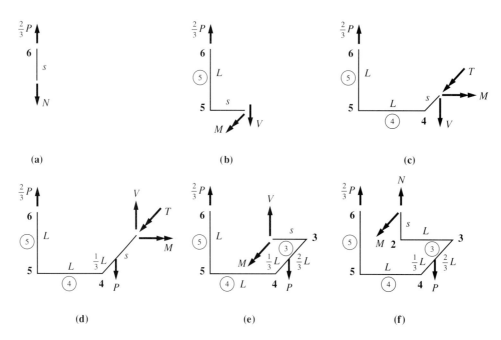

Figure 2.7.10 Free-body diagrams showing the internal forces acting on each transverse section of the frame in Figure 2.7.9.

Finally, Figure 2.7.10(f) reveals that member 1 has a uniform axial load $N = \frac{1}{3}P$ and a constant bending moment of $M = PL$.

The internal bending moment distribution is plotted in Figure 2.7.11. The bending moment produces compression on the side of the frame element on which the bending moment is drawn. Clearly, member 1 sustains the largest bending moment PL over its entire length.

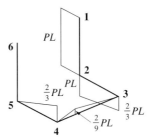

Figure 2.7.11   Bending moment distribution for the frame in Figure 2.7.9.

EXERCISES

2.1   The truss is symmetric about a vertical axis through node 3. Determine the range of values for P and ϕ for which the member forces do not exceed 15 kN tension and 1.5 kN compression.

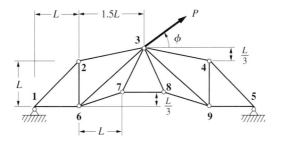

Exercise 2.1

2.2   Calculate the forces in rods 9–11 and 13–15 of the fuselage truss.

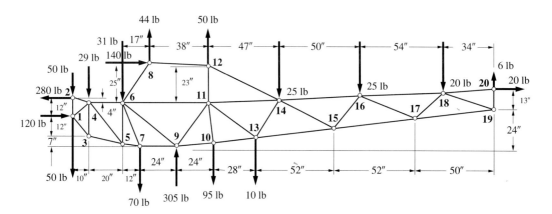

Exercise 2.2

2.3   Design a plane, statically determinate, pin-connected truss to transmit a vertical load P over the distance d to the rigid wall W, to which the structure is fastened. The load P ranges in value from P_d downward to P_u upward. Your objective is to define the geometry of the truss and the size of the members. The truss is required to be as light as possible and to have the minimum number of connections.

In the absence of any other instructions, let $d = 2$ m and -2.5 kN $\leq P \leq 5$ kN, where the minus sign means downward. Select a common steel tubing for all of the members, using (unless otherwise directed) the data of Table 2.2.2 and the same limits on axial load as prescribed in Example 2.2.1, including the buckling criterion.

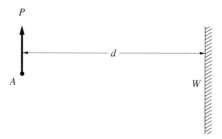

Exercise 2.3

2.4   Calculate the maximum load P that can be applied if the buckling load, Equation 2.2.4, must not be exceeded in any member of the truss. All members have a solid circular cross section.

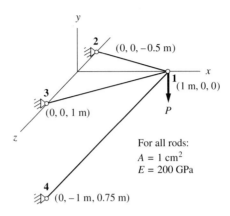

Exercise 2.4

2.5   Calculate the load in each member of the space truss.

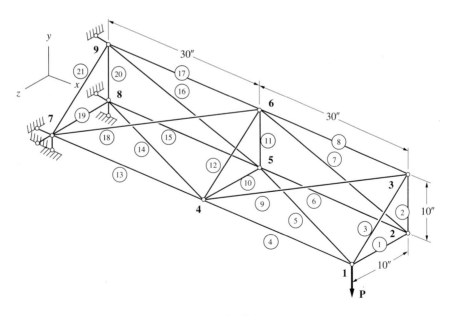

Exercise 2.5

2.6 Find the magnitude and location of the maximum bending moment in the pinned beam structure shown.

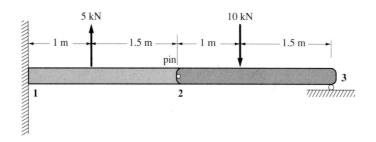

Exercise 2.6

2.7 Draw the shear and moment diagrams for both of the beam elements (elements 6 and 7) of the structure shown, all members of which are joined together by smooth pins. Note that element 6 has a uniformly distributed load of 10 lb/in., while a 5000 in.-lb point couple acts on element 7. A point load of 100 lb is applied to the *pin* at node 2. The support at node 1 acts on the pin at that location, whereas the roller support at the right acts directly on the beam.

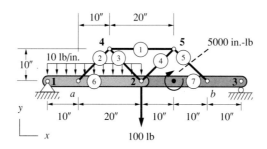

Exercise 2.7

2.8   Calculate the shear flow in each of the four panels.

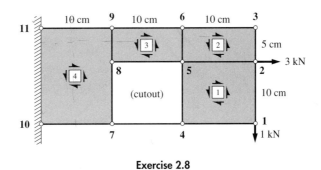

Exercise 2.8

2.9   Find the shear flow in each of the four webs and the maximum axial load in each stiffener.

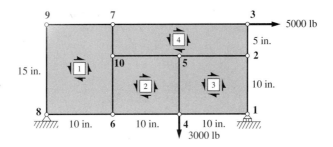

Exercise 2.9

2.10   Calculate the shear flow in each of the five webs and find the maximum axial load in each stiffener.

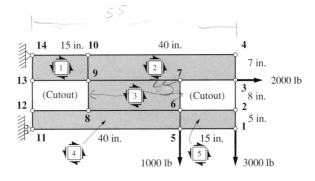

Exercise 2.10

2.11 Calculate the three panel shear flows, indicating the proper sign relative to the assumed directions shown.

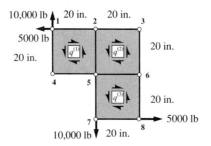

Exercise 2.11

2.12 Calculate the average shear flow in the panel. Remember that the axial members can take loads only along their length.

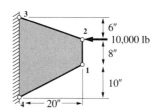

Exercise 2.12

2.13 Find the magnitude and location of the maximum tensile and compressive axial loads in the stiffeners of the tapered web structure.

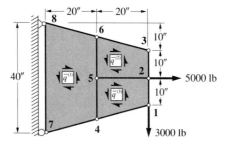

Exercise 2.13

2.14   In each panel, plot the variation of shear flow with distance x measured from section A of the torque box.

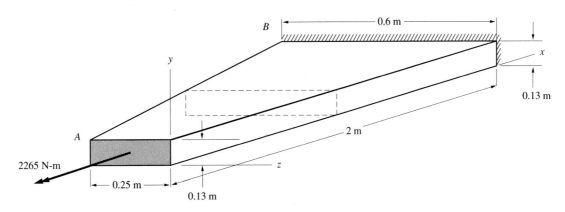

Exercise 2.14

2.15   Calculate the shear flows in the three webs of the section if they are statically equivalent to the 4 kN force acting as shown. The y–z axes are centered at O, and the y axis is an axis of symmetry.

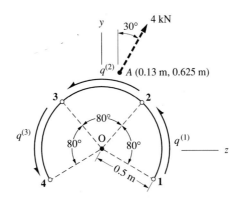

Exercise 2.15

2.16 A 1000 lb shear force acts on the beam cross section as shown. Calculate the magnitude of the shear flow $q^{(1)}$.

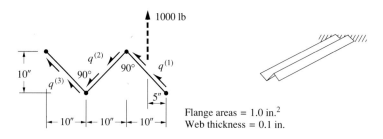

Flange areas = 1.0 in.^{2}
Web thickness = 0.1 in.

Exercise 2.16

2.17 A 10 kN load in the y–z plane, having the line of action shown, is applied to section A of the box beam. Calculate the shear flows in the three panels at section B, which is 5 meters away. The coordinates, in meters, of the endpoints of each of the three stringers are shown on the sketch.

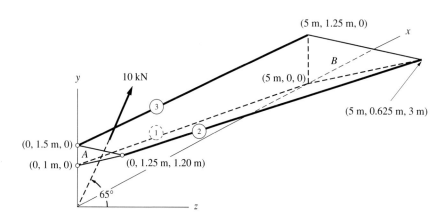

Exercise 2.17

2.18   Draw the shear and moment diagrams for the horizontal member of the frame. The connections at 2, 3, 6, and 7 are smooth pins.

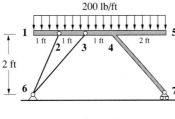

Exercise 2.18

2.19   Calculate the bending moment at point A of the circular beam

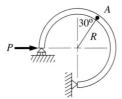

Exercise 2.19

2.20   The semicircular frame has a uniformly distributed load p on the horizontal portion and a uniformly distributed shear load p around the circular portion, as shown. Find the magnitude and location of the maximum bending moment in the frame.

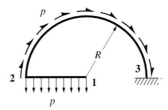

Exercise 2.20

2.21 Find the location and magnitude of the maximum bending moment in the frame, and specify where in the frame the bending moment is zero.

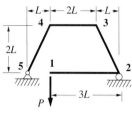

Exercise 2.21

2.22 Rods 1, 2, and 3 are two-force members, and the connection at 6 is a ball joint (no moment restraint). Find the reaction at 6 and the forces in members 1, 2, and 3 due to the applied force **P** at 1.

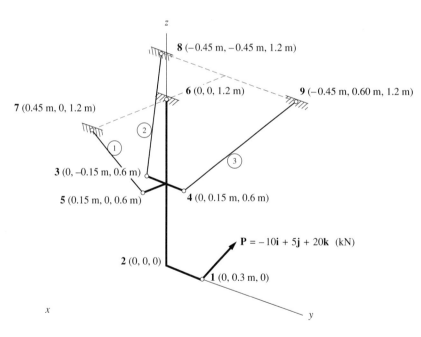

Exercise 2.22

2.23   The two-force members connect flotation devices to the fuselage of an amphibious aircraft. Points A, B, and C attach to the fuselage. Points D, E, F, and G attach to the floats. Each of the points D, E, F, and G receives a load of $(1400\mathbf{i} + 2900\mathbf{j})$ lb from the floats. Find the force in each member and the reactions at A, B, and C.

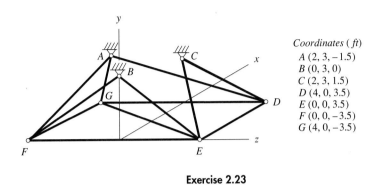

Coordinates (ft)
A (2, 3, −1.5)
B (0, 3, 0)
C (2, 3, 1.5)
D (4, 0, 3.5)
E (0, 0, 3.5)
F (0, 0, −3.5)
G (4, 0, −3.5)

Exercise 2.23

2.24   Forces act on the landing gear as shown. AC and AD are two-force members. The fixture at B can react to forces in all directions but can support only a twisting moment along the axis of the strut. Find all of the reactions at A, B, and C.

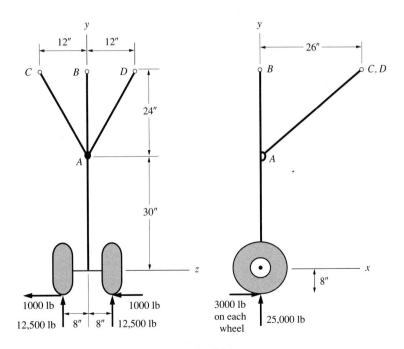

Exercise 2.24

2.25   The landing gear consists of the bent member *ABC* and the two supports *BD* and *BE*, both of which are two-force members. The connection at *C* can resist forces in all directions but only a twisting moment about the *BC* axis. Neglecting the weight of the landing gear, calculate (a) the forces in the struts *BE* and *BC*; (b) the reactions at *C*; (c) the magnitude and location of the maximum bending moment in *ABC*.

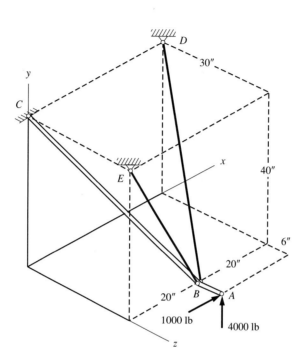

Exercise 2.25

2.26   The landing gear consists of the bent member *ABCD* and the two supports *CE* and *CF*, which are two-force members. Points *D*, *E*, and *F* connect to the fuselage. Point *A* receives the loads from the wheel. The connection at *D* can resist forces in all directions but only a torsion about *BCD* axis. Neglecting the weight of the landing gear, calculate (a) the forces in members *CE* and *CF*; (b) the reactions at *D*; (c) the magnitude and location of the maximum bending moment in *ABCD*.

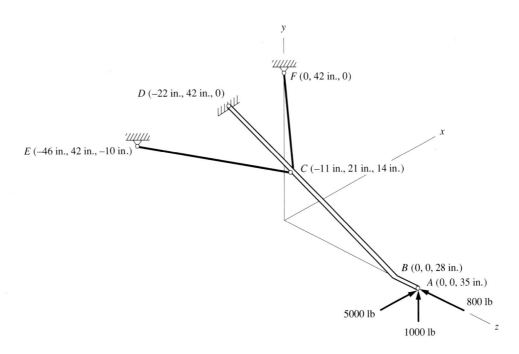

Exercise 2.26

Applied Elasticity: Fundamental Concepts

CHAPTER OUTLINE

3.1   INTRODUCTION

Elasticity is the ability to bounce back. An elastic structure once deformed by applied loads returns to its original shape when the loads are removed. Most engineering structures—aircraft being no exception—behave elastically within the range of the loads they are designed to support. They are also meant to retain their shape, deforming only slightly (in comparison to their overall size) under ordinary service conditions. Linear elasticity, in which the displacements are directly proportional to the applied loads, is the simplest and most pervasive kind of behavior in man-made structures. On the other hand, stretching a rubber band, inflating a balloon, buckling (bowing) a meter stick, bending a paper clip, and denting an automobile fender are familiar examples of nonlinear responses to applied load.

Structural analysis contains many time-tested formulas based on assumptions applicable to specific classifications of structures, materials, and load cases. We turn to the theory of elasticity to describe the behavior of elastic solids in precise detail and to assess the consequences of simplifying assumptions when they become necessary. This chapter reviews and explores the following topics:

- Stress and strain.
- Stress equilibrium and strain compatibility.
- Stress and strain transformations.
- Principal stress and strain.
- Generalized Hooke's Law.
- Saint-Venant's principle.
- Strain energy.
- Anisotropy.
- Failure theories for steady and fluctuating loads.
- Margins of safety.

Since plane elasticity provides much of the theoretical basis for treating structures typical of aircraft construction, parts of this chapter focus on plane stress and strain.

3.2   STRESS

Stress is a measure of force intensity. Its dimensions are force divided by area. Stress at a point must be defined carefully, because it is directionally dependent. We can apply a modest force to a rubber band by stretching it between our fingers. Because the force is distributed over the small cross section of the band, a sizeable stress develops in the material. In fact, we can easily pull hard enough to produce a load intensity on the cross section that overcomes the molecular bonding forces, causing the rubber band to break. Meanwhile, the lateral surface of the rubber band is free of stress (neglecting atmospheric pressure). Is the stress at a point on the surface of the band zero? Since the point lies on both the lateral surface and the cross section (which is only an imaginary surface unless the band actually breaks in two), the answer depends on which surface is being considered.

Figure 3.2.1a shows a three-dimensional solid body of arbitrary shape with a general set of boundary conditions. It has both point supports and supports that extend over portions of its surface (like the clamped edge of a plate). Acting on the *surface* of the body is a set of point loads **F**, point couples **C**, and distributed loads. Also shown is a body force **B**. Body forces act over the *volume* of the material. The weight of an object is a body force, as are inertial forces, such as centrifugal force in rotating machinery and maneuvering loads ("g forces") in aircraft.

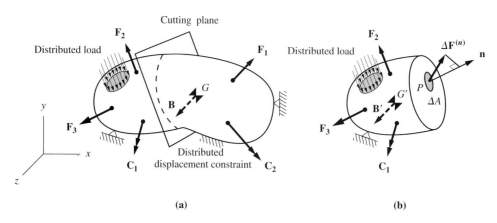

(a) **(b)**

Figure 3.2.1: (a) A solid body, loaded and constrained in an arbitrary fashion. (b) Force acting on a small area of a cutting plane through point *P* in the solid. G and G' are the centers of mass.

Pick a point *P* anywhere on or within the solid and pass a cutting plane with a unit normal **n** through *P*, thereby dividing the body into two parts, one on each side of the cutting plane. Figure 3.2.1b shows one side of the cut. The cut has revealed the small net force $\Delta\mathbf{F}^{(n)}$ acting at *P* and exerted on the particles within the small area ΔA surrounding *P* by molecular attraction of the neighboring points on the other side of the cut. The magnitude and direction of $\Delta\mathbf{F}^{(n)}$ depends on the orientation of the cutting plane. That is why the superscript *n* is attached to $\Delta\mathbf{F}$. The force $\Delta\mathbf{F}^{(n)}$ also depends on ΔA since changing the size of the area surrounding *P* increases or decreases the number of particles exerting the force across the cutting plane. Observe that, in general, $\Delta\mathbf{F}^{(n)}$ does not point in the direction of the unit normal **n**.

If we reduce the size of ΔA, then $\Delta\mathbf{F}^{(n)}$ will diminish in magnitude. If we divide the force $\Delta\mathbf{F}^{(n)}$ (a vector) by the area ΔA on which it acts (a scalar), we create another vector with the dimensions of force per unit area. In the limit as ΔA and $\Delta\mathbf{F}^{(n)}$ approach zero, the ratio $\Delta\mathbf{F}^{(n)}/\Delta A$ approaches the *traction* $\mathbf{T}^{(n)}$ at *P*, as follows:

$$\mathbf{T}^{(n)} = \lim_{\Delta A \to 0} \frac{\Delta\mathbf{F}^{(n)}}{\Delta A} \qquad\qquad \text{[3.2.1]}$$

This definition implies that we can subdivide a material indefinitely without ever encountering any discrete microstructure, let alone molecules and atoms. That is the *continuum hypothesis*, upon which all of solid and fluid mechanics rests. In continuum mechanics, the details of material microstructure are accounted for by the use of macroscopic quantities, such as mass density, modulus of elasticity, viscosity, etc.

It is useful to resolve the traction $\mathbf{T}^{(n)}$ into orthogonal components σ_n normal to the cutting plane and τ_n parallel to the plane, as shown in Figure 3.2.2:

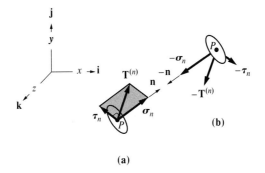

(a)

Figure 3.2.2   (a) Normal and shear components of the traction on an arbitrary plane through P. (b) Traction components on the opposite side of the cutting plane at P.

$$\mathbf{T}^{(n)} = \boldsymbol{\sigma}_n + \boldsymbol{\tau}_n \qquad \text{[3.2.2]}$$

where $\boldsymbol{\tau}_n$ is perpendicular to $\mathbf{n}$ and lies in the same plane as both $\mathbf{n}$ and $\mathbf{T}^{(n)}$. Observe, as illustrated in Figure 3.2.2, that the tractions on opposite sides of the cutting plane through the same point P form an action–reaction pair. Thus, if $\mathbf{T}^{(n)}$ is the traction on the surface, with unit normal $\mathbf{n}$, then the traction on the other side of the surface, with unit normal $-\mathbf{n}$, is $-\mathbf{T}^{(n)}$.

If, for the same point P, the cutting plane is chosen to be parallel to one of the three cartesian coordinate planes, then we have one of the three situations illustrated in Figure 3.2.3. The traction on the planes whose normals are in the positive x, y, and z directions, respectively, are $\mathbf{T}^{(x)}$, $\mathbf{T}^{(y)}$, and $\mathbf{T}^{(z)}$. Their components are

$$\begin{aligned}
\mathbf{T}^{(x)} &= \sigma_x \mathbf{i} + \tau_{xy} \mathbf{j} + \tau_{xz} \mathbf{k} \\
\mathbf{T}^{(y)} &= \tau_{yx} \mathbf{i} + \sigma_y \mathbf{j} + \tau_{yz} \mathbf{k} \\
\mathbf{T}^{(z)} &= \tau_{zx} \mathbf{i} + \tau_{zy} \mathbf{j} + \sigma_z \mathbf{k}
\end{aligned} \qquad \text{[3.2.3]}$$

where $\mathbf{i}$, $\mathbf{j}$, and $\mathbf{k}$ are the unit vectors in the direction of the positive x, y, and z axes, respectively. Note that each of the shear components of the traction has been further resolved into its two cartesian components. The first subscript on a shear component τ indicates the plane of the traction, and the second component specifies the direction

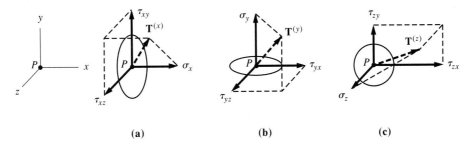

(a)   (b)   (c)

Figure 3.2.3   Positive stress components on positively-oriented cutting planes normal to: (a) the x axis, (b) the y axis, and (c) the z axis.

in which it acts. The nine components of traction on three orthogonal planes are called the *stresses*. Equations 3.2.3 give the stress components on *positively-oriented* coordinate surfaces, that is, those with unit normals pointing in the positive coordinate directions. On the opposite sides of those surfaces, which are oriented in the negative coordinate directions, the tractions are $-\mathbf{T}^{(x)}$, $-\mathbf{T}^{(y)}$, and $-\mathbf{T}^{(z)}$, which are the complements of the action–reaction pairs.

At a point P, we may relate the components of the tractions $\mathbf{T}^{(n)}$, $\mathbf{T}^{(x)}$, $\mathbf{T}^{(y)}$, and $\mathbf{T}^{(z)}$ by forming a small, tetrahedral, free body with the cutting plane through P as its base and the three coordinate planes as its other three surfaces, as illustrated in Figure 3.2.4. The tractions act on the centroids of the four outer faces of the tetrahedron, whereas the body force $\Delta\mathbf{B}$ acts at the center of mass G. The area of the inclined triangle pqr is ΔA_z, while the areas of the other three sides—Orp, Opq, and Oqr—are ΔA_x, ΔA_y, and ΔA_z, respectively. The surface tractions on the three negatively-oriented coordinate planes are shown with a negative sign in Figure 3.2.4, for the reasons previously discussed.

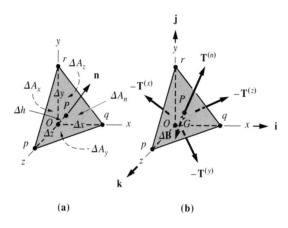

(a)   (b)

Figure 3.2.4   (a) Small tetrahedron $Opqr$ with the inclined cutting plane through P as its base. (b) Free-body diagram showing the surface tractions and body force density.

If the elemental tetrahedron in Figure 3.2.4(b) is to be in equilibrium, the net force on it must vanish, or

$$\mathbf{T}^{(n)}\Delta A_n - \mathbf{T}^{(x)}\Delta A_x - \mathbf{T}^{(y)}\Delta A_y - \mathbf{T}^{(z)}\Delta A_z + \Delta\mathbf{B} = 0 \qquad \text{[3.2.4]}$$

Each traction is multiplied by the area on which it acts. Here, ΔA_x, ΔA_y, and ΔA_z are the projections of ΔA_n onto the three coordinate planes. It is therefore easy to show that

$$\Delta A_x = n_x \Delta A_n \quad \Delta A_y = n_y \Delta A_n \quad \Delta A_z = n_z \Delta A_n \qquad \text{[3.2.5]}$$

where n_x, n_y, and n_z are the components of the unit normal $\mathbf{n}$ to the inclined surface pqr. The body force $\Delta\mathbf{B}$ is obtained by multiplying the *body force per unit volume* $\mathbf{b}$ by the volume of the tetrahedron, $\frac{1}{3}\Delta A_n\Delta h$, where Δh is its altitude, that is, the perpendicular distance from the base of pqr to the vertex O. Substituting Equations 3.2.5 into Equation 3.2.4 and canceling the common factor ΔA_n yields

$$\mathbf{T}^{(n)} - \mathbf{T}^{(x)}n_x - \mathbf{T}^{(y)}n_y - \mathbf{T}^{(z)}n_z + \mathbf{b}\frac{\Delta h}{3} = 0$$

Upon taking the limit as Δh goes to zero, so that the tetrahedron shrinks to point P, the body force term vanishes, and we are left with only the surface tractions at P:

$$\mathbf{T}^{(n)} = \mathbf{T}^{(x)}n_x + \mathbf{T}^{(y)}n_y + \mathbf{T}^{(z)}n_z \qquad [3.2.6]$$

The traction $\mathbf{T}^{(n)}$ may be resolved into its cartesian components:

$$\mathbf{T}^{(n)} = T_x^{(n)}\mathbf{i} + T_y^{(n)}\mathbf{j} + T_z^{(n)}\mathbf{k} \qquad [3.2.7]$$

Substituting Equation 3.2.7 together with Equations 3.2.3 into Equation 3.2.6, we get

$$T_x^{(n)}\mathbf{i} + T_y^{(n)}\mathbf{j} + T_z^{(n)}\mathbf{k} = (\sigma_x\mathbf{i} + \tau_{xy}\mathbf{j} + \tau_{xz}\mathbf{k})n_x + (\tau_{yx}\mathbf{i} + \sigma_y\mathbf{j} + \tau_{yz}\mathbf{k})n_y + (\tau_{zx}\mathbf{i} + \tau_{zy}\mathbf{j} + \sigma_z\mathbf{k})n_z$$

From this single vector equation, we extract three scalar equations by setting the x, y, and z components on each side equal to each other, as follows:

$$\boxed{\begin{aligned} T_x^{(n)} &= \sigma_x n_x + \tau_{yx} n_y + \tau_{zx} n_z \\ T_y^{(n)} &= \tau_{xy} n_x + \sigma_y n_y + \tau_{zy} n_z \\ T_z^{(n)} &= \tau_{xz} n_x + \tau_{yz} n_y + \sigma_z n_z \end{aligned}} \qquad [3.2.8]$$

These are *Cauchy's* formulas,[1] by means of which we can obtain the traction on any plane through a point if we know the nine cartesian components of stress. Thus, the *state of stress* at a point is given by specifying the three components of stress on each of the three cartesian coordinate planes, and these nine quantities are commonly expressed in the form of a square, three-by-three *stress matrix*, as follows:

$$[\sigma] = \begin{bmatrix} \sigma_x & \tau_{yx} & \tau_{zx} \\ \tau_{xy} & \sigma_y & \tau_{zy} \\ \tau_{xz} & \tau_{yz} & \sigma_z \end{bmatrix} \begin{array}{l} x \text{ plane} \\ y \text{ plane} \\ z \text{ plane} \end{array} \qquad [3.2.9]$$

(columns: x direction, y direction, z direction)

It is also convenient to represent the state of stress at a point by showing the nine components of stress acting on the sides of a differential cube, as illustrated in Figure 3.2.5. Observe that Figure 3.2.5a is a more compact version of Figure 3.2.3. Figure 3.2.5 also serves to illustrate *the sign convention for stresses*. A normal stress is positive if it is directed away from a surface, thereby exerting tension, whereas a negative normal stress acts towards the surface, producing compression. Positive shear stresses point in the positive coordinate directions on positively-oriented surfaces and in the negative coordinate directions on negatively–oriented surfaces.

Given the state of stress at a point, we can find the normal stress σ_n on a plane with unit normal $\mathbf{n}$ simply by projecting the traction $\mathbf{T}^{(n)}$ onto $\mathbf{n}$ by means of the dot product operation:

$$\sigma_n = \mathbf{T}^{(n)} \cdot \mathbf{n} = T_x^{(n)}n_x + T_y^{(n)}n_y + T_z^{(n)}n_z \qquad [3.2.10]$$

From Equation 3.2.2, the shear component of the traction is given by

$$\boldsymbol{\tau}_n = \mathbf{T}^{(n)} - \boldsymbol{\sigma}_n \qquad [3.2.11]$$

---

[1] Augustin Louis Cauchy (1789–1857), French mathematician.

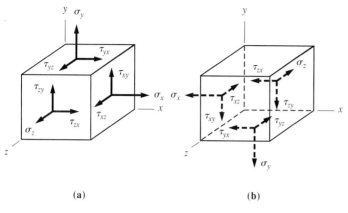

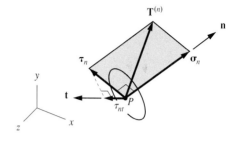

Figure 3.2.6   An in-plane component τ_{nt} of the shear traction on a plane.

(a)   (b)

Figure 3.2.5   State of stress on a differential cube at a point. Positive components of stress on: (a) the front, positively-oriented surfaces, and (b) the rear, negatively-oriented surfaces.

Let **t** be a unit vector lying in the plane normal to **n** (and therefore to $\boldsymbol{\sigma}_n$), and let τ_{nt} be the component of $\boldsymbol{\tau}_n$ in the direction of **t**, as illustrated in Figure 3.2.6. Then

$$\tau_{nt} = \boldsymbol{\tau}_n \cdot \mathbf{t} = (\mathbf{T}^{(n)} - \boldsymbol{\sigma}_n) \cdot \mathbf{t}$$

However, $\boldsymbol{\sigma}_n \cdot \mathbf{t} = 0$, therefore

$$\tau_{nt} = \mathbf{T}^{(n)} \cdot \mathbf{t} = T_x^{(n)} t_x + T_y^{(n)} t_y + T_z^{(n)} t_z \qquad [3.2.12]$$

The components of $\mathbf{T}^{(n)}$ are obtained from Cauchy's equations. Substituting those expressions into Equations 3.2.10 and 3.2.12 yields

$$\boxed{\begin{aligned} \sigma_n &= \sigma_x n_x^2 + \sigma_y n_y^2 + \sigma_z n_z^2 + 2\tau_{xy} n_x n_y + 2\tau_{xz} n_x n_z + 2\tau_{yz} n_y n_z \\ \tau_{nt} &= \sigma_x n_x t_x + \sigma_y n_y t_y + \sigma_z n_z t_z + \tau_{xy}(n_x t_y + t_x n_y) \\ &\quad + \tau_{xz}(n_x t_z + t_x n_z) + \tau_{yz}(n_y t_z + t_y n_z) \end{aligned}} \quad (\mathbf{n} \cdot \mathbf{t} = 0) \qquad [3.2.13]$$

These are formulas we can employ to find the component of normal stress on a plane and any in-plane component of shear stress at a point where the state of stress (Figure 3.2.5) is known. Remember that, in general, τ_{nt} is not the resultant shear, which is found by means of Equation 3.2.11, as follows:

$$\tau_n = \sqrt{\boldsymbol{\tau}_n \cdot \boldsymbol{\tau}_n} = \sqrt{T^{(n)^2} - \sigma_n^2} \qquad [3.2.14]$$

3.3   EQUILIBRIUM

The stress components in general vary from point to point within a solid, which means that the stress components differ slightly in magnitude on opposite faces of the differential element shown in Figure 3.3.1. To examine the variation of stress at a point, we can use the Taylor series expansion for a function of three variables. If $f(x, y, z)$

is the value of a function at the point with coordinates (x, y, z), the value of the function at a neighboring point with coordinates $(x + dx, y + dy, z + dz)$ is, to the first-order differentials in dx, dy, and dz,

$$f(x + dx, y + dy, z + dz) = f(x, y, z) + \frac{\partial f}{\partial x}dx + \frac{\partial f}{\partial y}dy + \frac{\partial f}{\partial z}dz \qquad [3.3.1]$$

Figure 3.3.1 is a free-body diagram of the stress element, with the surface tractions acting at the *centroids* of the faces and the body force $d\mathbf{B}$ acting at the center of mass G of the cube. Note that two of the three coordinates are held fixed as we move from the centroid of one face to that of the opposite face.

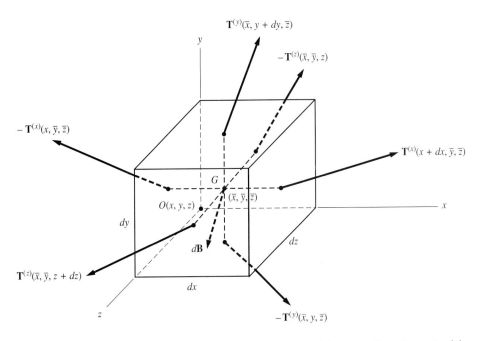

Figure 3.3.1   Variation of tractions at a point. The coordinates of the corner O are (x, y, z), while those of the center of mass G of the cube are $(\bar{x}, \bar{y}, \bar{z})$.

For the free body in Figure 3.3.1 to be in equilibrium, the net moment on the element must vanish. Multiply each traction by the surface area on which it acts to obtain the surface force, then sum the moments about the center of mass of the cube G (so that the moment of the body force is conveniently zero):

$$\left(\frac{dx}{2}\mathbf{i}\right) \times \mathbf{T}^{(x)}(x + dx, \bar{y}, \bar{z})dydz + \left(\frac{dy}{2}\mathbf{j}\right) \times \mathbf{T}^{(y)}(\bar{x}, y + dy, \bar{z})dxdz + \left(\frac{dz}{2}\mathbf{k}\right) \times \mathbf{T}^{(z)}(\bar{x}, \bar{y}, z + dz)dxdy$$

$$+ \left(-\frac{dx}{2}\mathbf{i}\right) \times [-\mathbf{T}^{(x)}(x, \bar{y}, \bar{z})dydz] + \left(-\frac{dy}{2}\mathbf{j}\right) \times [-\mathbf{T}^{(y)}(\bar{x}, y, \bar{z})dxdz]$$

$$+ \left(\frac{dz}{2}\mathbf{k}\right) \times [-\mathbf{T}^{(z)}(\bar{x}, \bar{y}, z)dxdy] = 0$$

Using Equation 3.3.1, we can write this as

$$\left(\frac{dx}{2}\mathbf{i}\right) \times \left(\mathbf{T}^{(x)} + \frac{1}{2}\frac{\partial \mathbf{T}^{(x)}}{\partial x}dx\right)dydz + \left(\frac{dy}{2}\mathbf{j}\right) \times \left(\mathbf{T}^{(y)} + \frac{1}{2}\frac{\partial \mathbf{T}^{(y)}}{\partial y}dy\right)dxdz + \left(\frac{dz}{2}\mathbf{k}\right) \times$$

$$\left(\mathbf{T}^{(z)} + \frac{\partial \mathbf{T}^{(z)}}{\partial z}dz\right)dxdy + \left(\frac{dx}{2}\mathbf{i}\right) \times \mathbf{T}^{(x)}dydz + \left(\frac{dy}{2}\mathbf{j}\right) \times \mathbf{T}^{(y)}dxdz + \left(\frac{dz}{2}\mathbf{k}\right) \times \mathbf{T}^{(z)}dxdy = 0$$

Collecting terms then leads to

$$\mathbf{i} \times \mathbf{T}^{(x)}dxdydz + \mathbf{j} \times \mathbf{T}^{(y)}dxdydz + \mathbf{k} \times \mathbf{T}^{(z)}dxdydz + \cdots = 0$$

where "$\cdots$" represents the terms which are of the fourth order in the differentials dx, dy, and dz and can therefore be neglected. Cancelling the common factor $dxdydz$ and introducing Equations 3.2.3 yields

$$\mathbf{i} \times (\sigma_x\mathbf{i} + \tau_{xy}\mathbf{j} + \tau_{xz}\mathbf{k}) + \mathbf{j} \times (\tau_{yx}\mathbf{i} + \sigma_y\mathbf{j} + \tau_{yz}\mathbf{k}) + \mathbf{k} \times (\tau_{zx}\mathbf{i} + \tau_{zy}\mathbf{j} + \sigma_z\mathbf{k}) = 0$$

or

$$\mathbf{i}(\tau_{yz} - \tau_{zy}) + \mathbf{j}(\tau_{zx} - \tau_{xz}) + \mathbf{k}(\tau_{xy} - \tau_{yx}) = 0$$

Each component on the left must vanish for the vector sum to be zero, so that

$$\boxed{\tau_{yx} = \tau_{xy} \qquad \tau_{zx} = \tau_{xz} \qquad \tau_{zy} = \tau_{yz}}$$ [3.3.2]

Therefore, the requirement of equilibrium of moments implies that *the shear stresses on orthogonal planes at a point are equal in magnitude*. This means, among other things, that only six components of stress—three normal stresses and three shear stresses—are required to specify the state of stress at a point. By virtue of Equation 3.3.2, the stress matrix in Equation 3.2.9 is symmetric.

Enforcing equilibrium also requires that the net *force* on the differential cube in Figure 3.3.1 be zero. Thus,

$$\mathbf{T}^{(x)}(x + dx, y, z)dydz - \mathbf{T}^{(x)}(x, y, z)dydz + \mathbf{T}^{(y)}(x, y + dy, z)dxdz - \mathbf{T}^{(y)}(x, y, z)dxdz$$
$$+ \mathbf{T}^{(z)}(x, y, z + dz)dxdy - \mathbf{T}^{(z)}(x, y, z)dxdy + d\mathbf{B} = 0$$

With the aid of Equation 3.3.1, this becomes

$$\frac{\partial \mathbf{T}^{(x)}}{\partial x}dxdydz + \frac{\partial \mathbf{T}^{(y)}}{\partial y}dydxdz + \frac{\partial \mathbf{T}^{(z)}}{\partial z}dzdxdy + d\mathbf{B} = 0$$

Dividing by the volume of the element, $dV = dxdydz$, and observing that $d\mathbf{B}/dV$ is equal to $\mathbf{b}$, which is the body force density, we get

$$\frac{\partial \mathbf{T}^{(x)}}{\partial x} + \frac{\partial \mathbf{T}^{(y)}}{\partial y} + \frac{\partial \mathbf{T}^{(z)}}{\partial z} + \mathbf{b} = 0$$ [3.3.3]

Substituting Equations 3.2.3 and then setting the scalar components of the resulting vector equation equal to zero yields the three *equilibrium equations*:

$$\boxed{\begin{array}{l} \dfrac{\partial \sigma_x}{\partial x} + \dfrac{\partial \tau_{yx}}{\partial y} + \dfrac{\partial \tau_{zx}}{\partial z} + b_x = 0 \\[2mm] \dfrac{\partial \tau_{xy}}{\partial x} + \dfrac{\partial \sigma_y}{\partial y} + \dfrac{\partial \tau_{zy}}{\partial z} + b_y = 0 \\[2mm] \dfrac{\partial \tau_{xz}}{\partial x} + \dfrac{\partial \tau_{yz}}{\partial y} + \dfrac{\partial \sigma_z}{\partial z} + b_z = 0 \end{array}}$$ [3.3.4]

Equations 3.3.4 are three partial differential equations governing the six stresses σ_x, σ_y, σ_z, τ_{xy} $(= \tau_{yx})$, τ_{xz} $(= \tau_{zx})$, and τ_{yz} $(= \tau_{zy})$. The body force densities b_x, b_y, and b_z, are often zero or negligible, but are nevertheless *prescribed* functions, whereas the stresses are the unknowns.

The equilibrium equations must be satisfied at all points within the material. On the boundary, the stresses must agree with the prescribed surface traction $\mathbf{T}^{(n)}$, that is, the *boundary conditions*, as given by the Cauchy formulas, Equations 3.2.8.

3.4 STRESS TRANSFORMATION

Figure 3.4.1 shows two cartesian coordinate systems: the unbarred system with axes xyz, and the barred system with axes $\bar{x}\,\bar{y}\,\bar{z}$. The orthogonal unit basis vectors for the unbarred system are $\mathbf{i}$, $\mathbf{j}$, and $\mathbf{k}$; for the barred system, they are $\bar{\mathbf{i}}$, $\bar{\mathbf{j}}$, and $\bar{\mathbf{k}}$. We can express the unit vectors of the barred system in terms of their components in the unbarred system, as follows:

$$\bar{\mathbf{i}} = R_{11}\mathbf{i} + R_{12}\mathbf{j} + R_{13}\mathbf{k}$$
$$\bar{\mathbf{j}} = R_{21}\mathbf{i} + R_{22}\mathbf{j} + R_{23}\mathbf{k} \qquad \text{[3.4.1]}$$
$$\bar{\mathbf{k}} = R_{31}\mathbf{i} + R_{32}\mathbf{j} + R_{33}\mathbf{k}$$

The R's in these expressions are the direction cosines of $\bar{\mathbf{i}}$, $\bar{\mathbf{j}}$, and $\bar{\mathbf{k}}$. Figure 3.4.1 illustrates the components of $\bar{\mathbf{k}}$, which are the projections of $\bar{\mathbf{k}}$ onto the x, y, and z axes. The unbarred unit vectors may be resolved into components along the barred system to obtain a set of equations similar to Equations 3.4.1:

$$\mathbf{i} = \bar{R}_{11}\bar{\mathbf{i}} + \bar{R}_{12}\bar{\mathbf{j}} + \bar{R}_{13}\bar{\mathbf{k}}$$
$$\mathbf{j} = \bar{R}_{21}\bar{\mathbf{i}} + \bar{R}_{22}\bar{\mathbf{j}} + \bar{R}_{23}\bar{\mathbf{k}} \qquad \text{[3.4.2]}$$
$$\mathbf{k} = \bar{R}_{31}\bar{\mathbf{i}} + \bar{R}_{32}\bar{\mathbf{j}} + \bar{R}_{33}\bar{\mathbf{k}}$$

However, $\bar{\mathbf{i}} \cdot \mathbf{i} = \mathbf{i} \cdot \bar{\mathbf{i}}$, so that $R_{11} = \bar{R}_{11}$; and $\bar{\mathbf{i}} \cdot \mathbf{j} = \mathbf{j} \cdot \bar{\mathbf{i}}$, which means that $R_{12} = \bar{R}_{21}$. Proceeding in this fashion, we can express the direction cosines in Equations 3.4.2 in terms of those in Equations 3.4.1. Therefore,

$$\mathbf{i} = R_{11}\bar{\mathbf{i}} + R_{21}\bar{\mathbf{j}} + R_{31}\bar{\mathbf{k}}$$
$$\mathbf{j} = R_{12}\bar{\mathbf{i}} + R_{22}\bar{\mathbf{j}} + R_{32}\bar{\mathbf{k}} \qquad \text{[3.4.3]}$$
$$\mathbf{k} = R_{13}\bar{\mathbf{i}} + R_{23}\bar{\mathbf{j}} + R_{33}\bar{\mathbf{k}}$$

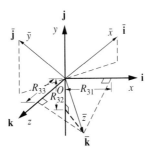

Figure 3.4.1   Two sets of cartesian reference axes, xyz and $\bar{x}\,\bar{y}\,\bar{z}$.

The fact that **i**, **j**, and **k** are unit vectors means that $\mathbf{i} \cdot \mathbf{i} = \mathbf{j} \cdot \mathbf{j} = \mathbf{k} \cdot \mathbf{k} = 1$; since they are mutually orthogonal, $\mathbf{i} \cdot \mathbf{j} = \mathbf{i} \cdot \mathbf{k} = \mathbf{j} \cdot \mathbf{k} = 0$. We therefore have six relations among the direction cosines in Equation 3.4.3[2]:

$$R_{11}^2 + R_{21}^2 + R_{31}^2 = R_{12}^2 + R_{22}^2 + R_{32}^2 = R_{13}^2 + R_{23}^2 + R_{33}^2 = 1$$

$$R_{11}R_{12} + R_{21}R_{22} + R_{31}R_{32} = R_{11}R_{13} + R_{21}R_{23} + R_{31}R_{33} = R_{12}R_{13} + R_{22}R_{23} + R_{32}R_{33} = 0$$

[3.4.4]

Equations 3.2.13 may be employed to find the three components of stress on planes normal to the $\bar{x}$, $\bar{y}$, and $\bar{z}$ axes, that is, to obtain the state of stress in the barred system:

$$[\bar{\boldsymbol{\sigma}}] = \begin{bmatrix} \bar{\sigma}_x & \bar{\tau}_{yx} & \bar{\tau}_{zx} \\ \bar{\tau}_{xy} & \bar{\sigma}_y & \bar{\tau}_{zy} \\ \bar{\tau}_{xz} & \bar{\tau}_{yz} & \bar{\sigma}_z \end{bmatrix}$$

To find the normal stresses, first set $\mathbf{n} = \bar{\mathbf{i}}$ (so that $n_x = R_{11}$, $n_y = R_{12}$, and $n_z = R_{13}$). Then Equation 3.2.13$_1$ yields $\bar{\sigma}_x$. With $\mathbf{n} = \bar{\mathbf{j}}$ and $\mathbf{n} = \bar{\mathbf{k}}$, we likewise obtain $\bar{\sigma}_y$ and $\bar{\sigma}_z$, respectively. To find the three shear stresses by means of Equation 3.2.13$_2$, we set $\mathbf{n} = \bar{\mathbf{i}}$ and $\mathbf{t} = \bar{\mathbf{j}}$ to obtain $\bar{\tau}_{xy}$. With $\mathbf{n} = \bar{\mathbf{i}}$ and $\mathbf{t} = \bar{\mathbf{k}}$, Equation 3.2.13$_2$ yields $\bar{\tau}_{xz}$; $\bar{\tau}_{yz}$ results by substituting $\bar{\mathbf{j}}$ and $\bar{\mathbf{k}}$ for $\mathbf{n}$ and $\mathbf{t}$. Then, taking advantage of the symmetry condition given by Equation 3.3.2, we are led to the *stress transformation equations*[3]

$$\begin{aligned}
\bar{\sigma}_x &= R_{11}^2\sigma_x + R_{12}^2\sigma_y + R_{13}^2\sigma_z + 2R_{11}R_{12}\tau_{xy} + 2R_{11}R_{13}\tau_{xz} + 2R_{12}R_{13}\tau_{yz} \\
\bar{\sigma}_y &= R_{21}^2\sigma_x + R_{22}^2\sigma_y + R_{23}^2\sigma_z + 2R_{21}R_{22}\tau_{xy} + 2R_{21}R_{23}\tau_{xz} + 2R_{22}R_{23}\tau_{yz} \\
\bar{\sigma}_z &= R_{31}^2\sigma_x + R_{32}^2\sigma_y + R_{33}^2\sigma_z + 2R_{31}R_{32}\tau_{xy} + 2R_{31}R_{33}\tau_{xz} + 2R_{32}R_{33}\tau_{yz} \\
\bar{\tau}_{xy} &= R_{11}R_{21}\sigma_x + R_{12}R_{22}\sigma_y + R_{13}R_{23}\sigma_z + (R_{11}R_{22} + R_{12}R_{21})\tau_{xy} + (R_{11}R_{23} + R_{13}R_{21})\tau_{xz} + (R_{12}R_{23} + R_{13}R_{22})\tau_{yz} \\
\bar{\tau}_{xz} &= R_{11}R_{31}\sigma_x + R_{12}R_{32}\sigma_y + R_{13}R_{33}\sigma_z + (R_{11}R_{32} + R_{12}R_{31})\tau_{xy} + (R_{11}R_{33} + R_{13}R_{31})\tau_{xz} + (R_{12}R_{33} + R_{13}R_{32})\tau_{yz} \\
\bar{\tau}_{yz} &= R_{21}R_{31}\sigma_x + R_{22}R_{32}\sigma_y + R_{23}R_{33}\sigma_z + (R_{21}R_{32} + R_{22}R_{31})\tau_{xy} + (R_{21}R_{33} + R_{23}R_{31})\tau_{xz} + (R_{22}R_{33} + R_{23}R_{32})\tau_{yz}
\end{aligned}$$

[3.4.5]

Because its components obey this transformation rule, the stress matrix $[\boldsymbol{\sigma}]$ is classified mathematically as a *tensor*. Furthermore, it is a *symmetric* tensor.

Example 3.4.1   Show that the *hydrostatic stress* $\frac{1}{3}(\sigma_x + \sigma_y + \sigma_z)$ is invariant under a coordinate transformation.
From the first three of Equations 3.4.5, we obtain

$$\begin{aligned}
\bar{\sigma}_x + \bar{\sigma}_y + \bar{\sigma}_z &= (R_{11}^2 + R_{21}^2 + R_{31}^2)\sigma_x + (R_{12}^2 + R_{22}^2 + R_{32}^2)\sigma_y + (R_{13}^2 + R_{23}^2 + R_{33}^2)\sigma_z \\
&\quad + 2(R_{11}R_{12} + R_{21}R_{22} + R_{31}R_{32})\tau_{xy} + 2(R_{11}R_{13} + R_{21}R_{23} + R_{31}R_{33})\tau_{xz} \\
&\quad + 2(R_{12}R_{13} + R_{22}R_{23} + R_{32}R_{33})\tau_{yz}
\end{aligned}$$

By virtue of Equations 3.4.4, this becomes

$$\bar{\sigma}_x + \bar{\sigma}_y + \bar{\sigma}_z = (1)\sigma_x + (1)\sigma_y + (1)\sigma_z + 2(0)\tau_{xy} + 2(0)\tau_{xz} + 2(0)\tau_{yz}$$

so that

$$\bar{\sigma}_x + \bar{\sigma}_y + \bar{\sigma}_z = \sigma_x + \sigma_y + \sigma_z$$

---

[2]Those familiar with matrices may recognize these equations in the form $[\mathbf{R}]^T[\mathbf{R}] = [\mathbf{I}]$, where $[\mathbf{I}]$ is the identity matrix. See Chapter 9.
[3]Using matrix notation, Equation 3.4.5 may be written compactly as $[\bar{\boldsymbol{\sigma}}] = [\mathbf{R}][\boldsymbol{\sigma}][\mathbf{R}]^T$. See Chapter 9.

The sum of the three normal stresses (which is the *trace* of the stress tensor) is the same in all coordinate systems, that is, it is a scalar invariant. Therefore, so is the hydrostatic stress.

Example 3.4.2   Let **m** and **n** be the unit normals to two planes through a point, and let $\mathbf{T}^{(m)}$ and $\mathbf{T}^{(n)}$ be the tractions on those planes. Show that $\mathbf{T}^{(n)} \cdot \mathbf{m} = \mathbf{T}^{(m)} \cdot \mathbf{n}$.

From Equations 3.2.7 and 3.2.8, we have

$$\mathbf{T}^{(n)} \cdot \mathbf{m} = T_x^{(n)} m_x + T_y^{(n)} m_y + T_z^{(n)} m_z = (\sigma_x n_x + \tau_{xy} n_y + \tau_{xz} n_z) m_x$$
$$+ (\tau_{yx} n_x + \sigma_y n_y + \tau_{yz} n_z) m_y + (\tau_{zx} n_x + \tau_{zy} n_y + \sigma_z n_z) m_z$$

Rearranging terms on the right yields

$$\mathbf{T}^{(n)} \cdot \mathbf{m} = (\sigma_x m_x + \tau_{yx} m_y + \tau_{zx} m_z) n_x$$
$$+ (\tau_{xy} m_x + \sigma_y m_y + \tau_{zy} m_z) n_y + (\tau_{xz} m_x + \tau_{yz} m_y + \sigma_z m_z) n_z$$

Using the symmetry of the stress tensor, Equations 3.3.2, this can be written

$$\mathbf{T}^{(n)} \cdot \mathbf{m} = (\sigma_x m_x + \tau_{xy} m_y + \tau_{xz} m_z) n_x$$
$$+ (\tau_{yx} m_x + \sigma_y m_y + \tau_{yz} m_z) n_y + (\tau_{zx} m_x + \tau_{zy} m_y + \sigma_z m_z) n_z$$

Then, with the Cauchy formulas, Equations 3.2.8, we are led to

$$\mathbf{T}^{(n)} \cdot \mathbf{m} = T_x^{(m)} n_x + T_y^{(m)} n_y + T_z^{(m)} n_z$$

However, on the right we have $\mathbf{T}^{(m)} \cdot \mathbf{n}$. Thus,

$$\mathbf{T}^{(n)} \cdot \mathbf{m} = \mathbf{T}^{(m)} \cdot \mathbf{n} \qquad \text{[3.4.6]}$$

3.5   PRINCIPAL STRESS

If the traction $\mathbf{T}^{(n)}$ on a surface acts entirely in the direction of the unit normal **n**, then $\mathbf{T}^{(n)} = \sigma \, \mathbf{n}$; that is,

$$T_x^{(n)} = \sigma n_x \qquad T_y^{(n)} = \sigma n_y \qquad T_z^{(n)} = \sigma n_z \qquad \text{[3.5.1]}$$

In that case, the shear stress on the surface is zero and σ is called a *principal stress*. The plane on which a principal stress acts is a *principal plane* and the normal **n** to that plane is a *principal normal* or *principal direction*. For a principal plane, Cauchy's Equations 3.2.8 become

$$\sigma_x n_x + \tau_{xy} n_y + \tau_{xz} n_z = \sigma n_x$$
$$\tau_{xy} n_x + \sigma_y n_y + \tau_{yz} n_z = \sigma n_y$$
$$\tau_{xz} n_x + \tau_{yz} n_y + \sigma_z n_z = \sigma n_z$$

or

$$(\sigma_x - \sigma) n_x + \tau_{xy} n_y + \tau_{xz} n_z = 0$$
$$\tau_{xy} n_x + (\sigma_y - \sigma) n_y + \tau_{yz} n_z = 0 \qquad \text{[3.5.2]}$$
$$\tau_{xz} n_x + \tau_{yz} n_y + (\sigma_z - \sigma) n_z = 0$$

These three equations have the trivial solution $n_x = n_y = n_z = 0$, which is not allowed, since n_x, n_y, and n_z are the components of a unit vector, satisfying the equation

$$n_x^2 + n_y^2 + n_z^2 = 1 \tag{3.5.3}$$

The only way Equations 3.5.2 possess a nontrivial solution is if the three equations are not independent of each other. According to linear algebra, this means that the determinant of the matrix of coefficients of n_x, n_y, and n_z must vanish, or:

$$\det \begin{bmatrix} \sigma_x - \sigma & \tau_{xy} & \tau_{xz} \\ \tau_{xy} & \sigma_y - \sigma & \tau_{yz} \\ \tau_{xz} & \tau_{yz} & \sigma_z - \sigma \end{bmatrix} = 0 \tag{3.5.4}$$

Expanding the determinant leads to the *characteristic equation* of the stress tensor:

$$\sigma^3 - I_1 \sigma^2 + I_2 \sigma - I_3 = 0 \tag{3.5.5}$$

where I_1, I_2, and I_3 are given by

$$
\begin{aligned}
I_1 &= \sigma_x + \sigma_y + \sigma_z \\
I_2 &= \det \begin{bmatrix} \sigma_x & \tau_{xy} \\ \tau_{xy} & \sigma_y \end{bmatrix} + \det \begin{bmatrix} \sigma_x & \tau_{xz} \\ \tau_{xz} & \sigma_z \end{bmatrix} + \det \begin{bmatrix} \sigma_y & \tau_{yz} \\ \tau_{yz} & \sigma_z \end{bmatrix} \\
I_3 &= \det \begin{bmatrix} \sigma_x & \tau_{xy} & \tau_{xz} \\ \tau_{xy} & \sigma_y & \tau_{yz} \\ \tau_{xz} & \tau_{yz} & \sigma_z \end{bmatrix}
\end{aligned}
\tag{3.5.6}
$$

The three I's are called the three *invariants* of the stress tensor, because they are unaltered by the coordinate transformation in Equation 3.4.5.

The three roots of the characteristic equation are the principal stresses σ_1, σ_2, and σ_3, all of which are *real* because the stress tensor is symmetric. From the theory of equations we find[4]

$$
\begin{aligned}
\sigma_1 &= \frac{I_1}{3} + \frac{2}{3}\sqrt{I_1^2 - 3I_2}\,\cos\left(\frac{\alpha}{3}\right) \\
\sigma_2 &= \frac{I_1}{3} + \frac{2}{3}\sqrt{I_1^2 - 3I_2}\,\cos\left(\frac{\alpha}{3} + \frac{2\pi}{3}\right) \\
\sigma_3 &= \frac{I_1}{3} + \frac{2}{3}\sqrt{I_1^2 - 3I_2}\,\cos\left(\frac{\alpha}{3} + \frac{4\pi}{3}\right)
\end{aligned}
\tag{3.5.7a}
$$

where

$$\alpha = \cos^{-1}\left[\frac{2I_1^3 - 9I_1 I_2 + 27 I_3}{2(I_1^2 - 3I_2)^{3/2}} \right] \tag{3.5.7b}$$

The principal stress direction $\mathbf{n}^{(p)}$ of principal stress σ_p ($p = 1, 2, 3$) is found by substituting σ_p into any two (but not all three) of Equations 3.5.2. In so doing, we obtain two of the three components of $\mathbf{n}^{(p)}$ in terms of the third. Equation 3.5.3 is then used to find numerical values of $n_x^{(p)}$, $n_y^{(p)}$, and $n_z^{(p)}$.

Let $\mathbf{n}^{(i)}$ and $\mathbf{n}^{(j)}$ be the principal normals for the principal stresses σ_i and σ_j, respectively. Then, according to Equation 3.4.6

| [4]See, for example, W. H. Beyer (ed.), *CRC Standard Mathematical Tables and Formulae*, 29th ed. Chapter II, CRC Press, 1991.

$$\mathbf{T}^{(i)} \cdot \mathbf{n}^{(j)} = \mathbf{T}^{(j)} \cdot \mathbf{n}^{(i)} \tag{3.5.8}$$

However, on the principal planes, we know that $\mathbf{T}^{(i)} = \sigma_i \mathbf{n}^{(i)}$ and $\mathbf{T}^{(j)} = \sigma_j \mathbf{n}^{(j)}$ (Equation 3.5.1). Thus, Equation 3.5.8 implies

$$\sigma_i \mathbf{n}^{(i)} \cdot \mathbf{n}^{(j)} = \sigma_j \mathbf{n}^{(j)} \cdot \mathbf{n}^{(i)} \qquad \text{or} \qquad (\sigma_i - \sigma_j)\mathbf{n}^{(i)} \cdot \mathbf{n}^{(j)} = 0 \tag{3.5.9}$$

If the two principal stresses are unique, that is, if $\sigma_i \neq \sigma_j$, then we see that $\mathbf{n}^{(i)} \cdot \mathbf{n}^{(j)} = 0$. In other words, *the principal directions corresponding to unique principal stresses are orthogonal.* If two principal stresses have the same value ($\sigma_i - \sigma_j = 0$), then Equation 3.5.9 places no restriction on $\mathbf{n}^{(i)} \cdot \mathbf{n}^{(j)}$, so we are free to set $\mathbf{n}^{(i)} \cdot \mathbf{n}^{(j)} = 0$ if we wish. Therefore, we can always imagine the principal directions to form an *orthogonal triad.* Observe that this conclusion follows from Equation 3.4.6, which in turn depends on the symmetry of the stress tensor.

It is a fairly simple matter to show that *the principal stresses are the extreme values of the normal stress.* To do so, let us assume that the three principal stresses are distinct and choose the three principal directions as the coordinate axes, so that the state of stress is

$$\begin{bmatrix} \sigma_1 & 0 & 0 \\ 0 & \sigma_2 & 0 \\ 0 & 0 & \sigma_3 \end{bmatrix}$$

Then, the normal stress on any plane is given by Equation 3.2.13, as follows:

$$\sigma_n = \sigma_1 n_x^2 + \sigma_2 n_y^2 + \sigma_3 n_z^2$$

Here, σ_n is a function of the three variables, n_x, n_y, and n_z. However, only two of them are independent, because they are related by Equation 3.5.3. Let us express n_x in terms of the other two components of $\mathbf{n}$, so that $n_x^2 = 1 - n_y^2 - n_z^2$. In that case, we have

$$\sigma_n = (\sigma_2 - \sigma_1)n_y^2 + (\sigma_3 - \sigma_1)n_z^2 + \sigma_1$$

Therefore,

$$\frac{\partial \sigma_n}{\partial n_y} = 2(\sigma_2 - \sigma_1)n_y \qquad \frac{\partial \sigma_n}{\partial n_z} = 2(\sigma_3 - \sigma_1)n_z$$

The extreme values of σ_n occur where both of these partial derivatives vanish. Since $\sigma_2 \neq \sigma_1$ and $\sigma_3 \neq \sigma_1$, the only way these two derivatives can both be zero is if $n_y = n_z = 0$, and that means $n_x = 1$. Thus, σ_1 is an extreme value of σ_n. Proceeding in a similar fashion, setting $n_y^2 = 1 - n_x^2 - n_z^2$ and then $n_z^2 = 1 - n_x^2 - n_y^2$, we find that σ_2 and σ_3 are the other two extrema of σ_n. For *all* orientations $\mathbf{n}$ of the cutting plane, the three principal stresses thus include the largest and the smallest normal stress (in the sense that a negative number of any magnitude is smaller than zero).

It turns out that a state of stress not only has three extreme values of normal stress, but also three extreme values of *shear* stress, which are related to the three principal stresses as follows[5]:

$$\tau_1 = \frac{|\sigma_1 - \sigma_2|}{2} \qquad \tau_2 = \frac{|\sigma_1 - \sigma_3|}{2} \qquad \tau_3 = \frac{|\sigma_2 - \sigma_3|}{2}$$

The three planes on which τ_1, τ_2, and τ_3 act bisect the angles between each of the three principal planes. Observe that *the absolute maximum shear stress at a point equals one-half the difference between the largest and the smallest principal stress,* or:

| [5]See, for example, S. Timoshenko, and J. N. Goodier, *Theory of Elasticity*, 2nd ed., Section 70, New York, McGraw-Hill, 1951.

$$\tau_{max} = \frac{1}{2}|\sigma_{max} - \sigma_{min}| \qquad \text{[3.5.10]}$$

Finally, it must be pointed out that whereas the shear stress, by definition, vanishes on planes of principal stress, the normal stress is generally not zero on planes where the shear stress acquires its extreme values.

Example 3.5.1 Find the principal stresses, principal normals, and the maximum shear stress if

$$[\boldsymbol{\sigma}] = \begin{bmatrix} -50 & 50 & -50 \\ 50 & 50 & 100 \\ -50 & 100 & 150 \end{bmatrix} \text{ (MPa)}$$

The invariants are

$$I_1 = (-50) + 50 + 150 = 150 \text{ MPa}$$

$$I_2 = \det\begin{bmatrix} -50 & 50 \\ 50 & 50 \end{bmatrix} + \det\begin{bmatrix} -50 & -50 \\ -50 & 150 \end{bmatrix} + \det\begin{bmatrix} 50 & 100 \\ 100 & 150 \end{bmatrix} = -17,500 \text{ MPa}^2$$

$$I_3 = \det\begin{bmatrix} -50 & 50 & -50 \\ 50 & 50 & 100 \\ -50 & 100 & 150 \end{bmatrix} = -875,000 \text{ MPa}^3$$

From Equation 3.5.7b, we have

$$\alpha = \cos^{-1}\left[\frac{2(150)^3 - 9(150)(-17,500) + 27(-875,000)}{2[150^2 - 3(-17,500)]^{3/2}}\right] = 80.54°$$

so that

$$\sigma_1 = 50 + 182.6\cos\frac{80.54°}{3} = 212.9 \text{ MPa}$$

$$\sigma_2 = 50 + 182.6\cos\left(\frac{80.54°}{3} + 120°\right) = -102.8 \text{ MPa}$$

$$\sigma_3 = 50 + 182.6\cos\left(\frac{80.54°}{3} + 240°\right) = 39.96 \text{ MPa}$$

From these, we find the maximum shear stress:

$$\tau_{max} = \frac{1}{2}(\sigma_{max} - \sigma_{min}) = \frac{1}{2}[212.9 - (-102.8)] = 157.5 \text{ MPa}$$

To find $\mathbf{n}^{(1)}$, the principal direction of σ_1, we substitute σ_1 into Equation 3.5.2:

$$-262.9n_x^{(1)} + 50n_y^{(1)} - 50n_z^{(1)} = 0 \qquad \text{[a]}$$

$$50n_x^{(1)} - 162.9n_y^{(1)} + 100n_z^{(1)} = 0 \qquad \text{[b]}$$

$$-50n_x^{(1)} + 100n_y^{(1)} - 62.90n_z^{(1)} = 0 \qquad \text{[c]}$$

From Equations a and b we obtain $n_y^{(1)}$ and $n_z^{(1)}$ in terms of $n_x^{(1)}$:

$$n_y^{(1)} = -7.565n_x^{(1)} \qquad n_z^{(1)} = -12.82n_x^{(1)} \qquad \text{[d]}$$

Substituting these expressions into Equation 3.5.3 yields

$$1 = n_x^{(1)^2} + n_y^{(1)^2} + n_z^{(1)^2}$$

$$= n_x^{(1)^2} + \left(-7.565 n_x^{(1)}\right)^2 + \left(-12.82 n_x^{(1)}\right)^2 = 222.6 n_x^{(1)^2}$$

so that $n_x^{(1)} = \pm 0.06702$. Choosing the positive sign (arbitrarily), we determine $\mathbf{n}^{(1)}$ using Equations d:

$$\sigma_1 = 212.9 \text{ MPa}: \qquad \mathbf{n}^{(1)} = 0.06702\mathbf{i} - 0.5070\mathbf{j} - 0.8594\mathbf{k}$$

We repeat this procedure to find the principal normals for the remaining two principal stresses:

$$\sigma_2 = -102.8 \text{ MPa}: \qquad \mathbf{n}^{(2)} = 0.7971\mathbf{i} - 0.4908\mathbf{j} + 0.3517\mathbf{k}$$

$$\sigma_3 = 39.96 \text{ MPa}: \qquad \mathbf{n}^{(3)} = 0.6001\mathbf{i} + 0.7086\mathbf{j} - 0.3712\mathbf{k}$$

Observe that the three principal normals are perpendicular to one another and that $\mathbf{n}^{(3)} = \mathbf{n}^{(1)} \times \mathbf{n}^{(2)}$.

3.6   PLANE STRESS

Aircraft structures are largely composed of material formed into thin sheets and plates of uniform thickness. For these structural elements, it is often appropriate to assume that the stress components are confined to a plane, which we will take to be the xy plane. In that case, $\sigma_z = \tau_{xz} = \tau_{yz} = 0$, so that for the case of plane stress, the general three-dimensional state of stress given by Equation 3.2.9 reduces to just three independent components:

$$\begin{bmatrix} \sigma_x & \tau_{xy} & 0 \\ \tau_{xy} & \sigma_y & 0 \\ 0 & 0 & 0 \end{bmatrix} \quad \text{plane stress} \qquad \textbf{[3.6.1]}$$

where Equations 3.3.2 have been invoked. Applying the plane stress assumptions to Equation 3.3.4, and assuming that the body force components lie in the plane, we are left with just two equilibrium equations:

$$\boxed{\begin{aligned} \frac{\partial \sigma_x}{\partial x} + \frac{\partial \tau_{xy}}{\partial y} + b_x = 0 \\ \frac{\partial \tau_{xy}}{\partial x} + \frac{\partial \sigma_y}{\partial y} + b_y = 0 \end{aligned}} \quad \text{plane stress} \qquad \textbf{[3.6.2]}$$

The Cauchy formulas, Equations 3.2.8, also simplify considerably, as follows:

$$\boxed{\begin{aligned} T_x^{(n)} = \sigma_x n_x + \tau_{xy} n_y \\ T_y^{(n)} = \tau_{xy} n_x + \sigma_y n_y \end{aligned}} \quad \text{plane stress} \qquad \textbf{[3.6.3]}$$

Figure 3.6.1 shows a state of plane stress relative to two cartesian coordinate systems. The relationships among the components of stress in the two frames are obtained from the three-dimensional stress transformations, Equations 3.4.5. We set all of the z components of stress equal to zero and observe that the direction cosines of the barred (rotated) axes are

$$\begin{array}{llll} \bar{x}: & R_{11} = \cos\theta & R_{12} = \sin\theta & R_{13} = 0 \\ \bar{y}: & R_{21} = -\sin\theta & R_{22} = \cos\theta & R_{23} = 0 \\ \bar{z}: & R_{31} = 0 & R_{32} = 0 & R_{33} = 1 \end{array}$$

Equations 3.4.5 then yield the *plane stress transformation formulas*

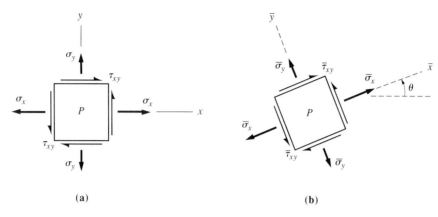

Figure 3.6.1   Positive components of plane stress at a point P in two different cartesian coordinate systems.

$$\bar{\sigma}_x = \sigma_x \cos^2 \theta + \sigma_y \sin^2 \theta + 2\tau_{xy} \sin \theta \cos \theta$$
$$\bar{\sigma}_y = \sigma_x \sin^2 \theta + \sigma_y \cos^2 \theta - 2\tau_{xy} \sin \theta \cos \theta \quad \text{plane stress} \qquad [3.6.4]$$
$$\bar{\tau}_{xy} = (\sigma_y - \sigma_x) \sin \theta \cos \theta + \tau_{xy} (\cos^2 \theta - \sin^2 \theta)$$

We then introduce the following trigonometric identities into these equations:

$$\sin^2 \theta = \frac{1 - \cos 2\theta}{2} \qquad \cos^2 \theta = \frac{1 + \cos 2\theta}{2} \qquad 2 \sin \theta \cos \theta = \sin 2\theta$$

In so doing, we obtain the alternative double-angle versions of Equations 3.6.4:

$$\bar{\sigma}_x = \frac{\sigma_x + \sigma_y}{2} + \frac{\sigma_x - \sigma_y}{2} \cos 2\theta + \tau_{xy} \sin 2\theta$$
$$\bar{\sigma}_y = \frac{\sigma_x + \sigma_y}{2} - \frac{\sigma_x - \sigma_y}{2} \cos 2\theta - \tau_{xy} \sin 2\theta \qquad [3.6.5]$$
$$\bar{\tau}_{xy} = -\frac{\sigma_x - \sigma_y}{2} \sin 2\theta + \tau_{xy} \cos 2\theta$$

In this form, the transformation equations have a convenient geometric graphical interpretation in the form of *Mohr's circle*,[6] which is plotted on the $\sigma\tau$ plane as shown in Figure 3.6.3. The coordinates of points on the circle are the normal and shear stresses on planes of different orientation θ. When working with Mohr's circle, we must convert to a sign convention for the shear stresses that is peculiar *only to Mohr's circle*. For Mohr's circle, the shear stress on the face of a plane stress element is positive (Figure 3.6.2a) if it exerts a clockwise moment about the center of the element and negative (Figure 3.6.2b) if it exerts a counterclockwise moment. According to this way of viewing the shear stresses, $\tau_{yx} = -\tau_{xy}$.

Given the plane stress components σ_x, σ_y, and τ_{xy} in Figure 3.6.1a, we construct Mohr's circle as follows, assuming for the sake of argument that $\sigma_x > \sigma_y > 0$ and $\tau_{xy} > 0$.

1. On Figure 3.6.3, plot point X, which represents the stress components on the x face. In view of the Mohr's circle sign convention for shear stress, τ_{xy}—assumed positive—plots downward, below the σ axis.

| [6]Otto Christian Mohr (1835–1918), German civil engineer.

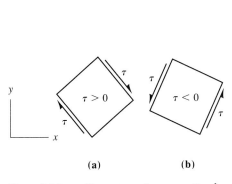

Figure 3.6.2   Shear stress sign convention for Mohr's circle only.

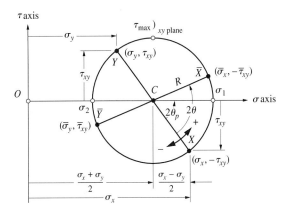

Figure 3.6.3   Mohr's circle for plane stress in the *xy* plane.

2. Plot point Y, which represents the stress components on the y face. If the shear stress plotted down in Step 1, then it must plot up in this step, as is the case in Figure 3.6.3.

3. Connect X and Y with a straight line XCY. This defines a diameter that crosses the σ axis at point C, the center of the circle. C is midway between the projections of X and Y onto the σ axis:

$$OC = \frac{\sigma_x + \sigma_y}{2} \qquad \textbf{[3.6.6]}$$

4. Sketch the circle and label the coordinates of $C, X,$ and Y.

From the Mohr circle it is clear that the angle $2\theta_p$, measured positive counterclockwise from line CX to the σ axis, may be calculated using the formula

$$\tan 2\theta_p = \frac{\tau_{xy}}{(\sigma_x - \sigma_y)/2} \qquad \textbf{[3.6.7]}$$

From the sketch, we also deduce that the radius of the circle is

$$R = \sqrt{\tau_{xy}^2 + \left(\frac{\sigma_x - \sigma_y}{2}\right)^2} \qquad \textbf{[3.6.8]}$$

The endpoints of the horizontal diameter of Mohr's circle are the principal stresses σ_1 and σ_2. Indeed, from Figure 3.6.3, we see that

$$\sigma_1 = OC + R \qquad \sigma_2 = OC - R \qquad \textbf{[3.6.9]}$$

Here, θ_p is the angle between the x axis and the normal to the plane on which the principal stress σ_1 acts.

We can use the sketch of Mohr's circle to find the stresses on an element rotated through a counterclockwise angle θ from the xy axes, as in Figure 3.6.1b. Note that the angles measured on the Mohr circle are *twice* those in the stress plane. Furthermore, they have the *same sense*, due to Mohr's sign convention for shear.

The stress components on the $\bar{x}$ face of the rotated element are represented by the point $\bar{X}$ in Figure 3.6.3. Reading directly off the Mohr circle, we see that

$$\bar{\sigma}_x = OC + R \cos(2\theta - 2\theta_p)$$

$$-\bar{\tau}_{xy} = R \sin(2\theta - 2\theta_p) \qquad \textbf{[3.6.10]}$$

From basic trigonometry, we have

$$\cos(2\theta-2\theta_p) = \cos 2\theta_p \cos 2\theta + \sin 2\theta_p \sin 2\theta$$

$$\sin(2\theta-2\theta_p) = \sin 2\theta_p \cos 2\theta - \cos 2\theta_p \sin 2\theta$$

so that Equations 3.6.10 become

$$\bar{\sigma}_x = OC + (R \cos 2\theta_p) \cos 2\theta + (R \sin 2\theta_p) \sin 2\theta$$

$$-\bar{\tau}_{xy} = (R \sin 2\theta_p) \cos 2\theta - (R \cos 2\theta_p) \sin 2\theta$$

[3.6.11]

The terms in parentheses may be read directly from Mohr's circle, as follows:

$$R \cos 2\theta_p = \frac{\sigma_x - \sigma_y}{2} \qquad R \sin \theta_p = \tau_{xy}$$

[3.6.12]

Substituting Equations 3.6.6 and 3.6.12 into Equations 3.6.11 yields Equations $3.6.5_1$ and $3.6.5_3$. Likewise, the stress coordinates of point $\bar{Y}$ in Figure 3.6.3 lead to Equation $3.6.5_2$. (We can also obtain Equation $3.6.5_2$ directly from Equation $3.6.5_1$ by replacing θ with $\theta + \pi/2$.) We have therefore verified that constructing Mohr's circle is graphically equivalent to establishing the plane stress transformation relations.

Example 3.6.1   A state of plane stress is represented on the element in Figure 3.6.4. Use a sketch of Mohr's circle to find: (a) the principal stresses and principal directions; (b) the state of maximum in-plane shear stress; and (c) the stress components on an element rotated 50 degrees counterclockwise.

(a)

The first step is to plot points X and Y on the $\sigma\tau$ plane. Since the 3000 psi shear stress on the x faces exerts a clockwise moment on the element, point X has a positive τ coordinate on Mohr's circle. On the other hand, the counterclockwise moment produced by τ_{yx} entitles Y to a negative τ coordinate. With X and Y plotted, we can sketch the diameter XCY, locating the center C at $\sigma = 9000$ psi, as shown in Figure 3.6.5.

Since

$$Ca = 14,000 - 9000 = 5000$$

[a]

it is clear from the sketch that the radius of the circle is

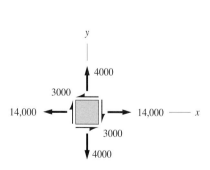

Figure 3.6.4   State of plane stress at a point: $\sigma_x = 14,000$ psi, $\sigma_y = 4,000$ psi, $\tau_{xy} = -3,000$ psi.

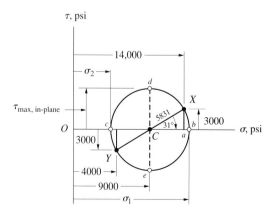

Figure 3.6.5   Mohr's circle for the state of stress in Figure 3.6.4.

$$R = \sqrt{5000^2 + 3000^2} = 5831 \text{ psi} \qquad \text{[b]}$$

Therefore, the principal stresses at points b and c on the circle are:

$$\sigma_1 = OC + R = 9000 + 5381 = 14,380 \text{ psi}$$
$$\sigma_2 = OC - R = 9000 - 5381 = 3619 \text{ psi} \qquad \text{[c]}$$

From Mohr's circle, we see that

$$2\theta_p = \tan^{-1}\frac{Xa}{Ca} = \tan^{-1}\frac{3000}{5000} = 30.96° \qquad \text{[d]}$$

It is clearly a clockwise rotation of 30.96 degrees around the Mohr circle from point X to point b where $\sigma = 14,380$ psi and the shear stress vanishes. It is therefore a clockwise rotation of 15.48 degrees from the x axis to the maximum principal stress direction. An additional 180 degrees clockwise rotation around the circle (90 degrees in the stress plane) takes us to point c, representing the other principal stress. The principal stresses are shown on the rotated element in Figure 3.6.6a.

(b)

Points d and e on the Mohr circle represent the state of maximum in-plane shear stress. The normal stress coordinate of these points is 9000 psi, and the shear stress equals the radius of the circle (5831 psi). From X we can arrive at point e by a clockwise rotation of $2\theta = 30.96° + 90°$, or $\theta = 60.48°$. The shear stress coordinate of point e is negative, so the shear stress on that face acts counterclockwise on the element. The opposite is true for point d, representing the orthogonal plane. The state of stress on the maximum shear planes is shown on the element in Figure 3.6.6b.

(c)

The state of stress on an element at 55 degrees counterclockwise is represented on the Mohr circle by points $\bar{X}$ and $\bar{Y}$ on a diameter rotated 110 degrees counterclockwise from the reference diameter XCY, as shown in Figure 3.6.7a. From the geometry of that figure, it is clear that

$$\bar{\sigma}_x = 9000 - 5831\cos 39° = 4468 \text{ psi} \qquad \bar{\sigma}_y = 9000 + 5831\cos 39° = 13,530 \text{ psi}$$

It is also evident that

$$|\tau_{xy}| = 5831\sin 39° = 3670 \text{ psi}$$

From the Mohr circle, we deduce that the moment of the shear is clockwise on the $\bar{x}$ faces and counterclockwise on the $\bar{y}$ faces. These and the normal stresses are sketched on the stress element in Figure 3.6.7b.

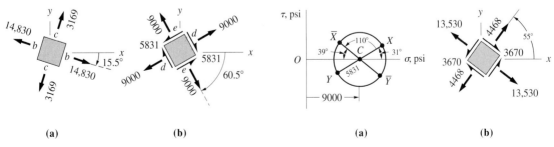

(a)   (b)   (a)   (b)

Figure 3.6.6   (a) State of principal stress (psi). (b) Planes of maximum shear (psi).

Figure 3.6.7   (a) Mohr's circle of Figure 3.6.5, used here to obtain the stress components in (b).

3.7   STRAIN

When loads are applied to a structure, it deforms. The deformation may produce a change in the dimensions of the body, as in a stretched rubber band (see Figure 3.7.1a). The deformation may also cause only a change in shape, as when the covers of a closed book are pushed in opposite directions (see Figure 3.7.1b) The deformation induces strains throughout the structure. *Strain* is a measure of the relative distortion of the material in the vicinity of a given point. Most often, we shall assume that the deformation is so small that the changes in an object's geometry can be ignored. That is, the stresses may be computed using the undeformed shape of the body, and although some movement of the applied loads accompanies the deformation, we assume the loads remain fixed in location and direction.

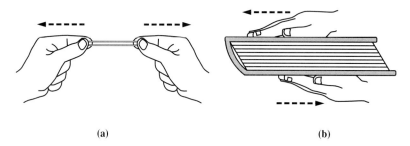

(a)   (b)

Figure 3.7.1   (a) Pulling a rubber band changes its length. (b) Shearing a book changes its shape.

When loads are applied to an initially unstressed body, each unconstrained material particle undergoes a small displacement, moving from its initial location to a new location a small distance away, as suggested in Figure 3.7.2. Consider a material particle occupying point P with coordinates x, y, and z. Imagine a differential line element $d\mathbf{r}$ joining an arbitrary material point P to a neighboring material point Q. The magnitude (length) of $d\mathbf{r}$ is ds, and its components are dx, dy, and dz. That is,

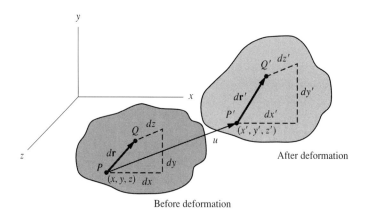

Figure 3.7.2   Directed material line segments of differential length, before and after deformation in the plane.

$$d\mathbf{r} = dx\mathbf{i} + dy\mathbf{j} + dz\mathbf{k} \qquad \text{[3.7.1]}$$

$$|d\mathbf{r}| = ds = \sqrt{dx^2 + dy^2 + dx^2} \qquad \text{[3.7.2]}$$

Let $\mathbf{n}$ be the unit vector in the direction of $d\mathbf{r}$, so that

$$\mathbf{n} = \frac{d\mathbf{r}}{ds} = \frac{dx}{ds}\mathbf{i} + \frac{dy}{ds}\mathbf{j} + \frac{dz}{ds}\mathbf{k} \qquad \text{[3.7.3]}$$

During deformation, point P undergoes a displacement $\mathbf{u}$, carrying it to P'. The *displacement vector* $\mathbf{u}$ is a function of position, varying from point to point within the body. In terms of its three components, it can be written:

$$\mathbf{u} = u(x, y, z)\mathbf{i} + v(x, y, z)\mathbf{j} + w(x, y, z)\mathbf{k} \qquad \text{[3.7.4]}$$

The displacement vector joins a point in the undeformed body to its new location in the deformed state. The coordinates of P' therefore are

$$x' = x + u(x, y, z) \quad y' = y + v(x, y, z) \quad z' = z + w(x, y, z) \qquad \text{[3.7.5]}$$

After deformation, the vector joining P' to Q' is $d\mathbf{r}'$ which is

$$d\mathbf{r}' = dx'\mathbf{i} + dy'\mathbf{j} + dz'\mathbf{k} \qquad \text{[3.7.6]}$$

Its magnitude is

$$|d\mathbf{r}'| = ds' = \sqrt{dx'^2 + dy'^2 + dz'^2} \qquad \text{[3.7.7]}$$

According to Equation 3.7.5 the components of $d\mathbf{r}'$ are

$$dx' = dx + du \qquad dy' = dy + dv \qquad dz' = dz + dw \qquad \text{[3.7.8]}$$

From differential calculus, the total derivatives in Equation 3.7.8 are obtained as follows:

$$du = \frac{\partial u}{\partial x}dx + \frac{\partial u}{\partial y}dy + \frac{\partial u}{\partial z}dz$$

$$dv = \frac{\partial v}{\partial x}dx + \frac{\partial v}{\partial y}dy + \frac{\partial v}{\partial z}dz$$

$$dw = \frac{\partial w}{\partial x}dx + \frac{\partial w}{\partial y}dy + \frac{\partial w}{\partial z}dz$$

Equation 3.7.8 may thus be written

$$dx' = dx + \frac{\partial u}{\partial x}dx + \frac{\partial u}{\partial y}dy + \frac{\partial u}{\partial z}dz$$

$$dy' = dy + \frac{\partial v}{\partial x}dx + \frac{\partial v}{\partial y}dy + \frac{\partial v}{\partial z}dz \qquad \text{[3.7.9]}$$

$$dz' = dy + \frac{\partial w}{\partial x}dx + \frac{\partial w}{\partial y}dy + \frac{\partial w}{\partial z}dz$$

We shall assume that each of the nine displacement derivatives in these expressions has a magnitude that is very much smaller than unity, so that in all of the following equations, we will retain displacement derivatives to the *first* power only.

3.7.1   Normal Strain

The *normal strain* ε_n of the line element PQ in Figure 3.7.2 is the ratio of the *change* of its length to its *original* length, or

$$\varepsilon_n = \frac{ds' - ds}{ds} = \frac{ds'}{ds} - 1 \qquad \text{[3.7.10]}$$

Strain is obviously a dimensionless quantity. A typical order of magnitude of normal strain is 10^{-3} (0.1 percent). This is written 1000μ and read "1000 microstrain," where $\mu = 10^{-6}$ length units/per length unit. In most engineering materials, strains on the order of 1 percent or more cause damage (e.g., fracture or permanent deformation), which is unacceptable in aircraft structures. The fact that, in applications, strains are indeed quite small compared to 1 justifies the assumption we have made regarding the magnitude of displacement gradients. That is good, because retaining only first-order terms keeps the equations as simple as possible. Elastomers (rubber-like materials) can endure strains in excess of 100 percent without damage. To characterize deformations of that magnitude requires retaining all terms, which makes the equations non-linear and beyond the scope of this text.

From Equations 3.7.7 and 3.7.9, neglecting powers of displacement derivatives higher than one, we have

$$ds' = \sqrt{ds^2 + 2\frac{\partial u}{\partial x}dx^2 + 2\frac{\partial v}{\partial y}dy^2 + 2\frac{\partial w}{\partial z}dz^2 + 2\left(\frac{\partial u}{\partial y} + \frac{\partial v}{\partial x}\right)dxdy + 2\left(\frac{\partial u}{\partial z} + \frac{\partial w}{\partial x}\right)dxdz + 2\left(\frac{\partial v}{\partial z} + \frac{\partial w}{\partial y}\right)dydz}$$

Factoring out ds^2 and recalling from Equation 3.7.3 that $n_x = dx/ds$, $n_y = dy/ds$, and $n_z = dz/ds$, we find that

$$\frac{ds'}{ds} = \sqrt{1 + 2\frac{\partial u}{\partial x}n_x^2 + 2\frac{\partial v}{\partial y}n_y^2 + 2\frac{\partial w}{\partial z}n_z^2 + 2\left(\frac{\partial u}{\partial y} + \frac{\partial v}{\partial x}\right)n_x n_y + 2\left(\frac{\partial u}{\partial z} + \frac{\partial w}{\partial x}\right)n_x n_z + 2\left(\frac{\partial v}{\partial z} + \frac{\partial w}{\partial y}\right)n_y n_z} \qquad \text{[3.7.11]}$$

Having assumed that the displacement derivatives are all very much smaller than 1, we can accurately replace the square-root operator on the right, as follows[7]:

$$\frac{ds'}{ds} = 1 + \frac{1}{2}\left[2\frac{\partial u}{\partial x}n_x^2 + 2\frac{\partial v}{\partial y}v + 2\frac{\partial w}{\partial z}n_z^2 + 2\left(\frac{\partial u}{\partial y} + \frac{\partial v}{\partial x}\right)n_x n_y + 2\left(\frac{\partial u}{\partial z} + \frac{\partial w}{\partial x}\right)n_x n_z + 2\left(\frac{\partial v}{\partial z} + \frac{\partial w}{\partial y}\right)n_y n_z\right]$$

Substituting this expression into Equation 3.7.10, we obtain a formula for *the normal strain of a line element in the direction of the unit vector* $\mathbf{n}$:

$$\varepsilon_n = \frac{\partial u}{\partial x}n_x^2 + \frac{\partial v}{\partial y}n_y^2 + \frac{\partial w}{\partial z}n_z^2 + \left(\frac{\partial u}{\partial y} + \frac{\partial v}{\partial x}\right)n_x n_y$$
$$+ \left(\frac{\partial u}{\partial z} + \frac{\partial w}{\partial x}\right)n_x n_z + \left(\frac{\partial v}{\partial z} + \frac{\partial w}{\partial y}\right)n_y n_z \qquad \text{[3.7.12]}$$

This relationship can be used to obtain the normal strain of an arbitrarily-oriented line element in terms of the displacement derivatives. The normal strain in the x direction is denoted ε_x. Equation 3.7.12 yields ε_x if $\mathbf{n}$ points in the x direction, that is, if $n_x = 1$ and $n_y = n_z = 0$. Making that substitution, we find that $\varepsilon_x = \partial u/\partial x$. In a similar fashion, the normal strain ε_y in the y direction ($\mathbf{n} = \mathbf{j}$) and ε_z in the z direction ($\mathbf{n} = \mathbf{k}$) are found, so that, in summary, we have

---

[7] $(1+\varepsilon)^{\frac{1}{2}} = 1 + \frac{\varepsilon}{2} - \frac{\varepsilon^2}{8} + \frac{\varepsilon^3}{16} + \cdots$

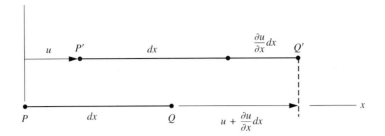

Figure 3.7.3   Normal strain in the x direction.

$$\varepsilon_x = \frac{\partial u}{\partial x} \qquad \varepsilon_y = \frac{\partial v}{\partial y} \qquad \varepsilon_y = \frac{\partial w}{\partial z} \qquad \text{[3.7.13]}$$

The normal strain describes the change in size or dimensions of a body. A physical interpretation of Equations 3.7.13 is as follows. First, consider a differential line element PQ aligned along the x axis, as shown in Figure 3.7.3. During deformation, P moves to P', a displacement u in the x direction, whereas Q, because of its separation dx from P, undergoes the same displacement plus a bit more $(\partial u/\partial x)dx$, to arrive at Q' on the x axis. Thus, the length of the line element $P'Q'$ after deformation is $dx + (\partial u/\partial x)dx$. According to Equation 3.7.10, the normal strain of the line element PQ is ε_x and is given by

$$\varepsilon_x = \frac{\left[dx + (\partial u/\partial x)dx\right] - dx}{dx} = \frac{\partial u}{\partial x}$$

in agreement with Equation 3.7.13$_1$. Similar arguments for ε_y and ε_z obviously follow, using line elements along the y and z directions.

3.7.2 Shear Strain

Let us now obtain the appropriate measure of the change in *shape* of a solid body. Figure 3.7.4 shows two *orthogonal* differential line elements $d\mathbf{r}^{(n)}$ and $d\mathbf{r}^{(t)}$ at point P, oriented in the direction of the unit vectors $\mathbf{n}$ and $\mathbf{t}$. In terms of their components along the x, y, and z axes,

$$d\mathbf{r}^{(n)} = dx^{(n)}\mathbf{i} + dy^{(n)}\mathbf{j} + dz^{(n)}\mathbf{k}$$
$$d\mathbf{r}^{(t)} = dx^{(t)}\mathbf{i} + dy^{(t)}\mathbf{j} + dz^{(t)}\mathbf{k} \qquad \text{[3.7.14]}$$

Let $ds^{(n)}$ be the magnitude of $d\mathbf{r}^{(n)}$ and let $ds^{(t)}$ be the magnitude of $d\mathbf{r}^{(t)}$. The unit vectors are therefore $\mathbf{n} = d\mathbf{r}^{(n)}/ds^{(n)}$ and $\mathbf{t} = d\mathbf{r}^{(t)}/ds^{(t)}$. We use Equation 3.7.14 to find the cartesian components of $\mathbf{n}$ and $\mathbf{t}$, as follows:

$$\mathbf{n} = n_x\mathbf{i} + n_y\mathbf{j} + n_z\mathbf{k} = \frac{dx^{(n)}}{ds^{(n)}}\mathbf{i} + \frac{dy^{(n)}}{ds^{(n)}}\mathbf{j} + \frac{dz^{(n)}}{ds^{(n)}}\mathbf{k}$$
$$\mathbf{t} = t_x\mathbf{i} + t_y\mathbf{j} + t_z\mathbf{k} = \frac{dx^{(t)}}{ds^{(t)}}\mathbf{i} + \frac{dy^{(t)}}{ds^{(t)}}\mathbf{j} + \frac{dz^{(t)}}{ds^{(t)}}\mathbf{k} \qquad \text{[3.7.15]}$$

Since the line elements are chosen to be orthogonal, $d\mathbf{r}^{(n)} \cdot d\mathbf{r}^{(t)} = 0$, that is,

$$dx^{(n)}dx^{(t)} + dy^{(n)}dy^{(t)} + dz^{(n)}dz^{(t)} = 0 \qquad \text{[3.7.16]}$$

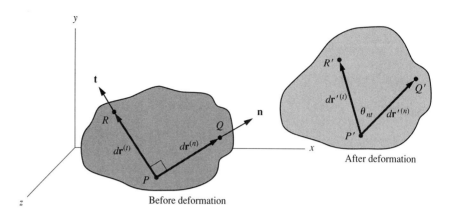

Figure 3.7.4 Change in angle between initially orthogonal directed differential line segments at a point.

After deformation, $d\mathbf{r}^{(n)}$ becomes $d\mathbf{r}'^{(n)}$ and $d\mathbf{r}^{(t)}$ becomes $d\mathbf{r}'^{(t)}$. Both the lengths and the orientations of these vectors change slightly during deformation, and afterwards, the angle θ_{nt} between them is likely to be something other than 90 degrees. Let γ_{nt} be the decrease in angle between $d\mathbf{r}^{(n)}$ and $d\mathbf{r}^{(t)}$, so that $\theta_{nt} = (\pi/2) - \gamma_{nt}$. The decrease in angle (in radians) between two orthogonal line elements is the *engineering shear strain*. From the definition of the dot product, we have

$$d\mathbf{r}'^{(n)} \cdot d\mathbf{r}'^{(t)} = ds'^{(n)}ds'^{(t)}\cos\left(\frac{\pi}{2} - \gamma_{nt}\right) = ds'^{(n)}ds'^{(t)}\sin\gamma_{nt} \qquad \text{[3.7.17]}$$

The deformations are assumed to be so small that $\sin\gamma_{nt} = \gamma_{nt}$ is valid to several significant figures.[8] In that case, we have

$$\gamma_{nt} = \frac{d\mathbf{r}'^{(n)} \cdot d\mathbf{r}'^{(t)}}{ds'^{(n)}ds'^{(t)}} \qquad \text{[3.7.18]}$$

In terms of the components of $d\mathbf{r}^{(n)}$ and $d\mathbf{r}^{(t)}$, the dot product in the numerator can be written

$$d\mathbf{r}'^{(n)} \cdot d\mathbf{r}'^{(t)} = (dx'^{(n)}\mathbf{i} + dy'^{(n)}\mathbf{j} + dz'^{(n)}\mathbf{k}) \cdot (dx'^{(t)}\mathbf{i} + dy'^{(t)}\mathbf{j} + dz'^{(t)}\mathbf{k})$$
$$= dx'^{(n)}dx'^{(t)} + dy'^{(n)}dy'^{(t)} + dz'^{(n)}dz'^{(t)}$$

We then use Equations 3.7.9 to express the right-hand side in terms of the displacement derivatives:

$$d\mathbf{r}'^{(n)} \cdot d\mathbf{r}'^{(t)} = \left(dx^{(n)} + \frac{\partial u}{\partial x}dx^{(n)} + \frac{\partial u}{\partial y}dy^{(n)} + \frac{\partial u}{\partial z}dz^{(n)}\right)\left(dx^{(t)} + \frac{\partial u}{\partial x}dx^{(t)} + \frac{\partial u}{\partial y}dy^{(t)} + \frac{\partial u}{\partial z}dz^{(t)}\right)$$
$$+ \left(dy^{(n)} + \frac{\partial v}{\partial x}dx^{(n)} + \frac{\partial v}{\partial y}dy^{(n)} + \frac{\partial v}{\partial z}dz^{(n)}\right)\left(dy^{(t)} + \frac{\partial v}{\partial x}dx^{(t)} + \frac{\partial v}{\partial y}dy^{(t)} + \frac{\partial v}{\partial z}dz^{(t)}\right)$$
$$+ \left(dz^{(n)} + \frac{\partial w}{\partial x}dx^{(n)} + \frac{\partial w}{\partial y}dy^{(n)} + \frac{\partial w}{\partial z}dz^{(n)}\right)\left(dz^{(t)} + \frac{\partial w}{\partial x}dx^{(t)} + \frac{\partial w}{\partial y}dy^{(t)} + \frac{\partial w}{\partial z}dz^{(t)}\right)$$

[8] $\sin x = x - \frac{x^3}{6} + \frac{x^5}{120} + \cdots$

Multiplying the brackets and then discarding terms with displacement derivatives to a power greater than 1, this becomes

$$d\mathbf{r}'^{(n)} \cdot d\mathbf{r}'^{(t)} = dx^{(n)}dx^{(t)} + dy^{(n)}dy^{(t)} + dz^{(n)}dz^{(t)} + 2\frac{\partial u}{\partial x}dx^{(n)}dx^{(t)} + 2\frac{\partial v}{\partial y}dy^{(n)}dy^{(t)}$$

$$+ 2\frac{\partial w}{\partial z}dz^{(n)}dz^{(t)} + \left(\frac{\partial u}{\partial y} + \frac{\partial v}{\partial x}\right)(dx^{(n)}dy^{(t)} + dx^{(t)}dy^{(n)})$$

$$+ \left(\frac{\partial u}{\partial z} + \frac{\partial w}{\partial x}\right)(dx^{(n)}dz^{(t)} + dx^{(t)}dz^{(n)}) + \left(\frac{\partial v}{\partial z} + \frac{\partial w}{\partial y}\right)(dy^{(n)}dz^{(t)} + dy^{(t)}dz^{(n)})$$

The first three terms on the right vanish according to Equation 3.7.16, so that substituting this expression into Equation 3.7.18 yields the following lengthy expression for the engineering shear strain:

$$\gamma_{nt} = 2\frac{\partial u}{\partial x}\frac{dx^{(n)}}{ds^{(n)}}\frac{dx^{(t)}}{ds^{(t)}} + 2\frac{\partial v}{\partial y}\frac{dy^{(n)}}{ds^{(n)}}\frac{dy^{(t)}}{ds^{(t)}} + 2\frac{\partial w}{\partial z}\frac{dz^{(n)}}{ds^{(n)}}\frac{dz^{(t)}}{ds^{(t)}} + \left(\frac{\partial u}{\partial y} + \frac{\partial v}{\partial x}\right)\left(\frac{dx^{(n)}}{ds^{(n)}}\frac{dy^{(t)}}{ds^{(t)}} + \frac{dx^{(t)}}{ds^{(t)}}\frac{dy^{(n)}}{ds^{(n)}}\right)$$

$$+ \left(\frac{\partial u}{\partial z} + \frac{\partial w}{\partial x}\right)\left(\frac{dx^{(n)}}{ds^{(n)}}\frac{dz^{(t)}}{ds^{(t)}} + \frac{dx^{(t)}}{ds^{(t)}}\frac{dz^{(n)}}{ds^{(n)}}\right) + \left(\frac{\partial v}{\partial z} + \frac{\partial w}{\partial y}\right)\left(\frac{dy^{(n)}}{ds^{(n)}}\frac{dz^{(t)}}{ds^{(t)}} + \frac{dy^{(t)}}{ds^{(t)}}\frac{dz^{(n)}}{ds^{(n)}}\right)$$

Using Equations 3.7.15, we introduce the components of the unit vectors $\mathbf{n}$ and $\mathbf{t}$ into this expression, thereby finally obtaining

$$\gamma_{nt} = 2\frac{\partial u}{\partial x}n_x t_x + 2\frac{\partial v}{\partial y}n_y t_y + 2\frac{\partial w}{\partial z}n_z t_z + \left(\frac{\partial u}{\partial y} + \frac{\partial v}{\partial x}\right)(n_x t_y + t_x n_y)$$

$$+ \left(\frac{\partial u}{\partial z} + \frac{\partial w}{\partial x}\right)(n_x t_z + t_x n_z) + \left(\frac{\partial v}{\partial z} + \frac{\partial w}{\partial y}\right)(n_y t_z + t_y n_z)$$

[3.7.19]

Equation 3.7.19 is a formula that allows us to find the engineering shear strain between any two orthogonal elements in terms of the displacement derivatives. The engineering shear strain between two line elements oriented in the x and y directions is denoted γ_{xy}. To find γ_{xy}, we set $\mathbf{n} = \mathbf{i}$ ($n_x = 1$, $n_y = n_z = 0$) and $\mathbf{t} = \mathbf{j}$ ($t_y = 1$, $t_x = t_z = 0$), whereupon Equation 3.7.19 yields $\gamma_{xy} = \partial u/\partial y + \partial v/\partial x$. Following this procedure, we can obtain similar relations for the other two shear strains γ_{xz} ($\mathbf{n} = \mathbf{i}$ and $\mathbf{t} = \mathbf{k}$) and γ_{yz} ($\mathbf{n} = \mathbf{j}$ and $\mathbf{t} = \mathbf{k}$). These three shear strain formulas, together with those obtained in Equation 3.7.13 for the normal strains, comprise the six *strain–displacement equations*

$$\boxed{\begin{array}{ccc} \varepsilon_x = \dfrac{\partial u}{\partial x} & \varepsilon_y = \dfrac{\partial v}{\partial y} & \varepsilon_z = \dfrac{\partial w}{\partial z} \\[2mm] \gamma_{xy} = \dfrac{\partial u}{\partial y} + \dfrac{\partial v}{\partial x} & \gamma_{xz} = \dfrac{\partial u}{\partial z} + \dfrac{\partial w}{\partial x} & \gamma_{yz} = \dfrac{\partial v}{\partial z} + \dfrac{\partial w}{\partial y} \end{array}}$$

[3.7.20]

Notice that to obtain the strain–displacement relation for γ_{yx} (instead of γ_{xy}) from Equation 3.7.19, we must set $\mathbf{n} = \mathbf{j}$ and $\mathbf{t} = \mathbf{i}$. Substituting the components of these vectors on the right of Equation 3.3.19, we calculate $\gamma_{yx} = \partial u/\partial y + \partial v/\partial x$. Clearly, $\gamma_{yx} = \gamma_{xy}$. This is true of all the shear strains: interchanging the order of the subscripts on the shear strains yields the same quantity. That is,

$$\gamma_{xy} = \gamma_{yx} \qquad \gamma_{xz} = \gamma_{zx} \qquad \gamma_{yz} = \gamma_{zy}$$

[3.7.21]

This set of relationships should be compared to Equations 3.3.2 for the shear stresses.

We can substitute the strain–displacement relationships into Equations 3.7.12 and 3.7.19 to obtain more compact forms of the *strain transformation formulas*, as follows:

$$\varepsilon_n = \varepsilon_x n_x^2 + \varepsilon_y n_y^2 + \varepsilon_z n_z^2 + 2\left(\frac{\gamma_{xy}}{2}\right) n_x n_y + 2\left(\frac{\gamma_{xz}}{2}\right) n_x n_z + 2\left(\frac{\gamma_{yz}}{2}\right) n_y n_z$$

$$\frac{\gamma_{nt}}{2} = \varepsilon_x n_x t_x + \varepsilon_y n_y t_y + \varepsilon_z n_z t_z + \left(\frac{\gamma_{xy}}{2}\right)\left(n_x t_y + t_x n_y\right)$$

$$+ \left(\frac{\gamma_{xz}}{2}\right)\left(n_x t_z + t_x n_z\right) + \left(\frac{\gamma_{yz}}{2}\right)\left(n_y t_z + t_y n_z\right)$$

$(\mathbf{n} \cdot \mathbf{t} = 0)$ [3.7.22]

The components of strain relative to any set of orthogonal axes are therefore known if we are given the *state of strain*

$$\begin{bmatrix} \varepsilon_x & \gamma_{xy} & \gamma_{xz} \\ \gamma_{xy} & \varepsilon_y & \gamma_{yz} \\ \gamma_{xz} & \gamma_{yz} & \varepsilon_z \end{bmatrix}$$ [3.7.23]

relative to a cartesian coordinate system.

A physical interpretation of the *shear* strain–displacement relationships is in order. Consider two orthogonal line elements, PQ along the x axis and PR along the y axis, as illustrated in Figure 3.7.5. Due to shearing action, PQ rotates through a small angle α, ending up in the orientation $P'Q'$ after deformation. Likewise, PR rotates through the angle β into the line element $P'R'$. From the figure we can see that

$$\alpha = \tan^{-1}\frac{(\partial v/\partial x)\, dx}{dx} \cong \frac{\partial v}{\partial x} \qquad \beta = \tan^{-1}\frac{(\partial u/\partial y)\, dy}{dy} \cong \frac{\partial u}{\partial y}$$

The angles α and β are very small—a few thousandths of a radian at most—so the indicated approximations are very accurate. The total decrease in the initially right angle between PQ and PR is $\alpha + \beta$. The engineering shear strain between the x and y directions is therefore

$$\gamma_{xy} = \alpha + \beta = \frac{\partial v}{\partial x} + \frac{\partial u}{\partial y}$$ [3.7.24]

in agreement with Equation 3.7.20$_4$. Interpretations of the other shear strain terms can be made in an identical fashion.

The transformation equations for strain, Equations 3.7.22, would be identical to those for stress, Equations 3.2.13, were it not for the factor of one-half attached to the engineering shear strain components. If we redefine the shear strain to be *one-half* the decrease in angle between initially orthogonal line elements and assign it the symbol ε_{nt}, then in place of Equation 3.7.22 we have

$$\varepsilon_n = \varepsilon_x n_x^2 + \varepsilon_y n_y^2 + \varepsilon_z n_z^2 + \varepsilon_{xy} n_x n_y + \varepsilon_{xz} n_x n_z + \varepsilon_{yz} n_y n_z$$

$$\varepsilon_{nt} = \varepsilon_x n_x t_x + \varepsilon_y n_y t_y + \varepsilon_z n_z t_z + \varepsilon_{xy}\left(n_x t_y + t_x n_y\right)$$

$$+ \varepsilon_{xz}\left(n_x t_z + t_x n_z\right) + \varepsilon_{yz}\left(n_y t_z + t_y n_z\right)$$

$(\mathbf{n} \cdot \mathbf{t} = 0)$ [3.7.25]

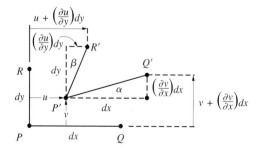

Figure 3.7.5   Shear strain between the x and y directions.

The one-half factor now shows up in the shear strain–displacement relations, as follows:

$$\varepsilon_{xy} = \frac{1}{2}\left(\frac{\partial u}{\partial y} + \frac{\partial v}{\partial x}\right) \qquad \varepsilon_{xz} = \frac{1}{2}\left(\frac{\partial u}{\partial z} + \frac{\partial w}{\partial x}\right) \qquad \varepsilon_{yz} = \frac{1}{2}\left(\frac{\partial v}{\partial z} + \frac{\partial w}{\partial y}\right) \qquad \textbf{[3.7.26]}$$

With this redefinition of strain, Equations 3.2.13 and 3.7.25 are identical, which means the stress and strain matrices

$$\begin{bmatrix} \sigma_x & \tau_{xy} & \tau_{xz} \\ \tau_{xy} & \sigma_y & \tau_{yz} \\ \tau_{xz} & \tau_{yz} & \sigma_z \end{bmatrix} \quad \text{and} \quad \begin{bmatrix} \varepsilon_x & \varepsilon_{xy} & \varepsilon_{xz} \\ \varepsilon_{xy} & \varepsilon_y & \varepsilon_{yz} \\ \varepsilon_{xz} & \varepsilon_{yz} & \varepsilon_z \end{bmatrix} \qquad \textbf{[3.7.27]}$$

are both tensors. Thus, for a given state of strain, there exist three values of principal strain, which we can order $\varepsilon_1 > \varepsilon_2 > \varepsilon_3$. The shear strain between each of the three principal strain directions is zero. Line elements aligned along the principal strain directions will stretch during deformation, but they will remain at right angles to each other. The three extreme values of shear strain are

$$\frac{\varepsilon_1 - \varepsilon_2}{2} \qquad \frac{\varepsilon_1 - \varepsilon_3}{2} \qquad \frac{\varepsilon_2 - \varepsilon_3}{2}$$

The maximum shear strain magnitude at a point is

$$\text{Max. shear strain} = \frac{|\varepsilon_{\max} - \varepsilon_{\min}|}{2}$$

We will continue to use the engineering shear strain γ, remembering to write it as $\frac{\gamma}{2}$ when transforming strains from one coordinate system to another. Therefore,

$$\gamma_{\max} = |\varepsilon_{\max} - \varepsilon_{\min}| \qquad \textbf{[3.7.28]}$$

3.8 VOLUMETRIC STRAIN

Consider a differential material cube whose volume is $dV_o = dx\,dy\,dz$, as in Figure 3.8.1. Due to the normal strain in the x, y, and z directions, the edges of this element increase in length by the amounts $\varepsilon_x dx$, $\varepsilon_y dy$, and $\varepsilon_z dz$. Therefore, the new volume is $dV = (1 + \varepsilon_x)(1 + \varepsilon_y)(1 + \varepsilon_z)dV_o$. The volumetric strain e is

$$e = \frac{\text{change in volume}}{\text{original volume}} = \frac{dV - dV_o}{dV_o} = (1 + \varepsilon_x)(1 + \varepsilon_y)(1 + \varepsilon_z) - 1$$

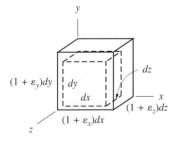

Figure 3.8.1   Change in volume of a differential element.

Neglecting the products of normal strain, we get

$$e = \varepsilon_x + \varepsilon_y + \varepsilon_z \qquad [3.8.1]$$

The quantity e is also referred to as the *dilatation*. The dilatation is useful when it is necessary to separate the total deformation into the portion causing only volume change and the portion causing only a change in shape.

3.9 COMPATIBILITY CONDITIONS

The strains in a solid must be consistent or *compatible* with the displacements of the solid. For example, if we hold the left end of a meter stick rigidly in place and move the other end one millimeter to the right, then the normal strain (Equation 3.7.10) in the meter stick is uniform and equals 0.001, or 1000μ. Now suppose *both* ends of the meter stick are rigidly fixed, yet we still insist that the strain throughout is 1000μ. Perhaps we allow the right end of the *left* half of the stick to move 0.5 mm to the right, while at the same time the left end of the *right* half moves 0.5 mm to the left. That way, the total length of the meter stick increases by one millimeter, the rigid constraints at each end remain enforced, and we obtain the required uniform strain of 1000μ. Of course, in so doing, the midpoint of the stick must move simultaneously to the right *and* to the left, which is absurd—unless we destroy the continuity of the body by physically cutting it in two at the middle. Compatibility, then, has to do with maintaining the continuity of a structure.

Compatibility between strains and displacements is expressed by Equations 3.7.20. Given the three displacements, we calculate the six strains, which are therefore compatible. On the other hand, given the strains, we have six equations from which to determine three displacements. We have more equations than unknowns. To ensure that Equations 3.7.20 will yield three physically reasonable (single-valued and continuous) displacements, some mathematical "gimmickery" is used to generate six auxiliary conditions that the strains alone must satisfy. The calculations depend on the fact that the order of taking partial derivatives can be interchanged without effect.

Starting with Equation 3.7.20₁, we take the partial derivative of both sides with respect to y and z, as follows:

$$\frac{\partial^2 \varepsilon_x}{\partial y \partial z} = \frac{\partial^2}{\partial y \partial z}\left(\frac{\partial u}{\partial x}\right) = \frac{\partial^3 u}{\partial y \partial z \partial x} = \frac{\partial^3 u}{\partial x \partial y \partial z} = \frac{\partial}{\partial x}\left(\frac{\partial^2 u}{\partial y \partial z}\right)$$

Since

$$\frac{\partial^2 u}{\partial y \partial z} = \left(\frac{1}{2}\right)\left(\frac{\partial^2 u}{\partial y \partial z} + \frac{\partial^2 u}{\partial z \partial y}\right)$$

this can be written

$$\frac{\partial^2 \varepsilon_x}{\partial y \partial z} = \frac{\partial}{\partial x}\left[\frac{1}{2}\left(\frac{\partial^2 u}{\partial y \partial z} + \frac{\partial^2 u}{\partial z \partial y}\right)\right] \qquad [3.9.1]$$

By adding and subtracting $\partial^2 v / \partial x \partial z$, we can write $\partial^2 u / \partial y \partial z$ as

$$\frac{\partial^2 u}{\partial y \partial z} = \frac{\partial^2 u}{\partial y \partial z} + \frac{\partial^2 v}{\partial x \partial z} - \frac{\partial^2 v}{\partial z \partial x}$$

$$= \frac{\partial}{\partial z}\left(\frac{\partial u}{\partial y} + \frac{\partial v}{\partial x}\right) - \frac{\partial^2 v}{\partial z \partial x} = \frac{\partial \gamma_{xy}}{\partial z} - \frac{\partial^2 v}{\partial z \partial x} \qquad [3.9.2]$$

where we used Equation 3.7.20$_4$, the formula for γ_{xy} in terms of the displacements. Likewise, using Equation 3.7.20$_5$, we have

$$\frac{\partial^2 u}{\partial z \partial y} = \frac{\partial^2 u}{\partial z \partial y} + \frac{\partial^2 w}{\partial x \partial y} - \frac{\partial^2 w}{\partial y \partial x}$$

$$= \frac{\partial}{\partial y}\left(\frac{\partial u}{\partial z} + \frac{\partial w}{\partial x}\right) - \frac{\partial^2 w}{\partial y \partial x} = \frac{\partial \gamma_{xz}}{\partial z} - \frac{\partial^2 w}{\partial y \partial x}$$

[3.9.3]

Adding Equations 3.9.2 and 3.9.3 and using Equation 3.7.20$_6$, we get

$$\frac{\partial^2 u}{\partial y \partial z} + \frac{\partial^2 u}{\partial z \partial y} = \frac{\partial \gamma_{xy}}{\partial z} + \frac{\partial \gamma_{xz}}{\partial z} - \frac{\partial}{\partial x}\left(\frac{\partial v}{\partial z} + \frac{\partial w}{\partial y}\right)$$

$$= \frac{\partial \gamma_{xy}}{\partial z} + \frac{\partial \gamma_{xz}}{\partial z} - \frac{\partial \gamma_{yz}}{\partial x}$$

[3.9.4]

Substituting Equation 3.9.4 into Equation 3.9.1 finally yields the first of the six *compatibility equations*,

$$2\frac{\partial^2 \varepsilon_x}{\partial y \partial z} = \frac{\partial}{\partial x}\left(\frac{\partial \gamma_{xy}}{\partial z} + \frac{\partial \gamma_{xz}}{\partial z} - \frac{\partial \gamma_{yz}}{\partial x}\right)$$

[3.9.5]

Starting with the other two normal strain–displacement relationships, Equations 3.7.20$_2$ and 3.7.20$_3$, and proceeding in a similar fashion leads to two more equations similar to Equation 3.9.5.

The second set of three strain compatibility equations is found by starting with the shear strain–displacement relationships. Taking the partial derivative of Equation 3.7.20$_4$ with respect to x and y and then reordering the partial differentiation yields

$$\frac{\partial^2 \gamma_{xy}}{\partial x \partial y} = \frac{\partial^2}{\partial x \partial y}\left(\frac{\partial u}{\partial y} + \frac{\partial v}{\partial x}\right) = \frac{\partial^3 u}{\partial x \partial y^2} + \frac{\partial^3 v}{\partial x \partial y \partial x}$$

$$= \frac{\partial^3 u}{\partial y^2 \partial x} + \frac{\partial^3 v}{\partial x^2 \partial y} = \frac{\partial^2}{\partial y^2}\left(\frac{\partial u}{\partial x}\right) + \frac{\partial^2}{\partial x^2}\left(\frac{\partial v}{\partial y}\right)$$

Substituting Equations 3.7.20$_1$ and 3.7.20$_2$, we find

$$\frac{\partial^2 \gamma_{xy}}{\partial x \partial y} = \frac{\partial^2 \varepsilon_x}{\partial y^2} + \frac{\partial^2 \varepsilon_y}{\partial x^2}$$

The two remaining equations are obtained by proceeding in a similar fashion by starting with Equations 3.7.20$_5$ and 3.7.20$_6$, respectively.

The six *compatibility equations* may be summarized as follows:

$$\frac{\partial^2 \gamma_{xy}}{\partial x \partial y} = \frac{\partial^2 \varepsilon_x}{\partial y^2} + \frac{\partial^2 \varepsilon_y}{\partial x^2} \qquad \frac{\partial^2 \gamma_{yz}}{\partial y \partial z} = \frac{\partial^2 \varepsilon_y}{\partial z^2} + \frac{\partial^2 \varepsilon_z}{\partial y^2} \qquad \frac{\partial^2 \gamma_{xz}}{\partial z \partial x} = \frac{\partial^2 \varepsilon_z}{\partial x^2} + \frac{\partial^2 \varepsilon_x}{\partial z^2}$$

$$2\frac{\partial^2 \varepsilon_x}{\partial y \partial z} = \frac{\partial}{\partial x}\left(-\frac{\partial \gamma_{yz}}{\partial x} + \frac{\partial \gamma_{xz}}{\partial y} + \frac{\partial \gamma_{xy}}{\partial z}\right) \qquad 2\frac{\partial^2 \varepsilon_y}{\partial z \partial x} = \frac{\partial}{\partial y}\left(-\frac{\partial \gamma_{xz}}{\partial y} + \frac{\partial \gamma_{xy}}{\partial z} + \frac{\partial \gamma_{yz}}{\partial x}\right)$$

$$2\frac{\partial^2 \varepsilon_z}{\partial x \partial y} = \frac{\partial}{\partial z}\left(-\frac{\partial \gamma_{xy}}{\partial z} + \frac{\partial \gamma_{yz}}{\partial x} + \frac{\partial \gamma_{xz}}{\partial y}\right)$$

[3.9.6]

The displacements that occur in a solid when loads are applied are unknown. They must be found as part of the solution of the problem. Ultimately, the strain–displacement equations are required. For example, to find the x component of displacement throughout the solid requires integrating Equation 3.7.20$_1$. That is,

$$u(x, y, z) = \int \varepsilon_x(x, y, z)dx + f(y, z)$$

where $f(y, z)$ is a function of integration, which depends on the specific problem. Similar expressions hold for $v(x, y, z)$ and $w(x, y, z)$. An admissible displacement field can be obtained in this fashion from a given strain field only if the strains satisfy the compatibility equations. Clearly, strains that are linear in x, y, and z are compatible.

3.10   PLANE STRAIN

Analogous to the state of plane stress is that of *plane strain*, in which the deformation is confined to the xy plane, so that the state of strain is given by the three components

$$\begin{bmatrix} \varepsilon_x & \gamma_{xy} & 0 \\ \gamma_{xy} & \varepsilon_y & 0 \\ 0 & 0 & 0 \end{bmatrix} \qquad \text{plane strain} \qquad \text{[3.10.1]}$$

and the strain–displacement relationships are

$$\varepsilon_x = \frac{\partial u}{\partial x} \qquad \varepsilon_y = \frac{\partial v}{\partial y} \qquad \gamma_{xy} = \frac{\partial u}{\partial y} + \frac{\partial v}{\partial x} \qquad \text{plane strain} \qquad \text{[3.10.2]}$$

The plane strain transformation relationships are the same as those for stress, Equations 3.6.4. Simply replace the normal stresses by the normal strains and substitute $\frac{\gamma_{xy}}{2}$ for τ_{xy}, to produce the following:

$$\bar{\varepsilon}_x = \varepsilon_x \cos^2 \theta + \varepsilon_y \sin^2 \theta + \gamma_{xy} \sin \theta \cos \theta$$
$$\bar{\varepsilon}_y = \varepsilon_x \sin^2 \theta + \varepsilon_y \cos^2 \theta - \gamma_{xy} \sin \theta \cos \theta$$
$$\bar{\gamma}_{xy} = 2\left(\varepsilon_y - \varepsilon_x\right) \sin \theta \cos \theta + \gamma_{xy} \left(\cos^2 \theta - \sin^2 \theta\right)$$
$$\text{plane strain} \qquad \text{[3.10.3]}$$

A Mohr circle for strain may be constructed similar to that for stress, with ε replacing σ and $\frac{\gamma}{2}$ replacing τ in Figure 3.6.3.

In plane strain, there is no dependence on z, and all of the z components of strain are zero. Hence, all of the compatibility equations are identities ($0 = 0$) except Equation 3.9.6$_1$, which is:

$$\frac{\partial^2 \gamma_{xy}}{\partial x \partial y} = \frac{\partial^2 \varepsilon_x}{\partial y^2} + \frac{\partial^2 \varepsilon_y}{\partial x^2} \qquad \text{plane strain} \qquad \text{[3.10.4]}$$

Example 3.10.1   Consider the following plane strain fields: (a) $\varepsilon_x = 0$, $\varepsilon_y = 0$, $\gamma_{xy} = 10^{-4}x$, and (b) $\varepsilon_x = 0$, $\varepsilon_y = 0$, $\gamma_{xy} = 10^{-4}xy$. For both fields, the units are centimeters. Sketch the accompanying displacement fields by applying the strains to the initially square elements of the coarse four-by-four grid of Figure 3.10.1.

(a)   The strains $\varepsilon_x = 0$, $\varepsilon_y = 0$, $\gamma_{xy} = 10^{-4}x$ clearly satisfy the compatibility requirement, Equation 3.10.4. Each small square of the grid must deform in such a way that its edges remain unchanged in length, since the normal strains are zero. The change in the right angle at the lower left corner of each element is given by the shear strain γ_{xy}, evaluated at the centroid of the element. Figure 3.10.2a shows that these strains yield continuous, single-valued displacements at all points of the grid.

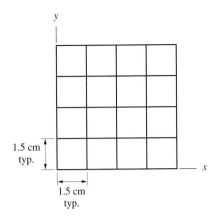

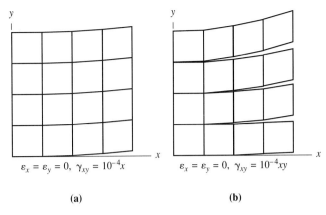

Figure 3.10.1   Square region divided into six-teen elements to which given strains are to be applied.

Figure 3.10.2   Deformation of the grid of initially square elements. (a) Compatible strain field. (b) Noncompatible strain field.

(b)   The strains $\varepsilon_x = 0$, $\varepsilon_y = 0$, $\gamma_{xy} = 10^{-4}xy$ do not satisfy the compatibility equation. If we sketch the elements deformed according to this set of strains, Figure 3.10.2b results. The displacements are clearly not single-valued, since gaps appear between each row of elements. These gaps are required in order to keep the edges of each element unchanged in length while accommodating the specified shear strain.

Example 3.10.2   Calculate the displacements corresponding to the following two-dimensional plane strain field:

$$\varepsilon_x = a\left(-3x^2 + 7y^2\right) \qquad \varepsilon_y = a\left(x^2 - 5y^2\right) \qquad \gamma_{xy} = 16axy$$

where a is a nonzero constant.

First, let us determine whether this set of strains satisfies the compatibility condition, Equation 3.10.4. To that end, we compute $\partial^2 \varepsilon_x / \partial y^2 = 14a$, $\partial^2 \varepsilon_y / \partial x^2 = 2a$, and $\partial^2 \gamma_{xy} / \partial x \partial y = 16a$, verifying that Equation 3.10.4 is satisfied. The strains are indeed compatible, which means they can be integrated to yield a set of continuous, single-valued displacements.

From the strain–displacement relation, Equation 3.10.2$_1$, we have

$$\frac{\partial u}{\partial x} = a\left(-3x^2 + 7y^2\right)$$

Integrating with respect to x yields

$$u = a\left(-x^3 + 7xy^2\right) + f(y) \qquad\qquad \text{[a]}$$

where $f(y)$ is an as yet undetermined function of y alone ($\partial f / \partial x = 0$). Integrating the strain–displacement equation 3.10.2$_2$ similarly yields

$$v = a\left(x^2 y - \tfrac{5}{3}y^3\right) + g(x) \qquad\qquad \text{[b]}$$

where $g(x)$ is a function of x only. To determine the specific form of $f(y)$ and $g(x)$ and thus complete our solution for the displacements, we substitute the expressions for u and v into the third strain–displacement equation 3.10.2$_3$ to find that

$$[14axy + \frac{df(y)}{dy}] + [2axy + \frac{dg(x)}{dx}] = 16axy$$

Combining terms, we get for this case

$$\frac{dg(x)}{dx} = -\frac{df(y)}{dy}$$

A function of x can equal a function of y for arbitrary x and y only if both functions are equal and constant. Thus,

$$\frac{dg(x)}{dx} = C \quad \text{and} \quad \frac{df(y)}{dy} = -C, \quad C \text{ is a constant}$$

Integrating both of these, we find that

$$g(x) = Cx + D \quad \text{and} \quad f(y) = -Cy + E, \quad D \text{ and } E \text{ are constants}$$

Finally, substituting the linear forms for f and g into Equations a and b yields

$$u = a\left(-x^3 + 7xy^2\right) - Cy + E$$
$$v = a\left(x^2 y - \tfrac{5}{3}y^3\right) + Cx + D$$

The constants E and D represent rigid-body translation in the x and y directions, respectively, whereas the terms involving C represent rigid-body rotation. None of the terms involving these three constants contribute to the strain, nor therefore to the stress. By imposing constraints, we can evaluate those constants. For example, if at the point $x = 0$, $y = 0$, we require that $u = v = \partial v / \partial x = 0$, then it follows that $C = D = E = 0$.

Example 3.10.3   Find the displacements corresponding to the following strains

$$\varepsilon_x = a\left(x^2 + y^2\right) \qquad \varepsilon_y = a\left(x^2 + y^2\right) \qquad \gamma_{xy} = -8axy$$

where a is a nonzero constant.

First, we check to see if these strains are compatible. Since, $\partial^2 \varepsilon_x / \partial y^2 = 2a$, $\partial^2 \varepsilon_y / \partial x^2 = 2a$, and $\partial^2 \gamma_{xy} / \partial x \partial y = -8a$, then Equation 3.10.4 is not satisfied, and the strains are not compatible. Therefore, a corresponding set of continuous, single-valued displacements cannot be found. Let us ignore this fact and attempt to integrate for the displacements in order to see what kind of difficulty occurs.

Following the steps of the previous example, we substitute ε_x into Equation 3.10.2$_1$ and ε_y into Equation 3.10.2$_2$ and integrate with respect to x and y, respectively, to find

$$u = a\left(\tfrac{1}{3}x^3 + xy^2\right) + f(y)$$
$$v = a\left(x^2 y + \tfrac{1}{3}y^3\right) + g(x)$$

which we substitute into Equation 3.10.2$_3$ to obtain

$$\frac{df(y)}{dy} + \frac{dg(x)}{dx} = 12axy$$

This is where we run into trouble and can proceed no further. We cannot organize this equation into terms that are a function of x alone, or of y alone, in order to integrate df/dy and dg/dx. Since we are unable to find the functional form of f and g, the displacements u and v remain undetermined.

3.11   STRESS–STRAIN EQUATIONS: ISOTROPIC ELASTIC MATERIALS

The deformation of a structure is clearly a function of the applied loads, and the resulting strains at a point are related to the local state of stress. A clue to the relationships between stress and strain can be found by subjecting a uniform rod to a tension test, as in Figure 3.11.1. An axial load P is applied to the rod, resulting in an axial deflection δ.

If the rod is elastic, the relationship between axial load P and the axial deflection δ is essentially linear for sufficiently small deflections, and it is the same whether the load is increasing or decreasing.

If the load P is applied at the centroid of the rod's cross section, then the axial stress and strain are uniform across all sections, except possibly in the vicinity of the ends. The uniform axial strain ε and stress σ are

$$\varepsilon = \frac{\delta}{L} \qquad \sigma = \frac{P}{A}$$

A plot of stress versus strain derived from Figure 3.11.2 will obviously be linear as well (Figure 3.11.3). The slope of the stress–strain diagram is the *modulus of elasticity*, or *Young's modulus*,[9] E.

All structural materials exhibit the *Poisson effect*.[10] That is, a tensile test specimen stretched in the axial direction contracts laterally; if axially compressed, it expands laterally. If the bar shown in Figure 3.11.4 has a circular cross section of unloaded diameter D, and if the contraction transverse to the pull direction is d, then the transverse strain $\varepsilon_\perp$ is

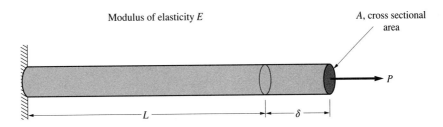

Figure 3.11.1   Deformation of a uniform elastic rod under axial load.

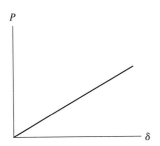

Figure 3.11.2   Linear load versus deflection behavior of an elastic rod.

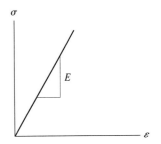

Figure 3.11.3   Stress-strain curve for a linearly elastic material.

[9]Thomas Young (1773–1829), English scientist.
[10]Siméon Denis Poisson (1781–1840), French mathematician.

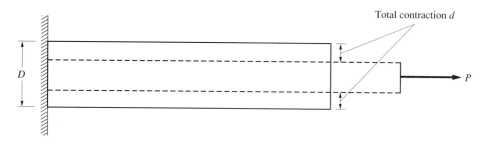

Figure 3.11.4   Poisson effect in an elastic rod.

$$\varepsilon_\perp = -\frac{d}{D}$$

Poisson's ratio ν is a material property that relates the axial strain to the transverse strain, as follows:

$$\nu = -\frac{\varepsilon_\perp}{\varepsilon} \qquad\qquad \textbf{[3.11.1]}$$

The minus sign in this equation is required because the axial strain and accompanying transverse strain are always opposite in sign.

The stress–strain relationships for a three-dimensional state of loading can be derived by calculating the strains that accompany the normal and shear stresses considered to act separately in each direction and then adding the results together. This is an application of the *principle of superposition*, which holds for linear elastic behavior, our sole concern in this text. Thus, if a plane differential element of material is subjected to stress σ_x, then the resulting normal strains will be as shown in Figure 3.11.5a. On the other hand, if only the stress σ_y or σ_z acts, the accompanying strains will be as shown in Figure 3.11.5b or c. If the three states of uniaxial stress in Figure 3.11.5 are combined using the principle of superposition, we obtain a triaxial state of stress in which the total normal strain in each direction is the sum of those shown in the figure (cf. Figure 3.11.6).

These results are valid only for *isotropic* materials, that is, materials whose stiffness, strength, and other properties are the same in all directions. Young's modulus is the same in both the x and y directions of a sheet of

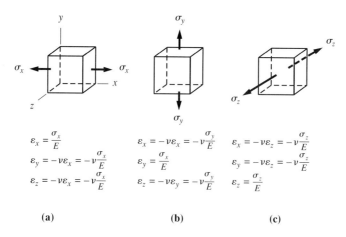

(a)   **(b)**   **(c)**

Figure 3.11.5   Normal strains accompanying uniaxial stress in the x, y, and z directions.

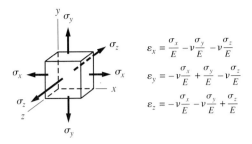

$$\varepsilon_x = \frac{\sigma_x}{E} - v\frac{\sigma_y}{E} - v\frac{\sigma_z}{E}$$

$$\varepsilon_y = -v\frac{\sigma_x}{E} + \frac{\sigma_y}{E} - v\frac{\sigma_z}{E}$$

$$\varepsilon_z = -v\frac{\sigma_x}{E} - v\frac{\sigma_y}{E} + \frac{\sigma_z}{E}$$

Figure 3.11.6   Normal strains accompanying a triaxial state of stress.

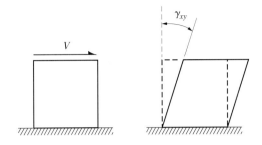

Figure 3.11.7   Pure shear loading.

isotropic material. On the other hand, material properties are direction dependent in *anisotropic* materials, which are discussed in section 3.17.

In isotropic materials, the shear stresses are independent of the normal strains. A shear load V applied to the plate of Figure 3.11.7 will induce only shear strain. A state of pure shear is also produced in thin-walled tubes subjected to pure torsion. Recall that the change in right angle is the shear strain γ_{xy}, which is linearly related to the shear stress τ_{xy}

$$\tau_{xy} = G\gamma_{xy}$$

The constant G is called the *shear modulus*.

Suppose we have a state of plane stress which *is* pure shear relative to the xy axes, so that

$$\sigma_x = \sigma_y = 0 \qquad \tau_{xy} = \tau$$

The state of in-plane strain is therefore

$$\varepsilon_x = \varepsilon_y = 0 \qquad \gamma_{xy} = \frac{\tau}{G}$$

From Equations 3.6.4 we know that at 45 degrees from the xy axes, the state of stress is

$$\bar{\sigma}_x = (0)\cos^2 45 + (0)\sin^2 45 + 2\tau\sin 45\cos 45 = \tau$$

$$\bar{\sigma}_y = (0)\sin^2 45 + (0)\cos^2 45 - 2\tau\sin 45\cos 45 = -\tau$$

$$\bar{\tau}_{xy} = (0)\sin 45\cos 45 + \tau(\cos^2 45 - \sin^2 45) = 0$$

Since the material is isotropic, we can use the expressions in Figure 3.11.6 to compute the corresponding strain components in the xy plane, as follows:

$$\bar{\varepsilon}_x = \frac{\tau}{E} - v\frac{(-\tau)}{E} = \frac{1+v}{E}\tau \qquad \bar{\varepsilon}_y = \frac{(-\tau)}{E} - v\frac{\tau}{E} = -\frac{1+v}{E}\tau \qquad \bar{\gamma}_{xy} = 0$$

On the other hand, Equations 3.10.3 require that

$$\bar{\varepsilon}_x = (0)\cos^2 45 + (0)\sin^2 45 + \left(\frac{\tau}{2G}\right)\sin 45\cos 45 = \frac{\tau}{2G}$$

$$\bar{\varepsilon}_y = (0)\sin^2 45 + (0)\cos^2 45 - \left(\frac{\tau}{2G}\right)\sin 45\cos 45 = -\frac{\tau}{2G}$$

$$\bar{\gamma}_{xy} = 2(0)\sin 45\cos 45 + \left(\frac{\tau}{2G}\right)(\cos^2 45 - \sin^2 45) = 0$$

The condition of material isotropy has thus led us to two independent equations for the same normal strain $\bar{\varepsilon}_x$, namely, $\bar{\varepsilon}_x = \tau/2G$ and $\bar{\varepsilon}_x = (1 + v)\tau/E$. For consistency, it must be true that

$$G = \frac{E}{2(1 + v)} \qquad \text{isotropic materials} \qquad \text{[3.11.2]}$$

We account for the fact that materials expand when heated and contract when cooled by adding the concept of *thermal strain*. The response of isotropic materials to temperature change T is characterized by the *coefficient of linear expansion* α. The thermal strain is given by

$$\varepsilon_T = \alpha T \qquad \text{[3.11.3]}$$

This normal strain at a point is the same in all directions in isotropic materials, in which the temperature change does not induce shear strain. The total strain is the mechanical strain plus the thermal strain. Therefore, the term αT is added to each of the normal strains in Figure 3.11.6. We will assume that the temperature throughout the body is determined by means of a separate *heat transfer* analysis.[11] This is *uncoupled thermoelasticity*, in which the temperature field T, like the body forces, is *prescribed* for a given stress problem. In *coupled thermoelasticity* the stress and temperature fields are interdependent and must be solved for simultaneously.

In summary, the strain–stress equations (generalized *Hooke's law*[12]) for an isotropic material are

$$
\begin{aligned}
\varepsilon_x &= \frac{\sigma_x}{E} - v\frac{\sigma_y}{E} - v\frac{\sigma_z}{E} + \alpha T & \gamma_{xy} &= \frac{\tau_{xy}}{G} \\
\varepsilon_y &= -v\frac{\sigma_x}{E} + \frac{\sigma_y}{E} - v\frac{\sigma_z}{E} + \alpha T & \gamma_{xz} &= \frac{\tau_{xz}}{G} \\
\varepsilon_z &= -v\frac{\sigma_x}{E} - v\frac{\sigma_y}{E} + \frac{\sigma_z}{E} + \alpha T & \gamma_{yz} &= \frac{\tau_{yz}}{G}
\end{aligned}
\qquad \text{[3.11.4]}
$$

If the strains are given, it is convenient to solve Equations 3.11.4 for the stresses in terms of the strains, obtaining the stress–strain relationships:

$$
\begin{aligned}
\sigma_x &= \frac{E}{(1+v)(1-2v)}\left[(1-v)\varepsilon_x + v\left(\varepsilon_y + \varepsilon_z\right)\right] - \frac{E\alpha T}{1-2v} & \tau_{xy} &= G\gamma_{xy} \\
\sigma_y &= \frac{E}{(1+v)(1-2v)}\left[(1-v)\varepsilon_y + v\left(\varepsilon_x + \varepsilon_z\right)\right] - \frac{E\alpha T}{1-2v} & \tau_{xz} &= G\gamma_{xz} \\
\sigma_z &= \frac{E}{(1+v)(1-2v)}\left[(1-v)\varepsilon_z + v\left(\varepsilon_x + \varepsilon_y\right)\right] - \frac{E\alpha T}{1-2v} & \tau_{yz} &= G\gamma_{yz}
\end{aligned}
\qquad \text{[3.11.5]}
$$

In section 3.8, the dilatation $e = \varepsilon_x + \varepsilon_y + \varepsilon_z$ was introduced as a measure of volumetric or bulk strain. Substituting the normal strains from Equations 3.11.4, we find that

$$e = \frac{1-2v}{E}(\sigma_x + \sigma_y + \sigma_z) + 3\alpha T$$

Let $p = \frac{1}{3}(\sigma_x + \sigma_y + \sigma_z)$ be the average normal stress, or the *hydrostatic stress*. Then

$$e = \frac{p}{K} + 3\alpha T \qquad p = K(e - 3\alpha T) \qquad \text{[3.11.6]}$$

where K is the *bulk modulus of elasticity*, defined as

---

[11] See, for example, J. P. Holman, *Heat Transfer*, 7th ed., New York, McGraw-Hill, 1990.
[12] Robert Hooke (1635–1703), English scientist.

$$K = \frac{E}{3(1-2v)} \tag{3.11.7}$$

In the absence of temperature change, it is clearly the hydrostatic stress p that causes a volume change (or density change) of a material. The bulk modulus is a measure of the material's resistance to density change under stress. Materials with a very large bulk modulus are nearly incompressible. Incompressible materials have a Poisson's ratio very close to 0.5, according to Equation 3.11.7. In fact, that equation places an upper limit of 0.5 on Poisson's ratio, for if that limit is exceeded, the bulk modulus becomes negative. In that case, a hydrostatic *pressure* ($p < 0$) would produce volumetric expansion instead of compression, contrary to known material behavior.

3.12   THE PLANE STRESS PROBLEM

For plane stress in the xy plane, $\sigma_z = \tau_{xz} = \tau_{yz} = 0$, so that Equations 3.11.4 reduce to

$$\boxed{\begin{aligned} \varepsilon_x &= \frac{\sigma_x}{E} - v\frac{\sigma_y}{E} + \alpha T \\ \varepsilon_y &= -v\frac{\sigma_x}{E} + \frac{\sigma_y}{E} + \alpha T \\ \varepsilon_z &= -\frac{v}{E}\left(\sigma_x + \sigma_y\right) + \alpha T \\ \gamma_{xy} &= \frac{\tau_{xy}}{G} \end{aligned}} \quad \text{plane stress} \tag{3.12.1}$$

Due to the Poisson effect (and to thermal strain, if present), plane stress is accompanied by normal strain in the z direction. For plane stress, Equations 3.11.5 yield

$$\varepsilon_z = \frac{-[v(\varepsilon_x + \varepsilon_y) + (1 + v)\alpha T]}{(1 - v)}$$

together with

$$\boxed{\begin{aligned} \sigma_x &= \frac{E}{1 - v^2}(\varepsilon_x + v\varepsilon_y) - \frac{E\alpha T}{1 - v} \\ \sigma_y &= \frac{E}{1 - v^2}(\varepsilon_y + v\varepsilon_x) - \frac{E\alpha T}{1 - v} \\ \tau_{xy} &= G\gamma_{xy} \end{aligned}} \quad \text{plane stress} \tag{3.12.2}$$

If the strain–displacement relations, Equations 3.10.2, are substituted into Equation 3.12.2, and those three equations are then substituted into Equations 3.6.2, the *equilibrium equations in terms of the displacements* are obtained, as follows:

$$2\frac{\partial^2 u}{\partial x^2} + (1 - v)\frac{\partial^2 u}{\partial y^2} + (1 + v)\frac{\partial^2 v}{\partial x \partial y} = 2\frac{1 - v^2}{E}b_x + 2\alpha\frac{\partial T}{\partial x}$$

$$2\frac{\partial^2 v}{\partial y^2} + (1 - v)\frac{\partial^2 v}{\partial x^2} + (1 + v)\frac{\partial^2 u}{\partial x \partial y} = 2\frac{1 - v^2}{E}b_y + 2\alpha\frac{\partial T}{\partial y} \tag{3.12.3}$$

This version of the equations of equilibrium is useful for problems in which the displacements instead of the stresses are the primary unknowns.

If the strain–stress equations, Equations 3.12.1, are substituted into the compatibility equation, Equation 3.10.4, we find

$$\frac{\partial^2 \sigma_x}{\partial y^2} + \frac{\partial^2 \sigma_y}{\partial x^2} - \nu \left(\frac{\partial^2 \sigma_x}{\partial x^2} + \frac{\partial^2 \sigma_y}{\partial y^2} \right) -2 \left(1 + \nu \right) \frac{\partial^2 \tau_{xy}}{\partial x \partial y} = -E\alpha \left(\frac{\partial^2 T}{\partial x^2} + \frac{\partial^2 T}{\partial y^2} \right) \qquad \textbf{[3.12.4]}$$

Plane stress fields in isotropic materials must satisfy this *stress compatibility equation*.

Stresses that are in equilibrium must also satisfy Equations 3.6.2, so that

$$\frac{\partial \tau_{xy}}{\partial y} = -\frac{\partial \sigma_x}{\partial x} - b_x \qquad \frac{\partial \tau_{xy}}{\partial x} = -\frac{\partial \sigma_y}{\partial y} - b_y$$

Taking the partial derivative of the first of these with respect to x and the second one with respect to y, then adding the results together yields

$$2\frac{\partial^2 \tau_{xy}}{\partial x \partial y} = -\frac{\partial^2 \sigma_x}{\partial x^2} - \frac{\partial^2 \sigma_y}{\partial y^2} - \frac{\partial b_x}{\partial x} - \frac{\partial b_y}{\partial y}$$

If this expression is substituted on the left of Equation 3.12.4, then combining terms and simplifying, we find that

$$\nabla^2 \left(\sigma_x + \sigma_y \right) = -E\alpha\nabla^2 T - (1 + \nu) \left(\frac{\partial b_x}{\partial x} + \frac{\partial b_y}{\partial y} \right) \qquad \textbf{[3.12.5]}$$

where $\nabla^2 = \partial^2/\partial x^2 + \partial^2/\partial y^2$. Compatible stress fields in isotropic elastic solids that are in equilibrium must satisfy this equation. It is interesting to note that $\nabla^2 T = 0$ for steady-state temperature distributions.[13]

Suppose we were faced with a plane elasticity problem in which the body force and temperature are known. In the absence of any justifiable simplifying assumptions, there are eight unknowns to be determined at every point within the body:

Three stresses: $\sigma_x,\ \sigma_y,\ \tau_{xy}$

Three strains: $\varepsilon_x,\ \varepsilon_y,\ \gamma_{xy}$

Two displacements: $u,\ v$

These eight unknown functions must satisfy the eight equations we have derived so far:

Two equilibrium equations :   Equations 3.6.2

Three strain–displacement equations: Equations 3.10.2

Three stress–strain equations: Equations 3.12.2

Furthermore, the stresses must satisfy the stress boundary conditions, Equations 3.6.3, and the displacements must match the prescribed displacements at the boundary.

One method of solving plane stress problems is the use of the *Airy stress function*, named for British astronomer G. B. Airy (1801–1892), who first observed that the equilibrium equations, Equations 3.6.2, with zero body forces are automatically satisfied by stresses defined in terms of an arbitrary function ϕ of x and y, as follows:

$$\sigma_x = \frac{\partial^2 \phi}{\partial y^2} \qquad \sigma_y = \frac{\partial^2 \phi}{\partial x^2} \qquad \tau_x = -\frac{\partial^2 \phi}{\partial x \partial y} \qquad \textbf{[3.12.6]}$$

| [13]J. P. Holman, *op cit,* Chapter 3.

It is easy to verify that Equation 3.6.2 is indeed satisfied by these stresses if $b_x = b_y = 0$. If, in addition, the temperature change T is steady state (or zero), the compatibility condition for isotropic materials, Equation 3.12.5, becomes,

$$\left(\frac{\partial^2}{\partial x^2} + \frac{\partial^2}{\partial y^2}\right)\left(\frac{\partial^2\phi}{\partial y^2} + \frac{\partial^2\phi}{\partial x^2}\right) = 0 \tag{3.12.7}$$

Substituting the stresses in terms of the Airy stress function into this equation, we find that

$$\frac{\partial^4\phi}{\partial x^4} + 2\frac{\partial^4\phi}{\partial x^2\partial y^2} + \frac{\partial^4\phi}{\partial y^4} = 0 \tag{3.12.8}$$

This *biharmonic* equation ($\nabla^4\phi = 0$) imposes the requirements of compatibility on the Airy stress function.

By selecting different polynomials, say, in x and y, subject to the compatibility condition, Equation 3.12.8, we can investigate the solutions to various plane stress problems, including some of practical interest.

Example 3.12.1 Consider the following plane stress field:

$$\sigma_x = -2\bar{\tau}h_1h_2\frac{g(x)}{h(y)^3} \qquad \sigma_y = 0 \qquad \tau_{xy} = \bar{\tau}\frac{h_1h_2}{h(y)^2} \tag{a}$$

where

$$g(x) = (h_2 - h_1)\frac{x}{l} - h_2\frac{x_1}{l} \qquad h(y) = h_2 - (h_2 - h_1)\frac{y}{l} \tag{b}$$

Show that these are the stress components within the trapezoidal shear panel of section 2.5, which is illustrated in Figure 3.12.1. The parallel bases of the trapezoid have the lengths h_1 and h_2, as shown, the altitude is l, and the width is $h(y)$ and is a function of position between its two parallel bases. The vertex P, with coordinates x_P and y_P, is the point where the nonparallel sides intersect, and $\bar{\tau}$ is a constant, with units of stress.

It is easy to show that the coordinates of the vertex P are

$$x_P = \frac{h_2x_1}{h_2 - h_1} \qquad y_P = \frac{h_2l}{h_2 - h_1} \tag{c}$$

The normal stress on the upper and lower (parallel) edges of the trapezoid is σ_y, which is zero throughout the region. Therefore, those segments of the boundary are in pure shear.

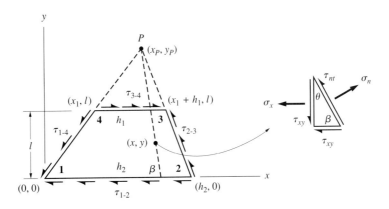

Figure 3.12.1 Trapezoidal plane stress region.

Choose within the trapezoid any point with coordinates x and y. Consider the plane containing that point and the vertex P; the edge of the plane, shown in Figure 3.12.1, makes an acute angle β with the x axis. The slope of this line is $dy/dx = -\tan\beta$; that is,

$$\tan\beta = -\frac{y - y_P}{x - x_P} \tag{d}$$

Substituting from Equations b and c, this can be written as

$$\tan\beta = \frac{h(y)}{g(x)} \tag{e}$$

A differential wedge surrounding the point is sketched in Figure 3.12.1, and the normal stress σ_n on the inclined edge, according to Equation 3.6.4$_1$—with σ_n playing the role of $\bar{\sigma}_x$—is

$$\sigma_n = \sigma_x \cos^2\theta + \sigma_y \sin^2\theta + 2\tau_{xy}\sin\theta\cos\theta$$

Since in this case $\sigma_y = 0$ and $\theta = 90° - \beta$, we have

$$\sigma_n = \sigma_x \sin^2\beta + 2\tau_{xy}\sin\beta\cos\beta \tag{f}$$

Using the trigonometric identities $\sin\beta = \tan\beta\big/\sqrt{1+\tan^2\beta}$ and $\cos\beta = 1\big/\sqrt{1+\tan^2\beta}$, we can write this as

$$\sigma_n = \frac{\tan\beta}{1+\tan^2\beta}(\sigma_x\tan\beta + 2\tau_{xy}) \tag{g}$$

Substituting Equations c and e into the parentheses on the right, we get

$$\sigma_x\tan\beta + 2\tau_{xy} = -2\bar{\tau}h_1h_2\frac{g(x)\,h(y)}{h(y)^3\,g(x)} + 2\bar{\tau}\frac{h_1h_2}{h(y)^2} = 0 \tag{h}$$

Therefore, $\sigma_n = 0$ on *any* plane passing through the vertex P of the trapezoid, including the edges 1–4 and 2–3.

We have shown that the stress field of Equation a satisfies the requirement that all four sides of the trapezoidal region have purely shear traction. According to Equation a$_2$, the shear stress is uniform over the length of both the upper edge ($y = l$) and the lower edge ($y = 0$), and has the values

$$\tau_{1-2} = \bar{\tau}\frac{h_1}{h_2} \qquad \tau_{3-4} = \bar{\tau}\frac{h_2}{h_1} \tag{i}$$

On the two inclined sides, the shear stress τ_{nt} is determined with the aid of Equation 3.6.4$_3$, as follows:

$$\tau_{nt} = -\sigma_x\sin\beta\cos\beta + \tau_{xy}(\sin^2\beta - \cos^2\beta)$$

Invoking the same trigonometric identities that led to Equation g, this becomes

$$\tau_{nt} = \frac{1}{1+\tan^2\beta}[-\sigma_x\tan\beta + \tau_{xy}(\tan^2\beta - 1)] \tag{j}$$

Substituting Equations a and e yields

$$\tau_{nt} = \frac{1}{1 + \frac{h(y)^2}{g(x)^2}}\left[2\bar{\tau}h_1h_2\frac{g(x)\,h(y)}{h(y)^3\,g(x)} + \bar{\tau}\frac{h_1h_2}{h(y)^2}\left(\frac{h(y)^2}{g(x)^2} - 1\right)\right]$$

$$= \frac{g(x)^2}{g(x)^2 + h(y)^2}\left\{\frac{\bar{\tau}h_1h_2[g(x)^2 + h(y)^2]}{h(y)^2\,g(x)^2}\right\}$$

which reduces to

$$\tau_{nt} = \frac{\bar{\tau} h_1 h_2}{h(y)^2}$$ [k]

Multiplying through by the thickness t of the panel yields the shear flow formula, Equation 2.5.4.

Example 3.12.2   Show that the plane stress field assumed for the trapezoidal shear panel of section 2.5 satisfies equilibrium but is compatible only if the panel is a parallelogram. Assume zero body forces and no thermal strain.

From the previous example, we know that, in terms of the cartesian coordinates of Figure 3.12.1, the components of stress are

$$\sigma_x = -2\bar{\tau} h_1 h_2 \frac{g(x)}{h(y)^3} \qquad \sigma_y = 0 \qquad \tau_{xy} = \bar{\tau} \frac{h_1 h_2}{h(y)^2}$$ [a]

Using Equation b of Example 3.12.1, we have

$$\frac{\partial \sigma_x}{\partial x} = \frac{\partial}{\partial x} \left[-2\bar{\tau} h_1 h_2 \frac{g(x)}{h(y)^3} \right] = -\frac{2\bar{\tau} h_1 h_2}{h(y)^3} \frac{dg(x)}{dx} = -\frac{2\bar{\tau} h_1 h_2}{h(y)^3} \frac{h_2 - h_1}{l}$$ [b]

$$\frac{\partial \tau_{xy}}{\partial y} = \frac{\partial}{\partial y} \left[\bar{\tau} \frac{h_1 h_2}{h(y)^2} \right] = -\frac{2\bar{\tau} h_1 h_2}{h(y)^3} \frac{dh(y)}{dy} = \frac{2\bar{\tau} h_1 h_2}{h(y)^3} \frac{h_2 - h_1}{l}$$ [c]

$$\frac{\partial \tau_{xy}}{\partial x} = \frac{\partial}{\partial x} \left[\bar{\tau} \frac{h_1 h_2}{h(y)^2} \right] = 0$$ [d]

$$\frac{\partial \sigma_y}{\partial y} = \frac{\partial [0]}{\partial y} = 0$$ [e]

From these it is clear that

$$\frac{\partial \sigma_x}{\partial x} + \frac{\partial \tau_{xy}}{\partial y} = 0 \qquad \frac{\partial \tau_{xy}}{\partial x} + \frac{\partial \sigma_y}{\partial y} = 0$$ [f]

Therefore, equilibrium is satisfied throughout the panel.

Since $\sigma_y = 0$, the compatibility equation, Equation 3.12.5, requires that

$$\frac{\partial^2 \sigma_x}{\partial x^2} + \frac{\partial^2 \sigma_x}{\partial y^2} = 0$$ [g]

From Equation b, it is true that $\partial^2 \sigma_x / \partial x^2 = 0$. However,

$$\frac{\partial^2 \sigma_x}{\partial y^2} = \frac{\partial^2}{\partial y^2} \left[-2\bar{\tau} h_1 h_2 \frac{g(x)}{h(y)^3} \right] = 24 \frac{\bar{\tau} h_1 h_2 (h_2 - h_1)}{l} \frac{g(x)}{h(y)^2}$$ [h]

This does not vanish unless $h_1 = h_2$, that is, unless the panel is a parallelogram.

Examples 3.12.1 and 3.12.2 show that the Garvey shear panel, with its appealingly simple, equilibrium shear flow distribution (which is why we will continue to employ it throughout most of this book), is not in strict accordance with elasticity theory and must be used with caution. A somewhat similar situation occurs in the *finite element method* of plane stress analysis, which employs triangular- or quadrilateral-shaped elements in which the form of the displacement field—not the stress field—is commonly assumed. The strain fields in these displacement-based elements are automatically compatibile, but they are not generally in equilibrium, which is just the reverse of force-based elements like the Garvey shear panel. The finite element method will be disussed further in Chapter 11.

Example 3.12.3   The materially isotropic beam of Figure 3.12.2 has a uniform shear traction applied to its upper surface. Use the Airy stress function to find the stresses in the beam and the displacements of point A at the free end. Assume $2h < 10L$.

Specifying $2h < 10L$ ensures that the geometry of the structure is that of a slender beam.

Let us try a complete fourth-degree polynomial for the Airy stress function:

$$\phi = (c_1 x^2 + c_2 xy + c_3 y^2) + (c_4 x^3 + c_5 x^2 y + c_6 xy^2 + c_7 y^3)$$
$$+ (c_8 x^4 + c_9 x^3 y + c_{10} x^2 y^2 + c_{11} xy^3 + c_{12} y^4)$$

Since the stresses are found as *second* derivatives of the Airy stress function, polynomial terms of degree less than two need not be included. Substituting ϕ into the compatibility equation, Equation 3.12.8, yields the following relationship among some of the polynomial coefficients:

$$3c_8 + c_{10} + 3c_{12} = 0 \qquad \text{[a]}$$

From Equations 3.12.6, we obtain the polynomial forms of the stress components, as follows:

$$\sigma_x = 2c_3 + 2c_6 x + 6c_7 y + 2c_{10} x^2 + 6c_{11} xy + 12c_{12} y^2 \qquad \text{[b]}$$

$$\sigma_y = 2c_1 + 6c_4 x + 2c_5 y + 12c_8 x^2 + 6c_9 xy + 2c_{10} y^2 \qquad \text{[c]}$$

$$\tau_{xy} = -c_2 - 2c_5 x - 2c_6 y - 3c_9 x^2 - 4c_{10} xy - 3c_{11} y^2 \qquad \text{[d]}$$

We next apply the stress boundary conditions on the top and bottom surface and on the free end. On the top surface, $y = h$, $\sigma_y = 0$, and $\tau_x = \tau_0$. Therefore, Equations c and d, evaluated at $y = h$, yield, respectively,

$$(2c_1 + 2c_5 h + 2c_{10} h^2) + (6c_4 + 6c_9 h)x + 12c_8 x^2 = 0$$
$$(c_2 + 2c_6 h + 3c_{11} h^2 + \tau_o) + (2c_5 + 4c_{10} h)x + 3c_9 x^2 = 0$$

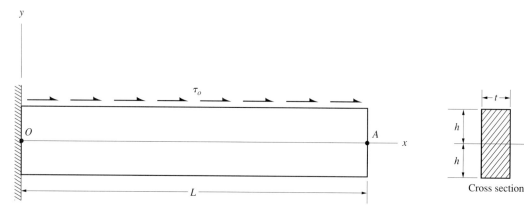

Figure 3.12.2   Beam with uniform shear traction on upper surface.

In order for these equations to be valid for all x, the coefficient of each power of x must vanish.[14] Thus, each equation produces three relationships among the coefficients of the stress function, namely,

$$c_1 + c_5 h + c_{10} h^2 = 0 \qquad c_4 + c_9 h = 0 \qquad c_8 = 0 \qquad \text{[e]}$$

and

$$c_2 + 2c_6 h + 3c_{11} h^2 = -\tau_o \qquad c_5 + 2c_{10} h = 0 \qquad c_9 = 0 \qquad \text{[f]}$$

On the bottom surface, $y = -h$, $\sigma_y = 0$, and $\tau_{xy} = 0$. Substituting $y = -h$ into Equations c and d, respectively, and requiring that both polynomials be zero for all values of x, again leads to six equations:

$$c_1 - c_5 h + c_{10} h^2 = 0 \qquad c_4 - c_9 h = 0 \qquad c_8 = 0 \qquad \text{[g]}$$

$$c_2 - 2c_6 h + 3c_{11} h^2 = 0 \qquad c_5 - 2c_{10} h = 0 \qquad c_9 = 0 \qquad \text{[h]}$$

From Equations f_2 and h, we get $c_5 = c_{10} = 0$. Then from Equation e_1 (or g_1), it follows that $c_1 = 0$. Equations f_3 and e_2 (or h_3 and g_2) then imply that $c_4 = c_9 = 0$. Using Equations e_3 (or g_3) and a, together with the fact that $c_{10} = 0$, we obtain $c_{12} = 0$. Finally, Equations f_1 and h_1 imply that $c_2 + 3c_{11} h^2 = \tau_o / 2$ and $c_6 = -\tau_o / 4h$. Therefore, up to this point, Equations b, c, and d have been reduced to the following:

$$\sigma_x = 2c_3 - \frac{\tau_o}{2h} x + 6c_7 y + 6c_{11} xy \qquad \text{[i]}$$

$$\sigma_y = 0 \qquad \text{[j]}$$

$$\tau_{xy} = \frac{\tau_o}{2} + 3c_{11} h^2 + \frac{\tau_o}{2h} y - 3c_{11} y^2 \qquad \text{[k]}$$

At the free end, $x = L$ and $\sigma_x = 0$. Therefore, Equation i requires that for all y between $\pm h$, we have

$$2c_3 - \frac{\tau_o}{2h} L + (6c_7 + 6c_{11} L) y = 0$$

The coefficient of y must therefore vanish, so that $c_7 = -c_{11} L$, which implies that $c_3 = \tau_o L / 4h$. Substituting these values of the coefficients into Equation i yields

$$\sigma_x = \frac{\tau_o L}{2h} - \frac{\tau_o L}{2h} x - 6c_{11} L y + 6c_{11} xy \qquad \text{[l]}$$

We might be tempted to impose the condition $\tau_{xy} = 0$ everywhere at the free end, for it appears entirely free of surface traction. Were we to do so, however, Equation k would require that $\tau_o = c_{11} = 0$, which means the beam would be unloaded. Therefore, we must instead require that the shear stress *resultant*, $\int \tau_{xy} dA$, vanish at $x = L$, that is,

$$\int_{-h}^{h} \left(\frac{\tau_o}{2} + 3c_{11} h^2 + \frac{\tau_o}{2h} y - 3c_{11} y^2 \right) t\, dy = 0$$

Carrying out the integration yields $\frac{\tau_o}{2} + 2c_{11} h^2 = 0$, or $c_{11} = -\tau_o / 4h^2$. Substituting this value of c_{11} into Equations k and l, together with Equation j, we finally obtain

$$\sigma_x = \frac{\tau_o}{4} \left(2\frac{L}{h} - 2\frac{x}{h} + 6\frac{Ly}{h^2} - 6\frac{xy}{h^2} \right) \qquad \text{[m]}$$

---

[14]Let the equation $a_0 + a_1 x + a_2 x^2 + \cdots + a_n x^n$ be valid for all x, where the a's are constants. Then, $a_0 = -x(a_1 + a_2 x + \cdots + a_n x^{n-1})$ Since a constant cannot equal a variable, this equation can be valid only if $a_0 = 0$ and $a_1 + a_2 x + \cdots + a_n x^{n-1} = 0$ for all x. Now we can repeat the argument to show that $a_1 = 0$, then that $a_2 = 0$, etc., and finally that $a_n = 0$.

$$\tau_{xy} = \frac{\tau_o}{4}\left(-1 + 2\frac{y}{h} + 3\frac{y^2}{h^2}\right)$$ [n]

$$\sigma_y = 0$$ [o]

These are the stresses throughout the beam.

To find the displacements, we must integrate the strain–displacement relations, Equations 3.10.2. From Equation 3.10.2$_1$, the strain–stress equation, Equation 3.12.1$_1$, and Equations m and o we have

$$\frac{\partial u}{\partial x} = \frac{\tau_o}{4E}\left(2\frac{L}{h} - 2\frac{x}{h} + 6\frac{Ly}{h^2} - 6\frac{xy}{h^2}\right)$$

Upon integrating, this yields

$$u = \frac{\tau_o}{4E}\left(2\frac{Lx}{h} - \frac{x^2}{h} + 6\frac{Lxy}{h^2} - 3\frac{x^2y}{h^2}\right) + f(y)$$ [p]

where $f(y)$ is an as yet arbitrary function of y. Similarly, from the strain–displacement equation, Equation 3.10.2$_2$, strain–stress equation, Equation 3.12.1$_2$ and Equations m and o, we have

$$\frac{\partial v}{\partial y} = -\frac{\nu\tau_o}{4E}\left(2\frac{L}{h} - 2\frac{x}{h} + 6\frac{Ly}{h^2} - 6\frac{xy}{h^2}\right)$$

which integrates to

$$v = -\frac{\nu\tau_o}{4E}\left(2\frac{Ly}{h} - 2\frac{xy}{h} + 3\frac{Ly^2}{h^2} - 3\frac{xy^2}{h^2}\right) + g(x)$$ [q]

From the strain–displacement equation, Equation 3.10.2$_3$, and the shear strain–shear stress equation, Equation 3.12.1$_3$, we have

$$\frac{\partial u}{\partial y} + \frac{\partial v}{\partial x} = \frac{2(1+\nu)}{E}\tau_{xy}$$

Substituting u and v from Equations p and q, and τ_{xy} from Equation n, then simplifying, we get,

$$\frac{dg(x)}{dx} + \frac{\tau_o}{4E}\left(-3\frac{x^2}{h^2} + 6\frac{Lx}{h^2}\right) = -\frac{df(y)}{dy} + (2+\nu)\frac{\tau_o}{4E}\left(2\frac{y}{h} + 3\frac{y^2}{h^2}\right) - \frac{\tau_o(1+\nu)}{2E}$$ [r]

The left side is a function of x alone, while the right side is a function of just y; therefore, both sides must equal a constant C. Taking the left side first, we have

$$\frac{dg(x)}{dx} + \frac{\tau_o}{4E}\left(-3\frac{x^2}{h^2} + 6\frac{Lx}{h^2}\right) = C$$

so that

$$g(x) = -\frac{\tau_o}{4E}\left(-\frac{x^3}{h^2} + 3\frac{Lx^2}{h^2}\right) + Cx + D$$ [s]

Setting the right side of Equation r equal to C and integrating with respect to y, we get

$$f(y) = (2+\nu)\frac{\tau_o}{4E}\left(2\frac{y^2}{h} + \frac{y^3}{h^2}\right) - \frac{\tau_o(1+\nu)}{2E}y - Cy + F$$ [t]

Finally, we substitute $f(y)$ and $g(x)$ back into Equations p and q. Then, to eliminate rigid-body translation, we can require $u = v = 0$ at point O ($x = y = 0$), from which it follows that $D = F = 0$. To remove the component of rigid-body rotation associated with the undetermined constant C, let us set $\partial v / \partial x$, the rotation of a horizontal line element, equal to zero at O. It is easy to verify that this forces C to be zero as well. Thus,

$$u = \frac{\tau_o L}{4E} \left\{ \left(2 - \frac{x}{L} \right) \frac{x}{h} \left(1 + 3\frac{y}{h} \right) + \left[(2+v) \left(1 + \frac{y}{h} \right) \frac{y^2}{h^2} - 2(1+v) \frac{y}{h} \right] \frac{h}{L} \right\} \qquad [u]$$

$$v = -\frac{\tau_o L}{4E} \left[v \left(1 - \frac{x}{L} \right) \left(2 + 3\frac{y}{h} \right) \frac{y}{h} + \left(3 - \frac{x}{L} \right) \frac{x^2}{h^2} \right] \qquad [v]$$

These two functions give the displacements everywhere in the beam. To find the displacements at point A on the free end, evaluate these expressions at $x = L$, $y = 0$, thereby obtaining

$$u = \frac{\tau_o L^2}{4Eh} \qquad v = -\frac{\tau_o L^3}{2Eh^2} \qquad \text{(Displacements of point A)} \qquad [w]$$

It should be observed that applying other displacement boundary conditions will alter the form of the displacement field. For example, if instead of setting $\partial v / \partial x$ equal to zero at O, we require that $\partial u / \partial y$ vanish at that point, then it turns out that $C = -\tau_o (1+v) / 2E$. This has the effect of superimposing a clockwise rigid-body rotation on the displacements u and v. It is a simple matter to verify that the horizontal displacement of point A remains unchanged, whereas the vertical component becomes

$$v = -\frac{\tau_o L^3}{2Eh^2} \left[1 + (1+v) \frac{h^2}{L^2} \right]$$

Exact solutions of the equations of elasticity can only be found in relatively few cases. Irregular boundaries, cutouts, complex loadings and constraints, variable material properties, etc., may force us to seek approximate solutions. One widely used approach is to abandon the effort to obtain an exact solution, that is, a single set of functions describing the state of stress and deformation at all of the infinite points comprising the plane elastic continuum. Instead, the region is subdivided into a number of small, but not *infinitesimally* small, elements. These subregions are called *finite elements*. The stresses or displacements within each element are assumed to be of an appropriate, relatively simple functional form, such as polynomials in x and y. The idea is to obtain a piecewise approximation of the overall solution for the stresses and displacements. Presumably, the approximation will become better as the region is more finely subdivided into finite elements. (Of course, subdividing indefinitely would lead us back to an infinite number of differential elements.) Basic aspects of the finite element method, which relies exclusively on digital computers to handle the typically massive amount of data and computations, is addressed in Chapter 11.

The *boundary element method*[15] is another alternative approximate method of analysis that requires only the boundary to be discretized into finite elements. The solution at any point within the solid is found in terms of values on the boundary, which are the primary unknowns.

3.13   SAINT-VENANT'S PRINCIPLE

The solution for the stresses and displacements in the beam of Example 3.12.3 is in accord with all of the equations of plane elasticity. However, the solution is valid only if the tractions on all four edges of the beam are distributed precisely according to Equations m, n, and o of the example and as sketched in Figure 3.13.1. Should the

---

[15]See, for example, P. K. Banerjee and R. Butterfield, *Boundary Element Methods in Engineering Science*, New York, McGraw-Hill, 1981; or C.A. Brebbia and J. Dominguez, *Boundary Elements: An Introductory Course*, New York, McGraw–Hill, 1989.

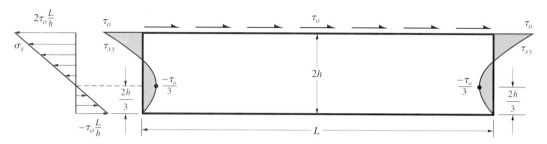

Figure 3.13.1   Surface tractions on the beam of Figure 4.6.1.

free right end of the beam not actually be in contact with any other body, there would be no means to enforce the parabolic shear stress distribution, and that surface would actually have to be completely free of shear stress, contrary to exact solution. Furthermore, the precise manner in which the left end of the beam is physically attached to the wall will determine the nature of the stress distribution on that surface; it is unlikely that those stresses would appear as shown in the figure. Although the actual, localized load systems at each end of the beam differ in their details, those that are *statically equivalent* to the "exact" stress distribution, that is, those that have the same force and moment resultants as the surface tractions accompanying the elasticity solution, will produce the same "exact" solution throughout the bulk of the material. Although the stresses and displacements from statically equivalent, locally applied loads may differ significantly *in the neighborhood* of the loads themselves, the discrepancies disappear at distances substantially greater than the linear dimension of the surface over which the load is applied.

This notion of the decay of localized load detail is called the *Saint-Venant principle*, after the nineteenth century French elastician[16] who first proposed it in 1855. According to the Saint-Venant principle, the stresses we computed in Example 3.12.3 for the cantilever beam are valid beyond a distance from the right end roughly equal to its depth $2h$ if the surface tractions at the right end are statically equivalent to zero shear, zero axial force, and zero couple. It is easy to show that the stresses shown acting on the left end of the beam in Figure 3.13.1 yield a zero shear force resultant and an axial load resultant to the left of $\tau_o t L$ acting at the *top* of the beam. Any load system having the same resultants will produce stresses equal to the exact solution in Example 3.12.3 beyond a distance of about $2h$ from the left end. On the other hand, if the traction on the top of the beam is not *precisely* that of a uniformly distributed shear force, the solution given in the example is void throughout the beam if $2h < 10L$, as assumed.

3.14   STRAIN ENERGY

In this section, we will ignore thermal effects by assuming that the solid remains at a constant, uniform temperature. Figure 3.14.1 shows a differential material element, with surface tractions at the centroids of each face and a body force at its center of mass. Also shown are the displacements of these points. The work done by a force **F** undergoing a differential displacement $d\mathbf{u}$ is $\mathbf{F} \cdot d\mathbf{u}$. Let us use this formula to calculate the work dW_o done on the element by the surface tractions and body force during a small increment of the loading. Multiplying the surface tractions by the areas on which they act, we have

| [16]Adéhmar Jean Claude Barré, Count de Saint-Venant (1797–1886).

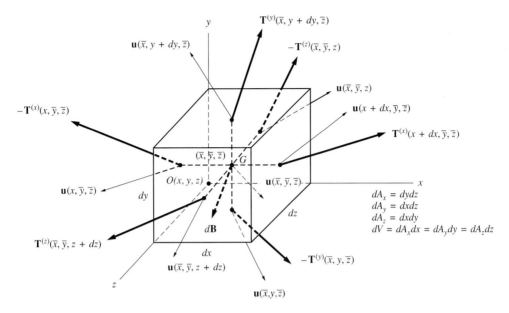

Figure 3.14.1   Surface tractions, body force, and the displacement vectors and their points of application on a differential material element.

The corner O has coordinates (x, y, z), and the center of mass G of the cube has coordinates $(\bar{x}, \bar{y}, \bar{z})$.

$$
\begin{aligned}
dW_o &= \mathbf{T}^{(x)}(x + dx, \bar{y}, \bar{z})dA_x \cdot d\mathbf{u}(x + dx, \bar{y}, \bar{z}) - \mathbf{T}^{(x)}(x, \bar{y}, \bar{z})dA_x \cdot d\mathbf{u}(x, \bar{y}, \bar{z}) \\
&+ \mathbf{T}^{(y)}(\bar{x}, y + dy, \bar{z})dA_y \cdot d\mathbf{u}(\bar{x}, y + dy, \bar{z}) - \mathbf{T}^{(y)}(\bar{x}, y, \bar{z})dA_y \cdot d\mathbf{u}(\bar{x}, y, \bar{z}) \\
&+ \mathbf{T}^{(z)}(\bar{x}, \bar{y}, z + dz)dA_z \cdot d\mathbf{u}(\bar{x}, \bar{y}, z + dz) - \mathbf{T}^{(z)}(\bar{x}, \bar{y}, z)dA_z \cdot d\mathbf{u}(\bar{x}, \bar{y}, z) + d\mathbf{B} \cdot d\mathbf{u}(\bar{x}, \bar{y}, \bar{z})
\end{aligned}
$$

Equation 3.3.1 is used to express the variation of the tractions and the displacements from face to face across the element, as follows:

$$
\begin{aligned}
dW_o &= \left(\mathbf{T}^{(x)} + \frac{\partial \mathbf{T}^{(x)}}{\partial x}dx \right)dA_x \cdot d\left(\mathbf{u} + \frac{\partial \mathbf{u}}{\partial x}dx \right) - \mathbf{T}^{(x)}dA_x \cdot d\mathbf{u} \\
&+ \left(\mathbf{T}^{(y)} + \frac{\partial \mathbf{T}^{(y)}}{\partial y}dy \right)dA_y \cdot d\left(\mathbf{u} + \frac{\partial \mathbf{u}}{\partial y}dy \right) - \mathbf{T}^{(y)}dA_y \cdot d\mathbf{u} \\
&+ \left(\mathbf{T}^{(z)} + \frac{\partial \mathbf{T}^{(z)}}{\partial z}dz \right)dA_z \cdot d\left(\mathbf{u} + \frac{\partial \mathbf{u}}{\partial z}dz \right) - \mathbf{T}^{(z)}dA_z \cdot d\mathbf{u} + d\mathbf{B} \cdot d\mathbf{u}
\end{aligned}
$$

Expanding the parentheses, cancelling like terms, and ignoring higher-order terms (those containing products of more than three of the differentials dx, dy, and dz), we are left with

$$
dW_o = \left(\frac{\partial \mathbf{T}^{(x)}}{\partial x} + \frac{\partial \mathbf{T}^{(y)}}{\partial y} + \frac{\partial \mathbf{T}^{(z)}}{\partial z} + \mathbf{b} \right) \cdot d\mathbf{u}dV + \left[\mathbf{T}^{(x)} \cdot d\left(\frac{\partial \mathbf{u}}{\partial x} \right) + \mathbf{T}^{(y)} \cdot d\left(\frac{\partial \mathbf{u}}{\partial y} \right) + \mathbf{T}^{(z)} \cdot d\left(\frac{\partial \mathbf{u}}{\partial z} \right) \right]dV + \cdots
$$

where "$\cdots$" represents the ignored terms, dV is the volume of the element, and $\mathbf{b}$ is the body force density ($\mathbf{b} = d\mathbf{B}/dV$). Assuming the element is in equilibrium, the first term on the right vanishes, according to Equation 3.3.3. The dot products in the square brackets are expanded in terms of the components of the displacement vector, $\mathbf{u} = u\mathbf{i} + v\mathbf{j} + w\mathbf{k}$, and the components of the tractions, as given in Equation 3.2.3. Thus,

$$dw_o = \left[\sigma_x d\left(\frac{\partial u}{\partial x}\right) + \tau_{xy} d\left(\frac{\partial v}{\partial x}\right) + \tau_{xz} d\left(\frac{\partial w}{\partial x}\right) \right]$$
$$+ \left[\tau_{yx} d\left(\frac{\partial u}{\partial y}\right) + \sigma_y d\left(\frac{\partial v}{\partial y}\right) + \tau_{yz} d\left(\frac{\partial w}{\partial y}\right) \right] + \left[\tau_{zx} d\left(\frac{\partial u}{\partial z}\right) + \tau_{zy} d\left(\frac{\partial v}{\partial z}\right) + \sigma_z d\left(\frac{\partial w}{\partial z}\right) \right]$$

where $dw_o = dW_o/dV$, is the work per unit volume. Using the symmetry of the stress tensor (Equations 3.3.2) and collecting terms, on this becomes

$$dw_o = \sigma_x d\left(\frac{\partial u}{\partial x}\right) + \sigma_y d\left(\frac{\partial v}{\partial y}\right) + \sigma_z d\left(\frac{\partial w}{\partial z}\right)$$
$$+ \tau_{xy} d\left(\frac{\partial u}{\partial x} + \frac{\partial v}{\partial x}\right) + \tau_{xz} d\left(\frac{\partial u}{\partial z} + \frac{\partial w}{\partial x}\right) + \tau_{yz} d\left(\frac{\partial v}{\partial z} + \frac{\partial w}{\partial y}\right)$$

Finally, recalling the strain–displacement relationships, Equations 3.7.20, we obtain

$$dw_o = \sigma_x d\varepsilon_x + \sigma_y d\varepsilon_y + \sigma_z d\varepsilon_z + \tau_{xy} d\gamma_{xy} + \tau_{xz} d\gamma_{xz} + \tau_{yz} d\gamma_{yz} \qquad \text{[3.14.1]}$$

This is a formula for the work done on a unit volume of material during an increment of a quasistatic loading process.[17] Observe that the derivation of this result did not involve the relation between stress and strain. Therefore, Equation 3.14.1 holds for any material, subject of course to the small strain assumption.

The increment of *complementary work* dw_o^* per unit volume is *defined* as

$$dw_o^* = \varepsilon_x d\sigma_x + \varepsilon_y d\sigma_y + \varepsilon_z d\sigma_z + \gamma_{xy} d\tau_{xy} + \gamma_{xz} d\tau_{xz} + \gamma_{yz} d\tau_{yz} \qquad \text{[3.14.2]}$$

This equation is obtained by interchanging the stresses and strains in Equation 3.14.1. Although it appears simply to be a mathematical gimmick, complementary work is a useful concept in structural mechanics, as we will see in Chapter 7.

By definition, an *elastic* solid has no internal dissipative mechanisms to convert work into lost energy. All of the work done on an elastic body is stored as internal energy, or *strain energy*, which is zero in the absence of deformation. The *strain energy density* u_o is a single-valued function of the state of deformation; it does *not* depend on how that state was arrived at. Therefore, in terms of the strains

$$u_o = u_o(\varepsilon_x, \varepsilon_y, \varepsilon_z, \gamma_{xy}, \gamma_{xz}, \gamma_{yz}) \qquad \text{[3.14.3]}$$

The strain energy U of an elastic solid is the integral of the strain energy density over the entire solid volume, that is:

$$U = \iiint_V u_o dV \qquad \text{[3.14.4]}$$

During a loading process, the increment in strain energy of a material element, $d(u_o dV)$, equals the work done on that element per unit volume, $dw_o dV$. Thus,

$$du_o = dw_o \qquad \text{[3.14.5]}$$

However, according to differential calculus, du_o is obtained from Equation 3.14.3 as

$$du_o = \frac{\partial u_o}{\partial \varepsilon_x} d\varepsilon_x + \frac{\partial u_o}{\partial \varepsilon_y} d\varepsilon_y + \frac{\partial u_o}{\partial \varepsilon_z} d\varepsilon_z + \frac{\partial u_o}{\partial \gamma_{xy}} d\gamma_{xy} + \frac{\partial u_o}{\partial \gamma_{xz}} d\gamma_{xz} + \frac{\partial u_o}{\partial \gamma_{yz}} d\gamma_{yz}$$

| [17]Quasistatic means "nearly in equilibrium." The process occurs slowly enough that kinetic energy and inertial effects may be neglected.

Substituting this and Equation 3.14.1 into Equation 3.14.3 and combining terms yields

$$\left(\sigma_x - \frac{\partial u_o}{\partial \varepsilon_x}\right)d\varepsilon_x + \left(\sigma_y - \frac{\partial u_o}{\partial \varepsilon_y}\right)d\varepsilon_y + \left(\sigma_z - \frac{\partial u_o}{\partial \varepsilon_z}\right)d\varepsilon_z$$

$$+ \left(\tau_{xy} - \frac{\partial u_o}{\partial \gamma_{xy}}\right)d\gamma_{xy} + \left(\tau_{xz} - \frac{\partial u_o}{\partial \gamma_{xz}}\right)d\gamma_{xz} + \left(\tau_{yz} - \frac{\partial u_o}{\partial \gamma_{yz}}\right)d\gamma_{yz} = 0$$

[3.14.6]

The coefficients of the strain differentials depend on the strains but not on the differential strains, which are independent of each other. Regardless of the choice of differentials, Equation 3.14.6 must remain true. Thus, for example, if we set all of the differentials except $d\varepsilon_x$ equal to zero, Equation 3.14.6 requires that $(\sigma_x - \partial u_o / \partial \varepsilon_x) = 0$. This argument may be made for each of the six terms in the equation to obtain

$$\sigma_x = \frac{\partial u_o}{\partial \varepsilon_x} \qquad \sigma_y = \frac{\partial u_o}{\partial \varepsilon_y} \qquad \sigma_z = \frac{\partial u_o}{\partial \varepsilon_z}$$

$$\tau_{xy} = \frac{\partial u_o}{\partial \gamma_{xy}} \qquad \tau_{xz} = \frac{\partial u_o}{\partial \gamma_{xz}} \qquad \tau_{yz} = \frac{\partial u_o}{\partial \gamma_{yz}}$$

[3.14.7]

Therefore, in an elastic material, the strain energy density is a potential from which the stresses are derived.

The *complementary strain energy density* u_o^* of an elastic solid is defined as follows:

$$u_o^* = \sigma_x \varepsilon_x + \sigma_y \varepsilon_y + \sigma_z \varepsilon_z + \tau_{xy} \gamma_{xy} + \tau_{xz} \gamma_{xz} + \tau_{yz} \gamma_{yz} - u_o$$

[3.14.8]

This relationship is illustrated in Figure 3.14.2. Clearly, u_o^*, like u_o, is a function only of the state of deformation,

$$u_o^* = u_o^*(\sigma_x, \sigma_y, \sigma_z, \tau_{xy}, \tau_{xz}, \tau_{yz})$$

[3.14.9]

Taking the differential of Equation 3.14.8 and using Equation 3.14.5 yields

$$du_o^* = d(\sigma_x \varepsilon_x + \sigma_y \varepsilon_y + \sigma_z \varepsilon_z + \tau_{xy} \gamma_{xy} + \tau_{xz} \gamma_{xz} + \tau_{yz} \gamma_{yz}) - dw_o$$

[3.14.10]

Noting that $d(\sigma_x \varepsilon_x) = \sigma_x d\varepsilon_x + \varepsilon_x d\sigma_x$, etc., and recalling Equations 3.14.1 and 3.14.2, we are led to the following conclusion:

$$du_o^* = dw_o^*$$

[3.14.11]

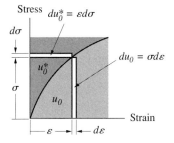

Figure 3.14.2   Elastic stress–strain relationship showing strain energy density (u_o) and complementary strain energy (u_o^*).

Thus, the same relationship exists between complementary work and complementary strain energy as exists between work and strain energy. In particular, following the same steps that led to Equation 3.14.7, we can show that

$$\varepsilon_x = \frac{\partial u_o^*}{\partial \sigma_x} \qquad \varepsilon_y = \frac{\partial u_o^*}{\partial \sigma_y} \qquad \varepsilon_z = \frac{\partial u_o^*}{\partial \sigma_z}$$

$$\gamma_{xy} = \frac{\partial u_o^*}{\partial \tau_{xy}} \qquad \gamma_{xz} = \frac{\partial u_o^*}{\partial \tau_{xz}} \qquad \gamma_{yz} = \frac{\partial u_o^*}{\partial \tau_{yz}}$$

[3.14.12]

The complementary strain energy density is a potential function for the strains. The total complementary energy U^* of an elastic solid is given by

$$U^* = \iiint_V u_o^* dV$$

For a linear elastic material, stress is proportional to strain, so the curve in Figure 3.14.2 is a straight line. For such materials, strain energy and complementary strain energy are equal. The distinction between the two is solely the choice of independent variables: strains for u_o, stresses for u_o^*. Thus, according to Equation 3.14.8, for *linear elastic materials*, we have

$$u_o = u_o^* = \frac{1}{2}(\sigma_x \varepsilon_x + \sigma_y \varepsilon_y + \sigma_z \varepsilon_z + \tau_{xy}\gamma_{xy} + \tau_{xz}\gamma_{xz} + \tau_{yz}\gamma_{yz})$$

[3.14.13]

Remember that the $\frac{1}{2}$ factor is due to the direct proportionality between stress and strain.

If we use Equations 3.11.4 to express the strains in Equation 3.14.13 in terms of the stresses, then for an *isotropic linear elastic material*, we obtain

$$u_o = u_o^* = \frac{1}{2E}\left[\sigma_x^2 + \sigma_y^2 + \sigma_z^2 - 2v(\sigma_x\sigma_y + \sigma_x\sigma_z + \sigma_y\sigma_z)\right] + \frac{1}{2G}\left(\tau_{xy}^2 + \tau_{yz}^2 + \tau_{zx}^2\right)$$

[3.14.14]

Alternatively, we can express the stresses in terms of the strains by means of Equation 3.11.5, in which case

$$u_o = u_o^* = \frac{1}{2}\frac{E}{(1+v)(1-2v)}\left[(1-v)(\varepsilon_x^2 + \varepsilon_y^2 + \varepsilon_z^2) + 2v(\varepsilon_x\varepsilon_y + \varepsilon_x\varepsilon_z + \varepsilon_y\varepsilon_z)\right]$$
$$+ \frac{1}{2}G(\gamma_{xy}^2 + \gamma_{xz}^2 + \gamma_{yz}^2)$$

[3.14.15]

Example 3.14.1   Show that the strain–stress equations, Equations 3.11.4 (with $T = 0$), follow from substituting the strain energy density, Equation 3.14.14, into Equations 3.14.12.

For ε_x, we have

$$\varepsilon_x = \frac{\partial u_o^*}{\partial \sigma_x} = \frac{\partial}{\partial \sigma_x}\left\{\frac{1}{2E}\left[\sigma_x^2 + \sigma_y^2 + \sigma_z^2 - 2v(\sigma_x\sigma_y + \sigma_x\sigma_z + \sigma_y\sigma_z)\right] + \frac{1}{2G}\left(\tau_{xy}^2 + \tau_{yz}^2 + \tau_{zx}^2\right)\right\}$$

Carrying out the partial derivatives yields

$$\varepsilon_x = \frac{1}{2E}\left[2\sigma_x - 2v(\sigma_y + \sigma_z)\right] = \frac{\sigma_x}{E} - \frac{v}{E}\sigma_y - \frac{v}{E}\sigma_z$$

which is Equation 3.11.4$_1$. Likewise,

$$\gamma_{xy} = \frac{\partial}{\partial \tau_{xy}} \left\{ \frac{1}{2E} \left[\sigma_x^2 + \sigma_y^2 + \sigma_z^2 - 2\nu(\sigma_x\sigma_y + \sigma_x\sigma_z + \sigma_y\sigma_z) \right] + \frac{1}{2G} \left(\tau_{xy}^2 + \tau_{yz}^2 + \tau_{zx}^2 \right) \right\} = \frac{\tau_{xy}}{G}$$

We proceed in this fashion to obtain all six of the strain–stress relationships from the strain energy density.

3.15   STATIC FAILURE THEORIES

A tension test is used to determine the basic mechanical properties of a material. A flat or round specimen of the material is placed in the grips of a tensile test machine and slowly pulled upon until it fails. The stress and strain throughout the test are recorded and plotted as a stress–strain curve, like that illustrated in Figure 3.15.1. The initial portion of the curve is a straight line, or very nearly so. The stress is directly proportional to the strain in this region, so Hooke's law is valid and the slope of the curve is Young's modulus, E. If the load is reduced, the stress–strain curve is retraced to the origin of the diagram, where both the stress and the strain are zero. With continued increase of the load, the curve straight-line behavior eventually ceases at a point called the *proportional limit*, where the stress is denoted σ_{pr} in Figure 3.15.1. The material may still be elastic beyond σ_{pr}, in which case the stress–strain curve is retraced if the load is removed.

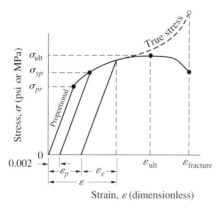

Figure 3.15.1   A typical tensile stress–strain diagram for ductile metal.

If the material is not brittle, that is, if it is ductile, then a further increase in the stress will induce *plastic* strain or *permanent set*. The onset of plastic behavior occurs at the *yield point*, at which the stress is denoted σ_{yp} in Figure 3.15.1. Some materials exhibit a definite yield point, but many do not. Therefore, the yield stress is commonly determined by the *0.2% offset method*. A straight line with a slope equal to the elastic modulus E is constructed through the point with zero stress and a strain of 0.002. The intersection of this line with the stress–strain curve determines the *yield point* σ_{yp}. Beyond the yield point, in the *plastic range*, the total strain in the test specimen consists of an elastic component ε_e and a plastic component ε_p, as shown in the figure. If the load is reduced in the plastic range, the unloading path on the stress–strain diagram follows a straight line of slope E, which intersects the strain axis at ε_p, the nonrecoverable plastic strain point. Upon reloading, the curve retraces the straight line until reaching the curved portion, which is thereafter followed by the material. Thus, inducing plastic strain raises the yield point, a phenomenon known as *work hardening*.

With a continued increase in the load, a maximum or *ultimate stress* σ_{ult} is reached. Thereafter, the test specimen begins to "neck down" considerably, with plastic flow pervading the material and the axial load dropping off as the elongation continues, until the specimen finally breaks. The strain $\varepsilon_{fracture}$ at fracture is a measure of the ductility of the material. It should be pointed out that the stress (*engineering stress* or *nominal stress*) is calculated using the *initial*, undeformed cross-sectional area of the test specimen. In the region where the material necks down, toward the end of the test, the actual transverse area decreases faster than the load drops off, so that the *true* stress at fracture is actually higher than the ultimate stress.

Brittle materials may be roughly classified as those for which $\varepsilon_{fracture} < 0.05$ (5 percent elongation of a specified gage length). They have no yield point, they endure little if any plastic flow, and they therefore fail (break) with little or no warning.

Of course, real engineering structures bear little resemblance to test specimens. In any case, they are subjected to combined loadings that produce biaxial or triaxial states of stress. So the question is how can we predict the failure of a structural member under static load, given information such as that in Figure 3.15.1—data commonly available.[18] Several theories have been proposed for isotropic materials, each based on the assumption that a particular combination of stresses or strains produces the critical condition. We will consider only three of those theories ordinarily used for design purposes.

3.15.1   Maximum Normal Stress Theory

Brittle materials tend to fail in a tension test by fracture (cleavage) across a plane normal to the pull direction. The *maximum normal stress theory* applies to brittle materials and states that failure occurs when the maximum principal stress at a point reaches the value at which failure occurred in a tensile test. All other stress components are ignored. Thus, according to this theory, the failure criterion is

$$\sigma_{max} = \sigma_{yp} \qquad \text{[3.15.1]}$$

3.15.2   Maximum Shear Stress Theory

The yield mechanism in crystalline materials is shear-induced slip on specific planes of atoms. The microstructure of a metal consists of numerous crystalline grains. Slip will occur within a grain when the shear stress parallel to one of its slip planes exceeds a critical value. In isotropic materials, the grains are randomly oriented, which means that yielding should depend on the magnitude of the maximum shear stress. That is the premise of the *maximum shear stress theory*, which states that yielding occurs at a point when the maximum shear stress τ_{max} at that point reaches the maximum shear stress at the onset of yield in a uniaxial tension test. The cross section of a uniaxial test specimen is a principal plane, and the normal stress on that plane is the only nonzero principal stress. Since that stress is σ_{yp} at yield, Equation 3.5.10 requires the corresponding maximum shear stress to be $\sigma_{yp}/2$. Therefore, the maximum shear stress theory lists the failure criterion as

$$2\tau_{max} = \sigma_{yp} \qquad \text{[3.15.2a]}$$

[18]*Military Handbook 5G, Metallic Materials and Elements for Aerospace Vehicle Structures*, 1 November 1994, Naval Publications and Forms Center, Philadelphia, PA. This two-volume handbook is maintained as a joint effort of the United States Department of Defense and the Federal Aviation Administration.

which can alternatively be expressed in terms of the maximum and minimum principal stresses, as follows:

$$\sigma_{max} - \sigma_{min} = \sigma_{yp}$$

$$[3.15.2b]$$

3.15.3   Maximum Distortion Energy Theory

In section 3.11, we pointed out that volume change is caused by the hydrostatic stress p. Shear stress, which does not cause volume change, has been found to be the dominant cause of yielding. Furthermore, very little change in the density of a material accompanies plastic deformation. It may therefore be reasoned that only that portion of the normal stress which causes shear distortion acts to promote yielding. These are called the stress deviations or *deviatoric stresses*, σ_x', σ_y', and σ_z'. They are defined such that

$$\sigma_x = \sigma_x' + p \qquad \sigma_y = \sigma_y' + p \qquad \sigma_z = \sigma_z' + p$$

$$[3.15.3]$$

Substituting these expressions for the normal stress in the strain energy density formula, Equation 3.14.14, yields

$$u_o = u_v + u_d$$

where

$$u_v = \frac{1}{2K}p^2 \qquad u_d = \frac{1}{2E}\left(\sigma_x'^2 + \sigma_y'^2 + \sigma_z'^2\right) - \frac{v}{E}\left(\sigma_x'\sigma_y' + \sigma_x'\sigma_z' + \sigma_y'\sigma_z'\right) + \frac{1}{2G}\left(\tau_{xy}^2 + \tau_{yz}^2 + \tau_{zx}^2\right)$$

Here, u_v is the portion of the strain energy density due to volume change, and u_d is the *distortion strain energy density*. From Equation 3.15.3 and the definition of hydrostatic stress, $p = \frac{1}{3}(\sigma_x + \sigma_y + \sigma_z)$, the deviatoric stresses may be written

$$\sigma_x' = \tfrac{2}{3}\sigma_x - \tfrac{1}{3}(\sigma_y + \sigma_z) \qquad \sigma_y' = \tfrac{2}{3}\sigma_y - \tfrac{1}{3}(\sigma_x + \sigma_z) \qquad \sigma_z' = \tfrac{2}{3}\sigma_z - \tfrac{1}{3}(\sigma_x + \sigma_y)$$

These, together with Equation 3.11.2, lead to the following expression for the distortion strain energy density:

$$u_d = \frac{1+v}{3E}\left\{ \frac{1}{2}\left[(\sigma_x - \sigma_y)^2 + (\sigma_x - \sigma_z)^2 + (\sigma_y - \sigma_z)^2\right] + 3\left(\tau_{xy}^2 + \tau_{xz}^2 + \tau_{yz}^2\right) \right\}$$

$$[3.15.4]$$

In a uniaxial tension test, the only nonzero stress is σ_x. Therefore, the distortion strain energy density when yield occurs ($\sigma_x = \sigma_{yp}$) is

$$u_d = \frac{1+v}{3E}\sigma_{yp}^2$$

According to the *maximum distortion energy theory* of failure, yield occurs when the distortion strain energy density equals the distortion strain energy density at the yield point of a tensile test specimen; that is,

$$\frac{1+v}{3E}\left\{ \frac{1}{2}\left[(\sigma_x - \sigma_y)^2 + (\sigma_x - \sigma_z)^2 + (\sigma_y - \sigma_z)^2\right] + 3\left(\tau_{xy}^2 + \tau_{xz}^2 + \tau_{yz}^2\right) \right\} = \frac{1+v}{3E}\sigma_{yp}^2$$

The von Mises stress, σ_{VM}, at a point is defined as[19]

$$\sigma_{VM} = \sqrt{\frac{1}{2}\left[(\sigma_x - \sigma_y)^2 + (\sigma_x - \sigma_z)^2 + (\sigma_y - \sigma_z)^2\right] + 3\left(\tau_{xy}^2 + \tau_{xz}^2 + \tau_{yz}^2\right)} \qquad \text{[3.15.5]}$$

Therefore, according to the maximum distortion energy theory, the failure criterion is

$$\sigma_{VM} = \sigma_{yp} \qquad \text{[3.15.6]}$$

Here, σ_{VM} is an invariant, having the same value regardless of the coordinate system used, so that in terms of the maximum, minimum, and intermediate values of the principal stresses,

$$\sigma_{VM} = \sqrt{\frac{1}{2}\left[(\sigma_{max} - \sigma_{min})^2 + (\sigma_{max} - \sigma_{int})^2 + (\sigma_{int} - \sigma_{min})^2\right]} \qquad \text{[3.15.7]}$$

or

$$\sigma_{VM} = \frac{1}{\sqrt{2}}(\sigma_{max} - \sigma_{min})\sqrt{1 + \left(\frac{\sigma_{max} - \sigma_{int}}{\sigma_{max} - \sigma_{min}}\right)^2 + \left(\frac{\sigma_{int} - \sigma_{min}}{\sigma_{max} - \sigma_{min}}\right)^2}$$

From this we can deduce that

$$0.866(\sigma_{max} - \sigma_{min}) \le \sigma_{VM} \le (\sigma_{max} - \sigma_{min}) \qquad \text{[3.15.8]}$$

Clearly, the von Mises stress is generally somewhat less than the maximum principal stress difference, which means that the maximum shear stress failure criterion is slightly more conservative than the distortion energy criterion. On the other hand, the distortion energy criterion is more convenient, because it does not require finding the principal stresses in order to calculate σ_{VM}.

Observe that in plane stress,

$$\sigma_{VM} = \sqrt{(\sigma_x^2 + \sigma_y^2 - \sigma_x\sigma_y) + 3\tau_{xy}^2} \qquad \text{[3.15.9]}$$

Example 3.15.1 The state of stress at a point is

$$\begin{bmatrix} p & 0 & 0 \\ 0 & p & 0 \\ 0 & 0 & p \end{bmatrix}$$

Find the maximum allowable value of p in terms of the yield stress σ_{yp}, according to each of the three failure theories presented.

The given state of stress is one of principal stress, so that, according to the maximum normal stress criterion,

$$p_{\text{max allowable}} = \sigma_{yp}$$

---

[19]Richard von Mises (1883–1953) was an Austrian-born American mathematician and philosopher. In a paper published in 1913, he proposed that yielding of a perfectly plastic material occurs at a constant value of I_2', the second invariant (see Equation 3.5.6) of the deviatoric stress. It can be shown that $\sigma_{VM} = \sqrt{3I_2'}$. It is also true that $\sigma_{VM} = \frac{3}{\sqrt{2}}\tau_0$, where τ_0 is the *octahedral shear stress*, that is, the resultant shear stress on a plane inclined at the same angle (54.7°) to all three principal directions.

This is a state of *pure hydrostatic stress*, since all of the principal stresses are equal. Therefore, both the maximum shear stress and the von Mises stress are zero. That is, the maximum shear stress and distortion energy theories place no restriction on p. This reflects the fact that a state of pure hydrostatic stress has little influence on plastic flow.

Example 3.15.2   The state of stress at a point is

$$\begin{bmatrix} -p & \tau & \tau \\ \tau & -p & \tau \\ \tau & \tau & -p \end{bmatrix} \qquad [a]$$

where $p > 0$ and $\tau > 0$. Find the maximum allowable values of p and τ, according to each of the three failure criteria. Assume the compressive and tensile yield stresses are identical.

The von Mises stress may be found immediately by substituting the stresses into Equation 3.15.5, as follows:

$$\sigma_{VM} = \sqrt{\frac{1}{2}\left[(-p+p)^2 + (-p+p)^2 + (-p+p)^2\right] + 3\left(\tau^2 + \tau^2 + \tau^2\right)} = 3\tau$$

Thus, the maximum distortion energy theory requires that

$$\tau \leq \frac{\sigma_{yp}}{3} \qquad [b]$$

For the other two criteria, we must find the principal stresses. According to Equation 3.5.6, the invariants of the stress tensor are

$$I_1 = \sigma_x + \sigma_y + \sigma_z = -3p$$
$$I_2 = \sigma_x \sigma_y + \sigma_x \sigma_z + \sigma_y \sigma_z - \tau_{xy}^2 - \tau_{xz}^2 - \tau_{yz}^2 = 3\left(p^2 - \tau^2\right)$$
$$I_3 = \sigma_x \sigma_y \sigma_z + 2\tau_{xy}\tau_{xz}\tau_{yz} - \sigma_x \tau_{yz}^2 - \sigma_y \tau_{xz}^2 - \sigma_z \tau_{xy}^2 = 2\tau^3 + 3p\tau^2 - p^3$$

Therefore, the characteristic equation, Equation 3.5.5, is

$$\sigma^3 + 3p\sigma^2 + 3(p^2 - \tau^2)\sigma - (2\tau^3 + 3p\tau^2 - p^3) = 0$$

We can use Equations 3.5.7 to find the roots of this equation, as follows:

$$\sigma_1 = 2\tau - p$$
$$\sigma_2 = \sigma_3 = -\tau - p$$

From these, we find that $\tau_{\max} = \frac{3}{2}\tau$, so that the maximum shear stress theory yields the same restriction on τ as the distortion energy criterion, Equation b. Neither places any restrictions on p.

Since it is required that $\tau > 0$ and $p > 0$, it is clear that σ_3 is compressive, and the maximum normal stress theory requires that the magnitude of σ_3 remain less than the compressive yield stress, or

$$\tau + p < \sigma_{yp} \qquad [c]$$

If $\tau < \frac{1}{2}p$, then σ_1 is compressive also, but σ_3 is more so; therefore, Equation c still governs. If $\tau > \frac{1}{2}p$, then $\sigma_1 > 0$, and the maximum normal stress theory requires that σ_1 be less than the tensile yield stress, or

$$2\tau - p < \sigma_{yp} \qquad [d]$$

For the inequalities of c and d to be satisfied simultaneously, τ and p must lie within the shaded region of the graph in Figure 3.15.2.

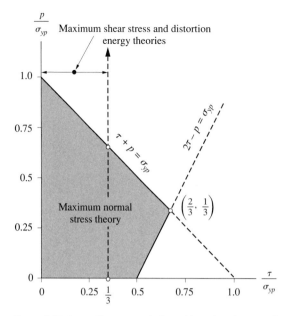

Figure 3.15.2   The range of allowable values for τ and p for the state of stress in Figure 3.15.1.

3.16   MARGIN OF SAFETY

A structure is intended to support a given maximum load. It must have the strength required to support that load. For a structure of a given size, the strength depends on the limits to which the material can be stressed without damage of one kind or another. Those limits are the *allowable stresses*. Safety factors may be applied to test data in order to come up with the allowable stresses. The working stresses required to support the loads must not exceed the allowable stresses. The excess strength of a structural component is the difference between its maximum strength (the allowable stress) and the required strength (the stress calculated for the maximum service load condition). The ratio of the excess strength to the required strength is the *margin of safety, MS*.

$$MS = \frac{\text{Excess strength}}{\text{Required strength}} = \frac{\sigma_{\text{allowable}} - \sigma_{\text{required}}}{\sigma_{\text{required}}}$$

$$MS = \frac{\sigma_{\text{allowable}}}{\sigma_{\text{required}}} - 1 \qquad\qquad \text{[3.16.1]}$$

An airplane must be able to operate and maneuver within its *flight envelope* without incurring structural damage. The flight envelope defines the *limit loads*, which are the worst possible loading conditions the airplane is expected to encounter during its service life. A safety factor of 1.5 is applied to the limit loads to define the *ultimate loads*. An airplane is designed (e.g., its various components are *sized*) to survive these ultimate load conditions.

A requirement of aircraft design is that all structural members satisfy three basic conditions. First, the stresses accompanying limit loads, $\sigma_{\text{limit loads}}$, must not cause plastic deformation, that is, they must not exceed the yield strength, σ_{yp}, of any part of the structure, or:

$$\sigma_{\text{limit loads}} < \sigma_{yp} \qquad\qquad [3.16.2]$$

Second, the stresses due to ultimate load conditions, $\sigma_{\text{ultimate loads}}$, must not exceed the *ultimate strength*, σ_{ult}, of any structural member. That is,

$$\sigma_{\text{ultimate loads}} < \sigma_{ult} \qquad \text{which implies that} \qquad \sigma_{\text{limit loads}} < \frac{\sigma_{ult}}{1.5} \qquad [3.16.3]$$

Notice that an ultimate load condition can result in plastic deformation of some parts of the structure, possibly requiring repair of the aircraft. Third, in addition to the above two conditions, the limit load stresses must not exceed the *buckling strength*, σ_{cr}, of any part,[20]

$$\sigma_{\text{limit loads}} < \sigma_{cr} \qquad\qquad [3.16.4]$$

The margin of safety with respect to plastic deformation, that is, the *yield* condition is

$$MS)_{\text{yield}} = \frac{\sigma_{yp} - \sigma_{\text{limit load}}}{\sigma_{\text{limit load}}} = \frac{\sigma_{yp}}{\sigma_{\text{limit load}}} - 1 \qquad [3.16.5]$$

For ultimate loads,

$$MS)_{\text{ult load}} = \frac{\sigma_{ult} - \sigma_{\text{ultimate load}}}{\sigma_{\text{ultimate load}}} = \frac{\sigma_{ult}}{\sigma_{\text{ultimate load}}} - 1 = \frac{\sigma_{ult}}{1.5\sigma_{\text{limit load}}} - 1 \qquad [3.16.6]$$

Finally, for buckling, we have

$$MS)_{\text{buckling}} = \frac{\sigma_{cr} - \sigma_{\text{limit load}}}{\sigma_{\text{limit load}}} = \frac{\sigma_{cr}}{\sigma_{\text{limit load}}} - 1 \qquad [3.16.7]$$

The margin of safety must be *positive*, but lightweight structures should be designed so that the margin is as small as possible.

Example 3.16.1   The state of stress at the most critical point of a structure is

$$\begin{bmatrix} 10,000 & 5000 & -6000 \\ 5000 & 15,000 & 8000 \\ -6000 & 8000 & 4000 \end{bmatrix} \text{(psi)}$$

The material yield stress is 25,000 psi. Calculate the margin of safety based on: (a) distortion energy theory, (b) maximum shear stress theory, and (c) maximum normal stress theory.

(a)

The von Mises stress is

$$\sigma_{VM} = \sqrt{\tfrac{1}{2}\left[(10,000 - 15,000)^2 + (10,000 - 4000)^2 + (15,000 - 4000)^2\right] + 3\left[5000^2 + (-6000)^2 + 8000^2\right]} = 21,590 \text{ psi}$$

According to the maximum distortion energy criterion, the margin of safety is

$$MS = \frac{25,000}{21,590} - 1 = 0.16$$

| [20]The term σ_{cr} is the *critical* value of compressive stress beyond which a part will buckle. Buckling is the subject of Chapter 12.

(b)

To apply the maximum shear stress theory, we must find the principal stresses. For the given stress tensor, the invariants are

$$I_1 = 10,000 + 15,000 + 4000 = 29,000 \text{ psi}$$

$$I_2 = \det \begin{bmatrix} 10,000 & 5000 \\ 5000 & 15,000 \end{bmatrix} = 125.0 \times 10^6 \text{ psi}^2$$

$$I_3 = \det \begin{bmatrix} 10,000 & 5000 & -6000 \\ 5000 & 15,000 & 8000 \\ -6000 & 8000 & 4000 \end{bmatrix} = 1.160 \times 10^{12} \text{ psi}^3$$

Therefore, the characteristic equation is

$$\sigma^3 - 29,000\sigma^2 + 125.0 \times 10^6 \sigma + 1.160 \times 10^{12} = 0$$

The three roots are

$$\sigma_1 = \sigma_{max} = 19,610 \text{ psi} \qquad \sigma_2 = 13,710 \text{ psi} \qquad \sigma_3 = \sigma_{min} = -4316 \text{ psi}$$

If follows that

$$\tau_{max} = \tfrac{1}{2}[19,610 - (-4316)] = 11,960 \text{ psi}$$

Applying the maximum shear stress criterion, we find that

$$MS = \frac{\tfrac{1}{2}(25,000)}{11,960} - 1 = 0.045$$

(c)

Using the maximum normal stress theory, we have

$$MS = \frac{25,000}{19,610} - 1 = 0.27$$

Of course, this margin of safety only applies if the material is brittle in nature. If the material is a ductile metal, then the results of either part (a) or (b) would be appropriate, with the maximum shear stress theory giving the more conservative answer (lowest excess strength), as previously discussed.

3.17   ANISOTROPIC ELASTIC MATERIALS

Anisotropic materials are those in which the physical properties are directionally dependent. If, as in Figure 3.17.1, we fabricate several tensile test specimens, taken at different angles from the same thin sheet of anisotropic material, we will find a different modulus of elasticity for each sample. Anisotropic materials require more than just the two independent mechanical properties—Young's modulus and Poisson's ratio—needed to characterize isotropic materials. For anisotropic elastic materials in plane stress, there appear to be nine *material flexibility coefficients* in the strain–stress relations:

$$\varepsilon_x = D_{11}\sigma_x + D_{12}\sigma_y + D_{13}\tau_{xy}$$
$$\varepsilon_y = D_{21}\sigma_x + D_{22}\sigma_y + D_{23}\tau_{xy} \qquad \text{[3.17.1]}$$
$$\gamma_{xy} = D_{31}\sigma_x + D_{32}\sigma_y + D_{33}\tau_{xy}$$

These coefficients have the following interpretation in terms of the *engineering elastic constants* measured relative to the *xy* reference system:

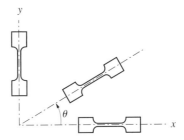

Figure 3.17.1   Several tensile test specimens prepared from the same thin sheet of material.

$$D_{11} = \frac{1}{E_x} \qquad D_{12} = -\frac{\nu_{yx}}{E_y} \qquad D_{13} = \frac{\eta_{x,xy}}{G_{xy}}$$

$$D_{21} = -\frac{\nu_{xy}}{E_x} \qquad D_{22} = \frac{1}{E_y} \qquad D_{23} = \frac{\eta_{y,xy}}{G_{xy}} \qquad [3.17.2]$$

$$D_{31} = \frac{\eta_{xy,x}}{E_x} \qquad D_{32} = \frac{\eta_{xy,y}}{E_y} \qquad D_{33} = \frac{1}{G_{xy}}$$

Here, E_x and E_y are the Young's moduli in the x and y directions, respectively, and ν_{xy} is a Poisson's ratio, namely, the magnitude of the ratio of the normal strain in the y direction—produced by a pure uniaxial tension in the x direction—to the strain in the x direction. Likewise, ν_{yx} relates the normal strain in the x direction to the normal strain in the y direction. Also, G_{xy} is the in-plane shear modulus. The other coefficients represent the coupling between normal and shear strains in the xy coordinate system: $\eta_{x,xy}$ is the ratio of the normal strain ε_x—due to a pure shear stress τ_{xy}—to the accompanying shear strain γ_{xy}; $\eta_{y,xy}$ is the ratio of normal strain ε_y to γ_{xy}; and $\eta_{xy,x}$ and $\eta_{xy,y}$ indicate the amount of shear strain γ_{xy} accompanying the normal strains ε_x and ε_y in a state of uniaxial stress in the x and y directions, respectively. The η quantities are referred to as *coefficients of mutual influence*.[21]

Since the material is linear elastic, there is a strain energy function, which means that relationships exist among the nine coefficients in Equations 3.17.1. Recall that the complementary strain energy density is a potential function for the strains, so that from Equations 3.14.12 we have

$$\varepsilon_x = \frac{\partial u_o^*}{\partial \sigma_x} \qquad \varepsilon_y = \frac{\partial u_o^*}{\partial \sigma_y} \qquad \gamma_{xy} = \frac{\partial u_o^*}{\partial \tau_{xy}} \qquad [3.17.3]$$

Observing, for example, that $D_{12} = \partial \varepsilon_x / \partial \sigma_y$, it follows from Equations 3.17.1 and 3.17.3 that

$$D_{12} = \frac{\partial \varepsilon_x}{\partial \sigma_y} = \frac{\partial}{\partial \sigma_y}\left(\frac{\partial u_o^*}{\partial \sigma_x}\right) = \frac{\partial}{\partial \sigma_x}\left(\frac{\partial u_o^*}{\partial \sigma_y}\right) = \frac{\partial \varepsilon_y}{\partial \sigma_x} = D_{21}$$

$$D_{13} = \frac{\partial \varepsilon_x}{\partial \tau_{xy}} = \frac{\partial}{\partial \tau_{xy}}\left(\frac{\partial u_o^*}{\partial \sigma_x}\right) = \frac{\partial}{\partial \sigma_x}\left(\frac{\partial u_o^*}{\partial \tau_{xy}}\right) = \frac{\partial \gamma_{xy}}{\partial \sigma_x} = D_{31} \qquad [3.17.4]$$

$$D_{23} = \frac{\partial \varepsilon_y}{\partial \tau_{xy}} = \frac{\partial}{\partial \tau_{xy}}\left(\frac{\partial u_o^*}{\partial \sigma_y}\right) = \frac{\partial}{\partial \sigma_y}\left(\frac{\partial u_o^*}{\partial \tau_{xy}}\right) = \frac{\partial \gamma_{xy}}{\partial \sigma_y} = D_{23}$$

| [21]R. M. Jones, *Mechanics of Composite Materials*, New York, Hemisphere Publishing, 1975.

This reduces the number of independent material-flexibility coefficients from nine to six. In fact, applying these three conditions to Equations 3.17.2, we are led to the conclusion that

$$\nu_{yx} = \frac{E_y}{E_x}\nu_{xy} \qquad \eta_{xy,x} = \frac{E_x}{G_{xy}}\eta_{x,xy} \qquad \eta_{xy,y} = \frac{E_y}{G_{xy}}\eta_{y,xy} \qquad \text{[3.17.5]}$$

Equations 3.17.1 can be solved for the stresses in terms of the strains to obtain

$$\sigma_x = E_{11}\varepsilon_x + E_{12}\varepsilon_y + E_{13}\gamma_{xy}$$
$$\sigma_y = E_{21}\varepsilon_x + E_{22}\varepsilon_y + E_{23}\gamma_{xy} \qquad \text{[3.17.6]}$$
$$\tau_{xy} = E_{31}\varepsilon_x + E_{32}\varepsilon_y + E_{33}\gamma_{xy}$$

in which the six *material stiffness coefficients* are found in terms of the flexibility terms by means of the following formulas:

$$E_{11} = \frac{1}{\Delta}(D_{22}D_{33} - D_{23}^2) \qquad E_{22} = \frac{1}{\Delta}(D_{33}D_{11} - D_{13}^2) \qquad E_{33} = \frac{1}{\Delta}(D_{11}D_{22} - D_{12}^2)$$

$$E_{12} = E_{21} = \frac{1}{\Delta}(D_{13}D_{23} - D_{33}D_{12}) \qquad E_{13} = E_{31} = \frac{1}{\Delta}(D_{12}D_{23} - D_{22}D_{13}) \qquad E_{23} = E_{32} = \frac{1}{\Delta}(D_{13}D_{12} - D_{11}D_{23})$$

$$\Delta = D_{11}D_{22}D_{33} - D_{11}D_{23}^2 - D_{22}D_{13}^2 - D_{33}D_{12}^2 + 2D_{12}D_{13}D_{23} \qquad \text{[3.17.7]}$$

On the other hand, if we have the nine stiffness coefficients, we may use Equations 3.17.7 to find the flexibility coefficients by replacing D throughout on the right by E, while at the same time replacing E everywhere on the left by D.

Since the properties of an anisotropic material vary with direction, we must be able to relate the stiffness and flexibility coefficients found in one coordinate system to those in another. For a given sheet of material, let $\bar{x}\bar{y}$ be a set of axes in the same plane as xy with a common z axis, but rotated through a counterclockwise angle θ, as illustrated in Figure 3.17.2. We seek the forms of Equations 3.17.1 and 3.17.6 in the rotated (barred) frame of reference. First, we recall that the strain components in the barred system are obtained from those in the unbarred system by means of Equations 3.10.3, as follows:

$$\bar{\varepsilon}_x = \varepsilon_x \cos^2\theta + \varepsilon_y \sin^2\theta + \gamma_{xy}\sin\theta\cos\theta$$
$$\bar{\varepsilon}_y = \varepsilon_x \sin^2\theta + \varepsilon_y \cos^2\theta - \gamma_{xy}\sin\theta\cos\theta \qquad \text{[3.17.8]}$$
$$\bar{\gamma}_{xy} = 2(\varepsilon_y - \varepsilon_x)\sin\theta\cos\theta + \gamma_{xy}(\cos^2\theta - \sin^2\theta)$$

Substituting Equations 3.17.1 on the right and using the abbreviations $s \equiv \sin\theta$ and $c \equiv \cos\theta$ yields

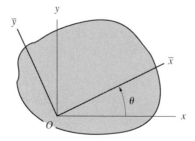

Figure 3.17.2   Two alternative cartesian coordinate systems for plane stress analysis.

$$\bar{\varepsilon}_x = (c^2 D_{11} + s^2 D_{12} + sc D_{13})\sigma_x + (c^2 D_{12} + s^2 D_{22} + sc D_{23})\sigma_y + (c^2 D_{13} + s^2 D_{23} + sc D_{33})\tau_{xy}$$

$$\bar{\varepsilon}_y = (s^2 D_{11} + c^2 D_{12} - sc D_{13})\sigma_x + (s^2 D_{12} + c^2 D_{22} - sc D_{23})\sigma_y + (s^2 D_{13} + c^2 D_{23} - sc D_{33})\tau_{xy}$$ [3.17.9]

$$\bar{\gamma}_{xy} = \left[-2sc D_{11} + 2sc D_{12} + (c^2 - s^2)D_{13}\right]\sigma_x + \left[-2sc D_{12} + 2sc D_{22} + (c^2 - s^2)D_{23}\right]\sigma_y + \left[-2sc D_{13} + 2sc D_{23} + (c^2 - s^2)D_{33}\right]\tau_{xy}$$

The unbarred stress components on the right must be replaced in terms of their counterparts in the barred system. This requires the opposite of the stress transformation given by Equation 3.6.4. However, we may use Equation 3.6.4 by simply interchanging the barred and unbarred symbols (so as to switch the roles of the xy and $\bar{x}\bar{y}$ frames) and changing θ to $-\theta$ so as to rotate clockwise instead of counterclockwise, consistent with a transformation *from $\bar{x}\bar{y}$ to xy* in Figure 3.17.2. In other words,

$$\sigma_x = \bar{\sigma}_x \cos^2\theta + \bar{\sigma}_y \sin^2\theta - 2\bar{\tau}_{xy}\sin\theta\cos\theta$$

$$\sigma_y = \bar{\sigma}_x \sin^2\theta + \bar{\sigma}_y \cos^2\theta + 2\bar{\tau}_{xy}\sin\theta\cos\theta$$ [3.17.10]

$$\tau_{xy} = (\bar{\sigma}_x - \bar{\sigma}_y)\sin\theta\cos\theta + \bar{\tau}_{xy}(\cos^2\theta - \sin^2\theta)$$

Substituting these equations into Equations 3.17.9 yields the material flexibility equations in the rotated reference frame:

$$\bar{\varepsilon}_x = \bar{D}_{11}\bar{\sigma}_x + \bar{D}_{12}\bar{\sigma}_y + \bar{D}_{13}\bar{\tau}_{xy}$$

$$\bar{\varepsilon}_y = \bar{D}_{12}\bar{\sigma}_x + \bar{D}_{22}\bar{\sigma}_y + \bar{D}_{23}\bar{\tau}_{xy}$$ [3.17.11]

$$\bar{\gamma}_{xy} = \bar{D}_{13}\bar{\sigma}_x + \bar{D}_{23}\bar{\sigma}_y + \bar{D}_{33}\bar{\tau}_{xy}$$

where the material flexibility coefficients in the barred system are given by the rather lengthy expressions listed here:

$$\bar{D}_{11} = \quad c^4 D_{11} \quad + s^4 D_{22} \quad + s^2 c^2 D_{33} \quad + 2s^2 c^2 D_{12} \quad + 2sc^3 D_{13} \quad + 2s^3 c D_{23}$$

$$\bar{D}_{22} = \quad s^4 D_{11} \quad + c^4 D_{22} \quad + s^2 c^2 D_{33} \quad + 2s^2 c^2 D_{12} \quad - 2s^3 c D_{13} \quad - 2sc^3 D_{23}$$

$$\bar{D}_{33} = 4s^2 c^2 D_{11} + 4s^2 c^2 D_{22} + (s^2 - c^2)^2 D_{33} \quad - 8s^2 c^2 D_{12} + 4sc(s^2 - c^2)D_{13} + 4sc(c^2 - s^2)D_{23}$$

$$\bar{D}_{12} = \quad s^2 c^2 D_{11} + s^2 c^2 D_{22} \quad - s^2 c^2 D_{33} \quad + (s^4 + c^4)D_{12} + sc(s^2 - c^2)D_{13} + sc(c^2 - s^2)D_{23}$$ [3.17.12]

$$\bar{D}_{13} = -2sc^3 D_{11} + 2s^3 c D_{22} + sc(c^2 - s^2)D_{33} + 2sc(c^2 - s^2)D_{12} + c^2(c^2 - 3s^2)D_{13} + s^2(3c^2 - s^2)D_{23}$$

$$\bar{D}_{23} = -2s^3 c D_{11} + 2sc^3 D_{22} + sc(s^2 - c^2)D_{33} + 2sc(s^2 - c^2)D_{12} + s^2(3c^2 - s^2)D_{13} + c^2(c^2 - 3s^2)D_{23}$$

In a similar fashion, we can obtain the material stiffness equations in the barred reference system:

$$\bar{\sigma}_x = \bar{E}_{11}\bar{\varepsilon}_x + \bar{E}_{12}\bar{\varepsilon}_y + \bar{E}_{13}\bar{\gamma}_{xy}$$

$$\bar{\sigma}_y = \bar{E}_{21}\bar{\varepsilon}_x + \bar{E}_{22}\bar{\varepsilon}_y + \bar{E}_{23}\bar{\gamma}_{xy}$$ [3.17.13]

$$\bar{\tau}_{xy} = \bar{E}_{31}\bar{\varepsilon}_x + \bar{E}_{32}\bar{\varepsilon}_y + \bar{E}_{33}\bar{\gamma}_{xy}$$

The transformation relationships for the material stiffness coefficients bear considerable resemblance to those for the flexibility coefficients, as shown here:

$$\bar{E}_{11} = \quad c^4 E_{11} \quad + s^4 E_{22} \quad + 4s^2 c^2 E_{33} \quad + 2s^2 c^2 E_{12} \quad + 4sc^3 E_{13} \quad + 4s^3 c E_{23}$$

$$\bar{E}_{22} = \quad s^4 E_{11} \quad + c^4 E_{22} \quad + 4s^2 c^2 E_{33} \quad + 2s^2 c^2 E_{12} \quad - 4s^3 c E_{13} \quad - 4sc^3 E_{23}$$

$$\bar{E}_{33} = s^2 c^2 E_{11} + s^2 c^2 E_{22} + (s^2 - c^2)^2 E_{33} \quad - 2s^2 c^2 E_{12} + 2sc(s^2 - c^2)E_{13} + 2sc(c^2 - s^2)E_{23}$$

$$\bar{E}_{12} = \quad s^2 c^2 E_{11} + s^2 c^2 E_{22} \quad - 4s^2 c^2 E_{33} \quad + (s^4 + c^4)E_{12} + 2sc(s^2 - c^2)E_{13} + 2sc(c^2 - s^2)E_{23}$$ [3.17.14]

$$\bar{E}_{13} = -sc^3 E_{11} + s^3 c E_{22} + 2sc(c^2 - s^2)E_{33} + sc(c^2 - s^2)E_{12} + c^2(c^2 - 3s^2)E_{13} + s^2(3c^2 - s^2)E_{23}$$

$$\bar{E}_{23} = -s^3 c E_{11} + sc^3 E_{22} + 2sc(s^2 - c^2)E_{33} + sc(s^2 - c^2)E_{12} + s^2(3c^2 - s^2)E_{13} + c^2(c^2 - 3s^2)E_{23}$$

Example 3.17.1   Show that the material flexibility coefficients of a plane isotropic material are unaltered by a rotation of the coordinate axes.

From Equations 3.12.1 (with $T = 0$) and Equation 3.11.2, the material flexibility coefficients of a plane isotropic material relative to the xy directions are

$$D_{11} = D_{22} = \frac{1}{E} \qquad D_{33} = \frac{2(1+\nu)}{E} \qquad D_{12} = -\frac{\nu}{E} \qquad D_{13} = D_{23} = 0 \qquad \text{[a]}$$

By substituting these into the right side of Equation 3.17.12, we get the material flexibility coefficients relative to the $\bar{x}\bar{y}$ axes at a counterclockwise angle θ relative to the xy directions (cf. Figure 3.17.2), as follows:

$$\bar{D}_{11} = c^4\left(\frac{1}{E}\right) + s^4\left(\frac{1}{E}\right) + s^2c^2\left[\frac{2(1+\nu)}{E}\right] + 2s^2c^2\left(-\frac{\nu}{E}\right) + 2sc^3(0) + 2s^3c(0)$$

$$\bar{D}_{22} = s^4\left(\frac{1}{E}\right) + c^4\left(\frac{1}{E}\right) + s^2c^2\left[\frac{2(1+\nu)}{E}\right] + 2s^2c^2\left(-\frac{\nu}{E}\right) - 2s^3c(0) - 2sc^3(0)$$

$$\bar{D}_{33} = 4s^2c^2\left(\frac{1}{E}\right) + 4s^2c^2\left(\frac{1}{E}\right) + (s^2 - c^2)^2\left[\frac{2(1+\nu)}{E}\right] - 8s^2c^2\left(-\frac{\nu}{E}\right) + 4sc(s^2 - c^2)(0) + 4sc(c^2 - s^2)(0)$$

$$\bar{D}_{12} = s^2c^2\left(\frac{1}{E}\right) + s^2c^2\left(\frac{1}{E}\right) - s^2c^2\left[\frac{2(1+\nu)}{E}\right] + (s^4 + c^4)\left(-\frac{\nu}{E}\right) + sc(s^2 - c^2)(0) + sc(c^2 - s^2)(0)$$

$$\bar{D}_{13} = -2sc^3\left(\frac{1}{E}\right) + 2s^3c\left(\frac{1}{E}\right) + sc(c^2 - s^2)\left[\frac{2(1+\nu)}{E}\right] + 2sc(c^2 - s^2)\left(-\frac{\nu}{E}\right) + c^2(c^2 - 3s^2)(0) + s^2(3c^2 - s^2)(0)$$

$$\bar{D}_{23} = -2s^3c\left(\frac{1}{E}\right) + 2sc^3\left(\frac{1}{E}\right) + sc(s^2 - c^2)\left[\frac{2(1+\nu)}{E}\right] + 2sc(s^2 - c^2)\left(-\frac{\nu}{E}\right) + s^2(3c^2 - s^2)(0) + c^2(c^2 - 3s^2)(0)$$

Carrying out the multiplications leads to

$$\bar{D}_{11} = \bar{D}_{22} = (s^2 + c^2)^2\left(\frac{1}{E}\right) = \frac{1}{E}$$

$$\bar{D}_{33} = (s^2 + c^2)^2\left[\frac{2(1+\nu)}{E}\right] = \frac{2(1+\nu)}{E}$$

$$\bar{D}_{12} = -(s^2 + c^2)^2\left(-\frac{\nu}{E}\right) = -\frac{\nu}{E} \qquad \text{[b]}$$

$$\bar{D}_{13} = \bar{D}_{23} = 0$$

These are identical to what we started out with in Equation a. Thus, for an isotropic material, the mechanical properties do not depend on the orientation of the coordinate axes.

An anisotropic material with three mutually orthogonal planes of material symmetry (three principal material axes) is known as an *orthotropic* material. These are commonly encountered in aircraft structures in the form of laminated composites. Let one of the principal material axes be the z axis, normal to the sheet of material itself. The other two principal axes in the plane of the sheet will be denoted as the $\bar{x}\bar{y}$ axes (Figure 3.17.2). In the principal material axes of an orthotropic material, the plane strain–stress relations are given by

$$\bar{\varepsilon}_x = \bar{D}_{11}\bar{\sigma}_x + \bar{D}_{12}\bar{\sigma}_y = \frac{1}{\bar{E}_x}\bar{\sigma}_x - \frac{\bar{\nu}_{yx}}{\bar{E}_y}\bar{\sigma}_y$$

$$\bar{\varepsilon}_y = \bar{D}_{21}\bar{\sigma}_x + \bar{D}_{22}\bar{\sigma}_y = -\frac{\bar{\nu}_{xy}}{\bar{E}_x}\bar{\sigma}_x + \frac{1}{\bar{E}_y}\bar{\sigma}_y \quad (\bar{D}_{13} = \bar{D}_{31} = \bar{D}_{23} = \bar{D}_{32} = 0) \qquad \text{[3.17.15a]}$$

$$\bar{\gamma}_{xy} = \bar{D}_{33}\bar{\tau}_{xy} = \frac{1}{\bar{G}_{xy}}\bar{\tau}_{xy}$$

There are just four independent, engineering elastic constants measured in the principal material frame, since we know that $\bar{D}_{12} = \bar{D}_{21}$, which implies

$$\frac{\bar{\nu}_{xy}}{\bar{E}_x} = \frac{\bar{\nu}_{yx}}{\bar{E}_y}$$

[3.17.15b]

Orthotropic materials are therefore characterized by the fact that in the principal material frame, *there is no coupling between shear strain and normal strain* (the coefficients of mutual influence vanish), so that pure shear strain induces zero normal strain and vice versa. Here, $\bar{E}_x$ and $\bar{E}_y$ are Young's moduli in the $\bar{x}$ and $\bar{y}$ directions, respectively; $\bar{\nu}_{xy}$ is Poisson's ratio of the magnitude of the normal strain produced in the $\bar{y}$ direction by a pure uniaxial tension in the $\bar{x}$ direction; $\bar{\nu}_{yx}$ relates the normal strain in the $\bar{x}$ direction to the normal strain induced in the $\bar{y}$ direction; and $\bar{G}_{xy}$ is the in-plane shear modulus relative to the principal material axes. Solving Equation 3.17.15a for the stresses in terms of the strains yields the material stiffness equations in the principal material axes, as follows:

$$\bar{\sigma}_x = \bar{E}_{11}\bar{\varepsilon}_x + \bar{E}_{12}\bar{\varepsilon}_y = \frac{\bar{E}_x}{1 - \bar{\nu}_{yx}\bar{\nu}_{xy}}\bar{\varepsilon}_x + \frac{\bar{\nu}_{xy}\bar{E}_y}{1 - \bar{\nu}_{yx}\bar{\nu}_{xy}}\bar{\varepsilon}_y$$

$$\bar{\sigma}_y = \bar{E}_{21}\bar{\varepsilon}_x + \bar{E}_{22}\bar{\varepsilon}_y = \frac{\bar{\nu}_{xy}\bar{E}_y}{1 - \bar{\nu}_{yx}\bar{\nu}_{xy}}\bar{\varepsilon}_x + \frac{\bar{E}_y}{1 - \bar{\nu}_{yx}\bar{\nu}_{xy}}\bar{\varepsilon}_y \qquad (\bar{E}_{13} = \bar{E}_{31} = \bar{E}_{23} = \bar{E}_{32} = 0)$$

[3.17.16]

$$\bar{\tau}_{xy} = \bar{E}_{33}\bar{\gamma}_{xy} \qquad\quad = \bar{G}_{xy}\bar{\gamma}_{xy}$$

The xy coordinate system used for plane stress analysis may not line up with the principal material $\bar{x}\bar{y}$ axes. To obtain the material stiffness and flexibility coefficients in orthogonal directions other than the barred principal material axes, we must carry out the transformations in Equations 3.17.12 and 3.17.14, using the material coefficients given in Equations 3.17.15 and 3.17.16. However, since we are transforming *from* the barred material axes *to* the unbarred analysis axes, we must shift the overbars on the symbols in Equations 3.17.12 and 3.17.14 from the left side of the equal sign to the right. We must also change the sign of the angle θ from $+\theta$ to $-\theta$ throughout the formulas. Thus, the material stiffness coefficients in an xy system oriented relative to the $\bar{x}\bar{y}$ principal material directions as shown in Figure 3.17.2 are:

$$E_{11} = \frac{2\bar{E}_x\cos^4\theta + \bar{\nu}_{xy}\bar{E}_y\sin^2 2\theta + 2\bar{E}_y\sin^4\theta}{2[1 - \bar{\nu}_{xy}^2(\bar{E}_y/\bar{E}_x)]} + \bar{G}_{xy}\sin^2 2\theta$$

$$E_{22} = \frac{2\bar{E}_x\sin^4\theta + \bar{\nu}_{xy}\bar{E}_y\sin^2 2\theta + 2\bar{E}_y\cos^4\theta}{2[1 - \bar{\nu}_{xy}^2(\bar{E}_y/\bar{E}_x)]} + \bar{G}_{xy}\sin^2 2\theta$$

$$E_{33} = \frac{(\bar{E}_x + \bar{E}_y - 2\bar{\nu}_{xy}\bar{E}_y)\sin^2 2\theta}{4[1 - \bar{\nu}_{xy}^2(\bar{E}_y/\bar{E}_x)]} + \bar{G}_{xy}\cos^2 2\theta$$

[3.17.17]

$$E_{12} = E_{21} = \frac{(\bar{E}_x + \bar{E}_y)\cos^2\theta\sin^2\theta + \bar{\nu}_{xy}\bar{E}_y(\sin^4\theta + \cos^4\theta)}{1 - \bar{\nu}_{xy}^2(\bar{E}_y/\bar{E}_x)} - \bar{G}_{xy}\sin^2 2\theta$$

$$E_{13} = E_{31} = \left[\frac{(\bar{E}_x\cos^2\theta - \bar{E}_y\sin^2\theta) - \bar{\nu}_{xy}\bar{E}_y\cos 2\theta}{1 - \bar{\nu}_{xy}^2(\bar{E}_y/\bar{E}_x)} - 2\bar{G}_{xy}\cos 2\theta\right]\frac{\sin 2\theta}{2}$$

$$E_{23} = E_{32} = \left[\frac{(\bar{E}_x\sin^2\theta - \bar{E}_y\cos^2\theta) + \bar{\nu}_{xy}\bar{E}_y\cos 2\theta}{1 - \bar{\nu}_{xy}^2(\bar{E}_y/\bar{E}_x)} + 2\bar{G}_{xy}\cos 2\theta\right]\frac{\sin 2\theta}{2}$$

Likewise, the material flexibility coefficients in terms of the engineering constants measured in the principal material frame are:

$$D_{11} = \frac{(\cos^2\theta - 2\sin^2\theta\,\bar{v}_{xy})\cos^2\theta}{\bar{E}_x} + \frac{\sin^4\theta}{\bar{E}_y} + \frac{\cos^2\theta\sin^2\theta}{\bar{G}_{xy}}$$

$$D_{22} = \frac{(\sin^2\theta - 2\cos^2\theta\,\bar{v}_{xy})\sin^2\theta}{\bar{E}_x} + \frac{\cos^4\theta}{\bar{E}_y} + \frac{\cos^2\theta\sin^2\theta}{\bar{G}_{xy}}$$

$$D_{33} = 4\left[\frac{(1+2\bar{v}_{xy})}{\bar{E}_x} + \frac{1}{\bar{E}_y}\right]\cos^2\theta\sin^2\theta + \frac{(\cos^2\theta - \sin^2\theta)^2}{\bar{G}_{xy}}$$

$$D_{12} = D_{21} = \frac{\cos^2\theta\sin^2\theta - (\sin^4\theta + \cos^4\theta)\bar{v}_{xy}}{\bar{E}_x} + \left(\frac{1}{\bar{E}_y} - \frac{1}{\bar{G}_{xy}}\right)\cos^2\theta\sin^2\theta$$

$$D_{13} = D_{31} = 2\frac{[(\cos^2\theta - \sin^2\theta)\bar{v}_{xy} + \cos^2\theta]\cos\theta\sin\theta}{\bar{E}_x} - 2\frac{\cos\theta\sin^3\theta}{\bar{E}_y} + \frac{(\sin^2\theta - \cos^2\theta)\cos\theta\sin\theta}{\bar{G}_{xy}}$$

$$D_{23} = D_{32} = 2\frac{[(\sin^2\theta - \cos^2\theta)\bar{v}_{xy} + \sin^2\theta]\cos\theta\sin\theta}{\bar{E}_x} - 2\frac{\cos^3\theta\sin\theta}{\bar{E}_y} + \frac{(\cos^2\theta - \sin^2\theta)\cos\theta\sin\theta}{\bar{G}_{xy}}$$

[3.17.18]

In Equations 3.17.2, the material flexibility coefficients are expressed in terms of the engineering material constants measured relative to the xy axes. Setting each of those components equal to its counterpart in Equations 3.17.18 leads to the following expressions for the material constants in the xy frame in terms of those (the barred quantities) measured in the principal material axes of the orthotropic material (Equations 3.7.14):

$$E_x = \frac{1}{\dfrac{(\cos^2\theta - 2\bar{v}_{xy}\sin^2\theta)\cos^2\theta}{\bar{E}_x} + \dfrac{\sin^4\theta}{\bar{E}_y} + \dfrac{\cos^2\theta\sin^2\theta}{\bar{G}_{xy}}}$$

$$E_y = \frac{1}{\dfrac{(\sin^2\theta - 2\bar{v}_{xy}\cos^2\theta)\sin^2\theta}{\bar{E}_x} + \dfrac{\cos^4\theta}{\bar{E}_y} + \dfrac{\cos^2\theta\sin^2\theta}{\bar{G}_{xy}}}$$

$$G_{xy} = \frac{1}{4\dfrac{(1+2\bar{v}_{xy})\sin^2\theta\cos^2\theta}{\bar{E}_x} + 4\dfrac{\sin^2\theta\cos^2\theta}{\bar{E}_y} + \dfrac{(\cos^2\theta - \sin^2\theta)^2}{\bar{G}_{xy}}}$$

[3.17.19]

$$v_{xy} = -E_x\left[\frac{\sin^2\theta\cos^2\theta - (\cos^4\theta + \sin^4\theta)\bar{v}_{xy}}{\bar{E}_x} + \left(\frac{1}{\bar{E}_y} - \frac{1}{\bar{G}_{xy}}\right)\sin^2\theta\cos^2\theta\right]$$

$$\eta_{x,xy} = G_{xy}\left\{2\frac{[\cos^2\theta + (\cos^2\theta - \sin^2\theta)\bar{v}_{xy}]\sin\theta\cos\theta}{\bar{E}_x} + \frac{(\sin^2\theta - \cos^2\theta)\sin\theta\cos\theta}{\bar{G}_{xy}} - 2\frac{\sin^3\theta\cos\theta}{\bar{E}_y}\right\}$$

$$\eta_{y,xy} = G_{xy}\left\{2\frac{[\sin^2\theta + (\sin^2\theta - \cos^2\theta)\bar{v}_{xy}]\sin\theta\cos\theta}{\bar{E}_x} + \frac{(\cos^2\theta - \sin^2\theta)\sin\theta\cos\theta}{\bar{G}_{xy}} - 2\frac{\sin\theta\cos^3\theta}{\bar{E}_y}\right\}$$

$$v_{yx} = \frac{E_y}{E_x}v_{xy} \qquad \eta_{xy,x} = \frac{E_x}{G_{xy}}\eta_{x,xy} \qquad \eta_{xy,y} = \frac{E_y}{G_{xy}}\eta_{y,xy}$$

Example 3.17.2   The material stiffness coefficients in the principal material axes of a plane orthotropic material are as follows:

$$\bar{E}_{11} = \bar{E}_{22} = 7.914 \times 10^6 \text{ psi} \qquad \bar{E}_{12} = 3.791 \times 10^6 \text{ psi} \qquad \bar{E}_{33} = 4.185 \times 10^6 \text{ psi}$$

Calculate the mechanical properties of the sheet relative to xy axes rotated 30 degrees clockwise from the principal directions.

First, we use Equations 3.17.7—interchanging the roles of D and E—to calculate the nonzero material flexibility coefficients, obtaining

$$\bar{D}_{11} = \bar{D}_{22} = 164.0 \times 10^{-9} \text{ psi}^{-1} \qquad \bar{D}_{12} = -78.54 \times 10^{-9} \text{ psi}^{-1} \qquad \bar{D}_{33} = 238.9 \times 10^{-9} \text{ psi}^{-1} \qquad \text{[a]}$$

From these constants, we find the principal engineering material constants, using Equations 3.7.14. Thus, the tensile elastic moduli are

$$\bar{E}_x = \frac{1}{\bar{D}_{11}} = \frac{1}{164.0 \times 10^{-9}} = 6.099 \times 10^6 \text{ psi} \qquad \text{[b]}$$

and

$$\bar{E}_y = \frac{1}{\bar{D}_{22}} = \frac{1}{164.0 \times 10^{-9}} = 6.099 \times 10^6 \text{ psi} \qquad \text{[c]}$$

For the shear modulus, we have

$$\bar{G}_{xy} = \frac{1}{\bar{D}_{33}} = \frac{1}{238.9 \times 10^{-9}} = 4.185 \times 10^6 \text{ psi} \qquad \text{[d]}$$

Finally, the Poisson's ratios are

$$\bar{v}_{xy} = -\bar{E}_x \bar{D}_{21} = \left(-6.099 \times 10^6\right)\left(-78.54 \times 10^{-9}\right) = 0.4790 \qquad \text{[e]}$$

and

$$\bar{v}_{yx} = -\bar{E}_y \bar{D}_{12} = \left(-6.099 \times 10^6\right)\left(-78.54 \times 10^{-9}\right) = 0.4790 \qquad \text{[f]}$$

Since the moduli of elasticity and the Poisson's ratios are the same in both the $\bar{x}$ and $\bar{y}$ directions, the material *appears* to be isotropic. But it is only *pseudo-isotropic* because, as we can readily verify,

$$\bar{G}_{xy} \neq \frac{\bar{E}_x}{2(1 + \bar{v}_{xy})}$$

According to Equation 3.11.2, the two sides of this relationship must be equal if the material is isotropic.

To find the material constants in the rotated coordinate system, we substitute Equations b through f and $\theta = 30°$ into Equations 3.17.19, which yields

$$E_x = E_y = 8.487 \times 10^6 \text{ psi}$$

$$G_{xy} = 2.361 \times 10^6 \text{ psi}$$

$$v_{xy} = v_{yx} = 0.2750$$

$$\eta_{x,xy} = -\eta_{y,xy} = 0.1258$$

$$\eta_{xy,x} = -\eta_{xy,y} = 0.4521$$

Considerations of anisotropic material behavior are usually not necessary if we restrict ourselves to structures composed of common metal alloys such as steel, aluminum and titanium. Some manufacturing processes (e.g., rolling, extruding and forging) may induce anisotropy in such metals, but we will neglect these effects. On the other hand, the anisotropy inherent in materials such as fiber-reinforced composites, laminates, and honeycomb sandwich structures cannot be ignored. The basic structural analysis methods developed in this text can be extended to anisotropic materials after their application to isotropic materials has been thoroughly understood. Therefore, although we recognize the increasing use of composites in modern aerospace structures, our focus will remain on isotropic materials.

3.18   STRESS CONCENTRATION AND FATIGUE

Figure 3.18.1 shows a straight, homogeneous, flat bar with a rectangular cross section of area $A = wt$, subjected to an axial load P directed along its longitudinal axis of symmetry (the x axis). Except for the regions in the vicinity of the load application points at each end, the state of stress throughout the bar is uniform, and is precisely

$$\sigma_x = \frac{P}{A} \qquad \sigma_y = \sigma_z = \tau_{xy} = \tau_{xz} = \tau_{yz} = 0 \qquad \text{[3.18.1]}$$

Suppose a circular hole of diameter d is drilled through the thickness of the bar at the center of its lateral surface ($x = y = 0$). Having thereby reduced the load-bearing area from wt to $(w - d)t$, we know that for the same load P, the axial stress σ_x at the middle of the bar must increase. On a transverse section through the center of the hole, the *nominal* stress, assumed to be uniformly distributed over the reduced or *net* area, is

$$\sigma_{\text{nom}} = \frac{P}{A_{\text{net}}} = \frac{P}{A(1 - \frac{d}{w})} \qquad \text{[3.18.2]}$$

As the diameter of the hole approaches the width of the bar, this stress becomes very large. This formula also implies that reducing the hole diameter to a very small but nonzero value would result in a stress σ_x that is nearly the same as if there were no hole at all. However, the theory of elasticity reveals that in the vicinity of the hole, the stress field is significantly more complex than that given by Equation 3.18.2. In reality, if the diameter of the hole is very small compared to the width of the bar ($d/w \ll 1$), the maximum stress in the bar is $\sigma_x = 3(P/A)$ at the top and bottom of the hole.[22] Moving away from the hole in the y direction, the normal stress σ_x drops rapidly toward the nominal value, as illustrated in Figure 3.18.2. The maximum stress varies with the size of the hole relative to the width of the bar. In general, $\sigma_{\text{max}} = K_t \sigma_{\text{nom}}$, or

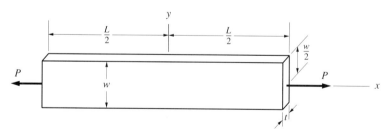

Figure 3.18.1   Flat bar of width w and thickness t in uniaxial tension.

[22]S. Timoshenko, and J. Goodier, *Theory of Elasticity*, 2nd ed. Chapter 4, New York, McGraw-Hill, 1951.

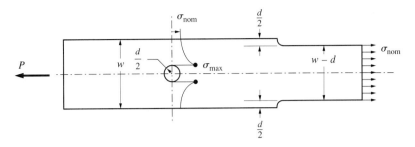

Figure 3.18.2   Small, central hole in a flat bar in uniaxial tension.

The ratio of the theoretical maximum stress at the hole to the nominal stress is the stress concentration factor.

$$K_t = \frac{\sigma_{max}}{\sigma_{nom}}$$ [3.18.3]

where K_t is the *stress concentration factor*. For our case, the variation in K_t with hole diameter is illustrated in Figure 3.18.3.[23]

Figure 3.18.4 shows a central elliptical hole in a plate that is very wide compared to the length $2a$ of the major axis of the ellipse. If a uniform tensile stress σ_{nom} is applied normal to the long axis of the ellipse, the maximum stress in the plate occurs at each end of the major axis. The stress concentration factor is[24]

$$K_t = 1 + 2\frac{a}{b}$$ [3.18.4a]

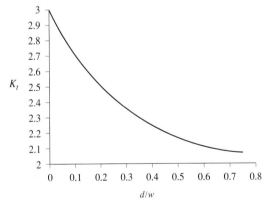

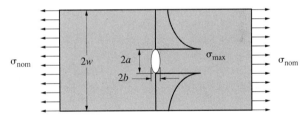

Figure 3.18.4   Stress concentration at the ends of an elliptical hole.

Figure 3.18.3   Stress concentration factor K_t for a central hole of diameter d in a thin, flat bar of width w in uniaxial tension (Figure 3.18.2).

[23]R. J. Roark, and W. C. Young, *Formulas for Stress and Strain*, 5th ed., New York, McGraw-Hill, 1975.
[24]S. Timoshenko, and J. Goodier, *op cit.*

This reduces to the K_t for a central circular hole if $a = b$. The radius of curvature ρ of an ellipse at the ends of its major axis is b^2/a (and a^2/b at the ends of the minor axis), so that Equation 3.18.4a may be written

$$K_t = 1 + 2\sqrt{\frac{a}{\rho}}$$

[3.18.4b]

Clearly, if a is held fixed while b is made very small, the curvature at each end of the major axis becomes extremely sharp ($\rho \to 0$), and the stress concentration increases without bound as the ellipse approaches the shape of a thin crack in the middle of the plate. The formation and presence of cracks are a matter of grave concern, which we will discuss further later in this section.

Stress concentration factors for elastic materials have been obtained mathematically and experimentally for a variety of geometries and load conditions encountered in machine design and structural design. The stress rise induced at intentional geometrical aberrations such as holes, notches, dimples, grooves, keyways, shoulders, etc., can be accounted for in design by reference to formulas or charts.[25] Consideration of stress concentrations depends on the type of material and the nature of the loading.

Ductile metal alloys under steadily increasing loads can undergo significant deformation after yielding and before reaching their ultimate stress. On the other hand, brittle materials experience very little (if any) plastic flow before fracturing. The two extreme cases are the elastic, perfectly-brittle material, with a stress–strain curve like that in Figure 3.18.5a, and the elastic, perfectly-plastic material of Figure 3.18.5b. Real engineering materials exhibit behavior in between these two idealized extremes, as illustrated in Figure 3.18.5c. Suppose the central hole drilled in a flat bar is very small, so that $K_t = 3$. Furthermore, let the bar be made of an elastic, perfectly-brittle material with an ultimate stress σ_{ult}. As the applied load is increased, the stress at the hole will reach σ_{ult} when the bulk of the material is at a nominal stress of $\frac{1}{3}\sigma_{\text{ult}}$, far below the breaking point. If the load is increased a bit more, the material in the vicinity of the hole will fracture, leading to complete fracture of the bar if the load is not removed. Thus, the stress riser dramatically reduces the load-carrying capability of the

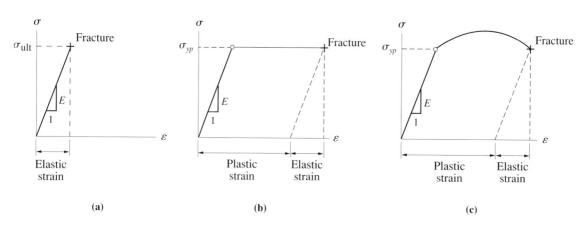

Figure 3.18.5   (a) Elastic, perfectly-brittle material. (b) Elastic, perfectly-plastic material. (E is Young's modulus.) (c) Ductile material exhibiting plastic work-hardening prior to fracture.

[25]For an extensive compilation, see R. E. Peterson, *Stress Concentration Factors*, New York, Wiley, 1974.

brittle bar. On the other hand, if the bar is made of an elastic, perfectly *plastic* material, then when the nominal stress reaches the yield stress σ_{yp}, plastic flow will begin in the neighborhood of the hole, confining the stress in that region to σ_{yp}. A further increase in the load will propagate the yield zone farther into the surrounding material. Ultimately, most of the cross section will be at the yield stress, with the bar retaining its structural integrity. Therefore, the presence of the hole will have little effect on the strength of the ductile bar. This simple example illustrates why the effects of stress concentration may be ignored in ductile materials subjected to static or steadily increasing loads. High-strength materials tend to have low ductility and are therefore sensitive to small cracks or flaws.

Aircraft are a clear example of structures that must endure nonsteady, repeated loading. A tensile test specimen in which the axial stress cycles continuously between a maximum and a minimum value may eventually break at a value of maximum stress well below the ultimate stress in a static test. Such a failure is known as a *fatigue* failure. Although the familiar desktop example of a fatigue test is bending a paper clip back and forth a number of times until it snaps in two, in engineering structures, fatigue failure occurs at ordinary working strain levels.

The fatigue properties of metallic materials used in aerospace vehicles appear in *MIL-HDBK-5*[26] in the form of *S–N diagrams* (stress versus number of cycles to failure) or *ε–N diagrams* (strain versus number of cycles to failure). The fatigue data presented in that document are for uniaxial test specimens (round or flat bars) subjected to constant-amplitude load cycling, which produces a stress history like that illustrated in Figure 3.18.6. The results of such tests can be presented as shown in Figure 3.18.7, which is for a common aluminum alloy. Clearly

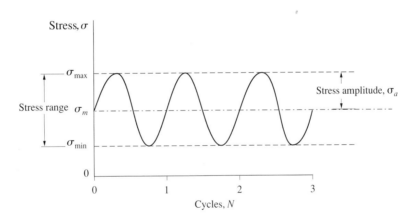

Figure 3.18.6   Typical fatigue loading (load control). σ_m is the mean stress.

and typically, the fatigue life diminishes at higher stress levels. Yet, even at relatively low stresses, the fatigue life in this particular alloy, though quite long, is still finite. Some ferrous alloys exhibit an *endurance limit,* a stress below which failure due to repeated loading will never occur. This endows a part with *infinite life.*

It is true of all metal alloys that a tensile mean stress reduces the fatigue life. As Figure 3.18.7 shows, the fatigue life at a given stress amplitude σ_a goes down as the average stress σ_m goes up. This is not surprising,

| [26]*Military Handbook 5G,* op. cit.

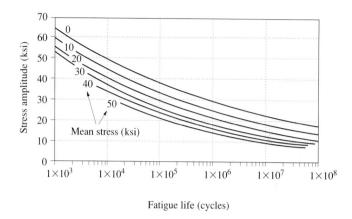

Figure 3.18.7   S–N curves for unnotched 2024-T4 aluminum tensile test specimens from correlated data presented in MIL-HDBK-5G.

since the material must endure higher peak stress when the mean stress is raised. On the other hand, *compressive* mean stress lengthens fatigue life.[27]

In actual service, a part may experience cyclic loading at a variety of stress amplitudes. One means of predicting fatigue life for variable amplitude loading is the *Palmgren-Miner* method.[28] This assumes that n_i cycles of a stress S_i for which the fatigue life is N_i uses up the fraction n_i/N_i of the total fatigue life of the part. n_i/N_i is called the *cycle ratio*. The damage associated with k stress levels is given by summing up the individual cycle ratios. The damage associated with k stress levels is given by

$$D = \sum_{i=1}^{k} \frac{n_i}{N_i}$$

A fatigue crack forms when $D = 1.0$.

Example 3.18.1   An unnotched, previously unstressed part made of 2024-T4 aluminum alloy is subjected to an alternating stress σ_a of 30 ksi and a variable mean stress as follows:

$$\sigma_m = \ \ 0 \, \text{ksi for } 30{,}000 \text{ cycles}$$
$$\sigma_m = 10 \, \text{ksi for } 20{,}000 \text{ cycles}$$
$$\sigma_m = 20 \, \text{ksi for } 10{,}000 \text{ cycles}$$
$$\sigma_m = 30 \, \text{ksi for} \ \ \ 7000 \text{ cycles}$$
$$\sigma_m = 40 \, \text{ksi for} \ \ \ 3000 \text{ cycles}$$
$$\sigma_m = 50 \, \text{ksi for} \ \ \ 1000 \text{ cycles}$$

[27]J. Bannantine, J. Comer, and J. Handrock, *Fundamentals of Metal Fatigue Analysis*, Englewood Cliffs, NJ: Prentice-Hall, 1990.
[28]M. C-Y. Niu, *Airframe Structural Design*, Chapter 15, Hong Kong, Conmilit Press, 1988.

According to Palmgren-Miner, what percentage of the fatigue life of the part remains? What is the remaining fatigue life for an alternating stress of 45 ksi together with a mean stress of 0 ksi?

Using the given data plus the fatigue life information presented in Figure 3.18.7, we can develop the following table:

Mean Stress, σ_m (ksi)	Stress Amplitude, σ_a (ksi)	Fatigue Life, N (cycles)	Number of Cycles, n	Cycle Ratio n/N
0	30	955,000	30,000	0.0314
10	30	272,000	20,000	0.0735
20	30	103,000	10,000	0.0971
30	30	46,400	7000	0.151
40	30	23,700	4000	0.169
50	30	13,200	1000	0.0756
			$\sum n/N$	= 0.598

Since $\sum_{i=1}^{6} n_i/N_i = 0.598$, nearly 60 percent of the part's fatigue life has been consumed. For continued use at $\sigma_a = 45$ ksi and $\sigma_m = 0$, Figure 3.18.7 yields a fatigue life of $N = 24{,}000$ cycles. The part will therefore presumably fail after

$$(1 - 0.598) \times 24{,}000 = 9650 \text{ cycles}$$

To account for uncertainties affecting the fatigue strength, a safety factor or *scatter factor* should be applied to this result. This lies in the range of 2 through 4, depending on the agency governing the design procedures. A scatter factor of 3 reduces the remaining predicted fatigue life of the part to 3200 cycles.

To determine the fatigue life of an entire airplane, one must first identify the principal components of the aircraft structure, which are the elements that carry the significant portions of the flight loads and are critical to the structural integrity of the airplane. The failure of any one component would be catastrophic. Examples of critical structural elements are the wing attachments fittings, spar caps and spar webs, door and window frames and the reinforcement around other cutouts, fuselage frames, and major splices. The critical substructures are analyzed to determine the cyclic stresses corresponding to load histories that the airplane will experience during its intended usage. The main types of unsteady loading are taxiing loads, gust and maneuvering loads during flight, and impact loads that occur during landing.

Figure 3.18.8a shows an airplane in unaccelerated, straight and level flight, so that its weight W and the net lifting force L act vertically through the center of mass G. If the lift is increased due to a gust or a pitching maneuver, the center of mass G will acquire an acceleration a_n normal to the level flight path. According to Newton's second law, $L - W = (W/g)a_n$, in which g is the acceleration of gravity. Solving for a_n yields

$$a_n = (n-1)g$$

where $n = L/W$, and n is called the *load factor*. In straight and level flight the load factor is 1.0, and the stresses in the airframe are called the "1.0 g" flight stresses. Let the critical 1.0 g stress at the wing root be denoted σ_0. Using formulas, we find σ_0 in terms of the internal shear V and the bending moment M, which are shown on the free-body diagram in Figure 3.18.8b.

Gust encounters and maneuvers cause the load factor to vary, as illustrated in Figure 3.18.9. This in turn superimposes a varying stress on the steady stress σ_0. Data gathered over many flight hours for a particular type

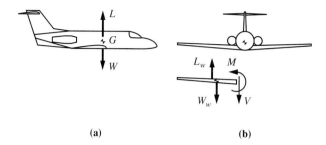

(a)   **(b)**

Figure 3.18.8   (a) Lift balances, weight in straight and level flight. (b) When $n = 1$, the shear and bending moment at the wing root are in equilibrium with the resultant lift L_w and weight W_w of the wing.

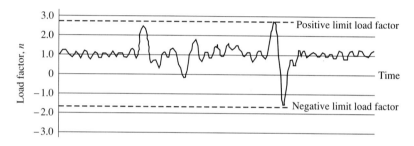

Figure 3.18.9   Load factor history.

Exceeding the limit load factors is likely to damage the airframe.

of airplane yield load spectra that can be presented in a format similar to that illustrated in Figure 3.18.10.[29, 30] The symbol n^+ stands for load factors greater than 1.0, and n^- represents load factors less than 1.0. The cumulative exceedance $C(n)$ is the expected total number of times that the load factor n will be exceeded per unit flight time interval (per hour in Figure 3.18.10). The part of the load spectrum labeled $C^+(n)$ is defined over the range of n^+, and $C^-(n)$ is defined over the range of n^-, which lies to the left of $n = 1.0$.

If the load factor fluctuates steadily between n^+ and n^-, the stress varies as shown in Figure 3.18.6, and the maximum and minimum stresses are found in terms of the load factors and the 1.0 g stress by means of the relations

$$\sigma_{\max} = n^+\sigma_0$$

$$\sigma_{\min} = n^-\sigma_0$$

It follows that the amplitude σ_a and the mean value σ_m of the cyclic stress are given by

---

[29]"Fatigue Evaluation of Wing and Associated Structures on Small Airplanes," *Report No. AFS–120–73–2,* Federal Aviation Administration, Washington, DC, May 1973

[30]J.E. Locke, H. W. Smith, E. A. Gabriel and T. DeFiore, "General Aviation Aircraft Normal Acceleration Data Analysis and Collection Project," *Report No. DOT/FAA/CT–91/20,* Federal Aviation Administration, Technical Center, Atlantic City, NJ, February 1993.

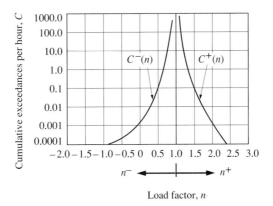

Figure 3.18.10   Typical maneuvering or gust load spectrum for a given type of aircraft, a specified cruise speed, and prescribed limit load factors.

$$\sigma_a = \frac{\sigma_{max} - \sigma_{min}}{2} = \frac{n^+ - n^-}{2}\sigma_0$$

[3.18.5]

$$\sigma_m = \frac{\sigma_{max} - \sigma_{min}}{2} = \frac{n^+ + n^-}{2}\sigma_0$$

To estimate the life of the part in question, the domain of the $C^+(n)$ cumulative exceedance curve is partitioned into k small intervals Δn_i, $i = 1, \cdots, k$. As illustrated in Figure 3.18.11a, the load factor at the center of the ith interval is denoted n_i^+. ΔC_i denotes the frequency of occurrence of load factors in the range Δn_i. The value of n_i^- corresponding to n_i^+ is that load factor (less than 1.0) for which $C^- = C^+$. Using n_i^+ and n_i^-,

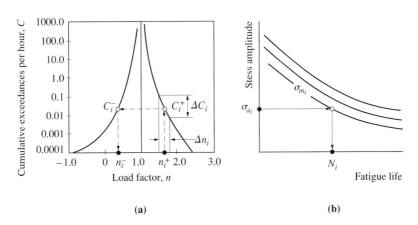

(a)   (b)

Figure 3.18.11   (a) Load spectrum, with the domain of C^+ partitioned into intervals Δn_i (b) S–N curves for the material of which the structural component is made.

Equation 3.18.5 yields the mean and varying stresses, σ_{m_i} and σ_{a_i}. These stresses are then used to obtain the fatigue life N_i from the $S–N$ curves for the material of which the part is made, as shown in Figure 3.18.11b.

If L is the fatigue life of the structural element under the type of loading represented by the spectrum in Figure 3.18.10, then $L\Delta C_i$ is the contribution, in number of cycles, of stress level i to the fatigue damage. Therefore, the cycle ratio at the ith factor is $L\Delta C_i/N_i$. According to Palmgren-Miner theory, for the complete load spectrum,

$$\sum_{i=1}^{k} \frac{L\Delta C_i}{N_i} = 1.0$$

Factoring out L, we obtain the fatigue life as

$$L = \frac{1.0}{\sum\limits_{i=1}^{k} \Delta C_i/N_i}$$

The units are hours if, as previously assumed, ΔC_i is cycles per hour.

The fatigue lives $L_1, L_2, L_3, \cdots$ for several different, relevant load spectra are computed as outlined above. The total fatigue life of the component is then found as

$$L_{\text{total}} = \frac{1}{\sum\limits_{i=1}^{m} \frac{1}{L_i}}$$

where m is the number of different types of fatigue loadings considered. The fatigue study must be repeated for every principal structural component of the aircraft. In the same way that a chain is only as strong as its weakest link, the safe life of the airplane is the shortest fatigue life of all the critical components. That computed life must be reduced by a scatter factor as high as 8.0 for civil aircraft in order to account for analytical uncertainty.

Assessing the fatigue life of an existing airplane is fairly straightforward following the procedure just described. The inverse problem—of designing an airplane to have a specified fatigue life—is much more difficult. In the conceptual design phase, nearly all of the information required to estimate the safe life of the airframe is nonexistent. The principal structural elements have yet to be identified and designed, so their size, shape, and material properties are unknown. Some approaches to safe-life aircraft design have been proposed.[31]

Fatigue failure, even in ductile metals, resembles brittle fracture. The failure begins as a microscopic flaw or crack in what might appear to be a notch-free part. If notches are present, fatigue failure initiates in regions of high stress concentration. Thus, the fatigue life of a member—ductile or not—is sensitive to stress risers. The *fatigue notch factor* K_f is a measure of this sensitivity, and is defined as

$$K_f = \frac{\sigma_f^{(\text{unnotched})}}{\sigma_f^{(\text{notched})}}$$

[3.18.6]

where, for a given fatigue life, $\sigma_f^{(\text{notched})}$ is the fatigue strength of a notched specimen and $\sigma_f^{(\text{unnotched})}$ is that of a specimen free of flaws.

---

[31]See, for example, J. G. Ladesic, *Life Cycle Design of Future General Aviation Airplanes*, AIAA/FAA/Mississippi State University Third Joint Symposium on General Aviation Systems, Starkville, Mississippi, May 1994.

Equation 3.18.6 bears some resemblance to the formula for stress concentration factor, Equation 3.18.3. However, whereas K_t depends only on the geometry and the nature of the applied load, the fatigue notch factor depends strongly on the material as well. This is evident from Figure 3.18.12, which illustrates that, aside from the fact that K_f increases with K_t, as we should expect, there is no clear relationship between the two factors. Evidently, notch sensitivity increases with the strength of an alloy.[32]

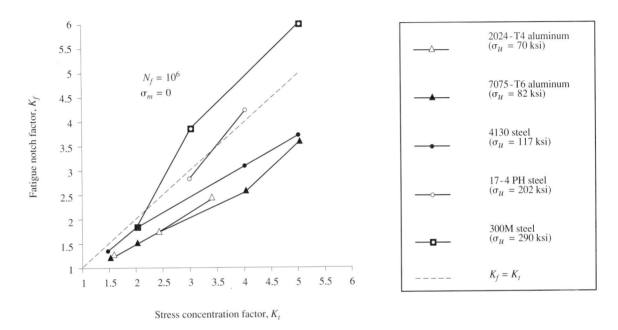

Figure 3.18.12   Fatigue notch factor versus stress concentration factor for several aluminum and steel alloys, using room-temperature data presented in *MIL-HDBK–5G* for fully-reversed loading and a fatigue life of 10^6 cycles. (σ_u is the ultimate static strength.)

For fatigue failure to occur, a crack must form at a notch or flaw and then propagate through the component, eventually becoming large enough to destroy its structural integrity. Fatigue life of a structural component therefore consists of crack initiation followed by crack growth. The study of crack propagation falls within the realm of *fracture mechanics*. Linear elastic fracture mechanics applies linear elasticity theory to the problem of solids with cracks or flaws. The load to which a crack is subjected may be considered a combination of one or more of the three modes illustrated in Figure 3.18.13.

Mode I is the *opening* mode in which tensile loads normal to the crack surfaces tend to pull those surfaces apart. Mode II is the *sliding* mode wherein shear force slides the crack surfaces over each other in the direction perpendicular to the leading edge of the crack. Mode III is the *tearing* mode, in which the crack surfaces slide over each other in the direction parallel to the leading edge. For each of these three loading modes, it can be shown,[33] using the theory of elasticity for isotropic linearly-elastic materials, that the stresses in the immediate neighborhood of a sharp crack tip are of the general form

[32]J. Bannantine et al., *op cit.*

[33]T. L. Anderson, *Fracture Mechanics: Fundamentals and Applications*, Boca Raton, FL, CRC Press, 1995.

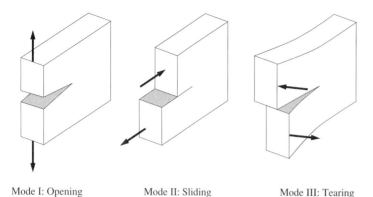

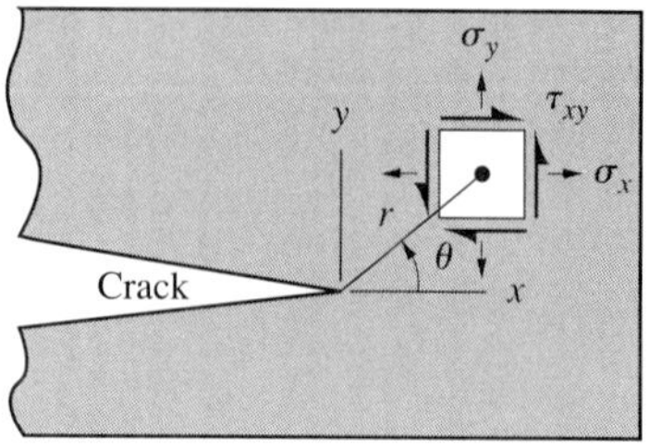

Figure 3.18.14   Stress field at a crack tip.

Mode I: Opening   Mode II: Sliding   Mode III: Tearing

Figure 3.18.13   The three basic modes of loading a crack.

$$\sigma_x = \frac{K}{\sqrt{2\pi r}}F(\theta) + \cdots \qquad \sigma_y = \frac{K}{\sqrt{2\pi r}}G(\theta) + \cdots \qquad \tau_{xy} = \frac{K}{\sqrt{2\pi r}}H(\theta) + \cdots \qquad \textbf{[3.18.7]}$$

where r and θ are polar coordinates centered at the crack tip, as illustrated in Figure 3.18.14.

In Equation 3.18.7 "$\cdots$" represents terms that depend on r and θ and the geometry, but they are finite at $r = 0$. Therefore, the $1/\sqrt{r}$ terms dominate the the stress field near the tip of the crack in what is called the *singularity zone*. Their magnitude is determined by K, the *stress intensity factor*, which is not to be confused with the stress concentration factor K_t and the fatigue notch factor K_f previously defined. Also, F, G, and H are dimensionless functions, the specific forms of which depend on the loading mode (I, II, or III). The stress intensity factor also depends on the type of loading, as well as the size and shape of the crack and the specimen geometry. For a given mode, the stress intensity factor may be expressed in the form

$$K = \sigma \sqrt{\pi a}Y \qquad \textbf{[3.18.8]}$$

where σ is the *remote* stress applied to the flawed section, a is a measure of the size of the crack or flaw, and Y is a nondimensional factor that relates the flaw size to the geometry of the specimen. Observe that the units of K are [stress $\times \sqrt{\text{length}}$]. Some analytical solutions are available for Y, but in recent years, the finite element method has become an important means of obtaining this information.

Figure 3.18.15 shows the stress intensity factors for several different crack geometries in Mode I loading.[34] For example, the stress intensity factor for a center-cracked plate in tension (Figure 3.18.4, with $b \ll a$) is

$$K_I = \sigma \sqrt{\pi a}\sqrt{\sec\left(\frac{\pi a}{2w}\right)} \qquad \text{CCT} \qquad \textbf{[3.18.9]}$$

According to linear elastic fracture mechanics, a crack begins to propagate when the stress intensity factor reaches a critical value K_c, known as the *fracture toughness*. The fracture toughness is therefore a measure of a material's resistance to fracture. For Mode I loading, the fracture toughness is denoted K_{Ic}.

Equations 3.18.7 would have us believe that near the tip of the crack, where r approaches zero, the stresses become infinite. However, these continuum-mechanics based formulas, like Equations 3.18.4, are not valid down at the level of molecular and atomic microstructure. Extending Equation 3.18.4b to that level would limit the

[34]T. L. Anderson, *op. cit.*

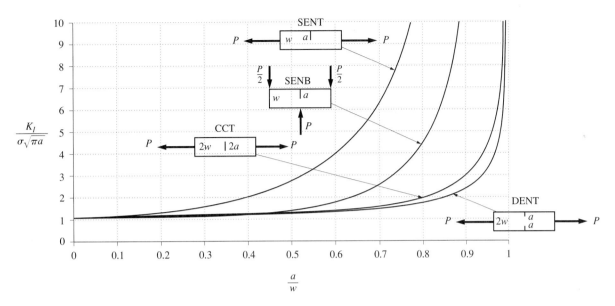

Figure 3.18.15   Mode I stress intensity factors for *single-edge notched tension* (SENT), *double-edge notched tension* (DENT), *single-edge notched bending* (SENB), and *center-cracked tension* (CCT) specimens. (*a* gives the flaw size and *w* relates to the specimen width.)

radius of curvature ρ at the crack tip to something on the order of an atomic radius, which, though quite small, is not zero. What we have near the tip of a crack is a region in which the large stresses produce a localized damage zone into which the crack tip may advance.

In metals, plastic flow within the damage zone causes a redistribution of the stresses, lowering the stress intensity and blunting the crack growth. Plastic flow occurs more readily in regions of plane stress than in those of plane strain. Some insight into this fact may be gained by applying a yield criterion—the maximum distortion energy theory—to plane stress and plane strain states. Recall that in plane stress (in the xy plane), we have

$$\sigma_z = \tau_{xz} = \tau_{yz} = 0$$

The normal strain ε_z perpendicular to the plane occurs freely, and Equations 3.12.1 reveal that for isotropic materials, in the absence of thermal effects, ε_z is related to the in-plane normal strains ε_x and ε_y by the formula

$$\varepsilon_z = -\frac{\nu}{1-\nu}(\varepsilon_x + \varepsilon_y)$$

Plane stress exists in thin sheets of material and on the traction-free surfaces of any solid. Plane stress is a biaxial stress state, since only the two in-plane normal stresses are present. On the other hand, the definition of plane strain is

$$\gamma_{xz} = \gamma_{yz} = \varepsilon_z = 0$$

The enforcement of zero normal strain in the z direction requires a nonzero value of the normal stress σ_z, giving rise to a *triaxial* state of stress. In fact, setting $\varepsilon_z = 0$ in Hooke's law, Equation 3.11.4$_3$, and ignoring thermal effects yields

$$\sigma_z = \nu(\sigma_x + \sigma_y)$$

Substituting this and $\tau_{xz} = \tau_{yz} = 0$ into the formula for von Mises stress, Equation 3.15.5, we find

$$\sigma_{VM})_{\text{plane strain}} = \sqrt{(\sigma_x - \sigma_y)^2 + \sigma_x\sigma_y + 3\tau_{xy}^2 - v(1-v)(\sigma_x + \sigma_y)^2} \qquad \text{plane strain} \qquad \textbf{[3.18.10a]}$$

The von Mises stress for plane stress, Equation 3.15.9, is

$$\sigma_{VM})_{\text{plane stress}} = \sqrt{(\sigma_x - \sigma_y)^2 + \sigma_x\sigma_y + 3\tau_{xy}^2} \qquad \text{plane strain} \qquad \textbf{[3.18.10b]}$$

Clearly, for the same set of in-plane stress components σ_x, σ_y, and τ_{xy}, we see that

$$\sigma_{VM})_{\text{plane strain}} \leq \sigma_{VM})_{\text{plane stress}}$$

In other words, according to the minimum distortion energy criterion, a state of stress that causes yielding under biaxial conditions will not generally give rise to plastic flow under plane strain (triaxial) conditions.

Figure 3.18.16 shows Mode I loading of two center-cracked tensile specimens, one of them thin ($t \ll 2w$), the other quite thick ($t \approx 2w$). Plane stress prevails near the tip of the crack in the thin sheet. In the thick plate, the bulk of the material in the interior resists strain normal to its lateral surfaces, giving rise to plane strain throughout most of the width of the crack ($A - A'$). In a thick specimen, the triaxial stress state at the crack tip confines plastic flow to a region that is much smaller than the plane stress plastic zone surrounding the crack tip in a thin sheet. Therefore, in the thick specimen, plastic flow presents less resistance to crack propagation than in the thin one. Propagation is due primarily to the rapid onset of brittle fracture, whereas in plane stress, the existence of a significant plastic zone permits slow, stable crack growth. Thin specimens therefore tend to have higher fracture toughness than do thick ones.

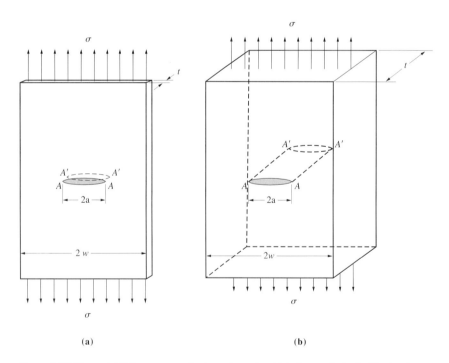

(a)   (b)

Figure 3.18.16   (a) Thin center-cracked specimen in which the crack tip A–A' is in plane stress throughout the thickness t ($t \ll 2w$). (b) Thick specimen in which the bulk of the crack tip A–A' is in triaxial stress (plane strain).

Since stress intensity factor—and therefore *critical* stress intensity factor K_c—is meaningful only if the material near the crack tip behaves elastically linear, data reported (e.g., in *MIL-HDBK–5*) for the fracture toughness K_{Ic} of a material in Mode I loading is for the critical *plane strain stress intensity factor*. The term K_{Ic} is called the *plane strain fracture toughness*. *MIL-HDBK–5G* lists measured K_{Ic} values for aluminum alloys (2000 and 7000 series) ranging from $18,000$ to $47,000 \text{ psi}\sqrt{\text{in.}}$, depending on the alloy, the heat treatment, and the product form (plate, forging, extrusion, etc.). For this same series of aluminum alloys, the yield strength ranges from 27,000 to 88,000 psi. A fracture toughness of $60,000 \text{ psi}\sqrt{\text{in.}}$ is typical of titanium alloy (Ti–6Al–4V), for which the yield strength is 130,000 psi. The yield strength of aircraft-grade steels exceeds 185,000 psi, and for these alloys, values of from $60,000 \text{ psi}\sqrt{\text{in.}}$ to $130,000 \text{ psi}\sqrt{\text{in.}}$ for K_{Ic} are reported.

Since aerospace vehicles comprise much in the way of thin sheets of material, a means of characterizing nonplane strain fracture is required. The crack growth in a thin center-cracked sheet is characterized by the crack growth curve shown in Figure 3.18.17, in which a_o is the half-length of the initial fatigue crack caused by cyclic loading. Steady load application results in slow, stable growth of the crack beyond point O. (In brittle fracture, rapid, unstable crack growth begins at once.) At any point between O and C, the specimen, though damaged by a crack of increasing length, is nevertheless capable of sustaining increased load without catastrophic failure. At point C, however, the crack reaches the critical length at which unstable, brittle-like fracture occurs. The nonlinear, inelastic behavior precludes defining a critical stress intensity factor; however, Equation 3.18.8 can be used as a basis for defining an *apparent* fracture toughness K_{app},

$$K_{app} = \sigma_c \sqrt{\pi a_o} Y_o \qquad \text{[3.18.11]}$$

where, for a center-cracked tension plate (cf. Equation 3.18.9),

$$Y_o = \sqrt{\sec\left(\frac{\pi a_o}{2w}\right)}$$

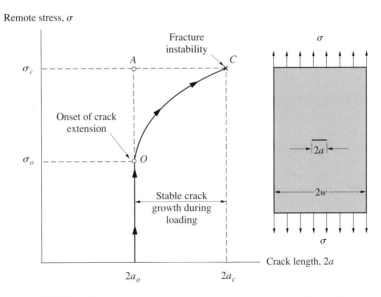

Figure 3.18.17 Crack growth curve for a thin sheet (plane stress fracture).

Source: *MIL-HDBK–5G.*

Here, a_o is the initial fatigue precrack length and σ_c is the stress at the end of stable crack growth and the onset of fracture.

The formation of a fatigue crack does not necessarily mean a part is in danger of imminent failure. Failure occurs when the crack attains a critical size, depending on K_{Ic}. Safe-life or *damage-tolerant* design allows for the presence of subcritical cracks that will not grow to critical length between periodic inspections.

Much of the fatigue life of a part may comprise the growth of a crack. Extending the service life of a structural component by allowing monitored crack propagation obviously requires data on the characteristics of crack growth for the constituent material. This information is obtained by subjecting specimens to constant-amplitude cyclic loading in which the remote stress σ varies as illustrated in Figure 3.18.6. Crack length a is measured as a function of the number of cycles N at various levels $\Delta\sigma$ of the applied stress, where $\Delta\sigma = \sigma_{max} - \sigma_{min}$. The results of such tests often appear in the form of crack growth rate curves, which are log-log plots of the fatigue crack propagation rate da/dN against the stress intensity factor range ΔK, where ΔK is obtained from Equation 3.18.8, as follows:

$$\Delta K = \Delta\sigma \sqrt{\pi a}\, Y \qquad\qquad \textbf{[3.18.12]}$$

These curves typically have the appearance illustrated in Figure 3.18.18. Region I is the threshold of fatigue crack growth. Below this stress intensity factor range, there is essentially no crack propagation. In Region II, the data may often be represented by a straight line on the log-log plot, which indicates a power-law relation known as the *Paris equation*, which is:

$$\frac{da}{dN} = C(\Delta K)^n$$

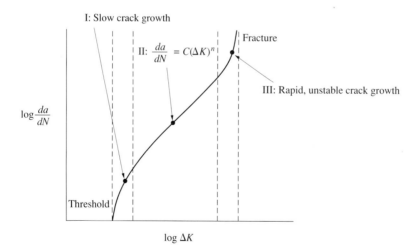

Figure 3.18.18   General form of crack growth rate curves, showing the three characteristic regions.

EXERCISES

3.1   If the surface traction on AB is $\mathbf{T}^{(n)} = 100\,\mathbf{i} + 300\,\mathbf{j}$ MPa, find the normal and shear components of stress on AB.

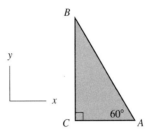

Exercise 3.1

3.2   If $\sigma_x = 100$ MPa, $\sigma_y = 200$ MPa, and $\tau_{xy} = 300$ MPa, find the normal stress and shear stress on surface AB of Exercise 3.1. Check your result using Mohr's circle.

3.3   The state of plane stress in the triangular sheet of thickness t is

$$\sigma_x = \frac{2\sigma_o}{L^2}xy \qquad \sigma_y = 0 \qquad \tau_{xy} = -\frac{\sigma_o}{L^2}y^2$$

where σ_o is a constant having units of stress. Calculate the resultant force normal to the inclined edge 1–2 and its line of action.

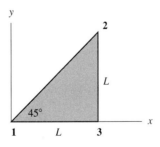

Exercise 3.3

3.4 Relative to an *xyz* cartesian coordinate system, the components of the stress tensor are

$$[\boldsymbol{\sigma}] = \begin{bmatrix} 10,000 & 5000 & -5000 \\ 5000 & 3000 & -2000 \\ -5000 & -2000 & 7000 \end{bmatrix} \text{(psi)}$$

Find the components of the stress tensor $[\bar{\boldsymbol{\sigma}}]$ in the $\bar{x}\,\bar{y}\,\bar{z}$ system whose axes point in the direction of the following three vectors, respectively:

$$\mathbf{X} = \mathbf{i} + \mathbf{j} + \mathbf{k} \qquad \mathbf{Y} = \mathbf{i} + \mathbf{j} - 2\mathbf{k} \qquad \mathbf{Z} = -\mathbf{i} + \mathbf{j}$$

Check your work by confirming that the stress invariants $I_1, I_2,$ and I_3 are the same for both $[\boldsymbol{\sigma}]$ and $[\bar{\boldsymbol{\sigma}}]$.

3.5 Calculate the principal stresses, principal stress directions, and the maximum shear stress for the following state of stress:

$$\begin{bmatrix} 100 & -200 & 300 \\ -200 & -100 & 400 \\ 300 & 400 & -500 \end{bmatrix} \text{(MPa)}$$

3.6 The state of plane stress at a point P is shown in part (a) of the figure.

a.   Determine the principal stresses and the principal stress directions. Sketch the in-plane principal stresses (i.e., the principal stress components lying in the *xy* plane) on a stress element properly oriented relative to the *xy* axes, as shown in part (b) of the figure.

b.   Determine the maximum shear stress τ_{max} and the maximum in-plane shear stress $\tau_{xy\,max}$. Sketch the state of maximum in-plane shear stress on a properly-oriented stress element, as suggested in part (c) of the figure.

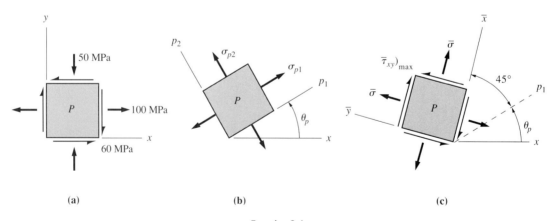

(a)   (b)   (c)

Exercise 3.6

3.7   Repeat Exercise 3.6 for the state of plane stress shown in the figure.

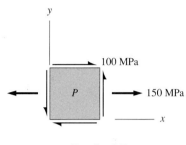

Exercise 3.7

3.8   The stresses on an equilateral triangular element of material in plane stress are as shown in the figure. Find σ_x, σ_y, and τ_{xy}.

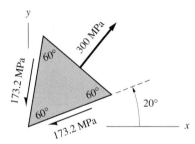

Exercise 3.8

3.9 A 100 mm by 75 mm rectangular sheet deforms as shown.

    a. Calculate the components of strain ε_x, ε_y, and γ_{xy}.

    b. Calculate principal strains, principal strain directions, and maximum shear strain.

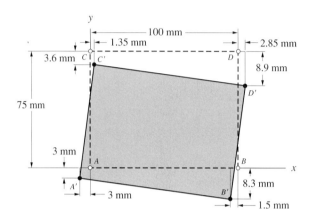

Exercise 3.9

3.10 The strain gage readings are as follows: $\varepsilon_1 = 300\mu$, $\varepsilon_2 = -750\mu$, and $\varepsilon_3 = 500\mu$.

    a. Calculate the components of strain ε_x, ε_y, and γ_{xy}.

    b. Calculate principal strains, principal strain directions, and maximum shear strain.

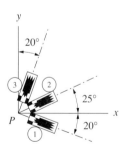

Exercise 3.10

3.11 Find the displacements $u(x, y, z)$, $v(x, y, z)$ and $w(x, y, z)$ corresponding to the following strains, if $u = v = w = 0$ and $\partial u/\partial y = \partial u/\partial z = \partial v/\partial z = 0$ at $x = y = z = 0$. (k is a nonzero constant.)

    a. $\varepsilon_x = ky^2$ $\varepsilon_y = kx^2$ $\gamma_{xy} = 4kxy$ $\varepsilon_z = \gamma_{xz} = \gamma_{yz} = 0$

    b. $\varepsilon_x = ky^2$ $\varepsilon_y = -kx^2$ $\gamma_{xy} = 4kxy$ $\varepsilon_z = \gamma_{xz} = \gamma_{yz} = 0$

3.12   On the surface of an isotropic aluminum plate ($E = 10 \times 10^6$ psi, $\nu = 0.25$), the state of plane stress in the vicinity of point P is $\sigma_x = 15,000$ psi, $\sigma_y = -10,000$ psi, and $\tau_{xy} = 5000$ psi. What is the strain in each gage of the strain gage rosette bonded to the plate at P?

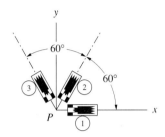

Exercise 3.12

3.13   Determine whether the following state of plane stress within an elastic solid is admissible. (A is a constant.)

$$\sigma_x = 2Ax^2 \qquad \sigma_y = 2A(4x^2 + y^2) \qquad \tau_{xy} = -4Axy$$

3.14   For the axially loaded bar, show that if the displacements are

$$u = \frac{Px}{AE} \qquad v = -\nu\frac{Py}{AE} \qquad w = -\nu\frac{Pz}{AE}$$

then the three equations of equilibrium and six compatibility equations are satisfied, as are the boundary conditions on the stress and the displacement on all six surfaces of the bar.

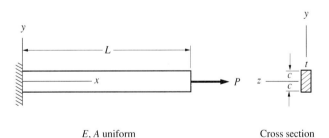

E, A uniform   Cross section

Exercise 3.14

3.15   The state of plane stress in the thin, square plate with Young's modulus E and Poisson's ratio v is

$$\sigma_x = 0 \qquad \sigma_y = 0 \qquad \tau_{xy} = \tau_o$$

where τ_o has the units of stress. Use elasticity theory to find the components of displacement at point A if $u = v = \partial v/\partial x = 0$ at O.

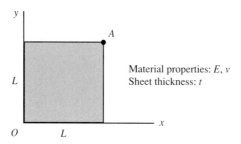

Exercise 3.15

3.16   The components of plane stress within the triangular sheet are

$$\sigma_x = \sigma_o \left[\left(\frac{x}{L} \right)^2 + \frac{y}{L} \right] \qquad \sigma_y = \sigma_o \left[-2 \left(\frac{x}{L} \right)^2 + \left(\frac{y}{L} \right)^2 \right] \qquad \tau_{xy} = -2\sigma_o \frac{xy}{L^2}$$

where σ_o is a constant with units of stress.

a.   Show that these stresses satisfy equilibrium and compatibility.

b.   At point P on the inclined edge, find the component of traction normal to the boundary.

c.   At that same point, find the shear component of traction.

d.   If all interior points of the sheet are in equilibrium [from part (a)], then the forces on its boundary must be in equilibrium. Calculate the surface traction resultants (including lines of action) on each of the three edges and show that, indeed, $\sum F_x = 0$, $\sum F_y = 0$, and $\sum M_o = 0$.

e.   If $u = v = \partial v/\partial x = 0$ at O, calculate the components of displacement at point A in terms of σ_o, v, E, L, and t.

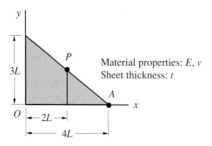

Exercise 3.16

3.17   For a cantilever beam with elastic modulus E loaded by a couple M at the free end, the following stress field is proposed: $\sigma_x = Dy$, $\sigma_y = \sigma_z = \tau_{xy} = \tau_{xz} = \tau_{yz} = 0$. ($D$ is a constant.)

    a.   Show that this stress field satisfies the three equations of equilibrium and the six compatibility equations.

    b.   Show that the top, bottom, front, and back surfaces are traction free for this set of stresses.

    c.   Show that the resultant normal and shear force are zero on the free end.

    d.   Evaluate the constant D such that the resultant couple on the free end equals the given couple M, that is, $M = - \int_{-c}^{c} \sigma_x t y \, dy$.

    e.   Evaluate the displacements of the beam if $u = v = w = 0$ and $\partial u/\partial y = \partial u/\partial z = \partial v/\partial z = 0$ at $x = y = z = 0$.

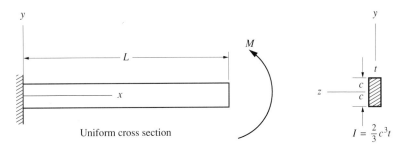

Uniform cross section

$I = \frac{2}{3} c^3 t$

Exercise 3.17

3.18   For a cantilever beam loaded by a shear force at the free end, the following state of plane stress is proposed: $\sigma_x = 2Ay(L - x)$, $\sigma_y = 0$, and $\tau_{xy} = A(y^2 - c^2)$, where A is a constant.

    a.   Show that this stress field satisfies equilibrium and compatibility.

    b.   Show that the stress boundary conditions are satisfied on the top, bottom, and free end, and determine the constant A such that the shear stress resultant on the free end equals the given load P, that is, $P = \int_{-c}^{c} \tau_{xy} t \, dy$.

    c.   Find the displacements $u(x, y)$ and $v(x, y)$ in the beam if $u = v = \partial v/\partial x = 0$ at $x = y = 0$.

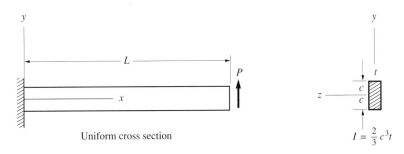

Uniform cross section

$I = \frac{2}{3} c^3 t$

Exercise 3.18

3.19 The state of plane stress in the thin, rectangular region is given by:

$$\sigma_x = \frac{3}{2}\frac{P}{tc^3}y^2 \qquad \sigma_y = -\frac{3}{2}\frac{P}{tc^3}x^2 \qquad \tau_{xy} = 0$$

Find the components of displacement of point A if $u = v = \partial v/\partial x = 0$ at $x = y = 0$.

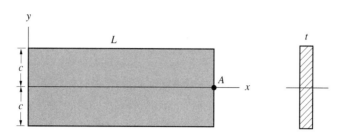

Material properties: E, v

Exercise 3.19

3.20 If the Airy stress function $\phi(x, y)$ is taken to be a complete fifth-degree bivariate polynomial satisfying Equation 3.12.8, then the stress components derived from Equations 3.12.6 may be written as:

$$\sigma_x = 2c_3 + 2c_6x + 6c_7y - 6c_8x^2 + 6c_{10}xy + c_{11}(12y^2 - 6x^2)$$
$$- 10c_{12}x^3 - 6c_{13}x^2y + c_{14}(12xy^2 - 2x^3) + c_{15}(20y^3 - 30x^2y)$$
$$\sigma_y = 2c_1 + 6c_4x + 2c_5y + c_8(12x^2 - 6y^2) + 6c_9xy - 6c_{11}y^2$$
$$+ c_{12}(20x^3 - 30xy^2) + c_{13}(12x^2y - 2y^3) - 6c_{14}xy^2 - 10c_{15}y^3$$
$$\tau_{xy} = -c_2 - 2c_5x - 2c_6y + 12c_8xy - 3c_9x^2 - 3c_{10}y^2 + 12c_{11}xy$$
$$+ 30c_{12}x^2y + c_{13}(-4x^3 + 6xy^2) + c_{14}(-4y^3 + 6x^2y) + 30c_{15}xy^2$$

a. Verify that in the absence of body forces and thermal effects, these third-degree polynomial stress fields satisfy equilibrium and compatibility (Equations 3.6.2 and 3.12.5).

b. Use these stress polynomials to obtain the exact solution of the problem of a simply-supported beam under a uniform load p per unit length, as illustrated in the figure. By symmetry, only one-half of the beam need be analyzed. Apply the following boundary conditions to determine the 15 constants (most of them turn out to be zero) in terms of p, l, c, and t:

   (1) For $0 \le x \le l$ and $y = -c$, $\sigma_y = \tau_{xy} = 0$. (The bottom surface is free of surface traction.)

   (2) For $0 \le x \le l$ and $y = +c$, $\sigma_y = -p/t$, $\tau_{xy} = 0$. (The top surface is free of shear, but has the compressive normal stress of the line load.)

   (3) At $x = l$, $\int_{-c}^{c} \sigma_x t\,dy = \int_{-c}^{c} \sigma_x yt\,dy = 0$. (The simply-supported ends of the beam are free of both axial load and bending moment.)

   (4) For $-c \le y \le c$ at $x = 0$, $\tau_{xy} = 0$ (by symmetry).

Uniform cross section

Exercise 3.20

3.21 The beam of Exercise 3.20 is made of isotropic material with Young's modulus E and Poisson's ratio v. Find the displacements $u(x, y)$ and $v(x, y)$, using the following boundary conditions:

    a.   At $x = y = 0$, $u = \partial v/\partial x = 0$ (by symmetry).

    b.   At $x = \pm l$ and $y = 0$, $v = 0$ (applying the support constraint at the ends to a point on the beam centerline).

3.22 The state of stress at a point is given by

$$\begin{bmatrix} 100 & 60 & \tau \\ 60 & -50 & 0 \\ \tau & 0 & 75 \end{bmatrix} (\text{MPa})$$

The yield stress in tension is 350 MPa.

    a.   Find the value of τ for a margin of safety of 0.12 according to the distortion energy theory of failure.

    b.   For the value of τ found in part (a), what is the margin of safety according to the maximum shear stress theory of failure?

4

Box Beam Stress Analysis

4.1 INTRODUCTION

In this chapter, we will calculate the stresses in three-dimensional, deformable, thin-walled beams of open and closed cross section, with and without taper. A beam is a slender structural member designed to carry transverse loads, that is, loads applied perpendicular to its axis. The axis is a line passing through the centroids of the beam's cross sections, which are perpendicular to that line. The cross-sectional dimensions are substantially smaller than the length of the beam. The transverse loads are transmitted from one end of the beam to the other by shear and bending. If the cross section is properly designed, the beam can also efficiently transmit torsion. Because of the shear, bending, and twisting loads, normal and shear stresses develop on the plane of the cross section. Normal stresses on planes perpendicular to the cross sections are small enough, if they exist, to be ignored.

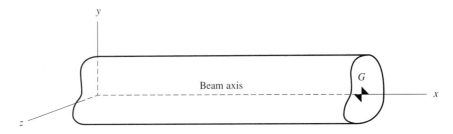

Figure 4.1.1   Three-dimensional beam of arbitrary cross section showing the beam axis passing through the centroid G of each cross section.

Figure 4.1.1 illustrates a cylindrical beam of arbitrary cross section. We will assume for simplicity that the elastic properties of the beam are uniform throughout.

The topics to be reviewed and explored in this chapter are:

- Area moments of inertia and their transformation properties.
- The right-hand sign convention, employed throughout this book.
- The theory of torsion due to Saint-Venant.
- Bernoulli-Euler flexure theory.
- Idealized beams (in which the walls carry only shear).
- Tapered idealized beams.

The Saint-Venant and Bernoulli-Euler displacement hypotheses lead, via Hooke's law, to the formulas required for computing stresses in terms of the applied loads and the geometrical properties of the cross section. That is the extent to which the topic of deformation analysis is covered in this chapter. We focus on stress analysis in otherwise statically determinate beams. Our coverage is therefore limited to single-cell closed sections. The analysis of multicell sections will be presented in Chapter 8, within the context of the general *force method* of analysis of redundant structures.

4.2   AREA MOMENTS OF INERTIA

Let us briefly review the geometric properties of plane areas, which are required for the formulas we will use to calculate stresses and deflections in beams. Consider the plane region of area A shown in Figure 4.2.1. The coordinates of the centroid G relative to the xy coordinate system are obtained from the formulas

$$x_G = \frac{1}{A} \iint_A x\, dA \qquad y_G = \frac{1}{A} \iint_A y\, dA \qquad\qquad \textbf{[4.2.1]}$$

The second moments of area A about the xy axes are given by

$$I_x = \iint_A y^2\, dA \qquad I_y = \iint_A x^2\, dA \qquad I_{xy} = \iint_A xy\, dA \qquad\qquad \textbf{[4.2.2]}$$

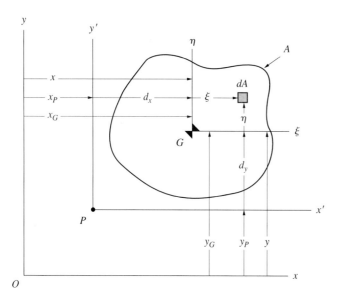

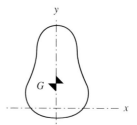

Figure 4.2.2   Plane area with a (vertical) axis of symmetry.

Figure 4.2.1   Plane region of arbitrary shape, of area A with centroid located at G.

Observe that under no conditions can I_x and I_y be negative or zero. On the other hand, I_{xy} may be positive or negative or zero. Within the context of the properties of plane areas, I_x and I_y are called the (area) *moments of inertia* and I_{xy} is the *product of inertia*. The *polar* moment of inertia J of area A is defined as

$$J = \iint_A r^2 dA$$ [4.2.3]

Since $r^2 = x^2 + y^2$, the polar moment of inertia is related to the moments of inertia I_x and I_y by the formula

$$J = I_x + I_y$$ [4.2.4]

If an area has an axis of symmetry, as in Figure 4.2.2, then the centroid must lie on that axis. Furthermore, if either the x axis or the y axis is an axis of symmetry, then $I_{xy} = 0$.

The second area moments $I_G)_x$, $I_G)_y$, and $I_G)_{xy}$, about the axes through the centroid G are found by means of Equation 4.2.2. Of course, x and y must be measured from G, as shown in Figure 4.2.1, in which the coordinates relative to G are denoted ξ and η.

The second area moment $I_P)_x$ is calculated by means of Equation 4.2.2$_1$ as follows:

$$I_P)_x = \iint_A (d_y + \eta)^2 dA = d_y^2 \iint_A dA + 2d_y \iint_A \eta dA + \iint_A \eta^2 dA$$

Also, $\iint_A dA = A$, $\iint_A \eta^2 dA = I_G)_x$,   and $\iint_A \eta dA = \eta_G A = 0$. This and similar calculations for $I_P)_y$, $I_P)_{xy}$, and J_P show that the second area moments about any point P are given in terms of the same quantities measured about *parallel* axes through G by the *parallel-axis formulas* (cf. Figure 4.2.1),

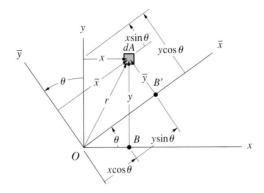

Figure 4.2.3   Relationship between the planar coordinates of a point in two different frames.

$$\boxed{\begin{aligned} I_P)_x &= I_G)_x + Ad_x^2 \qquad\quad I_P)_y = I_G)_y + Ad_y^2 \\ I_P)_{xy} &= I_G)_{xy} + Ad_x d_y \qquad J_P = J_G + A(d_x^2 + d_y^2) \end{aligned}}$$

[4.2.5a]

where

$$d_x = x_G - x_P \qquad d_y = y_G - y_P$$

[4.2.5b]

When computing products of inertia using Equation 4.2.5a$_3$, remember that the directed distances d_x and d_y need not be positive.

Now, consider two cartesian coordinate systems (xy) and ($\bar{x}\bar{y}$) that are rotated with respect to each other by the angle θ about their common z axis, as shown in Figure 4.2.3. The xy coordinates of dA are the components of the position vector of dA along the x and y axes. Similarly, the $\bar{x}\bar{y}$ coordinates of the same point in the rotated system are the projections of the position vector along the $\bar{x}$ and $\bar{y}$ axes. From the figure, we see that

$$\begin{aligned} \bar{x} &= x\cos\theta + y\sin\theta \\ \bar{y} &= -x\sin\theta + y\cos\theta \end{aligned}$$

[4.2.6]

The area moments of inertia of the area A about the $\bar{x}\bar{y}$ axes inclined to xy by the angle θ are given by

$$\bar{I}_x = \iint_A \bar{y}^2 dA \quad \bar{I}_y = \iint_A \bar{x}^2 dA \quad \bar{I}_{xy} = \iint_A \bar{x}\bar{y} dA$$

[4.2.7]

Substituting Equation 4.2.6 and using Equation 4.2.2 plus the double-angle trigonometric identities ultimately leads to

$$\boxed{\begin{aligned} \bar{I}_x &= \frac{I_x + I_y}{2} + \frac{I_x - I_y}{2}\cos 2\theta - I_{xy}\sin 2\theta \\ \bar{I}_y &= \frac{I_x + I_y}{2} - \frac{I_x - I_y}{2}\cos 2\theta + I_{xy}\sin 2\theta \\ \bar{I}_{xy} &= \frac{I_x - I_y}{2}\sin 2\theta + I_{xy}\cos 2\theta \end{aligned}}$$

[4.2.8]

Observe that $\bar{I}_x + \bar{I}_y = I_x + I_y$, so that the polar moment of inertia J is not affected by a rotation of the axes. Equations 4.2.8 are identical in form to the plane stress transformation formulas, Equations 3.6.5. Also, I_x and I_y play the role of the normal stresses σ_x and σ_y, and $(-I_{xy})$ plays that of τ_{xy}. Thus, the matrix

$$\begin{bmatrix} I_x & -I_{xy} \\ -I_{xy} & I_y \end{bmatrix}$$

is a symmetric tensor. Among its tensor properties are the existence of *principal values* relative to *principal axes of inertia*, having the orientation (cf. Equation 3.6.7)

$$\tan 2\theta = 2\frac{I_{xy}}{I_y - I_x} \qquad [4.2.9]$$

From Equation 4.2.8$_3$, we know that $\bar{I}_{xy} = 0$ for axes at this inclination. It follows that if area A has a symmetry axis, then it is also a principal axis of inertia.

Table 4.2.1 lists the area moments of inertia for several common shapes.

Table 4.2.1   Properties of selected shapes.

Section	Shape and Dimensions	Properties
Rectangle		$A = bh$ $I_x = \frac{1}{12}bh^3$ $I_y = \frac{1}{12}b^3h$ $I_{xy} = 0$
Triangle		$A = bh/2$ $I_x = \frac{1}{12}bh^3$ $I_y = \frac{1}{36}bh(a^2 - ab + b^2)$ $I_{xy} = \frac{1}{12}bh^2(b - 2a)$
Circle		$A = \pi r^2$ $I_x = I_y = \frac{\pi r^4}{4}$ $I_{xy} = 0$
Semicircle		$A = \frac{\pi}{2}r^2$ $I_x = \left(\frac{\pi}{8} - \frac{8}{9\pi}\right)r^4$ $I_y = \frac{\pi r^4}{8}$ $I_{xy} = 0$

4.3   INTERNAL FORCE SIGN CONVENTION

For the stress analysis of beams, the sign convention that will be employed consistently throughout the text is illustrated in Figure 4.3.1. It is consistent with the sign convention established in Chapter 3 for stress components. We will use a rectangular coordinate system whose x axis always coincides with or is parallel to the beam's axis. On any cross section of the beam, the internal stress distribution generally gives rise to a resultant force and a couple, each being vectors with three components. On the section whose outward normal points in the positive x direction (the positively-oriented surface), the components of these quantities are considered positive if they point in the positive coordinate directions. As can be seen from Figure 4.3.1, the components of force are the axial force P, which either pulls or pushes on the section (and is positive or negative accordingly), and a shear force acting in the plane of the section, with two components, V_y and V_z. The axial component of the couple moment is T, the *torque* or *twisting moment*. The components of the couple moment that lie in the plane of the section are the *bending moments*, M_y and M_z. To calculate the stresses at a section of a beam requires that we compute the internal forces and couples at the section due to the particular applied loading.

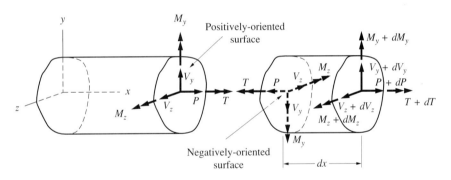

Figure 4.3.1   Sign convention for stress resultants on a beam cross section.

For example, consider the three-dimensional box beam in Figure 4.3.2. It has a uniform, rectangular cross section whose geometric details are such that the centroid G is located at the point shown in the figure. Let us place the origin of our coordinate system at the free end, centered at G. That means the x axis is the beam's axis, as previously defined. Suppose we are required to analyze the stress distribution at the cross section that is 150 in. from the free end, that is, at *station* 150. Just as with simple beams (cf. Chapter 2), we pass an imaginary transverse plane through the beam at that station, dividing the beam into two free bodies. The action of one part of the beam on the other is represented by the force resultants. We can select either one of the free bodies to do the equilibrium analysis in order to obtain values for these forces. Let us choose the free body to the left of the cut. (It is conventional to investigate the variation of internal loads by moving across the beam from left to right in the direction of increasing x, including an ever-increasing portion of the beam to the left of the cut.) Using the sign convention of Figure 4.3.1, the unknown forces on the section are drawn on the free body of Figure 4.3.3 such that they point in positive directions.

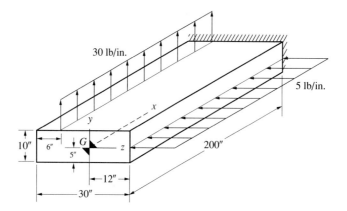

Figure 4.3.2   Cantilever beam loaded in two planes.

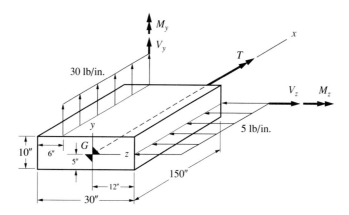

Figure 4.3.3   Free-body diagram of the beam segment between
x = 0 and x = 150 in.

The equilibrating internal stress resultants on the cut are shown positively
oriented, consistent with Figure 4.3.1.

Applying the equations of equilibrium, we compute each stress resultant in turn.

$$\sum F_y = 0 \quad \Rightarrow \quad V_y + (30\,\tfrac{\text{lb}}{\text{in.}} \times 150\,\text{in.}) = 0 \quad \Rightarrow \quad V_y = -4500\,\text{lb}$$

$$\sum F_z = 0 \quad \Rightarrow \quad V_z - (5\,\tfrac{\text{lb}}{\text{in.}} \times 150\,\text{in.}) = 0 \quad \Rightarrow \quad V_z = +750\,\text{lb}$$

Summing moments around the x, y, and z axes at the centroid G of the cut, we get

$$\sum M_{G_x} = 0 \quad \Rightarrow \quad T + (30\,\tfrac{\text{lb}}{\text{in.}} \times 150\,\text{in.}) \times 12\,\text{in.} = 0 \quad \Rightarrow \quad T = -54{,}000\,\text{in.-lb}$$

$$\sum M_{G_y} = 0 \quad \Rightarrow \quad M_y - (5\,\tfrac{\text{lb}}{\text{in.}} \times 150\,\text{in.}) \times \frac{150\,\text{in.}}{2} = 0 \quad \Rightarrow \quad M_y = +56{,}250\,\text{in.-lb}$$

$$\sum M_{G_z} = 0 \quad \Rightarrow \quad M_z - (30\,\tfrac{\text{lb}}{\text{in.}} \times 150\,\text{in.}) \times \frac{150\,\text{in.}}{2} = 0 \quad \Rightarrow \quad M_z = +337{,}500\,\text{in.-lb}$$

To show these load resultants on a sketch, we can redraw the three-dimensional free-body diagram so that it appears as shown in Figure 4.3.4. But this contains more detail than we need, and it is tedious during an analysis to redraw detailed three-dimensional figures. It is far simpler to show just the force and moment resultants on a *two*-dimensional sketch of the cross section alone. The question is, which way shall we view the section: in the direction of decreasing x, towards the positively-oriented side (back side) of the cut, as in Figure 4.3.5a, or in the direction of increasing x, towards the negatively-oriented side (front side) of the cut, as in Figure 4.3.5b? The choice is arbitrary, but we should stick with one or the other to avoid confusion. Therefore, throughout this text, we will always show the front side of the section, the negatively-oriented side of the cut, with the x axis running into the page, the z axis pointing to the right, and the y axis directed upwards.

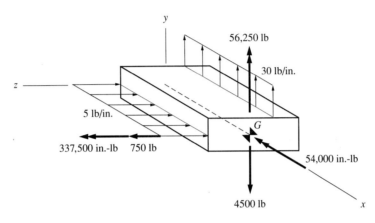

Figure 4.3.4 Free-body diagram of the beam segment between $x = 0$ and $x = 150$ in., showing the force and moment resultants on the cut section.

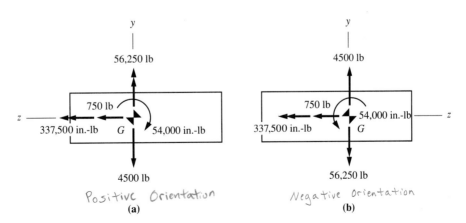

Figure 4.3.5 (a) Force and moment resultants in Figure 4.3.4 as viewed towards the negative x axis. (b) Force and moment resultants in Figure 4.3.4 as viewed towards the positive x axis, which is how they will henceforth be depicted in this text.

Once the side of the cut is chosen to show the section loads, the equilibrium analysis is finished. We proceed to the stress analysis; as such, we are henceforth dealing with *static equivalence,* that is, the integrated effect of the stress distribution on the section, which is the set of loads and moments obtained from equilibrium considerations.

For example, suppose we are analyzing a simple, straight bar of cross section A. Our equilibrium calculations have revealed that there is a tensile normal force P on the section in question, and, furthermore, it is the only nonzero load on that section. We can show this load as in Figure 4.3.6, acting through the centroid of the cross section. The stress giving rise to the axial load is calculated by requiring that the resultant of the stress distribution be statically equivalent to the single force P. That is,

$$\iint_A \sigma \, dA = P$$

Under the assumption that the stress σ is uniformly distributed over the section, we are then led to conclude that $\sigma = P/A$.

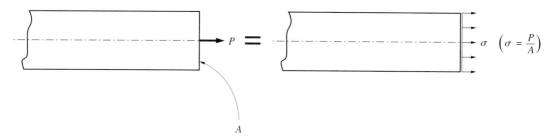

Figure 4.3.6   Static equivalence of the axial load resultant P to the integrated effect of the uniformly distributed stress σ:$P = \sigma A$.

The free-body diagram of a differential slice of the beam on the right of Figure 4.3.1 shows the variation of the forces and moments as we move from one section to another. Figure 4.3.7 presents two views of the slice. In Chapter 2, we analyzed the equilibrium of the free-body diagram in Figure 4.3.7a to show that

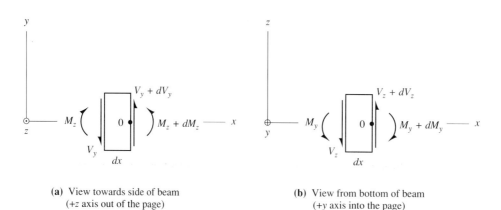

(a) View towards side of beam
(+z axis out of the page)

(b) View from bottom of beam
(+y axis into the page)

Figure 4.3.7   Stress resultants on each side of the differential slice shown in Figure 4.3.1.

$$\frac{dM_z}{dx} = -V_y \qquad\qquad [4.3.1]$$

We can obtain a similar formula involving M_y and V_z by setting the net moment about point O in Figure 4.3.7b equal to zero. Thus,

$$M_y + V_z dx - (M_y + dM_y) = 0$$

$$-dM_y + V_z dx = 0$$

$$\frac{dM_y}{dx} = V_z \qquad\qquad [4.3.2]$$

4.4   SAINT-VENANT TORSION THEORY

Consider a cylindrical shaft of arbitrary cross section subjected to pure torsion about the x axis, as illustrated in Figure 4.4.1. In response to the torsional loading, each cross section rotates as a rigid body through an angle θ about the x axis. Due to this rotation, a point in the cross section with position vector $\mathbf{r} = y\mathbf{j} + z\mathbf{k}$ undergoes the displacement $\theta\mathbf{i} \times \mathbf{r}$. According to the Saint-Venant theory of torsion,[1] the angle of twist varies linearly from one end of the rod to the other, so that

$$\theta = \phi x \qquad\qquad [4.4.1]$$

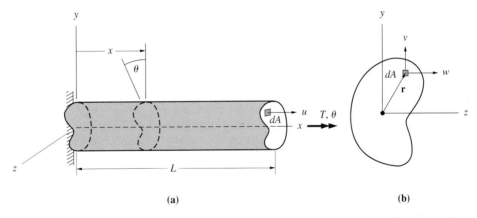

(a)   **(b)**

Figure 4.4.1   (a) Cylindrical bar of arbitrary cross section in pure torsion. (b) Displacements in the plane of the cross section.

where ϕ is the angle of twist per unit length. In general, the rotation is accompanied by a warping of the cross section, which is manifested by the axial displacement u of points on the initially plane section (cf. Figure 4.4.1a). The axial displacement u is assumed to be independent of x, proportional to the uniform rate of twist ϕ, and otherwise specified by a *warping function* $\psi(y, z)$. The displacement of a point on the cross section is

| [1] S. Timoshenko, and J. N. Goodier, *Theory of Elasticity*, Chapter 11, New York, McGraw-Hill, 1951.

$$\mathbf{u} = \overbrace{\phi\psi(y,z)\mathbf{i}}^{\text{warping}} + \overbrace{\theta\mathbf{i} \times (y\mathbf{j} + z\mathbf{k})}^{\text{rigid rotation}}$$

so that

$$u = \phi\psi(y,z) \quad v = -\phi xz \quad w = \phi xy \qquad \text{[4.4.2]}$$

With the displacements thus assumed, compatibility is assured, and the strains are found from the strain–displacement formulas, Equation 3.7.20:

$$\varepsilon_x = \frac{\partial u}{\partial x} = \frac{\partial}{\partial x}[\phi\psi(y,z)] = 0 \qquad \text{[4.4.7a]}$$

$$\varepsilon_y = \frac{\partial v}{\partial y} = \frac{\partial}{\partial y}(-\phi xz) = 0 \qquad \text{[4.4.7b]}$$

$$\varepsilon_z = \frac{\partial w}{\partial z} = \frac{\partial}{\partial z}(\phi xy) = 0 \qquad \text{[4.4.7c]}$$

$$\gamma_{xy} = \frac{\partial u}{\partial y} + \frac{\partial v}{\partial x} = \frac{\partial}{\partial y}[\phi\psi(y,z)] + \frac{\partial}{\partial x}(-\phi xz) = \phi\left[\frac{\partial\psi(y,z)}{\partial y} - z\right] \qquad \text{[4.4.7d]}$$

$$\gamma_{xz} = \frac{\partial u}{\partial z} + \frac{\partial w}{\partial x} = \frac{\partial}{\partial z}[\phi\psi(y,z)] + \frac{\partial}{\partial x}(-\phi xy) = \phi\left[\frac{\partial\psi(y,z)}{\partial z} + y\right] \qquad \text{[4.4.7e]}$$

$$\gamma_{yz} = \frac{\partial v}{\partial z} + \frac{\partial w}{\partial y} = \frac{\partial}{\partial z}(-\phi xz) + \frac{\partial}{\partial y}(\phi xy) = 0 \qquad \text{[4.4.7f]}$$

Substituting these strains into the stress–strain relations (with $T = 0$), Equation 3.11.5 yields

$$\sigma_x = \sigma_y = \sigma_z = \tau_{yz} = 0 \qquad \text{[4.4.8]}$$

and

$$\tau_{xy} = G\phi\left(\frac{\partial\psi}{\partial y} - z\right) \quad \tau_{xz} = G\phi\left(\frac{\partial\psi}{\partial z} + y\right) \qquad \text{[4.4.9]}$$

The stresses must satisfy Equations 3.3.4, the differential equations of equilibrium. Thus, prescribing the body force density to be equal to zero, Equation $3.3.4_1$ yields

$$\frac{\partial^2\psi}{\partial y^2} + \frac{\partial^2\psi}{\partial z^2} = 0 \qquad \text{[4.4.10]}$$

throughout the cross section. The other two equilibrium equations—Equation $3.3.4_2$ and $3.3.4_3$—are satisfied identically, that is, in both cases, $0 = 0$.

The stresses given by Equation 4.4.8 and 4.4.9 are required to comply with the boundary conditions. The lateral surface of the torsion bar is traction-free everywhere. That is, on the lateral surface, $T_x = T_y = T_z = 0$. Substituting these components of surface traction into Cauchy's formulas, Equation 3.2.8, we get

$$\sigma_x n_x + \tau_{yx} n_y + \tau_{zx} n_z = 0$$
$$\tau_{xy} n_x + \sigma_y n_y + \tau_{zy} n_z = 0 \qquad \text{[4.4.11]}$$
$$\tau_{xz} n_x + \tau_{yz} n_y + \sigma_z n_z = 0$$

However, the outward normal to the lateral surface is perpendicular to the x axis, so $n_x = 0$. Substituting this and the stress components into Equation 4.4.11, we find that the first equation yields

$$\left(\frac{\partial \psi}{\partial y} - z\right) n_y + \left(\frac{\partial \psi}{\partial z} + y\right) n_z = 0 \qquad \text{[4.4.12]}$$

everywhere on the boundary of the cross section, whereas the last two are identities $(0 = 0)$.

On transverse sections of the torsion bar, the resultant of the stress distribution must be a pure torque T in the x direction, as illustrated in Figure 4.4.2. That is,

$$\iint_A \mathbf{r} \times (-\tau_{xy} dA\mathbf{j} - \tau_{xz} dA\mathbf{k}) = -T\mathbf{i}$$

or

$$T = \iint_A (y\tau_{xz} - z\tau_{xy}) dA \qquad \text{[4.4.13]}$$

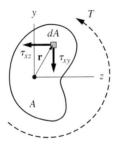

Figure 4.4.2

The shear stress distribution must be
statically equivalent to a pure torque T.

After substituting Equation 4.4.9 and simplifying, we obtain

$$T = GJ\phi \qquad \text{[4.4.14a]}$$

where J is the torsion constant,

$$J = \iint_A \left[(y^2 + z^2) + \left(y\frac{\partial \psi}{\partial z} - z\frac{\partial \psi}{\partial y}\right)\right] dA \qquad \text{[4.4.14b]}$$

GJ is called the *torsional rigidity,* which is a measure of the bar's resistance to twisting when subjected to an applied torque.

Since the stress distribution on the cross section must be statically equivalent to a pure torque about the x axis, it must be true that the y and z components of the resultant force vanish, or:

$$\iint_A \tau_{xy} dA = \iint_A \tau_{xz} dA = 0 \qquad \text{[4.4.15]}$$

Such is indeed the case, as can be shown[2] by substituting Equation 4.4.9 into these integrals and using Equations 4.4.10 and 4.4.12.

Example 4.4.1 Show that the stress distribution in a circular bar in pure torsion is obtained by setting the warping function $\psi = $ constant.

Here, $\psi = $ constant means the axial displacement of all points of the cross section is the same and that Equation 4.4.10 is trivially satisfied. There is no warping, and plane sections therefore remain plane. We may set $\psi = 0$ to eliminate rigid-body translation in the axial direction. Equation 4.4.12 becomes

$$(-z)n_y + (+y)n_z = 0$$

Since $n_y = -dz/ds$ and $n_z = dy/ds$ (see Figure 4.4.3), this implies that on the boundary of the cross section,

$$\frac{d}{ds}(y^2 + z^2) = 0$$

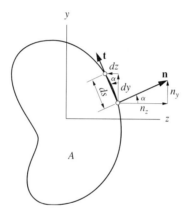

Figure 4.4.3 Relationship between the components of the unit normal **n** and unit tangent **t** on the boundary of the cross section.

That is,

$$y^2 + z^2 = \text{constant} \qquad \text{[a]}$$

which of course means that the boundary is a circle and the torsion bar is therefore a circular cylinder. From Equation 4.4.13b it is clear that $\psi = 0$ also means that

$$J = \iint_A (y^2 + z^2)\,dA \qquad \text{[b]}$$

| [2]C. T. Wang, *Applied Elasticity*, Chapter 5, New York, McGraw-Hill, 1950.

We recognize this integral to be the polar moment of inertia of the cross section, Equation 4.2.3. Thus, for circular sections, and *only* circular sections, the torsion constant is the polar moment of inertia, $J = \pi r^4/2$.

Equations 4.4.9 for the case at hand take the form

$$\tau_{xy} = -G\phi z \qquad \tau_{xz} = G\phi y \qquad \text{[c]}$$

From these two expressions, it is easy to see that

$$(\tau_{xy}\mathbf{j} + \tau_{xz}\mathbf{k}) \cdot (y\mathbf{j} + z\mathbf{k}) = 0 \qquad \text{[d]}$$

This means that the resultant shear stress, $\tau = \sqrt{\tau_{xy}^2 + \tau_{xz}^2}$, is perpendicular to the position vector drawn from the centerline of the bar. Equations c also imply that $\tau = G\phi r$, and Equation 4.4.14a yields $\phi = T/GJ$. Therefore,

$$\tau = \frac{Tr}{J} \qquad \text{[4.4.16]}$$

This is the well-known formula for the shear stress in a circular torsion bar. The variation is linear, from zero at the centerline to a maximum at the outer surface, and the stress around any circle is constant.

This formula remains valid if there is a concentric cylindrical hole through the length of the torsion bar, since Equation a is as true on the inner surface of radius r_i as it is on the outer one of radius r_o. Then, $J = \pi(r_o^4 - r_i^4)/2$ and Equation 4.4.16 holds for $r_i \le r \le r_o$.

If a hollow, circular torsion bar has a wall thickness t and a mean radius of $\bar{r}$, so that $r_i = \bar{r} - t/2$ and $r_o = \bar{r} + t/2$, then

$$J = 2\pi\bar{r}^3 t \left[1 + \left(\frac{t}{2\bar{r}} \right)^2 \right]$$

For thin-walled circular torsion tubes, $(t/2\bar{r})^2 \ll 1$, in which case

$$J = 2\pi\bar{r}^3 t = 2A\bar{r}t \qquad \text{[4.4.17]}$$

where A is the area enclosed by the mean radius. Equation 4.4.16 then becomes

$$\tau = \frac{T\bar{r}}{J} = \frac{T}{2At} \qquad \text{[4.4.18]}$$

Equation 4.4.18 is identical in form to Equation 2.5.11, if we recognize that τt is the uniform shear flow in the wall of the circular tube.

The torsion of noncircular sections is not as easy to determine. Consider the pure torsion of a bar having the rectangular cross section shown in Figure 4.4.4. The solution[3] of Equations 4.4.10 and 4.4.12 for the warping function ψ reveals that the displacements normal to the plane of the section are as illustrated in Figure 4.4.5. For thin rectangular sections ($h/t > 10$), analysis reveals that the contours of the resultant shear stress, $\tau = \sqrt{\tau_{xy}^2 + \tau_{xz}^2}$, are as indicated in Figure 4.4.6a. Except for a small region at each end of the long dimension, the shear stress varies linearly across the thickness, from zero at the midsection to a maximum at the outer surface, as was pointed out previously in Figure 2.6.3. For practical purposes, the maximum torsional shear stress may be visualized as shown in Figure 4.4.6b.

When $h/t > 10$, the torsion constant of a rectangular section can be satisfactorily approximated as

| [3]C. T. Wang, *ibid.*

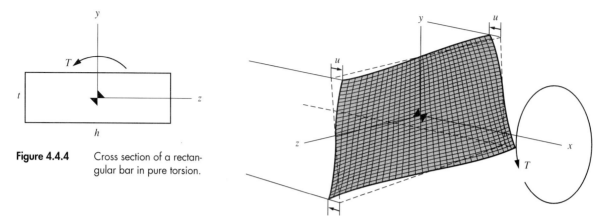

Figure 4.4.4   Cross section of a rectangular bar in pure torsion.

Figure 4.4.5   Warping of a rectangular bar in pure torsion $(h = 2t)$.

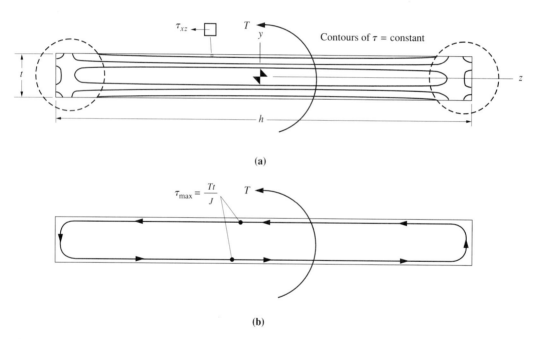

(a)

(b)

Figure 4.4.6   (a) Contours of constant shear stress in a rectangular bar in pure torsion $(h = 10t)$. (b) Approximate distribution of maximum shear stress in a thin rectangular section $(J \cong ht^3/3)$.

$$J = \frac{ht^3}{3} \qquad \left(\frac{h}{t} > 10\right) \qquad\qquad \text{[4.4.19]}$$

For an open section comprised of several narrow rectangles, J may be found by applying this formula to each rectangle and summing the results. This is illustrated in Figure 4.4.7 for an angle (b), channel (c), and I section (d).

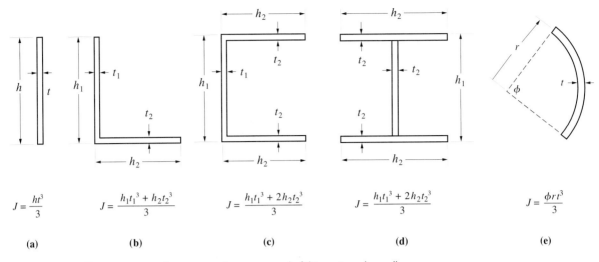

$$J = \frac{ht^3}{3}$$
(a)

$$J = \frac{h_1 t_1^3 + h_2 t_2^3}{3}$$
(b)

$$J = \frac{h_1 t_1^3 + 2 h_2 t_2^3}{3}$$
(c)

$$J = \frac{h_1 t_1^3 + 2 h_2 t_2^3}{3}$$
(d)

$$J = \frac{\phi r t^3}{3}$$
(e)

Figure 4.4.7   Torsion constants for open sections composed of thin rectangular walls.

For thin, curved sections, such as the circular arc in Figure 4.4.7e, J may be found by replacing the height of the rectangle by the developed length of the arc. For the circular arc, that length is equal to its radius times the angle ϕ in radians subtended by the arc. Away from sharp corners, which cause stress concentration, the maximum shear stress due to pure torsion in an open, composite thin-walled section is

$$\tau_{max} = \frac{Tt}{J} \tag{4.4.20}$$

The angle of twist per unit length remains as given in Equation 4.4.14a, or

$$\phi = \frac{T}{GJ} \tag{4.4.21}$$

Example 4.4.2   Compare the torsional shear stress and the torsional rigidity of two thin-walled tubes, one of them with a closed section (Figure 4.4.8a) and the other with a small longitudinal gap in the wall (Figure 4.4.8b).

For the closed section, $J = 2\pi r^3 t$; for the open section (cf. Figure 4.4.7e), $J = \frac{2}{3}\pi r t^3$. Therefore,

$$\frac{GJ)_{closed}}{GJ)_{open}} = \frac{2\pi r^3 t}{\frac{2}{3}\pi r t^3} = 3\left(\frac{r}{t}\right)^2 \tag{a}$$

Therefore, if $r/t > 10$, which must be true of a thin-walled tube, we see that the closed torque tube has over 300 times the resistance to twist than the one with the lengthwise gap.

According to Equation 4.4.18, the shear stress in the continuous-wall tube is $\tau_{closed} = T/2\pi r^2 t$. For the slit tube, Equation 4.4.20 yields

$$\tau_{open} = \frac{Tt}{\frac{2}{3}\pi r t^3} = \frac{3}{2}\frac{T}{\pi r t^2} \tag{b}$$

Thus,

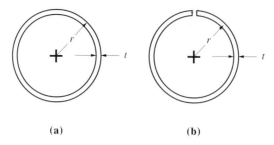

(a)   **(b)**

Figure 4.4.8   (a) Thin-walled closed tube. (b) Thin-walled open tube.

$$\frac{\tau_{\text{open}}}{\tau_{\text{closed}}} = \left(\frac{3}{2}\frac{T}{\pi r t^2}\right)\left(\frac{2\pi r^2 t}{T}\right) = 3\left(\frac{r}{t}\right)$$

[c]

For the same torque, the shear stress in the open tube is over thirty times greater than it is in the closed tube.

This discussion of torsional shear stress is based on the assumption that warping is free to occur. However, warping may be restrained by supports or attachments at either end of the torsion bar. If that is the case, then significant additional stresses develop due to the resistance to warping.[4]

4.5 BERNOULLI-EULER FLEXURE THEORY

If the force system on a beam consists of only a couple moment at each end, acting perpendicular to its axis, then the beam is in *pure bending*. Let the axis of the undeformed beam be the *x*-axis of a Cartesian coordinate system, as shown in Figure 4.5.1 According to what is generally referred to as the *Bernoulli-Euler hypothesis*,[5] a beam in pure bending deforms as follows. When the bending moments are applied, each point of the axis displaces in a

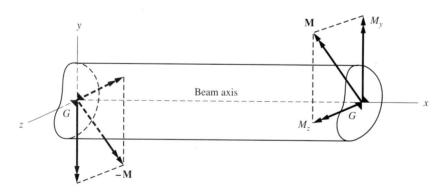

Figure 4.5.1   Beam in pure bending: *G* is the centroid of the cross section.

---

[4]See, for example, R. J. Roark, and W. C. Young, *Formulas for Stress and Strain*, Chapter 9, 5th ed. New York, McGraw-Hill, 1975.
[5]Jacob Bernoulli (1654–1705) and Leonhard Euler (1707–1783). Also attributed to Louis M. H. Navier (1785–1836).

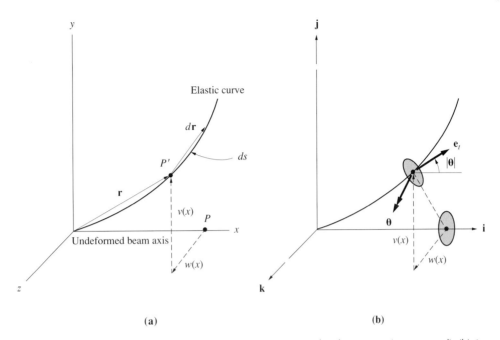

Figure 4.5.2   (a) Point *P* on the beam axis moves to point *P′* on the elastic curve (exaggerated). (b) A transverse section at point *P* on the beam axis undergoes a rigid-body translation to point *P′* followed by a rigid-body rotation.

direction perpendicular to the axis, which therefore bends into a curve, as shown in Figure 4.5.2a. Figure 4.5.2b illustrates the assumption that cross sections all remain perpendicular to this *elastic curve*, and each one undergoes an extremely small rigid-body rotation, $\boldsymbol{\theta} = \theta_y \mathbf{j} + \theta_z \mathbf{k}$, with no twist component ($\theta_x = 0$). There is no warping: "plane sections remain plane."

According to the Bernoulli-Euler assumption, a point on the beam axis moves to a location on the elastic curve with a position vector of

$$\mathbf{r} = x\mathbf{i} + v(x)\mathbf{j} + w(x)\mathbf{k} \tag{4.5.1}$$

where $v(x)$ and $u(x)$ are the y and z displacements of the point, which depend on x. The differential element of arc length ds measured along the elastic curve is given by $ds = \sqrt{d\mathbf{r} \cdot d\mathbf{r}}$, that is,

$$ds = \sqrt{dx^2 + dv^2 + dw^2} = dx\sqrt{1 + \left(\frac{dv}{dx}\right)^2 + \left(\frac{dw}{dx}\right)^2} \tag{4.5.2}$$

Consistent with the statement that the magnitude of vector rotation $\boldsymbol{\theta}$ is very small,[6] we will assume the displacement gradients are small compared to unity, so that Equation 4.5.2 yields $ds = dx$ as a satisfactory approximation. This means the elastic curve is shallow, with very small curvature.

By definition, the *unit tangent* $\mathbf{e}_t$ at a point on the elastic curve is the ratio of the differential change in the position vector to its magnitude, or:

---

[6]For the rotation vector $\boldsymbol{\theta}$, the rotation angle (in radians) is its magnitude and the normal to the plane of rotation is its direction, with the sense determined by the right-hand rule. Only infinitesimal rotations obey the vector rule of addition. Therefore, finite rotations, though possessing both magnitude and direction, are not truly vectors.

$$\mathbf{e}_t = \frac{d\mathbf{r}}{ds} \cong \frac{d\mathbf{r}}{dx} = \mathbf{i} + \frac{dv}{dx}\mathbf{j} + \frac{dw}{dx}\mathbf{k} \qquad\qquad \text{[4.5.3]}$$

where $\mathbf{e}_t$ is normal to the planar cross section of the deformed beam, just as the unit vector $\mathbf{i}$ was before deformation. We may therefore imagine that $\mathbf{e}_t$ results from subjecting the unit vector $\mathbf{i}$ to the small vector rotation $\boldsymbol{\theta}$ [7]:

$$\mathbf{e}_t = \mathbf{i} + \boldsymbol{\theta} \times \mathbf{i}$$

That is,

$$\mathbf{i} + \frac{dv}{dx}\mathbf{j} + \frac{dw}{dx}\mathbf{k} = \mathbf{i} + (\theta_y\mathbf{j} + \theta_z\mathbf{k}) \times \mathbf{i}$$

From this we obtain the components of $\boldsymbol{\theta}$ in terms of the derivatives of the displacements v and w, as follows:

$$\theta_y = -\frac{dw}{dx} \qquad \theta_z = \frac{dv}{dx} \qquad\qquad \text{[4.5.4]}$$

The position vector of a point in the cross section of the beam, relative to the beam axis, is $y\mathbf{j} + z\mathbf{k}$. The vector displacement of the point consists of the pure translation $v\mathbf{j} + w\mathbf{k}$ common to all points of the cross section, plus that due to the rigid-body rotation. That is,

$$\mathbf{u} = \overbrace{v(x)\mathbf{j} + w(x)\mathbf{k}}^{\text{pure translation}} + \overbrace{\boldsymbol{\theta}(x) \times (y\mathbf{j} + z\mathbf{k})}^{\text{rigid body rotation}}$$

After substituting Equations 4.5.4 we can write this in component form as

$$u = -y\frac{d}{dx}v(x) - z\frac{d}{dx}w(x) \qquad v = v(x) \qquad w = w(x) \qquad\qquad \text{[4.5.5]}$$

These are the beam displacements according to the Bernoulli-Euler hypothesis. With this information, we can calculate the strains using Equations 3.7.20.

$$\varepsilon_x = \frac{\partial u}{\partial x} = -y\frac{d^2v}{dx^2} - z\frac{d^2w}{dx^2} \qquad\qquad \text{[4.5.6a]}$$

$$\varepsilon_y = \frac{\partial v}{\partial y} = 0 \qquad\qquad \text{[4.5.6b]}$$

$$\varepsilon_z = \frac{\partial w}{\partial z} = 0 \qquad\qquad \text{[4.5.6c]}$$

$$\gamma_{xy} = \frac{\partial u}{\partial y} + \frac{\partial v}{\partial x} = -\frac{dv}{dx} + \frac{dv}{dx} = 0 \qquad\qquad \text{[4.5.6d]}$$

$$\gamma_{xz} = \frac{\partial u}{\partial z} + \frac{\partial w}{\partial x} = -\frac{dw}{dx} + \frac{dw}{dx} = 0 \qquad\qquad \text{[4.5.6e]}$$

$$\gamma_{yz} = \frac{\partial v}{\partial z} + \frac{\partial w}{\partial y} = 0 + 0 = 0 \qquad\qquad \text{[4.5.6f]}$$

The locus of points in a cross section for which $\varepsilon_x = 0$ is defined by Equation 4.5.6a and is called the *neutral axis*. The second derivatives d^2v/dx^2 and d^2w/dx^2 are the (shallow) *curvatures* of the deformed beam in the y and z directions, respectively.

| [7] The change $d\mathbf{V}$ of a vector $\mathbf{V}$ of constant magnitude when it undergoes a differential rotation $d\boldsymbol{\phi}$ is given by $d\mathbf{V} = d\boldsymbol{\phi} \times \mathbf{V}$.

4.6   ISOTROPIC BEAM FLEXURE STRESS

The Bernoulli-Euler hypothesis imposes a state of uniaxial strain in a beam, with ε_x being the only nonzero strain. But the Poisson effect, Equation 3.11.1, requires that $\varepsilon_y = \varepsilon_z = -\nu\varepsilon_x$. Therefore, in elementary beam theory, the Poisson effect is ignored. Setting $\nu = 0$ and substituting the strains given by Equations 4.5.6 into the generalized Hooke's law for isotropic materials in the absence of thermal strain, Equations 3.11.5, yields

$$\sigma_x = E\left(-y\frac{d^2v}{dx^2} - z\frac{d^2w}{dx^2}\right) \qquad [4.6.1]$$

All of the other stresses vanish.[8] On an area element dA like that shown in Figure 4.6.1, this gives rise to a differential force, $d\mathbf{F} = (\sigma_x dA)\mathbf{i}$, directed along the axis of the beam. The net force is found by integrating $d\mathbf{F}$ over the cross section A, as follows:

$$\iint_A E\left(-y\frac{d^2v}{dx^2} - z\frac{d^2w}{dx^2}\right)\mathbf{i}\,dA = -\mathbf{i}E\frac{d^2v}{dx^2}\iint_A y\,dA - \mathbf{i}E\frac{d^2w}{dx^2}\iint_A z\,dA = 0 \qquad [4.6.2]$$

The resultant is zero since the first area moments $\iint_A y\,dA$ and $\iint_A z\,dA$ are about the centroid G and are therefore zero.

The resultant moment is found by integrating $\mathbf{r} \times d\mathbf{F}$ over the cross section,

$$\mathbf{M} = \iint_A \mathbf{r} \times d\mathbf{F} = \iint_A (y\mathbf{j} + z\mathbf{k}) \times (\sigma_x\,dA\mathbf{i})$$

$$= \iint_A \left[(y\mathbf{j} + z\mathbf{k}) \times E\left(-\frac{d^2v}{dx^2}y - \frac{d^2w}{dx^2}z\right)\mathbf{i}\right]dA$$

That is,

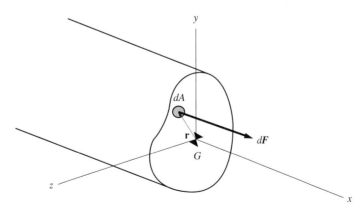

Figure 4.6.1   Normal force on a differential area of the cross section of a beam in pure bending, such as the one illustrated in Figure 4.5.1.

---

[8] The same result is obtained by retaining Poisson's ratio and setting $\epsilon_y = \epsilon_z = -\nu\epsilon_x$ in Equations 3.11.5, thereby ignoring Equations 4.5.6b and 4.5.6c.

$$M_y \mathbf{j} + M_z \mathbf{k} = -\mathbf{j}\left(\frac{d^2 w}{dx^2}\iint\limits_A Ez^2 \, dA + \frac{d^2 v}{dx^2}\iint\limits_A Eyz \, dA\right)$$

$$+ \mathbf{k}\left(\frac{d^2 v}{dx^2}\iint\limits_A Ey^2 \, dA + \frac{d^2 w}{dx^2}\iint\limits_A Eyz \, dA\right)$$

Under the assumption that the modulus of elasticity is the same for all elements of the cross section, equating the vector components on each side of this equation yields [9]

$$M_y = -E\frac{d^2 v}{dx^2}I_{yz} - E\frac{d^2 w}{dx^2}I_y$$
$$M_z = E\frac{d^2 v}{dx^2}I_z + E\frac{d^2 w}{dx^2}I_{yz}$$

$$[4.6.3]$$

where

$$I_y = \iint\limits_A z^2 \, dA, \text{ the moment of inertia of the cross section about the } y \text{ axis} \qquad [4.6.4a]$$

$$I_z = \iint\limits_A y^2 \, dA, \text{ the moment of inertia of the cross section about the } z \text{ axis} \qquad [4.6.4b]$$

$$I_{yz} = \iint\limits_A yz \, dA, \text{ the product of inertia of the cross section relative to the centroid} \qquad [4.6.4c]$$

Solving Equations 4.6.3 for the curvatures $d^2 v/dx^2$ and $d^2 w/dx^2$, we get

$$\frac{d^2 v}{dx^2} = \frac{1}{E(I_y I_z - I_{yz}^2)}(M_z I_y + M_y I_{yz}) \qquad [4.6.5a]$$

$$\frac{d^2 w}{dx^2} = -\frac{1}{E(I_y I_z - I_{yz}^2)}(M_y I_z + M_z I_{yz}) \qquad [4.6.5b]$$

Substituting these equations into Equation 4.6.1 yields the *flexure formula* for the normal stress, in terms of the bending moments and the geometry of the beam cross section, as follows:

$$\boxed{\sigma_x = \frac{1}{I_y I_z - I_{yz}^2}[-(M_z I_y + M_y I_{yz})y + (M_y I_z + M_z I_{yz})z]} \qquad [4.6.6]$$

It must be remembered that the origin of the yz axes coincides with the centroid of the section. If the yz axes are also principal axes of inertia of the cross section, then $I_{yz} = 0$ and the flexure stress formula reduces to

$$\sigma_x = -\frac{M_z y}{I_z} + \frac{M_y z}{I_y} \qquad [4.6.7]$$

An axis of symmetry is a principal axis of inertia.

---

[9]*Modulus-weighted* areas and moments of inertia are employed in beams with nonhomogeneous cross section. See, for example, B.K. Donaldson, *Analysis of Aircraft Structures: An Introduction*, Kappa Chapter, New York, McGraw-Hill, 1993

The flexure stress formula results from the assumption that plane sections remain plane. That is true only if the bending moments are caused by pure couples. These applied couples must have no axial (torsional) component to produce twist about the beam's longitudinal axis. Torsion causes rotation and warping of the cross section, which is contrary to the Bernoulli-Euler hypothesis. Bending moments are also produced by shear loads acting at a distance from the cross section. Using the flexure formula to calculate the normal stress due to transverse shear implies that the shear force is positioned precisely within the plane of the cross section so as not to induce torsional deformation along with the bending. The *shear center* is that point in the plane of the cross section through which the resultant shear force on the section must act to produce only bending, with no accompanying twist. If a beam is loaded in shear and all of the stress calculations are based exclusively on Equation 4.6.6, then, by definition, the shear force acts at the shear center. To locate the shear center of a section, we use the fact that the resultant shear on the section must pass through that point if there is to be no twist; that is, the moment of the resultant shear force about the shear center must vanish. Finding the shear center thus awaits the introduction of a formula for calculating shear stress. For idealized open section beams, that is a simple procedure and was illustrated in Example 2.6.1.

Even in the presence of nontorsional shear loading, the initially plane cross sections of a beam warp into slightly curved surfaces. The shear load induces shear stresses, which are accompanied by shear strains (in violation of the Bernoulli-Euler assumption), which in turn produce the warping. If the shear force in the beam is constant, as, for instance, in a cantilever beam with a point load at its free end (cf. Figure 2.4.1), then the warping will be the same at every section, so that—in the absence of restraints—no axial strain will be induced and the axial stress will be given by the flexure formula. If the shear varies along the span, then the warping will vary from section to section, inducing axial strain in addition to that due to bending alone. However, the additional stress will be negligible compared to the maximum bending stress *provided that the span of the beam is large compared to its depth*. For example, in a uniformly loaded, simply supported beam (cf. Figure 2.4.2), the ratio of the maximum stress due to shear to the maximum bending stress is about one-tenth of the *square* of the depth-to-span ratio.[10] In practical applications, the depth of a beam is less, often *much* less, than 10 percent of the span, so that the transverse shear correction amounts to a tenth of a percent or less.

If a beam carries an axial load P *directed along its axis* (the axis of centroids), then the accompanying stress is uniform over the cross section and equal to P/A. We can superimpose this direct stress on the bending stress, in which case Equation 4.6.6 becomes

$$\sigma_x = \frac{P}{A} + \frac{1}{I_y I_z - I_{yz}^2} [-(M_z I_y + M_y I_{yz})y + (M_y I_z + M_z I_{yz})z] \qquad \textbf{[4.6.8]}$$

Example 4.6.1   Figure 4.6.2a shows the first 50 inches of a 200-inch long cantilever beam. Applied loads act as shown through the centroid G (which in this case coincides with the shear center) of the free end. Calculate the bending stress distribution on the section at 50 inches from the free end.

We must calculate the area moments of inertia about the centroid G. To do so, we divide the cross section into three rectangles (Figure 4.6.2b) and use the parallel axis theorems. Thus,

$$
\begin{aligned}
I_y &= I_{y_1} + I_{y_2} + I_{y_3} \\
&= [\tfrac{1}{12}(8)(1)^3] + (2)[\tfrac{1}{12}(1)(8)^3 + (8 \times 1)(1.5)^2] \\
&= 0.6667 + 165.3 \\
&= 122.0 \text{ in.}^4
\end{aligned}
$$

| [10]S. P. Timoshenko, and J. N. Goodier, *Theory of Elasticity*, 3rd ed., New York, McGraw-Hill, 1970, p. 48.

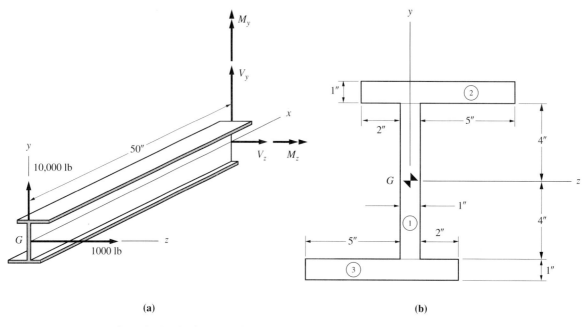

(a)   (b)

Figure 4.6.2   (a) Loads applied to the free end of a cantilever beam 200 in. long, and the internal reactions at station $x = 50$ in. (b) Details of the beam cross section.

$$I_z = I_{z_1} + I_{z_2} + I_{z_3}$$
$$= [\tfrac{1}{12}(1)(8)^3] + (2)[\tfrac{1}{12}(8)(1)^3 + (8 \times 1)(4.5)^2]$$
$$= 42.67 + 325.3$$
$$= 368 \text{ in.}^4$$

$$I_{yz} = I_{yz_1} + I_{yz_2} + I_{yz_3}$$
$$= [0] + [0 + (8)(4.5)(1.5)] + [0 + (8)(-4.5)(-1.5)]$$
$$= 108 \text{ in.}^4$$

The shear forces and bending moments on the section are shown as positive on Figure 4.6.2a, according to our sign convention (Figure 4.3.1). For equilibrium of the portion of the beam shown, $M_y + (1000 \text{ lb})(50 \text{ in.}) = 0$, or $M_y = -50,000$ in.-lb. Similarly, $M_z = +50,000$ in.-lb.

Substituting the bending moments into Equation 4.6.6, we get

$$\sigma_x = \frac{1}{[(122)(368) - 108^2]} \left\{ -[(5 \times 10^5)(122) + (-5 \times 10^4)(108)] y + [(-5 \times 10^4)(368) + (5 \times 10^5)(108)] z \right\}$$

$$= -1673y + 1071z$$

The maximum tensile stress on the section is 11,040 psi at $y = -5$ in., $z = 2.5$ in.

The *neutral axis* of this cross section is defined by those points for which $\sigma_x = 0$, namely, $y = 0.6403z$.

4.7   THIN-WALLED BEAM SHEAR STRESS

In this text, we are primarily concerned with thin-walled, lightweight structures across which the shear stress is essentially constant. In lightweight beams, the thin walls carry all or part of the bending load, which causes the shear stress to vary *along* the walls of transverse sections. It is convenient to deal with shear flow instead of shear stress, which at any point along a wall is the shear flow divided by the wall thickness. The variation of shear flow along a wall is found as described in the following discussion.

To isolate a differential element of a thin-walled beam such as that shown in Figure 4.7.1, first pass two planes spaced dx apart through the beam perpendicular to the x axis. Then pass two planes parallel to the x axis and perpendicular to the inner and outer wall surfaces. The distance between the latter two planes, measured along the wall *in the direction of increasing shear flow*, is ds. For equilibrium in the axial direction, we have

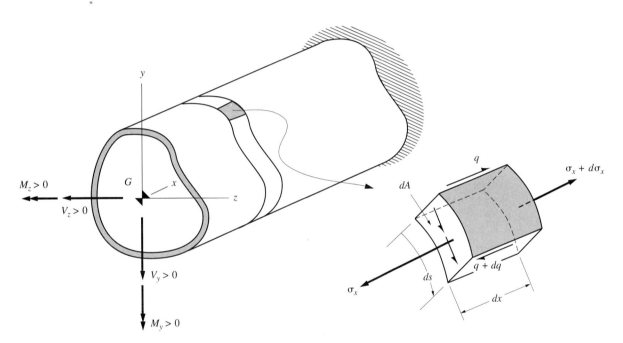

Figure 4.7.1   Differential element of a thin-walled beam showing shear flow and bending stress.

$$(\sigma_x + d\sigma_x)dA + q\,dx - \sigma_x dA - (q + dq)dx = 0$$

$$dq\,dx = d\sigma_x dA$$

$$dq = \frac{d\sigma_x}{dx}dA$$

or

$$\Delta q = \iint_A \frac{d\sigma_x}{dx}\,dA \qquad\qquad \textbf{[4.7.1]}$$

Here, $\Delta q = q_2 - q_1$ is the change in the shear flow between points 1 and 2 on a wall of the cross section, as illustrated in Figure 4.7.2. Also, A is the area of the wall cross section between 1 and 2. If there is no axial load resultant, the axial stress σ_x is given by Equation 4.6.6, as follows:

$$\sigma_x = \frac{1}{(I_y I_z - I_{yz}^2)}[-(I_y M_z + I_{yz} M_y)y + (I_z M_y + I_{yz} M_z)z]$$

From section 4.3, we know that $dM_z/dx = -V_y$ and $dM_y/dx = V_z$. Differentiating σ_x with respect to x and assuming a constant cross section, we obtain

$$\frac{d\sigma_x}{dx} = \frac{1}{(I_y I_z - I_{yz}^2)}[(I_y V_y - I_{yz} V_z)y + (I_z V_z - I_{yz} V_y)z] \qquad \textbf{[4.7.2]}$$

Substituting this result into Equation 4.7.1 yields the following expression for the change in shear flow between points 1 and 2 of a wall (Figure 4.7.2):

$$\Delta q = q_2 - q_1 = \frac{1}{(I_y I_z - I_{yz}^2)}[(I_y V_y - I_{yz} V_z)Q_z + (I_z V_z - I_{yz} V_y)Q_y] \qquad \textbf{[4.7.3]}$$

where $Q_z = \iint_A y \, dA$ and $Q_y = \iint_A z \, dA$ are the first area moments of the wall segment about the centroidal z axis and y axis, respectively, of the cross section.

Figure 4.7.3 illustrates the common special case of a simple beam with a symmetric, open, thin-walled cross section, loaded in shear in one plane only. If the bending plane, which must contain the shear center, is parallel to the xy plane, then $V_z = 0$ and, by symmetry, $I_{yz} = 0$. In such an application, Equation 4.7.3 reduces to the simpler form

$$q_2 - q_1 = \frac{Q_z V_y}{I_z}$$

Let point 1 lie on a stress-free external surface, so that $q_1 = 0$. Then the shear stress τ at an interior point, where the thickness of the section is t, is given by

$$\tau = \frac{q}{t} = \frac{Q_z V_y}{I_z t} \qquad \textbf{[4.7.4]}$$

where Q_z is the first area moment of the cross section between the free surface and the interior point, that is, between points 1 and 2 in Figure 4.7.3.

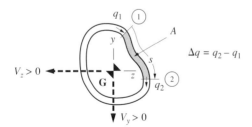

Figure 4.7.2   Change in shear flow between two points on a wall of the cross section.

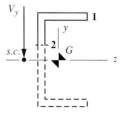

Figure 4.7.3   Open section of a beam with a bending load in one plane only.

In thin-walled beams, it is common for several webs to join along a common line. Figure 4.7.4 illustrates the case of three webs coming together. Let us assume that positive shear flows are directed away from the intersection on the near side, as shown in the figure. The shear flow $q^{(i)}$ ($i = 1, 2, \cdots$) increases by $dq^{(i)}$ over the differential length dx. So the average shear flow in each web is $q^{(i)} + \frac{1}{2}dq^{(i)}$. The net force in the direction of the line of intersection (the x direction) is equal to zero, so that $\sum (q^{(i)} + \frac{1}{2}dq^{(i)}) = 0$. The differential terms may be neglected; therefore, where n webs of a beam cross section join together, we have

$$\sum_{i=1}^{n} q^{(i)} = 0 \qquad\qquad [4.7.6]$$

That is, *the net shear flow out of (or into) a web junction is zero.*

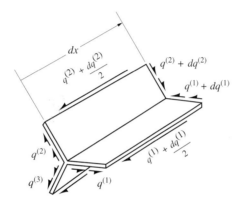

Figure 4.7.4   Shear flows on a differential portion of a multiweb junction.

Example 4.7.1   Find the shear flow distribution in the thin-walled open section of Figure 4.7.5 if the 1000 lb vertical force acts through the shear center. Also, locate the shear center relative to the vertical web. All of the walls have the same thickness of 0.1 inch. All dimensions are to the midplanes of the walls.

First, we locate the centroid G of the cross section. By symmetry, it lies midway between the top and bottom flanges. Its distance z_G from the vertical web is found in the usual way. After subdividing the section into five rectangles, we have

$$\left(\sum A_i\right) z_G = \sum A_i z_{G_i}$$

so that

$$z_G = \frac{2[(3 \times 0.1)(-1.5)] + 2[(5 \times 0.1)(2.5)] + (10 \times 0.1)(0)}{2(3 \times 0.1) + 2(5 \times 0.1) + (10 \times 0.1)}$$

$$= \frac{16}{26} = 0.6154 \text{ in.}$$

Place the origin of the yz coordinate system at G and orient it as shown in the figure. Since the z axis is an axis of symmetry of the section, the yz axes are principal axes of inertia. Therefore, the product of inertia I_{yz} vanishes. Noting also that $V_z = 0$, Equation 4.7.3 reduces to

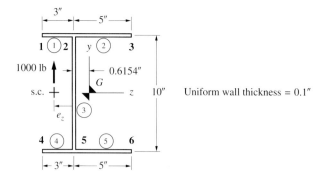

Figure 4.7.5   Thin-walled open section with webs effective in bending, as well as in shear.

$$\Delta q = \frac{V_y Q_z}{I_z}$$

Here, I_z is found by summing the area moments of all of the subregions and using the parallel axis theorem, as follows:

$$I_z = \sum I_{z_i}$$

$$= 2\left[\tfrac{1}{12}(3)(0.1)^3 + (3 \times 0.1)(5)^2\right] + 2\left[\tfrac{1}{12}(5)(0.1)^3 + (5 \times 0.1)(5)^2\right] + \tfrac{1}{12}(0.1)(10)^3$$

$$= 15 + 25 + 8.333$$

$$= 48.33 \text{ in.}^4$$

Since the 1000 lb load acts *up* on the section, and the x axis is *into* the page, it follows from the sign conventions illustrated in Figure 4.3.1 that $V_y = -1000$ lb. Thus,

$$\Delta q = \frac{V_y Q_z}{I_z} = \frac{-1000}{48.33} Q_z = -20.69 Q_z \tag{a}$$

To obtain the shear flow $q^{(i)}$ ($i = 1, \ldots, 5$) in each of the five walls of the section, we must evaluate the first area moments $Q_z^{(i)}$ as a function of position within each wall.

Wall 1

Let s be the distance measured along the wall, starting at a point where we know the shear flow. For wall 1 that is point 1, at a free surface where the shear flow is zero. To find $Q_z^{(1)}$ for the rectangular segment (cf. Figure 4.7.6a), we determine the y coordinate of its centroid (paying close attention to sign) relative to the neutral axis of the cross section (the z axis) and multiply that number by the area of the segment. Therefore,

$$Q_z^{(1)} = 5(0.1 \times s) = 0.5s$$

For this wall, Equation a becomes

$$\Delta q^{(1)} = q^{(1)} - 0 = -10.35s$$

or

$$q^{(1)} = -10.35s \tag{b}$$

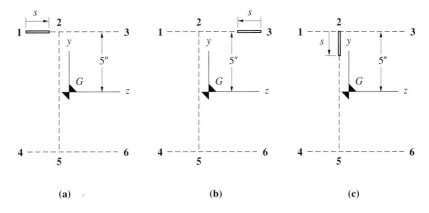

Figure 4.7.6   Segment of (a) wall 1, (b) wall 2, and (c) wall 3.

The shear flow varies linearly from zero at point 1 ($q_1^{(1)} = 0$) to $q_2^{(1)} = -31.05$ lb/in. at the junction with the vertical web. The minus sign signifies that the shear flow points opposite to the direction of increasing s. For this wall, that means the shear flow is *out* of point 2 and *towards* point 1.

Wall 2

The expression for the shear flow in this wall, measuring the distance from point 3, is the same as that for wall 1, or

$$q^{(2)} = -10.35s \tag{c}$$

The shear flow is *away* from the vertical web, where its value is $q_2^{(2)} = -51.75$ lb/in., and it decreases linearly to zero.

Wall 3

The net shear flow out of the junction at 2 is zero, where *out* is positive. Thus,

$$q_2^{(1)} + q_2^{(2)} + q_2^{(3)} = 0$$

From our previous calculations, we know that shear flows in walls 1 and 2 are *away* from the vertical web. Therefore,

$$(+31.05) + (+51.75) + q_2^{(3)} = 0$$

or

$$q_2^{(3)} = -82.80 \text{ lb/in.}$$

This means that the shear flow in the vertical wall is *into* point 2 (and consequently *out* of point 5).

From Figure 4.7.6c, the first area moment for the segment of the vertical wall is

$$Q_z^{(3)} = (5 - s/2)(0.1 \times s) = 0.5s - 0.05s^2$$

Substituting this into Equation a yields

$$\Delta q^{(3)} = q^{(3)} - q_2^{(3)} = -20.69(0.5s - 0.05s^2)$$

or

$$q^{(3)} = -82.80 - 20.69(0.5s - 0.05s^2) \tag{d}$$

Here we have a parabolic distribution, with the magnitude of the shear flow increasing from 82.8 lb/in. at either end of the web to a maximum value of 108.7 lb/in. at the neutral axis.

Walls 4 and 5

These walls are identical to their counterparts in the top flange, but the y coordinate of their centerline is opposite in sign to that of the top flange. Therefore, the shear flows in the bottom flanges will be equal but opposite in direction to those in walls 1 and 2.

Figure 4.7.7 illustrates the shear flow distribution over the section.

The resultant shear force R_V in each wall is found by integrating the shear flow over the length of the wall, that is, $R_V = \int q\,ds$.

$$\text{Wall 1}: \quad R_V^{(1)} = \int_0^3 (-10.35s)\,ds = -46.68 \text{ lb}$$

$$\text{Wall 2}: \quad R_V^{(2)} = \int_0^5 (-10.35s)\,ds = -129.5 \text{ lb}$$

$$\text{Wall 3}: \quad R_V^{(3)} = \int_0^{10} [-82.80 - 20.69(0.5s - 0.05s^2)]\,ds = -1000 \text{ lb}$$

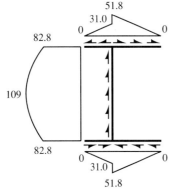

Figure 4.7.7   Shear flow distribution (lb/in.) for the loading shown in Figure 4.7.5.

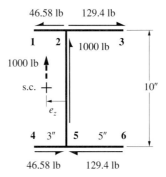

Figure 4.7.8   Shear flow resultants.

These shear flow resultants, as well as those for the bottom flange, are shown in Figure 4.7.8. The shear flows were computed using only the flexure formula. Therefore, the resultant force system is statically equivalent to the system comprised of the single, 1000 lb vertical force acting through the point, which *by definition* is the shear center. The moment of the two systems about any point must therefore be equal. Choosing point 5, that implies

$$1000 \times e_z = (129.4 - 46.58) \times 10$$

$$e_z = 0.8282 \text{ in.}$$

The shear center, which always lies on an axis of symmetry (the z axis, in this case), is 0.828 inch to the *left* of the web.

If the shear force does not pass through the shear center, there will be torsional shear stresses, as well as those due to flexure alone. We can break a thin-walled open section problem of this kind into two parts. First, the shear load is shifted to the shear center, and the flexure shear stress distribution is found in the manner illustrated in the previous example. Then, the moment of the shear force in its actual location about the shear center is used as the applied torque to find the torsional shear stresses, using the procedure outlined in section 4.4. The two solutions—pure flexure and pure torsion—are finally superimposed to yield the *net* shear flow distribution throughout the section.

Example 4.7.2   Calculate the maximum shear stress in the section of Example 4.7.1 if the shear force acts through the vertical web instead of through the shear center, as shown in Figure 4.7.9a.

First, the shear force is shifted to the shear center (Figure 4.7.9b), which produces the shear stress distribution already calculated in the previous example. As illustrated in part (c) of the figure, the moment of the shear force about the shear center is 828.2 in.-lb, which is applied as a pure torque to the open section. The torsion constant J of the section, composed of three 0.1 in. thick rectangles, is

$$J = \sum \frac{1}{3} h_i t_i^3 = 2 \times \left[\frac{1}{3} (8) (0.1)^3 \right] + \frac{1}{3} (10) (0.1)^3 = 0.008666 \text{ in.}^4$$

Therefore, according to Equation 4.4.20,

$$\tau_{max} = \frac{T t}{J} = \frac{(828.2)(0.1)}{0.008666} = 9556 \text{ psi}$$

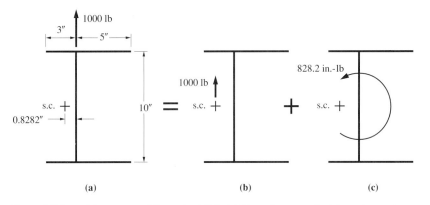

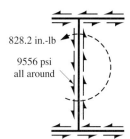

Figure 4.7.10   Maximum shear stress on the open section due to pure torsion.

Figure 4.7.9   The section of Example 4.7.1. (a) Shear force applied through a point other than the shear center. (b) Load at the shear center (per Example 4.7.1). (c) Moment of the applied load about the shear center.

This shear stress acts all around the outer boundary of the section (except for stress concentrations in the corners), as suggested in Figure 4.7.10. It is significantly larger than the maximum flexure stress, which, upon referring to Figure 4.7.7, we see occurs at the midpoint of the vertical web and has the value $109.0/0.1 = 1090$ psi. (Remember that the shear flow is uniformly distributed across the wall thickness.) Superimposing the flexural and torsional shear stresses around the outer boundary of the section reveals that the maximum shear stress is $9556 + 1090 = 10{,}650$ psi at the midpoint of the vertical web, on the right side, where the torsional and flexural shear stresses are in the same direction.

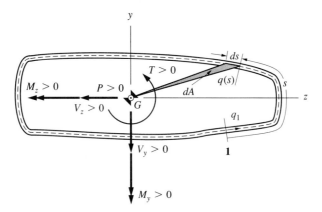

Figure 4.7.11   A closed, single-cell section carrying shear, torsion, bending, and axial load. The symbol represents a force directed out of the page.

Consider a beam with a closed section, such as that pictured in Figure 4.7.11. The negatively-oriented section (x axis into the page) is shown with a full complement of six load resultants, all located at the centroid G and all oriented in the conventionally positive directions (Figure 4.3.1). If the axial load P is actually offset from the beam's axis, then its moments about the y and z axes are included in the values of the bending moments M_y and M_z. Likewise, if the shear forces V_y and V_z do not really pass through G, then they contribute to the twisting moment T. The bending and twisting couples are free vectors and may be shown acting at any point on the section.

The normal stress distribution on the section is calculated by substituting the given axial load and bending moments, along with the geometric properties of the section, into the flexure formula, Equation 4.6.8. The shear flow caused by the pure torque T is obtained using Equation 2.5.11, where A is the area enclosed by the midline of the wall (shown dashed in Figure 4.7.11). Torsional shear stress in thin-walled closed sections, like flexural shear stress, may be considered uniform across the wall thickness. Finally, the shear flow due to the shear forces V_y and V_z are calculated using the flexural shear flow formula, Equation 4.7.3.

That formula gives the *difference* between the shear flows at two points in a wall. Therefore, we must know the shear flow at one of the points in order to find the shear flow at the other one. In an open section, there are at least two points where the shear flow is known to be zero. Starting at these locations, the interior shear flows are obtained from Equation 4.7.3 in the straightforward manner illustrated in Example 4.7.2.

On the other hand, the cross section in Figure 4.7.11 is closed. There is no point on the wall at which the value of the shear flow is known. Therefore, we select an arbitrary and convenient point, say point 1, and label the shear flow there as the unknown q_1. The shear flow anywhere else on the wall is found in terms of q_1 by means of Equation 4.7.3 as follows:

$$q(s) = q_1 + V_y f(s) + V_z g(s) \qquad \text{[4.7.7a]}$$

where

$$f(s) = \frac{I_y Q_z(s) - I_{yz} Q_y(s)}{I_y I_z - I_{yz}^2} \qquad g(s) = \frac{I_z Q_y(s) - I_{yz} Q_z(s)}{I_y I_z - I_{yz}^2} \qquad \text{[4.7.7b]}$$

To find q_1, we require that the shear flow distribution be statically equivalent to the loads on the section shown in Figure 4.7.11. The use of the flexure formula, in the form of Equation 4.7.3, to calculate the shear flows guarantees their static equivalence to the transverse loads V_y and V_z. *Moment equivalence* requires that the moment of the shear flow about any point in the cross section equals the moment of the section loads about that same point. If we choose the centroid G in Figure 4.7.11, then for moment equivalence, we must have $\sum M_G = \oint 2q(s)dA$. The integral is carried out over one complete circuit of the closed section. As illustrated in Figure 4.7.11, dA is the area swept by the line joining G to the wall. Since $\sum M_G = T$, we have, upon substituting Equation 4.7.7a,

$$T = 2Aq_1 + 2 \oint [V_y f(s) + V_z g(s)]dA \qquad \text{[4.7.8]}$$

This equation is then solved for q_1.

Equation 4.7.8 may be interpreted as follows. The term $2Aq_1$ is the moment of the constant shear flow q_1 directed around the entire closed section, that is, the shear flow due to a pure torque. The second term on the right is the moment of the *variable* shear flow produced by applying the shear forces V_y and V_z through the shear center of the open section created by an imaginary "cut" at 1. The shear flow in a closed section is then the superposition of the pure torsional shear flow onto the open-section shear flow produced by an infinitesimal gap placed anywhere around the wall.

Example 4.7.3   Calculate the shear flow in the walls of the box beam with the cross section shown in Figure 4.7.12 if the webs are effective in bending as well as in shear.

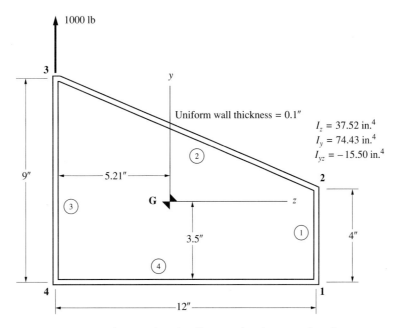

Figure 4.7.12   Box beam with webs effective in bending as well as shear.

The location of the centroid G shown in Figure 4.7.12 is found by considering the section to be composed of four thin, rectangular walls. Before we can use Equation 4.7.3, the second area moments I_y, I_z, and I_{yz} about the centroid must be calculated. For walls 1, 3, and 4, the parallel axis theorem may be employed, to find

$$I_z^{(1)} = 1.433 \text{ in.}^4 \quad I_y^{(1)} = 18.44 \text{ in.}^4 \quad I_{yz}^{(1)} = -4.074 \text{ in.}^4$$

$$I_z^{(3)} = 6.975 \text{ in.}^4 \quad I_y^{(3)} = 24.43 \text{ in.}^4 \quad I_{yz}^{(3)} = -4.689 \text{ in.}^4$$

$$I_z^{(4)} = 14.70 \text{ in.}^4 \quad I_y^{(4)} = 15.15 \text{ in.}^4 \quad I_{yz}^{(4)} = -3.318 \text{ in.}^4$$

Since wall 2 is not parallel to either the y or z axis, we must first use Equation 4.2.8 for transforming area moments between inclined axes. For the particular case of a thin rectangle, those relationships are listed in Figure 4.7.13. When coupled with the parallel axis theorem, they yield the following for wall 2:

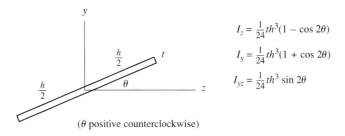

$$I_z = \frac{1}{24}th^3(1 - \cos 2\theta)$$

$$I_y = \frac{1}{24}th^3(1 + \cos 2\theta)$$

$$I_{yz} = \frac{1}{24}th^3 \sin 2\theta$$

(θ positive counterclockwise)

Figure 4.7.13 Second area moments for an inclined, thin rectangle about axes through its centroid.

$$I_z^{(2)} = 14.41 \text{ in.}^4 \qquad I_y^{(2)} = 16.41 \text{ in.}^4 \qquad I_{yz}^{(2)} = -3.419 \text{ in.}^4$$

Thus, for the entire section, we have

$$I_z = I_z^{(1)} + I_z^{(2)} + I_z^{(3)} + I_z^{(4)} = 37.52 \text{ in.}^4$$

$$I_y = I_y^{(1)} + I_y^{(2)} + I_y^{(3)} + I_y^{(4)} = 74.43 \text{ in.}^4$$

$$I_{yz} = I_{yz}^{(1)} + I_{yz}^{(2)} + I_{yz}^{(3)} + I_{yz}^{(4)} = -15.50 \text{ in.}^4$$

Substituting these results into Equation 4.7.3, and remembering that $V_z = 0$, we find

$$\Delta q = -29.16Q_z - 6.073Q_y \qquad \text{[a]}$$

To obtain the shear flow in each wall, we must evaluate the first area moments Q_z and Q_y as functions of position within the wall. As in the previous example, let us use a superscript to identify the shear flow with the wall to which it belongs. Subscripts denote the corner points around the cross section. Unlike the beam in the previous example, this section is closed, so at the outset, there is no point at which we know the shear flow to be zero.

Wall 1

Let s be the distance measured along the wall starting at point 1, just inside the wall, where the shear flow is $q_1^{(1)}$, as shown in Figure 4.7.14. Initially, $q_1^{(1)}$ is unknown; its value will be determined later when we invoke moment equivalence near the end of the solution. In the meantime, the shear flows all around the section will be expressed in terms of $q_1^{(1)}$. To find the first area moment $Q_z^{(1)}$ for the rectangular segment of this wall, we determine the y coordinate of its centroid relative to the neutral axis of the cross section (the z axis) and multiply that number by the area of the segment. For $Q_y^{(1)}$, we locate the segment's centroid relative to the y axis. Thus, from Figure 4.7.14, we see that

$$Q_z^{(1)} = \left[-\left(3.5 - \frac{s}{2}\right)\right](0.1 \times s) = -0.05(7s - s^2)$$
$$Q_y^{(1)} = 6.790(0.1 \times s) = 0.6790s$$

Substituting $Q_y^{(1)}$ and $Q_z^{(1)}$ into Equation a, noting that $\Delta q^{(1)} = q^{(1)}(s) - q_1^{(1)}$, yields

$$q^{(1)}(s) = q_1^{(1)} - 29.16\left[-0.05\left(7s - s^2\right)\right] - 6.073\left[0.6790s\right]$$

or

$$q^{(1)}(s) = q_1^{(1)} + 6.082s - 1.458s^2 \qquad [b]$$

which shows that the shear flow in wall 1 varies parabolically starting at $q_1^{(1)}$ and ending up at $q_2^{(1)} = q_1^{(1)} + 1.000$ lb/in. ($s = 4$ in.).

Wall 2

The centroid of the wall segment shown in Figure 4.7.15 is located at $s/2$. From the geometry, we see that

$$\frac{6.790 - \bar{z}}{s/2} = \frac{12}{13} \qquad \Rightarrow \qquad \bar{z} = 6.790 - \frac{6}{13}s$$

and

$$\frac{\bar{y} - 0.5}{s/2} = \frac{5}{13} \qquad \Rightarrow \qquad \bar{y} = 0.5 + \frac{5s}{26}$$

Therefore, the area moments are

$$Q_y^{(2)} = \left(6.790 - \frac{6}{13}s\right)(0.1 \times s) = 0.6790s - 0.04615s^2$$
$$Q_z^{(2)} = \left(0.5 + \frac{5s}{26}\right)(0.1 \times s) = 0.05s + 0.01923s^2$$

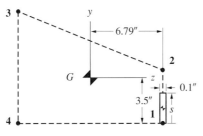

Figure 4.7.14   Segment of wall 1, starting with point 1 at the bottom.

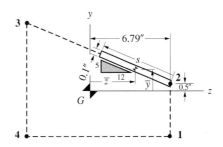

Figure 4.7.15   Segment of wall 2, starting with point 2 at its right end.

Now, the shear flow $q_2^{(1)}$ at the end of wall 1 equals the shear flow $q_1^{(2)}$ at the beginning of wall 2. According to Equation b, $q_2^{(1)} = q_1^{(1)} + 1.000$ lb/in. Consequently, from Equation a we get

$$q^{(2)}(s) = (q_1^{(1)} + 1.000) - 29.16(0.05s + 0.01923s^2) - 6.073(0.6790s - 0.04615s^2)$$

or

$$q^{(2)}(s) = q_1^{(1)} + 1.000 - 5.580s - 0.2802s^2 \qquad \text{[c]}$$

From this we obtain $q_3^{(2)} = q_1^{(1)} - 118.9$ lb/in.

Wall 3

Again, since the shear flow is continuous, $q_3^{(3)} = q_3^{(2)} = q_1^{(1)} - 118.9$ lb/in. From Figure 4.7.16, we deduce that

$$Q_z^{(3)} = \left(5.5 - \frac{s}{2}\right)(0.1 \times s) = 0.55s - 0.05s^2$$
$$Q_y^{(3)} = (-5.210)(0.1 \times s) = -0.5210s$$

Therefore,

$$q^{(3)}(s) = (q_1^{(1)} - 118.9) - 29.16(0.55s - 0.05s^2) - 6.073(-0.5210s)$$

or

$$q^{(3)}(s) = q_1^{(1)} - 118.9 - 12.87s + 1.458s^2 \qquad \text{[d]}$$

At the bottom of wall 3, this equation yields

$$q_4^{(3)} = q_1^{(1)} - 116.6 \text{ lb/in.}$$

Wall 4

From Figure 4.7.17, we have

$$Q_z^{(4)} = (-3.5)(0.1 \times s) = -0.35s$$
$$Q_y^{(4)} = \left[-\left(5.21 - \frac{s}{2}\right)\right](0.1 \times s) = -0.521s + 0.05s^2$$

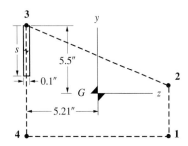

Figure 4.7.16   Segment of wall 3, starting with point 3 at its top end.

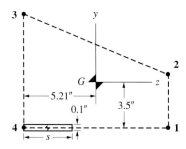

Figure 4.7.17   Segment of wall 4, beginning with point 4 at its left end.

Since $q_4^{(4)} = q_4^{(3)} = q_1^{(1)} - 116.6$ lb/in., Equation a gives

$$q^{(4)}(s) = (q_1^{(1)} - 116.6) - 29.16(-0.35s) - 6.073(-0.521s + 0.05s^2)$$

so that

$$q^{(4)}(s) = q_1^{(1)} - 116.6 + 13.37s - 0.3036s^2 \qquad \text{[e]}$$

We now have, in terms of $q_1^{(1)}$, the shear flow distribution around the closed cross section (counterclockwise flow positive) due to the 1000 lb shear force. The resultant shear force R_V in each wall is found by integrating the shear flow over its length.

$$\text{Wall 1}: \quad R_V^{(1)} = \int_0^4 \left(q_1^{(1)} + 6.082s - 1.458s^2 \right) ds = 4q_1^{(1)} + 17.55 \text{ lb}$$

$$\text{Wall 2}: \quad R_V^{(2)} = \int_0^{13} \left(q_1^{(1)} + 1.000 - 5.580s - 0.2802s^2 \right) ds = 13q_1^{(1)} - 663.7 \text{ lb}$$

$$\text{Wall 3}: \quad R_V^{(3)} = \int_0^9 \left(q_1^{(1)} - 118.9 - 12.87s + 1.458s^2 \right) ds = 9q_1^{(1)} - 1237 \text{ lb}$$

$$\text{Wall 4}: \quad R_V^{(4)} = \int_0^{12} \left(q_1^{(1)} - 116.6 + 13.37s - 0.3036s^2 \right) ds = 12q_1^{(1)} - 611.4 \text{ lb}$$

These resultant shear forces are shown in Figure 4.7.18. The moment of the shear flow resultants about 3 must equal the moment of the 1000 lb shear force about 3. Taking positive moments counterclockwise,

$$\left(12q_1^{(1)} - 611.4 \right) \times 9 + \left(4q_1^{(1)} + 17.55 \right) \times 12 = 0$$

so that $q_1^{(1)} = 33.92$ lb/in. Substituting this shear flow into Equation b through e yields

$$\text{Wall 1}: \quad q^{(1)}(s) = 33.9 + 6.08s - 1.46s^2$$
$$\text{Wall 2}: \quad q^{(2)}(s) = 34.9 - 5.58s - 0.208s^2$$
$$\text{Wall 3}: \quad q^{(3)}(s) = -85.0 - 12.9s + 1.46s^2$$
$$\text{Wall 4}: \quad q^{(4)}(s) = -82.7 + 13.4s - 0.304s^2$$

This parabolic shear flow distribution is sketched in Figure 4.7.19.

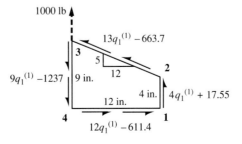

Figure 4.7.18   Shear flow resultants on the section.

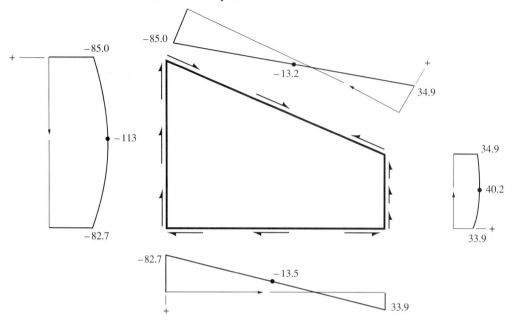

Figure 4.7.19 Parabolic shear flow distribution around the closed section of Figure 4.7.12.

Counterclockwise shear flow is considered positive. Numerical values (lb/in.) are shown at the ends and the midpoints of each wall.

4.8 IDEALIZED THIN-WALLED SECTIONS

The cross section of an *idealized* beam consists of thin webs and concentrated flange areas. The flanges carry all of the bending load, while the webs transmit only shear between adjacent flanges. As suggested in Figure 4.8.1, the process of idealizing a beam cross section involves lumping the area of the webs into discrete points. These concentrated areas include the areas of actual stiffeners (e.g., stringers, longerons, or spar caps) to which the skin is attached. Some concentrated areas may just be an outcome of the modeling and may not represent actual physical stiffeners. For example, the small black dots in the upper and lower skins between the vertical webs (spars) of Figure 4.8.1b are lumped skin areas, not physical stiffeners. The larger black dots on the other hand include the area of the spar caps. Since the webs of idealized beams are not active in bending, the shear flow in each wall of a section is constant. This greatly simplifies the shear flow analysis.

In Chapter 2, our analysis of idealized beams was limited to those that could be solved without regard to their deformability, that is, using static equilibrium considerations alone. As a consequence of the Bernoulli-

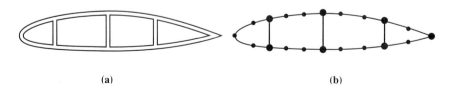

(a)   (b)

Figure 4.8.1   (a) Actual thin-walled section. (b) Idealized section.

Euler assumption on the deformation accompanying bending, we now have Equation 4.6.6, which is the flexure formula for isotropic, homogeneous beams:

$$\sigma_x = \frac{1}{I_y I_z - I_{yz}^2}[-(M_z I_y + M_y I_{yz})y + (M_y I_z + M_z I_{yz})z]$$

[handwritten: Flexure formula, σ_x is normal STRESS, used for bending stress]

With this formula, we can obtain the flange loads in idealized beams of open or closed section, with any number of flanges. The normal stress σ_x is assumed to be uniformly distributed on the concentrated flange area A_f. Therefore, the axial load in the flange is given simply by $P_x^{(f)} = \sigma_x A_f$, where

[handwritten: axial loads of stringers, $P_x^{(f)} = \sigma_x A_f$ where $P_x^{(f)} =$]

$$P_x^{(f)} = \frac{1}{I_y I_z - I_{yz}^2}[-(M_z I_y + M_y I_{yz})y_f + (M_y I_z + M_z I_{yz})z_f]A_f$$ [4.8.1]

[handwritten: (used for stringer loads)]

The moments of inertia and the flange coordinates are measured relative to axes through the centroid of the cross section. If the flange area does not vary with the span, it follows that $dP_x^{(f)}/dx = A_f(d\sigma_x/dx)$. Letting $P_x'^{(f)} = dP_x^{(f)}/dx$ and substituting Equation (4.7.2) yields

[handwritten: Flange load gradient]

$$P_x'^{(f)} = \frac{1}{I_y I_z - I_{yz}^2}[(I_y V_y - I_{yz} V_z)y_f + (I_z V_z - I_{yz} V_y)z_f]A_f$$ [4.8.2]

for a beam of constant cross section. The relationship between the flange load gradient $P_x'^{(f)}$ and the shear flows in the adjacent webs is found by adding a flange to the free-body diagram of the web juncture in Figure 4.7.4, as illustrated in Figure 4.8.2. Requiring equilibrium in the flange direction yields

$$P_x'^{(f)} = \sum_{i=1}^{\text{No. of webs}} q^{(i)}$$

[handwritten: $\Rightarrow \Sigma q_{out} = \Sigma q_{in} + P_x'^{(f)}$]   [4.8.3]

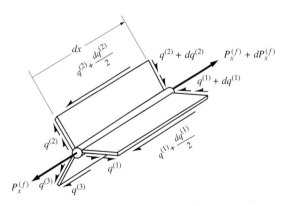

Figure 4.8.2   Shear flows and flange loads at a differential portion of a flange–web juncture.

To solve idealized constant-cross-section beam problems, we use Equation 4.8.2 to find the flange load gradients, and then employ Equation 4.8.3 to obtain the shear flows in the webs. This is illustrated in the next example.

Example 4.8.1 Find the stresses in the stringers and the shear flow in the webs at station $x = 100$ in. of the box beam shown in Figure 4.8.3. The webs are ineffective in bending, and their areas have been lumped at the points indicated in the table in Figure 4.8.3. The skin cross sections are straight lines, except for the curved leading edge, which has an enclosed area of 35 in.[2]

Step 1

Locate the centroid G of the cross section, using only the areas concentrated at the stringers.

$$z_G = \frac{\sum A_i z_i}{A} = \frac{0.75 \times 0 + 1.0 \times 0 + 0.25 \times (-8) + 0.6 \times (-16) + 0.5 \times (-16)}{0.75 + 1.0 + 0.25 + 0.6 + 0.5} = \frac{-19.6}{3.1} = -6.323 \text{ in.}$$

$$y_G = \frac{\sum A_i y_i}{A} = \frac{0.75 \times 0 + 1.0 \times 12 + 0.25 \times 9 + 0.6 \times 6 + 0.5 \times 0}{3.1} = \frac{17.9}{3.1} = 5.758 \text{ in.}$$

Figure 4.8.4 shows the coordinates of the stringers relative to $\bar{x}\bar{y}\bar{z}$ axes located at G.

Step 2

Calculate the moments of inertia I_{Gy}, I_{Gz}, and I_{Gyz} about the axes through the centroid G.

$$I_{Gy} = \sum A_i (z_i - z_G)^2 = 0.75(6.323)^2 + 1.0(6.323)^2 + 0.25(-1.677)^2$$
$$+ 0.6(-9.677)^2 + 0.5(-9.677)^2 = 173.7 \text{ in.}^4$$

$$I_{Gz} = \sum A_i (y_i - y_G)^2 = 0.75(-5.758)^2 + 1.0(6.242)^2 + 0.25(3.242)^2$$
$$+ 0.6(0.2419)^2 + 0.5(-5.758)^2 = 83.07 \text{ in.}^4$$

$$I_{Gyz} = \sum A_i (y_i - y_G)(z_i - z_G) = 0.75(6.323)(-5.758) + 1.0(6.323)(6.242) + 0.25(-1.677)(3.242)$$
$$+ 0.6(-9.677)(0.2419) + 0.5(-9.677)(-5.758) = 37.26 \text{ in.}^4$$

Step 3:

Find the shear and bending moment at the section 100 inches from the free end.

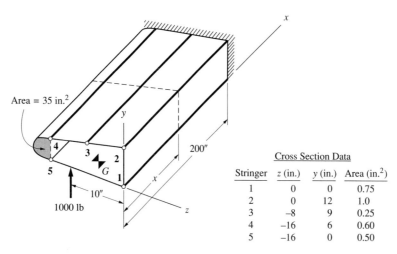

Figure 4.8.3 Single-cell cantilevered box beam with a shear force applied at the free end.

Cross Section Data			
Stringer	z (in.)	y (in.)	Area (in.²)
1	0	0	0.75
2	0	12	1.0
3	-8	9	0.25
4	-16	6	0.60
5	-16	0	0.50

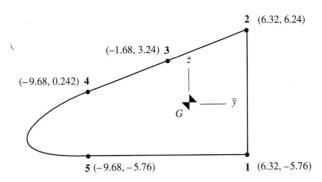

Figure 4.8.4   Beam cross section showing the stringer coordinates (in inches) relative to the centroid G.

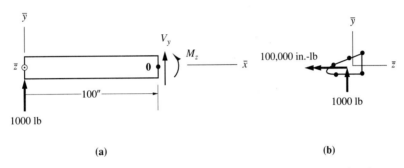

(a)   (b)

Figure 4.8.5   (a) Free-body diagram of the beam in Figure 4.8.3, revealing the shear and bending moment at $x = 100$ in. (b) Loads on the section at $x = 100$ in.

In this case, bending occurs in the xy plane. Figure 4.8.5 shows a free-body diagram of the first 100 inches of the beam. The shear V_y and bending moment M_z on the cross section of the cut are shown in their conventionally positive directions (cf. Figure 4.3.1). Equilibrium requires that

$$\sum F_y = 0: \qquad \Rightarrow \qquad V_y + 1000 = 0 \qquad \Rightarrow \qquad V_y = -1000\,\text{lb}$$

$$\sum M_o = 0: \qquad \Rightarrow \qquad M_z - 100 \times 1000 = 0 \qquad \Rightarrow \qquad M_z = 100,000\,\text{in.-lb}$$

Step 4:

Calculate the bending stresses in the stringers.

From Equation 4.6.8 we have the flexure stress formula

$$\sigma_x = \frac{P}{A} + \frac{1}{I_{Gy}I_{Gz} - I_{Gyz}^2}\left[-\left(I_{Gy}M_z + I_{Gyz}M_y\right)(y - y_G) + \left(I_{Gz}M_y + I_{Gyz}M_z\right)(z - z_G)\right]$$

In our case, $P = 0$, $V_z = 0$, $M_y = 0$ and the moments of inertia are as previously computed. Thus,

$$\sigma_x = \frac{1}{(173.7)(83.07) - 37.26^2}\left[-(173.7 \times 100,000)(y - 5.758) + (37.26 \times 100,000)(z + 6.323)\right]$$

$$= -1332y + 285.7z + 9476$$

where y and z are measured from flange 1, as indicated in Figure 4.8.3. Substituting the ith stringer's coordinates and multiplying by the area gives the stringer loads, $P_x^{(i)}$. The stringer loads on the section are listed in Table 4.8.1 and sketched in Figure 4.8.6.

Table 4.8.1   Stringer stresses and loads at station $x = 100$ in., for the beam of Figure 4.8.3.

Stringer	σ_x (psi)	$P_x^{(i)}$ (lb)
1	9476	7107
2	−6508	−6508
3	−4798	−1199
4	−3088	−1852
5	4904	2452

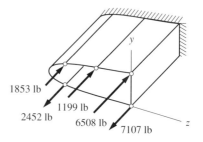

Figure 4.8.6   Stringer loads at station $x = 100$ in.

Step 5:

Calculate the flange load gradients at $x = 100$ in.

We compute the flange load gradients $P_x'^{(f)}$ using Equation (4.8.2), as follows:

$$P_x'^{(f)} = \frac{1}{I_{Gy}I_{Gz} - I_{Gyz}^2}[(I_{Gy}V_y - I_{Gyz}V_z)(y - y_G) + (I_{Gz}V_z - I_{Gyz}V_y)(z - z_G)]A_f$$

Recalling that $V_y = -1000$ lb and $V_z = 0$, this gives

$$P_x'^{(f)} = \frac{1}{173.7 \times 83.07 - (37.26)^2}\{[173.7 \times (-1000)](y - 5.758) + [-(37.26) \times (-1000)](z + 6.323)\}A_f$$

so that for the ith flange,

$$P_x'^{(i)} = (-13.32y_i + 2.857z_i + 94.76)A_i$$

This expression yields the flange load gradients listed in Table 4.8.2. They are shown in parentheses alongside each stringer in Figure 4.8.7.

Step 6:

Starting with any one of the webs, calculate the shear flows around the section in terms of the shear flow in that web. Assume all positive shear flows are directed counterclockwise around the cell.

Let us arbitrarily begin with the web between stringers 1 and 2 in which the shear flow is $q^{(1-2)}$. Then, starting at stringer 2, we move counterclockwise around the section, and at each stringer, we use Equation 4.8.3 to write

$$\sum q_{out} = \sum q_{in} + P_x'^{(f)}$$

thereby expressing the shear flows in the webs in terms of $q^{(1-2)}$. For example, at stringers 2 and 3, we get

$$q^{(2-3)} = q^{(1-2)} + P_x'^{(2)} = q^{(1-2)} - 65.08 \quad \text{(lb/in.)}$$
$$q^{(3-4)} = q^{(2-3)} + P_x'^{(3)} = (q^{(1-2)} - 65.08) + (-11.99) = q^{(1-2)} - 77.07 \quad \text{(lb/in.)}$$

Table 4.8.2   Flange load gradients in the beam of Figure 4.8.3.

Stringer	$P_x^{\prime(f)}$ (lb/in.)
1	71.07
2	−65.08
3	−11.99
4	−18.53
5	24.52

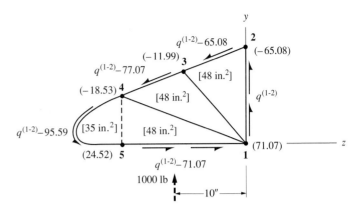

Figure 4.8.7   Shear flows (lb/in.) in terms of the shear flow in web 1–2. Flange load gradients (lb/in.) are shown in parentheses. Enclosed areas are shown in brackets. (The x axis is *into* the page, just as in Figure 4.8.4.)

Continuing around the section, we obtain all of the shear flows for the section, as illustrated in Figure 4.8.7.

Step 7:

Invoke moment equivalence at the section.

The shear flows acting on the section must be statically equivalent to a 1000 lb force directed upwards, with a line of action 10 inches to the left of flange 1. In particular, the moment of the shear flows about a given point, say flange 1, must equal the moment of the 1000 lb applied load. That is, referring to Figure 4.8.7 and taking counterclockwise moments as positive, we require that

$$(2 \times 48)\left(q^{(1-2)}-65.08\right) + (2 \times 48)\left(q^{(1-2)}-77.07\right) + [2 \times (48+35)]\left(q^{(1-2)}-95.59\right) = -10 \times 1000$$

$$q^{(1-2)} = 54.51 \text{ lb/in.}$$

Thus, moment equivalence yields the shear flow in the web where we started. Substituting this counterclockwise shear flow into those shown in Figure 4.8.7 yields the shear flows in the remaining webs, as illustrated in Figure 4.8.8.

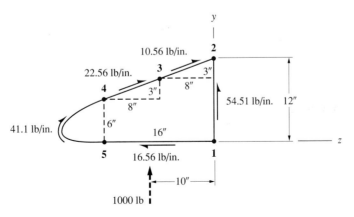

Figure 4.8.8   Shear flow distribution that is statically equivalent to the 1000 lb load.

Example 4.8.2   Find the shear flows in the webs at station $x = 1.3$ m of the box beam shown in Figure 4.8.9.
Since the stringer loads are not required, bending moments are not an issue in this problem. The shear loading is constant from one end of the beam to the other, which means the shear flows will be constant throughout each of the six rectangular shear panels. Thus, the section at 1.3 m is equivalent to any other section, as far as shear flows are concerned.

No σ_x or P_i^s

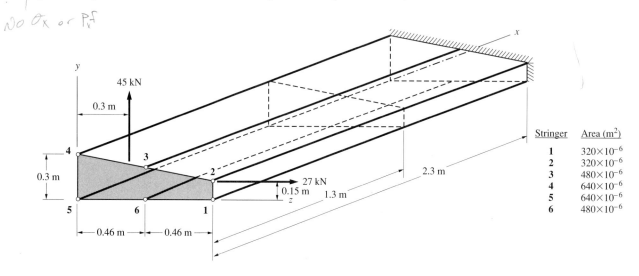

Stringer	Area (m²)
1	320×10^{-6}
2	320×10^{-6}
3	480×10^{-6}
4	640×10^{-6}
5	640×10^{-6}
6	480×10^{-6}

Figure 4.8.9   Cantilevered, idealized box beam with shear loads applied at the free end.
The cross-sectional areas of the stringers are listed.

The centroid of the concentrated flange areas of the beam's uniform trapezoidal cross section is found to have the coordinates $y_G = 0.1208$ m and $z_G = 0.3578$ m, as shown in Figure 4.8.10. The moments of inertia of the beam cross section with respect to axes through the centroid, parallel to y and z, are then

$$I_{Gz} = 47.05 \times 10^{-6} \text{ m}^4 \qquad I_{Gy} = 376.2 \times 10^{-6} \text{ m}^4 \qquad I_{Gyz} = -30.67 \times 10^{-6} \text{ m}^4$$

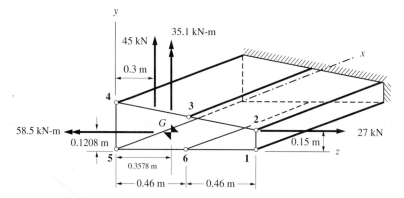

Figure 4.8.10   Negatively-oriented cross section of the beam at station 1.3m,
showing the location of the centroid G and the shear and bending
moments of the loads applied to the free end.

Since $V_z = -27$ kN and $V_y = -45$ kN, the flange load gradient $P_x'^{(i)}$ at the ith stringer, according to Equation 4.8.2, is

$$P_x'^{(i)} = \frac{10^{-6}}{\left[47.05 \cdot 376.2 - (-30.67)^2\right] \cdot 10^{-12}} \times$$

$$\{[(376.2)(-45) - (-30.67)(-27)](y_i - y_G) + [(47.05)(-27) - (-30.67)(-45)](z_i - z_G)\} A_i$$

$$= \left[(-1060 y_i - 158.2 z_i + 184.6) \times 10^3\right] A_i \ \ (\text{kN/m})$$

where y_i and z_i are the coordinates, in meters, of the ith concentrated flange area relative to the coordinate system shown in Figure 4.8.9. Substituting these and the stringer areas (in m^2) listed in Figure 4.8.9, we obtain the flange load gradients shown in parentheses in Figure 4.8.11a.

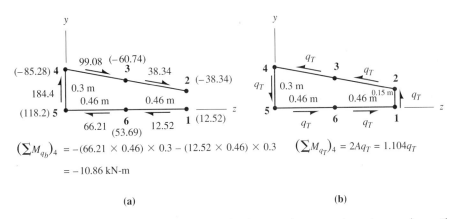

$$\left(\sum M_{q_b}\right)_4 = -(66.21 \times 0.46) \times 0.3 - (12.52 \times 0.46) \times 0.3 \qquad \left(\sum M_{q_T}\right)_4 = 2A q_T = 1.104 q_T$$

$$= -10.86 \text{ kN-m}$$

(a)   (b)

Figure 4.8.11   (a) The flange load gradients (kN/m) at each stringer, shown in parentheses. The indicated purely flexural shear flows (kN/m) arise if the applied shear loads in Figure 4.8.10 act at the shear center of the *open* section (web 1–2 removed). (b) The constant torsional shear flow q_T due to a pure counterclockwise torque applied to the *closed* section (web 1–2 restored). Moments are assumed positive counterclockwise.

We can now start with, say web 1–2 and calculate the shear flows in all of the webs in terms of $q^{(1-2)}$, in the same way as in the previous example. An alternative, equivalent approach is to imagine that web 1–2 is cut, eliminating it from the section, thereby producing the open section shown in Figure 4.8.11a. Of course, any other web could have been chosen. With web 1–2 absent, we can start with flange 2 and use Equation 4.8.3 (shear flow out equals shear flow in plus flange load gradient) at each stringer to obtain the shear flows shown in Figure 4.8.11a. The web was not physically removed; what we did was shift the lines of action of the applied shear loads so that they would pass through the shear center of the open section with web 1–2 absent, thereby reducing the shear flow in that web to zero.

To bring the lines of action of the 45 kN and 27 kN loads from the open section's shear center back to where they were to begin with, we superimpose a pure torsional counterclockwise shear flow q_T onto the closed section with web 1–2 in place, as illustrated in Figure 4.8.11b. The value of q_T is obtained by requiring static equivalence between the torsional moment of the net shear flows (bending plus torsional) and that of the externally applied loads carried by the section. As shown in Figure 4.8.11a, the moment M_{q_b} of the shear flows about flange 4 due to pure bending is -10.86 kN-m. The moment M_{q_T} of the purely torsional shear flow q_T about flange 4 (or any other point) is $1.104 q_T$, according to Figure 4.8.11b. The sum of these two moments must be equal to the moment of the applied shear loads about flange 4. That is, taking positive moments counterclockwise around point 4, we have

$$\left(\sum M_{q_b}\right)_4 + \left(\sum M_{q_T}\right)_4 = 45 \times 0.3 + 27 \times 0.15$$

$$-10.86 + 1.104q_T = 17.55$$
$$q_T = 68.63 \text{ kN/m}$$

Superimposing this counterclockwise torsional shear flow onto the flexural shear flows in Figure 4.8.11a yields the net shear flows in the webs, illustrated in Figure 4.8.12.

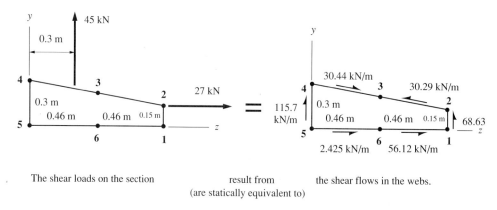

The shear loads on the section   result from   the shear flows in the webs.
(are statically equivalent to)

Figure 4.8.12   Shear force resultants and their statically equivalent shear flow distribution.

Although the methods of shear flow analysis illustrated in the previous two examples are nearly identical, we will use the approach illustrated in Example 4.8.1 throughout the remainder of this text.

Idealized beams are relatively simple to analyze, for several reasons. Since the cross-sectional area is lumped into a finite number of points, computing the location of the centroid and the area moments of inertia of a section does not involve integration or the use of the transformation rules for the area moments of inertia. Also, since the shear flow in the webs is constant, integration is not required to find the force resultants of the shear flow distribution.

Suppose we wish to solve Example 4.7.3 by treating it as an idealized beam problem, rather than one in which the walls carry bending as well as shear. In that case, we assume that the walls act as shear webs and that all of the bending load is carried by longitudinal stiffeners (stringers). Yet, there are no stringers in Figure 4.7.12. Therefore, to approximately represent the bending-load carrying capability of the walls, which was accurately accounted for in Example 4.7.3, we must create some concentrated stringer areas by lumping portions of the wall areas at selected points around the section. Let us consider three methods of doing this.

Method 1

The simplest approach is to apportion a wall's area equally to two concentrated areas located at each end of the wall. This method preserves the location of the centroid of the section. Subdividing each wall into two or more wall segments increases the accuracy. Figures 4.8.13a and 4.8.13b show the results of applying this method to the cross section of Figure 4.7.12, using one and three wall segments, respectively. Notice that the location of the centroid remains the same as the original section, but the area moments of inertia are different. The methods illustrated previously in examples 4.8.1 and 4.8.2 were employed to compute the constant shear flows shown for each web of Figure 4.8.13. These shear flows should be compared with the exact solution pictured in Figure 4.7.19.

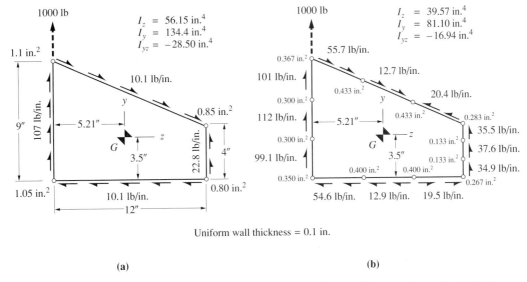

Figure 4.8.13   Shear flows in the section of Figure 4.7.12 if the webs are effective in shear only. (a) Wall areas lumped equally to the corners. (b) Accuracy is improved if each wall is subdivided into three equal lengths, and the area of each segment is lumped equally to its endpoints.

Method 2

In this approach,[11] each wall (or wall segment) is idealized as two stringer areas, each having half the wall area and spaced to preserve the centroidal area moment of inertia, as illustrated in Figure 4.8.14. The results of applying this method to the beam of Figure 4.7.12 are shown in Figure 4.8.15. By comparing Figure 4.8.15 with Figure 4.7.19, we see that the second area moments of the section are indeed identical and that the shear flows around the idealized section are very close to the exact values.

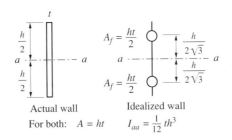

Figure 4.8.14   Idealization of a thin rectangular wall into two concentrated areas for shear flow calculations.

| [11]R. M. Rivello, *Theory and Analysis of Flight Structures*, Section 7.6, New York, McGraw-Hill, 1969.

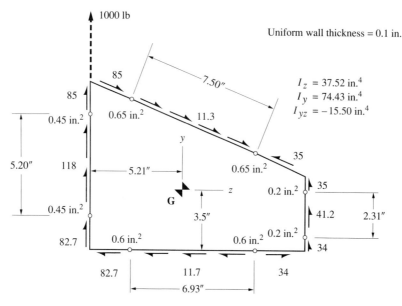

Figure 4.8.15   Shear flows (lb/in.) on the section of Figure 4.7.10, idealized by lumping wall areas equally to concentrated areas located according to
Figure 4.8.14. The webs are effective in shear only.

Method 3

Consider any thin rectangular wall of a beam cross section, and denote its endpoints as 1 and 2, respectively. According to Equation 4.6.6, the normal stress on the wall varies linearly from σ_1 to σ_2, as illustrated in Figure 4.8.16. That is,

$$\sigma_x = \left(1 - \frac{s}{h}\right)\sigma_1 + \frac{s}{h}\sigma_2, \qquad 0 \le s \le h \qquad \text{[4.8.4]}$$

where h is the length of the wall, and s is the distance measured along the wall from point 1. Let the concentrated areas A_1 and A_2 at each end of the idealized wall be selected so that the force systems on the wall and its ideal-

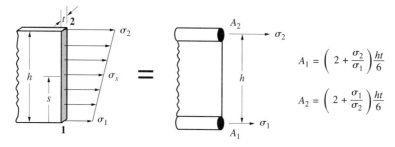

Figure 4.8.16   For static equivalence of the true and idealized stress systems, the
concentrated areas must be related to the wall geometry and end-
point stresses by the formulas shown.

ized counterpart are statically equivalent.[12] That means the resultant force and the resultant moment about any point (such as point 1) must be identical. In particular,

$$\sigma_1 A_1 + \sigma_2 A_2 = \int_0^h \sigma_x (t\,ds)$$

$$\sigma_2 A_2 h = \int_0^h \sigma_x s (t\,ds)$$

where t is the wall thickness. Substituting Equation 4.8.4, integrating, and solving the two equations for A_1 and A_2 yields

$$A_1 = \left(2 + \frac{\sigma_2}{\sigma_1} \right) \frac{ht}{6}$$

$$A_2 = \left(2 + \frac{\sigma_1}{\sigma_2} \right) \frac{ht}{6}$$

[4.8.5]

If the normal stress is uniform across the wall, then $\sigma_1 = \sigma_2$, and we see that $A_1 = A_2 = ht/2$, which coincides with Method 1. If, on the other hand, the wall is in pure flexure, then $\sigma_1 = -\sigma_2$ and Equations 4.8.5 yield $A_1 = A_2 = ht/6$. Clearly, using this method to idealize thin-walled beams results in different concentrated areas for different types of loadings. Since the areas depend on the flexure stresses, and the flexure stresses depend on the areas through the moment of inertia terms in Equation 4.6.6, an iterative procedure is generally required.

Figure 4.8.17 shows the results of applying Method 3 to the solution of the problem illustrated in Figure 4.7.12. We assume that the 1000 lb shear force on the section is accompanied by a bending moment $M_z = 1000$ in.-lb. (Observe that since only ratios of stresses appear in Equation 4.8.5, the magnitude of the bending moment may be chosen arbitrarily.) The wall areas are initially lumped to the corners of the section, just as in Method 1, and cross-sectional geometry is used to calculate the flexure stresses by means of Equation 4.6.6. Equation 4.8.5 are then

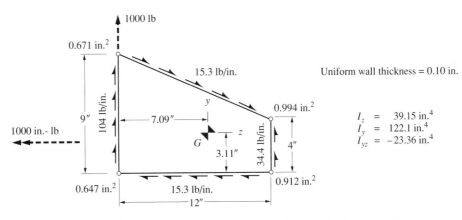

Figure 4.8.17   Shear flows on the section of Figure 4.7.10, idealized by lumping wall areas according to Equation 4.8.5. The 1000 in.-lb bending moment was used to compute the normal stress ratios in those formulas.

| [12]T. H. G. Megson, *Aircraft Structures for Engineering Students*, Second Ed., Section 8.8. New York, Halstead Press, 1990.

employed to update the concentrated areas, for which a new set of flexure stresses is computed, from which a new set of concentrated areas is obtained, and so on. After several repetitions of this procedure, the concentrated areas converge to equal the values shown in Figure 4.8.17. We use those areas to find the shear flows by the usual procedure.

If stringers are present in the beam cross section, their areas are concentrated at the stringer locations and added to the array of lumped skin areas, calculated by methods such as those just described.

4.9   TAPERED IDEALIZED BEAM SHEAR FLOW

In the interest of aerodynamic and structural efficiency, the major components of aircraft are tapered. This is particularly evident in wings and tail surfaces, which are thicker at the root than at the tip and which also taper chordwise. Taper also occurs in the fuselage, in the forward portion and usually aft as well. Up to now, our analysis has been limited to beams of constant cross section. However, the formulas and procedures presented thus far can be applied to tapered beams, provided we limit ourselves to situations in which the taper is slight.[13] Fortunately, that is often the case in aircraft structures, which means there are practical problems we can solve—designs we can check—reasonably accurately by means of familiar beam theory. In approaching any problem, however, we must remind ourselves of the assumptions on which that theory is based and proceed with caution if they do not all apply. We can take some clever liberties in first-pass, order-of-magnitude load path analyses using hand calculations. However, the availability of structural analysis computer codes for even the most modest computing facilities—including modern desktop computers—obviates the need to stretch the limits of applicability of stress formulas in order to keep hand calculations tractable.

Figure 4.9.1 shows examples of plane, tapered, idealized cantilever beams. The component $P_x^{(i)}$ of the axial load in stringer i normal to the cross section (in the direction of the x axis) is given by Equation 4.8.1, which in this situation takes the form

$$P_x^{(i)} = -\frac{M_z(y_i - y_G)}{I_{Gz}} A_i \qquad [4.9.1]$$

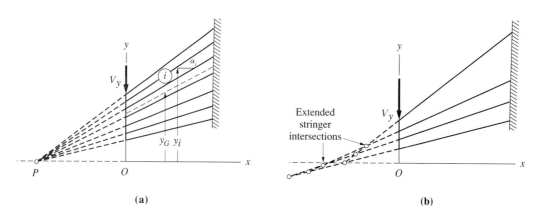

Figure 4.9.1   Multistringer, plane, tapered idealized beams. (a) Shear webs sharing a common vertex P. (b) Multiple stringer intersects.

[13]See also J. M. Gere and S. P. Timoshenko, *Mechanics of Materials,* Second Ed., Section 5.9. Belmont, CA, Brooks/Cole Engineering Division, Wadsworth, Inc., 1984.

To find the net shear flow at a given flange in a beam of uniform cross section, we have previously employed Equation 4.8.3, which requires computing the flange load gradient, $dP_x^{(i)}/dx$ (cf. Equation 4.8.2). For a beam of variable cross section, both the bending moment M_z and, in general, all of the quantities in Equation 4.9.1 are functions of x, including the stringer areas. Thus, recalling that $dM_z/dx = -V_y$, we have the lengthy formula

$$\frac{dP_x^{(i)}}{dx} = \left[\left(V_y + \frac{M_z}{I_{G_z}}\frac{dI_{G_z}}{dx}\right)\frac{A_i}{I_{G_z}} - \frac{M_z}{I_{G_z}}\frac{dA_i}{dx}\right](y_i - y_G) - \frac{M_z A_i}{I_{G_z}}\frac{d}{dx}(y_i - y_G)$$ [4.9.2]

Compare this expression with the far simpler version applicable to uniform cross section beams:

$$\frac{dP_x^{(i)}}{dx} = \frac{V_y(y_i - y_G)A_i}{I_{G_z}}$$ [4.9.3]

If all of the flange areas are constant across the span and if all of the stringers taper towards a common point of intersection (Figure 4.9.1a), then—and only then—does Equation 4.9.2 simplify to the following expression:

$$\frac{dP_x^{(i)}}{dx} = \frac{V_{y,\text{webs}}(y_i - y_G)A_i}{I_{G_z}}$$ [4.9.4]

This is identical to Equation 4.9.3 with the total shear V_y replaced by $V_{y,\text{webs}}$, which is the portion of the total shear carried by the webs. The remainder is borne by the transverse y components of the axial loads in the inclined stringers. Assuming the flange load $P_x^{(i)}$ is tensile, the transverse load component in stringer i is

$$P_y^{(i)} = P_x^{(i)} \tan\alpha_i$$ [4.9.5]

where α_i is the angle of inclination of the stringer, as indicated in Figure 4.9.2. Thus,

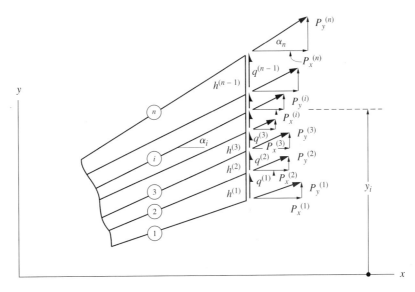

Figure 4.9.2   Flange loads and shear flows at a section of a multistringer, plane, tapered idealized beam.

$$V_y = V_{y,\text{webs}} + \sum_{i=1}^{\text{No. stringers}} P_x^{(i)} \tan \alpha_i \qquad\qquad \textbf{[4.9.6]}$$

We know from the discussion in Section 2.5 that equilibrium requires the shear flow in tapered pure-shear webs to be given by Equation 2.5.4:

$$q(x) = \frac{\bar{q} h_o h_L}{h(x)^2}$$

Do the shear flows calculated using Equation 4.8.3 ($\sum q^{(i)} = P_x'^{(i)}$) and Equation 4.9.2 satisfy this equation? They do, but *only if the stringers all share a common point of intersection*. Granted, this requirement is usually met in plane idealized beams. However, in three-dimensional, tapered box beams, the side, top, and bottom panels do not generally have a common vertex.

The method of analyzing tapered structures so as to yield shear flows that are in equilibrium was first addressed in section 2.6, in the discussion of tapered torque boxes. The analysis must focus from the outset on computing the *average* shear flow in each shear web of the structure. The panel equilibrium condition, Equation 2.5.4, can be invoked (if and when required) to evaluate the shear flow anywhere *within* a panel in terms of its average shear flow. We will henceforth consistently attack the solution of tapered beams in the same way. Just as the flange loads in tapered torque boxes are assumed to be zero at each end of the structure, we will always assume that the flange loads at each end of a tapered beam are given by the flexure formula. We will then set all of the stringers in equilibrium under the action of the computed flange endloads and the average shear flows in the panels to which they are attached.

This *average shear flow method* will be used in this text to analyze the stresses in tapered two- and three-dimensional idealized beams. The procedure differs little from that used in the analysis of beams of uniform cross section. The flexure stress formula is used to compute the axial components of flange loads at each end of the beam. Each stringer is then required to be in equilibrium under the action of these computed endloads and the average shear flows in adjacent shear panels. Thus, for a stringer such as that illustrated in Figure 4.9.3, for which the number of adjoining shear webs is arbitrary, we have

$$P_x^{(i)}(L) - P_x^{(i)}(0) - \bar{q}^{(j)}L - \bar{q}^{(k)}L - \bar{q}^{(l)}L - \cdots = 0 \qquad\qquad \textbf{[4.9.7]}$$

The assumed direction of the average shear flows is such that positive shear flows in the plane of the near section, $x = 0$, are *out* of the flange, consistent with our previous notation (cf. Figure 4.8.2). The *average flange load gradient* for the stringer is defined as

$$\boxed{\overline{P_x'^{(i)}} = \frac{P_x^{(i)}(L) - P_x^{(i)}(0)}{L}} \qquad\qquad \textbf{[4.9.8]}$$

Substituting this into Equation 4.9.7 and rearranging terms, we obtain

$$\boxed{\sum_{\text{adjoining webs}} \bar{q}_{out} = \overline{P_x'^{(i)}}} \qquad\qquad \textbf{[4.9.9]}$$

which is analogous to Equation 4.8.3. Equation 4.9.9 relates the average shear flows to the average flange load gradients in the same way that Equation 4.8.3 relates shear flows to flange load gradients at any section of a constant-cross-section beam.

Equation 4.9.9 is not written for any specific cross section. Instead, it applies to the beam as a whole and can therefore be imagined to apply at a generic section, as will be made clear in the following examples.

As in constant-cross-section box beam analysis, it is necessary to invoke moment equivalence of the shear flows in tapered beams. Unlike constant-cross-section beams, the section geometry and the shear flows vary

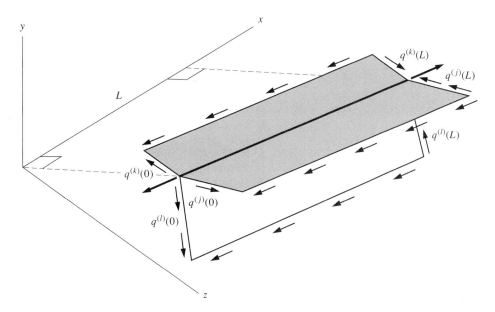

Figure 4.9.3   Typical stringer of a tapered box beam.

It is joined to several shear panels (in this case, three are shown).

along the span, and we must account for these variations when selecting the section at which moment about the beam axis is to be computed. The flange loads in stringers inclined to the beam's axis (the x direction) have components in transverse sections (the yz plane), and these components not only carry a portion of the section shear loads, but also exert twisting moments.

Using the flexure formula, Equation 4.8.1, to compute flange loads in the stringers produces the components normal to the section of a tapered beam, that is,

$$P_x^{(f)} = \frac{A_f}{I_y I_z - I_{yz}^2}[-(M_z I_y + M_y I_{yz})y_f + (M_y I_z + M_z I_{yz})z_f]$$

Strictly speaking, A_f is actually A_{fx}, which is the projection of the stringer cross-sectional area onto the beam cross section. In practical applications, however, $A_{fx} \cong A_f$. To find the x, y, and z components of the flange load of an arbitrarily oriented stringer in a three-dimensional beam, we project the flange load $P^{(f)}$ in the direction of the stringer onto the coordinate axes, or

$$P_x^{(f)} = lP^{(f)} \qquad P_y^{(f)} = mP^{(f)} \qquad P_z^{(f)} = nP^{(f)} \qquad\qquad \textbf{[4.9.10]}$$

where l, m, and n are the direction cosines of the stringer, that is, the components of the unit vector directed along the stringer in the direction of the beam axis (increasing x). If the difference between the coordinates of any two points of the stringer are Δx, Δy, and Δz, where $\Delta x > 0$, then the length ΔL of the line joining the points is $\Delta L = \sqrt{\Delta x^2 + \Delta y^2 + \Delta z^2}$. The direction cosines are $l = \Delta x/\Delta L$, $m = \Delta y/\Delta L$, and $n = \Delta z/\Delta L$. Substituting these expressions into Equation 4.9.10 and solving for $P_y^{(f)}$ and $P_z^{(f)}$ in terms of $P_x^{(f)}$ yields

$$P_y^{(f)} = \frac{\Delta y}{\Delta x}P_x^{(f)} \qquad P_z^{(f)} = \frac{\Delta z}{\Delta x}P_x^{(f)} \qquad\qquad \textbf{[4.9.11]}$$

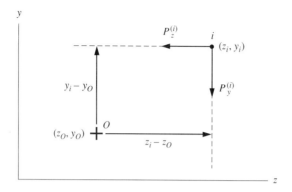

Figure 4.9.4   Components of a tapered beam's ith stringer flange loads in the plane of the beam cross section, with the x axis running into the page.

The flange load components shown are positive since the negatively oriented side of the section is pictured (cf. Figure 4.3.1).

The moment of the in-plane components of the flange load of the ith stringer of a tapered beam are calculated with the aid of Figure 4.9.4. The right-handed xyz coordinate system shown is the one arbitrarily chosen at the outset to define the beam geometry, and it may or may not pass through the centroid of the cross section. Point O is the arbitrarily chosen point in the plane of the section about which moment equivalence is to be established. The in-plane coordinates of point O and the concentrated flange area are shown in parentheses. The flange load components are assumed to be positive. Since they act on the negatively oriented side of the section, they are shown pointing in the *negative* in-plane coordinate directions. The net moment of the flange loads about O is

$$M_{Oi} = P_z^{(i)}(y_i - y_O) - P_y^{(i)}(z_i - z_O)$$

where moments are considered positive if directed counterclockwise in the plane, consistent with the sign convention in Figure 4.3.1. Using Equation 4.9.11, we can write this as

$$M_{Oi} = P_x^{(i)}\left[(y_i - y_O)\left(\frac{\Delta z}{\Delta x}\right)_i - (z_i - z_O)\left(\frac{\Delta y}{\Delta x}\right)_i\right]$$

Thus, for all of the stringers, we obtain

$$M_{O,\text{in–plane stringer loads}} = \sum_{i=1}^{\text{No. of stringers}} P_x^{(i)}[(y_i - y_O)(\frac{\Delta z}{\Delta x})_i - (z_i - z_O)(\frac{\Delta y}{\Delta x})_i] \qquad [4.9.12]$$

This moment, plus the moment about O of the shear flows in the webs of the section, must be equal to the moment about O of the applied loads.

In general, a beam carries shear loads V_y and V_z in both the y and z directions, so that the net shear carried by the webs in each direction is given by

$$V_{y,\text{webs}} = V_y - \sum_{i=1}^{\text{No. stringers}} P_y^{(i)} \qquad V_{z,\text{webs}} = V_z - \sum_{i=1}^{\text{No. stringers}} P_z^{(i)} \qquad [4.9.13]$$

Example 4.9.1   Use the average shear flow method to calculate the shear flow in the tapered cantilever beam of Figure 4.9.5. The area of the top and bottom flanges is 1 in.^{2}, while that of the two inner ones is 0.5 in.^{2}

By symmetry, the centroid of any section of this beam lies midway between the top and bottom flanges. The centroidal moment of inertia of the rightmost section is

$$I_z(70) = 2(0.5 \times 3^2 + 1 \times 9^2) = 171 \text{ in.}^4$$

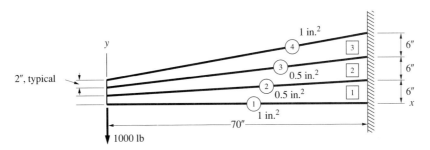

Figure 4.9.5   Plane tapered idealized beam.

Using Equation 4.9.1, the flange loads at that section are

$$P_x^{(1)}(70) = \frac{(1000 \times 70) \cdot (-9) \cdot 1}{171} = -3684 \text{ lb} \qquad P_x^{(4)}(70) = +3684 \text{ lb} \qquad \text{[a]}$$

$$P_x^{(2)}(70) = \frac{(1000 \times 70) \cdot (-3) \cdot 0.5}{171} = -614.0 \text{ lb} \qquad P_x^{(3)}(70) = +614.0 \text{ lb}$$

Since the flange loads at the free end of the beam are zero, dividing each of the quantities in Equation a by the length of the beam—70 inches—yields the average flange load gradient in each stringer. These numbers are shown alongside each stringer on the generic cross section shown in Figure 4.9.6a. Equation 4.9.9 (average shear flow out equals average shear flow in plus

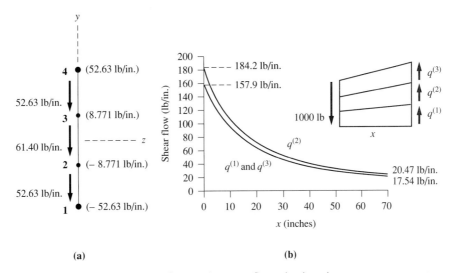

(a)   **(b)**

Figure 4.9.6   (a) Average shear flows and average flange load gradients on a generic section of the tapered beam in Figure 4.9.4. (b) Shear flow variation from end to end.

the average flange load gradient) delivers the average shear flows shown by each web in Figure 4.9.6a.

With the average shear flow $\bar{q}^{(i)}$ for each web available, we use Equation 2.5.4 to calculate the shear flow throughout each panel, as follows:

$$q^{(i)}(x) = \bar{q}^{(i)} \frac{h^{(i)}(0)h^{(i)}(70)}{h^{(i)}(x)^2}$$

The shear flow variation in the beam is plotted in Figure 4.9.6b.

Example 4.9.2   Use the average shear flow method to calculate the shear flows in the tapered idealized box beam shown in Figure 4.9.7. All cross sections are symmetric about the z axis, which passes through the centroid of the left end of the beam ($x = 0$).

The centroid of the cross section at the wall is located at $y_G = 0$ mm and $z_G = -312.5$ mm. The centroidal moments of inertia of that cross section are

$$I_{Gz} = 2\left[(1000 + 500 + 500) \times 312.5^2\right] = 390.6 \times 10^6 \text{ mm}^4$$
$$I_{Gy} = 2\left[1000 \times 187.5^2 + 500 \times 62.5^2 + 500 \times 312.5^2\right] = 171.9 \times 10^6 \text{ mm}^4$$
$$I_{Gyz} = 0 \text{ (by symmetry)}\}$$

The bending moments at the wall are

$$M_y = 15 \text{ kN-m} \qquad M_z = -90 \text{ kN-m}$$

According to Equation 4.8.1, the flange load in the ith stringer, in kN, is

$$P_x^{(i)} = \frac{1}{(171.9 \times 10^6)(390.6 \times 10^6) - 0}\left\{-\left[\left(-90 \times 10^3\right)\left(171.9 \times 10^6\right) + 0\right]y_i + \left[\left(15 \times 10^3\right)\left(3.906 \times 10^6\right) + 0\right](z_i + 312.5)\right\} A_i$$

where the coordinates y_i and z_i, in millimeters, are relative to the system shown in Figure 4.9.7. Substituting the coordinates and the area of each stringer into this expression, we obtain the following flange loads at $x = 3$ m:

$$P_x^{(1)} = -22.36 \text{ kN} \quad P_x^{(2)} = 49.64 \text{ kN} \quad P_x^{(3)} = 38.73 \text{ kN}$$
$$P_x^{(4)} = 55.64 \text{ kN} \quad P_x^{(5)} = -88.36 \text{ kN} \quad P_x^{(6)} = -33.27 \text{ kN}$$

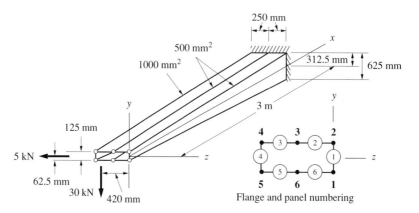

Figure 4.9.7   Box beam, tapered in depth only.

All of the flange loads are, of course, zero at the free end of the beam where the bending moment is zero. The average flange load gradients over the 3 meter span are obtained from Equation 4.9.8 and are shown in parentheses alongside the flanges in Figure 4.9.8, which illustrates a generic section of the beam. The average shear flows are related to the average flange load gradients by the stringer equilibrium relations, Equation 4.9.9. Thus, starting at flange 1, for example, and proceeding counter-clockwise around the section, we get

$$\bar{q}^{(1)} = \bar{q}^{(6)} - 7.454$$
$$\bar{q}^{(2)} = \bar{q}^{(1)} + 16.54 = \bar{q}^{(6)} + 9.091$$
$$\bar{q}^{(3)} = \bar{q}^{(2)} + 12.91 = \bar{q}^{(6)} + 22.00 \qquad \text{[a]}$$
$$\bar{q}^{(4)} = \bar{q}^{(3)} + 18.54 = \bar{q}^{(6)} + 40.54$$
$$\bar{q}^{(5)} = \bar{q}^{(4)} - 29.45 = \bar{q}^{(6)} + 11.09$$

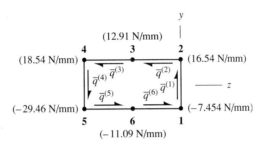

Figure 4.9.8   A generic section of the beam of Figure 4.9.7, showing the average flange load gradients (in parentheses) and the assumed directions of the average panel shear flows.

The average shear flows in panels 1 through 5 have thus been found in terms of $\bar{q}^{(6)}$, the average shear flow in panel 6. To find $\bar{q}^{(6)}$, we need to apply moment equivalence, just as we would if the beam were not tapered. Since the cross-section geometry varies from end to end, we must decide at which station, the near or far end of the beam, to sum torsional moments about the beam x axis. The clear choice is the near end, that is, the section at which the bending moment is zero, because the flange loads are zero and we consequently do not have to compute the moments of their projections in the plane of the section.

Figure 4.9.9 shows the shear flows acting at the free end of the beam. The shear flow directions are consistent with those in Figure 4.9.8. Setting the moments of the shear flows about flange 1 equal to the moments of the 30 kN and 5 kN shear loads, we have

$$125 \times 250 q^{(2)}(0) + 125 \times 250 q^{(3)}(0) + 500 \times 125 q^{(4)}(0) = 62.5 \times 5000 + 420 \times 30,000 \qquad \text{[b]}$$

These shear flows are related to the average shear flows by the shear panel equilibrium relation, Equation 2.5.4,

$$q^{(i)}(0) = \bar{q}^{(i)} \frac{h^{(i)}(100)}{h^{(i)}(0)} \qquad \text{[c]}$$

where $h^{(i)}(0)$ and $h^{(i)}(100)$ are the widths of the ith panel at the near and far ends of the beam, respectively. Substituting this expression into Equation b yields

$$31,250 \times \bar{q}^{(2)} + 31,250 \times \bar{q}^{(3)} + 62,500 \times 5\bar{q}^{(4)} = 12.91 \times 10^6$$

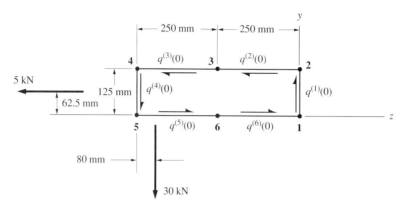

Figure 4.9.9   Shear flows on the section at station $x = 0$.

Since $\bar{q}^{(2)}$, $\bar{q}^{(3)}$, and $\bar{q}^{(4)}$ are related to $\bar{q}^{(6)}$ by Equation a, this equation can be written

$$31,250 \times (\bar{q}^{(6)} + 9.091) + 31,250 \times (\bar{q}^{(6)} + 22.00) + 62,500 \times 5(\bar{q}^{(6)} + 40.54) = 12.91 \times 10^6$$

from which we obtain

$$\bar{q}^{(6)} = -1.946 \text{ N/mm} \qquad \textbf{[d1]}$$

From Equations a it follows that the average shear flows in the other panels are

$$\bar{q}^{(1)} = -9.400 \text{ N/mm} \quad \bar{q}^{(2)} = 7.146 \text{ N/mm} \quad \bar{q}^{(3)} = 20.06 \text{ N/mm} \quad \bar{q}^{(4)} = 38.60 \text{ N/mm} \quad \bar{q}^{(5)} = 9.146 \text{ N/mm} \quad \textbf{[d2]}$$

The upper and lower shear panels (2, 3, 5, and 6) are rectangular, so the shear flow in each of them is constant throughout and equals the average shear flow. In the trapezoidal side panels (1 and 4), the shear flow varies according to Equation 2.5.4,

$$q^{(i)}(x) = \bar{q}^{(i)} \frac{h^{(i)}(0)h^{(i)}(L)}{h^{(i)}(x)^2}$$

where $h^{(i)}(x)$ is the linearly varying width of the panel. Panels 1 and 4 have the same shape. For both of them, $h = 125$ mm at $x = 0$; $h = 625$ mm at the wall; and $h(x) = \frac{x}{6} + 125$ (x in mm). The shear flows in each of the shear panels are plotted in Figure 4.9.10.

The analysis of tapered beams commonly uses the section-by-section method, as opposed to the average shear flow method. In the section-by-section approach, the shear flow is computed at several selected intermediate stations, using the same procedure employed in nontapered idealized beams. However, care is taken to account for the fact that some or all of the stringers are inclined to the beam axis. Thus, at each station, the shear flow calculation must be preceded by a computation of the bending loads in the inclined stringers. Using Equation 4.9.13, the resultant of the components of these stringer loads lying in the plane of the beam cross section are then subtracted from the net shear on the section, thereby yielding the portion of the transverse shear load that is picked up by the webs. (The webs, of course, carry all of the shear in constant-cross-section idealized beams.) Furthermore, the torsional moment of the in-plane stringer load components must be added to the moment of the shear flows when the moment equivalence at each station is established (see Equation 4.9.12).

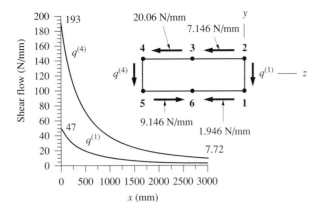

Figure 4.9.10   Shear flows on any section of the beam of Figure 4.9.7. The shear flow is uniform in the rectangular panels (top and bottom).

Unless all of the stringers of an idealized beam taper towards a common point of intersection, the section-by-section procedure will yield shear flows that are different from those predicted by the average shear flow method. Recall that only the average shear flow method guarantees that shear flow in each panel, tapered or not, is self-equilibrating.

Example 4.9.3   Use the section-by-section method to calculate the spanwise shear flow variation in the tapered box beam of the previous example.

Let us compute the shear flow at stations 300 mm apart, starting with the free end of the beam. Using the same xyz reference as before (Figure 4.9.7), we note that the cross-section centroidal coordinates are constant from one end of the beam to the other and are

$$y_G = 0 \qquad z_G = -312.5 \text{ mm} \qquad \text{[a]}$$

Because the taper is in the y direction only, the z coordinates of the stringers do not vary with span; therefore,

$$z_1 = z_2 = 0 \qquad z_3 = z_6 = -250 \text{ mm} \qquad z_4 = z_5 = -500 \text{ mm} \qquad \text{[b]}$$

On the other hand, for the y coordinates, we have

$$y_1 = y_5 = y_6 = -\frac{1}{12}x - 62.5 \text{ (mm)} \qquad y_2 = y_3 = y_4 = \frac{1}{12}x + 62.5 \text{ (mm)} \qquad \text{[c]}$$

The centroidal area moments of inertia across the span are therefore

$$I_{Gz} = \frac{250}{9}x^2 + (41.667 \times 10^3)x + 15.625 \times 10^6 \text{ mm}^4 \qquad I_{Gy} = 171.9 \times 10^6 \text{ mm}^4 \qquad I_{Gyz} = 0 \text{ (by symmetry)} \qquad \text{[d]}$$

Consistent with the sign convention of Figure 4.3.1, the section shear loads anywhere along the span are

$$V_y = 30 \times 10^3 \text{N} \qquad V_z = 5 \times 10^3 \text{ N} \qquad \text{[e]}$$

whereas the bending moments (M_y and M_z) and torsional moment (M_x) around the centroidal axes at each section are

$$M_x = 3.225 \times 10^6 \qquad M_y = 5000x \qquad M_z = -(30 \times 10^3)x \text{ (N-mm)} \tag{f}$$

We can now substitute the data in Equations a, b, c, d, and e, together with the flange areas, into Equation 4.8.1 to get

$$P_x^{(i)} = \frac{1}{I_{Gy}I_{Gz} - I_{Gyz}^2}[-(M_z I_{Gy} + M_y I_{Gyz})(y_i - y_G) + (M_y I_{Gz} + M_z I_{Gyz})(z_i - z_G)]A_i \quad i = 1, \cdots, 6 \tag{g}$$

from which we can compute the flange loads at the selected stations. These are listed in Table 4.9.1.

Table 4.9.1   Flange loads (N) at the selected stations.

Station, x (mm)	$P_x^{(1)}$	$P_x^{(2)}$	$P_x^{(3)}$	$P_x^{(4)}$	$P_x^{(5)}$	$P_x^{(6)}$
0	0	0	0	0	0	0
300	−11490	14220	13130	24080	−27350	−12580
600	−17270	22730	20540	36730	−43270	−19460
900	−20460	28640	25360	44180	−54000	−23730
1200	−22240	33150	28780	48840	−61930	−26600
1500	−23180	36820	31360	51820	−68180	−28640
1800	−23580	39950	33400	53710	−73350	−30130
2100	−23610	42700	35070	54860	−77770	−31250
2400	−23380	45200	36470	55480	−81660	−32100
2700	−22940	47490	37670	55710	−85160	−32760
3000	−22360	49640	38730	55640	−88360	−33270

Next, we can project each of the flange loads in Table 4.9.1 onto the beam cross section (the yz plane), as indicated in Equation 4.9.11. Since there is no taper in the z direction, $P_z^{(f)} = 0$ for each of the stringers at every section. On the other hand, $\Delta y/\Delta x = -1/12$ for stringers 1, 5, and 6, and $\Delta y/\Delta x = +1/12$ for stringers 2, 3, and 4. Therefore, Equation 4.9.13, Equation d, and Table 4.9.1 yield the shears borne by the webs, which are listed in Table 4.9.2.

At each section of the beam, we substitute the web shear loads in Table 4.9.2 into Equation 4.8.2,

$$P_x'^{(i)} = \frac{1}{I_{Gy}I_{Gz} - I_{Gyz}^2} \left\{ \left[(I_{Gy} V_y)_{\text{webs}} - I_{Gyz} V_z)_{\text{webs}} \right](y_i - y_G) + \left[I_{Gz} V_z)_{\text{webs}} - I_{Gyz} V_y)_{\text{webs}} \right](z_i - z_G) \right\} A_i \quad i = 1, \cdots, 6 \tag{h}$$

thereby obtaining the flange load gradients presented in Table 4.9.3.

We are now in a position to find the shear flows at any section of the box beam. To illustrate, let us choose the section at $x = 1500$ mm, illustrated in Figure 4.9.11. Applying Equation 4.8.3 to stringers 1 through 5, in turn, yields the expressions for all of the shear flows (N/mm) in terms of just q_6:

$$
\begin{aligned}
q^{(1)} &= q^{(6)} - 2.121 \\
q^{(2)} &= q^{(1)} + 11.21 = q^{(6)} + 9.091 \\
q^{(3)} &= q^{(2)} + 7.576 = q^{(6)} + 16.67 \\
q^{(4)} &= q^{(3)} + 7.879 = q^{(6)} + 24.55 \\
q^{(5)} &= q^{(4)} - 18.79 = q^{(6)} + 5.758
\end{aligned}
\tag{i}
$$

Table 4.9.2   Section shear loads carried by the webs.

Station, x (mm)	$V_y)_{webs}$ (N)	$V_z)_{webs}$ (N)
0	30000	5000
300	21430	5000
600	16670	5000
900	13640	5000
1200	11540	5000
1500	10000	5000
1800	8824	5000
2100	7895	5000
2400	7143	5000
2700	6522	5000
3000	6000	5000

Table 4.9.3:   Flange load gradients (N/mm) at the selected stations.

Station, x (mm)	$P_x'^{(1)}$	$P_x'^{(2)}$	$P_x'^{(3)}$	$P_x'^{(4)}$	$P_x'^{(5)}$	$P_x'^{(6)}$
0	−55.45	64.54	60.91	114.6	−125.4	−59.09
300	−26.07	35.16	31.52	55.77	−66.68	−29.70
600	−13.97	23.06	19.43	31.58	−42.49	−17.61
900	−7.851	16.94	13.31	19.34	−30.25	−11.49
1200	−4.330	13.42	9.785	12.30	−23.21	−7.967
1500	−2.121	11.21	7.576	7.879	−18.79	−5.758
1800	−0.6449	9.736	6.099	4.926	−15.84	−4.281
2100	0.3903	8.700	5.064	2.856	−13.76	−3.246
2400	1.144	7.947	4.310	1.348	−12.26	−2.492
2700	1.710	7.381	3.745	0.2165	−11.13	−1.926
3000	2.145	6.945	3.309	−0.6545	−10.25	−1.491

Summing positive counterclockwise moments about stringer 1 at the lower right, we have, from Figure 4.9.11a and Equations i that the moment of the shear flows is

$$M_1)_{\text{shear flow}} = (250q^{(2)}) \times 375 + (250q^{(3)}) \times 375 + (375q^{(4)}) \times 500$$
$$= [250(q^{(6)} + 9.091)] \times 375 + [250(q^{(6)} + 16.67)] \times 375 + [375(q^{(6)} + 24.55)] \times 500 \qquad \text{[j]}$$
$$= 375{,}000q^{(6)} + 7.018 \times 10^6 \text{ N-mm}$$

From Figure 4.9.11a, we also obtain the net moment of the in-plane components of flange load about stringer 1:

$$M_1)_{\text{flange loads}} = (4318 + 5682) \times 500 + (2614 + 2386) \times 250 = 6.250 \times 10^6 \text{ N-mm} \qquad \text{[k]}$$

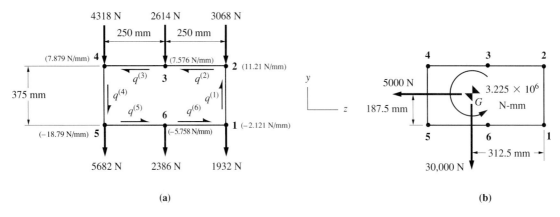

Figure 4.9.11 Station $x = 1500$ mm of the beam in Figure 4.9.7. (a) Flange load gradients (in parentheses) from Table 4.9.3; in-plane components of flange loads, from Table 4.9.1 and Equation 4.9.11; and the assumed shear flow directions. (b) In-plane reactions to the external loads.

Finally, Figure 4.9.11b reveals the moments of the external loads about flange 1, as follows:

$$M_1)_{\text{applied loads}} = 5000 \times 187.5 + 30,000 \times 312.5 + 3.225 \times 10^6 = 13.54 \times 10^6 \text{ N} - \text{mm} \qquad \text{[l]}$$

Moment equivalence then requires that

$$M_1)_{\text{shear flow}} + M_1)_{\text{flange loads}} = M_1)_{\text{applied loads}} \qquad \text{[m]}$$

Thus,

$$(375,000q^{(6)} + 7.018 \times 10^6) + 6.250 \times 10^6 = 13.54 \times 10^6$$

from which it follows that

$$q^{(6)} = 0.7212 \text{ N/mm}$$

Substituting q_6 into Equation i yields the five remaining shear flows around the section at $x = 1500$ mm :

$$q^{(1)} = -1.400 \text{ N/mm} \qquad q^{(2)} = 9.812 \text{ N/mm} \qquad q^{(3)} = 17.39 \text{ N/mm} \qquad q^{(4)} = 25.56 \text{ N/mm} \qquad q^{(5)} = 6.479 \text{ N/mm}$$

Repeating this series of calculations at each of the other ten beam stations provides the data required to plot the shear flow variation across the span. These curves are shown in Figure 4.9.12.

As is evident from Figures 4.9.10 and 4.9.12, for the box beam of Figure 4.9.7, the average shear flow method (Example 4.9.2) and the section-by-section method (Example 4.9.3) yield fairly similar shear flow distributions ($q^{(1)}$ and $q^{(4)}$) in the tapered side panels. However, according to the average shear flow method, the shear flow is constant in the four rectangular cover panels; in the section-by-section approach, there are spanwise variations, which are obvious in Figure 4.9.12, including a reversal in direction of both $q^{(1)}$ (at $x = 1866$ mm) and $q^{(6)}$ (at $x = 474$ and 2283 mm). The two examples also reveal that the spanwise shear flow distribution obtained by the average shear flow method requires much less computational effort than the section-by-section procedure.

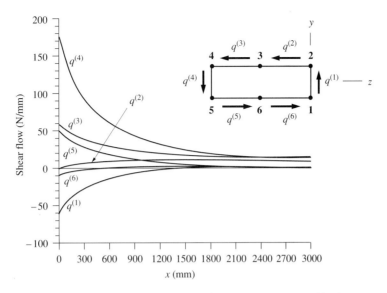

Figure 4.9.12   Shear flows in the beam of Figure 4.9.7, computed by the section-by-section method. Positive shear flows have the directions indicated.

Example 4.9.4   Use the average shear flow method to calculate the shear flow throughout each of the panels in the idealized tapered box beam pictured in Figure 4.9.13. Table 4.9.4 lists the area and endpoint coordinates relative to the global xyz coordinate system shown, for each of the six stringers.

The first step is to calculate the flange loads at the far end of the beam, station 200, and use those results in Equation 4.9.8 to calculate the average flange load gradients $\overline{P_x'^{(i)}}$ for each stringer. The cross section of the beam at the 200 in. station is shown in Figure 4.9.14, which also shows the location of its centroid, computed in the usual fashion. Relative to the global coordinates, the centroid G is located at

$$x_G(200) = 200 \text{ in.} \qquad y_G(200) = 5 \text{ in.} \qquad z_G(200) = -93.33 \text{ in.} \qquad \text{[a]}$$

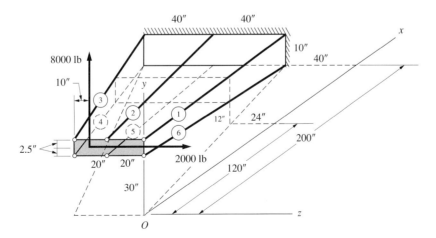

Figure 4.9.13   Tapered wing box with sweep and dihedral.

Table 4.9.4: Stringer area (in.²) and coordinates (in.) at each end of the box beam in Figure 4.9.13.

Stringer	Area	Station 0			Station 200		
		x	y	z	x	y	z
1	0.5	0	35	0	200	10	−40
2	1.0	0	35	−20	200	10	−80
3	1.5	0	35	−40	200	10	−120
4	1.5	0	30	−40	200	0	−120
5	1.0	0	30	−20	200	0	−80
6	0.5	0	30	0	200	0	−40

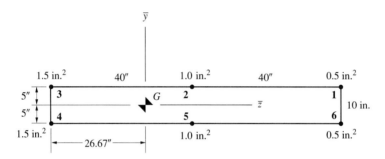

Figure 4.9.14 Beam cross section at station x = 200 in.

From Figure 4.9.14 we see that, at $x = 200$ in., the area moments of inertia about the $\bar{y}\bar{z}$ coordinate axes through the centroid G are

$$I_{\bar{z}}(200) = 2 \times [5^2 \times (0.5 + 1.0 + 1.5)] = 150 \text{ in.}^4$$

and

$$I_{\bar{y}}(200) = 2 \times [(53.33^2 \times 0.5) + (13.33^2 \times 1.0) + (26.67^2 \times 1.5)] = 5333 \text{ in.}^4$$

By symmetry, then, we have $I_{\bar{y}\bar{z}} = 0$.

We shift the loads acting at the near end of the beam to the centroid G at $x = 200$ in. and calculate the moments of the loads about G. The loads act at the point with global coordinates $x = 0$, $y = 32.5$ in., and $z = -30$ in. The position vector from G, whose global coordinates are listed in Equation a, to the load application point is

$$\mathbf{r} = (0 - 200)\mathbf{i} + (32.5 - 5)\mathbf{j} + [-30 - (-93.33)]\mathbf{k} = -200\mathbf{i} + 27.5\mathbf{j} + 63.33\mathbf{k} \text{ in.}$$

Therefore, the vector moment of the loads about G is

$$\mathbf{M}_G = \mathbf{r} \times \mathbf{F} = \begin{vmatrix} \mathbf{i} & \mathbf{j} & \mathbf{k} \\ -200 & 27.5 & 63.33 \\ 0 & 8000 & 2000 \end{vmatrix}$$

$$= (27.5 \times 2000 - 63.33 \times 8000)\,\mathbf{i} - (-200 \times 2000)\,\mathbf{j} + (-200 \times 8000)\,\mathbf{k}$$

$$= -451{,}700\mathbf{i} + 400{,}000\mathbf{j} - 1{,}600{,}000\mathbf{k} \text{ lb-in.}$$

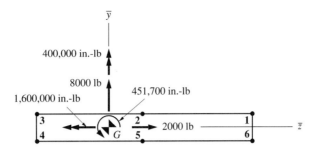

Figure 4.9.15   Forces and moments on the section at x = 200 in.

The forces and moments applied to the section at $x = 200$ in. are illustrated in Figure 4.9.15. According to our sign convention (Figure 4.3.1), we have

$$V_{\bar{y}}(200) = -8000\,\text{lb}$$

$$V_{\bar{z}}(200) = -2000\,\text{lb}$$

$$T(200) = +451,700\,\text{in.} - \text{lb}$$

$$M_{\bar{y}}(200) = -400,000\,\text{in.} - \text{lb}$$

$$M_{\bar{z}}(200) = +1,600,000\,\text{in.} - \text{lb}$$

The flange load in stringer i normal to the section is found using Equation 4.8.1 with $I_{yz} = 0$,

$$P_x^{(i)}(200) = A_i \left\{ -\frac{M_z(200)[y_i(200) - y_G(200)]}{I_z(200)} + \frac{M_y(200)[z_i(200) - z_G(200)]}{I_y(200)} \right\}$$

Using the numbers computed, this equation becomes

$$P_x^{(i)} = A_i \left\{ -10,670[y_i(200) - y_G(200)] - 75.00[z_i(200) - z_G(200)] \right\}$$

Substituting the flange areas and coordinates from Table 4.9.4 and the centroid coordinates from Equation a yields

$$P_x^{(1)}(200) = -26,670\,\text{lb} \qquad P_x^{(2)}(200) = -54,330\,\text{lb} \qquad P_x^{(3)}(200) = -77,000\,\text{lb}$$
$$P_x^{(4)}(200) = 83,000\,\text{lb} \qquad P_x^{(5)}(200) = 52,330\,\text{lb} \qquad P_x^{(6)}(200) = 24,670\,\text{lb}$$

[b]

Since the flange loads are zero at $x = 0$, the average flange load gradients over the 200 in. span are

$$\overline{P_x'^{(i)}} = \frac{P_x^{(i)}(200) - P_x^{(i)}(0)}{200} = \frac{P_x^{(i)}(200)}{200}$$

We have thus obtained the flange load gradients shown in parentheses alongside each flange of the *generic* section pictured in Figure 4.9.16, which also shows the assumed directions of the average shear flows in the panels. Since the average shear flows are related to the average flange load gradients by the stringer equilibrium relations, Equation 4.9.9, we can start at, say, flange 2 and proceed counterclockwise around the section, to obtain the following in terms of $\bar{q}^{(1)}$ only:

$$\bar{q}^{(2)} = \bar{q}^{(1)} - 271.7$$
$$\bar{q}^{(3)} = \bar{q}^{(2)} - 385 = \bar{q}^{(1)} - 656.7$$
$$\bar{q}^{(4)} = \bar{q}^{(3)} + 415 = \bar{q}_1^{(1)} - 241.7$$
$$\bar{q}^{(5)} = \bar{q}^{(4)} + 261.7 = \bar{q}^{(1)} + 20$$
$$\bar{q}^{(6)} = \bar{q}^{(5)} + 123.3 = \bar{q}^{(1)} + 143.3$$

[c]

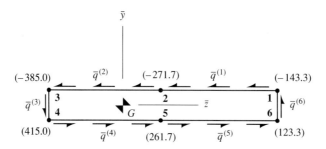

Figure 4.9.16   Average flange load gradients (lb/in.) and the assumed directions of the average shear flows, for a generic cross section of the beam in Figure 4.9.13.

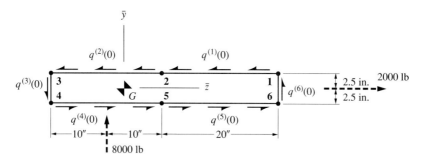

Figure 4.9.17   Section loads and shear flows at station $x = 0$.

The next step is to require that the twisting moment of the shear flows at $x = 0$ be statically equivalent to the torsional moment of the applied loads. The cross section at $x = 0$ is shown in Figure 4.9.17. Invoking this moment equivalence around flange 4, we get

$$5q^{(6)}(0) \times 40 + 20q^{(1)}(0) \times 5 + 20q^{(2)}(0) \times 5 = 8000 \times 10 - 2000 \times 2.5 \qquad \text{[d]}$$

The shear flows at $x = 0$ are related to the average shear flows by the shear panel equilibrium relation, Equation 2.5.4,

$$q^{(i)}(0) = \bar{q}^{(i)} \frac{h^{(i)}(200)}{h^{(i)}(0)} \qquad \text{[e]}$$

where $h^{(i)}(0)$ and $h^{(i)}(200)$ are the widths of the ith panel at the near and far ends of the beam, respectively. Substituting this expression into Equation d yields

$$5\bar{q}^{(6)} \frac{10}{5} \times 40 + 20\bar{q}^{(1)} \frac{40}{20} \times 5 + 20\bar{q}^{(2)} \frac{40}{20} \times 5 = 75,000 \qquad \text{[f]}$$

Here, $\bar{q}^{(2)}$ and $\bar{q}^{(6)}$ are related to $\bar{q}^{(1)}$ in Equation c, by means of which Equation f can be written

$$400(\bar{q}^{(1)} + 143.3) + 200\bar{q}^{(1)} + 200(\bar{q}^{(1)} - 271.7) = 75,000$$

Solving this equation for $\bar{q}_1$ and substituting the result into Equation c yields all of the average shear flows:

$$\bar{q}^{(1)} = 90 \text{ lb/in.} \qquad \bar{q}^{(2)} = -181.7 \text{ lb/in.} \qquad \bar{q}^{(3)} = -566.7 \text{ lb/in.}$$
$$\bar{q}^{(4)} = -151.7 \text{ lb/in.} \qquad \bar{q}^{(5)} = 110.0 \text{ lb/in.} \qquad \bar{q}^{(6)} = 233.3 \text{ lb/in.}$$

[g]

The shear flow at any point within a given shear panel is found in terms of the panel's average shear flow by means of Equation 2.5.4, which, for this beam, appears as

$$q^{(i)}(x) = \bar{q}^{(i)} \frac{h^{(i)}(0)h^{(i)}(200)}{h^{(i)}(x)^2}$$

[h]

where $h^{(i)}(x)$ is the width of the trapezoidal panel i at station x. Plotting Equation h for each of the six panels yields the curves shown in Figure 4.9.18. A negative value indicates that the shear flow direction is opposite to that initially assumed. For example, the shear flows at station $x = 120$ in. are sketched in Figure 4.9.19.

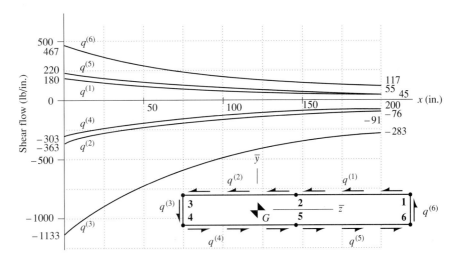

Figure 4.9.18   Shear flow variation with span in each of the six panels of the tapered beam in Figure 4.9.13.

A generic cross section is shown at the bottom of the figure.

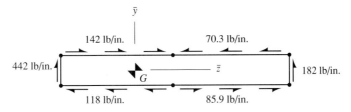

Figure 4.9.19   Shear flows on the section at station $x = 120$ in.

EXERCISES

4.1   Find the maximum tensile and compressive values of axial stress σ_x at station $x = 50$ cm.

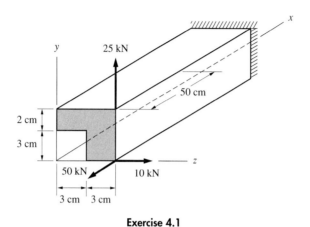

Exercise 4.1

4.2   Calculate the maximum compressive and maximum tensile values of axial stress σ_x at the built-in end of the cantilever beam. A detail of the hollow section is shown.

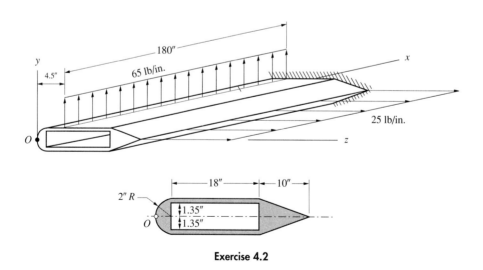

Exercise 4.2

4.3   Find the y and z coordinates of the shear center of the thin-walled open section.

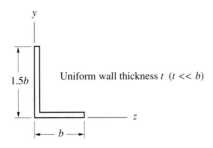

Exercise 4.3

4.4   Find the location of the shear center of the thin-walled section shown. Assume the wall thickness t is uniform and that $t \ll r$.

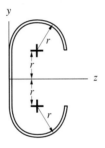

Exercise 4.4

4.5   Find the shear flow distribution in the rectangular section shown, given that the walls are effective in bending as well as shear.

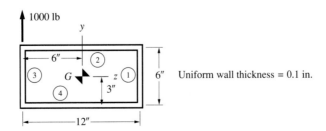

Exercise 4.5

4.6   A thin-walled square section has a small slit at the midpoint of the vertical wall. Find the maximum shear stress if the section transmits a load P applied as shown. Compare that to the maximum shear stress if there were no slit. The walls are effective in bending as well as shear.

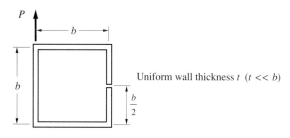

Exercise 4.6

4.7   An idealized torque box having the section shown transmits a torque of 10 kN-m. The nose abc of the section is one-half of an ellipse. The dimensions are measured to the median line of the skin. Calculate the shear flow around the section.

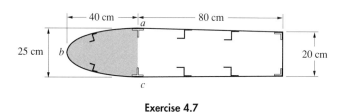

Exercise 4.7

4.8   Find the shear flow distribution in the idealized beam section shown.

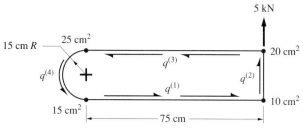

Exercise 4.8

4.9   Repeat Exercise 4.5 assuming the walls are effective in shear only, lumping each wall area equally to its ends, as illustrated.

Exercise 4.9

4.10   Repeat Exercise 4.5 assuming the walls are effective in shear only, lumping the wall areas according to Figure 4.8.14, as illustrated.

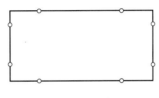

Exercise 4.10

4.11   Verify the shear flows shown in Figure 4.8.13.

4.12   Verify the shear flows shown in Figure 4.8.15.

4.13 a. Locate the centroid of the idealized beam cross section.

   b. Calculate the centroidal area moments of inertia for the beam.

   c. Calculate the normal stresses in the flanges at station $x = 150$ in.

   d. Calculate the shear flows in the beam.

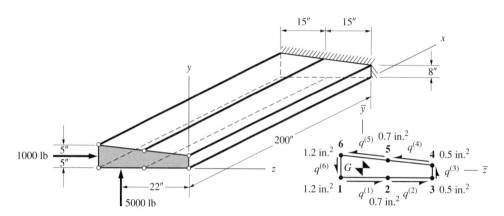

Exercise 4.13

4.14 Calculate the shear flow distribution around the section of the cantilevered box beam located 30 in. from the wall (70 in. from the free end). The panels (webs) are effective in shear only.

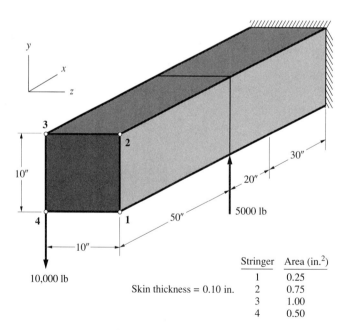

Skin thickness = 0.10 in.

Stringer	Area (in.^{2})
1	0.25
2	0.75
3	1.00
4	0.50

Exercise 4.14

4.15   The idealized cantilevered box beam shown has an upward-directed, spanwise-distributed load acting along one of the top stringers and an aft-directed distributed load acting along the middle of the leading edge, which is a semicircular cylinder centered at O, midway between stringers 1 and 7. The stringer coordinates with respect to the axes shown are listed in the table, along with their areas. Find the load in each stringer and the shear flow in the webs at $x = 150$ in. (i.e., 150 in. from the free end.)

Stringer	z coordinate (in.)	y coordinate (in.)	Area (in.^{2})
1	0.0	0.0	2.0
2	12.0	0.0	1.0
3	24.0	0.0	0.5
4	24.0	6.0	0.5
5	16.0	7.0	0.75
6	8.0	8.0	0.75
7	0.0	9.0	2.0

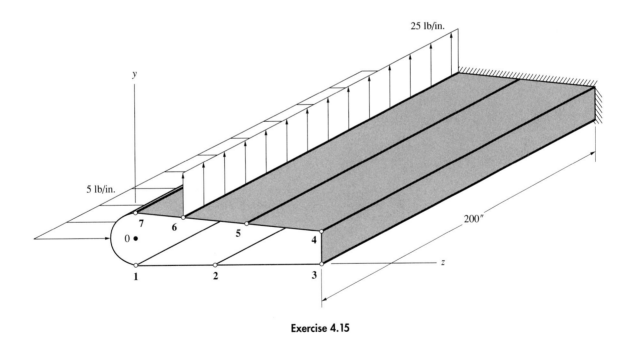

Exercise 4.15

4.16 Solve Example 4.7.1 as an idealized beam using the area lumping scheme of Figure 4.8.14.

4.17 Calculate the shear flow in the panel at a distance of 15 in. from the free end. The panel is effective in shear only.

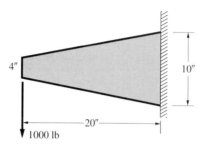

Exercise 4.17

4.18 Calculate the shear flow in the panels at a distance of 15 in. from the wall, if the three stringer areas are 1 in.^{2}. The panels are effective in shear only.

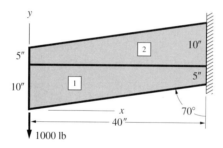

Exercise 4.18

4.19   Calculate the average shear flow in each of the three shear panels of the idealized tapered beam. Also, tabulate the shear flow in each panel at stations 0, 5, 10, 15, 20, 25, and 30. What percentage of the 1000 lb shear load is carried by the stringers at the wall?

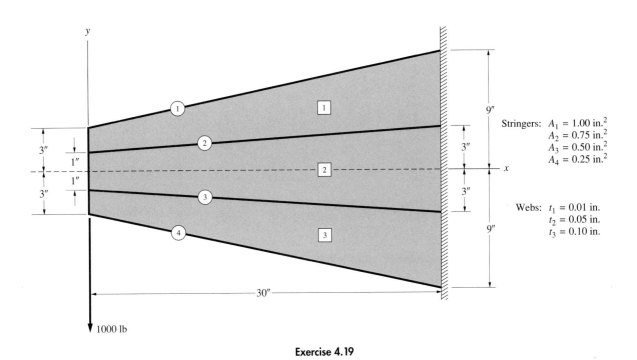

Stringers: $A_1 = 1.00$ in.^{2}
$A_2 = 0.75$ in.^{2}
$A_3 = 0.50$ in.^{2}
$A_4 = 0.25$ in.^{2}

Webs: $t_1 = 0.01$ in.
$t_2 = 0.05$ in.
$t_3 = 0.10$ in.

Exercise 4.19

4.20   The near end ($x = 0$) of the idealized torque box is in pure torsion, as shown. The far end ($x = 150$ in.) is fixed. Find the shear flow in each of the six panels at the near and far ends. (Stringer areas are not required since there is no bending.)

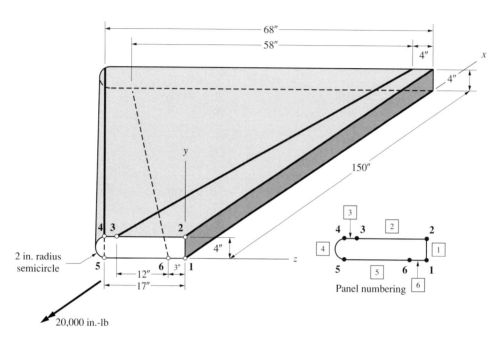

Exercise 4.20

4.21   The top and front views of an idealized box beam are shown, including a 1000 lb applied load. Calculate the shear flow in each of the panels at a distance of 20 in. from the free end of the beam.

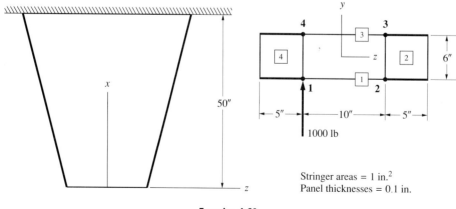

Exercise 4.21

4.22   For the loading shown, the computed flange loads, in pounds, are as shown in the table.

Stringer	$x = 0$	$x = 100$ in.
1	0	17,361 lb
2	0	11,111 lb
3	0	4861.1 lb
4	0	−17,361 lb
5	0	−11,111 lb
6	0	−4861.1 lb

At the free end of the idealized box beam, the shear flow in the panel between stringers 1 and 2 has been found to be $q_{1-2}(0) = 188.5$ lb/in., as shown. Using this information, calculate the shear flow in each of the six panels at $x = 75$ in. (from the free end).

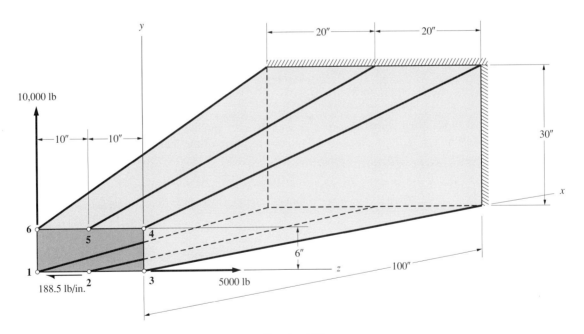

Exercise 4.22

4.23 Calculate the shear flows $q^{(1)}$ through $q^{(6)}$ at the midspan ($x = 1$ m) of the tapered box beam. Notice that the areas of the stringers adjacent to the tapered panels increase linearly from one end of the beam to the other.

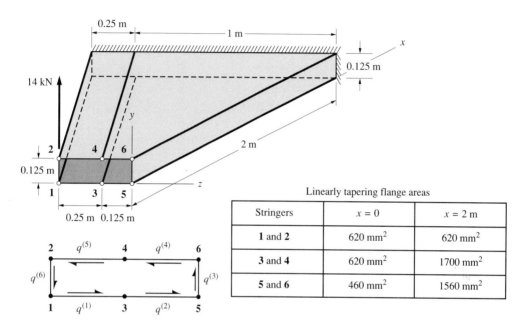

Linearly tapering flange areas

Stringers	$x = 0$	$x = 2$ m
1 and 2	620 mm^2	620 mm^2
3 and 4	620 mm^2	1700 mm^2
5 and 6	460 mm^2	1560 mm^2

Exercise 4.23

4.24 The cantilevered idealized wing box carries a 5000 lb load applied at the tip, as shown. Calculate the shear flow in each panel at $x = 0$, 20 in., and 40 in. What percentage of the 5000 lb shear load is taken out by the shear panels at the root?

Stringer	x (in.)	y (in.)	z (in.)	Area (in.²)
1	0	2	22	0.5
	40	0	0	0.5
2	0	2.5	26	0.2
	40	1.25	10	0.2
3	0	3	30	0.3
	40	2.5	20	0.3
4	0	5	30	0.3
	40	5.5	20	0.3
5	0	5.5	26	0.2
	40	6.75	10	0.2
6	0	6	22	0.5
	40	8	0	0.5

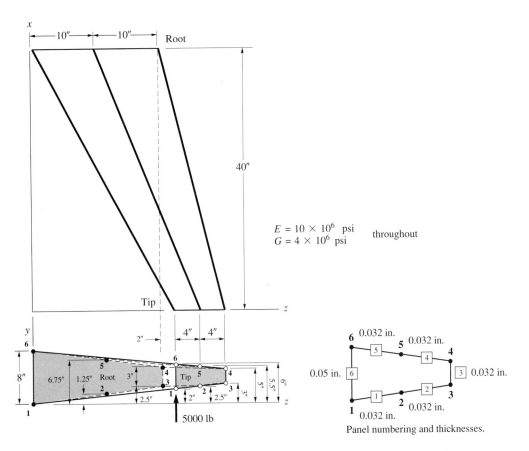

$E = 10 \times 10^6$ psi   throughout
$G = 4 \times 10^6$ psi

Panel numbering and thicknesses.

Exercise 4.24

5

Load Transfer in Stiffened Panel Structures

5.1 INTRODUCTION

Point loads acting on idealized wing and fuselage structures cannot be applied directly to the shear panels, which are capable of supporting only pure shear exerted in their plane, along and in the direction of their boundaries. In section 2.5, we studied planar assemblies of rods and shear webs. In that type of structure, the rods act as stiffeners to which direct loads are applied and by means of which the loads are diffused to the shear panels. For wing and fuselage structures, the analogous stiffening members are the rib and frame, respectively.

Suppose the cantilevered thin-walled cylinder of radius r in Figure 5.1.1a is required to support a point load applied to its free end, as shown. The load P cannot be applied directly to the fragile skin. A "hard point" must be provided, which in this case is a sturdy, circular ring, bonded to the skin of the cylinder (by rivets, for example) around its periphery. In this way, the concentrated load P is applied directly to the ring by means of a lug, as indicated in Figure 5.1.1c, and distributed to the skin in the form of the continuous shear flow distribution, illustrated in Figure 5.1.1b. As we shall see, the shear flow q varies with the angle ϕ, measured from the line of action of P, according to the formula

$$q = \frac{P}{\pi r} \sin \phi \qquad \text{[5.1.1]}$$

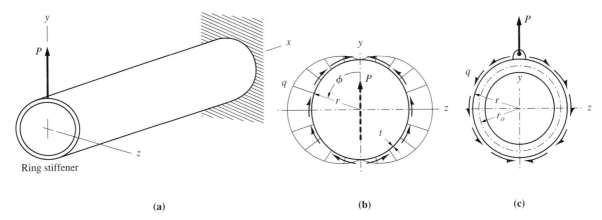

Figure 5.1.1   (a) Thin-walled cylinder with a point load applied to a ring stiffener. (b) Shear flow applied to the skin by the stiffener. (c) Free-body diagram of the stiffener, with the shear flow applied by the skin in equilibrium with the load P applied at the lug.

Part (c) of the figure shows the action of the skin on the ring, in the form of this same shear flow transferred across the bond line from the skin to the frame, where it is reversed in direction by the action–reaction principle.

Figure 5.1.2 illustrates a similar situation in which concentrated loads are applied to an intermediate section of a box beam. In this case, a rib is employed to transfer the point loads into the shear flows along the bond line *abcd*. Ribs are considered rigid in their plane, while possessing negligible out-of-plane stiffness. We will show

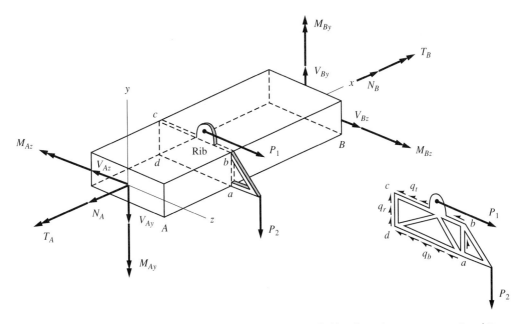

Figure 5.1.2   Concentrated forces applied to the rib of a thin-walled box beam between stations A and B.

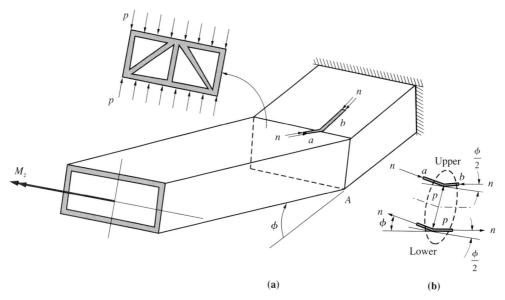

Figure 5.1.3   (a) Pure bending of a thin-walled box beam with a junction at section *A*. (b) At *A*, the membrane forces per unit length *n* in the upper and lower walls must be reacted to by the rib force *p*.

how to compute the shear flows indicated on the free-body diagram of the rib. Once these are determined, the load distribution within the truss-like rib itself may be analyzed. Treated as a truss, the rib of Figure 5.1.2 is statically determinate, so that finding the loads within it is a straightforward statics problem. However, the ring stiffener of Figure 5.1.1 is statically *indeterminate*, and the methods discussed in Chapter 7 must be used to determine the internal shear and bending moment distribution.

Another example of the use of ribs is shown in Figure 5.1.3. The constant-cross-section, thin-walled box beam of Figure 5.1.3a consists of two nonaligned, straight portions that intersect at section *A*. The angle between the two segments is ϕ. The pure bending load shown acting on the structure induces compression in the top panels and tension in the bottom ones. The total force in the top (or bottom) panel, divided by the width of the beam, is the force per unit width *n*. This load intensity acts on small, longitudinal strips like those pictured at points *a* and *b* of the upper surface in Figure 5.1.3a. These internal loads *n* in both the upper and lower panels are viewed from the side in Figure 5.1.3b at section *A*, where they are clearly misaligned. Since the individual thin webs of the beam, like a sheet of paper, are ineffective in bending, an external force intensity *p* must be applied along the panel intersection to provide equilibrium. This force intensity is

$$p = 2n \sin \frac{\phi}{2}$$

This load intensity is exerted by a rib, which is shown in compression in Figure 5.1.3. Reversing the bending moment would change this to a tensile loading.

This chapter presents a method for quantitatively determining how loads applied to ribs and frames of *semi-monocoque structures* are transferred to the attached skin. We will also consider a means of assessing the influence of cutouts on the stress distribution in idealized structures. However, we will focus on problems we can solve using just the equations of statics, supplemented where applicable by beam theory.

5.2   RIB AND BULKHEAD SHEAR FLOW

Consider a transverse stiffener (a rib or a fuselage frame or bulkhead) together with the attached skin. Let loads be applied to the stiffener, in its plane, as illustrated in Figure 5.2.1.

The difference in the section shear loads on each side of the stiffener is due to the load applied directly to the stiffener. That is,

$$P_y = V_y^+ - V_y^- = \Delta V_y$$
$$P_z = V_z^+ - V_z^- = \Delta V_z$$

[5.2.1]

where the superscripts "−" and "+" represent the positions just fore and aft of the stiffener, respectively.

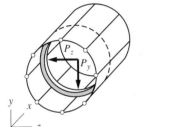
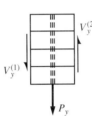
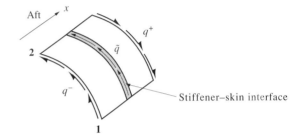

Figure 5.2.1   Loads applied to a fuselage frame, and the shear force resultants in the adjacent skin.

Figure 5.2.2   Shear flows immediately fore and aft of a stiffener and along the bond line.

Figure 5.2.2 shows the shear flows at some point on the box beam perimeter. We are interested in the difference between the shear flows fore and aft of the stiffener, because that difference is the shear flow $\tilde{q}$, which acts on the skin along the stiffener–skin interface, where the skin is bonded to the frame. Here, $\tilde{q}$ is the *jump* in the shear flow as we move aft from one side of the stiffener to the other,

$$\tilde{q} = q^+ - q^-$$

The change in $\tilde{q}$ between points 1 and 2 along the *bond line* is

$$\Delta\tilde{q} = \Delta q^+ - \Delta q^-$$

Assuming the geometry of the beam is the same on either side of the stiffener, Δq^+ and Δq^- are obtained from Equation 4.7.3:

$$\Delta\tilde{q} = \frac{1}{(I_y I_z - I_{yz}^2)}\{[I_y(V_y^+ - V_y^-) - I_{yz}(V_z^+ - V_z^-)]Q_z + [I_z(V_z^+ - V_z^-) - I_{yz}(V_y^+ - V_y^-)]Q_y\}$$

Substituting Equation 5.2.1 into this expression yields

$$\Delta\tilde{q} = \tilde{q}_2 - \tilde{q}_1 = \frac{1}{(I_y I_z - I_{yz}^2)}\left[(I_y P_y - I_{yz} P_z)Q_z + (I_z P_z - I_{yz} P_y)Q_y\right]$$

[5.2.2]

Remember that $\tilde{q}$ is the shear flow acting on the skin along the stiffener–skin interface. Equation 5.2.2 shows how this *interface shear flow* varies as we move around the perimeter of the box beam. Note carefully that

P_y and P_z are the loads applied directly to the stiffener. Also note that P_y is positive if it acts downward, that is, in the negative y direction, and P_z is positive if it acts in the negative z direction. Finally, observe that Equation 5.2.2 is identical in form to Equation 4.7.3, which gives the shear flow between stiffeners in terms of the section shear loads, V_y and V_z. Thus, the shear flow distribution around a frame or rib is calculated using the approach described in Chapter 4, except we use the applied frame loads P_y and P_z (the jumps in the section shear loads) in place of the section shear loads.

It follows that for idealized beams, for which, in the walls joining longitudinal stiffeners, the shear flow is constant, Equation 4.8.2 and 4.8.3—for computing section shear flows—take the following form for calculating the bond line shear flows:

$$\sum_{i=1}^{\text{No. of webs}} \tilde{q}^{(i)} = \tilde{P}_x'^{(f)} \qquad [5.2.3]$$

where

$$\tilde{P}_x'^{(f)} = \frac{1}{I_y I_z - I_{yz}^2}\left[\left(I_y P_y - I_{yz} P_z\right)y_f + \left(I_z P_z - I_{yz} P_y\right)z_f\right]A_f \qquad [5.2.4]$$

in which $\tilde{P}_x'^{(f)}$ is the jump in the flange load gradient across a stiffener. That is, $\tilde{P}_x'^{(f)} = (P_x'^{(f)})^+ - (P_x'^{(f)})^-$.

Briefly, the procedure for calculating the shear flow around a transverse stiffener, such as a frame, is as follows. "Remove" (figuratively) the frame from the skin–stringer structure. Transfer the frame loads to the skin–stringer structure, and calculate the *statically equivalent* shear flow distribution around the periphery, using the methods in Chapter 4. Then, reverse the computed shear flow directions to show them acting on the frame. The steps are illustrated in Figure 5.2.3.

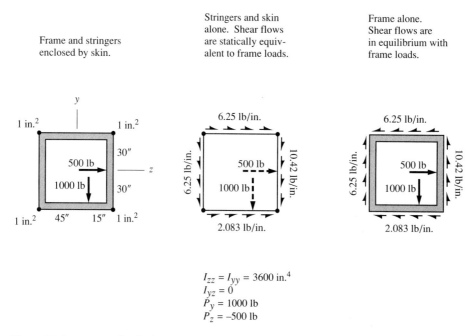

Figure 5.2.3 Example of a frame-to-skin load transfer calculation.

Only the rightmost sketch is a free-body diagram.

Example 5.2.1   Verify Equation 5.1.1 for the ring stiffener in Figure 5.1.1. There are no longitudinal stiffeners; the skin is effective in both shear and bending. Assume the thickness t of the skin is very small compared to the radius r.

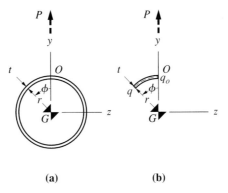

(a)   **(b)**

Figure 5.2.4   (a) Cross section of the skin of the cylinder in Figure 5.1.1.
(b) An arc of the cross section.

The cross section of the skin is a thin ring as shown in Figure 5.2.4. The area moment of inertia about the z axis is

$$I_z = \iint_A y^2 dA = \int_0^{2\pi} (r\cos\phi)^2 (tr\,d\phi) = \pi r^3 t \qquad \text{[a]}$$

By symmetry, I_y is the same as I_z, and $I_{yz} = 0$. Also, in this case $P_y = -P$ and $P_z = 0$. Thus, Equation 5.2.2 reduces to

$$\Delta\tilde{q} = \frac{P_y Q_z}{I_z} = -\frac{P}{\pi r^3 t} Q_z \qquad \text{[b]}$$

Let us set the shear flow at O (the top of the cross section) equal to $\tilde{q}_o$, in terms of which the shear flow at any point of the section may then be obtained using Equation (b). The moment of the area of an arc of the circle, starting at O and subtended by the angle ϕ, is

$$Q_z = \iint_A y\,dA = \int_0^{\phi} (r\cos\phi')(tr\,d\phi') = r^2 t \sin\phi \qquad \text{[c]}$$

Substituting Equation c into b, we get

$$\tilde{q} = \tilde{q}_o - \frac{P}{\pi r}\sin\phi \qquad \text{[d]}$$

The moment of the applied load about G is zero. Therefore, the resultant moment of this shear flow about G must also be zero, or

$$\int_0^{2\pi} r \times (\tilde{q}\,r\,d\phi) = 0$$

Thus, since $\int_0^{2\pi} \sin\phi\, d\phi = 0$, we find that

$$r^2 \int_0^{2\pi} \left(\tilde{q}_o - \frac{P}{\pi r} \sin\phi \right) d\phi = 2\pi r^2 \tilde{q}_o = 0$$

It follows that $\tilde{q}_o = 0$ (a fact we could much more readily have established by noting the symmetry about the vertical diameter), so that Equation d becomes

$$\tilde{q} = -\frac{P}{\pi r} \sin\phi \qquad \text{[e]}$$

The minus sign means that starting at O, the bond line shear flow on the skin is opposite to our assumed direction, which was counterclockwise. It should therefore be reversed and shown acting clockwise around the left half of the skin (Figure 5.1.1b), which means it must be counterclockwise around the left half of the ring (Figure 5.1.1c).

Observe in Figure 5.1.1 that the shear flow on the ring stiffener depends only on the loads applied directly to the stiffener. Consider the more complex situation illustrated in Figure 5.2.5 in which there are five stiffeners, A through E. The stiffener at D is loaded precisely as the one in Figure 5.1.1, so its peripheral shear flow distribution will be identical. By the same token, in spite of the fact that loads are applied to the structure on either side of the stiffener at C, since C itself has no concentrated load, the shear flow around its perimeter is zero. This point is further illustrated in the following example.

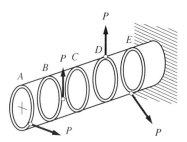

Figure 5.2.5   Stiffened cylinder with several concentrated loads.

Example 5.2.2   Consider the cantilever box beam of square cross section illustrated in Figure 5.2.6. Point loads are applied to the frames at sections A, B, C, and D, as shown. Calculate (a) the panel shear flows in each of the two bays adjacent to section C, and (b) the shear flow around the perimeter of frame C.

(a)

This beam has a uniform cross section that is symmetric, so that relative to axes through the centroid G, $I_{yz} = 0$. Since each of the stringers has a 1 in.^{2} cross section, the moments of inertia of the cross section are

$$I_y = I_z = 4[5^2(1)] = 100 \text{ in.}^4$$

According to Equation 4.8.2, the flange load gradients at any section of the beam are given by

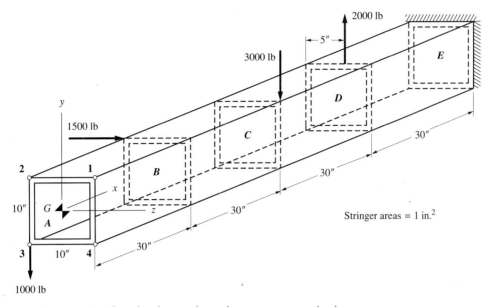

Figure 5.2.6   Cantilever box beam subjected to transverse point loads.

$$P_x'^{(f)} = \left(\frac{V_y y}{I_z} + \frac{V_z z}{I_y}\right) A_f = 0.01(V_y y + V_z z)$$ [a]

where y and z are measured from the centroid of the square cross section. In the bay between sections B and C, $V_y = +1000$ lb and $V_z = -1500$ lb. Therefore, the flange load gradient for this bay is $P_x'^{(f)} = 10y - 15z$, and its values at the four stringers are shown in parentheses in Figure 5.2.7. Using Equation 4.8.3, we have

$$\sum q_{out} = \sum q_{in} + P_x'^{(f)}$$ [b]

at each flange. Starting with flange 1, the shear flows in each web are determined and are shown in Figure 5.2.7. Summing moments about flange 2 yields

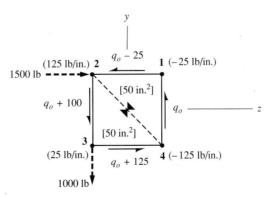

Figure 5.2.7   Flange load gradients (in parentheses) and shear flows around any section between B and C of the box beam of Figure 5.2.6.

$$2 \times 50 \times (q_o + 125) + 2 \times 50 \times q_o = 0$$

so that

$$q_o = -62.5 \ \text{lb/in.} \tag{c}$$

Between sections C and D, the section shear loads are $V_y = 1000 \ \text{lb} + 3000 \ \text{lb} = 4000 \ \text{lb}$ and $V_z = -1500 \ \text{lb}$. Therefore, from Equation a, $P_x'^{(f)} = 40y - 15z$. Using this to calculate the flange load gradients and using Equation b to obtain the shear flows yield the data shown in Figure 5.2.8. Summing moments around flange 2, as before, we get

$$2 \times 50 \times (q_o + 275) + 2 \times 50 \times q_o = -(10 \times 3000)$$

which means

$$q_o = -287.5 \ \text{lb/in.} \tag{d}$$

Substituting Equations c and d, respectively, into Figures 5.2.7 and 5.2.8 yields the shear flows on each side of section C, as in Figure 5.2.9.

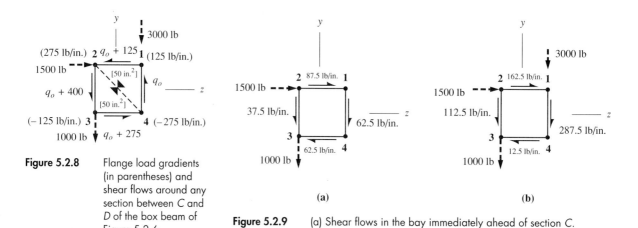

Figure 5.2.8 Flange load gradients (in parentheses) and shear flows around any section between C and D of the box beam of Figure 5.2.6.

Figure 5.2.9 (a) Shear flows in the bay immediately ahead of section C.
(b) Shear flows in the bay immediately aft of section C.

(b)

To compute the shear flows acting on the frame at section C, we focus on the skin bonded to the frame periphery. The shear flow around the section must be statically equivalent to the 3000 lb load applied directly to the frame. That is, in Equation 5.2.4, $P_y = +3000 \ \text{lb}$ and $P_z = 0$. Therefore,

$$\tilde{P}_x'^{(f)} = \frac{P_y y}{I_z} A_f = \frac{3000y}{100}(1) = 30y$$

This is used to calculate the jump in the flange load gradients in each stringer. Equation 5.2.3 yields the bond line shear flows between each stringer, as illustrated in Figure 5.2.10. Summing moments about flange 2, we find

$$2 \times 50 \times (\tilde{q}_o + 150) + 2 \times 50 \times \tilde{q}_o = -(10 \times 3000)$$

from which

$$\tilde{q}_o = -225 \ \text{lb/in.} \tag{e}$$

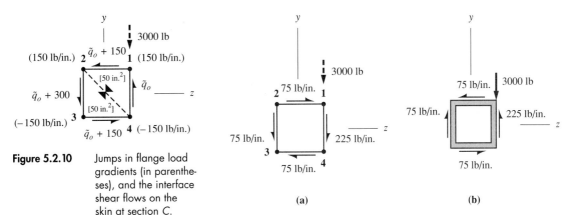

Figure 5.2.10 Jumps in flange load gradients (in parentheses), and the interface shear flows on the skin at section C.

Figure 5.2.11 (a) Shear flows exerted by the frame on the skin around the frame–skin interface perimeter at section C. (b) Shear flows exerted by the skin on the frame around the frame-skin interface perimeter.

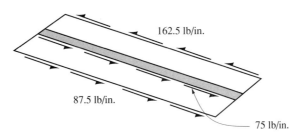

Figure 5.2.12 Portion of the web between stringers 1 and 2, showing that the computed 75 lb/in. shear flow along the frame–skin interface is required for equilibrium.

Therefore, the shear flows on the skin around the frame–skin interface are as sketched in Figure 5.2.11a (which is *not* a free-body diagram). By the action–reaction principle, the shear flows around the frame are as pictured in part (b) of that figure (which *is* a free-body diagram).

We can readily confirm that the interface shear flows just calculated are indeed the change in the panel shear flows fore and aft of the frame. For example, consider the web between stringers 1 and 2. In Figure 5.2.9, we see that the shear flow jumps from 87.5 lb/in. on the fore side of the frame to 162.5 lb/in. in the opposite direction on the aft side of the frame. Figure 5.2.12 clearly shows that the 75 lb/in. shear flow exerted by the frame on the skin in Figure 5.2.11 is required for equilibrium. The reader should check the other three sides of the section as well.

Example 5.2.3 Figure 5.2.13 shows a 100 in. diameter circular fuselage frame with a 1000 lb vertical load applied to the floor beam. The 24 equally spaced stringers all have the same 0.2 in.^{2} area. Calculate the shear flow distribution around the frame.

To begin, we "remove" the frame and show just the stringer–skin structure, in Figure 5.2.14. Number the stringers counterclockwise from 1 to 24, as shown. The stringer areas are all the same, so the centroid G of the section lies at the center of the circle. Since $P_z = 0$ and, by symmetry, $I_{yz} = 0$, the jump in the flange load gradients as given by Equation 5.2.4 is

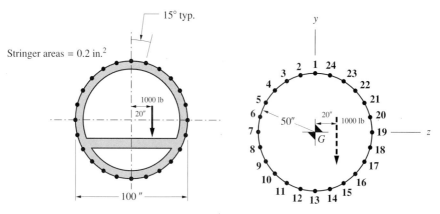

Figure 5.2.13   Circular fuselage frame with a load applied to the floor beam.

Figure 5.2.14   Skin and stringers around the circular frame.

$$\tilde{P}_x^{\prime(i)} = \frac{P_y\, y_i}{I_z} A_i \qquad\qquad [a]$$

To calculate the area moment of inertia, we use $I_z = \sum\limits_{i=1}^{24} y_i^2 A_i = A_f \sum\limits_{i=1}^{24} y_i^2$ since the flanges all have a common area, $A_f = 0.2$ in.² Taking advantage of the symmetry and the fact that $y_7 = y_{19} = 0$, we can write this as

$$I_z = A_f[2y_1^2 + 4(y_2^2 + y_3^2 + y_4^2 + y_5^2 + y_6^2)]$$

If θ_i is the angle from the negative z axis to stringer i, then $y_i = 50 \sin\theta_i$. Thus,

$$I_z = 0.2[2(50^2) + 4(50^2)(\sin^2 75° + \sin^2 60° + \sin^2 45° + \sin^2 30° + \sin^2 15°)] = 6000 \text{ in.}^4$$

Substituting this into equation a, together with $P_y = +1000$ lb and $A_f = 0.2$ in.², yields

$$\tilde{P}_x^{\prime(i)} = 0.03333 y_i \qquad\qquad [b]$$

We use this to calculate the jump in the flange load gradients at each stringer. The results are shown in parentheses in Figure 5.2.15a. Assuming that the shear flow q_0 in the web joining flanges 1 and 24 is counterclockwise, then starting at flange 1, we can use Equation 5.2.3 to find each of the web shear flows around the section in terms of q_0. These are shown in Figure 5.2.15. We then require that the moment of the shear flows equals the moment of the vertical 1000 lb load applied 20 inches to the right of the center of the frame. Thus, if A is the area of the 100 in. diameter circle,

$$2Aq_o + 2\left(\frac{A}{24}\right)[2(1.667 + 3.277 + 4.720 + 5.898 + 6.731 + 7.163)]$$

$$- 2\left(\frac{A}{24}\right)[2(1.610 + 3.053 + 4.231 + 5.064 + 5.496)] = -(20 \times 1000)$$

Introducing the area, $A = \pi(50 \text{ in.})^2$, and solving for q_o yields

$$q_o = -2.107 \text{ lb/in.}$$

We substitute this uniform clockwise shear flow into the expressions for the individual web shear flows in Figure 5.2.15 to get the net shear flow acting on the skin around the frame–skin interface. These are shown in Figure 5.2.16a. In Figure 5.2.16b the direction of each of these shear flows is reversed to reveal the action of the skin on the frame. The bending moments in the frame due to the applied load and the reaction shear flows cannot be calculated until we develop the tools for handling statically indeterminate structures, in Chapter 7.

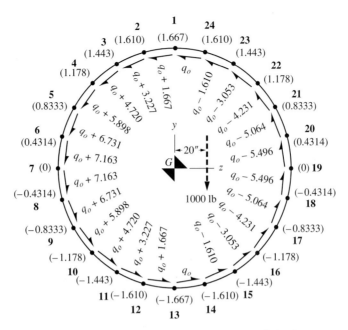

Figure 5.2.15   Interface shear flows due to the 1000 lb load, in terms of the shear flow q_o in webs 1–24.

The flange load gradients are shown in parentheses.

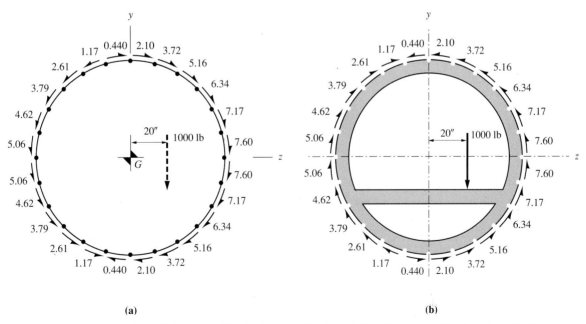

(a)   **(b)**

Figure 5.2.16   (a) Shear flows (lb/in.) exerted by the frame on the skin. (b) Shear flows exerted by the skin on the frame.

Example 5.2.4   A concentrated load of 5000 lb is applied to the wing rib in Figure 5.2.17 at the point where the left-most vertical stiffener *ef* joins the bottom rib flange. The area of the front and rear spar caps (6, 7, 1, and 12) is 0.3 in.², and the area of the eight numbered stringers is 0.1 in.² Calculate the average shear flow in each of the four rib webs and the axial loads in the rib flanges at the location just to the right of the vertical stiffeners *ef*, *cd*, and *ab*. (The rib is bonded to the front and rear spar webs and to the wing skin along the rib flanges, whereas the spanwise stringers are bonded to the wing skin.)

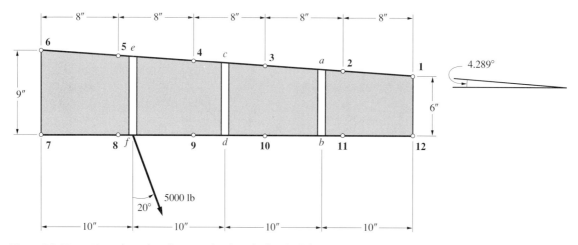

Figure 5.2.17   Wing box rib with a point load applied to the leftmost vertical stiffener.

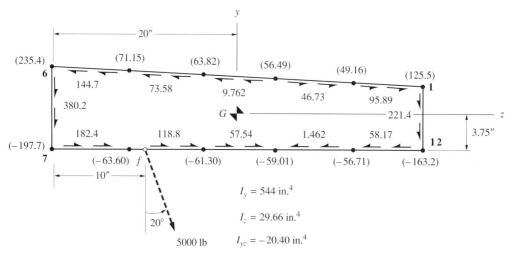

Figure 5.2.18   Shear flows (lb/in.) on the skin around the rib–skin interface. The jumps in flange load gradients (lb/in.) are in parentheses.

We begin by imagining that the rib is removed and replaced by the shear flows acting around the bond line between the rib and the wing skin and spar webs. Figure 5.2.18 shows just the webs and the concentrated flange areas together with the results of a series of straightforward computations, as follows:

1. Calculate the location of the centroid G of the flange area.

2. Compute the area moments of inertia I_y, I_z, and I_{yz} relative to axes through the centroid.

3. Substitute the area moments of inertia, along with $P_y = 5000 \cos 20°$ and $P_z = -5000 \sin 20°$, into Equation 5.2.4 to obtain the jump in flange load gradients, which are shown in parentheses in the figure.

4. Set the shear flow equal to q_o in one of the webs, say 12–1, and starting with flange 1, work your way counterclockwise around the section, applying Equation 5.2.3, "shear flow out equals shear flow in plus flange load gradient," at each flange, assuming all shear flows are counterclockwise, to calculate each of the web shear flows in terms of q_o.

5. Set the moment of the shear flow distribution about a point, say f, equal to the moment of the applied load about that same point. Solve the equation for the only unknown, namely q_o.

6. For each of the twelve skins, substitute q_o into the expression for shear flow computed in Step 4, finally yielding the shear flows shown in Figure 5.2.18.

We now apply these shear flows to the rib itself, as in Figure 5.2.19, which—unlike Figure 5.2.18—is a free-body diagram. By sectioning this free body, we can reveal the internal loads carried by the rib webs, flanges, and stiffeners.

It is clear from Figure 5.2.19 that the shear flow on the left edge of the leftmost rib web is 380.2 lb/in. Equation 2.5.4 relates the average shear flow in a trapezoidal shear panel to the shear flow on its parallel sides. For our case, we therefore have

$$\bar{q}_1 = 380.2 \times \frac{9}{8.25} = 414.8 \text{ lb/in.} \tag{a}$$

Moving to the right into the next web, we section the rib just to the right of the vertical stiffener ef to reveal the shear flow q'_{ef} shown in Figure 5.2.20, as well as the rib flange loads P_e and P_f. Equilibrium analysis of this free body yields the following three equations for the three unknowns P_e, P_f, and q'_{ef}:

$$\sum M_e = 0: \quad 8.25 P_f + 8.25 \times 5000 \sin 20° - 10 \times (380.2 \times 9)$$
$$- 8.25 \times (182.4 \times 8) - 8.25 \times (118.8 \times 2) = 0$$

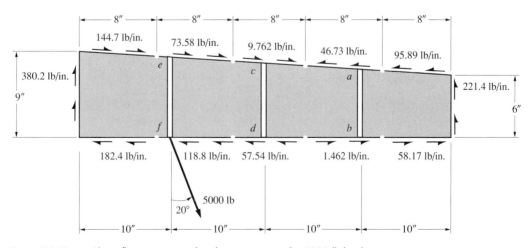

Figure 5.2.19 Shear flows acting on the rib in reaction to the 5000 lb load.

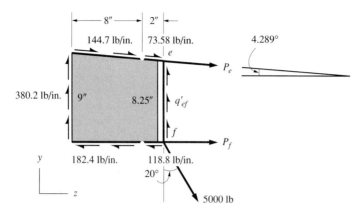

Figure 5.2.20 Free-body diagram of a portion of the rib extending from the extreme left to just past the vertical stiffener ef.

$$\sum F_z = 0: \quad P_e \cos 4.289° + P_f + 5000 \sin 20° - 182.4 \times 8$$
$$- 118.8 \times 2 + 144.7 \times 8 + 73.58 \times 2 = 0$$
$$\sum F_y = 0: \quad - P_e \sin 4.289 + 8.25 q'_{ef} - 5000 \cos 20° + 380.2 \times 9$$
$$- 144.7 \times (8 \tan 4.289°) - 73.58 \times (2 \tan 4.289°) = 0$$

The solution of this system is

$$P_f = 4134 \text{ lb} \qquad P_e = -5468 \text{ lb} \qquad q'_{ef} = 117.0 \text{ lb/in.} \qquad \text{[b]}$$

The average shear flow in this second web, according to Equation 2.5.4, is

$$\bar{q}_2 = 117.0 \times \frac{8.25}{7.5} = 128.7 \text{ lb/in.} \qquad \text{[c]}$$

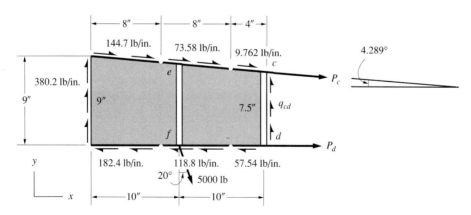

Figure 5.2.21 Free-body diagram of a portion of the rib extending from the extreme left to just past the vertical stiffener cd.

The free-body diagram in Figure 5.2.21 shows the rib flange loads and web shear flow immediately to the right of the vertical stiffener de. The equilibrium equations in this case are

$$\sum M_c = 0: \quad 7.5P_d + 7.5 \times 5000 \sin 20° + 10 \times 5000 \cos 20° - 20 \times (380.2 \times 9)$$
$$- 7.5 \times (182.4 \times 8 + 118.8 \times 8 + 57.54 \times 4) = 0$$

$$\sum F_z = 0: \quad P_c \cos 4.289° + P_d + 5000 \sin 20° + (144.7 + 73.58 - 182.4 - 118.8) \times 8$$
$$+ (9.762 - 57.54) \times 4 = 0$$

$$\sum F_y = 0: \quad - P_c \sin 4.289° + 7.5q_{cd} - 5000 \cos 20° + 380.2 \times 9$$
$$- (144.7 + 73.58) \times (8 \tan 4.289°) - 9.762 \times (4 \tan 4.289°) = 0$$

Solving for the three unknowns P_d, P_c, and q_{cd}, we obtain

$$P_d = 3790 \text{ lb} \qquad P_c = -4658 \text{ lb} \qquad q_{cd} = 141.6 \text{ lb/in.} \qquad \text{[d]}$$

and the average shear flow is

$$\bar{q}_3 = 141.6 \times \frac{7.5}{6.75} = 157.3 \text{ lb/in.} \qquad \text{[e]}$$

Finally, the free-body diagram in Figure 5.2.22 may be used to find the flange loads and shear flow at section ab. Equilibrium requires that

$$\sum M_a = 0: \quad -6.75P_b + 10 \times (221.4 \times 6) + 6.75 \times (1.462 \times 2 + 58.17 \times 8) = 0$$
$$\sum F_z = 0: \quad -P_a \cos 4.289° - P_b + (1.462 - 46.73) \times 2 + (58.17 - 95.89) \times 8 = 0$$
$$\sum F_y = 0: \quad +P_a \sin 4.28° - 6.75q_{ab} + 221.4 \times 6 + 46.73 \times (2 \tan 4.289°)$$
$$+ 95.89 \times (8 \tan 4.289°) = 0$$

and these yield

$$P_b = 2436 \text{ lb} \qquad P_a = -2836 \text{ lb} \qquad q_{ab} = 174.9 \text{ lb/in.} \qquad \text{[e]}$$

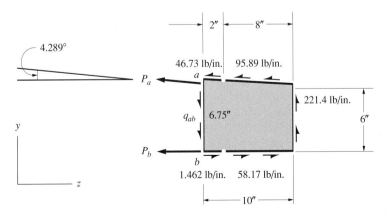

Figure 5.2.22   Free-body diagram of the rib segment between vertical stiffener ab and the extreme right end.

The average shear flow in the rightmost web is obtained from Equation 2.5.4 and the shear flow on either of the parallel edges, as follows:

$$\bar{q}_4 = 174.9 \times \frac{6.75}{6} = 221.4 \times \frac{6}{6.75} = 196.8 \text{ lb/in.} \qquad \text{[f]}$$

Figure 5.2.23 presents a summary of the shear flow distribution throughout the rib. The jump in the shear flow at section ef is due to the concentrated load applied to the stiffener at that location.

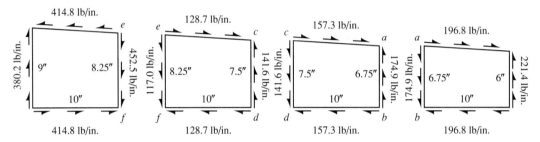

Figure 5.2.23   Shear flows around the boundaries of the rib webs.

The average shear flows are shown on the nonparallel edges.

5.3   SHEAR FLOW AROUND CUTOUTS

Figure 5.3.1 shows a stiffened panel structure with evenly spaced cutouts. The width and height of the cutouts are w and h, respectively, d_f is the space between windows, and d_s is the stringer spacing above and below the cutouts. The vertical and horizontal dotted lines represent the stiffeners (e.g., frames and stringers), which surround the rectangular shear panels. This might serve as a model of an airplane fuselage with windows, in which case the assembly is likely to be curved rather than flat.

The edges of the structure support a uniform shear flow q_0. This shear flow would be uniform throughout were it not for the presence of the evenly spaced windows. The question is, What effect do the cutouts have on the shear flow distribution? Because of the symmetry of the structure and the simple nature of the applied load, this question can be answered using statics alone, if we make some reasonable simplifying assumptions. One of them is that the effect of the windows on the shear flow pattern does not extend more than one shear panel beyond the top and bottom of the cutouts. Thus, the top and bottom rows of shear panels will all have the same shear flow, q_0. Also, by symmetry, there are only three different shear flows: q_1 is the shear flow in the panels between the cutouts; q_2 is its value in the panels directly above and below the cutouts; and q_3 is the shear flow in the panel directly above that which separates the cutouts.

Since there are, by assumption, only three unknown shear flows, the three equations of static equilibrium in the plane should yield their values in terms of q_0. Figure 5.3.2 shows three free-body diagrams, which we can use to evaluate the three shear flows. Notice that axial loads develop in a stiffener for which the shear flows differ in adjacent panels. Although we are not concerned here with those loads, they must be shown on the diagrams for completeness.

For the free-body diagram in Figure 5.3.2a, the equilibrium of forces in the horizontal direction requires that

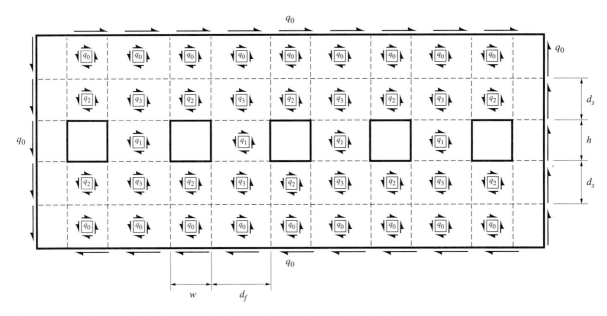

Figure 5.3.1   Portion of a fuselage panel with regularly spaced cutouts (windows).

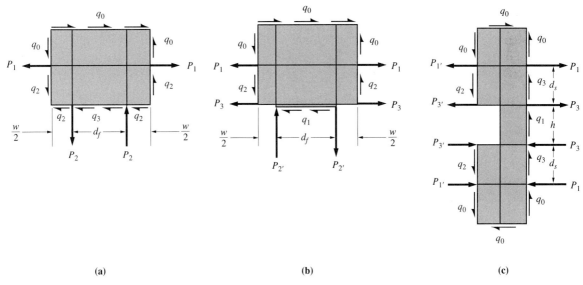

(a)   **(b)**   **(c)**

Figure 5.3.2   Free-body diagrams of portions of the panel: (a) Spans the midpoints of the stringer above two adjacent cutouts, the top edge extending into the uniform shear flow region and the bottom edge lying just above the top of the window line; (b) same as in (a), except the bottom edge extends into the panel between the windows; (c) the top and bottom edges extend to the uniform shear flow region; the left edge cuts through the midpoint of a window, and the right edge extends into the region between windows.

$$q_2 w + q_3 d_f = q_0 \left(d_f + w \right) \qquad \text{[5.3.1]}$$

From part (b) of the figure, we deduce similarly that

$$q_1 d_f = q_0 \left(d_f + w \right) \qquad \text{[5.3.2]}$$

Finally, summing forces in the vertical direction in Figure 5.3.2c yields

$$q_1 h - 2q_2 d_s + 2q_3 d_s = 0 \qquad \text{[5.3.3]}$$

The solution of these three equations for q_1, q_2, and q_3 is

$$q_1 = q_0 \left(1 + \frac{w}{d_f} \right) \qquad q_2 = q_0 \left(1 + \frac{h}{2d_s} \right) \qquad q_3 = q_0 \left(1 - \frac{hw}{2d_f d_s} \right) \qquad \text{[5.3.4]}$$

For example, if $w = h = d_f = d_s$, then $q_1 = 2q_0$, $q_2 = 1.5q_0$ and $q_3 = 0.5q_0$.

Example 5.3.1 The flat, stiffened, panel structure in Figure 5.3.3a is loaded by a uniform shear flow $q_o = 50\,\text{kN/m}$ on its boundary, as shown. (a) Calculate the independent shear flows q_1 and q_2. (b) Find the location and value of the maximum load within the stiffeners.

(a)

The fact that there are only two unknown shear flows, located as shown in Figure 5.3.3a, is justified by symmetry. We can find these shear flows by writing the lengthwise equilibrium equation for any two of the eight stiffeners isolated as free bodies. Figure 5.3.3b is a free-body diagram of the topmost stiffener. Summing forces to the right, we have (using $q_o = 50\,\text{kN/m}$)

$$-0.250q_1 - 2 \times 0.375q_2 + 50 = 0$$

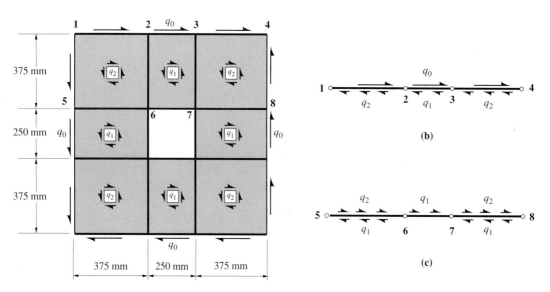

Figure 5.3.3 (a) Stiffened flat panel with a central cutout. (b) and (c) Free-body diagrams of the topmost two horizontal stiffeners, which, by symmetry, represent all of the stiffeners.

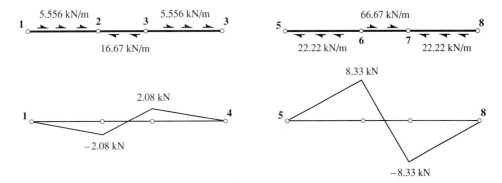

Figure 5.3.4 Axial load distributions in the top two horizontal stiffeners (which represent all of them).

or

$$0.250q_1 + 0.750q_2 = 50 \qquad \text{[a]}$$

Likewise, from Figure 5.3.3c, we have

$$-2 \times 0.375q_1 + 0.250q_1 + 2 \times 0.375q_2 = 0$$

or

$$-0.500q_1 + 0.75q_2 = 0 \qquad \text{[b]}$$

Solving Equations a and b yields

$$q_1 = 66.67 \ \text{kN/m} \quad q_2 = 44.44 \ \text{kN/m} \qquad \text{[c]}$$

Without the cutout, the shear flow would be 50 kN/m throughout. Thus, in those panels having an edge in common with the cutout, the shear flow is increased 33 percent, whereas in the remaining panels, it is reduced by 11 percent.

(b)

Applying the shear flows in Equation c to the free-body diagrams in Figures 5.3.3b and c yields the net shear flows, shown at the top of Figure 5.3.4. Moving from left to right, passing transverse cutting planes through each stiffener, we generate the internal axial load diagrams shown (with compression negative, as usual). From these figures, we see that the maximum axial load (8.33 kN, or 224 lb) occurs in the stiffeners adjacent to the hole, at its corners. If there were no cutout, the axial load would be zero throughout all of the stiffeners.

5.4 BOX BEAM CUTOUTS

For a given external load condition, we can determine the effect of a cutout on an idealized wing box or fuselage structure by using the following procedure. First, use idealized beam theory to calculate the stress distribution throughout the structure, assuming there is no cutout. We thereby ascertain the stresses on the interface between the main structure and the substructure to be cut away. These surface tractions represent the action of the substructure on the main structure. Then, the presence of the substructure is effectively eliminated by reversing the direction of the surface tractions, calculating the resulting stresses throughout the structure, and superimposing them, point by point, onto the stresses originally computed. Within and on the boundary of the substructure, the stresses are thereby cancelled out. The state of zero stress represents the cutout, that is, the absence of the substructure.

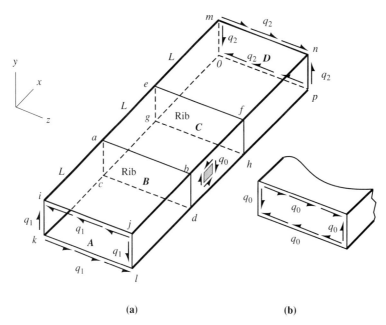

(a) **(b)**

Figure 5.4.1 (a) Rectangular box beam with shear flow q_o applied as shown to the vertical panel *bdhf* of the central bay. (b) Shear flow around a section through the central bay.

The system of internal loads imposed to simulate the cutout must be self-equilibrating. No reactions arise at the supports, and the stresses that develop throughout the structure in response to the imposed load are such that the resultant force is zero on any cross section. Furthermore, the effect of a cutout is assumed to be localized. Beyond a distance roughly equal to the size of the cutout, its effect on the stress distribution is considered to be negligible.

We will restrict our attention to structures for which the technique described here can be applied through the use of statics. For statically indeterminate situations, the methods of Chapter 8 must be employed.

Figure 5.4.1a illustrates a three-bay box beam with shear flow q_o applied to a vertical web of the central bay, in the direction indicated on the shaded element. Ribs at sections B and C separate the central bay from the fore and aft bays, which, like the center one, are of length L. There are no externally applied loads, so the net shear force and twisting moment on any section are zero. Absence of net shear means that the shear flows in opposite walls of the rectangular cross section have the same magnitude but opposite directions. This, together with the fact that the torque is zero, leads to the conclusion that the shear flows in adjacent walls are equal. It follows that the shear flow directions in the central bay are as pictured in Figure 5.4.1b. The same reasoning yields the shear flows q_1 and q_2 shown in the adjacent bays.

In order to obtain q_1 and q_2 in terms of the given shear flow q_0, we isolate any one of the four stringers as a free body, as illustrated in Figure 5.4.2. The effect of q_0 is assumed to extend no further than one bay length L from the central bay; therefore, the flange loads at each end of the stringer (points j and n) vanish. The net force on the stringer must be zero, so that

$$-2(q_1 L) + 2(q_0 L) - 2(q_2 L) = 0$$

or

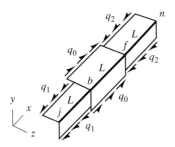

Figure 5.4.2   Free-body diagram of a stringer, through which the shear flow q_o is related to the shear flow in the adjacent bays.

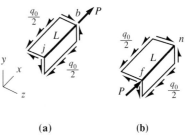

(a)   **(b)**

Figure 5.4.3   Free-body diagrams of segments of the stringer spanning (a) the fore bays and (b) the aft bays.

$$q_1 + q_2 = q_0$$

Since, by symmetry, $q_1 = q_2$, we see that

$$q_1 = q_2 = \frac{q_o}{2}$$

In the stringer the axial load's maximum value P occurs at each end of the center bay. From Figure 5.4.3, we see that

$$P = 2 \times \left(\frac{q_o}{2}L\right) = q_o L$$

This force is alternately tensile and compressive from stringer to stringer around the box. The axial load $q_o L$ is tensile at one end of the stringer segments spanning the central bay and compressive at the other, as illustrated in Figure 5.4.4. Therefore, the flange load in each stringer is zero at the center of the bay (Figure 5.4.4b). Figure 5.4.5 presents a summary of the analysis up to this point.

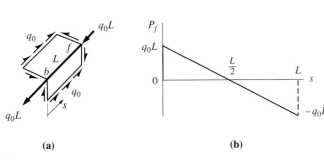

(a)   **(b)**

Figure 5.4.4   (a) Free-body diagram of a stringer segment spanning the central bay. (b) Flange load versus position within the central bay.

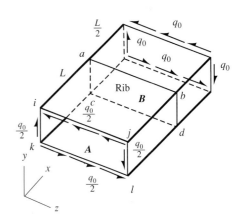

Figure 5.4.5   Free-body diagram of one-half of the structure, showing the shear flows in the central and forward bays.

The stringer loads are zero at each end of the figure.

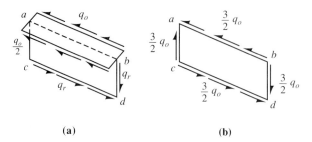

(a)   **(b)**

Figure 5.4.6   (a) Free-body diagram of the rib at B and a portion of the top panels. (b) The shear flows required for equilibrium of the rib.

The ribs separating the bays pick up the jump in the shear flow between each one. By considering the equilibrium of a free-body diagram of a rib, such as that shown in Figure 5.4.6a, we obtain the shear flow distribution shown in Figure 5.4.6b. This is the information we need to compute the stresses within the rib, as discussed in section 5.2.

Example 5.4.1   Calculate the shear flows in the spar webs and cover skins of the idealized wing box shown in Figure 5.4.7. The geometric properties of the constant cross section are listed. The uniformly distributed airloads (15 lb/in. lift, 3 lb/in. drag) on the wing have been lumped at the ribs, which are spaced 20 inches apart.

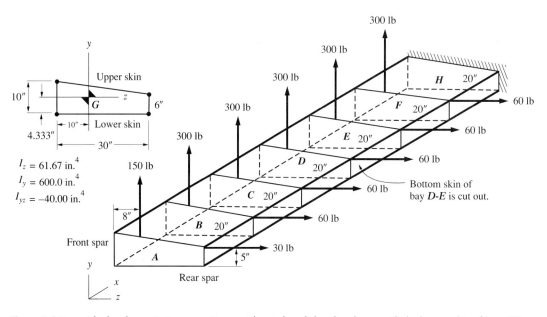

Figure 5.4.7   Idealized, constant-cross-section, cantilevered, multi-bay box beam with the bottom skin of bay *D-E* removed.

At the upper left, the location of the centroid is shown and the centroidal moments of inertia are listed. Lift and drag loads have been lumped at the ribs.

The first step is to analyze the structure as though there were no cutout, using beam theory and the methods of Chapter 4. Since the distributed airloads have been lumped into concentrated shear loads at the ribs, the shear flows are constant throughout each of the bays and undergo a jump at the ribs. Figure 5.4.8 shows the shear flows calculated for each bay if the bottom skin of bay D-E is present.

To simulate the cutout in bay D-E, we apply a shear flow of 10.8 lb/in. to the bottom panel, in the direction opposite to that shown in Figure 5.4.8, and we assume that the accompanying self-equilibrating internal loads extend no further than the two bays adjacent to bay D-E. This means that the axial loads due to the applied shear flow in the spar caps vanish at sections C and F. Figure 5.4.9 shows the seven shear flows that arise in response to the applied reversed shear flow in the affected bays. Notice that the shear flows around bays C-D and E-F are the same, by symmetry. As pointed out prior to this example, it follows that the axial load in the spar caps is zero at the midsection of bay D-E.

The shear flows in bay D-E are reproduced in Figure 5.4.10. They must be statically equivalent to a net zero force and torque resultant on the section. For the moment about, say, the upper left spar cap to be zero, recalling the dimensions of the cross section from Figure 5.4.7, we have $(6q_1) \times 30 - (30 \times 10.8) \times 10 = 0$, or

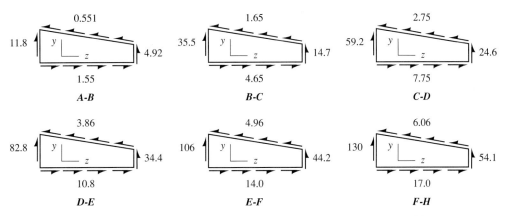

Figure 5.4.8   Shear flows (lb/in.) around each of the bays of the beam in Figure 5.4.7, if there is no cutout.

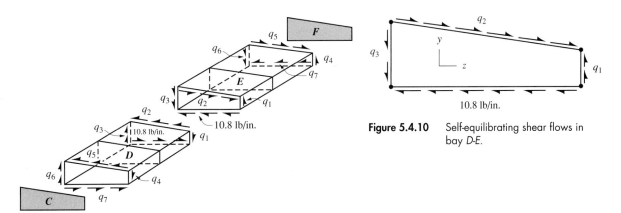

Figure 5.4.9   Free-body diagrams of a portion of bay D-E and the adjacent bays C-D and E-F, with a shear flow of 10.8 lb/in. applied as shown to the bottom panel of bay D-E.

Figure 5.4.10   Self-equilibrating shear flows in bay D-E.

$$q_1 = 18.1 \text{ lb/in.} \tag{a}$$

A zero force resultant in the z direction requires

$$q_2 = 10.8 \text{ lb/in.} \tag{b}$$

Finally, the resultant force in the y direction will be zero if $6q_1 - 4q_2 - 10q_3 = 0$, which, together with Equations a and b, implies that

$$q_3 = 6.52 \text{ lb/in.} \tag{c}$$

The shear flows in the bays adjacent to D-E are related to those just computed, as illustrated in the free-body diagram in Figure 5.4.11. This figure shows the portion of the upper rear spar cap starting at rib C and extending to the middle of bay D-E. The axial load at each end of the spar cap is zero. Therefore, setting the net force in the longitudinal direction equal to zero yields

$$20q_4 + 20q_5 - 10 \times 18.1 - 10 \times 10.8 = 0$$

or

$$q_4 + q_5 = 14.5 \text{ lb/in.} \tag{d}$$

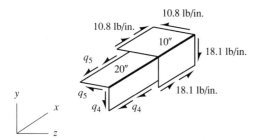

Figure 5.4.11   Free-body diagram of the portion of the upper rear spar cap between rib C and the middle of bay D-E.

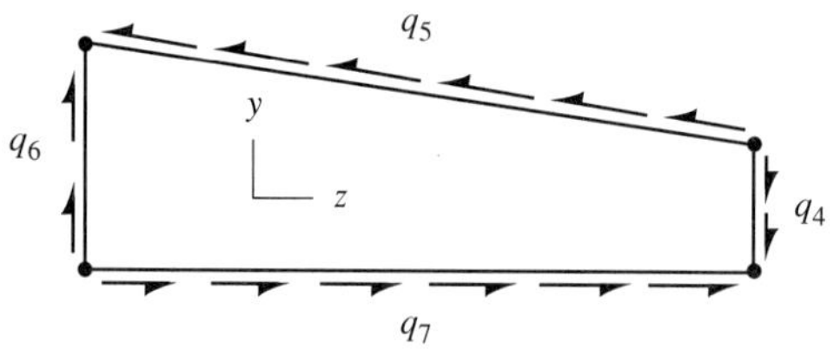

Figure 5.4.12   Self-equilibrating shear flows around bay C-D.

Since the external load on the beam is zero, the shear flows around bay C-D, pictured in Figure 5.4.12, must yield zero force and moment resultants. For the z direction, this requires that

$$-q_5 + q_7 = 0 \tag{e}$$

whereas for the y direction,

$$-6q_4 + 4q_5 + 10q_6 = 0 \tag{f}$$

The moment of the shear flows about the upper left spar cap must vanish. Therefore,

$$(30q_7) \times 10 - (6q_4) \times 30 = 0$$

or

$$-6q_4 + 10q_7 = 0 \tag{g}$$

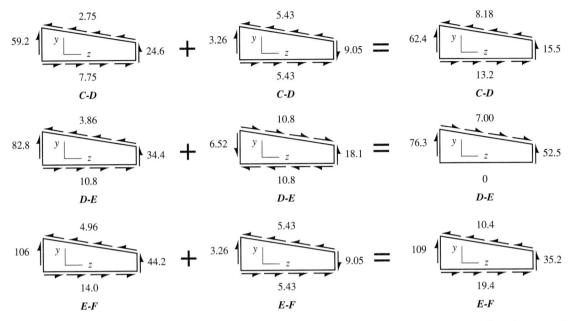

Figure 5.4.13   Superposition of the shear flows due to the external loads (left) and those (center) arising from the reversed shear flow applied to bay *D-E*, yielding the shear flows modified by the presence of the cutout. All shear flow units are in lb/in.

Solving Equations d through g yields the four remaining shear flows:

$$q_4 = 9.05 \text{ lb/in.} \qquad q_5 = q_7 = 5.43 \text{ lb/in.} \qquad q_6 = 3.26 \text{ lb/in.} \tag{h}$$

The shear flow distribution due to the external loads in Figure 5.4.7 and the presence of the cutout are found by superimposing the shear flows in Equations a, b, c, and h onto those in Figure 5.4.8. The shear flows in bays *A-B, B-C* and *F-H* are unaltered, whereas the shear flows in the three bays affected by the cutout are shown at the right in Figure 5.4.13.

EXERCISES

5.1   Loads of 6000 lb and 4000 lb are applied as shown to the rib truss of a box beam. (The diagonal member is used instead of a web to provide rigidity.) Calculate the force in the diagonal member.

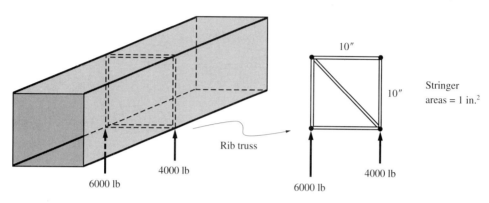

Exercise 5.1

5.2   The split, square frame of a thin-web box beam has a tiny gap at the center of the bottom span, at the location of stringer 8. A 1000 lb load is applied to the frame as shown, on the right side of the gap. Find the magnitude and location of the maximum bending moment in the frame.

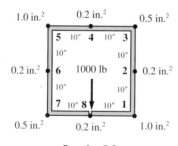

Exercise 5.2

5.3   Calculate and sketch the shear flow around the fuselage ring, to which point loads are applied as shown. The area of stringers 1 through 5 is 0.6 cm², and stringers 6 through 8 have an area of 1.5 cm². The fuselage is symmetric about the *xy* plane.

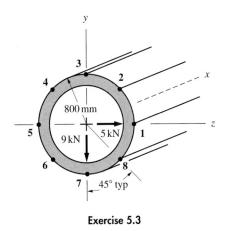

Exercise 5.3

5.4   The point loads shown are applied directly to the circular frame of the 20 in. diameter fuselage. Find the skin reactions on the frame.

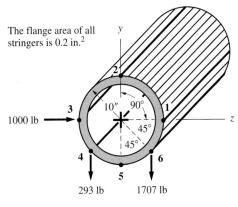

Exercise 5.4

5.5   Calculate the shear flow in the rib web just to the left of the vertical stiffener 5-2. (The shear flow around the rib is shown.)

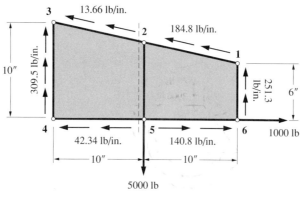

Exercise 5.5

5.6   Calculate the shear flow in the web of rib *B*.

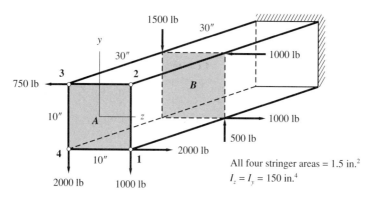

Exercise 5.6

5.7 Point loads are applied to the ribs of a box beam, as shown. All four stringer areas are 1 in.^{2} For rib C, calculate:

a. The shear flow in the rib webs adjacent to the vertical stiffeners ef, dc, and ab.

b. The maximum force in the upper rib flange (fbd).

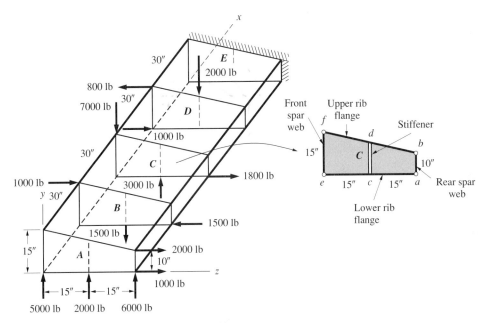

Exercise 5.7

5.8 Loads are applied as shown to the three ribs of a box beam. Calculate the shear stress in the web of the center rib.

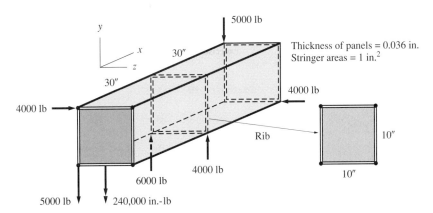

Exercise 5.8

5.9   A 3000 lb point load is applied directly to the square frame of an idealized box beam, as shown. Calculate the skin reactions on the frame, and label and indicate their directions.

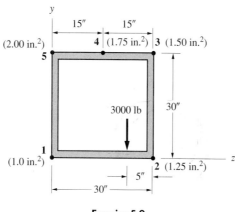

Exercise 5.9

5.10   The skin reactions on the rib due to the inclined 5000 lb load are shown. Calculate the shear flow in the rib web just to the right of vertical stiffener *ab*.

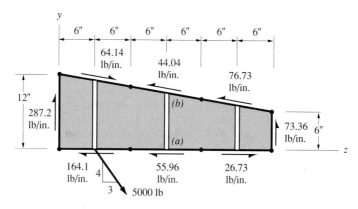

Exercise 5.10

5.11   Loads are applied directly to point 7 of a wing box rib, as shown. Find the shear flows around the periphery of the rib. Then calculate the rib web shear flows and axial loads in the rib flanges at the five sections indicated, adjacent to the front and rear spars and the intermediate vertical stiffeners. The areas of both front spar caps (9 and 10) are 20 cm^2, while the rear spar cap areas (1 and 2) are 10 cm^2.

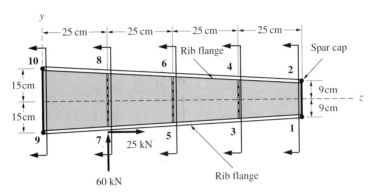

Exercise 5.11

5.12   A 2000 lb force is applied to the truss-like rib of an idealized box beam as shown. Calculate the shear flow around the periphery of the rib, and use that information to find the axial load in each of the diagonal members. The areas of the beam's four stringers 1, 2, 7 and 8 are given in the figure.

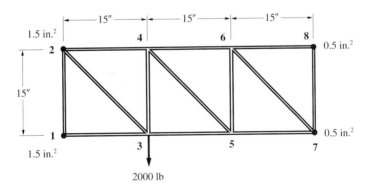

Exercise 5.12

5.13 Calculate the shear flows $q^{(1)}$ through $q^{(5)}$ in the webs of the idealized wing rib, to which point loads are applied as shown. The areas of stringers 1 and 3 are 1 in.^{2}, and the areas of stringers 9 and 10 are 2 in.^{2}

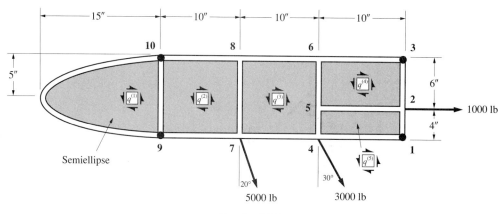

Exercise 5.13

5.14 If front panel 2 is removed from the torque box, as shown in the bottom figure, calculate the shear flow in panel 1.

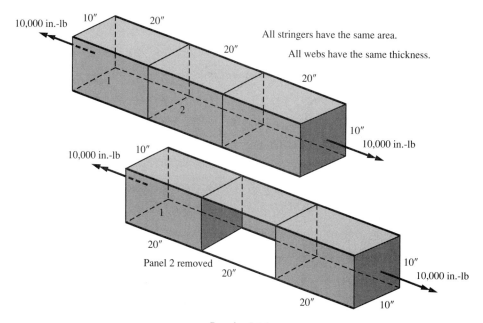

Exercise 5.14

5.15   Calculate the shear flow around the periphery of the ribs adjacent to the 10 in. cutout in the box structure loaded as shown.

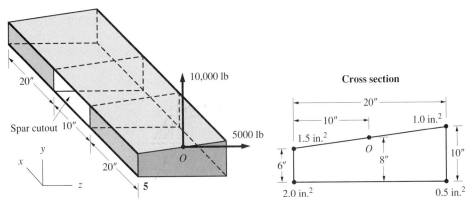

Exercise 5.15

5.16   Calculate the shear flow around the periphery of the circular frames on each side of the cutout in the semimonocoque structure shown. Assume it is subjected to a pure axial torque of 100,000 in.-lbs.

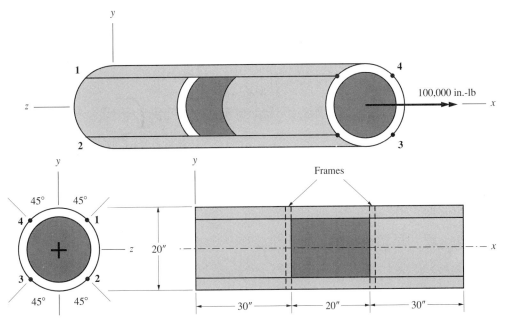

Exercise 5.16

5.17 In Exercise 5.7, calculate the shear flow around rib *C* if:

a. The bottom panel in bay *CD* is removed.

b. The rear spar web in bay *BC* is removed.

5.18 The window size and spacing are as shown. If the allowable shear stress is 10,000 psi, calculate the minimum allowable fuselage skin thickness for the loading shown.

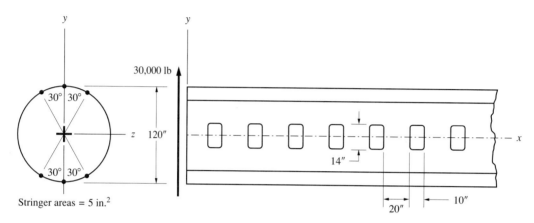

Exercise 5.18

5.19 Calculate the shear flow at point *A* between a pair of windows of the fuselage panel, on the periphery of which a uniform 1000 lb/in. shear flow acts.

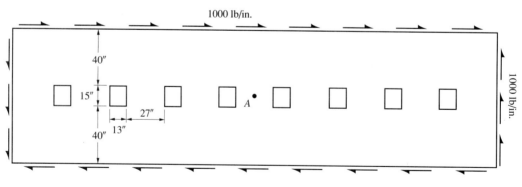

Exercise 5.19

chapter

6

Work-Energy Principles

CHAPTER OUTLINE

6.1 INTRODUCTION

Virtual work methods, now widely implemented on computers, are the practical means of solving for the loads and deflections in complex structures. Virtual work principles convey the requirements of equilibrium and compatibility as *integral* equations, instead of the partial differential equations presented in Chapter 3 (Equations 3.3.4 and 3.9.6). The virtual work principles are mathematical alternatives to—not approximations of—the differential equations of elasticity.

This chapter develops the *principle of virtual work* and the *principle of complementary virtual work* from the basic concepts of vector statics. These principles are then extended to deformable continua, cast in general terms from which formulas for specific structural elements will be obtained in subsequent chapters. *Castigliano's theorems* and the *theorems of minimum potential energy* are consequences of applying the principles of virtual work to linear elastic structures. Since they underlie the Castigliano and minimum energy methods, virtual work methods will be used nearly exclusively in this text.

The principle of virtual work (*ergo* Castigliano's first theorem and the principle of minimum potential energy) is the cornerstone of the *displacement method* of structural analysis (Chapters 9, 10, and 11). In this procedure, the displacements are the primary unknowns, and the principle of virtual work is used to enforce overall equilibrium of the structure. Consider the truss in Figure 6.1.1a, which is statically indeterminate. It is easy to show that simply by requiring the three members to remain pinned together at B, the axial loads in the rods will depend on the small components of displacement at B, as shown in Figure 6.1.1b. Requiring the rods to remain pinned together is an enforcement of *displacement compatibility,* which keeps the truss a continuous structure. By varying u and v, we generate an infinite number of values for the set of member loads F_1, F_2, and F_3. Only one of these sets is admissible, namely, the one for which F_1, F_2, and F_3 are in equilibrium with the applied load P. That set uniquely determines the equilibrium values of u and v, which are delivered by the virtual work procedure. It turns out that $u = 0.01879Pl/AE$ and $v = -1.755Pl/AE$. The rod element loads are calculated from these by means of the equations in Figure 6.1.1b.

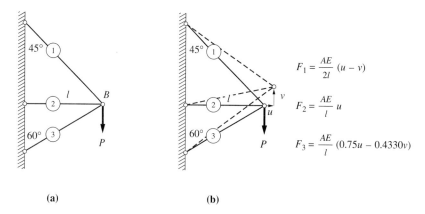

$$F_1 = \frac{AE}{2l}(u - v)$$

$$F_2 = \frac{AE}{l}u$$

$$F_3 = \frac{AE}{l}(0.75u - 0.4330v)$$

(a)   (b)

Figure 6.1.1   (a) Truss with an applied load P at joint A. (b) Member forces as a function of the displacements of point A.

All members have the same axial rigidity AE.

The principle of complementary virtual work (*ergo* Castigliano's second theorem and the principle of minimum complementary potential energy) forms the basis of the classical *force method* of analysis (Chapters 7 and 8). As the name suggests, the forces are sought first and then used thereafter to compute the displacements. The complementary virtual work principle ensures that the forces are consistent with the continuity of the structure. Figure 6.1.2a is a free-body diagram of joint B of the truss in Figure 6.1.1. Unlike the displacement method, the force method enforces equilibrium from the outset. Since the truss has three members and there are only two equations of equilibrium, we cannot find the member loads F_1, F_2, and F_3 uniquely in terms of P. The most we can do is use equilibrium to obtain, say, F_1 and F_2 in terms of F_3. Those expressions are shown in Figure 6.1.2a. (Observe that according to the free-body diagram, a positive value of F_3 means *compression.*)

Having thus determined the equilibrium relationship among the member loads, we can show that the amount of stretch s in each of the rods is as shown in Figure 6.1.2b. There is exactly one choice of F_3 that will produce just the right amount of stretch in each rod to permit their joining together at a common point, thereby maintaining displacement compatibility. This value of F_3 is an automatic outcome of the complementary virtual work procedure.

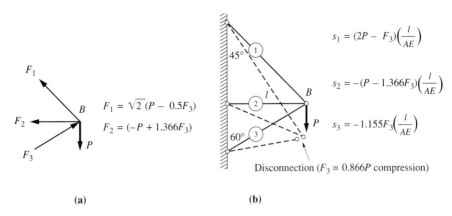

$$s_1 = (2P - F_3)\left(\frac{l}{AE}\right)$$

$$s_2 = -(P - 1.366F_3)\left(\frac{l}{AE}\right)$$

$$s_3 = -1.155F_3\left(\frac{l}{AE}\right)$$

$$F_1 = \sqrt{2}\,(P - 0.5F_3)$$

$$F_2 = (-P + 1.366F_3)$$

Disconnection ($F_3 = 0.866P$ compression)

(a)   (b)

Figure 6.1.2   (a) Free-body diagram and the equilibrium equations for joint B of the truss in Figure 6.1.1a. (b) Example of a failed attempt to select F_3 such that displacement compatibility is maintained at B.

Figure 6.1.2b illustrates what happens when the wrong value of F_3 is chosen. For example, if $F_3 = 0.866P$ (compression), then the stretch of each rod is such that we can bring members 2 and 3 together at a common point, but it cannot be reached by rod 3. The truss "comes apart" at the joint. The correct solution is $F_3 = 0.7458P$ (compression), $F_1 = 0.8868P$, and $F_2 = 0.01879P$ (both tensile).

6.2   SYSTEM EQUILIBRIUM AND COMPATIBILITY

Consider a system of N particles. Figure 6.2.1 illustrates the case for which $N = 3$ (for simplicity and without loss of generality). Particle i is acted upon by a net external force $\mathbf{Q}_i$ coming from outside the system (such as gravity or contact with an adjacent system) and a net internal force $\mathbf{f}_i$ coming from the other particles within the system. Let $\mathbf{f}_{ij}$ be the force on particle i due to particle j. We can then write

$$\mathbf{f}_i = \sum_{\substack{j=1 \\ j \neq i}}^{N} \mathbf{f}_{ij}$$ [6.2.1]

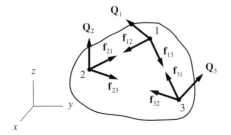

Figure 6.2.1   System of particles, showing external and internal forces.

where $j \neq i$ since a particle cannot exert an external force on itself. By the action–reaction principle, we know that

$$\mathbf{f}_{ij} = -\mathbf{f}_{ji} \qquad [6.2.2]$$

For the three-particle system of Figure 6.2.1, Equations 6.2.1 and 6.2.2 imply that

$$\mathbf{f}_1 = \mathbf{f}_{12} + \mathbf{f}_{13}$$
$$\mathbf{f}_2 = \mathbf{f}_{21} + \mathbf{f}_{23} = -\mathbf{f}_{12} + \mathbf{f}_{23} \qquad [6.2.3]$$
$$\mathbf{f}_2 = \mathbf{f}_{31} + \mathbf{f}_{32} = -\mathbf{f}_{13} - \mathbf{f}_{23}$$

If the system as a whole is in equilibrium, then every particle of the system must be in equilibrium. This implies that the net force on each particle must vanish; that is,

$$\mathbf{Q}_i + \mathbf{f}_i = \mathbf{0}, \quad i = 1, \cdots, N \qquad [6.2.4]$$

This together with Equations 6.2.3 means that for the three-particle system,

$$\mathbf{Q}_1 = -\mathbf{f}_{12} - \mathbf{f}_{13}$$
$$\mathbf{Q}_2 = \mathbf{f}_{12} - \mathbf{f}_{23} \qquad [6.2.5]$$
$$\mathbf{Q}_3 = \mathbf{f}_{13} + \mathbf{f}_{23}$$

Suppose that in the equilibrium state, the system is deformed such that each particle undergoes a small displacement $\mathbf{q}$ from the undeformed state.[1] Before the displacement occurs, the position vector of particle i is $\mathbf{r}_i$ and that of particle j is $\mathbf{r}_j$. The position vector $\mathbf{r}_{ij}$ of point j relative to point i before deformation is

$$\mathbf{r}_{ij} = \mathbf{r}_j - \mathbf{r}_i \qquad [6.2.6]$$

as illustrated in Figure 6.2.2. The magnitude $|\mathbf{r}_{ij}|$ of the relative position vector $\mathbf{r}_{ij}$ is the distance between the two particles before deformation. After the deformation, the relative position vector becomes $\mathbf{r}'_{ij}$. From Figure 6.2.2, we see that

$$\mathbf{r}'_{\mathbf{ij}} = (\mathbf{r}_j + \mathbf{q}_j) - (\mathbf{r}_i + \mathbf{q}_i) = \mathbf{r}_{ij} + \mathbf{q}_j - \mathbf{q}_i$$

That is,

$$\Delta \mathbf{r}_{ij} = \mathbf{q}_j - \mathbf{q}_i \qquad [6.2.7]$$

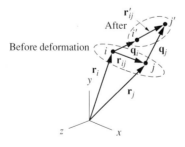

Figure 6.2.2   Relative position vectors of a pair of particles i and j before and after deformation.

[1] Note that $\mathbf{q}$ is used for generalized displacement in the present context, and elsewhere the symbol q is used to denote shear flow.

where $\Delta\mathbf{r}_{ij} = \mathbf{r}'_{ij} - \mathbf{r}_{ij}$ is the change in the relative position vector due to the deformation. Equation 6.2.7 is a *compatibility condition* which states that: *the change in relative position between two particles of the system cannot be prescribed independently of the displacements of those points.*

Let us take the dot product of the equilibrium equation, Equation 6.2.4 for particle i with that particle's displacement vector $\mathbf{q}_i$. In so doing, we obtain the scalar expression

$$\mathbf{Q}_i \cdot \mathbf{q}_i = -\mathbf{f}_i \cdot \mathbf{q}_i$$

Summing this equation over all the particles of the system yields

$$\sum_{i=1}^{N} \mathbf{Q}_i \cdot \mathbf{q}_i = -\sum_{i=1}^{N} \mathbf{f}_i \cdot \mathbf{q}_i \qquad \text{[6.2.8]}$$

For the special case of the three-particle system of Figure 6.2.1, the summation on the right is

$$\sum_{i=1}^{N} \mathbf{f}_i \cdot \mathbf{q}_i = \mathbf{f}_1 \cdot \mathbf{q}_1 + \mathbf{f}_2 \cdot \mathbf{q}_2 + \mathbf{f}_3 \cdot \mathbf{q}_3 \qquad (N = 3)$$

Substituting Equations 6.2.3 leads to

$$\sum_{i=1}^{N} \mathbf{f}_i \cdot \mathbf{q}_i = (\mathbf{f}_{12} + \mathbf{f}_{13}) \cdot \mathbf{q}_1 + (-\mathbf{f}_{12} + \mathbf{f}_{23}) \cdot \mathbf{q}_2 + (-\mathbf{f}_{13} - \mathbf{f}_{23}) \cdot \mathbf{q}_3$$

$$= \mathbf{f}_{12} \cdot (\mathbf{q}_1 - \mathbf{q}_2) + \mathbf{f}_{13} \cdot (\mathbf{q}_1 - \mathbf{q}_3) + \mathbf{f}_{23} \cdot (\mathbf{q}_2 - \mathbf{q}_3)$$

Using the compatibility relation, Equation 6.2.7, this becomes

$$\sum_{i=1}^{N} \mathbf{f}_i \cdot \mathbf{q}_i = \mathbf{f}_{12} \cdot (-\Delta\mathbf{r}_{12}) + \mathbf{f}_{13} \cdot (-\Delta\mathbf{r}_{13}) + \mathbf{f}_{23} \cdot (-\Delta\mathbf{r}_{23})$$

Substituting this back into Equation 6.2.8, we may conclude that *if a system is in equilibrium and the relative displacements are compatible, then*[2]

$$\sum_{i=1}^{N} \mathbf{Q}_i \cdot \mathbf{q}_i = \sum_{\substack{i,j=1 \\ i<j}}^{N} \mathbf{f}_{ij} \cdot \Delta\mathbf{r}_{ij} \qquad \text{[6.2.9]}$$

It must be emphasized that Equation 6.2.9 is valid for an arbitrary self-equilibrating load system and an independently chosen, arbitrary displacement field. No relationship was assumed to exist between the forces and the displacements; that is, the forces and displacements in Equation 6.2.9 are independent of each other.

Now, suppose that nothing is said about the forces, but Equation 6.2.9 is required to hold for *any* arbitrary but compatible set of displacements and relative displacements. For the three-particle system, Equation 6.2.9 is

$$\mathbf{Q}_1 \cdot \mathbf{q}_1 + \mathbf{Q}_2 \cdot \mathbf{q}_2 + \mathbf{Q}_3 \cdot \mathbf{q}_3 = \mathbf{f}_{12} \cdot \Delta\mathbf{r}_{12} + \mathbf{f}_{13} \cdot \Delta\mathbf{r}_{13} + \mathbf{f}_{23} \cdot \Delta\mathbf{r}_{23}$$

We can use the compatibility equation, Equation 6.2.7, which means that

$$\mathbf{Q}_1 \cdot \mathbf{q}_1 + \mathbf{Q}_2 \cdot \mathbf{q}_2 + \mathbf{Q}_3 \cdot \mathbf{q}_3 = \mathbf{f}_{12} \cdot (\mathbf{q}_2 - \mathbf{q}_1) + \mathbf{f}_{13} \cdot (\mathbf{q}_3 - \mathbf{q}_1) + \mathbf{f}_{23} \cdot (\mathbf{q}_3 - \mathbf{q}_2)$$

---

[2] $\sum_{\substack{i,j=1 \\ i<j}}^{N} a_{ij}b_{ij}$ is shorthand for $\sum_{i=1}^{j-1}\left(\sum_{j=1}^{N} a_{ij}b_{ij}\right)$, which contains only those terms of the double summation $\sum_{i=1}^{N}\left(\sum_{j=1}^{N} a_{ij}b_{ij}\right)$ in which $i < j$.

Substituting Equation 6.2.3 on the right, simplifying, and collecting terms leads to

$$(\mathbf{Q}_1 + \mathbf{f}_1) \cdot \mathbf{q}_1 + (\mathbf{Q}_2 + \mathbf{f}_2) \cdot \mathbf{q}_2 + (\mathbf{Q}_3 + \mathbf{f}_3) \cdot \mathbf{q}_3 = 0 \qquad \text{[6.2.10]}$$

The displacements $\mathbf{q}_1$, $\mathbf{q}_2$, and $\mathbf{q}_3$ are arbitrary and may be chosen independently of each other. Regardless, Equation 6.2.10 must remain valid. Let us select $\mathbf{q}_1 \neq \mathbf{0}$ and $\mathbf{q}_2 = \mathbf{q}_3 = \mathbf{0}$. Then,

$$(\mathbf{Q}_1 + \mathbf{f}_1) \cdot \mathbf{q}_1 = 0 \qquad \text{for any } \mathbf{q}_1 \neq \mathbf{0}$$

Since $\mathbf{q}_1$ is arbitrary, we can choose it to lie in the direction of the vector $\mathbf{Q}_1 + \mathbf{f}_1$ and to have any magnitude except zero. It follows then that $|\mathbf{Q}_1 + \mathbf{f}||\mathbf{q}_1| = 0$, while at the same time $|\mathbf{q}_1| \neq 0$, which means that $\mathbf{Q}_1 + \mathbf{f}_1 = 0$. Repeating this argument for each of the N particles, we can conclude that

$$\mathbf{Q}_i + \mathbf{f}_i = 0, \qquad i = 1, \cdots, N$$

Therefore, the set of external and internal forces is self-equilibrating: the system is in equilibrium. In summary:

> *A system is in equilibrium if and only if Equation 6.2.9 is valid for any compatible deformation.*   **[6.2.11]**

Next, let us place no restrictions on the displacements, and at the same time, require Equation 6.2.9 to be valid for *any* self-equilibrating load system. For the system in Figure 6.2.1 this implies that

$$\mathbf{Q}_1 \cdot \mathbf{q}_1 + \mathbf{Q}_2 \cdot \mathbf{q}_2 + \mathbf{Q}_3 \cdot \mathbf{q}_3 = \mathbf{f}_{12} \cdot \Delta\mathbf{r}_{12} + \mathbf{f}_{13} \cdot \Delta\mathbf{r}_{13} + \mathbf{f}_{23} \cdot \Delta\mathbf{r}_{23}$$

Since the loads must be in equilibrium, we can substitute Equation 6.2.5 into this equation and collect terms to obtain

$$[\Delta\mathbf{r}_{12} - (\mathbf{q}_2 - \mathbf{q}_1)] \cdot \mathbf{f}_{12} + [\Delta\mathbf{r}_{13} - (\mathbf{q}_3 - \mathbf{q}_1)] \cdot \mathbf{f}_{13} + [\Delta\mathbf{r}_{23} - (\mathbf{q}_3 - \mathbf{q}_2)] \cdot \mathbf{f}_{23} = 0 \qquad \text{[6.2.12]}$$

The forces are arbitrary and independent, so we can set $\mathbf{f}_{13} = \mathbf{f}_{23} = 0$ and choose $\mathbf{f}_{12}$ to lie in the direction of $\Delta\mathbf{r}_{12} - (\mathbf{q}_2 - \mathbf{q}_1)$. Equation 6.2.12 then reduces to

$$|\Delta\mathbf{r}_{12} - (\mathbf{q}_2 - \mathbf{q}_1)||\mathbf{f}_{12}| = 0 \qquad \text{for any value of } |\mathbf{f}_{12}|$$

It follows that $|\Delta\mathbf{r}_{12} - (\mathbf{q}_2 - \mathbf{q}_1)|$ must vanish, which means that

$$\Delta\mathbf{r}_{12} - (\mathbf{q}_2 - \mathbf{q}_1) = 0 \quad \text{or} \quad \Delta\mathbf{r}_{12} = \mathbf{q}_2 - \mathbf{q}_1$$

This argument can be repeated for each of the forces, showing that all of the bracketed terms in Equation 6.2.12 must be zero, which proves that compatibility, represented by Equation 6.2.7, is indeed satisfied.

> *The deformation of a system is compatible if and only if Equation 6.2.9 is valid for any self-equilibrating load system.*   **[6.2.13]**

6.3   THE VIRTUAL WORK PRINCIPLE

Consider a particle acted on by N forces, $\mathbf{Q}_i$, $i = 1, \ldots, N$. The resultant force on the particle is $\mathbf{Q}_R$, where $\mathbf{Q}_R = \sum\limits_{i=1}^{N} \mathbf{Q}_i$. If the particle undergoes a real, infinitesimal displacement $d\mathbf{q}$, then the incremental work done by the forces on the particle is $dW = \mathbf{Q}_R \cdot d\mathbf{q}$. If instead we imagine that the particle is given a small but fictitious, or *virtual*, displacement $\delta\mathbf{q}$, *while the forces are held constant*, then the total *virtual work* δW done on the particle is

$$\delta W = \mathbf{Q}_R \cdot \delta\mathbf{q}$$

If the particle is in equilibrium, that is, if $\mathbf{Q}_R = 0$, it is clear that $\delta W = 0$ for any choice of the virtual displacement vector $\delta\mathbf{q}$.

Conversely, suppose we place no conditions on $\mathbf{Q}_R$, but require that its virtual work be zero for all possible virtual displacements of the particle. That is, let us demand that $\mathbf{Q}_R \cdot \delta\mathbf{q} = 0$ no matter what the value of $\delta\mathbf{q}$. One of the infinite possible choices of $\delta\mathbf{q}$ is a nonzero displacement in the direction of the net force $\mathbf{Q}_R$, so that θ

Figure 6.3.1   Particle undergoing a virtual displacement while acted on by the net force $\mathbf{Q}_R$.

in Figure 6.3.1 is zero. In this case, $\mathbf{Q}_R \cdot \delta\mathbf{q} = |\mathbf{Q}_R||\delta\mathbf{q}|$. Since $|\delta\mathbf{q}| \neq 0$ and $|\mathbf{Q}_R||\delta\mathbf{q}| = 0$, it follows that $|\mathbf{Q}_R| = 0$. That is, the net force on the particle must be zero, and this means that the particle is in equilibrium.

Thus, the *principle of virtual work for a particle* is as follows:

> *A particle is in equilibrium if and only if the virtual work done on the particle is zero for any virtual displacement.*

Let a system of N particles, such as that in Figure 6.2.1, undergo a small virtual distortion, so that a virtual displacement is imparted to each particle. The virtual displacements must be small so that the geometry of the system is essentially unchanged by the virtual deformation. During this virtual distortion, all forces—the *true* forces that exist in the system—are held fixed. In this case, the virtual work done by the external forces is

$$\delta W_{\text{ext}} = \sum_{i=1}^{N} \mathbf{Q}_i \cdot \delta\mathbf{q}_i \qquad \text{[6.3.1]}$$

The internal virtual work δW_{int} is given by

$$\delta W_{\text{int}} = \sum_{\substack{i,j=1 \\ i<j}}^{N} \mathbf{f}_{ij} \cdot \delta\mathbf{r}_{ij} \qquad \text{[6.3.2]}$$

Observe that the dot product $\mathbf{f}_{ij} \cdot \delta\mathbf{r}_{ij}$ is the product of the true force mutually applied to points i and j times the virtual change in the distance between those points. Thus, δW_{int} is the virtual work done on the material joining the points of the system. For example, if you imagine each particle to be joined to all of the others by elastic springs, then δW_{int} is the virtual work done on all of those springs during the virtual distortion of the system.

The term δW_{int} is always zero for a rigid body since, by definition, the distance between points on a rigid body cannot change.

If the actual displacements $\mathbf{q}_i$ and $\Delta\mathbf{r}_{ij}$ in Equation 6.2.9 are replaced by the virtual displacements $\delta\mathbf{q}_i$ and $\delta\mathbf{r}_{ij}$, where $\delta\mathbf{r}_{ij} = \delta\mathbf{q}_j - \delta\mathbf{q}_i$ (cf. Equation 6.2.7), we obtain

$$\sum_{i=1}^{N} \mathbf{Q}_i \cdot \delta\mathbf{q}_i = \sum_{\substack{i,j=1 \\ i<j}}^{N} \mathbf{f}_{ij} \cdot \delta\mathbf{r}_{ij}$$

That is,

$$\delta W_{\text{ext}} = \delta W_{\text{int}}$$

Equation 6.2.11 therefore leads to the general *principle of virtual work*, which is:

> A system is in equilibrium if and only if $\delta W_{\text{ext}} = \delta W_{\text{int}}$ for any compatible virtual deformation.

The principle of virtual work is an *exact*, alternative statement of equilibrium. Observe that the principle does not depend on material properties.

Example 6.3.1   Figure 6.3.2 shows a weight W supported by two cables. Use the principle of virtual work to find the tension in each cable.

This problem is statically determinate. The forces T_1 and T_2 therefore do not depend on the specific nature of the two-force members joining point C to points A and B. In particular, they are unaffected by the displacement of point C. Therefore, we may remove the members themselves (the cables) and replace them by the forces they exert.

If u and v are the x and y components of the displacement of point C, then the small virtual displacement of the point is $\delta\mathbf{q} = \delta u\mathbf{i} + \delta v\mathbf{j}$. Resolving the forces on C into components, the total virtual work of the weight and the cable tensions is

$$\delta W = (-W\mathbf{j}) \cdot (\delta u\mathbf{i} + \delta v\mathbf{j}) + (-T_1 \cos 45°\mathbf{i} + T_1 \sin 45°\mathbf{j}) \cdot (\delta u\mathbf{i} + \delta v\mathbf{j}) + (T_2 \cos 60°\mathbf{i} + T_2 \sin 60°\mathbf{j}) \cdot (\delta u\mathbf{i} + \delta v\mathbf{j})$$

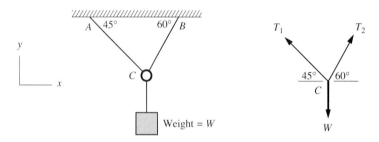

Figure 6.3.2   Supported weight and the free-body diagram.

Carrying out the dot products, collecting terms, and setting $\delta W = 0$, we get

$$(-T_1 \cos 45° + T_2 \cos 60°)\delta u + (-W + T_1 \sin 45° + T_2 \sin 60°)\delta v = 0$$

According to the principle of virtual work, this equality must hold for arbitrary values of δu and δv if point C is to be in equilibrium. The coefficients of δu and δv must therefore vanish, yielding two equations:

$$-T_1 \cos 45° + T_2 \cos 60° = 0$$
$$T_1 \sin 45° + T_2 \sin 60° = W$$

These are the same equilibrium equations we would find for point C by using elementary statics. The solution of these equations is $T_1 = 0.518W$ and $T_2 = 0.732W$.

Example 6.3.2   The figure shows three springs rigidly attached to the wall at points B, C, and D, and attached to each other at point A, where the external load P is applied. Each spring has a unique spring constant. Use the principle of virtual work to find the spring forces F_1, F_2, and F_3.

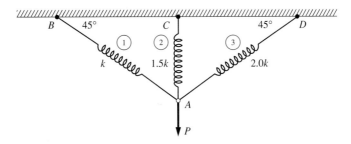

Figure 6.3.3   Three springs in equilibrium under the point load P.

Let the displacement of point A be $\mathbf{q}_A$, so that in terms of its horizontal and vertical components, $\mathbf{q}_A = u_A\mathbf{i} + v_A\mathbf{j}$. As always, we assume these displacements are small, so that the geometry of the system is left essentially unchanged after the load is applied. The virtual displacement of point A is

$$\delta\mathbf{q}_A = \delta u_A\mathbf{i} + \delta v_A\mathbf{j}$$

Assuming that the springs are undeformed before the load is applied, let the extension or *stretch* of each spring due to the load be s_i, $i = 1, 2, 3$. These extensions are found by projecting the displacement $\mathbf{q}_A$ onto each spring. Thus, if $\mathbf{n}_{BA}$, $\mathbf{n}_{CA}$, and $\mathbf{n}_{DA}$ are the unit vectors drawn from the attachment points of each spring to the common point A, we have

$$
\begin{aligned}
s_1 &= \mathbf{q}_A \cdot \mathbf{n}_{BA} = u_A \cos 45^\circ - v_A \sin 45^\circ \\
s_2 &= \mathbf{q}_A \cdot \mathbf{n}_{CA} = -v_A \\
s_3 &= \mathbf{q}_A \cdot \mathbf{n}_{DA} = -u_A \cos 45^\circ - v_A \sin 45^\circ
\end{aligned}
$$ [a]

The internal virtual work is the work done by the actual spring forces under the virtual stretch δs_i, $i = 1, 2, 3$, where

$$
\begin{aligned}
\delta s_1 &= \delta\mathbf{q}_A \cdot \mathbf{n}_{BA} = \delta u_A \cos 45^\circ - \delta v_A \sin 45^\circ \\
\delta s_2 &= \delta\mathbf{q}_A \cdot \mathbf{n}_{CA} = -\delta v_A \\
\delta s_3 &= \delta\mathbf{q}_A \cdot \mathbf{n}_{DA} = -\delta u_A \cos 45^\circ - \delta v_A \sin 45^\circ
\end{aligned}
$$

Thus,

$$
\begin{aligned}
\delta W_{\text{int}} &= \delta W_{\text{int,spring 1}} + \delta W_{\text{int,spring 2}} + \delta W_{\text{int,spring 3}} \\
&= k s_1 \delta s_1 + (1.5k)\, s_2 \delta s_2 + (2k)\, s_3 \delta s_3 \\
&= k(u_A \cos 45^\circ - v_A \sin 45^\circ)(\delta u_A \cos 45^\circ - \delta v_A \sin 45^\circ) + (1.5k)(-v_A)(-\delta v_A) \\
&\quad + (2k)\,(-u_A \cos 45^\circ - v_A \sin 45^\circ)(-\delta u_A \cos 45^\circ - \delta v_A \sin 45^\circ)
\end{aligned}
$$

Collecting terms on the right-hand side, we get

$$\delta W_{\text{int}} = (3ku_A \cos^2 45^\circ + kv_A \sin 45^\circ \cos 45^\circ)\delta u_A + (ku_A \cos 45^\circ \sin 45^\circ + 3kv_A \sin^2 45^\circ + 1.5kv_A)\delta v_A$$

or

$$\delta W_{\text{int}} = (1.5ku_A + 0.5kv_A)\delta u_A + (0.5ku_A + 3.0kv_A)\delta v_A$$

Next, we need the external virtual work of the applied load. This is obtained as follows:

$$\delta W_{\text{ext}} = (-P\mathbf{j}) \cdot \delta\mathbf{q}_A = -P\delta v_A$$

Equating the external and internal virtual works, and recalling that δu_A and δv_A are arbitrary, yields two equations:

$$1.5ku_A + 0.5kv_A = 0$$

$$0.5ku_A + 3.0kv_A = -P$$ [b]

The solution of this system is therefore $u_A = 0.1176(P/k)$ and $v_A = -0.3529(P/k)$. These are the components of the displacement of point A. We substitute them into Equation a to find the extensions of springs 1, 2, and 3, respectively, as follows:

$$s_1 = 0.3328(P/k) \qquad s_2 = 0.3529(P/k) \qquad s_3 = 0.1664(P/k)$$

Finally, multiplying each of these by the appropriate spring constant yields the spring forces themselves, which is what we set out to find:

$$F_1 = k\left[0.3328\left(\frac{P}{k}\right)\right] = 0.3328\,P \qquad F_2 = 1.5k\left[0.3529\left(\frac{P}{k}\right)\right] = 0.5294\,P \qquad F_3 = 2.0k\left[0.1664\left(\frac{P}{k}\right)\right] = 0.3328\,P$$

Observe that this problem was statically indeterminate. There were three unknown forces acting on A, but only two equations of equilibrium: $\Sigma F_x = 0$ and $\Sigma F_y = 0$. Statically indeterminate problems always require consideration of the structural deformations. Using the principle of virtual work was our way of doing so. We know that the principle guarantees equilibrium as long as we maintain compatibility, which we did by invoking Equation a.

The expression for internal virtual work given by Equation 6.3.2 applies to a system of particles. To extend that expression to deformable continuous structures, we need to imagine the number of particles in the system going to infinity, while the finite mass of the system becomes distributed continuously throughout the volume it occupies. We can then subdivide the system into differential elements. The summation sign in Equation 6.3.2 becomes an integral, and $\mathbf{f}_{ij} \cdot \delta\mathbf{r}_{ij}$ can be written in a form appropriate to a continuum, that is, in terms of stresses and strains.

From Equation 3.14.1, we know that in a quasistatic loading process, the work done within a solid by the true stresses during an increment of the true strains is

$$dW = \iiint_V dw_o dV = \iiint_V \left(\sigma_x d\varepsilon_x + \sigma_y d\varepsilon_y + \sigma_z d\varepsilon_z + \tau_{xy} d\gamma_{xy} + \tau_{xz} d\gamma_{xz} + \tau_{yz} d\gamma_{yz}\right) dV$$ [6.3.3]

regardless of the specific relationship between stress and strain. Therefore, the internal virtual work of the true stresses (held fixed) acting through the virtual strains is

$$\delta W_{\text{int}} = \iiint_V \left(\sigma_x \delta\varepsilon_x + \sigma_y \delta\varepsilon_y + \sigma_z \delta\varepsilon_z + \tau_{xy} \delta\gamma_{xy} + \tau_{xz} \delta\gamma_{xz} + \tau_{yz} \delta\gamma_{yz}\right) dV$$ [6.3.4]

This is the continuum version of Equation 6.3.2. We will use Equation 6.3.4 as a formula to find the internal virtual work for specific structural elements.

For a continuous medium, the external virtual work, Equation 6.3.1, must be generalized to include the work of surface tractions and body forces. Thus, for a solid such as that pictured in Figure 6.3.4, we have

$$\delta W_{\text{ext}} = \sum_{i=1}^{n} \mathbf{Q}_i \cdot \delta\mathbf{q}_i + \iint_S \mathbf{T}^{(n)} \cdot \delta\mathbf{u}\, dS + \iiint_V \mathbf{b} \cdot \delta\mathbf{u}\, dV$$ [6.3.5]

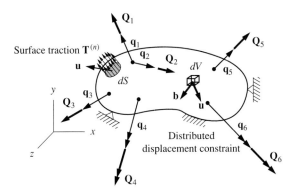

Figure 6.3.4   A solid body, constrained in an arbitrary fashion and acted upon by generalized point loads (**Q**) plus surface (**T**$^{(n)}$) and volume (**b**) force fields.

where S is the closed surface bounding the volume V, $\mathbf{Q}_i$ is the ith *generalized* force (either a point force or a point couple), and $\mathbf{q}_i$ is the generalized displacement (either a translation or a rotation) *in the direction of* $\mathbf{Q}_i$, so that, for example, $\mathbf{Q}_i \cdot \delta\mathbf{q}_i = Q_i\delta q_i$. Thus, $\sum_{i=1}^{n} Q_i\delta q_i$ is the virtual work of the n generalized point loads.

Whereas $\mathbf{q}$ (or $\delta\mathbf{q}$) is a displacement at a specific point, $\mathbf{u}$ (or $\delta\mathbf{u}$) is a displacement *field*, defined as a continuous function of position throughout the solid. To calculate the virtual work of the traction $\mathbf{T}^{(n)}$ at a point on the surface S of the solid where the virtual displacement is $\delta\mathbf{u}$, we multiply $\mathbf{T}^{(n)}$ (force per unit area) by the element of surface area dS on which it acts, and take the dot product of the differential force vector $\mathbf{T}^{(n)}dS$ into $\delta\mathbf{u}$. Integrating this differential work over the entire surface of the solid (including regions where $\mathbf{T}^{(n)} = 0$ and $\delta\mathbf{u} = 0$) yields the finite virtual work of the distributed surface forces, which is the second term on the right of Equation 6.3.5. Likewise, the last term in the equation accounts for the virtual work of the body forces per unit volume acting on the bulk of the solid material.

6.4   MINIMUM POTENTIAL ENERGY AND CASTIGLIANO'S FIRST THEOREM[3]

If a solid is linearly elastic, then according to section 3.14, the internal work associated with a quasistatic loading process equals the strain energy U, or

$$W_{\text{int}} = U$$

where, by virtue of Equations 3.14.3 and 3.14.4, U can be viewed as a function of the displacements:

$$U = U(q_1, q_2, \cdots, q_n) \tag{6.4.1}$$

As before, q_i, $i = 1, \cdots, n$, are the generalized displacements in the direction of the n generalized point loads Q_i. For a virtual deformation, it follows that

$$\delta W_{\text{int}} = \delta U \tag{6.4.2}$$

| [3]Carlo Alberto Pio Castigliano (1847–1884), Italian civil engineer.

The virtual displacements may be viewed as arbitrary small changes, or variations, of the actual displacements. Therefore, the accompanying variation δU of the strain energy is calculated by expanding the total differential:

$$\delta U = \sum_{i=1}^{n} \frac{\partial U}{\partial q_i} \delta q_i \tag{6.4.3}$$

Since Q_i and δq_i are colinear,

$$\delta W_{\text{ext}} = \sum_{i=1}^{n} Q_i \delta q_i$$

and the principle of virtual work for an elastic solid is

$$\sum_{i=1}^{n} Q_i \delta q_i = \sum_{i=1}^{n} \frac{\partial U}{\partial q_i} \delta q_i \tag{6.4.4}$$

This must be true for any choice of the virtual displacements $\delta q_1, \delta q_2, \cdots, \delta q_n$. If we set all of them except δq_1 equal to zero, then Equation 6.4.4 becomes

$$Q_1 \delta q_1 = \left(\frac{\partial U}{\partial q_1} \right) \delta q_1,$$

or

$$Q_1 = \frac{\partial U}{\partial q_1}$$

Repeating the argument for the remaining virtual displacements yields

$$Q_i = \frac{\partial U}{\partial q_i}, \ i = 1, 2, \cdots, n \tag{6.4.5}$$

This is *Castigliano's first theorem*. For an elastic structure, each generalized load is the derivative of the strain energy with respect to the corresponding displacement. Note that Castigliano's first theorem was deduced from the more general principle of virtual work, which does not require the structure to be elastic.

Using the undeformed configuration of a structure as its datum, the potential energy V of the external loads may be written

$$V = -\sum_{i=1}^{n} Q_i q_i \tag{6.4.6}$$

The constant force Q_i, having moved through the parallel displacement q_i, has lost its capacity to do the work $Q_i q_i$. That is the reason for the minus sign. With the loads Q_i held fixed, we therefore have

$$\frac{\partial V}{\partial q_i} = -Q_i \quad i = 1, \cdots, n \tag{6.4.7}$$

Castigliano's first theorem may thus be written

$$\frac{\partial \Pi}{\partial q_i} = 0 \quad i = 1, \cdots, n \tag{6.4.8}$$

Here, $\Pi = U + V$ is the total potential energy of the structure, and Equation 6.4.8 is therefore a statement of the *theorem of minimum potential energy*. Actually, Equation 6.4.8 only implies that the potential energy is *stationary*. Proof that it is indeed a minimum for elastic structures in stable equilibrium may be found elsewhere.[4] Again, the reader is reminded that the source of Equation 6.4.8 is the principle of virtual work.

Example 6.4.1   Solve the problem in Example 6.3.2 using the principle of minimum potential energy.

The elastic potential energy of each spring is $ks^2/2$, where k is the spring constant and s is the spring extension. Therefore, the total strain energy of the three-spring assembly in Figure 6.3.3 is

$$U = \frac{1}{2}k_1 s_1^2 + \frac{1}{2}k_2 s_2^2 + \frac{1}{2}k_3 s_3^2 = \frac{1}{2}k s_1^2 + \frac{1}{2}(1.5k)s_2^2 + \frac{1}{2}(2k)s_3^2$$

Using Equation a of Example 6.3.2 to express the individual spring extensions in terms of the displacements u_A and v_A leads to

$$U = \frac{1}{2}k(u_A \cos 45° - v_A \sin 45°)^2 + \frac{1}{2}(1.5k)(-v_A)^2 + \frac{1}{2}(2k)(-u_A \cos 45° - v_A \sin 45°)^2$$

$$= \frac{3}{4}k u_A^2 + \frac{3}{2}k v_A^2 + \frac{1}{2}k u_A v_A \qquad\qquad \textbf{[a]}$$

From Equation 6.4.6, the potential energy of the load P is

$$V = -(-P)v_A \qquad\qquad \textbf{[b]}$$

since P is directed downwards, which is opposite to the direction of positive displacement (up). The total potential energy is $\Pi = U + V$, so that in this case, we have

$$\Pi = \frac{3}{4}k u_A^2 + \frac{3}{2}k v_A^2 + \frac{1}{2}k u_A v_A + P v_A \qquad\qquad \textbf{[c]}$$

With the applied load P held fixed, Π is a function of the two displacement components u_A and v_A. From Equation 6.4.8 we obtain

$$\frac{\partial \Pi}{\partial u_A} = 0: \qquad \frac{3}{2}k u_A + \frac{1}{2}k v_A \quad\;\; = 0$$

$$\frac{\partial \Pi}{\partial v_A} = 0: \qquad \frac{1}{2}k u_A + 3k v_A + P = 0 \qquad\qquad \textbf{[d]}$$

These two equations for u_A and v_A are identical to Equation b in Example 6.3.2, yielding

$$u_A = 0.1176\left(\frac{P}{k}\right) \qquad v_A = -0.3529\left(\frac{P}{k}\right) \qquad\qquad \textbf{[e]}$$

From this point on, we follow the same procedure as before to complete the problem.

Let us calculate the value of the potential energy Π at equilibrium by substituting the displacements in Equation e into Equation c:

$$\Pi = \frac{3}{4}k\left(0.1176\frac{P}{k}\right)^2 + \frac{3}{2}k\left(-0.3529\frac{P}{k}\right)^2 + \frac{1}{2}k\left(0.1176\frac{P}{k}\right)\left(-0.3529\frac{P}{k}\right)$$

$$+ P\left(-0.3529\frac{P}{k}\right) = -0.1765\frac{P^2}{k} \qquad\qquad \textbf{[f]}$$

According to the calculus of functions of several variables, this is a *minimum* value of Π if, for the same values of u_A and v_A,

---

| [4]See, for example, Y. C. Fung, *Foundations of Solid Mechanics*, Chapter 10, Englewood Cliffs, NJ, Prentice-Hall, 1965.

$$\frac{\partial^2 \Pi}{\partial u_A^2} \cdot \frac{\partial^2 \Pi}{\partial v_A^2} - \left(\frac{\partial^2 \Pi}{\partial u_A \partial v_A}\right)^2 > 0 \qquad \frac{\partial^2 \Pi}{\partial u_A^2} > 0 \qquad \text{[g]}$$

Calculating the second partial derivatives of Π yields

$$\frac{\partial^2 \Pi}{\partial u_A^2} = \frac{3}{2}k \qquad \frac{\partial^2 \Pi}{\partial v_A^2} = 3k \qquad \frac{\partial^2 \Pi}{\partial u_A \partial v_A} = \frac{1}{2}k$$

Since k is positive, we can see that both conditions in Equation g are satisfied: Π is indeed a minimum.

6.5   STIFFNESS MATRIX

If a structure is not only elastic but *linearly elastic*, then, by definition, the generalized loads Q_i and the generalized displacements q_i in the direction of the loads are directly proportional to each other. That is,

$$Q_i = \sum_{l=1}^{n} k_{il} q_l, \qquad i = 1, 2, \cdots, n$$

[6.5.1]

In this set of n equations, the coefficients k_{il} form a square matrix[5]—the *stiffness matrix*—with n rows and n columns (an "n by n" matrix), as follows:

$$\begin{bmatrix} k_{11} & k_{12} & \cdot & \cdot & \cdot & k_{1n} \\ k_{21} & k_{22} & \cdot & \cdot & \cdot & \cdot \\ \cdot & & \cdot & \cdot & \cdot & \cdot \\ \cdot & & & \cdot & \cdot & \cdot \\ \cdot & & & & \cdot & \cdot \\ k_{n1} & k_{n2} & \cdot & \cdot & \cdot & k_{nn} \end{bmatrix}$$

The number n of generalized displacements is the number of *degrees of freedom* which the structure has been assigned for purposes of analysis. For example, the linearly elastic cantilever beam in Figure 6.5.1a has just two degrees of freedom: the transverse displacement v_1, and the rotation θ_{z_1} at the free end. These are linearly related to the transverse force Y_1 and couple M_{z_1} by a two-by-two stiffness matrix,

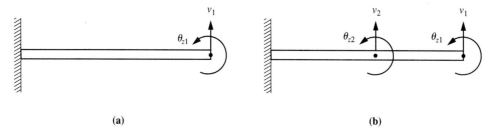

(a)   (b)

Figure 6.5.1   (a) Cantilever beam with two degrees of freedom. (b) The same beam with four degrees of freedom.

| [5]See Chapter 9.

$$Y_1 = k_{11}v_1 + k_{12}\theta_{z_1}$$
$$M_{z_1} = k_{21}v_1 + k_{22}\theta_{z_1}$$

If we wish to include the displacements of the midpoint of the beam, then the number of degrees of freedom doubles to four and the number of stiffness coefficients quadruples to 16. Generally speaking, the more complex the structure, the larger the number of degrees of freedom required for acceptable accuracy.

Observe that from Equation 6.5.1, we have

$$\frac{\partial Q_i}{\partial q_j} = \frac{\partial}{\partial q_j} \sum_{l=1}^{n} k_{il}q_l = \sum_{l=1}^{n} k_{il} \frac{\partial q_l}{\partial q_j} \qquad [6.5.2]$$

Since the generalized displacements are independent of each other by definition, it follows that $\partial q_l / \partial q_j$ is zero unless $l = j$, in which case $\partial q_j / \partial q_j = 1$. Therefore, the only nonzero term in the summation on the right of Equation 6.5.2 is the one for which $l = j$:

$$\frac{\partial Q_i}{\partial q_j} = k_{ij} \qquad [6.5.3]$$

Likewise

$$\frac{\partial Q_j}{\partial q_i} = k_{ji} \qquad [6.5.4]$$

However, from Castigliano's first theorem, we have

$$\frac{\partial Q_i}{\partial q_j} = \frac{\partial}{\partial q_j} \frac{\partial U}{\partial q_i} = \frac{\partial}{\partial q_i} \frac{\partial U}{\partial q_j} = \frac{\partial Q_j}{\partial q_i} \qquad [6.5.5]$$

Therefore, it follows from Equations 6.5.3 and 6.5.4 that

$$k_{ij} = k_{ji}, \qquad i, j = 1, 2, \cdots, n \qquad [6.5.6]$$

This means that the coefficients in *row i* of the stiffness matrix are identical to those in *column i. The stiffness matrix of an elastic structure is symmetric.* This reduces the number of independent stiffness coefficients. In general, then

> The number of independent components of an n by n symmetric matrix is $\dfrac{n(n+1)}{2}$.   [6.5.7]

Suppose all of the generalized displacements are zero except q_1. Then the corresponding generalized loads are found from Equation 6.5.1

$$Q_i = k_{i1}q_1, \quad i = 1, \cdots, n$$

In particular, $Q_1 = k_{11}q_1 \cdot Q_1$ is the nonzero force exerted at the same location and in the same direction as q_1. On physical grounds alone, Q_1 and q_1 must have the same sign. It follows that $k_{11} > 0$. Repeating this argument, in turn, for degrees of freedom $2, \cdots, n$ leads to the conslusion that

$$k_{ii} > 0 \quad i = 1, \cdots, n \qquad [6.5.8]$$

From Equation b of Example 6.3.2 we see that the stiffness matrix of the two degree of freedom system in Figure 6.3.3 is

$$\begin{bmatrix} 1.5k & 0.5k \\ 0.5k & 3.0k \end{bmatrix}$$

Clearly, this matrix is symmetric, and both k_{11} and k_{22} are positive.

6.6 THE COMPLEMENTARY VIRTUAL WORK PRINCIPLE

In an analogy to the definition of internal virtual work given in Equation 6.3.2, the *internal complementary virtual work* for a system of particles such as that of Figure 6.2.1 is defined as

$$\delta W_{int}^* = \sum_{\substack{i,j=1 \\ i<j}}^{N} \Delta \mathbf{r}_{ij} \cdot \delta \mathbf{f}_{ij} \qquad \text{[6.6.1]}$$

Here, we have the product of the *actual* change in the relative displacement vector with the *virtual* internal forces, whereas before, we had the *virtual* relative position change times the *actual* internal forces. Likewise, in an analogy with Equation 6.3.1, we define the *external complementary virtual work*, where the virtual quantities are the loads instead of the displacements, as follows:

$$\delta W_{ext}^* = \sum_{i=1}^{N} \mathbf{q}_i \cdot \delta \mathbf{Q}_i \qquad \text{[6.6.2]}$$

If the actual forces $\mathbf{Q}_i$ and $\mathbf{f}_{ij}$ in Equation 6.2.9 are replaced by the virtual forces $\delta\mathbf{Q}_i$ and $\delta\mathbf{f}_{ij}$, we obtain

$$\sum_{i=1}^{N} \mathbf{q}_i \cdot \delta \mathbf{Q}_i = \sum_{\substack{i,j=1 \\ i<j}}^{N} \Delta \mathbf{r}_{ij} \cdot \delta \mathbf{f}_{ij}$$

That is,

$$\delta W_{ext}^* = \delta W_{int}^*$$

Equation 6.2.13 therefore results in the *principle of complementary virtual work:*

> *The displacements of a system satisfy compatibility if and only if $\delta W_{ext}^* = \delta W_{int}^*$*   **[6.6.3]**
> *for any self-equilibrating virtual loading.*

Example 6.6.1 Figure 6.6.1a shows a load W supported by two springs with identical spring constants k. The picture is similar to that for Example 6.3.1, in which the spring loads were found to have the values shown. Find the horizontal displacement of point P, using the principle of complementary virtual work.

First, we remove the actual loads from the system and place an externally applied virtual load $\delta\mathbf{Q}$ at P in the direction of the desired displacement, as shown in Figure 6.6.1b. We then use statics to find the values of the internal virtual loads, which are shown alongside each spring. The external virtual complementary work is the product of the *virtual* horizontal load δQ times the *actual* horizontal displacement u of point P, that is,

$$\delta W_{ext}^* = u\delta Q$$

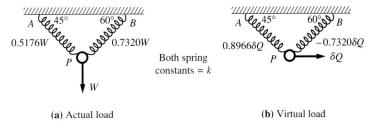

(a) Actual load   **(b)** Virtual load

Figure 6.6.1   Point load supported by two springs.

To calculate the internal complementary virtual work, we use Equation 6.6.1. In general, if points i and j are joined by a spring, then

$$\Delta \mathbf{r}_{ij} \cdot \delta \mathbf{f}_{ij} = \Delta s_{ij} \delta f_{ij}$$

where Δs_{ij} is the change in the distance between the points caused by the actual loading, and δf_{ij} is the signed magnitude (positive for tension, negative for compression) of the virtual force in the spring. If the spring is elastic with spring rate k, then

$$\Delta s_{ij} = \frac{f_{ij}}{k}$$

where f_{ij} is the actual load in the spring.

For our problem, then,

$$\delta W^*_{\text{int}} = \left(\frac{f_{AP}}{k} \right) \delta f_{AP} + \left(\frac{f_{BP}}{k} \right) \delta f_{BP}$$

$$= \left(\frac{0.5176W}{k} \right) (0.8966 \delta Q) + \left(\frac{0.7320W}{k} \right) (-0.7320 \delta Q)$$

$$= \frac{0.4641W}{k} \delta Q - \frac{0.5358W}{k} \delta Q$$

$$= -\frac{0.0717W}{k} \delta Q$$

Setting $\delta W^*_{\text{ext}} = \delta W^*_{\text{int}}$, we have

$$u \delta Q = -0.0717 \left(\frac{W}{k} \right) \delta Q$$

Finally, cancelling δQ on both sides yields

$$u = -0.0717 \left(\frac{W}{k} \right)$$

The negative sign means that the displacement is in the direction opposite to that of δQ, that is, point P moves to the left when the downward load W is applied.

Example 6.6.2   The structure shown in Figure 6.6.2a supports a vertical load at A. Use the principle of complementary virtual work to find (a) the horizontal displacement u of point C, and (b) the rotation θ_{AD} of member AD, due to the load P.

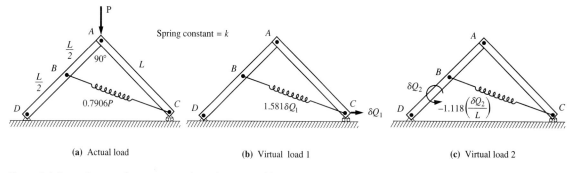

(a) Actual load   (b) Virtual load 1   (c) Virtual load 2

Figure 6.6.2   Two rigid pin-connected members joined by a spring.

(a)

The force in the spring due to P is found using statics, and the result is shown in Figure 6.6.2a. To find the horizontal displacement at C, we apply a horizontal virtual force δQ_1 at that point and use statics to calculate the resulting virtual spring force, the value of which is shown in Figure 6.6.2b. The external complementary virtual work of δQ_1 is

$$\delta W_{\text{ext}}^* = u\delta Q_1$$

The internal complementary virtual work is that of the spring alone, since the other two members of the structure are rigid. Therefore,

$$\delta W_{\text{int}}^* = \left(\frac{f_{BC}}{k}\right)\delta f_{BC}$$
$$= \left(\frac{0.7906P}{k}\right)(1.581\delta Q_1)$$
$$= 1.25\left(\frac{P}{k}\right)\delta Q_1$$

Equating δW_{ext}^* and δW_{int}^* yields the displacement of point C:

$$u = 1.25\left(\frac{P}{k}\right) \qquad \text{(to the right)}$$

(b)

To calculate the rotation of member AD, we must apply a virtual couple to AD. The associated external complementary virtual work is the product of the actual rotation of AD—due to the actual load P—times the virtual couple, denoted δQ_2 in Figure 6.6.2c. Thus,

$$\delta W_{\text{ext}}^* = \theta_{AD}\delta Q_2$$

Since AD is a rigid body, every point on it has the same rotation. Therefore, δQ_2 can be placed anywhere on the beam, which is why a specific location of the virtual couple is not shown in Figure 6.6.2c. (If the beam were deformable, the location of δQ_2 would determine the rotation *at that point*.) From static equilibrium, we can find the force in the spring caused by δQ_2 acting alone, and that value is shown on the sketch. Notice that the virtual spring force in this case is compressive.

As before, internal complementary virtual work is done only on the elastic spring, so that

$$\delta W_{int}^* = \left(\frac{f_{BC}}{k}\right)\delta f_{BC}$$

$$= \left(\frac{0.7906P}{k}\right)\left(-1.118\frac{\delta Q_2}{L}\right)$$

$$= -0.884\left(\frac{P}{kL}\right)\delta Q_2$$

Finally, we equate the internal and external complementary virtual work expressions to obtain

$$\theta_{AD} = -0.884\left(\frac{P}{kL}\right)$$

The negative sign means that beam AD rotates clockwise (opposite to the direction of the virtual couple) under the influence of the applied load P.

For a continuous medium, the internal complementary virtual work of the true strains (held fixed) acting through the virtual stresses is inferred from Equation 6.3.4 by analogy:

$$\delta W_{int}^* = \iiint_V (\varepsilon_x\delta\sigma_x + \varepsilon_y\delta\sigma_y + \varepsilon_z\delta\sigma_z + \gamma_{xy}\delta\tau_{xy} + \gamma_{xz}\delta\tau_{xz} + \gamma_{yz}\delta\tau_{yz})dV \qquad \text{[6.6.4]}$$

Likewise, the external complementary virtual work for a solid continuum is obtained from Equation 6.3.5 by shifting the virtual symbol δ from the displacement variables to the loads:

$$\delta W_{ext}^* = \sum_{i=1}^{n}\mathbf{q}_i \cdot \delta\mathbf{Q}_i + \iint_S \mathbf{u} \cdot \delta\mathbf{T}^{(n)}dS + \iiint_V \mathbf{u} \cdot \delta\mathbf{b}dV \qquad \text{[6.6.5]}$$

Here, $\mathbf{q}_i$ is the generalized displacement *in the direction of* the virtual generalized load $\delta\mathbf{Q}_i$, so that, for example, $\mathbf{q}_i \cdot \delta\mathbf{Q}_i = q_i\delta Q_i$. Thus $\sum_{i=1}^{n} q_i\delta Q_i$ is the complementary virtual work of the n generalized point displacements.

6.7   MINIMUM COMPLEMENTARY POTENTIAL ENERGY AND CASTIGLIANO'S SECOND THEOREM

From section 3.14, we know that in an elastic solid, the internal complementary work of a quasistatic process from the undeformed to the deformed state equals the complementary strain energy U^*, or

$$W_{int}^* = U^*$$

The complementary strain energy U^* is a function of the applied loads; that is,

$$U^* = U^*(Q_1, Q_2, \cdots, Q_n) \qquad \text{[6.7.1]}$$

Therefore, for a virtual deformation, we have

$$\delta W_{int}^* = \delta U^* = \sum_{i=1}^{n}\frac{\partial U^*}{\partial Q_i}\delta Q_i \qquad \text{[6.7.2]}$$

Since the displacements and the virtual loads are colinear,

$$\delta W_{\text{ext}}^* = \sum_{i=1}^{n} q_i \delta Q_i$$

and the principle of complementary virtual work—$\delta W_{\text{ext}}^* = \delta W_{\text{int}}^*$—for an elastic solid requires that

$$\sum_{i=1}^{n} q_i \delta Q_i = \sum_{i=1}^{n} \frac{\partial U^*}{\partial Q_i} \delta Q_i \qquad \text{[6.7.3]}$$

for any choice of the virtual loads δQ_1, δQ_2, $\cdots$, δQ_n. This means that the coefficients of δQ_i on each side of the equation must be the same:

$$q_i = \frac{\partial U^*}{\partial Q_i} \qquad i = 1, 2, \cdots, n \qquad \text{[6.7.4]}$$

This is *Castigliano's second theorem*: For an elastic structure, the displacements are found as derivatives of the complementary strain energy with respect to the corresponding loads. Let us emphasize again that Castigliano's theorems, which apply only to elastic materials, are a consequence of the more general virtual work principles, which are independent of material properties.

In a manner completely analogous to that of section 6.4, we are led at once to another consequence of the principle of complementary virtual work, namely the *theorem of minimum complementary potential energy*,

$$\frac{\partial \Pi^*}{\partial Q_i} = 0 \qquad i = 1, \cdots, n \qquad \text{[6.7.5]}$$

where $\Pi^* = U^* + V^*$ is the *total complementary potential energy*, and

$$V^* = V = -\sum_{i=1}^{n} Q_i q_i \qquad \text{[6.7.6]}$$

Example 6.7.1   Solve the problem of Example 6.3.2 using the principle of minimum complementary potential energy as an alternative to using the principle of minimum potential energy, which was done in Example 6.4.1. The sketch for that problem is reproduced in Figure 6.7.1a for convenience.

The (complementary) strain energy of a spring, in terms of its axial load F and spring constant k, is

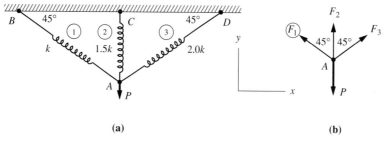

(a)   (b)

Figure 6.7.1   (a) The system of Example 6.3.2. (b) Free-body diagram of point A.

F_1 is circled to highlight its selection as the redundant force.

$$U_s^* = \frac{1}{2}Fs = \frac{1}{2}F\left(\frac{F}{k}\right) = \frac{F^2}{2k}$$

Therefore, the total complementary strain energy of the system in Figure 6.7.1a is

$$U^* = \frac{F_1^2}{2k_1} + \frac{F_2^2}{2k_2} + \frac{F_3^2}{2k_3} \tag{a}$$

The potential energy of the applied load is

$$V^* = -(-P)v_A = Pv_A$$

It follows that the total complementary potential energy, $\Pi^* = U^* + V^*$, is

$$\Pi^* = \frac{F_1^2}{2k_1} + \frac{F_2^2}{2k_2} + \frac{F_3^2}{2k_3} + Pv_A \tag{b}$$

Before applying Equation 6.7.5, we must require that the forces F_1, F_2, F_3, and P be in equilibrium. (This was not required in Examples 6.3.2 and 6.4.1, where instead we had to enforce displacement compatibility, which is not an issue here.) Using the free-body diagram in Figure 6.7.1b, we obtain the equations of equilibrium for point A:

$$x: \quad -F_1 \cos 45° + F_3 \cos 45° = 0$$
$$y: \quad F_1 \sin 45° + F_2 + F_3 \sin 45° + P = 0 \tag{c}$$

Here, P is considered given, but F_1, F_2, and F_3 are unknowns. Using the equilibrium equations, Equations c, we can express any two of these variables in terms of the remaining one and the load P. Let us arbitrarily choose to find F_2 and F_3 in terms of F_1 and P. So doing yields

$$F_2 = -\sqrt{2}F_1 + P \quad F_3 = F_1 \tag{d}$$

In Figure 6.7.1b, F_1 is circled to highlight its role as the chosen independent unknown. It is referred to as a *redundant* load, since it is not required for equilibrium. That is, removing spring 1 leaves an assembly fully capable of supporting the load P, while F_1 is replaced by zero in Equations d.

Substituting Equations d, together with the spring constants $k_1 = k$, $k_2 = 1.5k$, and $k_3 = 2k$, into Equation b leads to

$$\Pi^* = 1.417\frac{F_1^2}{k} + 0.3333\frac{P^2}{k} - 0.9428\frac{F_1 P}{k} + Pv_A \tag{e}$$

The displacement v_A, like k, is a constant (the load P was fixed in Example 6.4.1). Since F_1 and P are independent variables, the partial derivatives of Π^* are

$$\frac{\partial \Pi^*}{\partial F_1} = 2.833\frac{F_1}{k} - 0.9428\frac{P}{k} \quad \frac{\partial \Pi^*}{\partial P} = -0.9428\frac{F_1}{k} + 0.6667\frac{P}{k} + v_A \tag{f}$$

According to Equation 6.7.5, the two partial derivatives in Equation f must vanish, and we thereby obtain the two equations

$$2.833\frac{F_1}{k} = 0.9428\frac{P}{k}$$
$$0.9428\frac{F_1}{k} + v_A = -0.6667\frac{P}{k} \tag{g}$$

Solving these for F_1 and v_A yields

$$F_1 = 0.3328P \quad v_A = -0.3529\frac{P}{k} \tag{h}$$

These are in agreement with Example 6.3.2, as are the values of F_2 and F_3, which are found by substituting F_1 into the equilibrium equations, Equations d.

Finally, using the expressions obtained for F_1 and v_A in Equation h to evaluate the complementary potential Π^* in Equation e yields a value identical to that found for potential Π in Equation f of Example 6.4.1. For linear elastic solids, it is always true that $\Pi^* = \Pi$.

6.8 FLEXIBILITY MATRIX

The *flexibility matrix* of a linearly elastic structure is the set of coefficients that relate the generalized displacements to the loads, or

$$q_i = \sum_{l=1}^{n} c_{il} Q_l \qquad i = 1, 2, \cdots, n \qquad \text{[6.8.1]}$$

The flexibility matrix is important when the loads, rather than the displacements, are considered to be the independent variables. Just as the stiffness matrix was shown to be symmetric in section 6.5, Castigliano's second theorem leads to the conclusion that the flexibility matrix is also symmetric. In fact, the two matrices are the inverse[6] of each other. In analogy with Equation 6.5.8, it follows that

$$c_{ii} > 0 \ i = 1, \cdots, n \qquad \text{[6.8.2]}$$

As an important example of the use of the symmetry property, consider Figure 6.8.1, which shows an elastic beam cross section with two points highlighted. Point 1 is the shear center, with which we are familiar (cf. section 4.6). Point 2 denotes the *center of twist*, which is the point of the cross section that undergoes no displacement when pure torsion is applied to the beam. Let a vertical shear force P_1 act through point 1, and let a point couple M_2 act at point 2. If u_1 is the displacement at point 1 due to P_1 and M_2, and θ_2 is the rotation at 2, then,

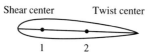

Shear center   Twist center

1   2

Figure 6.8.1   Shear center versus twist center.

$$u_1 = c_{11} P_1 + c_{12} M_2$$

$$\theta_2 = c_{21} P_1 + c_{22} M_2$$

where $c_{12} = c_{21}$. Therefore, if we apply a load just to point 1, as in Figure 6.8.2a, the displacements at points 1 and 2 are:

$$u_1 = c_{11} P_1$$

$$\theta_2 = c_{21} P_1$$

| [6]See Chapter 9.

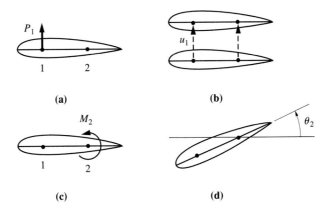

Figure 6.8.2:   (a) and (b) A load through the shear center produces no twist. (c) and (d) A pure torque produces no displacement at the center of twist.

Since point 1 is the shear center, then, by definition, a load acting through it, as in Figure 6.8.2a, produces no twist of the cross section. Therefore, $c_{21}P_1 = 0$, which means that $c_{21} = 0$. If we apply just a point couple M_2 to point 2, as in Figure 6.8.2c, then,

$$u_1 = c_{12}M_2$$

$$\theta_2 = c_{22}M_2$$

However, since c_{21} is zero, so is c_{12}. Therefore, $u_1 = 0$. That is, point 1 undergoes no displacement if a pure torque is applied to the section. This means that point 1 is the center of twist. Yet, point 1 was defined to be the shear center. We have therefore shown that *in an elastic structure, the shear center and the center of twist coincide.*

The locus of centers of twist along the wing span is called the *elastic axis*, which should not be confused with the beam axis that we have previously defined to be the locus of the cross-sectional centroids.

EXERCISES

6.1   Each member of the linkage in the figure is rigid. Use the principle of virtual work to calculate the angle θ for equilibrium.

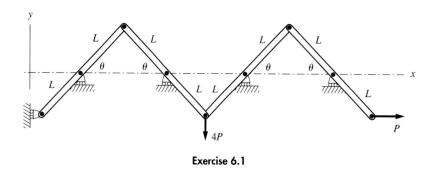

Exercise 6.1

6.2   Using the principle of virtual work, find the magnitude of the force P required for equilibrium.

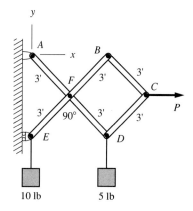

Exercise 6.2

6.3   Use the principle of virtual work to find the relationship between the forces P and Q.

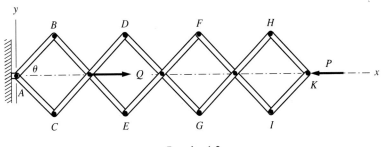

Exercise 6.3

6.4   Find the forces in each of the four springs, in terms of P and θ, using the principle of virtual work.

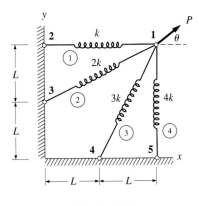

Exercise 6.4

6.5   Use the principle of virtual work to calculate the load P. The eight rigid, weightless links are connected by smooth pins. Six of the links are of length $2l$, while the two at the far right are of length l.

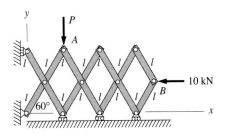

Exercise 6.5

6.6   Use the principle of virtual work to find the angle θ. The three rigid, weightless rods are connected by smooth pins, except that inclined member 5–2 is pinned to a collar that slides freely on the horizontal rod 1–3, as shown.

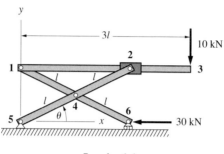

Exercise 6.6

6.7   Use the principle of complementary virtual work to find the displacement of point A due to the weight W of the uniform rod. Neglect friction.

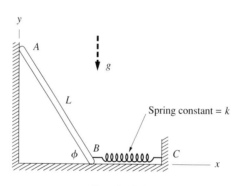

Exercise 6.7

6.8   Except for the spring, the members of the structure are rigid. Use the principle of complementary virtual work to calculate the vertical displacement of point A.

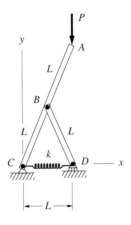

Exercise 6.8

Force Method: Trusses, Beams, and Frames

CHAPTER OUTLINE

7.1   INTRODUCTION

The *force method* of analyzing a structure begins with the use of statics to obtain the equilibrium equations, which relate the unknown forces to the known forces. The known forces are the applied loads. The unknowns are the reactions at the supports and the internal member loads. If the number of unknowns equals the number of equilibrium equations, then the problem is statically determinate, which means that as far as the forces are concerned, the problem is finished once the equations are solved.

Before seeking the solution of the equilibrium equations, we must be sure that the structure is stable; if it is not, a solution is not possible. A properly designed

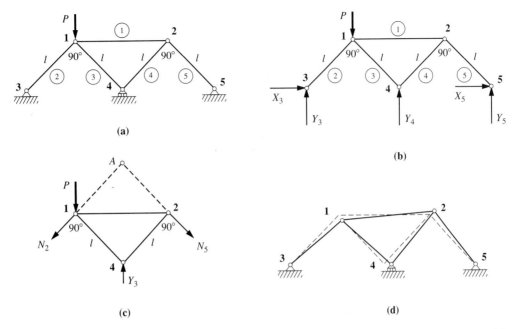

Figure 7.1.1 (a) Plane truss. (b) Free-body diagram showing the support reactions. (c) Free-body diagram of the portion of the truss spanning nodes 1, 2, and 4. (d) Illustration of the kinematic instability.

structure is stable. An example of one that is not is the plane truss in Figure 7.1.1. It has five member forces and five support reactions [see part (b)], for a total of ten unknown loads. Also, there are five joints with two equilibrium equations each, so the number of equations matches the number of unknowns. However, the structure is unstable, and an attempt to solve the equations of equilibrium will fail. The free-body diagram in Figure 7.1.1c reveals that the lines of action of the sole reaction at point 4, as well as the two member forces N_2 and N_5, all intersect at point A above the truss. The moment about A of the applied force P has no equilibrant. Figure 7.1.1d shows the unconstrained rigid-body motion responsible for the instability.

Once the forces have been determined, we can find the displacements at selected points on the structure using the principle of complementary virtual work. This requires bringing the structure's material properties and cross-sectional details into the picture. All real structures are flexible to one degree or another; there is no such thing as a perfectly rigid body. The degree of flexibility allowed is part of the design process, and deflection analysis is required to ensure that static displacements remain within the limits prescribed. Calculating displacements is therefore fundamental to the analysis of structural dynamics and structural stability.

Statically indeterminate structures are those for which the methods of statics alone are not sufficient for calculating the internal loads and external reactions. A structure of this type has more than the minimum number of members and/or supports required for it to sustain a given load without collapsing or moving off as a rigid body. The excess members and supports are called *redundants*. Highly redundant structures, such as aircraft, provide a variety of internal load path options. Should a given redundant member fail for some reason, the remaining ones will continue to provide the means of carrying the load.

Figure 7.1.2 is a simple example of a redundant structure. The two parallel elastic springs, with spring rates k_1 and k_2, respectively, share in supporting the applied load P. One of the springs is redundant, because only a single spring is needed to transmit the load P to the wall. Let us arbitrarily name spring 2 as the redundant member.

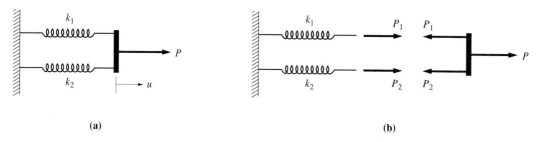

Figure 7.1.2 (a) Parallel (redundant) elastic spring structure. (b) Free-body diagrams showing internal loads.

Applying statics to the free-body diagram in part (b) of the figure, we find the internal load P_1 in terms of both the externally applied load P and the redundant internal load P_2, as follows.

$$P_1 = P - P_2 \qquad [7.1.1]$$

Here, P and P_2 are the independent force variables, while statics (and our choice of redundant) makes P_1 the dependent variable. Typical of any statically indeterminate problem, we must also take into consideration the *deformation* of the structure and invoke a compatibility condition—a requirement that the structure hang together. In this case, the requirement is that the right end of each spring must have a common displacement u in the direction of the applied load. If s_1 is the stretch of spring 1 and s_2 that of spring 2, then compatibility simply means that

$$s_1 = s_2 \qquad [7.1.2]$$

Next, we impose Hooke's Law, which brings in the material properties of the springs:

$$s_1 = \frac{P_1}{k_1} \qquad s_2 = \frac{P_2}{k_2} \qquad [7.1.3]$$

Substituting these two expressions into Equation 7.1.2 and solving for P_2 yields

$$P_2 = \frac{k_2}{k_1} P_1 \qquad [7.1.4]$$

Equation 7.1.3 (statics) and Equation 7.1.4 (compatibility) are two conditions from which we can uniquely find P_1 and P_2 in terms of the applied load P:

$$P_1 = \frac{k_1}{k_1 + k_2} P \qquad P_2 = \frac{k_2}{k_1 + k_2} P \qquad [7.1.5]$$

From this we see that *the portion of the applied load carried by a spring depends on its relative stiffness.* The larger part of the load is borne by the stiffest spring. Obviously, as the spring rate k_1 goes to zero, so does P_1, and the other spring must absorb all of the load. If $k_1 = k_2$, then $P_1 = P_2 = P/2$: both springs share the load equally.

Adding more parallel springs to the assembly would not limit our ability to find all of the internal loads using this procedure, which is a simple example of the force method. In more complex structures, the principle of complementary virtual work is used to enforce compatibility.

In this chapter, we will apply the force method to the analysis of skeletal or "stick-like" structures, using the principle of complementary virtual work to find displacements and forces in statically determinate and indeterminate trusses, beams, and frames.

7.2 RODS: COMPLEMENTARY VIRTUAL WORK

The rod is a slender bar, possibly slightly tapered, which transmits loads parallel to its long axis. The load resultant N at any section must pass through the centroid of that section. The rod may carry an intermediate load p distributed along its axis, such as when it is attached to shear panels in a stiffened web structure. There may also be a variable temperature change T from ambient along the rod, producing thermal strain. The length of the element is L, its cross-sectional area (which may vary along the rod) is A, the uniform modulus of elasticity is E, and the thermal expansion coefficient is α. Since the axial load N acts at the centroid, the stress σ_x is uniform across each section and is the only nonzero stress component; likewise $\delta\sigma_x$ is the only virtual stress component, as shown in Figure 7.2.1b. Thus, from Equation 6.6.4, the internal complementary virtual work for the rod, using Hooke's Law (isotropic materials), is

$$\delta W_{int}^* = \iiint_V \delta\sigma_x \varepsilon_x dV$$

$$= \int_0^L \delta\sigma_x \left(\frac{\sigma_x}{E} + \alpha T\right)(Adx)$$

$$= \int_0^L \left(\frac{\delta N}{A}\right)\left(\frac{N}{AE}\right)(Adx) + \int_0^L \left(\frac{\delta N}{A}\right)(\alpha T)(Adx)$$

or

$$\delta W_{int}^* = \int_0^L \frac{N\delta N}{AE} dx + \int_0^L (\alpha T)\,\delta P dx \qquad [7.2.1]$$

The product AE is called the *axial rigidity*.

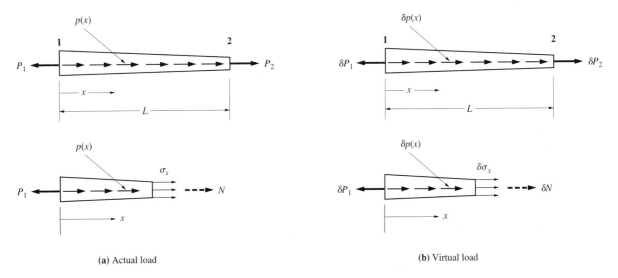

(a) Actual load   (b) Virtual load

Figure 7.2.1   The rod element.

The rods in a truss structure usually have a constant cross section and carry no distributed axial load. Therefore, the loads N and δN are constant, so that for the *constant stress rod*, Equation 7.7.1 simplifies to

$$\delta W_{int}^* = \frac{NL}{AE}\delta N + \left(\int_0^L (\alpha T)\,dx\right)\delta P \qquad \text{truss member} \qquad [7.2.2]$$

internal complementary virtual work due to axial loading *internal complementary virtual work due to thermal strain*

The variation of temperature change $T(x)$ along the rod must be known before the integral can be computed. Of course, the integral vanishes in the absence of a temperature change.

7.3 TRUSS DEFLECTIONS USING COMPLEMENTARY VIRTUAL WORK

To find the displacement at a given node of a statically determinate truss, using the force method, we first calculate the internal forces $N^{(e)}$ in each truss member e due to the actual applied loads. Then, we remove all of the true loads and apply a fictitious force δQ to the given node, in the direction of the desired displacement component. We then solve for the resultant virtual forces $\delta N^{(e)}$ throughout the truss. Since the virtual load need not bear any resemblance to the true load, an actual load need not be present where virtual load is applied. The internal complementary virtual work of the complete truss is the sum of the individual complementary virtual works given by Equation 7.2.2 over all of its members, or

$$\delta W_{int}^* = \sum_{e=1}^{\text{no. of rods}} \left\{ \frac{L^{(e)}N^{(e)}}{A^{(e)}E^{(e)}} + \int_0^{L^{(e)}} \left(\alpha^{(e)}T^{(e)}\right)ds \right\}\delta N^{(e)} \qquad [7.3.1]$$

Here, s is used to represent position along a rod, since not all truss members line up with a common x axis. Also, $T^{(e)}(s)$ is the actual temperature change in rod e.

Recall that the complementary virtual work of the virtual force δQ is the product of δQ and the actual displacement q in the direction of the virtual force, or

$$\delta W_{ext}^* = q\delta Q$$

Finally, we equate the internal and external complementary virtual work, as follows:

$$\delta W_{ext}^* = \delta W_{int}^*$$

which must hold for arbitrary values of the virtual applied load. This yields the desired displacement, as illustrated in the following examples.

Example 7.3.1 For the statically determinate truss in Figure 7.3.1, calculate the horizontal displacement u_4 at node 4 due to a vertical load P at node 1, using the principle of complementary virtual work. The axial rigidity AE is the same for all members of the structure.

We can determine the member forces resulting from the vertical force P at node 1 using statics, and the results are presented alongside each member in Figure 8.3.4. Since we are seeking the horizontal movement of node 4, we remove the load from node 1 and apply a virtual force δQ acting horizontally at node 4. The corresponding member loads are shown in parentheses in Figure 7.3.2.

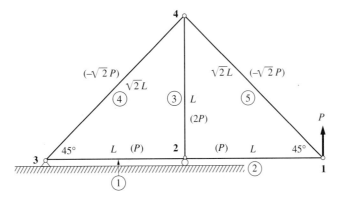

Figure 7.3.1   True load P at node 1 and the corresponding member forces (in parentheses).

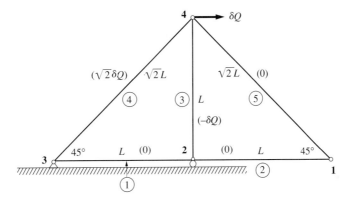

Figure 7.3.2   Virtual applied load and corresponding virtual member loads (in parentheses).

Equating the external and internal complementary virtual work, we get

$$
u_4 \delta Q = \overbrace{\frac{(P)\,L}{AE}}^{\text{rod 1}}(0) + \overbrace{\frac{(P)\,L}{AE}}^{\text{rod 2}}(0) + \overbrace{\frac{(2P)\,L}{AE}}^{\text{rod 3}}(-\delta Q) + \overbrace{\frac{\left(-\sqrt{2}P\right)\sqrt{2}L}{AE}}^{\text{rod 4}}\left(\sqrt{2}\delta Q\right) + \overbrace{\frac{\left(-\sqrt{2}P\right)\sqrt{2}L}{AE}}^{\text{rod 5}}(0)
$$

This reduces to

$$
u_4 = -2\left(1 + \sqrt{2}\right)\frac{PL}{AE}
$$

The minus sign means that the displacement is to the left, in the direction opposite to that of δQ.

Example 7.3.2   For the truss of the previous example, loaded as shown in Figure 7.3.1, calculate the rotation of member 5.
    The complementary virtual work associated with a (small) rotation $\theta^{(5)}$ of member 5 is $\theta^{(5)}\delta C$, where δC is a virtual couple in the plane of rotation. We cannot apply a point couple to rod 5, because it is a two-force member, and loads can only be applied

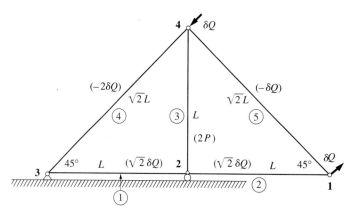

Figure 7.3.3   Virtual loads required to find the rotation of member 5.

to it at its ends. Therefore, we apply a pair of equal but opposite virtual forces δQ to nodes 1 and 4, as shown in Figure 7.3.3, which also shows the resulting member loads, found by using statics. The moment of the couple formed by the virtual forces is $\delta Q \times \sqrt{2}L$, counterclockwise. Therefore, from the principle of complementary virtual work, we have

$$
\theta^{(5)} \left(\sqrt{2}L\delta Q \right) = \overbrace{\frac{(P)L}{AE} \left(\sqrt{2}\delta Q \right)}^{\text{rod 1}} + \overbrace{\frac{(P)L}{AE} \left(\sqrt{2}\delta Q \right)}^{\text{rod 2}} + \overbrace{\frac{(2P)L}{AE} \left(\sqrt{2}\delta Q \right)}^{\text{rod 3}}
$$
$$
+ \overbrace{\frac{\left(-\sqrt{2}P\right)\left(\sqrt{2}L\right)}{AE}}^{\text{rod 4}} (-2\delta Q) + \overbrace{\frac{\left(-\sqrt{2}P\right)\left(\sqrt{2}L\right)}{AE}}^{\text{rod 5}} (-\delta Q)
$$

Upon simplification, we get

$$
\theta^{(5)} \left(\sqrt{2}L\delta Q \right) = \left(6 + 4\sqrt{2} \right) \frac{PL}{AE} \delta Q
$$

so that

$$
\theta^{(5)} = 8.243 \frac{P}{AE}
$$

The rotation of the rod is counterclockwise, in the same direction as the virtual couple.

Example 7.3.3   The truss in Figure 7.3.4a is not loaded, but member 1 is heated to a uniform temperature T above ambient and member 2 is heated to $2T$. The temperature in the other rods increases linearly from T to $2T$. If the axial rigidity AE and the thermal expansion coefficient α are the same for all members, calculate the displacement u_1 of node 1, using the principle of complementary virtual work.

The linear temperature variation in rods 3 and 4 is given by $T \left[1 + \left(s/\sqrt{2}L \right) \right]$, while that in rod 5 is $T \left[1 + (s/L) \right]$.

To find the displacement at node 1, we remove the thermal loading and apply the virtual force δQ, which is shown in Figure 7.3.4b along with the corresponding member loads, found by using statics. Even though the true loads are zero, we still have a thermal strain term in the virtual work expression, Equation 7.3.1. Therefore, by the principle of complementary virtual work, we have

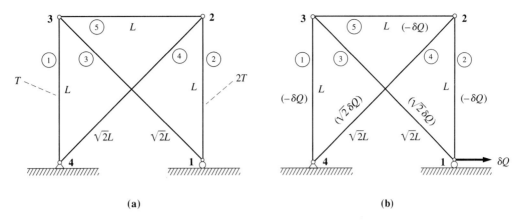

Figure 7.3.4   (a) Unloaded truss with thermal strain. (b) Virtual load at node 1 and the corresponding member forces.

$$\overset{\text{rod 1}}{u_1\delta Q = \overbrace{(\alpha TL)\,(-\delta Q)}} + \overset{\text{rod 2}}{\overbrace{(2\alpha TL)\,(-\delta Q)}}$$

$$+ \left[\overset{\text{rod 3}}{\overbrace{\int_0^{\sqrt{2}L} \alpha T\left(1 + \frac{s}{\sqrt{2}L}\right)ds}}\right](\sqrt{2}\delta Q) + \left[\overset{\text{rod 4}}{\overbrace{\int_0^{\sqrt{2}L} \alpha T\left(1 + \frac{s}{\sqrt{2}L}\right)ds}}\right](\sqrt{2}\delta Q)$$

$$+ \left[\overset{\text{rod 5}}{\overbrace{\int_0^{L} \alpha T\left(1 + \frac{s}{L}\right)ds}}\right](-\delta Q)$$

This reduces to

$$u_1\delta Q = -\alpha TL\delta Q - 2\alpha TL\delta Q + 3\alpha TL\delta Q + 3\alpha TL\delta Q - \frac{3}{2}\alpha TL\delta Q$$

so that we finally obtain a rightward displacement of node 1 in the following amount

$$u_1 = \frac{3}{2}\alpha TL$$

7.4   STATICALLY INDETERMINATE TRUSSES

In statically determinate structures, the number of unknown forces equals the number of equations of static equilibrium, which was true for the examples in the previous section. For indeterminate structures, the number of unknown forces exceeds the number of equations available from statics, by the number of redundant forces.

In the force method, the first step towards calculating the forces in an indeterminate structure—in particular, a truss—is to single out the redundant members and/or supports and, on a sketch, show their effect on the structure through the loads they exert. In other words, we imagine that we cut through each redundant member, revealing the force within it and applying that force to the structure as though it were an external load. As an example, consider the truss in Figure 7.4.1. It has two degrees of redundancy: an extra member plus an extra support. If we take away member 4, the truss remains stable. If we also take away the support at node 4, the truss is still unable to undergo rigid-body motion. Upon removal of both the member and the support, the truss becomes statically determinate. Of course we cannot physically remove the member and support, for we would then have a different structure. Rather, we imagine the two elements to be cut away, and we replace them by the forces they apply to the structure, as shown in Figure 7.4.2a. The statically determinate structure that remains after cutting away the redundants is called the *base structure*.

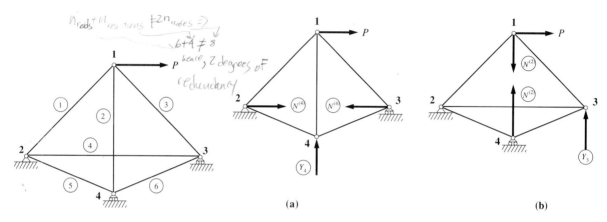

Figure 7.4.1   Doubly-redundant truss.   **Figure 7.4.2**   Possible choices of redundants in the truss of Figure 7.4.1.

Assuming member 4 is in tension, it exerts the force $N^{(4)}$ on nodes 2 and 3, as shown. The roller support at node 4 applies a vertical force Y_4 to the node. We assume this force to be directed upwards. The truss in Figure 7.4.2a appears to be a statically determinate truss acted on by four external forces. We circle the forces coming from the redundants to distinguish them from the given, true externally applied load P. Using statics, we can now solve for the forces throughout the base structure, in terms of P, $N^{(4)}$, and Y_4.

Figure 7.4.2b shows an alternate choice of base structure. Indeed, there are several other possibilities, all of them equally valid. In any case, the base structure we end up with must be stable and properly supported. The one in Figure 7.4.3a is stable, but "cutting away" the horizontal constraint at node 2 removes the only means of preventing horizontal rigid-body motion. In other words, there is no way of reacting to the external loads P and X_2. Part (b) shows the base structure that would be obtained by cutting away rods 2 and 4; it would collapse.

After settling on a proper base structure, say that of Figure 7.4.2a, we draw its free-body diagram and calculate the axial load $N^{(e)}$ in each member e in terms of the applied load P and the redundants $N^{(4)}$ and Y_4. We then remove the true loads from the truss and replace the redundant loads by virtual loads, as shown in Figure 7.4.4. Using statics, we calculate the resulting virtual internal load $\delta N^{(4)}$ in each member of the truss. With the real and virtual loads thus determined, we can write the complementary internal virtual work $\delta W_{int}^{*\,(e)}$ for each member of the original structure in Figure 7.4.1, using Equation 7.2.2 (with $T = 0$, in this case). We sum them up to obtain δW_{int}^{*} for the whole truss, as follows:

$$\delta W_{int}^{*} = \sum_{e} \delta W_{int}^{*(e)} = \sum_{e} \left(\frac{L}{AE}\right)^{(e)} N^{(e)}\delta N^{(e)} \qquad [7.4.1]$$

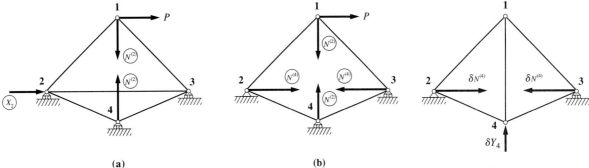

(a)   **(b)**

Figure 7.4.3   Invalid choices of base structure for the indeterminate truss in Figure 7.4.1.

Figure 7.4.4   The base truss of Figure 7.4.2a subjected only to the virtual redundant loads.

According to the principle of complementary virtual work, the external and internal complementary virtual works must always be the same. Therefore, for the truss, we have

$$\delta W^*_{\text{ext}} = \sum_e \left(\frac{L}{AE}\right)^{(e)} N^{(e)} \delta N^{(e)} \qquad [7.4.2]$$

The external complementary virtual work δW^*_{ext} is that of the virtual forces in Figure 7.4.4 acting through the real displacements of their points of application. The complementary virtual work of δY_4 is $\delta Y_4 \times v_4$, where v_4 is the true vertical component of displacement of node 4. But $v_4 = 0$ because the support at node 4 presumably prevents vertical motion. The complementary virtual work of $\delta N^{(4)}$ is also zero, as can be seen by studying Figure 7.4.5.

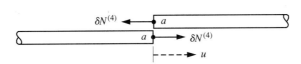

Figure 7.4.5   Virtual member loads and true displacement at the point of the imaginary cut.

Remember that we do not physically cut the redundant members out of the structure. The redundant forces shown on the base structure in Figure 7.4.4 occur at an imaginary cut taken at some point along the member, such as point a in Figure 7.4.5. The complementary virtual work of these virtual forces is

$$\delta W^* = \delta N^{(4)} \times u_{\text{left}} + \delta N^{(4)} \times u_{\text{right}}$$

where u_{left} and u_{right} are the true displacements of the points just to the left and right of the cut, respectively. But the member is actually continuous at point a. If the displacement u of point a is to the right, we would have

$$\delta W^* = \delta N^{(4)} \times u + \delta N^{(4)} \times (-u) = 0$$

This same argument can be used to show that the complementary virtual work of redundant member forces is zero not only for truss elements but for members of any other type of structure, as well.

Since the external virtual work is zero, we have from Equation 7.4.2,

$$\sum_e \left(\frac{L}{AE}\right)^{(e)} N^{(e)} \delta N^{(e)} = 0 \qquad\qquad \textbf{[7.4.3]}$$

This equation must be valid for *any* choice of the two virtual loads δY_4 and $\delta N^{(4)}$. Therefore, Equation 7.4.3 will yield the two additional equations needed to solve for all of the forces in the indeterminate truss. We will illustrate this in several examples.

It should be pointed out that although the displacements at a structure's supports are usually zero, this need not be the case. For example, node 4 of the truss in Figure 7.4.1 might well have been given a specified upward displacement v_4. In that case, the complementary virtual work of δY_4 would be $\delta Y_4 \times v_4$ instead of zero, which would appear on the right of Equation 7.4.3. The point is that v_4 is a *known* nonzero quantity, so that the complementary virtual work equality,

$$\sum_e \left(\frac{L}{AE}\right)^{(e)} N^{(e)} \delta N^{(e)} = v_4 \delta Y_4$$

still yields the equations we need.

At supports, the displacements are always known (usually zero), whereas the applied loads (reactions) are unknown. Where known loads (including those that are zero) are applied, the displacements are unknown. We cannot simultaneously prescribe a load *and* a displacement in the same direction at a given point of a structure.

Example 7.4.1   Calculate the internal forces in the truss in Figure 7.4.6, using the principle of complementary virtual work.

Let us choose the horizontal support at node 4 as the redundant, so that our base structure is as shown in Figure 7.4.7. The support reactions are immediately obtained from statics. Using the joint method, we can find the member loads, as shown in the figure, in terms of the applied load P and the redundant reaction X_4. In Figure 7.4.8, we remove all of the true loads on the truss, apply only a virtual redundant load δX_4, and calculate the virtual internal loads shown.

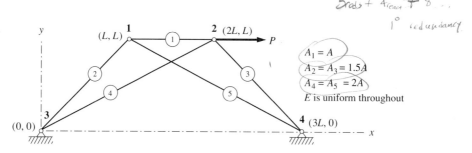

$A_1 = A$
$A_2 = A_3 = 1.5A$
$A_4 = A_5 = 2A$
E is uniform throughout

Figure 7.4.6   Truss with single degree of indeterminacy.

According to the principle of complementary virtual work, $\delta W^*_{\text{int}} = \delta W^*_{\text{ext}}$, so that in this case,

$$\left(\frac{L}{AE}\right)^{(e)} N^{(e)} \delta N^{(e)} = u_4 \delta X_4 = (0)\,\delta X_4 = 0$$

since the horizontal displacement u_4 at node 4 is constrained to zero by the pin support. Expanding the summation, using the data shown in Figure 7.4.7, and noting that $L^{(1)} = L$, $L^{(2)} = L^{(3)} = 1.414L$, and $L^{(4)} = L^{(5)} = 2.236L$, we get

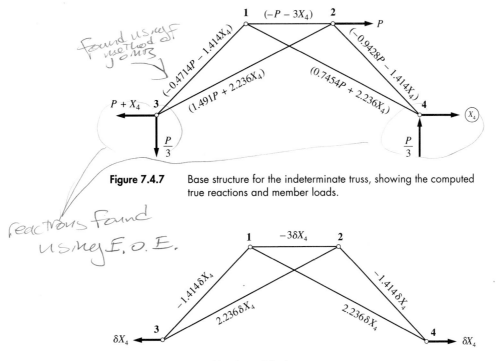

Figure 7.4.7 Base structure for the indeterminate truss, showing the computed true reactions and member loads.

reactions found using E.O.E.

Figure 7.4.8 Virtual loading of the base truss.

$$\overbrace{\frac{L}{AE}(P-3X_4)(-3\delta X_4)}^{\text{element 1}} + \overbrace{\frac{1.414L}{1.5AE}(-0.4714P-1.414X_4)(-1.414\delta X_4)}^{\text{element 2}} + \overbrace{\frac{1.414L}{1.5AE}(-0.9428P-1.414X_4)(-1.414\delta X_4)}^{\text{element 3}}$$

$$+\overbrace{\frac{2.236L}{2AE}(1.491P+2.236X_4)(2.236\delta X_4)}^{\text{element 4}} + \overbrace{\frac{2.236L}{2AE}(0.7454P+2.236X_4)(2.236\delta X_4)}^{\text{element 5}} = 0$$

Collecting terms and simplifying, we get

$$\left(10.47\frac{PL}{AE} + 23.95\frac{X_4L}{AE}\right)\delta X_4 = 0 \qquad \text{[a]}$$

which means that

$$X_4 = -\frac{10.47}{23.95}P = -0.4374P$$

The redundant load X_4 has thus been found in terms of the applied load P. Substituting X_4 into Figure 7.4.7, we obtain the results shown in Figure 7.4.9. Observe that the actual direction of X_3 turned out to be opposite to what we assumed.

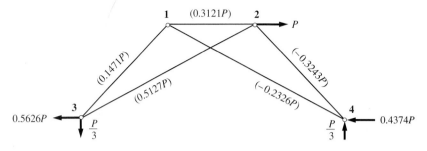

Figure 7.4.9 Reactions and member forces for the truss of Figure 7.4.6

In the previous example, one of the horizontal constraints was chosen as the redundant. That is not the only choice. All of the supports may be retained and any one of the five truss members selected as the redundant. The following example illustrates this alternative.

Example 7.4.2   Solve the problem in the previous example by selecting member 1 of the truss as the redundant.

With the horizontal member 1 of the truss in Figure 7.4.6 chosen as the redundant, the base structure becomes that shown in Figure 7.4.10, and statics yields the true reactions and member loads, also shown. Removing the true loads and applying just a virtual redundant load yields the situation illustrated in Figure 7.4.11.

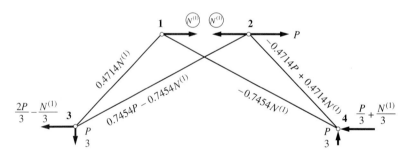

Figure 7.4.10   Base structure of the truss in Figure 7.4.6 when member 1 is selected as the redundant.

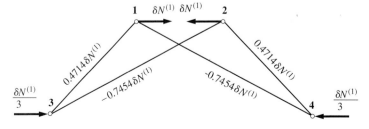

Figure 7.4.11   Virtual loads due to virtual redundant $\delta N^{(1)}$.

Since the redundant is an internal force, the external complementary virtual work is zero. Therefore, according to the principle of complementary virtual work, the internal complementary virtual work δW_{int}^* must also be zero. That is,

$$\overbrace{\frac{L}{AE}N^{(1)}\delta N^{(1)}}^{\text{element 1}} + \overbrace{\frac{1.414L}{1.5AE}(0.4714N^{(1)})(0.4714\delta N^{(1)})}^{\text{element 2}} + \overbrace{\frac{1.414L}{1.5AE}(-0.4714P + 0.4714N^{(1)})(0.4714\delta N^{(1)})}^{\text{element 3}}$$

$$+ \overbrace{\frac{2.236L}{2AE}(0.7454P - 0.7454N^{(1)})(-0.7454\delta N^{(1)})}^{\text{element 4}} + \overbrace{\frac{2.236L}{2AE}(-0.7454N^{(1)})(-0.7454\delta N^{(1)})}^{\text{element 5}} = 0$$

or

$$\frac{L}{AE}(2.661N^{(1)} - 0.8307P)\delta N^{(1)} = 0$$

so that the redundant load, in terms of P, is

$$N^{(1)} = \frac{0.8307}{2.661}P = -0.3121P$$

We can substitute $N^{(1)}$ into Figure 7.4.10 to obtain the values for all the other initially unknown loads, which are identical to those in Figure 7.4.9, as the reader can verify.

After all of the forces in an indeterminate structure have been determined using the principle of complementary virtual work, the method can be used to calculate the displacements of selected points. We do so just as in section 7.3, by applying a virtual point force in the direction of the desired displacement. Recall that in section 7.3, we dealt exclusively with statically determinate structures, so that the internal virtual force distribution due to the external virtual load was determined by a static equilibrium analysis of the *complete* structure. Although the virtual load system selected for computing the internal complementary virtual work δW_{int}^* must be self-equilibrating, it need bear no resemblance to the true load distribution.

Because there are no redundant load paths in a statically determinate structure, there is only one equilibrium set of internal loads corresponding to a given applied virtual load. However, the redundant load paths in indeterminate structures yield the possibility of a variety of different internal virtual load distributions for the same applied virtual load. According to the principle of complementary virtual work, $\delta W_{\text{int}}^* = \delta W_{\text{ext}}^*$ for *any* self-equilibrating virtual load system. This means that any stable, statically determinate base structure may be selected to support the virtual load in the direction of the desired displacement. This is best illustrated by an example.

Example 7.4.3   Calculate the vertical component of the displacement of node 1 of the statically indeterminate truss of examples 7.4.1 and 7.4.2.

For convenience, the structure and its computed loads are reproduced in Figure 7.4.12.

To calculate the vertical displacement of node 1, we apply a vertical virtual force to that point of the truss. That virtual load can be supported by any stable, statically determinate substructure of the complete truss. For example, we can imagine δP to be supported entirely by members 2 and 5, as illustrated in Figure 7.4.13, which also shows the rod loads required for node 1 to be in equilibrium. The virtual forces throughout the rest of the truss may be assumed to be zero. Therefore, according to the principle of complementary virtual work, we have

$$v_1\delta P = \frac{N^{(2)}L^{(2)}}{A^{(2)}E^{(2)}}\delta N^{(2)} + \frac{N^{(5)}L^{(5)}}{A^{(5)}E^{(5)}}\delta N^{(5)} = \frac{(0.1471P)(1.414L)}{1.5AE}(0.9428\delta P) + \frac{(-0.2327P)(2.236L)}{2AE}(0.7454\delta P)$$

$$= -0.06310\frac{PL}{AE}\delta P$$

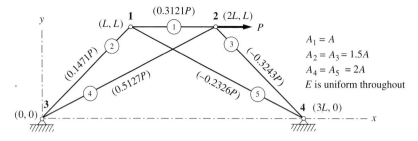

Figure 7.4.12 Computed member loads (in parentheses) in the statically indeterminate truss of the previous two examples.

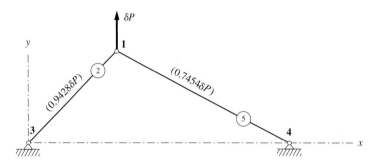

Figure 7.4.13 Statically determinate substructure of the truss containing node 1.

The vertical displacement of the node is then

$$v_1 = -0.06310 \frac{PL}{AE}$$ [a]

The minus sign means that the displacement is downward, in the direction opposite to that of the virtual load δP.

To illustrate the fact that *any* statically determinate substructure containing node 1 may be used, let us choose the one shown in Figure 7.4.14. Analysis of the equilibrium of nodes 1 and 2 yields the internal virtual forces written alongside each rod. The virtual forces throughout the remainder of the complete truss are zero. Once again, the principle of complementary virtual work requires that

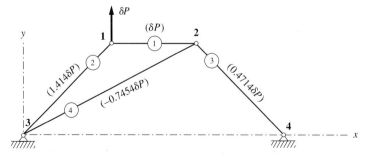

Figure 7.4.14 Alternative choice of statically determinate substructure to support the virtual load at node 1.

$$v_1 \delta P = \frac{N^{(1)} L^{(1)}}{A^{(1)} E^{(1)}} \delta N^{(1)} + \frac{N^{(2)} L^{(2)}}{A^{(2)} E^{(2)}} \delta N^{(2)} + \frac{N^{(3)} L^{(3)}}{A^{(3)} E^{(3)}} \delta N^{(3)} + \frac{N^{(4)} L^{(4)}}{A^{(4)} E^{(4)}} \delta N^{(4)}$$

$$= \overbrace{\frac{(0.3121 P)(L)}{AE}}^{\text{rod 1}} (\delta P) + \overbrace{\frac{(0.1471 P)(1.414 L)}{(1.5 A) E}}^{\text{rod 2}} (1.414 \delta P)$$

$$+ \overbrace{\frac{(-0.3243 P)(1.414 L)}{(1.5 A) E}}^{\text{rod 3}} (0.4714 \delta P) + \overbrace{\frac{(0.5127 P)(2.236 L)}{(2A) E}}^{\text{rod 4}} (-0.7454 \delta P)$$

$$= -0.06308 \frac{PL}{AE} \delta P$$

so that, as before,

$$v_1 = -0.06310 \frac{PL}{AE} \tag{b}$$

The internal loads of statically determinate trusses are unaffected by the presence of thermal strain. Statics alone suffices to determine the member loads. However, in statically indeterminate trusses, nonuniform heating produces stresses—thermal stresses—even in the absence of externally applied loads. This point is simply illustrated in the following example .

Example 7.4.4   Each member of the indeterminate truss in Figure 7.4.15 undergoes a uniform temperature rise from ambient, in which state the truss is unstressed. Assuming the axial rigidity AE and the thermal expansion coefficient α are the same for all of the members, find the internal loads.

Select member 2 as the redundant, so that the base structure is as shown in Figure 7.4.16a, along with the true loads. Removing the true loads and applying a virtual redundant load yields the same picture, Figure 7.4.16b. Since the supports are immobile, the external complementary virtual work is zero. Therefore, $\delta W_{\text{int}}^* = 0$, and from Equation 7.2.2,

$$\delta W_{\text{int}}^* = \sum_{e=1}^{3} \delta W_{\text{int}}^{*(e)} = \sum_{e=1}^{3} \left(\frac{N^{(e)}}{A^{(e)} E^{(e)}} + \alpha^{(e)} T^{(e)} \right) L^{(e)} \delta N^{(e)} = 0$$

$$\overbrace{\left(\frac{-\frac{\sqrt{2}}{2} N^{(2)}}{AE} + \alpha T^{(1)} \right) L \left(-\frac{\sqrt{2}}{2} \delta N^{(2)} \right)}^{\text{element 1}} + \overbrace{\left(\frac{N^{(2)}}{AE} + \alpha T^{(2)} \right) \sqrt{2} L \delta N^{(2)}}^{\text{element 2}} + \overbrace{\left(\frac{-\frac{\sqrt{2}}{2} N^{(2)}}{AE} + \alpha T^{(3)} \right) L \left(-\frac{\sqrt{2}}{2} \delta N^{(2)} \right)}^{\text{element 3}} = 0$$

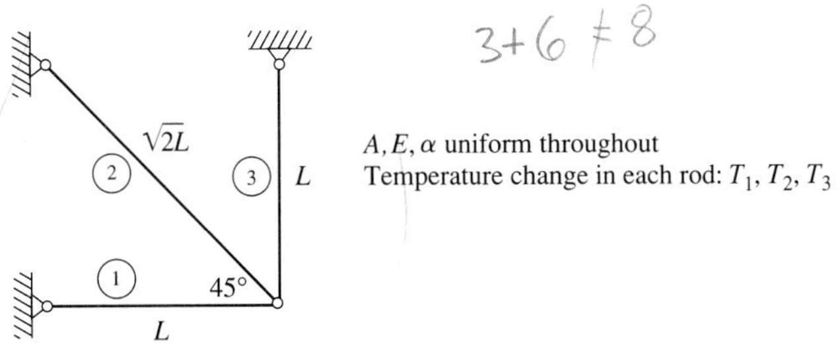

$3 + 6 \neq 8$

A, E, α uniform throughout
Temperature change in each rod: T_1, T_2, T_3

Figure 7.4.15   Thermally-loaded indeterminate truss.

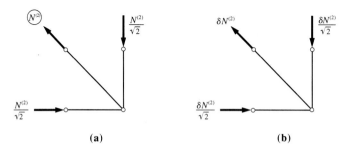

Figure 7.4.16   (a) Base truss showing the redundant load and the reactions to it. (b) The virtual loads.

This simplifies to

$$\left\{\left(1+\sqrt{2}\right)\frac{N^{(2)}L}{AE}+\alpha L\left[\sqrt{2}T^{(2)}-\frac{\sqrt{2}}{2}\left(T^{(1)}+T^{(3)}\right)\right]\right\}\delta N^{(2)}=0$$

Solving for $N^{(2)}$, we have

$$N^{(2)}=\frac{\sqrt{2}}{1+\sqrt{2}}AE\alpha\left[\frac{1}{2}\left(T^{(1)}+T^{(3)}\right)-T^{(2)}\right]$$

Then, from Figure 7.4.16a, we get

$$N^{(1)}=N^{(3)}=-\frac{\sqrt{2}}{2}N^{(2)}=-\frac{1}{1+\sqrt{2}}AE\alpha\left[\frac{1}{2}\left(T^{(1)}+T^{(3)}\right)-T^{(2)}\right]$$

If the temperature change is the same for all of the rods ($T^{(1)}=T^{(2)}=T^{(3)}=T$), then $N^{(1)}=N^{(2)}=N^{(3)}=0$, that is, no loads develop within the structure if the heating is uniform. Otherwise, axial loads will accompany temperature changes. For example, if $T^{(2)}=T$ and $T^{(1)}=T^{(3)}=0$, then $P^{(1)}=P^{(3)}=0.4142AE\alpha T$ and $P^{(2)}=-0.5858AE\alpha T$; if $T^{(1)}=T$ and $T^{(2)}=T^{(3)}=0$, then $P^{(1)}=P^{(3)}=-0.2071AE\alpha T$ and $P^{(2)}=-0.2929AE\alpha T$.

As the degree of static indeterminacy increases, so does the effort required to solve the problem. The following example illustrates the force method for a doubly-redundant truss.

Example 7.4.5   Find the forces throughout the truss in Figure 7.4.17, using the principle of complementary virtual work. The axial rigidity AE is the same for all of the members.

This truss has two redundants, which we can select as members 8 and 9, in which case the base structure is as shown in Figure 7.4.18. The reactions shown in the figure are obtained from statics, and the joint method of equilibrium analysis yields the member loads in terms of the applied loads P and the redundants $N^{(8)}$ and $N^{(9)}$. These are listed here in Equation a. The virtual loads on the base structure are shown in Figure 7.4.19. The corresponding virtual member loads can be obtained from the true member loads by setting the true applied load P equal to zero and replacing the true redundant member loads $N^{(8)}$ and $N^{(9)}$ by their virtual counterparts, $\delta N^{(8)}$ and $\delta N^{(9)}$, respectively. In Equation a all of the self-equilibrating virtual member loads are listed alongside the true loads.

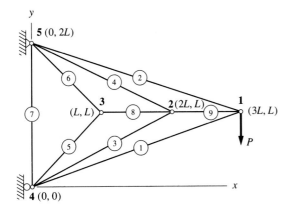

Figure 7.4.17   A truss with two degrees of static indeterminacy.

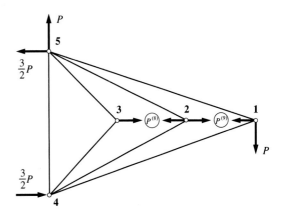

Figure 7.4.18   Statically determinate base structure and the true loading for the truss in Figure 7.4.17.

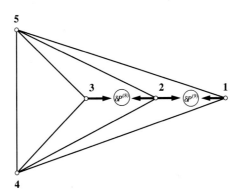

Figure 7.4.19   Virtual loads on the base truss.

True Loads	Virtual Loads

$$N^{(1)} = -1.581P - 0.5270N^{(9)} \qquad\qquad \delta N^{(1)} = -0.5270\delta N^{(9)}$$

$$N^{(2)} = 1.581P - 0.5270N^{(9)} \qquad\qquad \delta N^{(2)} = -0.5270\delta N^{(9)}$$

$$N^{(3)} = -0.5590N^{(8)} + 0.5590N^{(9)} \qquad \delta N^{(3)} = -0.5590\delta N^{(8)} + 0.5590\delta N^{(9)}$$

$$N^{(4)} = -0.5590N^{(8)} + 0.5590N^{(9)} \qquad \delta N^{(4)} = -0.5590\delta N^{(8)} - 0.5590\delta N^{(9)} \qquad \text{[a]}$$

$$N^{(5)} = 0.7071N^{(8)} \qquad\qquad\qquad \delta N^{(5)} = 0.7071\delta N^{(8)}$$

$$N^{(6)} = 0.7071N^{(8)} \qquad\qquad\qquad \delta N^{(6)} = 0.7071\delta N^{(8)}$$

$$N^{(7)} = 0.5P - 0.25N^{(8)} - 0.08333N^{(9)} \qquad \delta N^{(7)} = -0.25\delta N^{(8)} - 0.08333\delta N^{(9)}$$

The internal complementary virtual work of the truss is as follows:

$$\delta W_{int}^* = \sum_{e=1}^{12} \frac{N^{(e)} L^{(e)}}{A^{(e)} E^{(e)}} \delta N^{(e)}$$

$$= \overbrace{\frac{\left(-1.581P - 0.5270N^{(9)}\right)(3.162L)}{AE}}^{(1)} \left(-0.5270\delta N^{(9)}\right) + \overbrace{\frac{\left(1.581P - 0.5270N^{(9)}\right)(3.162L)}{AE}}^{(2)} \left(-0.5270\delta N^{(9)}\right)$$

$$+ \overbrace{\frac{\left(-0.5590N^{(8)} + 0.5590N^{(9)}\right)(2.236L)}{AE}}^{(3)} \left(-0.5590\delta N^{(8)} + 0.5590\delta N^{(9)}\right)$$

$$+ \overbrace{\frac{\left(-0.5590N^{(8)} + 0.5590N^{(9)}\right)(2.236L)}{AE}}^{(4)} \left(-0.5590\delta N^{(8)} + 0.5590\delta N^{(9)}\right)$$

$$+ \overbrace{\frac{\left(0.7071N^{(8)}\right)(1.414L)}{AE}}^{(5)} \left(0.7071\delta N^{(8)}\right) + \overbrace{\frac{\left(0.7071N^{(8)}\right)(1.414L)}{AE}}^{(6)} \left(0.7071\delta N^{(8)}\right)$$

$$+ \overbrace{\frac{\left(0.5P - 0.25N^{(8)} - 0.08333N^{(9)}\right)(2L)}{AE}}^{(7)} \left(-0.25\delta N^{(8)} - 0.08333\delta N^{(9)}\right) + \overbrace{\frac{N^{(8)} L}{AE} \delta N^{(8)}}^{(8)} + \overbrace{\frac{N^{(9)} L}{AE} \delta N^{(9)}}^{(9)}$$

Collecting terms and simplifying, we arrive at

$$\delta W_{int}^* = \frac{L}{AE} [(3.937N^{(8)} - 1.356N^{(9)} - 0.25P)\delta N^{(8)} + (-1.356N^{(8)} + 4.168N^{(9)} - 0.08333P)\delta N^{(9)}] \qquad \text{[b]}$$

The external complementary virtual work δW_{ext}^* is zero, since both of the redundants are internal loads. Therefore, δW_{int}^* must also be zero, for any choice of $\delta N^{(8)}$ and $\delta N^{(9)}$. This implies that the coefficients of these two virtual loads must vanish. The two equations for the two unknown redundants are therefore

$$3.937N^{(8)} - 1.356N^{(9)} = 0.25P$$
$$-1.356N^{(8)} + 4.168N^{(9)} = 0.08333P \qquad \text{[c]}$$

The solution of this system is

$$N^{(8)} = 0.07927P \qquad N^{(9)} = 0.04578P$$

Substituting these two redundant forces into Equation a gives the loads in all of the members, as summarized in Figure 7.4.20.

The following example reemphasizes the fact that after having used the force method to find the internal forces in a redundant structure, the procedure for obtaining its displacements does not require another lengthy, statically indeterminate analysis.

Example 7.4.6   Use the principle of complementary virtual work to calculate the horizontal component of the displacement at node 1 of the previous example.

Figure 7.4.21a shows the results of the analysis in Example 7.4.5. Using the principle of complementary virtual work to find the horizontal component of the displacement at node 1 requires applying a horizontal virtual load at that point, as shown in Figure 7.4.21b. Since the virtual loads need not satisfy compatibility, we can select any stable, statically determinate sub-

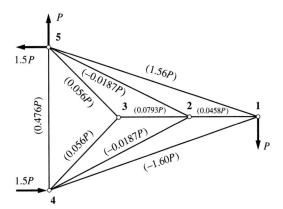

Figure 7.4.20   Computed loads in the truss of Figure 7.4.17.

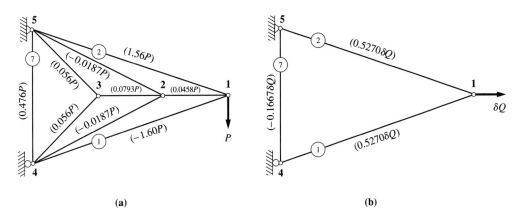

(a)   (b)

Figure 7.4.21   (a) The truss of Example 7.4.5 and the computed member loads. (b) Statically determinate substructure and the member loads due to a virtual load δQ applied to find the horizontal displacement of node 1.

structure of the truss to pick up the virtual load δQ. One of several such substructures is shown in Figure 7.4.21b and is comprised of only members 1, 2 and 7, the virtual loads in which are readily found from statics. Since the virtual loads in the rest of the truss are zero, the internal complementary virtual work is given by

$$
\begin{aligned}
\delta W_{\text{int}}^* &= \frac{N^{(1)}L^{(1)}}{A^{(1)}E^{(1)}}\delta N^{(1)} + \frac{N^{(2)}L^{(2)}}{A^{(2)}E^{(2)}}\delta N^{(2)} + \frac{N^{(7)}L^{(7)}}{A^{(7)}E^{(7)}}\delta N^{(7)} \\
&= \frac{(-1.6053P)(3.162L)}{AE}(0.5270\delta Q) + \frac{(1.557P)(3.162L)}{AE}(0.5270\delta Q) + \frac{(0.4764P)(2L)}{AE}(-0.1667\delta Q)
\end{aligned}
$$

so that

$$
\delta W_{\text{int}}^* = -0.2392\frac{PL}{AE}\delta Q
$$

In terms of the true horizontal displacement u_1 and the virtual load δQ, the external complementary virtual work is

$$\delta W_{ext}^* = u_1 \delta Q$$

Since $\delta W_{ext}^* = \delta W_{int}^*$, we have

$$u_1 = -0.2392 \frac{PL}{AE}$$

and the displacement is to the left.

It is not uncommon for a structure to have a prescribed, nonzero displacement in addition to applied loads and fixed supports. The external load Q required to produce the specified displacement q is added to the list of unknown external reactions. The prescribed displacement q appears in the external complementary virtual work term $q\delta Q$, where δQ is the virtual load applied in the direction of q. The following example illustrates this procedure as applied to the indeterminate truss studied in the previous two examples.

Example 7.4.7   The truss in Figure 7.4.23 is identical to that in Example 7.4.5. However, in addition to the applied load P, the vertical displacement at the roller support 4 is prescribed to be d. Calculate the member loads.

The truss has two degrees of static indeterminacy, and in Example 7.4.5, the loads in members 8 and 9 were selected as the redundants. Since the load at 4 required to produce the specified displacement d is unknown, the degree of static indeterminacy increases from two to three. We will treat the vertical reaction Y_4 at node 4 as the additional redundant, although several other choices are apparent. Figure 7.3.24a shows the member loads, obtained from statics, as functions of the applied load P and the redundants $N^{(8)}$, $N^{(9)}$, and Y_4. Removing the true load P and applying virtual loads in place of the redundants yields the system illustrated in Figure 7.4.24b. Except for member 7, the loads in the rod elements are identical to those found in Example 7.4.5. Therefore, the only difference in the internal complementary virtual work will be that contributed by element 7. Thus,

$$\delta W_{int}^* = \sum_{i=1}^{9} \delta W_{int}^{*\,(i)} = \left(\sum_{i=1}^{6} \delta W_{int}^{*\,(i)} + \delta W_{int}^{*\,(8)} + \delta W_{int}^{*\,(9)} \right) + \delta W_{int}^{*\,(7)} \qquad \text{[a]}$$

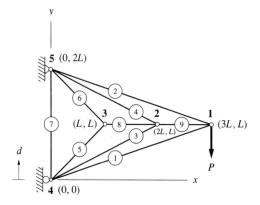

Figure 7.4.23   The truss of Figure 7.4.17, with a prescribed displacement at node 4.

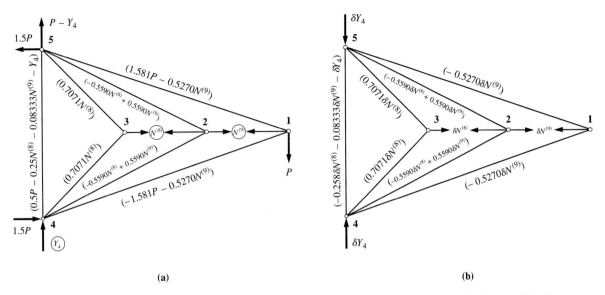

Figure 7.4.24   (a) Member loads in terms of the applied loads and the redundant loads (circled). (b) The virtual loading.

Using the calculations presented in Example 7.4.5, we get

$$\delta W^*_{\text{int}} = \frac{L}{AE} \left[\left(3.812N^{(8)} - 1.398N^{(9)}\right) \delta N^{(8)} + \left(-1.398N^{(8)} + 4.154N^{(9)}\right) \delta N^{(8)} \right] + \delta W^{*\,(7)}_{\text{int}} \qquad \textbf{[b]}$$

Since, in this case, we have

$$\delta W^{*(7)}_{\text{int}} = \frac{N^{(7)}L^{(7)}}{A^{(7)}E^{(7)}} \delta N^{(7)} = \frac{\left(0.5P - 0.25N^{(8)} - 0.08333N^{(9)} - Y_4\right)(2L)}{AE}\left(-0.25\delta N^{(8)} - 0.08333\delta N^{(9)} - \delta Y_4\right)$$

Equation b yields, after substitution,

$$\delta W^*_{\text{int}} = \frac{L}{AE}\Big[\left(3.937N^{(8)} - 1.356N^{(9)} + 0.5Y_4 - 0.25P\right)\delta N^{(8)}$$
$$+ \left(-1.356N^{(8)} + 4.168N^{(9)} + 0.1667Y_4 - 0.08333P\right)\delta N^{(9)} + \left(0.5N^{(8)} + 0.1667N^{(9)} + 2Y_4 - P\right)\delta Y_4\Big]$$

This expression must be equated to the external complementary virtual work, which is the product of the true prescribed displacement d at node 4 and the external virtual load in the direction of d. That is,

$$\delta W^*_{\text{ext}} = \delta Y_4 \times d$$

Since $\delta W^*_{\text{ext}} = \delta W^*_{\text{int}}$, after rearranging terms, we therefore obtain

$$\frac{L}{AE}\left(3.937N^{(8)} - 1.356N^{(9)} + 0.5Y_4 - 0.25P\right)\delta N^{(8)}$$
$$+ \frac{L}{AE}\left(-1.356N^{(8)} + 4.168N^{(9)} + 0.1667Y_4 - 0.08333P\right)\delta N^{(9)}$$
$$+ \frac{L}{AE}\left(0.5N^{(8)} + 0.1667N^{(9)} + 2Y_4 - P - \frac{AE}{L}d\right)\delta Y_4 = 0$$

Requiring this equality to hold for any choice of $\delta P^{(8)}$, $\delta P^{(9)}$, and δY_4 yields three equations for $N^{(8)}$, $N^{(9)}$, and Y_4, as follows:

$$3.937 N^{(8)} - 1.356 N^{(9)} + 0.5 Y_4 = 0.25 P$$

$$-1.356 N^{(8)} + 4.168 N^{(9)} + 0.1667 Y_4 = 0.08333 P$$

$$0.5 N^{(8)} + 0.1667 N^{(9)} + 2 Y_4 = P + \frac{AE}{L} d$$

The solution of this system of equations is

$$N^{(8)} = -0.08320 \frac{AEd}{L} \qquad N^{(9)} = 0.04805 \frac{AEd}{L} \qquad Y_4 = 0.5248 \frac{AEd}{L} + 0.5P$$

After substituting these values of $N^{(8)}$, $N^{(9)}$, and Y_4 into Figure 7.4.24a, we find the truss loads to be as illustrated in Figure 7.4.25, where $F = AEd/L$.

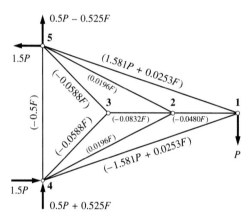

Figure 7.4.25 Member loads and reactions for the truss of Figure 7.4.23. $(F=AEd/L.)$

7.5 SIMPLE BEAMS: COMPLEMENTARY VIRTUAL WORK

At any point in the cross section of a beam loaded in the xy plane, as illustrated in Figure 7.5.1, there is a normal stress σ_x due to the bending moment M_z and a shear stress τ_{xy} due to shear load V_y on the section. Assuming, for simplicity, that the xy plane is a plane of symmetry of the beam cross section, it follows that $I_{yz} = 0$; and since $M_y = 0$, we know from Chapter 4 that in the absence of thermal strain, the normal stress at a distance y from the neutral axis is given by

$$\sigma_x = -\frac{M_z}{I_z} y \qquad\qquad [7.5.1]$$

For a virtual load, this becomes

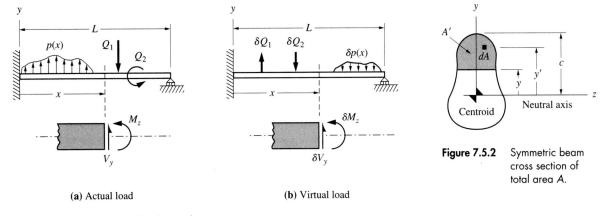

(a) Actual load

(b) Virtual load

Figure 7.5.1   Beam loaded in the xy plane.

Figure 7.5.2   Symmetric beam cross section of total area A.

$$\delta\sigma_x = -\frac{\delta M_z}{I_z}y \qquad [7.5.2]$$

The shear stress at a distance y from the neutral axis, as given in Chapter 4, is

$$\tau_{xy} = \frac{Q_z V_y}{I_z t} \qquad [7.5.3a]$$

where Q_z is the first moment about the neutral axis of the shaded area A' shown in Figure 7.5.2, that is, the area of the cross section enclosed by the horizontal line a distance y from the neutral axis and the boundary of the cross section. Specifically,

$$Q_z = \iint\limits_{A'} y'\,dA \qquad [7.5.3b]$$

The shear stress arising from a virtual load is given by the same formula,

$$\delta\tau_{xy} = \frac{Q_z \delta V_y}{I_z t} \qquad [7.5.4]$$

Using Equations 7.5.1 through 7.5.4, as well as Hooke's Law for isotropic materials, Equation 6.6.4 for the beam becomes

$$\delta W_{int}^* = \iiint\limits_V \left(\delta\sigma_x \varepsilon_x + \delta\tau_{xy}\gamma_{xy}\right) dV$$

$$= \iiint\limits_V \left[\delta\sigma_x \left(\frac{\sigma_x}{E}\right) + \delta\tau_{xy}\left(\frac{\tau_{xy}}{G}\right)\right] dV$$

$$= \int_0^L \left(\iint\limits_A \frac{\sigma_x \delta\sigma_x}{E}\,dA\right) dx + \int_0^L \left(\iint\limits_A \frac{\tau_{xy}\delta\tau_{xy}}{G}\,dA\right) dx$$

$$= \int_0^L \iint_A \frac{M_z \delta M_z}{E I_z^2} y^2 \, dA \, dx + \int_0^L \iint_A \frac{V_y \delta V_y}{G} \frac{Q_z^2}{I_z^2 t^2} \, dA \, dx$$

$$= \int_0^L \left(\frac{M_z \delta M_z}{E I_z^2} \right) \left(\iint_A y^2 \, dA \right) dx + \int_0^L \left(\frac{V_y \delta V_y}{G I_z^2} \right) \left(\iint_A \frac{Q_z^2}{t^2} \, dA \right) dx$$

or

$$\delta W_{\text{int}}^* = \underbrace{\int_0^L \frac{M_z \delta M_z}{E I_z} \, dx}_{\delta W_{\text{int,bending}}^*} + \underbrace{\int_0^L \frac{V_y \delta V_y}{G A_s} \, dx}_{\delta W_{\text{int,shear}}^*} \qquad \text{simple beam} \qquad \text{[7.5.5]}$$

Here, the familiar definition of area moment of inertia, $I_z = \int_A y^2 \, dA$, was substituted and a new quantity called the *area effective in shear*, A_s, was introduced into the last integral. The term A_s is defined by the formula

$$A_s = \frac{I_z^2}{\int_A \left(\frac{Q_z}{t} \right)^2 dA} \qquad \text{[7.5.6]}$$

The product EI_z is the *flexural rigidity* and GA_s is the *shear rigidity*.

The interpretation of A_s is as follows. Recall that the shear stress distribution over a cross section, as given by Equation 7.5.3, is not uniform. In the case of a rectangular section, it varies parabolically; for other sections, the distribution is more complex. Suppose we define a *nominal* shear stress $\tau = V_y / A_s$, which is uniformly distributed over the cross section A_s. Let the corresponding shear strain be denoted γ. Using these quantities to calculate the internal complementary virtual work due to shear would yield the following:

$$\delta W_{\text{int,shear}}^* = \iiint_V \gamma \delta \tau \, dV = \iiint_V \left(\frac{\tau}{G} \right) \delta \tau \, dV = \int_0^L \iint_{A_s} \left(\frac{V_y}{G A_s} \right) \left(\frac{\delta V_y}{A_s} \right) dA \, dx = \int_0^L \frac{V_y \delta V_y}{G A_s} \, dx$$

This agrees with Equation 7.5.5. Thus, for purposes of analyzing beam deflections A_s is the area over which the shear stress may be considered uniformly distributed.

We can now introduce the *form factor*

$$k = \frac{A_s}{A} \qquad \text{[7.5.7]}$$

in terms of which the shear component of the complementary internal virtual work can be written as follows:

$$\delta W_{\text{int}}^* = \int_0^L \frac{V_y \delta V_y}{k G A} \, dx \qquad \text{[7.5.8]}$$

From Equations 7.5.6 and 7.5.7, the form factor for a given section is therefore:

$$k = \frac{\frac{I_z^2}{A}}{\iint\limits_{A} \left(\frac{Q_z}{t}\right)^2 dA}$$

[7.5.9]

Example 7.5.1   Use Equation 7.5.9 to calculate the form factor for the rectangular section illustrated in Figure 7.5.6.

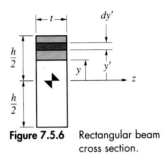

Figure 7.5.6   Rectangular beam cross section.

The area is $A = ht$, the centroidal area moment of inertia is $I_z = \frac{1}{12}th^3$, and Q_z is the first area moment of the cross-hatched area about the z axis, which is

$$Q_z(y) = \iint\limits_{A} y' dA = \int_{y}^{\frac{h}{2}} y'(t\,dy') = \frac{1}{2}ty'^2\Big|_{y}^{\frac{h}{2}} = \frac{t}{2}\left(\frac{h^2}{4} - y^2\right)$$

For the denominator of Equation 7.5.9, we thus have

$$\iint\limits_{A} \left(\frac{Q_z}{t}\right)^2 dA = \int_{-\frac{h}{2}}^{\frac{h}{2}} \left(\frac{Q_z}{t}\right)^2 t\,dy = \frac{t}{4}\int_{-\frac{h}{2}}^{\frac{h}{2}} \left(\frac{h^2}{4} - y^2\right)^2 dy$$

$$= \frac{t}{2}\int_{0}^{\frac{h}{2}} \left(\frac{h^4}{16} - \frac{h^2 y^2}{2} + y^4\right) dy = \frac{t}{2}\left(\frac{h^4 y}{16} - \frac{h^2 y^3}{6} + \frac{y^5}{5}\right)\Bigg|_{0}^{\frac{h}{2}} = \frac{th^5}{120}$$

Substituting this result into Equation 7.5.9 yields

$$k = \frac{I_z^2}{A}\left(\frac{1}{\iint\limits_{A} \left(\frac{Q_z}{t}\right)^2 dA}\right) = \frac{\left(\frac{1}{12}th^3\right)^2}{th}\left(\frac{1}{\frac{th^5}{120}}\right) = \frac{120}{144} = \frac{5}{6}$$

The form factor for a rectangular section is $5/6$. That is, for purposes of deflection analysis the shear force may be imagined to be uniformly distributed over 83 percent of a rectangular cross section.

The form factor for a solid circular section is $9/10$; for a wide flange section, it is A_{web}/A_{total}.

7.6   TORSION: COMPLEMENTARY VIRTUAL WORK

Recall from section 4.4 that the state of stress in an unrestrained torsion bar, such as the one illustrated in Figure 7.6.1, is

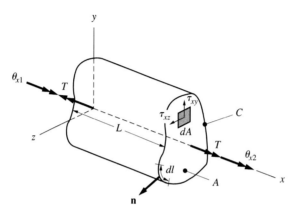

Figure 7.6.1   Prismatic torsion bar. C is the closed curve bounding the arbitrarily shaped cross section A, which may be multiply connected (e.g., hollow).

$$\sigma_x = \sigma_y = \sigma_z = \tau_{yz} = 0 \qquad \tau_{xy} = \frac{T}{J}\left(\frac{\partial \psi}{\partial y} - z\right) \qquad \tau_{xz} = \frac{T}{J}\left(\frac{\partial \psi}{\partial y} + y\right) \qquad [7.6.1]$$

where T is the torque, J is the torsion constant (Equation 4.4.14b), and $\psi = \psi(y, z)$ is the warping function. The internal complementary virtual work for isotropic materials, according to Equation 6.6.4 and including only the nonzero stress terms, is

$$\delta W^*_{int} = \iiint_V \left(\delta\tau_{xy}\gamma_{xy} + \delta\tau_{xz}\gamma_{xz}\right) dV = \iiint_V \left(\delta\tau_{xy}\frac{\tau_{xy}}{G} + \delta\tau_{xz}\frac{\tau_{xz}}{G}\right) dV = \frac{1}{G}\iiint_V \left(\tau_{xy}\delta\tau_{xy} + \tau_{xz}\delta\tau_{xz}\right) dV \quad [7.6.2]$$

where

$$\delta\tau_{xy} = \frac{\delta T}{J}\left(\frac{\partial \psi}{\partial y} - z\right) \qquad \delta\tau_{xz} = \frac{\delta T}{J}\left(\frac{\partial \psi}{\partial z} + y\right) \qquad [7.6.3]$$

Substituting Equations 7.6.1 and 7.6.3 into Equation 7.6.2 yields

$$\delta W^*_{int} = \int_0^L \frac{T\,\delta T}{GJ^2}\left\{\iint_A \left[\left(\frac{\partial \psi}{\partial y} - z\right)^2 + \left(\frac{\partial \psi}{\partial z} + y\right)^2\right] dA\right\} dx \qquad [7.6.4]$$

Let us simplify the right-hand side. Observe that, after expanding the terms of the integrand within the curly brackets and using the formula for the torsion constant J given by Equation 4.4.14b, we obtain

$$\iint_A \left[\left(\frac{\partial \psi}{\partial y} - z \right)^2 + \left(\frac{\partial \psi}{\partial z} + y \right)^2 \right] dA = J + \iint_A \left[\left(\frac{\partial \psi}{\partial y} \right)^2 + \left(\frac{\partial \psi}{\partial z} \right)^2 - z \frac{\partial \psi}{\partial y} + y \frac{\partial \psi}{\partial z} \right] dA$$

Solving for J and making use of Equation 4.4.10 leads to the expression

$$J = \iint_A \left[\left(\frac{\partial \psi}{\partial y} - z \right)^2 + \left(\frac{\partial \psi}{\partial z} + y \right)^2 \right] dA - \iint_A \left\{ \frac{\partial}{\partial y} \left[\psi \left(\frac{\partial \psi}{\partial y} - z \right) \right] + \frac{\partial}{\partial z} \left[\psi \left(\frac{\partial \psi}{\partial z} + y \right) \right] \right\} dA$$

By applying the *divergence theorem*[1] for the plane, the second integral over the arbitrarily shaped cross section A of the bar can be converted into a line integral around the boundary C of A (see Figure 7.6.1), so that

$$J = \iint_A \left[\left(\frac{\partial \psi}{\partial y} - z \right)^2 + \left(\frac{\partial \psi}{\partial z} + y \right)^2 \right] dA - \oint_C \left[\psi \left(\frac{\partial \psi}{\partial y} - z \right) n_y + \psi \left(\frac{\partial \psi}{\partial z} + y \right) n_z \right] dl$$

According to Equation 4.4.12, the line integral vanishes and we are left with the following formula for the torsion constant as an alternative to equation 4.4.14b:

$$J = \iint_A \left[\left(\frac{\partial \psi}{\partial y} - z \right)^2 + \left(\frac{\partial \psi}{\partial z} + y \right)^2 \right] dA$$

Substituting this into Equation 7.6.4 yields

$$\delta W_{\text{int}}^* = \int_0^L \frac{T \delta T}{GJ} dx \qquad \text{torsion bar} \qquad \text{[7.6.5]}$$

Obviously, if the torque is constant over the length of the torsion member, then Equation 7.6.5 becomes

$$\delta W_{\text{int,torsion}}^* = \frac{TL}{GJ} \delta T \qquad \text{[7.6.6]}$$

7.7   BEAM AND FRAME DEFLECTIONS USING COMPLEMENTARY VIRTUAL WORK

The basic procedure for calculating beam deflections using the principle of complementary virtual work is identical to that for trusses and, indeed, for any elastic structure. The difference lies in computing the internal loads, which are more complex in a beam than in a rod. We will illustrate the process by a series of examples, the first of which will investigate the relative significance of the second integral in Equation 7.5.5, that is, the magnitude of the shear distortion.

[1] Variously attributed to Karl F. Gauss (1777–1855), George Green (1793–1855) and Mikhail V. Ostrogradsky (1801–1862). See any textbook on vector calculus.

Example 7.7.1   Calculate the vertical displacement and the rotation at the left end of a cantilever beam under a uniformly-distributed load, as in Figure 7.7.1, and assess the magnitude of the deformation due to shear.

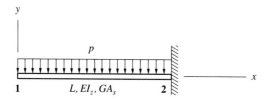

Figure 7.7.1   Uniformly-loaded simple cantilever beam.

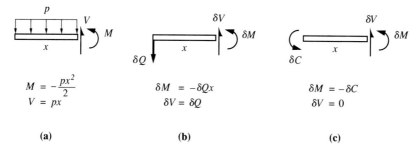

$$M = -\frac{px^2}{2}$$
$$V = px$$

$$\delta M = -\delta Q x$$
$$\delta V = \delta Q$$

$$\delta M = -\delta C$$
$$\delta V = 0$$

(a)   **(b)**   **(c)**

Figure 7.7.2   (a) Shear and bending moment due to the actual load on the beam.
(b) and (c) Virtual shear and bending moments due to a virtual point load and point couple, respectively, applied at the free end.

The shear and moment, as a function of the distance s from the free end of the beam, are found using statics and are shown in Figure 7.7.1a. To find the tip displacement, we take away the actual, uniformly-distributed load and replace it with a virtual point load δQ at the end. We then calculate the virtual shear and bending moment, which are shown in Figure 7.7.2a. From the principle of complementary virtual work and Equation 7.5.5, it follows that

$$v_1 \delta Q = \frac{1}{EI_z} \int_0^L \left(-\frac{ps^2}{2}\right)(-\delta Q \, s)ds + \frac{1}{GA_s} \int_0^L (ps)(\delta Q)ds$$

Carrying out the integrals, we get

$$v_1 \delta Q = \frac{pL^4}{8EI_z}\delta Q + \frac{pL^2}{2GA_s}\delta Q$$

Therefore, the transverse displacement due to the uniform load is

$$v_1 = \underbrace{\frac{pL^4}{8EI_z}}_{\text{due to bending}} + \underbrace{\frac{pL^2}{2GA_s}}_{\text{due to shear}}$$

To assess the contribution of shear to the total deformation, let us substitute $G = \frac{1}{2}E/(1 + v)$ for the shear modulus (cf. Chapter 3), set Poisson's ratio equal to 0.25, let the effective area in shear be equal to the total cross-sectional area A, and write the area as $A = I_z/\rho_z^2$, where ρ_z is the radius of gyration of the cross section. The expression for v_1 can then be written

$$v_1 = \frac{pL^4}{8EI_z}\left[1 + 10\left(\frac{\rho_z}{L}\right)^2\right]$$

The ratio of the shear portion of the displacement to the total displacement is therefore given by

$$\frac{v_{1,\text{shear}}}{v_{1,\text{total}}} = \frac{10\left(\frac{\rho_z}{L}\right)^2}{1 + 10\left(\frac{\rho_z}{L}\right)^2} = \frac{1}{1 + 0.1\left(\frac{L}{\rho_z}\right)^2}$$

Here, L/ρ_z is the *slenderness ratio*. As this relationship shows, the influence of shear on the displacement decreases rapidly with increasing slenderness ratio. Shear accounts for less than 5 percent of the deflection when the slenderness ratio is greater than about 15. Since the radius of gyration is less than about one-half of the depth of the beam's cross section, then, as a rule of thumb, we say that the shear distortion is negligible in simple beams whose length-to-depth ratio exceeds 10.

To find the rotation of the end of the beam, we apply a virtual point couple, as in Figure 8.5.2c. We then set the external complementary virtual work equal to the internal complementary virtual work,

$$\theta_1\delta C = \frac{1}{EI_z}\int_0^L\left(-\frac{ps^2}{2}\right)(-\delta C)\,ds + \frac{1}{GA_s}\int_0^L(ps)(0)\,ds$$

which means that

$$\theta_1 = \frac{pL^3}{6EI_z}$$

Example 7.7.2   Calculate the rotation of point 1 of the cantilever beam in Figure 7.7.3, using the principle of complementary virtual work. Neglect shear deformation.

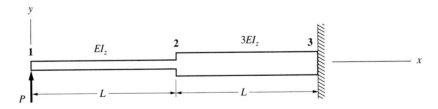

Figure 7.7.3   Stepped cantilever beam.

The actual shear and bending moments are shown in Figure 7.7.4a, along with the virtual shear and moments due to the virtual point couple δQ. For the external complementary virtual work, we have

$$\delta W^*_{\text{ext}} = \theta_1\delta Q$$

The internal complementary virtual work without the shear component, paying attention to sign conventions, is

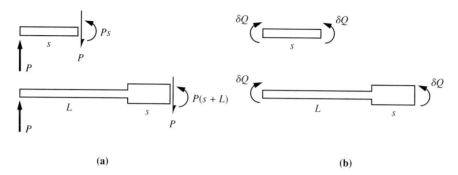

Figure 7.7.4   (a) Actual loading and internal loads. (b) Virtual loading and virtual internal loads.

$$\delta W_{int}^* = \int \frac{M_z \delta M_z}{E I_z} ds = \frac{1}{E I_z} \int_0^L (Ps)(\delta Q) \, ds + \frac{1}{3 E I_z} \int_0^L P(L+s) \delta Q \, ds$$

$$= \frac{P \delta Q}{E I_z} \frac{L^2}{2} + \frac{P \delta Q}{3 E I_z} \frac{3 L^2}{2}$$

$$= \frac{P L^2}{E I_z} \delta Q$$

Finally, equating the two virtual work expressions yields

$$\theta_1 = \frac{P L^2}{E I_z} \qquad \text{(clockwise)}$$

Frames are composed of slender, possibly curved, beam-like elements capable of carrying shear and bending loads, as well as axial loads like the members of a truss. Even though bending deflections usually dominate, we may wish to consider the contribution of axial deformation, as well as shear deformation, in the frame members. To do so, we must combine Equations 7.2.2 and 7.5.5, so that the internal complementary virtual work of a plane frame member (neglecting thermal strain) is

$$\delta W_{int}^* = \int_0^L \frac{M_z \delta M_z}{E I_z} ds + \int_0^L \frac{V_y \delta V_y}{G A_s} ds + \int_0^L \frac{N \delta N}{A E} ds \qquad \text{[7.7.1]}$$

where N is the force normal to the cross section of the member. In a curved frame like that of Figure 7.7.5, in which the depth h of the cross section is much smaller than the radius of curvature R, we can use Equation 7.7.1, replacing ds with $R d\phi$

Example 7.7.3   Find the horizontal displacement of the free end of the statically determinate, thin circular frame (curved beam) shown in Figure 7.7.6.

Figure 7.7.6 also shows a free-body diagram of a segment of the frame, starting at the free end. Setting both the net force normal and parallel to the cut and the net moment about the cut equal to zero, we find that

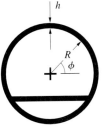

Figure 7.7.5 Circular frame in which $R \gg h$.

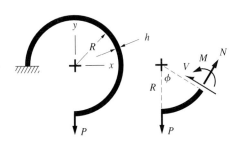

Figure 7.7.6 Curved beam.

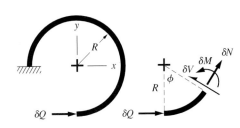

Figure 7.7.7 Virtual load on the circular frame.

$$M = -PR\sin\phi$$
$$V = P\cos\phi \qquad \text{[a]}$$
$$N = P\sin\phi$$

To find the horizontal displacement at the free end, we remove the actual load and apply a virtual load δQ, as shown in Figure 7.7.7. Equilibrium then requires that

$$\delta M = -\delta Q\, R(1 - \cos\phi)$$
$$\delta V = \delta Q\, \sin\phi \qquad \text{[b]}$$
$$\delta N = -\delta Q\, \cos\phi$$

Substituting Equations a and b into Equation 7.7.1 yields

$$\delta W_{\text{int}}^{*} = \frac{1}{EI_z} \int_0^{\frac{3\pi}{2}} (-PR\sin\phi)\,[-\delta Q\, R\,(1 - \cos\phi)]\, Rd\phi$$

$$+ \frac{1}{GA_s} \int_0^{\frac{3\pi}{2}} (P\cos\phi)\,(\delta Q\, \sin\phi)\, Rd\phi + \frac{1}{AE} \int_0^{\frac{3\pi}{2}} (P\sin\phi)\,(-\delta Q\, \cos\phi)\, Rd\phi$$

$$= \frac{PR^3}{EI_z}\delta Q \int_0^{\frac{3\pi}{2}} \sin\phi\, d\phi + \left(-\frac{PR^3}{EI_z} + \frac{PR}{GA_s} - \frac{PR}{AE}\right)\delta Q \left(\frac{1}{2} \int_0^{\frac{3\pi}{2}} \sin 2\phi\, d\phi\right)$$

$$= \frac{PR^3}{EI_z}\delta Q\,(1) + \left(-\frac{PR^3}{EI_z} + \frac{PR}{GA_s} - \frac{PR}{AE}\right)\delta Q \left(\frac{1}{2}\right)$$

so that

$$\delta W_{\text{int}}^{*} = \frac{\delta Q}{2}\left(\frac{PR^3}{EI_z} + \frac{PR}{GA_s} - \frac{PR}{AE}\right)$$

Setting this equal to the external complementary virtual work, $\delta W_{\text{ext}}^{*} = u\delta Q$, substituting $G = \frac{1}{2}E/(1 + v)$, and simplifying, we get

$$u = \frac{PR^3}{2EI_z} \left\{ 1 + \left[\frac{2(1+v)}{k} - 1 \right] \left(\frac{\rho_z}{R} \right)^2 \right\}$$

where k is the form factor ($A_s = kA$), and ρ_z is the radius of gyration of the cross section ($I_z = A\rho_z^2$). Since $\rho_z \ll R$, then for all practical purposes, the curly brackets can be set equal to 1; that is, the deflection is dominated by bending.

In the analysis of the deflections of complex frames, special attention must be paid to the computation of the internal loads, as illustrated in the following example.

Example 7.7.4   Use the principle of complementary virtual work to calculate the horizontal displacement of point 1 of the statically determinate frame in Figure 7.7.8a. The area, moment of inertia, and material properties are uniform throughout.

Figure 7.7.8b shows the internal force sign convention for a horizontal beam. This sign convention, which will be used throughout the frame, requires that we specify which side of a vertical element is the "top," and that is indicated in Figure 7.7.9a. To obtain the shear, normal force, and moment in each portion of the structure, we write the equations of equilibrium for the free-body diagram of that portion of the frame between the free end and a transverse section through each of the five beam members in succession. The transverse section is taken at a distance s from the beginning of each beam element. As we move from element 1 to element 5, the free bodies include more and more of the frame. The results of this equilibrium analysis are shown alongside each member of the frame, for both the actual and virtual load, in Figure 7.7.9. The quantities are positive or negative according to the sign convention in Figure 7.7.8.

Writing Equation 7.7.1 for each of the five members of the frame and summing them up, we obtain

$$\delta W^*_{int} = \overbrace{\frac{1}{EI_z} \int_0^L (-Ps)(0)ds + \frac{1}{AE} \int_0^L (0)(\delta Q)ds + \frac{1}{GA_s} \int_0^L (-P)(0)ds}^{\text{beam 1}}$$

$$+ \overbrace{\frac{1}{EI_z} \int_0^L (PL)(-\delta Q\, s)ds + \frac{1}{AE} \int_0^L (P)(0)ds + \frac{1}{GA_s} \int_0^L (0)(\delta Q)ds}^{\text{beam 2}}$$

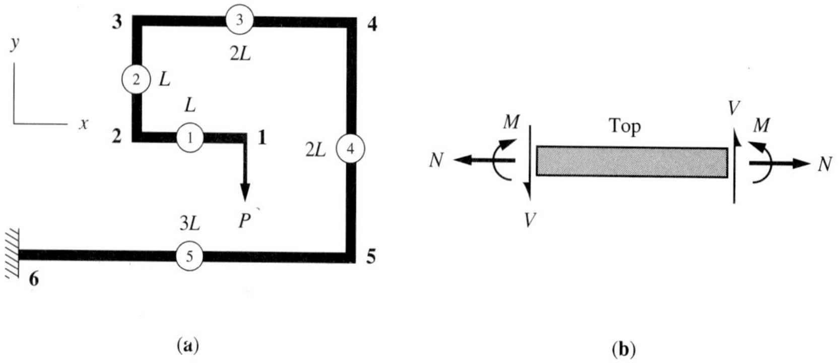

Figure 7.7.8   (a) Frame with load at point 1. (b) Sign convention for internal forces.

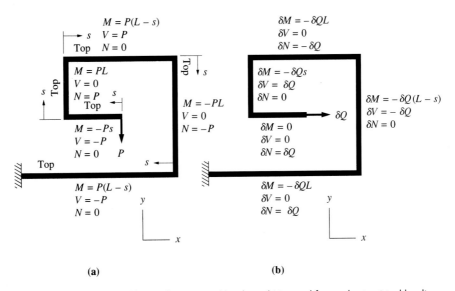

Figure 7.7.9   (a) Internal forces due to actual loading. (b) Internal forces due to virtual loading.

$$\overbrace{+ \frac{1}{EI_z} \int_0^{2L} [P(L-s)](-\delta Q L)ds + \frac{1}{AE} \int_0^{2L} (0)(-\delta Q)ds + \frac{1}{GA_s} \int_0^{2L} (P)(0)ds}^{\text{beam 3}}$$

$$\overbrace{+ \frac{1}{EI_z} \int_0^{2L} (-PL)[-\delta Q(L-s)]ds + \frac{1}{AE} \int_0^{2L} (-P)(0)ds + \frac{1}{GA_s} \int_0^{L} (0)(-\delta Q)ds}^{\text{beam 4}}$$

$$\overbrace{+ \frac{1}{EI_z} \int_0^{3L} [P(L-s)][-\delta Q L]ds + \frac{1}{AE} \int_0^{3L} (0)(\delta Q)ds + \frac{1}{GA_s} \int_0^{3L} (-P)(0)ds}^{\text{beam 5}}$$

which reduces to

$$\delta W_{\text{int}}^* = \delta Q \left[0 + \left(-\frac{PL^3}{2EI_z}\right) + (0) + (0) + \left(\frac{3PL^3}{2EI_z}\right) \right]$$

Equating this to the external complementary virtual work, $u_1 \delta Q$, we get the horizontal displacement of point 1,

$$u_1 = \frac{PL^3}{EI_z}$$

Example 7.7.5   Use the principle of complementary virtual work to calculate the rotation at point 4 of the frame of the previous example.

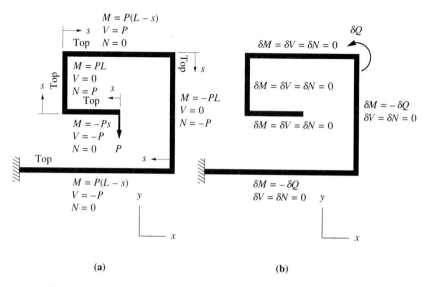

Figure 7.7.10 (a) Internal forces due to the actual loading. (b) Internal forces due to the virtual couple at point 4.

Figure 7.7.10 is the same as Figure 7.7.9, except for the virtual loading. To find the rotation θ_4 at point 4, we need to apply a virtual point couple δQ, as shown in Figure 7.7.10 b. The corresponding virtual internal loads are indicated on the figure. Notice that no virtual internal loads are induced in that portion of the frame between the point where δQ is applied and the free end (where the true applied load P exists). Since the virtual shear and axial loads are zero everywhere, we have

$$\delta W_{int}^* = \overbrace{0}^{\text{beam 1}} + \overbrace{0}^{\text{beam 2}} + \overbrace{0}^{\text{beam 3}} + \frac{1}{EI_z} \overbrace{\int_0^{2L} (-PL)(-\delta Q)\,ds}^{\text{beam 4}} + \frac{1}{EI_z} \overbrace{\int_0^{3L} [P(L-s)](-\delta Q)\,ds}^{\text{beam 5}}$$

so that

$$\delta W_{int}^* = \frac{2PL^2}{EI_z}\delta Q + \frac{3PL^2}{2EI_z}\delta Q = \frac{7PL^2}{2EI_z}\delta Q$$

Since the external complementary virtual work is $\theta_{x_4}\delta Q$, then $\delta W_{ext}^* = \delta W_{int}^*$ implies that

$$\theta_{x_4} = \frac{7PL^2}{2EI_z}$$

The positive sign of this result means that the rotation is in the same direction as δQ, that is, counterclockwise.

7.8   STATICALLY INDETERMINATE BEAMS AND FRAMES

In this section, we will use the principle of complementary virtual work to analyze simple, statically indeterminate beams and frames. The procedure for redundant beams is identical to that of section 7.4 for trusses. First, we identify the redundants and treat them as external loads applied to a statically determinate base structure. Then we remove the true loads from the structure and apply virtual loads in the direction of each redundant. Finally, we equate the external and internal complementary virtual work expressions and solve the resulting equations for the redundant loads, after which all the other loads follow from the analysis of the statically determinate base structure.

Example 7.8.1   Use the principle of complementary virtual work to calculate the reaction at the left end of the beam in Figure 7.8.1. The flexural rigidity EI_z is constant. Consider bending only.

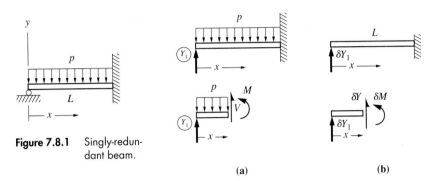

Figure 7.8.1   Singly-redundant beam.

(a)   **(b)**

Figure 7.8.2   (a) Statically determinate base structure
with the actual applied loads. (b) Base
structure with the virtual load only.

The roller support on the left is redundant. Figure 7.8.2a shows the statically determinate base structure, with the reaction at the roller treated as an externally applied load. An equilibrium analysis of the free-body diagram in part (a) of the figure yields the true bending moment in terms of the true loads, as follows:

$$M = Y_1 x - \frac{px^2}{2} \qquad\qquad \text{[a]}$$

In Figure 7.8.2b, the actual loads have been removed and replaced by the virtual load in the direction of the redundant reaction. From that free-body diagram, we get the virtual bending moment in terms of the virtual load:

$$\delta M = \delta Y_1 x \qquad\qquad \text{[b]}$$

The internal complementary virtual work, neglecting shear, is given by Equation 7.5.5,

$$\delta W_{\text{int}}^* = \int_0^L \frac{M_z \delta M_z}{EI_z} = \frac{1}{EI_z} \int_0^L \left(Y_1 x - \frac{px^2}{2} \right) (\delta Y_1 x)\, dx = \frac{1}{EI_z} \left(\frac{Y_1 L^3}{3} - \frac{pL^4}{8} \right) \delta Y_1 \qquad\qquad \text{[c]}$$

The principle of complementary virtual work requires that $\delta W_{\text{int}}^* = \delta W_{\text{ext}}^*$; however, since the displacement in the direction of the redundant load is zero, then $\delta W_{\text{ext}}^* = 0$. Therefore, $\delta W_{\text{int}}^* = 0$, that is,

$$\frac{1}{EI_z} \left(\frac{Y_1 L^3}{3} - \frac{pL^4}{8} \right) \delta Y_1 = 0$$

This means that the redundant support reaction is

$$Y_1 = \frac{3}{8}pL \qquad\qquad \text{[d]}$$

Example 7.8.2   Using the principle of complementary virtual work, find the location and magnitude of the maximum bending moment in the clamped–clamped simple beam in Figure 7.8.3. The flexural rigidity EI_z is uniform. Neglect shear.

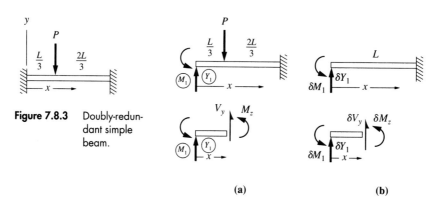

Figure 7.8.3   Doubly-redundant simple beam.

(a)   **(b)**

Figure 7.8.4   (a) Statically determinate base structure with the actual applied loads. (b) Base structure with virtual load only.

Let us select the two reactions Y_1 and M_1 at the left wall as the redundants. On the base structure of Figure 7.8.4a, they are circled. Equilibrium of the base beam under the true applied loads requires that

$$M_z = Y_1 x - M_1 \qquad\qquad\qquad\qquad 0 \le x \le \frac{L}{3}$$

$$M_z = Y_1 x - P\left(x - \frac{L}{3}\right) - M_1 = (Y_1 - P)x + \frac{PL}{3} - M_1 \qquad \frac{L}{3} \le x \le L \qquad \text{[a]}$$

Figure 7.8.4b shows the base beam with all of the true loads removed and replaced by virtual redundants. For equilibrium,

$$\delta M_z = \delta Y_1 x - \delta M_1 \qquad 0 \le x \le L \qquad\qquad \text{[b]}$$

In this case, the internal complementary virtual work, neglecting shear, yields

$$\delta W_{int}^* = \int_0^L \frac{M_z \delta M_z}{EI_z}\,dx = \left(\frac{1}{EI_z}\right)\left\{ \int_0^{L/3} (Y_1 x - M_1)(\delta Y_1 x - \delta M_1)\,dx + \int_{L/3}^L \left[(Y_1 - P)x + \frac{PL}{3} - M_1\right](\delta Y_1 x - \delta M_1)\,dx \right\}$$

After carrying out the integrations and simplifying, we have

$$\delta W_{int}^* = \frac{1}{EI_z}\left(M_1 L - \frac{1}{2}Y_1 L^2 + \frac{2}{9}PL^2 \right)\delta M_1 + \frac{1}{EI_z}\left(-\frac{1}{2}M_1 L^2 + \frac{1}{3}Y_1 L^3 - \frac{14}{81}PL^3 \right)\delta Y \qquad \text{[c]}$$

The external complementary virtual work δW_{ext}^* is that of the redundant loads δM_1 and δY_1 acting through the actual rotation and transverse displacement, respectively, at the left end of the beam. But the beam is clamped there, so both of these displacements are zero. It follows that $\delta W_{ext}^* = 0$. According to the principle of complementary virtual work, $\delta W_{int}^* = \delta W_{ext}^*$ for arbitrary choice of the virtual loads. Thus, from Equation c we have

$$E I_z \left(M_1 L - \frac{1}{2} Y_1 L^2 + \frac{2}{9} P L^2 \right) \delta M_1 + E I_z \left(-\frac{1}{2} M_1 L^2 + \frac{1}{3} Y_1 L^3 - \frac{14}{81} P L^3 \right) \delta Y_1 = 0$$

for any choice of δM_1 and δY_1. This means that the coefficients of δM_1 and δY_1 must vanish, yielding the two equations,

$$L M_1 - \frac{1}{2} L^2 Y_1 = -\frac{2}{9} P L$$

$$-\frac{1}{2} L^2 M_1 + \frac{1}{3} L^3 Y_1 = \frac{14}{81} P L^3$$

[d]

We solve this set of equations to obtain the redundant reactions in terms of the applied load P,

$$M_1 = \frac{4}{27} P L \qquad Y_1 = \frac{20}{27} P$$

[e]

We can now substitute these expressions into Equation a to find the bending moment distribution across the beam. The bending moment diagram is plotted in Figure 7.8.5. The maximum value occurs at the left wall and has a magnitude of $4PL/27$.

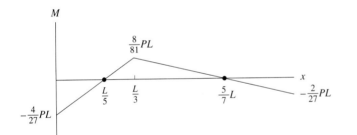

Figure 7.8.5 Bending moment diagram for the beam in Figure 7.8.3.

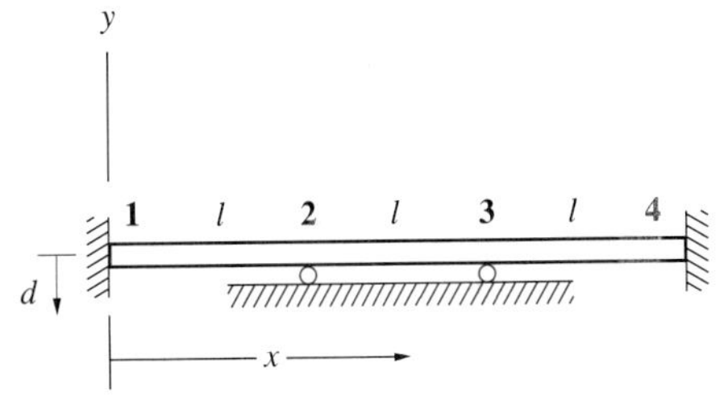

Figure 7.8.6 Simple beam of redundancy four.

Example 7.8.3 The simple beam in Figure 7.8.6 is built in at both ends, and there are two intermediate roller supports. The left wall is displaced downwards a prescribed amount d but remains vertical. Neglecting shear and assuming EI is uniform, use the principle of complementary virtual work to calculate the six reactions at the four supports.

This beam has four redundants, which we will choose as the reactions at 1, 2 and 3. In Figure 7.8.7a the statically determinate base beam is shown with the true, applied redundant loads. Part (b) of the figure shows the virtual load pattern, which in this case is identical to that of the true loads. For equilibrium in both the true and virtual load cases, we have:

For $0 \leq x \leq l,$ $M_z = -M_1 + Y_1 x$ $\delta M_z = -\delta M_1 + \delta Y_1 x$

For $l \leq x \leq 2l,$ $M_z = -M_1 + Y_1 x + Y_2 (x - l)$ $\delta M_z = -\delta M_1 + \delta Y_1 x + \delta Y_2 (x - l)$

For $2l \leq x \leq 3l,$ $M_z = -M_1 + Y_1 x + Y_2 (x - l) + Y_3 (x - 2l)$ $\delta M_z = -\delta M_1 + \delta Y_1 x + \delta Y_2 (x - l) + \delta Y_3 (x - 2l)$

Therefore, the internal complementary virtual work is

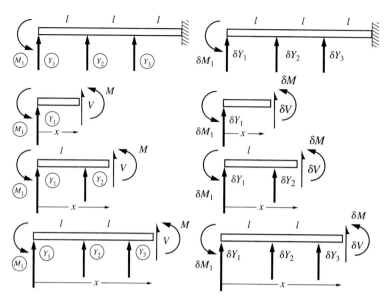

Figure 7.8.7   (a) Statically determinate base beam with the true applied loads, and the intermediate free-body diagrams. (b) Base structure with the four applied virtual loads, and the intermediate free-body diagrams.

$$\delta W_{\text{int}}^* = \frac{1}{EI_z}\left[\int_0^l \{[-M_1 + Y_1 x][-\delta M_1 + \delta Y_1 x]\}\, dx + \int_l^{2l} \{[-M_1 + Y_1 x + Y_2(x-l)][-\delta M_1 + \delta Y_1 x + \delta Y_2(x-l)]\}\, dx\right.$$

$$\left. + \int_{2l}^{3l} \{[-M_1 + Y_1 x + Y_2(x-l) + Y_3(x-2l)][-\delta M_1 + \delta Y_1 x + \delta Y_2(x-l) + \delta Y_3(x-2l)]\}\, dx\right]$$

Expanding the integrands, carrying out the integrals, and simplifying yields

$$\delta W_{\text{int}}^* = \left(\frac{1}{EI_z}\right)\left[\left(3l M_1 - \tfrac{9}{2}l^2 Y_1 - 2l^2 Y_2 - \tfrac{1}{2}l^2 Y_3\right)\delta M_1 + \left(-\tfrac{9}{2}l^2 M_1 + 9l^3 Y_1 + \tfrac{14}{3}l^3 Y_2 + \tfrac{4}{3}l^3 Y_3\right)\delta Y_1\right.$$

$$\left. + \left(-2l^2 M_1 + \tfrac{14}{3}l^3 Y_1 + \tfrac{8}{3}l^3 Y_2 + \tfrac{5}{6}l^3 Y_3\right)\delta Y_2 + \left(-\tfrac{1}{2}l^2 M_1 + \tfrac{4}{3}l^3 Y_1 + \tfrac{5}{6}l^3 Y_2 + \tfrac{1}{3}l^3 Y_3\right)\delta Y_3\right]$$ [a]

The external complementary virtual work is that of the virtual load δY_1 acting through the true, prescribed displacement d, or

$$\delta W_{\text{ext}}^* = -v\delta Y_1$$ [b]

The minus sign is required because we assume δY_1 acts upwards, whereas the displacement d is specified as downwards.

Setting Equation a equal to Equation b and equating the coefficients of δM_1, δY_1, δY_2, and δY_3 on the left side of the equation to those on the right yields four equations for the four unknown redundant reactions:

$$3l M_1 - \tfrac{9}{2}l Y_1 - 2l^2 Y_2 - \tfrac{1}{2}l^2 Y_3 = 0$$

$$-\tfrac{9}{2}l M_1 + 9l^3 Y_1 + \tfrac{14}{3}l^3 Y_2 + \tfrac{4}{3}l^3 Y_3 = -EI_z d$$

$$-2l^2 M_1 + \tfrac{14}{3}l^3 Y_1 + \tfrac{8}{3}l^3 Y_2 + \tfrac{5}{6}l^3 Y_3 = 0$$

$$-\tfrac{1}{2}l^2 M_1 + \tfrac{4}{3}l^3 Y_1 + \tfrac{5}{6}l^3 Y_2 + \tfrac{1}{3}l^3 Y_3 = 0$$ [c]

This system has the following solution:

$$M_1 = -\frac{22EI_z}{5l^2}d \qquad Y_1 = -\frac{36EI_z}{5l^3}d \qquad Y_2 = \frac{54EI_z}{5l^3}d \qquad Y_3 = -\frac{24EI_z}{5l^3}d \qquad \text{[d]}$$

With this information, we can find the force and couple at the right wall, using statics. The results are summarized in Figure 7.8.8.

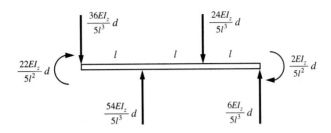

Figure 7.8.8   Support reactions on the beam of Figure 7.8.6.

Example 7.8.4   Calculate the transverse displacement at the midpoint of the indeterminate beam in the previous example.
    To find the displacement, we apply a virtual force δP upwards at the beam midpoint $x = 1.5l$ in Figure 7.8.6. The virtual load system arising in response to δP need not satisfy all of the constraints placed on the true load system. That is, it does not have to satisfy compatibility, but it must be self-equilibrating. Therefore, let us assume that the virtual load is counteracted by just the fixed support at the right end, as shown in Figure 7.8.9b. Since the virtual bending moment is zero for $x \le 1.5l$, the internal complementary virtual work is

$$\delta W_{\text{int}}^* = \frac{1}{EI_z}\int_0^{3l} M\delta M\, dx = \frac{1}{EI_z}\int_{1.5l}^{3l} M\delta P(x - 1.5l)\, dx \qquad \text{[a]}$$

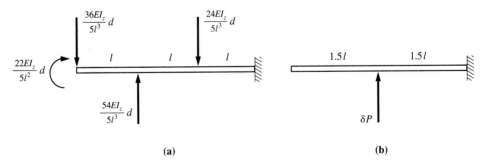

(a)   **(b)**

Figure 7.8.9   (a) Actual loads, computed in Example 7.8.3. (b) Virtual load and the selected statically determinate base structure.

The true bending moment is found by using the calculated true loads shown in Figure 7.8.9a. Thus,

$$\delta W^*_{int} = \frac{1}{EI_z} \int_{\frac{3l}{2}}^{2l} \left[\frac{22EI_z d}{5l^2} - \frac{36EI_z d}{5l^3}x + \frac{54EI_z d}{5l^3}(x-l) \right] \delta P \left(x - \frac{l}{2} \right) dx$$

$$+ \frac{1}{EI_z} \int_{2l}^{3l} \left[\frac{22EI_z d}{5l^2} - \frac{36EI_z d}{5l^3}x + \frac{54EI_z d}{5l^3}(x-l) - \frac{24EI_z d}{5l^3}(x-2l) \right] \delta P \left(x - \frac{l}{2} \right) dx$$

The integrands simplify to become

$$\delta W^*_{int} = \frac{\delta P d}{5l^3} \int_{\frac{3l}{2}}^{2l} \left(18x^2 - 59lx + 48l^2 \right) dx + \frac{\delta P d}{l^3} \int_{2l}^{3l} \left(-6x^2 + 25lx - 24l^2 \right) dx$$

and the integrations yield

$$\delta W^*_{int} = 0.025\delta P d + 0.100\delta P d = 0.125\delta P d \qquad \text{[b]}$$

The external complementary virtual work is $\delta W^*_{ext} = v\delta P$, where v is the actual transverse displacement at the virtual load. Since it must be true that $\delta W^*_{ext} = \delta W^*_{int}$, it follows that $v = 0.125d$. Thus, a downward displacement d at the free end of the beam produces an upward displacement $d/8$ at its midpoint.

Using the force method to solve redundant frame problems follows the same steps used for trusses and beams and is best demonstrated through several examples. We will confine ourselves to frames composed of slender members, that is, those with length-to-depth ratios large enough that they can be treated as simple beams, with bending dominating both shearing and stretching.

Example 7.8.5   Calculate the reaction at point 1 of the frame in Figure 7.8.10, using the principle of complementary virtual work. Assess the effects of shear and stretching on the result. Assume that the material properties and section properties are uniform throughout.

The frame has one degree of static indeterminacy, and we will treat the vertical reaction Y_1 at support 1 as the redundant, which is circled in the free-body diagrams of Figure 7.8.11. An equilibrium analysis of the free-body diagrams shows that in the horizontal member of the frame, we have

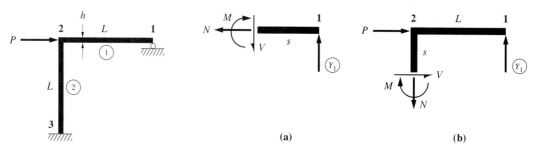

Figure 7.8.10   Singly-redundant frame.

Figure 7.8.11   Free-body diagrams of the statically determinate base structure, showing the redundant reaction Y_1 as an externally applied load.

$$M = Y_1 s$$
$$V = Y_1 \qquad \text{Element 1} \qquad \text{[a]}$$
$$N = 0$$

In the vertical member, we have

$$M = Y_1 L - P s$$
$$V = -P \qquad \text{Element 2} \qquad \text{[b]}$$
$$N = Y_1$$

We next remove the true load P from the frame and apply the virtual load δY_1 to the base structure in the direction of the redundant at 1, as shown in Figure 7.8.12. An equilibrium analysis reveals the following internal virtual loads:

$$\delta M = \delta Y_1 s$$
$$\delta V = \delta Y_1 \qquad \text{Element 1} \qquad \text{[c]}$$
$$\delta N = 0$$

$$\delta M = \delta Y_1 L$$
$$\delta V = 0 \qquad \text{Element 2} \qquad \text{[d]}$$
$$\delta N = \delta Y_1$$

The complementary virtual work expression, accounting for bending, shear, and stretching, is found by Equation 7.7.1,

$$\delta W_{\text{int}}^* = \delta W_{\text{int,bending}}^* + \delta W_{\text{int,stretching}}^* + \delta W_{\text{int,shear}}^* \qquad \text{[e]}$$

where

$$\delta W_{\text{int,bending}}^* = \sum_{e=1}^{2} \int_0^{L^{(e)}} \left(\frac{M \delta M}{E I_z} \right)^{(e)} ds \qquad \text{[f]}$$

$$\delta W_{\text{int,shear}}^* = \sum_{e=1}^{2} \int_0^{L^{(e)}} \left(\frac{V \delta V}{G A_s} \right)^{(e)} ds \qquad \text{[g]}$$

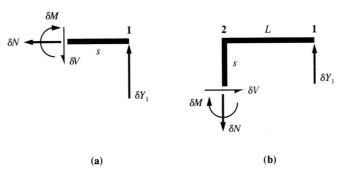

(a)   (b)

Figure 7.8.12   Free-body diagrams of the base frame with a virtual load applied at 1 in the direction of the redundant.

$$\delta W^*_{int,stretching} = \sum_{e=1}^{2} \int_0^{L^{(e)}} \left(\frac{N \delta N}{AE} \right)^{(e)} ds \tag{h}$$

After substituting the true and virtual bending moments from Equations a, b, and c, we have from Equation f

$$\delta W^*_{int,bending} = \int_0^{L^{(1)}} \left(\frac{M \delta M}{EI_z} \right)^{(1)} ds + \int_0^{L^{(2)}} \left(\frac{M \delta M}{EI_z} \right)^{(2)} ds = \frac{1}{EI_z} \int_0^L (Y_1 s)(\delta Y_1 s) \, ds + \frac{1}{EI_z} \int_0^L (Y_1 L - Ps)(\delta Y_1 L) \, ds$$

so that

$$\delta W^*_{int,bending} = \left(\frac{4}{3} \frac{Y_1 L^3}{EI_z} - \frac{1}{2} \frac{PL^3}{EI_z} \right) \delta Y_1 \tag{i}$$

Combining Equations g, c, and d, we find

$$\delta W^*_{int,shear} = \int_0^{L^{(1)}} \left(\frac{V \delta V}{GA_s} \right)^{(1)} ds + \int_0^{L^{(2)}} \left(\frac{V \delta V}{GA_s} \right)^{(2)} ds = \frac{1}{GA_s} \int_0^L (Y_1)(\delta Y_1) \, ds + \frac{1}{GA_s} \int_0^L (-P)(0) \, ds$$

or

$$\delta W^*_{int,shear} = \frac{Y_1 L}{GA_s} \delta Y_1 \tag{j}$$

Finally, the complementary virtual work due to stretching, Equation h, is

$$\delta W^*_{int,stretching} = \int_0^{L^{(1)}} \left(\frac{N \delta N}{AE} \right)^{(1)} ds + \int_0^{L^{(2)}} \left(\frac{N \delta N}{AE} \right)^{(2)} ds = \frac{1}{AE} \int_0^L (0)(0) \, ds + \frac{1}{AE} \int_0^L (Y_1)(\delta Y_1) \, ds$$

which yields

$$\delta W^*_{int,stretching} = \frac{Y_1 L}{AE} \delta Y_1 \tag{k}$$

Substituting Equations i, j, and k into Equation e gives the expression for the internal complementary virtual work for the frame, in terms of the true and virtual loads:

$$\delta W^*_{int} = \left(\frac{4}{3} \frac{Y_1 L^3}{EI_z} - \frac{1}{2} \frac{PL^3}{EI_z} \right) \delta Y_1 + \frac{Y_1 L}{GA_s} \delta Y_1 + \frac{Y_1 L}{EA} \delta Y_1 \tag{l}$$

Since δY_1 acts through zero displacement at roller support 1, the external complementary virtual work is zero. Therefore, $\delta W^*_{int} = \delta W^*_{ext} = 0$, so that regardless of δY_1, Equation l yields

$$\left(\frac{4}{3} \frac{Y_1 L^3}{EI_z} - \frac{1}{2} \frac{PL^3}{EI_z} \right) + \frac{Y_1 L}{GA_s} + \frac{Y_1 L}{EA} = 0 \tag{m}$$

Solving this for the redundant load, Y_1, we get

$$Y_1 = \frac{3P}{2} \frac{GAA_s L^2}{4GAA_s L^2 + 3GI_z A_s + 3EI_z A}$$

We then substitute the relationship between shear modulus, Young's modulus, and Poisson's ratio (cf. Chapter 3), $G = \frac{1}{2}E/(1+v)$; the moment of inertia in terms of the area and the radius of gyration, $I_z = A\rho_z^2$; and the effective shear area in terms of the actual area, through the form factor, $A_s = kA$. The expression for Y_1 can then be written as

$$Y_1 = \frac{3P}{8} \frac{1}{1 + \frac{3}{4}(\frac{\rho_z}{L})^2[1 + \frac{2(1+v)}{k}]} \qquad \text{[n]}$$

Observe that as ρ_z/L becomes smaller, the denominator approaches 1, that is,

$$\lim_{(\frac{\rho_z}{L}) \to 0} \frac{1}{1 + \frac{3}{4}(\frac{\rho_z}{L})^2\left[1 + \frac{2(1+v)}{k}\right]} = 1$$

in which case $Y_1 = 3P/8$, which is the value we would have obtained for the redundant load had we considered only bending and ignored shear and stretching altogether. Poisson's ratio v is typically around 0.3, the form factor k is on the order of 1, and $\rho_z \cong h/2$, where h is the depth of the beam. Thus, if $h/L \cong 0.1$, then

$$\frac{1}{1 + \frac{3}{4}\left(\frac{\rho_z}{L}\right)^2\left[1 + \frac{2(1+v)}{k}\right]} \cong 0.99$$

Taking $h/L = 0.1$ as the upper limit for the slender beam assumption, it appears that for such structures, a negligible loss in accuracy is incurred by ignoring shear and stretch in favor of bending alone.

Example 7.8.6   A pin-supported semicircular frame supports a horizontal load at point 2 on the top, as shown in Figure 7.8.13. Assuming that the frame can be treated as a curved, slender beam of constant cross section, use the principle of complementary virtual work to calculate the reactions at the supports and the magnitude and locations of the maximum bending moment. Neglect shear and axial deformation.

Let us treat the horizontal reaction at node 1 as the redundant. We easily find that the vertical reaction at 1 is $P/2$. Figure 7.8.14 shows free-body diagrams of the base structure under the true loads. From them, we use statics to deduce the section bending moment M as a function of ϕ measured positive counterclockwise from the horizontal x axis, as shown.

$$\text{For } 0 \le \phi \le \frac{\pi}{2}: \qquad M = X_1 R \sin\phi + \frac{PR}{2}(1 - \cos\phi) \qquad \text{[a]}$$

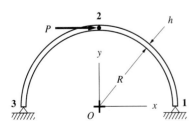

Figure 7.8.13   Semicircular frame joined to the floor by pin supports.

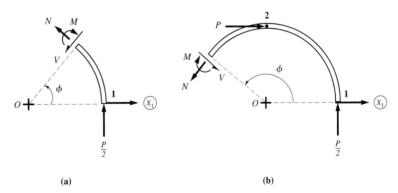

Figure 7.8.14   Free-body diagrams of the statically determinate base frame, showing the horizontal reaction at 1 as the redundant.

For $\dfrac{\pi}{2} \le \phi \le \pi$: $M = (X_1 + P)\, R \sin \phi - \dfrac{PR}{2}(1 + \cos \phi)$ **[b]**

Removing the actual load P from the base frame, we apply a virtual load in the direction of the redundant, as shown in Figure 7.8.15. Using that free-body diagram, we obtain a virtual bending moment expression that applies all around the frame, namely,

$$\delta M = \delta X_1 R \sin \phi \qquad 0 \le \phi \le \pi$$ **[c]**

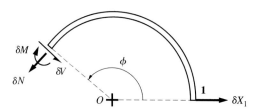

Figure 7.8.15   Free-body diagram of the base frame with true loads removed and a virtual load applied in the direction of the redundant.

The internal complementary virtual work expression, accounting for bending alone, is

$$\delta W^*_{\text{int}} = \int_0^\pi \frac{M \delta M}{E I_z}(R d\phi)$$ **[d]**

Note that $R d\phi$ replaces ds as the element of length along the curved beam axis.

Substituting the true and virtual bending moments from Equations a, b, and c into Equation d, we have

$$\delta W^*_{\text{int}} = \int_0^{\pi/2} \frac{M \delta M}{E I_z}(R d\phi) + \int_{\pi/2}^\pi \frac{M \delta M}{E I_z}(R d\phi)$$

$$= \frac{R}{E I_z} \int_0^{\pi/2} \left[X_1 R \sin \phi + \frac{PR}{2}(1 - \cos \phi) \right] [\delta X_1 R \sin \phi]\, d\phi + \frac{R}{E I_z} \int_{\pi/2}^\pi \left[(X_1 + P) R \sin \phi - \frac{PR}{2}(1 + \cos \phi) \right] (\delta X_1 R \sin \phi)\, d\phi$$

Expanding the integrands and then integrating,[2] we get

[2] In carrying out the integrals, we use the following formulas:

$$\int_0^{\pi/2} \cos^2 \theta\, d\theta = \int_0^{\pi/2} \sin^2 \theta\, d\theta = \int_{\pi/2}^\pi \cos^2 \theta\, d\theta = \int_{\pi/2}^\pi \sin^2 \theta\, d\theta = \frac{\pi}{4}$$

$$\int_0^{\pi/2} \sin \theta \cos \theta\, d\theta = -\int_{\pi/2}^\pi \sin \theta \cos \theta\, d\theta = \frac{1}{2}$$

$$\int_0^{\pi/2} \sin \theta\, d\theta = \int_{\pi/2}^\pi \sin \theta\, d\theta = 1$$

$$\delta W_{\text{int}}^* = \frac{R^3}{4EI_z}\{[(\pi+1)P + 2X_1\pi]\}\,\delta X_1 \tag{e}$$

The external complementary virtual work is zero, since δX_1 acts at a fixed support. Therefore, $\delta W_{\text{int}}^* = \delta W_{\text{ext}}^* = 0$, so that Equation e becomes

$$X_1 = -\frac{P}{2}(1 + \frac{1}{\pi}) = -0.659P \tag{f}$$

Substituting Equation f into the expressions for the bending moment in Equations a and b, we can compute the bending moment around the frame. The results are most easily seen in the polar plot shown in Figure 7.8.16. The value of the bending moment at each point is plotted normal to the frame, that is, along a radial and on the side of the beam that is in compression. The maximum bending moment is 1.31PR, at the point 52.8 degrees away from the right support, and it produces compression at the inner surface. A lesser extreme value, 0.42PR, 34.2 degrees from the left support, yields tension at the inner surface. The least stressed point of the frame occurs where the bending moment is zero, 69 degrees from the left support.

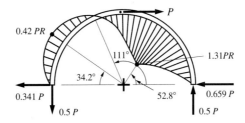

Figure 7.8.16   Computed reactions and bending moment distribution for the frame in Figure 7.8.13.

The bending moment is plotted normal to the frame, on the side in compression.

In three-dimensional frames, the elements of the structure are likely to be subjected to torsion as well as bending. Although we can ignore shear and extension, we cannot neglect torsion, as we will demonstrate in the next example.

Example 7.8.7   Use the principle of complementary virtual work to calculate the maximum bending moment in the frame (*grillage*) of Figure 7.8.17. Assume the member cross sections are solid circles of radius r and that the material properties are uniform throughout, with Poisson's ratio being 0.25. Neglect shear and stretching.

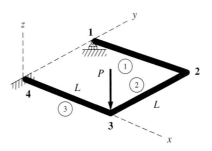

Figure 7.8.17   Singly-redundant grillage.

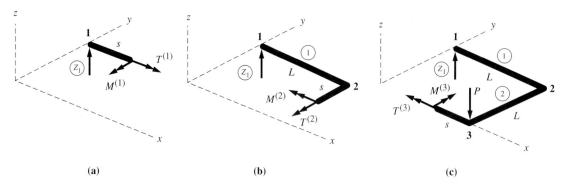

Figure 7.8.18   Free-body diagrams of the true loads on the statically determinate base structure.

Since the effects of shear and extension are neglected, only the bending and twisting couples on each section are shown, for simplicity.

The frame is built-in at point 4, which is the only support required to restrain the structure fully. The vertical reaction of the roller support at point 1 will be taken as the redundant and treated as an externally applied load, as shown in Figure 7.8.18. The figure shows free-body diagrams revealing the bending and twisting couples in each of the three members of the frame. The diagrams are incomplete because they do not show the shear and normal forces on the sections. These forces are absent because we are to neglect their effect, so they are omitted for simplicity. Equilibrium requires that

$$M^{(1)} = Z_1 s$$
$$T^{(1)} = 0$$
Element 1   [a]

$$M^{(2)} = Z_1 s$$
$$T^{(2)} = Z_1 L$$
Element 2   [b]

$$M^{(3)} = (Z_1 - P) s - Z_1 L$$
$$T^{(3)} = Z_1 L$$
Element 3   [c]

Using Figure 7.8.19, we infer the equilibrium equations for the virtually-loaded base structure:

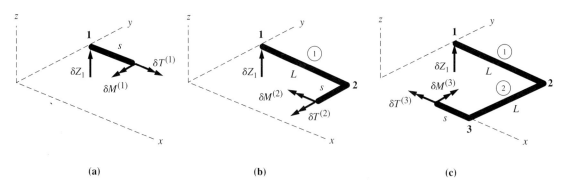

Figure 7.8.19   Virtual load of the base structure.

$$\delta M^{(1)} = \delta Z_1 s$$
$$\delta T^{(1)} = 0$$

Element 1   [d]

$$\delta M^{(2)} = \delta Z_1 s$$
$$\delta T^{(2)} = \delta Z_1 L$$

Element 2   [e]

$$\delta M^{(3)} = \delta Z_1 (s - L)$$
$$\delta T^{(3)} = \delta Z_1 L$$

Element 3   [f]

The total internal complementary virtual work equals the work due to bending plus the work due to torsion, or

$$\delta W^*_{\text{int}} = \delta W^*_{\text{int,bending}} + \delta W^*_{\text{int,torsion}} \qquad \text{[g]}$$

For the bending component, we have

$$\delta W^*_{\text{int,bending}} = \sum_{e=1}^{3} \int_{0}^{L^{(e)}} \left(\frac{M \delta M}{EI} \right)^{(e)} ds$$

where $I = I_z = I_y$ for the circular cross section. After substituting the true bending moments from Equations a through c and the virtual bending moments from Equations d through f, we get

$$\delta W^*_{\text{int,bending}} = \frac{1}{EI} \int_{0}^{L} (Z_1 s)(\delta Z_1 s)\, ds + \frac{1}{EI} \int_{0}^{L} (Z_1 s)(\delta Z_1 s)\, ds$$
$$+ \frac{1}{EI} \int_{0}^{L} [(Z_1 - P)s - Z_1 L][\delta Z_1 (s - L)]\, ds$$

Carrying out the integrals and simplifying terms leads to the following result:

$$\delta W^*_{\text{int,bending}} = \frac{L^3}{EI} \left(\frac{Z_1}{3} + \frac{P}{6} \right) \delta Z_1 \qquad \text{[h]}$$

According to the internal complementary virtual work expression presented in Equation 7.6.5, the total contribution from torsion is

$$\delta W^*_{\text{int,torsion}} = \sum_{e=1}^{3} \int_{0}^{L^{(e)}} \left(\frac{T \delta T}{GJ} \right)^{(e)} ds$$

Using the true and virtual torques listed in Equations a through f, we have

$$\delta W^*_{\text{int,torsion}} = \frac{1}{GJ} \int_{0}^{L} (0)(0)\, ds + \frac{1}{GJ} \int_{0}^{L} (Z_1 L)(\delta Z_1 L)\, ds + \frac{1}{GJ} \int_{0}^{L} (Z_1 L)(\delta Z_1 L)\, ds$$

or

$$\delta W^*_{\text{int,torsion}} = \frac{2Z_1 L^3}{GJ} \delta Z_1 \qquad \text{[i]}$$

Substituting this term into Equation g, along with that due to bending (Equation h), yields

$$\delta W^*_{\text{int}} = \frac{2Z_1 L^3}{GJ} \delta Z_1 + \frac{L^3}{EI} \left(\frac{Z_1}{3} + \frac{P}{6} \right) \delta Z_1 \qquad \text{[j]}$$

Since the displacement in the direction of the redundant load δZ_1 is zero, the external complementary virtual work is zero, so that from the principle of complementary virtual work and Equation j (with $\delta W^*_{int} = 0$), we can solve for Z_1, as follows:

$$Z_1 = -\frac{P}{2}\frac{GJ}{GJ + 6EI} = -\frac{P}{2}\frac{\left[\frac{E}{2(1+\nu)}\right]J}{\left[\frac{E}{2(1+\nu)}\right]J + 6EI} = -\frac{P}{2}\frac{1}{1 + 12(1+\nu)\frac{I}{J}} \qquad [k]$$

For a solid circle, $I = \pi r^4/4$ and $J = \pi r^4/2$. Substituting these and a typical Poisson's ratio of $\nu = 0.25$ into Equation k, we find that

$$Z_1 = -\frac{P}{17} \qquad [l]$$

The reaction at 1 is therefore downward, in the same direction as the applied load P. Using this value of Z_1 in the bending moment expressions Equations a, b, and c, we can sketch the bending moment along each of the three frame elements. This is done in Figure 7.8.20, where the curves are plotted on the side of the element in compression. Obviously, the bending moments throughout the structure are very small compared to the moment of the couple that develops at the fixed support 4.

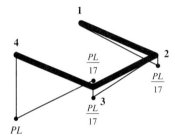

Figure 7.8.20 Bending moment distribution in the grillage of Figure 7.8.17.

The bending moment diagram is plotted on the side of the element in compression.

If a structure has geometric symmetry and is supported and loaded in a symmetric fashion, then we should take advantage of those symmetries to reduce the effort required to obtain the internal loads. This is illustrated in the next example.

Example 7.8.8   Use the principle of complementary virtual work to find the magnitude and location of the maximum bending moment in the circular frame in Figure 7.8.21. The material and section properties are uniform throughout. Neglect shear and stretching.

To calculate the internal loads, we must cut the ring at some point around its circumference. In so doing, we reveal the internal bending moment M and the internal shear and normal forces, V and N, at that section, as shown in Figure 7.8.21b. These are the three unknown redundant loads for this problem. Once we find M, V, and N, statics can be used to obtain the bending moment, in terms of those quantities, at any other point in the ring, and the externally applied load P.

Suppose we divide the circular frame into two semicircles by cutting across the vertical diameter through points 1 and 2 at the top and bottom of the ring, as in Figure 7.8.22. By symmetry, the free-body diagram in Figure 7.8.22a should be the mirror image of that in part (b) of the figure. This means that if V_1 is assumed to act upwards on the left free-body, it should be shown acting upwards on the right one as well. But we have drawn it downwards in part (b) to abide by the action–reaction principle.

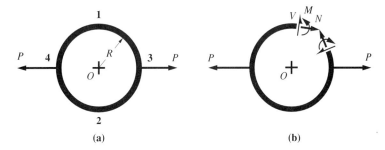

Figure 7.8.21   Circular frame loaded in tension along a diameter.

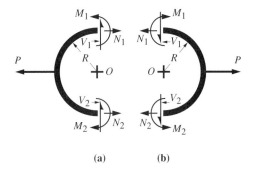

Figure 7.8.22   Free-body diagrams of the frame divided in two along the vertical diameter.

Since we cannot have it both ways, we must conclude that $V_1 = 0$. The same argument holds to conclude that $V_2 = 0$. After the shear forces from parts (a) and (b) of Figure 7.8.22 are deleted, the two halves become mirror images.

Each of the semicircular free bodies is also symmetric about a horizontal diameter. If we draw a free-body diagram of the quarter circle defined by diameters through points 1 and 3, we obtain Figure 7.8.23a. By the mirror image argument, we can deduce that the internal shear force V_3 at point 3 is zero. Furthermore, by symmetry, we can apportion one-half of the external

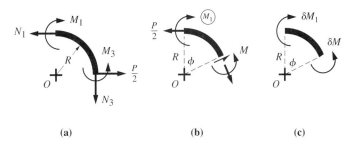

Figure 7.8.23   (a) Free-body diagram of a quarter circle of the ring. (b) Free-body diagram of the sectioned quarter circle showing M_1 as an applied load. (c) Free-body diagram of the sectioned quarter circle with the true loading removed and replaced by the virtual couple δM_1.

load P to the quarter circle above and below 3, as shown. We are then left with four unknowns in Figure 7.8.23a: M_1, N_1, M_3, and N_3. Using statics, we can easily conclude that $N_1 = P/2$, $N_3 = 0$, and $M_3 = M_1 - PR$. Only M_1 remains as the redundant load.
From the free-body diagram of Figure 7.8.23b, we find that

$$M = M_1 - \frac{P}{2}R(1 - \cos\phi) \qquad\qquad [a]$$

and from Figure 7.8.23c, we have

$$\delta M = \delta M_1 \qquad\qquad [b]$$

Applying the principle of complementary virtual work to the complete circular frame, using symmetry and recognizing that the complementary virtual work of the internal virtual load δM_1 is zero, we have, for bending alone,

$$\delta W_{int}^* = 4\left[\frac{1}{EI_z}\int_0^{\pi/2}(M\delta M)\,Rd\phi\right] = \frac{4R}{EI_z}\int_0^{\pi/2}\left[M_1 - \frac{PR}{2}(1 - \cos\phi)\right]\delta M\,d\phi = 0$$

which reduces to

$$\frac{4R}{EI_z}\left[\frac{\pi}{2}M_1 - \frac{\pi}{4}PR + \frac{1}{2}PR\right] = 0$$

so that

$$M_1 = \frac{\pi - 2}{2\pi}PR = 0.1817PR \qquad\qquad [c]$$

Substituting M_1 into Equation a yields the bending moment in terms of ϕ and the applied load P. This is plotted in Figure 7.8.24, which shows that the maximum bending moment occurs at point 3 (and, by symmetry, point 4).

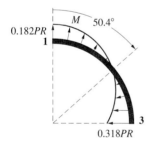

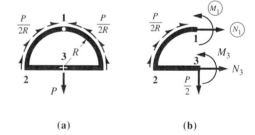

Figure 7.8.24 Symmetrical bending moment distribution over a quadrant of the circular frame of Figure 7.8.21.

The bending moment is plotted on the side of the frame in compression.

Figure 7.8.25 (a) Statically indeterminate frame. (b) Free-body diagram showing internal loads at 1 and 3.

Example 7.8.9 A frame composed of a semicircular beam and a horizontal floor beam has a point load P applied to the midspan of the floor beam, as shown in Figure 7.8.25a. This load is equilibrated by a uniform shear flow $\frac{1}{2}P/R$ acting around the periphery of the circular beam. Using the principle of complementary virtual work, find the maximum bending moment in the frame, as well as the point where the moment is zero. Consider bending only, and assume that all material and sectional properties are uniform.

By symmetry, the shear force on a section through points 1 and 3 is zero. Hence, the free-body diagram in Figure 7.8.25b shows only the normal force and the bending moment on the cuts. Using statics, we can determine M_3 and N_3 in terms of M_1

and N_1, but that is all. Hence, we will take M_1 and N_1 as the redundant loads for this analysis, so that the base frame has an imaginary cut at 1.

To find the true and virtual bending moment distribution, we start with the free-body diagrams in Figure 7.8.26. From part (a) of the figure, using Equation 2.5.11 to compute the moment of the shear flow, summing the moments about the cut at an angle ϕ from 1, and equating this to zero yields

$$-M + M_1 - N_1 R \left(1 - \cos\phi\right) - 2A_1 \left(\frac{P}{2R}\right) = 0 \qquad [a]$$

where A_1, the area of the shaded segment of the circle, is the area $R^2\phi/2$ of the pie-shaped sector subtended by the angle ϕ minus the area $R^2 \sin\phi/2$ of the isosceles triangle with legs R and vertex angle ϕ. Thus,

$$A_1 = \frac{R^2}{2} \left(\phi - \sin\phi\right) \qquad [b]$$

Substituting this into Equation a and solving for the true bending moment M, we get

$$M = M_1 - N_1 R \left(1 - \cos\phi\right) - \frac{PR}{2} \left(\phi - \sin\phi\right) \qquad [c]$$

Similarly, an analysis of Figure 7.8.26b yields the following for the virtual bending moment:

$$\delta M = \delta M_1 - \delta N_1 R \left(1 - \cos\phi\right) \qquad [d]$$

Moving into the horizontal beam, we have the free-body diagrams shown in Figure 7.8.27. From part (a) of the figure, summing the moments about the bottom cut, which is a distance s from point 2, we obtain

$$M + M_1 - N_1 R - 2 \left(A_1 + A_2\right) \left(\frac{P}{2R}\right) = 0 \qquad [e]$$

where A_1 is found from Equation b by setting $\phi = \pi/2$; that is,

$$A_1 = \frac{R^2}{2} \left(\frac{\pi}{2} - 1\right)$$

It can be seen from the figure that A_2 is the area of the triangle with base s and height R, or $A_2 = Rs/2$. We substitute A_1 and A_2 into Equation e and solve for the true bending moment in this portion of the frame, as follows:

$$M = -M_1 + N_1 R + \frac{Ps}{2} + \frac{PR}{2} \left(\frac{\pi}{2} - 1\right) \qquad [f]$$

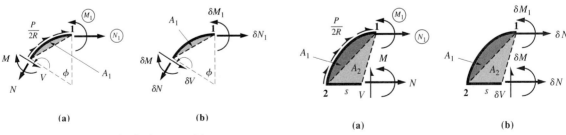

(a)   (b)   (a)   (b)

Figure 7.8.26   (a) Free-body diagram of the semicircular portion of the statically determinate base frame acted on by the true loads. (b) The same free-body diagram acted on by the virtual loads applied at 1.

Figure 7.8.27   (a) Free-body diagram of the frame for a cut through the floor beam, showing the actual applied loads. (b) The same free-body diagram with only the virtual loads applied.

A similar analysis of the free-body diagram in Figure 7.8.27b yields the virtual bending moment,

$$\delta M = -\delta M_1 + \delta N_1 R \qquad \text{[g]}$$

We are now in a position to determine the internal complementary virtual work for the structure in Figure 7.8.25a. Considering only bending and using symmetry, we have

$$\delta W_{int}^* = 2\left(\frac{1}{EI_z}\int_1^2 M\delta M ds + \frac{1}{EI_z}\int_2^3 M\delta M ds\right)$$

where $ds = Rd\phi$ in the first integral. Substituting Equations c, d, f, and g into the integrands yields

$$\delta W_{int}^* = \frac{2}{EI_z}\int_0^{\pi/2}\left[M_1 - N_1 R(1-\cos\phi) - \frac{PR}{2}(\phi-\sin\phi)\right][\delta M_1 - \delta N_1(1-\cos\phi)]Rd\phi$$

$$\qquad \text{[h]}$$

$$+\frac{2}{EI_z}\int_0^R\left[-M_1 + N_1 R + \frac{Ps}{2} + \frac{PR}{2}\left(\frac{\pi}{2}-1\right)\right][-\delta M_1 + \delta N_1 R]ds$$

Expanding the integrands and collecting coefficients of the virtual loads δM_1 and δN_1, we get the following lengthy expression:

$$\delta W_{int}^* = \frac{2\delta M_1}{EI_z}\left\{\int_0^{\pi/2}\left[M_1 - N_1 R(1-\cos\phi) - \frac{PR}{2}(\phi-\sin\phi)\right]Rd\phi + \int_0^R\left[M_1 - N_1 R - \frac{Ps}{2} - \frac{PR}{2}\left(\frac{\pi}{2}-1\right)\right]ds\right\}$$

$$+\frac{2\delta N_1}{EI_z}\left\{\int_0^{\pi/2}\left[M_1 R(\cos\phi-1) + N_1 R^2\left(1-2\cos\phi+\cos^2\phi\right) + \frac{PR^2}{2}(\phi-\sin\phi-\phi\cos\phi+\sin\phi\cos\phi)\right]Rd\phi\right.$$

$$\left.+\int_0^R\left[-RM_1 + N_1 R^2 + \frac{RPs}{2} + \frac{PR^2}{2}\left(\frac{\pi}{2}-1\right)\right]ds\right\}$$

Carrying out the simple integrations leads ultimately to

$$\delta W_{int}^* = \frac{2\delta M_1}{EI_z}\left[\left(\frac{\pi}{2}+1\right)RM_1 - \frac{\pi}{2}R^2 N_1 + \left(\frac{3}{4}-\frac{\pi^2}{16}-\frac{\pi}{4}\right)PR^2\right]$$

$$\qquad \text{[i]}$$

$$+\frac{2\delta N_1}{EI_z}\left[-\frac{\pi}{2}R^2 M_1 + \left(\frac{3\pi}{4}-1\right)R^3 N_1 + \frac{\pi^2}{16}PR^3\right]$$

According to the principle of complementary virtual work, $\delta W_{int}^* = \delta W_{ext}^*$ for all choices of virtual loads δM_1 and δN_1. In this case, however, $\delta W_{ext}^* = 0$, since the virtual loads are internal and not externally applied. Thus we require δW_{int}^* in Equation i to be zero for all δM_1 and δN_1, which means their coefficients must vanish, resulting in the following two equations:

$$\left(\frac{\pi}{2}+1\right)RM_1 \quad -\frac{\pi}{2}R^2 N_1 = \left(\frac{3}{4}-\frac{\pi^2}{16}-\frac{\pi}{4}\right)PR^2$$

$$\qquad \text{[j]}$$

$$-\frac{\pi}{2}R^2 M_1 + \left(\frac{3\pi}{4}-1\right)R^3 N_1 = -\frac{\pi^2}{16}PR^3$$

Solving for M_1 and N_1, we find that

$$M_1 = -0.08279PR \qquad N_1 = -0.5507P \qquad \text{[k]}$$

Substituting these values into Equations c and f yields the bending moment equations, which can be plotted on the frame, as shown in Figure 7.8.28. As usual, the bending moment is plotted on the side of the frame in compression. As we can see, the maximum bending moment occurs under the point load P at the middle of the floor beam.

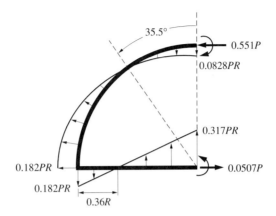

Figure 7.8.28   Symmetric bending moment distribution around the frame of Figure 7.8.25.

Example 7.8.10   Using the principle of complementary virtual work, find the bending moment distribution in the portal frame of Figure 7.8.29a. Consider bending only, and assume the flexural rigidity EI_z is uniform throughout.

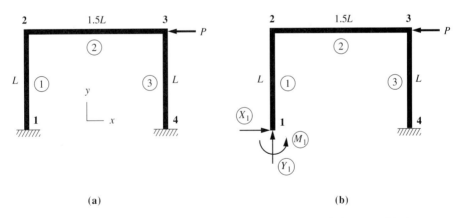

(a)   (b)

Figure 7.8.29   (a) Portal frame built in at each support. (b) Free-body diagram of the base structure, showing the three support reactions chosen as the redundants.

This frame is triply redundant. The support at 1 is not required for stability of the structure. Hence, it can theoretically be removed, thereby creating a statically determinate base structure for which the three reactions X_1, Y_1, and M_1 are viewed as externally applied loads, as indicated in the free-body diagram in Figure 7.8.29b.

In Figure 7.8.30, starting at point 1, we work our way around the frame clockwise, sectioning through each of the three beam elements in turn, to expose the true internal shear V, true normal force N, and true bending moment M. Since we can neglect the effects of shear and normal force, we need only compute the bending moments. Summing to zero the moment

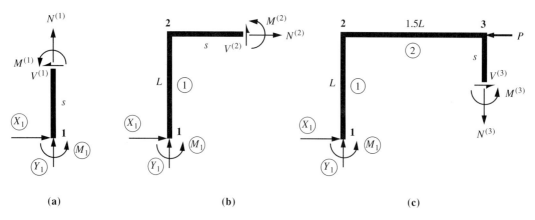

Figure 7.8.30   Free-body diagrams of portions of the base frame, revealing the true section loads in each beam element.

about each of the three interior cuts, we then use statics to obtain the following three expressions for the true bending moment as a function of distance from the beginning of each beam element:

$$\text{Element 1}: \quad M^{(1)} = -X_1 s - M_1 \tag{a}$$

$$\text{Element 2}: \quad M^{(2)} = -X_1 L + Y_1 s - M_1 \tag{b}$$

$$\text{Element 3}: \quad M^{(3)} = -X_1(L - s) + 1.5 L Y_1 - M_1 - P s \tag{c}$$

Likewise, from the free-body diagrams of the virtual load, Figure 7.8.31, we find

$$\text{Element 1}: \quad \delta M^{(1)} = -\delta X_1 s - \delta M_1 \tag{d}$$

$$\text{Element 2}: \quad \delta M^{(2)} = -\delta X_1 L + \delta Y_1 s - \delta M_1 \tag{e}$$

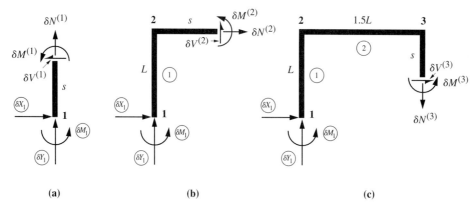

Figure 7.8.31   Free-body diagrams of portions of the base frame, revealing the virtual section loads in each beam element.

$$\text{Element 3}: \quad \delta M^{(3)} = -\delta X_1(L-s) + 1.5L\delta Y_1 - \delta M_1 \qquad \text{[f]}$$

The internal complementary virtual work for each element can be written in turn using Equations a through f, as follows:

$$
\delta W_{\text{int}}^{*(1)} = \frac{1}{EI_z} \int_0^{L^{(1)}} M^{(1)} \delta M^{(1)} ds = \frac{1}{EI_z} \int_0^{L} (-X_1 s - M_1)(-\delta X_1 s - \delta M_1)\, ds
$$

$$
= \frac{1}{EI_z}\left[\left(\tfrac{1}{3}L^3 X_1 + \tfrac{1}{2}L^2 M_1\right)\delta X_1 + \left(\tfrac{1}{2}L^2 X_1 + L M_1\right)\delta M_1 \right] \qquad \text{[g]}
$$

$$
\delta W_{\text{int}}^{*(2)} = \frac{1}{EI_z}\int_0^{L^{(2)}} M^{(2)} \delta M^{(2)} ds = \frac{1}{EI_z}\int_0^{1.5L}(-X_1 L + Y_1 s - M_1)(-\delta X_1 L + \delta Y_1 s - \delta M_1)\, ds
$$

$$
= \frac{1}{EI_z}\left[\left(\tfrac{3}{2}L^3 X_1 - \tfrac{9}{8}L^3 Y_1 + \tfrac{3}{2}L^2 M_1\right)\delta X_1 + \left(-\tfrac{9}{8}L^3 X_1 + \tfrac{9}{8}L^3 Y_1 - \tfrac{9}{8}L^2 M_1\right)\delta Y_1 \right.
$$
$$
\left. + \left(\tfrac{3}{2}L^2 X_1 - \tfrac{9}{8}L^2 Y_1 + \tfrac{3}{2}L M_1\right)\delta M_1 \right] \qquad \text{[h]}
$$

$$
\delta W_{\text{int}}^{*(3)} = \frac{1}{EI_z}\int_0^{L^{(3)}} M^{(3)} \delta M^{(3)} ds = \frac{1}{EI_z}\int_0^{L}[-X_1(L-s) + 1.5LY_1 - M_1 - Ps][-\delta X_1(L-s) + 1.5L\delta Y_1 - \delta M_1]\, ds
$$

$$
= \frac{1}{EI_z}\left[\left(\tfrac{1}{3}L^3 X_1 - \tfrac{3}{4}L^3 Y_1 + \tfrac{1}{2}L^2 M_1 + \tfrac{1}{6}L^3 P\right)\delta X_1 + \left(-\tfrac{3}{4}L^3 X_1 + \tfrac{9}{4}L^3 Y_1 - \tfrac{3}{2}L^2 M_1 - \tfrac{3}{4}L^3 P\right)\delta Y_1 \right.
$$
$$
\left. + \left(\tfrac{1}{2}L^2 X_1 - \tfrac{3}{2}L^2 Y_1 + L M_1 + \tfrac{1}{2}L^2 P\right)\delta M_1 \right] \qquad \text{[i]}
$$

The total internal complementary virtual work of the frame is the sum of the values for the three beam elements, or

$$
\delta W_{\text{int}}^* = \delta W_{\text{int}}^{*\,(1)} + \delta W_{\text{int}}^{*\,(2)} + \delta W_{\text{int}}^{*\,(3)}
$$

$$
= \frac{1}{EI_z}\left[\left(\tfrac{13}{6}L^3 X_1 - \tfrac{15}{8}L^3 Y_1 + \tfrac{5}{2}L^2 M_1 + \tfrac{1}{6}L^3 P\right)\delta X_1 + \left(-\tfrac{15}{8}L^3 X_1 + \tfrac{27}{8}L^3 Y_1 - \tfrac{21}{8}L^2 M_1 - \tfrac{3}{4}L^3 P\right)\delta Y_1 \right. \qquad \text{[j]}
$$
$$
\left. + \left(\tfrac{5}{2}L^2 X_1 - \tfrac{21}{8}L^2 Y_1 + \tfrac{7}{2}L M_1 + \tfrac{1}{2}L^2 P\right)\delta M_1 \right]
$$

The external complementary virtual work is zero, since the true displacement in the direction of each of the three redundants is zero. It follows then that $\delta W_{\text{int}}^* = 0$ for an arbitrary choice of δX_1, δY_1, and δM_1. Therefore, their coefficients in Equation j must each be zero, yielding the following three equations,

$$
\tfrac{13}{6}L^3 X_1 - \tfrac{15}{8}L^3 Y_1 + \tfrac{5}{2}L^2 M_1 = -\tfrac{1}{6}PL
$$
$$
-\tfrac{15}{8}L^3 X_1 + \tfrac{27}{8}L^3 Y_1 - \tfrac{21}{8}L^2 M_1 = \tfrac{3}{4}PL^3 \qquad \text{[k]}
$$
$$
\tfrac{5}{2}L^2 X_1 - \tfrac{21}{8}L^2 Y_1 + \tfrac{7}{2}L M_1 = -\tfrac{1}{2}PL^2
$$

The solution to this system is

$$
X_1 = \tfrac{1}{2}P \qquad Y_1 = \tfrac{4}{15}P \qquad M_1 = -\tfrac{3}{10}PL \qquad \text{[l]}
$$

We can now substitute these values for the redundant loads into Equations a, b, and c to obtain the bending moments throughout the frame in terms of the applied load P. These are plotted in Figure 7.8.32.

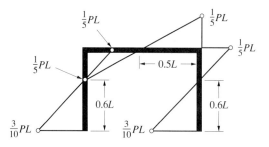

Figure 7.8.32 Bending moment distribution around the frame of Figure 7.8.29.

The moment is plotted on the side of the beam in compression.

Circular frames are a common component of aircraft structures. Normally, many stringers are distributed around the periphery of such frames. The shear flow applied to a frame between each pair of stringers varies with position, as illustrated in Example 5.2.3. To determine the contribution of a given portion of the shear flow to the bending moment at another point of the frame, a formula is useful. First, let A be the shaded area abc subtended by the uniform shear flow q in Figure 7.8.33. Then, the moment of that shear flow about point c is

$$M_c = 2Aq$$

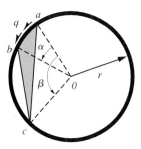

Figure 7.8.33 Shaded area subtended by the constant shear flow q.

However, A is the area of segment abc of the circle minus the area of segment bc (see Equation b of Example 7.8.9). Using that equation to calculate the areas of the two segments in question and then subtracting the larger from the smaller, we get

$$A = \frac{r^2}{2}(\beta - \sin\beta) - \frac{r^2}{2}[(\beta - \alpha) - \sin(\beta - \alpha)] = \frac{r^2}{2}[\alpha - \sin\beta + \sin(\beta - \alpha)]$$

The moment of the shear flow q about point c is therefore

$$M_c = qr^2[\alpha - \sin\beta + \sin(\beta - \alpha)] = qr^2[\alpha - (1 - \cos\alpha)\sin\beta - \sin\alpha\cos\beta] \qquad [7.8.1]$$

Example 7.8.11   Use the principle of complementary virtual work to find the bending moment distribution in the circular fuselage frame of Figure 7.8.34. Assume that the flexural rigidity EI_z is uniform throughout both the ring and the floor beam, and neglect shear and stretching. The area of all the stringers is 0.4 in.²

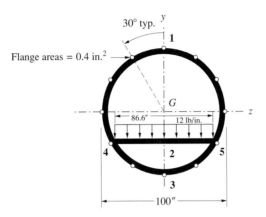

Figure 7.8.34   Circular frame and floor beam combination.

The floor beam carries a uniformly distributed line load.

The load on the floor beam is equilibrated by shear flows applied by the skin to the periphery of the frame. The method of determining these shear flows is described in Chapter 5. Following a procedure similar to that of Example 5.2.3, we obtain the shear flows shown in Figure 7.8.35.

To reveal the bending moments within the frame, we must section the structure appropriately. The vertical diameter through points 1 and 3 is a symmetry axis, so we will cut through the frame along that line, dividing it into two free bodies that are mirror images. The left one is shown in Figure 7.8.36. There is no shear force on the sections at 1, 2, and 3, since they lie on the axis of symmetry. Two of the three pairs of normal force and bending moments in Figure 7.8.36 are redundant. Our choice will be those at 1 and 2, which are circled for emphasis.

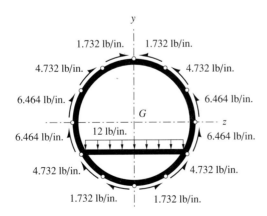

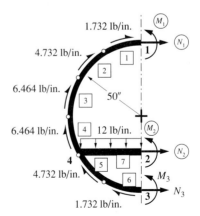

Figure 7.8.35   Shear flows acting around the periphery of the circular fuselage frame in response to the load on the floor beam.

Figure 7.8.36   Free-body diagram of one-half of the symmetric frame, with the chosen redundant loads circled.

It is convenient to use a number to identify the portions of the circular frame lying between two stringers. Therefore, including the half-span of floor beam, we can consider the symmetric half of the structure in Figure 7.8.36 to be composed of seven elements. Starting with point 1, at the top of the structure, we work our way around the frame counterclockwise, sectioning each of the six curved elements in turn and drawing the free-body diagrams shown in Figure 7.8.37. Observe that we have identified the shear flows symbolically ($q^{(1)}$, $q^{(2)}$, etc.) and denoted the 12 lb/in. load on the floor beam as p.

From Figure 7.8.37a, using Equation 7.8.1, we find that

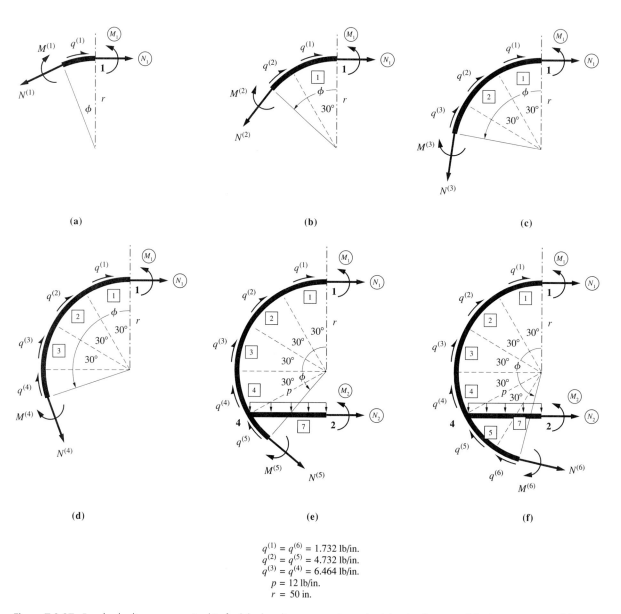

$$q^{(1)} = q^{(6)} = 1.732 \text{ lb/in.}$$
$$q^{(2)} = q^{(5)} = 4.732 \text{ lb/in.}$$
$$q^{(3)} = q^{(4)} = 6.464 \text{ lb/in.}$$
$$p = 12 \text{ lb/in.}$$
$$r = 50 \text{ in.}$$

Figure 7.8.37 Free-body diagrams required to find the bending moment in each of the six elements of the frame in terms of the redundants.

$$M^{(1)} = M_1 - N_1 r(1 - \cos\phi) - q^{(1)}r^2(\phi - \sin\phi) \tag{a1}$$

The virtual bending moment is found by removing the shear flow and replacing M_1 and N_1 by δM_1 and δN_1, respectively, so that

$$\delta M^{(1)} = \delta M_1 - \delta N_1 r(1 - \cos\phi) \tag{a2}$$

The true and virtual bending moments in elements 2 through 4 are found in a similar fashion by using the free-body diagrams in parts (b) through (d) of Figure 7.8.37. The number of terms in the equations increases as we move from element to element, picking up more and more of the applied loads.

$$M^{(2)} = M_1 - N_1 r\left(1 - \cos\phi\right) - q^{(1)}r^2\left[\tfrac{\pi}{6} + \left(\cos\tfrac{\pi}{6} - 1\right)\sin\phi - \sin\tfrac{\pi}{6}\cos\phi\right] - q^{(2)}r^2\left[\phi - \tfrac{\pi}{6} - \sin\left(\phi - \tfrac{\pi}{6}\right)\right] \tag{b1}$$

$$\delta M^{(2)} = \delta M_1 - \delta N_1 r(1 - \cos\phi) \tag{b2}$$

$$M^{(3)} = M_1 - N_1 r\left(1 - \cos\phi\right) - q^{(1)}r^2\left[\tfrac{\pi}{6} + \left(\cos\tfrac{\pi}{6} - 1\right)\sin\phi - \sin\tfrac{\pi}{6}\cos\phi\right]$$
$$- q^{(2)}r^2\left[\tfrac{\pi}{6} + \left(\cos\tfrac{\pi}{6} - 1\right)\sin\left(\phi - \tfrac{\pi}{6}\right) - \sin\tfrac{\pi}{6}\cos\left(\phi - \tfrac{\pi}{6}\right)\right] - q^{(3)}r^2\left[\phi - \tfrac{\pi}{3} - \sin\left(\phi - \tfrac{\pi}{3}\right)\right] \tag{c1}$$

$$\delta M^{(3)} = \delta M_1 - \delta N_1 r(1 - \cos\phi) \tag{c2}$$

$$M^{(4)} = M_1 - N_1 r\left(1 - \cos\phi\right) - q^{(1)}r^2\left[\tfrac{\pi}{6} + \left(\cos\tfrac{\pi}{6} - 1\right)\sin\phi - \sin\tfrac{\pi}{6}\cos\phi\right]$$
$$- q^{(2)}r^2\left[\tfrac{\pi}{6} + \left(\cos\tfrac{\pi}{6} - 1\right)\sin\left(\phi - \tfrac{\pi}{6}\right) - \sin\tfrac{\pi}{6}\cos\left(\phi - \tfrac{\pi}{6}\right)\right]$$
$$- q^{(3)}r^2\left[\tfrac{\pi}{6} + \left(\cos\tfrac{\pi}{6} - 1\right)\sin\left(\phi - \tfrac{\pi}{3}\right) - \sin\tfrac{\pi}{6}\cos\left(\phi - \tfrac{\pi}{3}\right)\right] - q^{(4)}r^2\left[\phi - \tfrac{\pi}{2} - \sin\left(\phi - \tfrac{\pi}{2}\right)\right] \tag{d1}$$

$$\delta M^{(4)} = \delta M_1 - \delta N_1 r(1 - \cos\phi) \tag{d2}$$

The free-body diagrams in parts (e) and (f) of Figure 7.8.37 reveal the distributed load p on the floor beam and the redundant loads M_2 and N_2 acting on the cut at point 2. The corresponding free-body diagrams for the virtual load would show all of the shear flows removed and the normal load and bending moment at points 1 and 2 replaced, as usual, by their virtual counterparts. Therefore, for elements 5 and 6, moment equilibrium implies

$$M^{(5)} = M_1 - N_1 r\left(1 - \cos\phi\right) - q^{(1)}r^2\left[\tfrac{\pi}{6} + \left(\cos\tfrac{\pi}{6} - 1\right)\sin\phi - \sin\tfrac{\pi}{6}\cos\phi\right]$$
$$- q^{(2)}r^2\left[\tfrac{\pi}{6} + \left(\cos\tfrac{\pi}{6} - 1\right)\sin\left(\phi - \tfrac{\pi}{6}\right) - \sin\tfrac{\pi}{6}\cos\left(\phi - \tfrac{\pi}{6}\right)\right]$$
$$- q^{(3)}r^2\left[\tfrac{\pi}{6} + \left(\cos\tfrac{\pi}{6} - 1\right)\sin\left(\phi - \tfrac{\pi}{3}\right) - \sin\tfrac{\pi}{6}\cos\left(\phi - \tfrac{\pi}{3}\right)\right]$$
$$- q^{(4)}r^2\left[\tfrac{\pi}{6} + \left(\cos\tfrac{\pi}{6} - 1\right)\sin\left(\phi - \tfrac{\pi}{2}\right) - \sin\tfrac{\pi}{6}\cos\left(\phi - \tfrac{\pi}{2}\right)\right]$$
$$- q^{(5)}r^2\left[\phi - \tfrac{2\pi}{3} - \sin\left(\phi - \tfrac{2\pi}{3}\right)\right] + M_2 + N_2 r\left(\cos\tfrac{\pi}{3} + \cos\phi\right) + pr^2\sin\tfrac{\pi}{3}\left(\tfrac{1}{2}\sin\tfrac{\pi}{3} - \sin\phi\right) \tag{e1}$$

$$\delta M^{(5)} = \delta M_1 - \delta N_1 r(1 - \cos\phi) + \delta M_2 + \delta N_2 r(\cos\tfrac{\pi}{3} + \cos\phi) \tag{e2}$$

$$M^{(6)} = M_1 - N_1 r\left(1 - \cos\phi\right) - q^{(1)}r^2\left[\tfrac{\pi}{6} + \left(\cos\tfrac{\pi}{6} - 1\right)\sin\phi - \sin\tfrac{\pi}{6}\cos\phi\right]$$
$$- q^{(2)}r^2\left[\tfrac{\pi}{6} + \left(\cos\tfrac{\pi}{6} - 1\right)\sin\left(\phi - \tfrac{\pi}{6}\right) - \sin\tfrac{\pi}{6}\cos\left(\phi - \tfrac{\pi}{6}\right)\right]$$
$$- q^{(3)}r^2\left[\tfrac{\pi}{6} + \left(\cos\tfrac{\pi}{6} - 1\right)\sin\left(\phi - \tfrac{\pi}{3}\right) - \sin\tfrac{\pi}{6}\cos\left(\phi - \tfrac{\pi}{3}\right)\right]$$
$$- q^{(4)}r^2\left[\tfrac{\pi}{6} + \left(\cos\tfrac{\pi}{6} - 1\right)\sin\left(\phi - \tfrac{\pi}{2}\right) - \sin\tfrac{\pi}{6}\cos\left(\phi - \tfrac{\pi}{2}\right)\right]$$
$$- q^{(5)}r^2\left[\tfrac{\pi}{6} + \left(\cos\tfrac{\pi}{6} - 1\right)\sin\left(\phi - \tfrac{2\pi}{3}\right) - \sin\tfrac{\pi}{6}\cos\left(\phi - \tfrac{2\pi}{3}\right)\right]$$
$$- q^{(6)}r^2\left[\phi - \tfrac{5\pi}{6} - \sin\left(\phi - \tfrac{5\pi}{6}\right)\right] + M_2 + N_2 r\left(\cos\tfrac{\pi}{3} + \cos\phi\right) + pr^2\sin\tfrac{\pi}{3}\left(\tfrac{1}{2}\sin\tfrac{\pi}{3} - \sin\phi\right) \tag{f1}$$

$$\delta M^{(6)} = \delta M_1 - \delta N_1 r \left(1 - \cos\phi\right) + \delta M_2 + \delta N_2 r \left(\cos\tfrac{\pi}{3} + \cos\phi\right) \qquad \text{[f2]}$$

For the floor beam between points 2 and 4, the true and virtual bending moments are simply

$$M^{(7)} = M_2 - \frac{ps^2}{2} \qquad \text{[g1]}$$

$$\delta M^{(7)} = \delta M_2 \qquad \text{[g2]}$$

The internal complementary virtual work for the frame is the sum of the individual contributions from each of the seven elements (multiplied by 2 to account for both symmetric halves of the complete frame):

$$\delta W_{int}^* = \frac{2}{EI_z} \left(\int_0^{\pi/6} M^{(1)} \delta M^{(1)} r d\phi + \int_{\pi/6}^{\pi/3} M^{(2)} \delta M^{(2)} r d\phi + \int_{\pi/3}^{\pi/2} M^{(3)} \delta M^{(3)} r d\phi + \int_{\pi/2}^{2\pi/3} M^{(4)} \delta M^{(4)} r d\phi \right.$$

$$\left. + \int_{2\pi/3}^{5\pi/6} M^{(5)} \delta M^{(5)} r d\phi + \int_{5\pi/6}^{\pi} M^{(6)} \delta M^{(6)} r d\phi + \int_0^{r \sin\frac{\pi}{3}} M^{(7)} \delta M^{(7)} ds \right) \qquad \text{[h]}$$

Substituting Equations a through g into this equation, inserting the numerical values for the shear flows, distributed load, and frame radius, collecting terms, simplifying, and doing the integrals finally yields

$$\delta W_{int}^* = \frac{1}{EI_z} \left[\left(157.1M_1 - 7854N_1 + 52.36M_2 - 856.1N_2 - 1.420 \times 10^6\right) \delta M_1 \right.$$

$$+ \left(-7854M_1 + 589.0 \times 10^3 N_1 - 4783M_2 + 81190N_2 + 121.2 \times 10^6\right) \delta N_1 \qquad \text{[i]}$$

$$+ \left(52.36M_1 - 4783N_1 + 95.66M_2 - 856.1N_2 - 1.334 \times 10^6\right) \delta M_2$$

$$\left. + \left(-856.1M_1 + 81190N_1 - 856.1M_2 + 16990N_2 + 19.25 \times 10^6\right) \delta N_2 \right]$$

All of the redundants are internal loads, so $\delta W_{ext}^* = 0$. Therefore, $\delta W_{int}^* = 0$ and, by the usual argument, the principle of complementary virtual work requires the four coefficients of the virtual loads in Equation i to vanish. The result is the four equations in the four unknowns M_1, N_1, M_2, and N_2, as follows:

$$
\begin{aligned}
157.1M_1 \quad &- 7854N_1 + 52.36M_2 - 856.1N_2 = 1.420 \times 10^6 EI_z \\
-7854M_1 + 589.0 \times 10^3 N_1 &- 4783M_2 + 81190N_2 = -121.2 \times 10^6 EI_z \\
52.36M_1 \quad &- 4783N_1 + 95.66M_2 - 856.1N_2 = 1.334 \times 10^6 EI_z \\
-856.1M_1 \quad &+ 81190N_1 - 856.1M_2 + 16990N_2 = -19.25 \times 10^6 EI_z
\end{aligned}
\qquad \text{[j]}
$$

The solution of this system is

$$M_1 = -1788 \text{ in.-lb} \qquad N_1 = -161 \text{ lb} \qquad M_2 = 5115 \text{ in.-lb} \qquad N_2 = -196 \text{ in.-lb}$$

The bending moment distribution around the frame is found by substituting these redundant load values into Equations a1 through g1. The bending moment is plotted in Figure 7.8.38. We can see that the maximum value occurs at the junction of the circular frame and the floor beam, and the next largest value occurs at the midpoint of the floor beam.

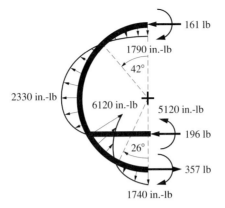

Figure 7.8.38   Bending moment distribution around the circular portion of the frame.

The moment is plotted on the side of the frame in compression.

EXERCISES

Use the force method to solve the following problems. Neglect shear and stretching deformation in beam and frame problems, unless otherwise directed.

7.1   Calculate the displacement of point C of the truss in the direction of the incline.

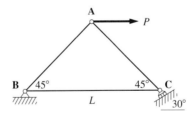

A, E uniform throughout

Exercise 7.1

7.2   Find the displacements u_3 and v_5 of the truss in terms of $P, L, A,$ and E.

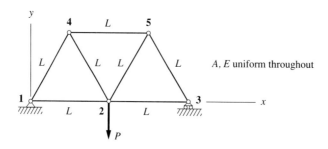

Exercise 7.2

7.3   Each member of the truss undergoes a temperature change, as follows: rod 1-2: +50° F; rod 1-4: +40° F; rod 2-3: +30° F; rod 2-4: +20° F; rod 3-4: +10° F; rod 3-5: 0° F; rod 3-6: –10° F; rod 4-5: –20° F; rod 4-6: –30° F. Calculate the vertical component of displacement at node 1. For all members, $A = 1$ in.^{2}, $E = 10^7$ psi, and $\alpha = 10^{-5}$ °F^{-1}. Also, $l = 15$ in.

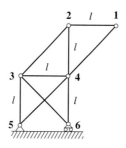

Exercise 7.3

7.4   In the truss, the axial loads due to the external forces P and Q at nodes 1 and 6 are shown alongside each member. Calculate the magnitude of the horizontal component of displacement of node 1 if $P = 1000$ lb and $Q = 0$.

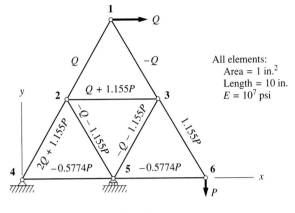

Exercise 7.4

7.5   The member loads (in pounds) due to the 15,000 lb load are shown in parentheses. If the 15,000 lb load is removed and the temperature of member 5-4 (only) is increased 100° F, find the horizontal displacement of node 2.

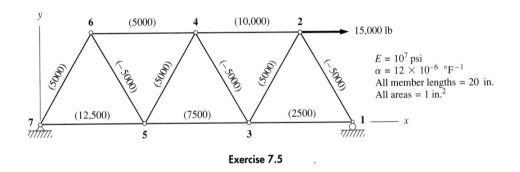

Exercise 7.5

7.6   All members of the truss have the same E and A. Calculate the displacement in the direction of the load P.

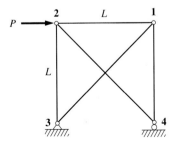

Exercise 7.6

7.7   All members of the truss have the same E and A. Calculate the horizontal displacement at node 4 due to the horizontal load P at node 2.

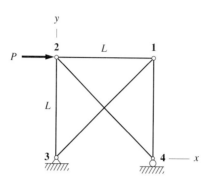

Exercise 7.7

7.8   If all members of the truss have the same axial rigidity AE, compute the horizontal component of the displacement of node 1.

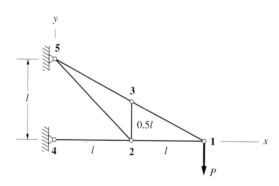

Exercise 7.8

7.9   All members of the hexagonal truss have a common length (20 in.), cross-sectional area (0.300 in.^{2}), and modulus of elasticity (30,000,000 psi). Alongside each member is the axial force due to P_1 and P_3, applied as shown. Calculate the horizontal displacement of node 3 if $P_1 = 10,000$ lb and $P_3 = 0$.

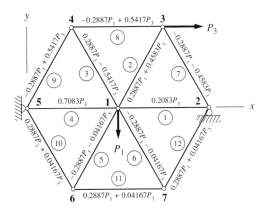

Exercise 7.9

7.10   In the truss on the left, the temperature of rod 1 rises ΔT, while the other two rods remain at the initial temperature. (A free-body diagram of the truss is shown on the right.) Calculate the force P_2 in member 2 due to the thermal strain.

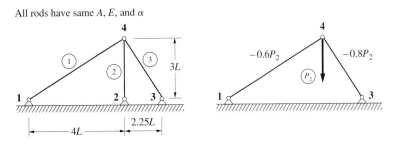

Exercise 7.10

7.11 The statically indeterminate truss in (a) has a horizontal load P applied to node 4, as shown. Take the vertical reaction Y_2 at node 2 and the axial load P_5 in rod 5 as the redundants. Figures b, c, and d show the member loads (alongside each member, with proper sign) as functions of P, Y_2, and P_5, respectively. Compute the values of the redundants Y_2 and P_5, in terms of P.

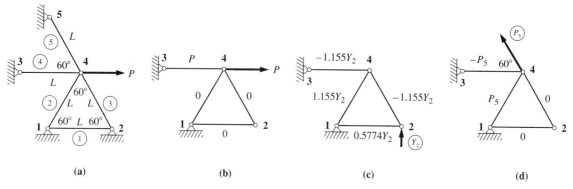

(a) (b) (c) (d)

All rods have the same A, E, and L

Exercise 7.11

7.12 Find the axial load in each rod of the truss.

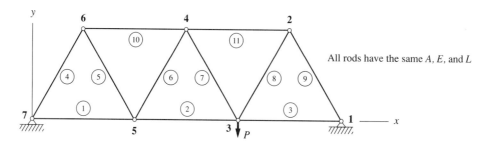

All rods have the same A, E, and L

Exercise 7.12

7.13   The temperature of rods 1 and 2 increases by ΔT. Calculate the axial load in each rod of the truss.

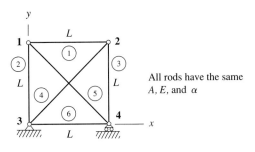

All rods have the same
A, E, and α

Exercise 7.13

7.14   Given the axial loads in each member of the truss, calculate the horizontal displacement at node 1.

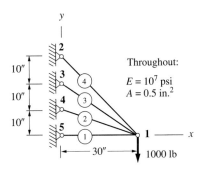

Throughout:

$E = 10^7$ psi
$A = 0.5$ in.^{2}

Rod	Axial load
1	−1116.0 lb
2	−92.74 lb
3	+630.0 lb
4	+961.5 lb

Exercise 7.14

7.15 Calculate the member loads in the triply-redundant truss. Use the base structure and equilibrium load information shown in the remaining figures. All members have the same cross-sectional area and modulus of elasticity.

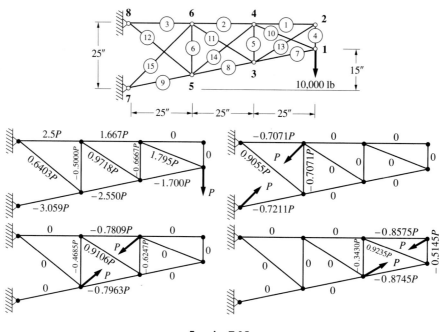

Exercise 7.15

7.16 Calculate the rotation θ at node 1 of the beam.

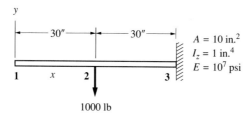

Exercise 7.16

7.17   The computed reactions on the statically indeterminate beam are shown. The properties are $A = 4$ in.^{2}, $I_z = 5$ in.^{4}, and $E = 10^7$ psi. Calculate the rotation at node 1.

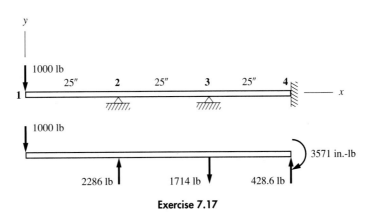

Exercise 7.17

7.18   Find the deflection at the midspan of the beam, in terms of p, L, E, and I_z.

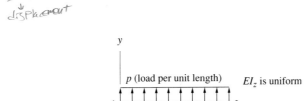

Exercise 7.18

7.19   For the cantilever beam of length L and flexural rigidity EI_z, find the rotation at point 1 due to the applied couple M.

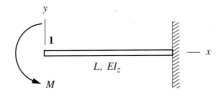

Exercise 7.19

7.20   The simply-supported beam of length L has a uniform flexural rigidity EI_z. Find the rotation at point 2 in terms of the point couple at point 1.

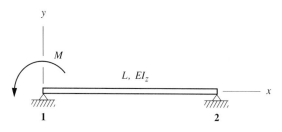

Exercise 7.20

7.21   If the center support moves downward by the amount $0.01\,pL^4/EI_z$ and remains attached to the beam, what will be the reactions at the left and right supports?

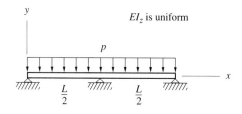

Exercise 7.21

7.22   Calculate the vertical reaction at point 1 of the beam under the uniform distributed load p.

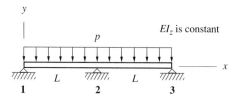

Exercise 7.22

7.23   Calculate the reaction at the left support for the beam under uniform load p.

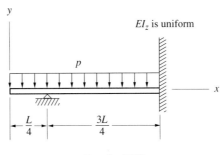

Exercise 7.23

7.24   Find the vertical reaction at point 1 if the rotations at points 1 and 2 are restrained while point 1 is given an upward displacement d, as shown.

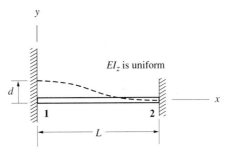

Exercise 7.24

7.25   Taking advantage of symmetry, find the moment at point 1.

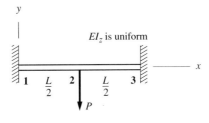

Exercise 7.25

7.26   Calculate the reactions and find those points where the bending moment is zero.

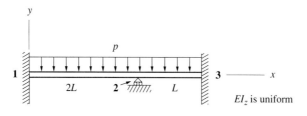

Exercise 7.26

7.27   The structure consists of six rods and a beam. Calculate the axial load in rod 2-5.

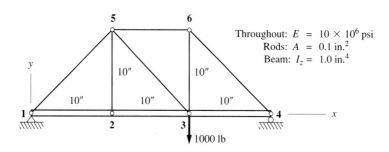

Exercise 7.27

7.28   Including shear deformation, calculate the rotation of the beam at pin B .

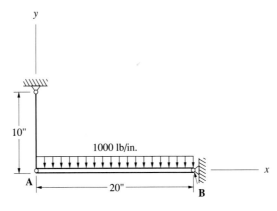

Rod: $A = 1$ in.^{2}
Beam: $A_s = 1$ in.^{2}, $I_z = 100$ in.^{4}
Throughout: $E = 10 \times 10^6$ psi
$G = 4 \times 10^6$ psi

Exercise 7.28

7.29   Member BC is joined to the horizontal beam by a smooth pin at B. Find the rotation of the beam at pin D.

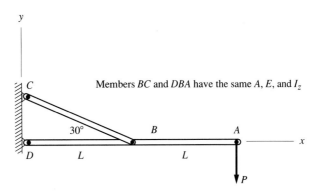

Exercise 7.29

7.30   Find: (a) the rotation θ_3 of point 3, and (b) the vertical displacement v_5 of point 5 of the frame.

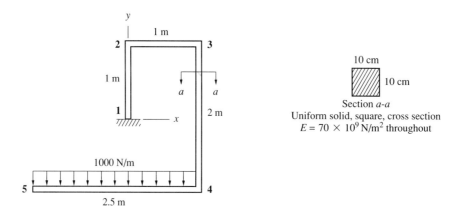

Exercise 7.30

7.31   Point 2 of the split ring is rigidly fixed, while the adjacent point 1 is forced radially inward. Find the displacement u_1 in the direction of the force P, in terms of P, R, E, and I_z.

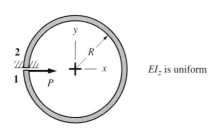

Exercise 7.31

7.32   Find the horizontal displacement at point 1. EI_z is uniform.

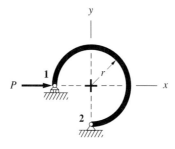

Exercise 7.32

7.33   A vertical force P is applied to point 6 of the frame. Calculate the horizontal component of displacement u_6 at that point. Assume EI_z is uniform throughout.

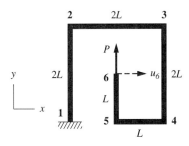

Exercise 7.33

7.34   A clockwise point couple C is applied to point 1 of the frame, which is built in at point 6. Calculate the rotation at point 5. Assume EI_z is uniform throughout.

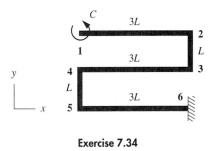

Exercise 7.34

7.35   A counterclockwise point couple C is applied to point 1 of the frame, which is built in at point 3. Calculate the vertical displacement v_1 at point 1. Assume EI_z is uniform throughout.

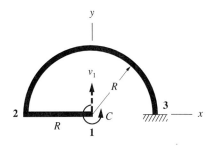

Exercise 7.35

7.36   The square frame has a uniformly distributed load p along opposite sides. Find the bending moment at point 1, which is located at the midpoint of the top of the frame. Take advantage of symmetry.

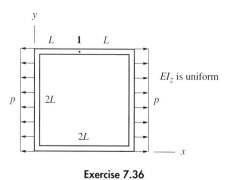

Exercise 7.36

7.37   The portal frame is supported by smooth pins as shown. Calculate the reactions at the supports. Include axial deformation, but ignore shear.

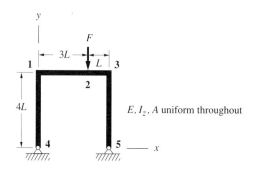

Exercise 7.37

7.38   The structure is composed of rigidly-joined beams and pin-jointed rods. Find the location and value of the maximum bending moment in the frame and the location of those interior points where the bending moment is zero.

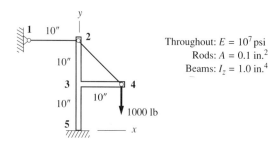

Exercise 7.38

7.39   Find the horizontal reaction at point 1.

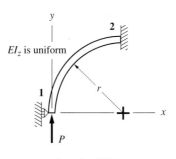

Exercise 7.39

7.40   The square frame is in equilibrium under the combination of the point load P applied as shown and the given shear flow distribution around its periphery.

    a.   Calculate the bending moment at the point where P is applied.

    b.   Locate the points where the bending moment is zero.

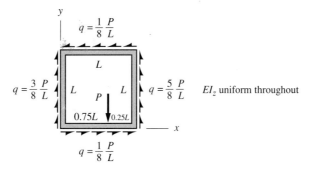

Exercise 7.40

7.41   The semicircular frame of radius R is supported by smooth pins at both ends (points 1 and 3). A downward load P is applied to point 2 at the top. Find the value and location of the maximum bending moment in the frame, in terms of P and R. Assume EI_z is constant.

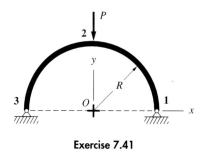

Exercise 7.41

7.42   The circular ring is supported at points 2 and 3 by frictionless surfaces. If a load P is applied as shown to point 1 on top, calculate the internal bending moment at that point. Assume EI_z is uniform throughout. Choose the internal loads at point 1 as the redundants.

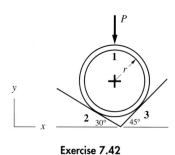

Exercise 7.42

7.43   A circular ring is in equilibrium under the action of a point load P and a symmetric uniform shear flow $P/4r$. Find the bending moment at the point where the load P is applied. Choose the internal loads at point 1 on top as the redundants. Take advantage of symmetry.

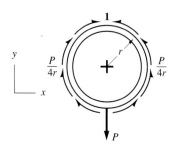

Exercise 7.43

7.44   Calculate the horizontal reaction at the smooth roller support at point 3.

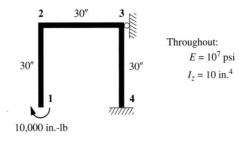

Throughout:
$E = 10^7$ psi
$I_z = 10$ in.^{4}

10,000 in.-lb

Exercise 7.44

7.45   Calculate the bending moment at point 1 at the top of the frame. Assume EI_z is uniform.

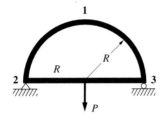

Exercise 7.45

7.46   For the semicircular frame of Example 7.8.9, calculate the displacement in the direction of the load P at point 3, at the midpoint of the horizontal portion.

7.47 Calculate the x displacement at point B of the L-shaped rod.

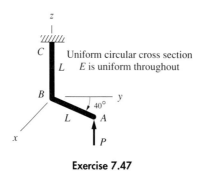

Exercise 7.47

7.48 The wire frame consists of two right-angle bends, with point 4 built in. The solid circular cross section has a uniform diameter d. A load P in the x direction acts at point 1. Find the three components of translational displacement (u_1, v_1, w_1) of point 1 in terms of P, L, d, and E. Assume $G = 0.4E$.

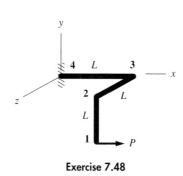

Exercise 7.48

7.49 The structure is made of a 1 in. diameter, solid circular steel rod. The piece with two right-angle bends is built into the wall at point 4, while its other end, point 1, is fastened to the wall by means of a 10 in. rod and smooth pins, as shown. Calculate the force in rod 1-5.

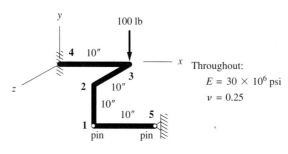

Exercise 7.49

Force Method: Idealized Thin-Walled Structures

8.1   INTRODUCTION

This chapter applies the force method based on the complementary virtual work principle to the analysis of assemblies of thin shear panels and stiffeners; deflections of box beams, with and without taper; shear flows in multicell box beams; the unrestrained warping of beam cross sections due to the torsional component of loading; and the effects on shear flows of warping restraints, as occurs near supports. The chapter concludes with a disscussion of shear lag, which is not so much an aspect of the force method as it is a means of assessing the influence of deformation restraints on shear flow distribution.

8.2 SHEAR PANELS AND STIFFENERS

The rod element discussed in section 7.2 commonly plays the role of a constant-area stiffener attached to one or more constant shear flow panels, as in a box beam. This situation is depicted in Figure 8.2.1. The constant shear flow q in the webs attached to the stiffeners gives rise to a linear axial load variation along the rod, as shown in the figure. From the free-body diagram of the stiffener–web combination at the top of Figure 8.2.1a we find that

$$P_2 + qL - P_1 = 0$$

or

$$q = \frac{P_1 - P_2}{L} \qquad [8.2.1]$$

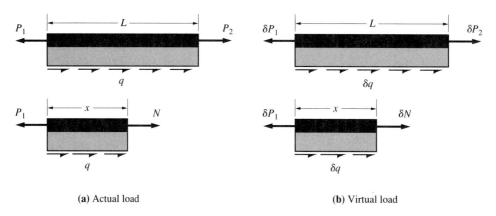

(a) Actual load   (b) Virtual load

Figure 8.2.1   Rod and adjacent shear web.

Using the free-body diagram at the bottom of Figure 8.2.1a, which reveals the varying internal axial load N, we obtain $N = P_1 - qx$. Substituting Equation 8.2.1 into this expression yields

$$N = P_1\left(1 - \frac{x}{L}\right) + P_2\left(\frac{x}{L}\right) \qquad [8.2.2]$$

Similarly, for a virtual load, we have

$$\delta N = \delta P_1\left(1 - \frac{x}{L}\right) + \delta P_2\left(\frac{x}{L}\right) \qquad [8.2.3]$$

From these two results, it is clear that

$$N\delta N = P_1\delta P_1\left(1 - \frac{x}{L}\right)^2 + P_2\delta P_2\left(\frac{x}{L}\right)^2 + (P_1\delta P_2 + P_2\delta P_1)\left(1 - \frac{x}{L}\right)\left(\frac{x}{L}\right) \qquad [8.2.4]$$

After substituting this expression into Equation 7.2.1, and doing the first integral (recalling that AE is constant), we find, for the *linear stress rod*, that

$$\delta W_{int}^* = \frac{L}{3AE} \left[\left(P_1 + \tfrac{1}{2} P_2\right) \delta P_1 + \left(P_2 + \tfrac{1}{2} P_1\right) \delta P_2 \right] + \int_0^L (\alpha T) \left[\delta P_1 \left(1 - \frac{x}{L}\right) + \delta P_2 \left(\frac{x}{L}\right) \right] dx \quad \text{web stiffener} \qquad \textbf{[8.2.5]}$$

As before, the integral can be computed only after the temperature variation is known, and it vanishes if there is no temperature change. Observe that if $P_1 = P_2 = N$, that is, if there is no shear flow along the rod, then Equation 8.2.5 and 7.2.2 are identical.

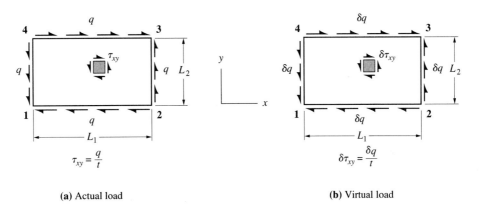

(a) Actual load

(b) Virtual load

Figure 8.2.2   Rectangular, constant shear flow panel.

The shear panel is used along with the rod element to model stiffened thin-walled structures. For a rectangular panel, a state of pure shear stress exists with respect to the edges of the element, and it is uniform throughout. The relationship between the shear stress and shear flow for a panel of thickness t is given in Figure 8.2.2, for both the real and virtual loads. Since τ_{xy} is the only nonzero stress, for materially isotropic panels Equation 6.6.4 reduces to

$$\delta W_{int}^* = \iiint_V \delta \tau_{xy} \gamma_{xy} dV$$

$$= \iint_A \delta \tau_{xy} \left(\frac{\tau_{xy}}{G}\right) t dA$$

$$= \iint_A \left(\frac{\delta q}{t}\right) \left(\frac{q}{Gt}\right) t dA$$

$$= \frac{q \delta q}{Gt} \iint_A dA$$

Thus, for a *rectangular shear panel*,

$$\delta W_{int}^* = \frac{A}{Gt} q \delta q \qquad \textbf{[8.2.6]}$$

where $A = L_1 \times L_2$ is the area of the rectangle. This same equation can be used for curved cylindrical panels, like the one illustrated in Figure 8.2.3. In this case, the surface area is $L \times s$, where s is the length of the generating curve in the plane of the cross section.

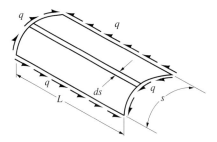

Figure 8.2.3   Cylindrical shear panel.

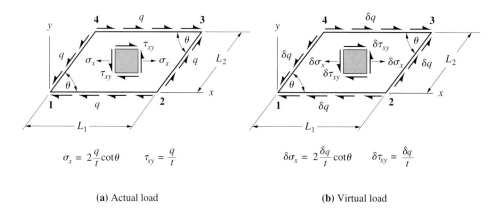

$$\sigma_x = 2\frac{q}{t}\cot\theta \qquad \tau_{xy} = \frac{q}{t}$$

$$\delta\sigma_x = 2\frac{\delta q}{t}\cot\theta \qquad \delta\tau_{xy} = \frac{\delta q}{t}$$

(a) Actual load   (b) Virtual load

Figure 8.2.4   Parallelogram shear panel.

A constant shear flow q applied to the edges of the parallelogram-shaped shear panel in Figure 8.2.4 is self-equilibrating. The state of stress shown in the figure is constant throughout the panel. Relative to the given xy axes, the only nonzero stress components anywhere inside the parallelogram are

$$\sigma_x = 2\frac{q}{t}\cot\theta \qquad \tau_{xy} = \frac{q}{t} \qquad [8.2.7]$$

where t is the panel thickness and θ is the acute included angle of the parallelogram. Substituting these stresses into Equation 6.6.4 along with their virtual counterparts, $\delta\sigma_x$ and $\delta\tau_{xy}$, yields

$$
\begin{aligned}
\delta W^*_{\text{int}} &= \iiint_V \delta\sigma_x \varepsilon_x \, dV + \iiint_V \delta\tau_{xy}\gamma_{xy} \, dV \\
&= \iint_A \delta\sigma_x \frac{\sigma_x}{E} \, t\,dA + \iint_A \delta\tau_{xy}\frac{\tau_{xy}}{G}\, t\,dA \\
&= \iint_A \left(2\frac{\delta q}{t}\cot\theta\right)\left[\frac{2(q/t)\cot\theta}{2\,(1+v)\,G}\right]t\,dA + \iint_A \left(\frac{\delta q}{t}\right)\left(\frac{q/t}{G}\right)t\,dA
\end{aligned}
$$

so that for a *parallelogram shear panel* of area A and acute included angle θ, we have

$$\delta W_{\text{int}}^{*} = \left[1 + \frac{2\cot^2\theta}{1+\nu}\right] \frac{A}{Gt} q\,\delta q \qquad \text{parallelogram shear panel} \qquad \textbf{[8.2.8]}$$

where A is the area of the parallelogram. When $\theta = 90°$, this expression reduces to that for a rectangular panel, Equation 8.2.6.

Figure 8.2.5 shows a flat *trapezoidal* panel, two edges of which are parallel while the other two (extended) intersect at the *vertex P* through which a *baseline* parallel to the parallel edges is drawn. The angles α and γ are the two included angles of the trapezoid *farthest* from the baseline. As discussed in section 2.5, such a panel, if subjected to pure shear forces on its edges, is in overall equilibrium if, on any differential parallelogram (shaded in the figure) formed by sections through P and sections parallel to the baseline, the shear flow q is given by Equation 2.5.3, which is

$$q = \frac{\beta}{p^2} \qquad \textbf{[8.2.9]}$$

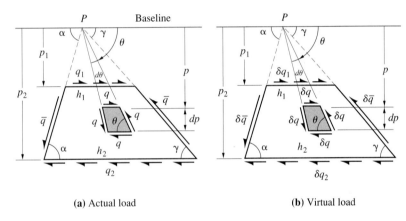

(a) Actual load   **(b)** Virtual load

Figure 8.2.5   Trapezoidal shear panel, where h_1 and h_2 are the lengths of the parallel sides.

p is the perpendicular distance from the vertex to the differential element and β is a constant related to the average shear flow in the panel by the formula $\beta = p_1\,p_2\bar{q}$. In Example 3.12.2, it was shown that the trapezoidal-shaped pure shear panel does not satisfy compatibility and is therefore not in full compliance with the theory of elasticity.

To calculate the internal complementary virtual work of the trapezoidal shear panel, we integrate the complementary virtual work of the differential parallelogram over the trapezoid, as follows:

$$\delta W_{\text{int}}^{*} = \int_{\text{trapezoid}} \delta W_{\text{int,differential parallelogram}}^{*}$$

Substituting Equation 8.2.9 into Equation 8.2.8 and replacing A by dA yields

$$\delta W_{\text{int}}^{*} = \iint_{A} \frac{1}{Gt}\left[1 + \frac{2\cot^2\theta}{1+\nu}\right] \frac{\beta\,\delta\beta}{p^4}\,dA \qquad \textbf{[8.2.10]}$$

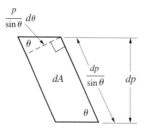

Figure 8.2.6   Detail of the shaded differential parallelogram in Figure 8.2.5.

From Figure 8.2.6, it is easily seen that the area of the differential parallelogram is given by

$$dA = \left(\frac{p}{\sin\theta}d\theta\right)\left(\frac{dp}{\sin\theta}\right) = \frac{p\,dp\,d\theta}{\sin^2\theta}$$

[8.2.11]

Substituting this into Equation 8.2.10 yields

$$\delta W_{int}^* = \frac{\beta\ \delta\beta}{G\,t}\left[\int_{\gamma}^{180-\alpha}\left(1 + \frac{2\cot^2\theta}{1+\nu}\right)\frac{d\theta}{\sin^2\theta}\right]\left[\int_{p_1}^{p_2}\frac{dp}{p^3}\right]$$

[8.2.12]

We can simplify the first integral by making the substitution $u = \cot\theta$ so that $du = -d\theta/\sin^2\theta$, which leads to

$$\delta W_{int}^* = -\frac{\beta\delta\beta}{Gt}\left[\int_{\cot\gamma}^{-\cot\alpha}\left(1 + \frac{2u^2}{1+\nu}\right)du\right]\left[\int_{p_1}^{p_2}\frac{dp}{p^3}\right]$$

After carrying out the integrals, we substitute $\beta = p_1\,p_2\bar{q}$ and $\delta\beta = p_1\,p_2\,\delta\bar{q}$ to obtain, for a *trapezoidal shear panel*,

$$\delta W_{int}^* = \left[1 + \frac{2}{3(1+\nu)}\left(\cot^2\gamma - \cot\alpha\ \cot\gamma + \cot^2\alpha\right)\right]\frac{A}{Gt}\bar{q}\delta\bar{q}$$

[8.2.13]

where A is the trapezoid area, which is

$$A = \frac{1}{2}(h_1 + h_2)(p_2 - p_1)$$

For a quadrilateral shear panel no two sides of which are parallel, the expression for δW_{int}^* is even more complicated and will not be given here.

8.3   STATICALLY INDETERMINATE STIFFENED WEBS

Section 2.5 presents the procedure for finding the internal loads in planar, statically determinate, stiffened web structures. The elements of these structures are shear panels, the edges of which are bonded to rods or stiffeners. The individual stiffeners are joined together at the nodes (cf. Figure 2.5.11). Point loads are applied to the structure at the stiffeners, which transmit the load to the panels by shear action alone. According to Equation 2.5.7, a plane stiffened panel is statically determinate if

$$\text{(no. rods)} + \text{(no. panels)} + \text{(no. reactions)} = 2 \times \text{(no. nodes)}$$

Therefore,

$$\text{(no. rods)} + \text{(no. panels)} + \text{(no. reactions)} - 2 \times \text{(no. nodes)} = \text{degree of static indeterminancy}\quad\textbf{[8.3.1]}$$

The structure in Figure 8.3.1a is statically determinate. For stiffened panels with redundancies, we must supplement statics by considering the structure's deflections. If the structure—such as the singly-redundant one in Figure 8.3.1b—resembles a beam, we may be able to assume that the deflections are those of a beam in bending, and we can then use idealized beam theory (Chapter 4) to calculate the stress distribution. However, for situations like that illustrated in Figure 8.3.1c, beam theory is clearly not appropriate and we must appeal to a virtual work principle. We will use the principle of complementary virtual work.

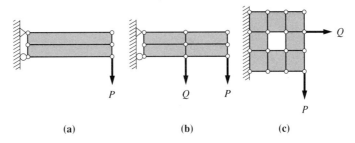

(a)   **(b)**   **(c)**

Figure 8.3.1   (a) Statically determinate, stiffened panel assembly.
(b) and (c) Statically indeterminate structures.

Example 8.3.1   Use the principle of complementary virtual work to calculate the shear flows in the stiffened web structure in Figure 8.3.2. All of the stiffeners have the same cross-sectional area A, and all of the panels have the same thickness t. The material properties are uniform throughout.

Since there are twelve rods, four panels, three reactions, and nine nodes, Equation 8.3.1 indicates that this structure has one degree of static indeterminacy. The structure is also symmetric, as is the loading. Therefore, we can infer the reactions and symmetric shear flow distribution indicated in Figure 8.3.3. The element numbering scheme is also shown. A relationship between the two unknown shear flows, $q^{(1)}$ and $q^{(2)}$, and the applied load P can be obtained by analyzing stiffener 1–2–3 as a free body. Thus, from Figure 8.3.4a, we have

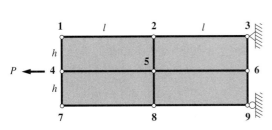

Figure 8.3.2   Singly-redundant stiffened panel structure.

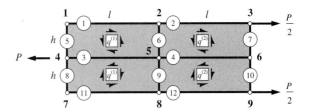

Figure 8.3.3   Free-body diagram of the symmetric structure of Figure 8.3.2, showing the shear flows and rod element numbering scheme.

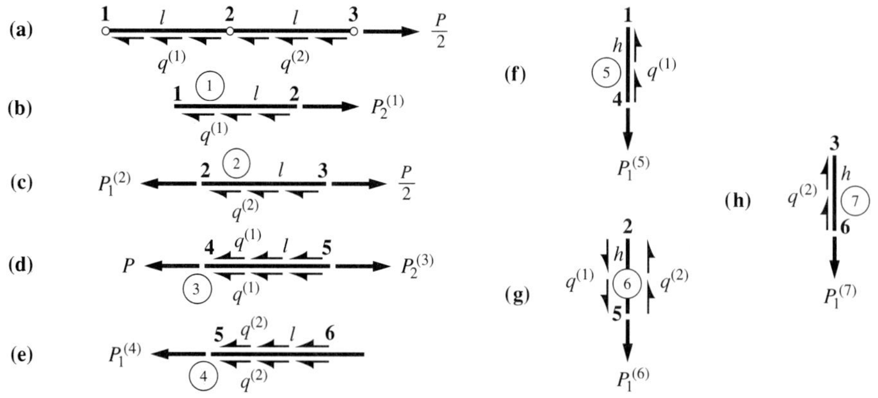

Figure 8.3.4 Free-body diagrams of individual stiffener elements of the structure in Figure 8.3.2.

$$q^{(1)}l + q^{(2)}l = \frac{P}{2}$$

Let us choose $q^{(1)}$ as the redundant shear flow, since the two shear panels in the leftmost bay could be removed and the structure would remain stable for *this* loading. Then, $q^{(2)}$ in terms of $q^{(1)}$ is

$$q^{(2)} = -q^{(1)} + \frac{P}{2l} \qquad \text{[a]}$$

which implies that

$$\delta q^{(2)} = -\delta q^{(1)} \qquad \text{[b]}$$

From Figure 8.3.4b we see that $P_2^{(1)} - q^{(1)}l = 0$, or $P_2^{(1)} = q^{(1)}l$. It follows that $\delta P_2^{(1)} = \delta q^{(1)}l$. Thus, for rod element 1, we have, using Equation 8.2.5,

$$\delta W_{int}^{*(1)} = \frac{l}{3AE}\left[\left(P_1^{(1)} + \tfrac{1}{2}P_2^{(1)}\right)\delta P_1^{(1)} + \left(P_2^{(1)} + \tfrac{1}{2}P_1^{(1)}\right)\delta P_2^{(1)}\right]$$

$$= \frac{l}{3AE}\left[\left(0 + \tfrac{1}{2}q^{(1)}l\right)0 + \left(q^{(1)}l + \tfrac{1}{2}\times 0\right)\delta q^{(1)}l\right]$$

that is,

$$\delta W_{int}^{*(1)} = \frac{l^3}{3AE}q^{(1)}\delta q^{(1)} \qquad \text{[c]}$$

For rod element 2, $P_1^{(2)} = P_2^{(1)} = q^{(1)}l$, and Figure 8.3.4c shows that the force on the right end of rod 2 is $P_2^{(2)} = P/2$. Thus, $\delta P_1^{(2)} = \delta q^{(1)}l$ and $\delta P_2^{(2)} = 0$, so that

$$\delta W_{int}^{*(2)} = \frac{l}{3AE}\left[\left(\delta q^{(1)}l + \tfrac{1}{2}\times\frac{P}{2}\right)\delta q^{(1)}l + \left(\frac{P}{2} + \tfrac{1}{2}\delta q^{(1)}l\right)\times 0\right] = \frac{l^2}{3AE}\left(\frac{P}{4} + lq^{(1)}\right)\delta q^{(1)} \qquad \text{[d]}$$

From Figure 8.3.4d, we see that $P_1^{(3)} = P$ and $P_2^{(3)} = P - 2q^{(1)}l$, which means that $\delta P_1^{(3)} = 0$ and $\delta P_2^{(3)} = -2\delta q^{(1)}l$. Therefore,

$$\delta W_{int}^{*(3)} = \frac{l}{3AE}\left\{\left[P + \tfrac{1}{2}\left(P - 2q^{(1)}l\right)\right]\times 0 + \left[\left(P - 2q^{(1)}l\right) + \tfrac{1}{2}P\right]\left(-2\delta q^{(1)}l\right)\right\} = \frac{2l^2}{3AE}\left(\tfrac{3}{2}P - 2q^{(1)}l\right)\delta q^{(1)} \qquad \text{[e]}$$

Using the remaining free-body diagrams in Figure 8.3.4 and proceeding as before leads to the following complementary virtual work expressions for the remaining rods:

$$\delta W_{\text{int}}^{*(4)} = \frac{2l^2}{3AE} \left(P - 2q^{(1)}l \right) \delta q^{(1)} \tag{f}$$

$$\delta W_{\text{int}}^{*(5)} = \frac{h^3}{3AE} q^{(1)} \delta q^{(1)} \tag{g}$$

$$\delta W_{\text{int}}^{*(6)} = \frac{h^3}{3AE} \left(4q^{(1)} - \frac{P}{l} \right) \delta q^{(1)} \tag{h}$$

$$\delta W_{\text{int}}^{*(7)} = \frac{h^3}{3AE} \left(q^{(1)} - \frac{P}{2l} \right) \delta q^{(1)} \tag{i}$$

The total internal complementary virtual work for the structure, including all 12 rods, is

$$\delta W_{\text{int,stiffeners}}^{*} = 2\delta W_{\text{int}}^{*(1)} + 2\delta W_{\text{int}}^{*(2)} + \delta W_{\text{int}}^{*(3)} + \delta W_{\text{int}}^{*(4)} + 2\delta W_{\text{int}}^{*(5)} + 2\delta W_{\text{int}}^{*(6)} + 2\delta W_{\text{int}}^{*(7)}$$

Substituting Equations c through i into this expression yields

$$\delta W_{\text{int,stiffeners}}^{*} = \frac{l^2}{AE} \left\{ 4q^{(1)}l \left[1 + \left(\frac{h}{l} \right)^3 \right] - P \left[\frac{3}{2} + \left(\frac{h}{l} \right)^3 \right] \right\} \delta q^{(1)} \tag{j}$$

For the complementary internal virtual work of the webs, we use Equation 8.2.6,

$$\delta W_{\text{int,panels}}^{*} = 2 \times \left(\frac{hl}{Gt} q^{(1)} \delta q^{(1)} + \frac{hl}{Gt} q^{(2)} \delta q^{(2)} \right)$$

which includes all four panels. Substituting Equations a and b, this can be expressed in terms of the sole redundant $q^{(1)}$ as

$$\delta W_{\text{int,panels}}^{*} = \frac{h}{Gt} \left(4q^{(1)}l - P \right) \delta q^{(1)} \tag{k}$$

The total complementary internal virtual work for the structure is

$$\delta W_{\text{int}}^{*} = \delta W_{\text{int,stiffeners}}^{*} + \delta W_{\text{int,panels}}^{*}$$

Substituting Equations j and k, we obtain

$$\delta W_{\text{int}}^{*} = \frac{1}{2AEGlt} \left\{ 8 \left[AEhl^2 + Glt \left(l^3 + h^3 \right) \right] q^{(1)} - \left[2AEl + Gt \left(2h^3 + 3l^3 \right) \right] P \right\} \delta q^{(1)} \tag{l}$$

Since $q^{(1)}$, the redundant shear flow, is an internal load, the external complementary virtual work is zero. The principle of complementary virtual work therefore requires $\delta W_{\text{int}}^{*} = 0$. Setting the right side of Equation l equal to zero and solving for $q^{(1)}$ yields

$$q^{(1)} = \frac{\frac{3}{2} + \frac{AE}{Glt} \frac{h}{l} + \frac{h^3}{l^3}}{4 \left(1 + \frac{AE}{Glt} \frac{h}{l} + \frac{h^3}{l^3} \right)} \frac{P}{l} \tag{m}$$

Substituting this into Equation a shows that the other shear flow is

$$q^{(2)} = \frac{\frac{1}{2} + \frac{AE}{Glt} \frac{h}{l} + \frac{h^3}{l^3}}{4 \left(1 + \frac{AE}{Glt} \frac{h}{l} + \frac{h^3}{l^3} \right)} \frac{P}{l} \tag{n}$$

The shear flows are clearly influenced by the relative stiffness of the rods and the webs, through the ratio AE/Glt. Further insight might be gained by substituting some typical numerical values into these equations. For example, let $G = 0.4E$ ($\nu = 0.25$), $l = 15$ in., $h = 5$ in., $t = 0.05$ in., $A = 0.1$ in.^{2} and $P = 1000$ lb. Then Equations m and n yield the shear flows $q^{(1)} = 23.9$ lb/in and $q^{(2)} = 9.41$ lb/in. The free-body diagrams of the individual elements of the stiffened web structure are shown in Figure 8.3.5. Notice that the shear flows in the left and right panels influence the stiffeners as linear variations of axial load.

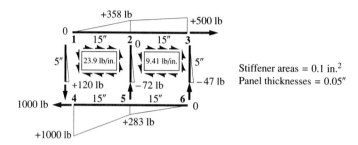

Figure 8.3.5 Panel shear flows and stiffener axial load distributions in the top half of the symmetric structure of Figure 8.3.2 corresponding to the indicated numerical data.

The results of the analysis in Example 8.3.1 are only as good as the model used to represent the actual structure depicted in Figure 8.3.2. We used constant shear flow panels to solve this statically indeterminate problem; a shear lag approach could also have been used (cf. section 8.10). Such an analysis reveals that the loads applied to the stiffeners diffuse into the shear webs in a more complex, exponentially-varying fashion, rather than at the constant rate predicted by our shear panels. In fact, using the same numerical data as Example 8.3.1, a shear lag analysis yields the shear flow distribution shown in Figure 8.3.6. The associated average shear flows computed over the first and second half of the structure, respectively, are 22.2 lb/in. and 11.2 lb/in. The shear flows computed in the example are 23.9 lb/in. and 9.41 lb/in. These results show that the shear panels are oblivious to the dramatically varying shear flow pattern within the webs, but they fairly portray the overall effect of the shear flows in terms of their average values. This is not too surprising, since the shear lag analysis, like the force method used to get our results, assumes the structure is in equilibrium throughout.

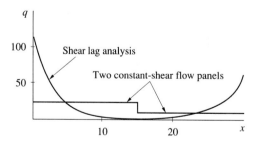

Figure 8.3.6 Shear flow q (lb/in.) vs. station x (in.) for the stiffened web structure of Figure 8.3.2.

Shear lag results are compared to those in the previous example, using the same numerical data.

Example 8.3.2   Use the principle of complementary virtual work to find the shear flows in each of the six panels of the stiffened web structure in Figure 8.3.7.

This structure has 15 rods, 6 panels, 6 reactions, and 12 nodes, so that according to Equation 8.3.1, the degree of static indeterminacy is

$$(15 + 6 + 6) - 2 \times 12 = 3$$

We can remove one shear panel from each of the three bays and the structure will remain stable. Therefore, let us choose the shear flows $q^{(4)}$, $q^{(5)}$, and $q^{(6)}$ to be our redundants. In Figure 8.3.8 we have isolated the three horizontal and three vertical stiffeners as free bodies, showing on each: the applied loads F_1 and F_2; the reactions at the wall, X_4, X_8, and X_{12}; and the six

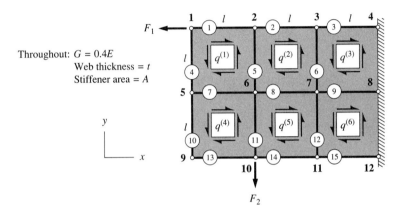

Figure 8.3.7   Stiffened web structure with three degrees of static indeterminacy.

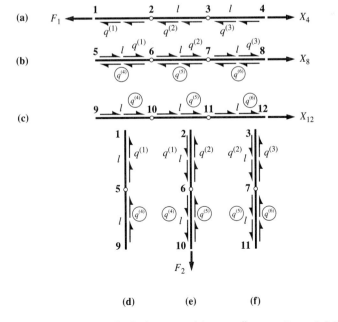

Figure 8.3.8   Free-body diagrams of the six stiffeners in Figure 8.3.7.

shear flows, with the redundants circled. We require the net force on and in the direction of each stiffener to be zero. Thus, from parts (a) through (f) of the figure, we obtain, in turn,

$$-lq^{(1)} - lq^{(2)} - lq^{(3)} + X_4 = F_1 \tag{a}$$

$$lq^{(1)} + lq^{(2)} + lq^{(3)} + X_8 = lq^{(4)} + lq^{(5)} + lq^{(6)} \tag{b}$$

$$X_{12} = -lq^{(4)} - lq^{(5)} - lq^{(6)} \tag{c}$$

$$lq^{(1)} = -lq^{(4)} \tag{d}$$

$$-lq^{(1)} + lq^{(2)} = lq^{(4)} - lq^{(5)} + F_2 \tag{e}$$

$$-lq^{(2)} + lq^{(3)} = lq^{(5)} - lq^{(6)} \tag{f}$$

Solving for the unknowns on the left, we get

$$q^{(1)} = -q^{(4)} \tag{g1}$$

$$q^{(2)} = -q^{(5)} + \frac{F_2}{l} \tag{h1}$$

$$q^{(3)} = -q^{(6)} + \frac{F_2}{l} \tag{i1}$$

$$X_4 = -lq^{(4)} - lq^{(5)} - lq^{(6)} + F_1 + 2F_2 \tag{j1}$$

$$X_8 = 2lq^{(4)} + 2lq^{(5)} + 2lq^{(6)} - 2F_2 \tag{k1}$$

$$X_{12} = -lq^{(4)} - lq^{(5)} - lq^{(6)} \tag{l1}$$

To obtain the virtual load counterparts of these true loads, we set F_1 and F_2 equal to zero and replace the shear flows on the right by $\delta q^{(4)}$, $\delta q^{(5)}$, and $\delta q^{(6)}$:

$$\delta q^{(1)} = -\delta q^{(4)} \tag{g2}$$

$$\delta q^{(2)} = -\delta q^{(5)} \tag{h2}$$

$$\delta q^{(3)} = -\delta q^{(6)} \tag{i2}$$

$$\delta X_4 = -l\delta q^{(4)} - l\delta q^{(5)} - l\delta q^{(6)} \tag{j2}$$

$$\delta X_8 = 2l\delta q^{(4)} + 2l\delta q^{(5)} + 2l\delta q^{(6)} \tag{k2}$$

$$\delta X_{12} = -l\delta q^{(4)} - l\delta q^{(5)} - l\delta q^{(6)} \tag{l2}$$

We then use this information to compute the internal complementary virtual work δW_{int}^* in terms of the applied loads F_1 and F_2 and the three redundant shear flows. The complementary virtual work of a stiffener element is found in Equation 8.2.5,

$$\delta W_{int}^{*(e)} = \frac{L^{(e)}}{3A^{(e)}E^{(e)}} \left[\left(P_1^{(e)} + \tfrac{1}{2}P_2^{(e)} \right) \delta P_1^{(e)} + \left(P_2^{(e)} + \tfrac{1}{2}P_1^{(e)} \right) \delta P_2^{(e)} \right] \tag{m}$$

To determine the axial loads $P_1^{(e)}$ and $P_2^{(e)}$ at the ends of element e, we must treat the element as a free body and then substitute as required from the stiffener equilibrium data. For example, for stiffener element 1, we see from Figure 8.3.9 that the force at the left end is the applied force tensile load F_1; therefore, $P_1^{(1)} = F_1$. Then equilibrium requires that

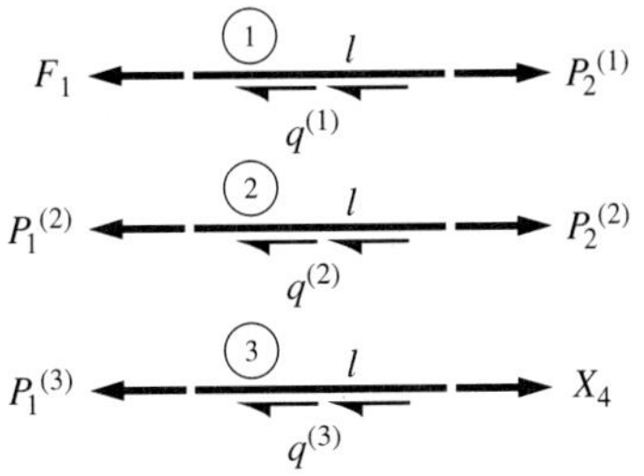

Figure 8.3.9   Free-body diagrams of the individual elements comprising the topmost stiffener.

$$P_2^{(1)} = F_1 + q^{(1)}l$$

But according to Equation (g1), $q^{(1)} = -q^{(4)}$. It follows that

$$P_2^{(1)} = F_1 - q^{(4)}l$$

The virtual end loads are then

$$\delta P_1^{(1)} = \delta F_1 = 0$$

and

$$\delta P_2^{(1)} = \delta F_1 - \delta q^{(4)}l = -\delta q^{(4)}l$$

Thus, for element 1, we have

$$\delta W_{\text{int}}^{*(1)} = \frac{l}{3AE} \left\{ \left[F_1 + \tfrac{1}{2}\left(F_1 - q^{(4)}l \right) \right](0) + \left(F_1 - q^{(4)}l + \tfrac{1}{2}F_1 \right)\left(-\delta q^{(4)}l \right) \right\} = \frac{l}{3AE}\left(q^{(4)}l^2 - \frac{3}{2}F_1 l \right)$$

Moving on to stiffener element 2, we note that

$$P_1^{(2)} = P_2^{(1)} = F_1 - q^{(4)}l$$

For equilibrium, we need

$$P_2^{(2)} = P_2^{(1)} + q^{(2)}l = \left(F_1 - q^{(4)}l \right) + q^{(2)}l$$

Substituting Equation h1, we get

$$P_2^{(2)} = -q^{(4)}l - q^{(5)}l + F_1 + F_2$$

From these relationshps, we find that

$$\delta P_2^{(1)} = -\delta q^{(4)}l$$

and

$$\delta P_2^{(2)} = -\delta q^{(4)}l - \delta q^{(5)}l$$

From Equation m, we therefore obtain, for element 2,

$$\delta W_{int}^{*(2)} = \frac{l}{3AE}\left[\left(3q^{(4)}l^2 + \tfrac{3}{2}q^{(5)}l^2 - 3F_1l - \tfrac{3}{2}F_2l\right)\delta q^{(4)} + \left(\tfrac{3}{2}q^{(4)}l^2 + q^{(5)}l^2 - \tfrac{3}{2}F_1l - F_2l\right)\delta q^{(5)}\right]$$

For stiffener element 3, we have

$$P_1^{(3)} = P_2^{(2)} = -q^{(4)}l - q^{(5)}l + F_1 + F_2$$

so that

$$\delta P_1^{(3)} = -\delta q^{(4)}l - \delta q^{(5)}l_2$$

At the other end, we see that $P_2^{(3)} = X_4$. X_4 and δX_4, in terms of the redundant and applied loads, were found in Equations j1 and j2. Therefore, the complementary virtual work of element 3, from Equation m, is

$$\delta W_{int}^{*(3)} = \frac{l}{3AE}\left[\left(3q^{(4)}l^2 + 3q^{(5)}l^2 + \tfrac{3}{2}q^{(6)}l^2 - 3F_1l - \tfrac{9}{2}F_2l\right)\delta q^{(4)} + \left(3q^{(4)}l^2 + 3q^{(5)}l^2 + \tfrac{3}{2}q^{(6)}l^2 - 3F_1l - \tfrac{9}{2}F_2l\right)\delta q^{(5)}\right.$$
$$\left. + \left(\tfrac{3}{2}q^{(4)}l^2 + \tfrac{3}{2}q^{(5)}l^2 + q^{(6)}l^2 - \tfrac{3}{2}F_1l - \tfrac{5}{2}F_2l\right)\delta q^{(6)}\right]$$

Proceeding in this fashion through the remaining 12 rods, we finally obtain the total complementary virtual work of all the stiffeners, as follows:

$$\delta W_{int,stiffeners}^* = \sum_{e=1}^{15}\delta W_{int}^{*(e)}$$

$$= \frac{l^2}{6AE}[(92q^{(4)}l + 50q^{(5)}l + 18q^{(6)}l - 15F_1 - 31F_2)\delta q^{(4)} + \tag{n}$$
$$+ (50q^{(4)}l + 56q^{(5)}l + 14q^{(6)}l - 15F_1 - 31F_2)\delta q^{(5)}$$
$$+ .(18q^{(4)}l + 14q^{(5)}l + 16q^{(6)}l - 3F_1 - 15F_2)\delta q^{(6)}]$$

To calculate the complementary virtual work of the shear panels, Equation 8.2.6 applies. Summing over all six panels, we have

$$\delta W_{int,panels}^* = \frac{l^2}{Gt}\sum_{e=1}^{6}q^{(e)}\delta q^{(e)}$$

Substituting the true and virtual shear flows in terms of the redundants and the applied loads, using Equations g1 through h1 and g2 through h2, yields

$$\delta W_{int,panels}^* = \frac{l^2}{Gt}\left[2q^{(4)}\delta q^{(4)} + \left(2q^{(5)} - \frac{F_2}{l}\right)\delta q^{(5)} + \left(2q^{(6)} - \frac{F_2}{l}\right)\delta q^{(6)}\right] \tag{o}$$

The total internal complementary virtual work for the structure is that of the stiffeners, Equation n, plus that of the panels, Equation o,

$$\delta W_{int}^* = \frac{1}{6AEGt}\left[\left(12AEl^2 + 92Gl^3t\right)q^{(4)} + 50Gl^3tq^{(5)} + 18Gl^3tq^{(6)} - 15Gl^2tF_1 - 31Gl^2tF_2\right]\delta q^{(4)}$$
$$+ \frac{1}{6AEGt}\left[50Gl^3tq^{(4)} + \left(12AEl^2 + 56Gl^3t\right)q^{(5)} + 14Gl^3tq^{(6)} - 9Gl^2tF_1 - \left(6AEl + 38Gl^2t\right)F_2\right]\delta q^{(5)}$$
$$+ \frac{1}{6AEGt}\left[18Gl^3tq^{(4)} + 14Gl^3tq^{(5)} + \left(12AEl^2 + 16Gl^3t\right)q^{(6)} - 3Gl^2tF_1 - \left(6AEi + 15Gl^2t\right)F_2\right]\delta q^{(6)}$$

The external complementary virtual work is zero because all of the redundants—the three shear flows $q^{(4)}$, $q^{(5)}$, and $q^{(6)}$—are internal loads. Therefore, δW_{int}^* must be zero for arbitrary choice of the virtual redundants, and this leads as usual to three equations for the true redundants, as follows:

$$\left(12AEl^2 + 92Gl^3t\right)q^{(4)} \qquad + 50Gl^3tq^{(5)} \qquad + 18Gl^3q^{(6)} = 15Gl^2tF_1 + 31Gl^2tF_2$$

$$50Gl^3tq^{(4)} + \left(12AEl^2 + 56Gl^3t\right)q^{(5)} \qquad + 14Gl^3tq^{(6)} = \ 9Gl^2tF_1 + \left(6AEl + 38Gl^2t\right)F_2$$

$$18Gl^3q^{(4)} \qquad + 14Gl^3tq^{(5)} + \left(12AEl^2 + 16Gl^3t\right)q^{(6)} = \ 3Gl^2tF_1 + \left(6AEl + 15Gl^2t\right)F_2$$

We solve this system for $q^{(4)}$, $q^{(5)}$, and $q^{(6)}$, and then substitute the results into Equations g1, h1, and i1 to obtain the remaining shear flows $q^{(1)}$, $q^{(2)}$, and $q^{(3)}$. This leads to the following expressions for the six shear flows in terms of the two applied loads F_1 and F_2:

$$q^{(1)} = \frac{1}{\Delta}\left\{-\left[2160\left(\frac{AE}{Glt}\right)^2 + 6912\frac{AE}{Glt} + 4644\right]\frac{F_1}{l} + \left[432\left(\frac{AE}{Glt}\right)^2 + 4392\frac{AE}{Glt} + 3744\right]\frac{F_2}{l}\right\} \qquad \text{[p]}$$

$$q^{(2)} = \frac{1}{\Delta}\left\{-\left[1296\left(\frac{AE}{Glt}\right)^2 + 2160\frac{AE}{Glt} + 948\right]\frac{F_1}{l} + \left[864\left(\frac{AE}{Glt}\right)^3 + 11,376\left(\frac{AE}{Glt}\right)^2 + 21,312\frac{AE}{Glt} + 10,640\right]\frac{F_2}{l}\right\} \qquad \text{[q]}$$

$$q^{(3)} = \frac{1}{\Delta}\left\{\left[-432\left(\frac{AE}{Glt}\right)^2 - 576\frac{AE}{Glt} + 156\right]\frac{F_1}{l} + \left[864\left(\frac{AE}{Glt}\right)^3 + 11,808\left(\frac{AE}{Glt}\right)^2 + 26,856\frac{AE}{Glt} + 15,968\right]\frac{F_2}{l}\right\} \qquad \text{[r]}$$

$$q^{(4)} = \frac{1}{\Delta}\left\{\left[2160\left(\frac{AE}{Glt}\right)^2 + 6912\frac{AE}{Glt} + 4644\right]\frac{F_1}{l} - \left[432\left(\frac{AE}{Glt}\right)^2 + 4392\frac{AE}{Glt} + 3744\right]\frac{F_2}{l}\right\} \qquad \text{[s]}$$

$$q^{(5)} = \frac{1}{\Delta}\left\{\left[1296\left(\frac{AE}{Glt}\right)^2 + 2160\frac{AE}{Glt} + 948\right]\frac{F_1}{l} + \left[864\left(\frac{AE}{Glt}\right)^3 + 12,240\left(\frac{AE}{Glt}\right)^2 + 32,688\frac{AE}{Glt} + 20,816\right]\frac{F_2}{l}\right\} \qquad \text{[t]}$$

$$q^{(6)} = \frac{1}{\Delta}\left\{\left[432\left(\frac{AE}{Glt}\right)^2 + 576\frac{AE}{Glt} - 156\right]\frac{F_1}{l} + \left[864\left(\frac{AE}{Glt}\right)^3 + 11,808\left(\frac{AE}{Glt}\right)^2 + 27,144\frac{AE}{Glt} + 15,488\right]\frac{F_2}{l}\right\} \qquad \text{[u]}$$

where

$$\Delta = 1728\left(\frac{AE}{Glt}\right)^3 + 23,616\left(\frac{AE}{Glt}\right)^2 + 54,000\frac{AE}{Glt} + 31,456 \qquad \text{[v]}$$

The ratio AE/Glt, which appears throughout these shear flow formulas, as well as those in the previous example, is a measure of the rigidity of the stiffeners relative to that of the shear webs. The ratio clearly influences the way in which the external load is transferred through the structure. At one extreme, in which the panels are very rigid compared to the relatively flexible stiffeners, we have $AE/Glt \to 0$. In this case Equation p through v imply in the limit that

$$q^{(1)} = -0.1476\frac{F_1}{l} + 0.1190\frac{F_2}{l}$$

$$q^{(2)} = -0.03014\frac{F_1}{l} + 0.3382\frac{F_2}{l}$$

$$q^{(3)} = 0.004959\frac{F_1}{l} + 0.5076\frac{F_2}{l} \qquad \text{[w]}$$

$$q^{(4)} = 0.1476\frac{F_1}{l} - 0.1190\frac{F_2}{l}$$

$$q^{(5)} = 0.03014\frac{F_1}{l} + 0.6617\frac{F_2}{l}$$

$$q^{(6)} = -0.004959\frac{F_1}{l} + 0.4924\frac{F_2}{l}$$

At the other extreme, in which the stiffeners are much more rigid than the panels, the shear flows are

$$q^{(1)} = q^{(4)} = 0$$

$$q^{(2)} = q^{(3)} = q^{(5)} = q^{(6)} = \frac{1}{2}\frac{F_2}{l}$$ [x]

Notice that in this case, all of the load F_1, applied to the top stiffener, is transferred directly to the wall by that member alone.

For a more typical situation, set $A = 0.5$ in.^{2}, $t = 0.05$ in., $E = 10^7$ lb/in.^{2} and $G = 0.4E$. If $l = 15$ in., then $AE/Glt = 1.666$, so that

$$q^{(1)} = -0.1136\frac{F_1}{l} + 0.06287\frac{F_2}{l}$$

$$q^{(2)} = -0.04177\frac{F_1}{l} + 0.4192\frac{F_2}{l}$$

$$q^{(3)} = 0.01027\frac{F_1}{l} + 0.5000\frac{F_2}{l}$$

$$q^{(4)} = 0.1136\frac{F_1}{l} - 0.06287\frac{F_2}{l}$$ [y]

$$q^{(5)} = 0.04177\frac{F_1}{l} + 0.5808\frac{F_2}{l}$$

$$q^{(6)} = -0.01027\frac{F_1}{l} + 0.5000\frac{F_2}{l}$$

8.4 THIN-WALLED BEAMS

In three-dimensional idealized beams of arbitrary cross section and lateral dimensions, the internal virtual work is divided between bending and shear, as follows:

$$\delta W^*_{int} = \delta W^*_{int, bending} + \delta W^*_{int, shear}$$ [8.4.1]

where

$$\delta W^*_{int, bending} = \iiint_V \delta\sigma_x \varepsilon_x dV = \iiint_V \frac{\sigma_x \delta\sigma_x}{E} dAdx$$ [8.4.2]

and

$$\delta W^*_{int, shear} = \iiint_V \delta\tau_{xs}\gamma_{xs}dV = \iiint_V \frac{\tau_{xs}\delta\tau_{xs}}{G} dAdx$$ [8.4.3]

where τ_{xs} is the shear stress, directed along the tangent to the middle surface of the wall (Figure 8.4.1). Since $\tau_{xs} = q/t$, where q is the shear flow and t is the wall thickness, Equation 8.4.3 may be written

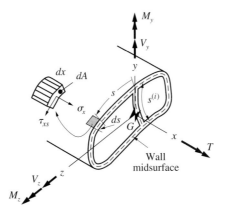

Figure 8.4.1   Section of a thin-walled box beam.

$$\delta W^*_{\text{int, shear}} = \iiint_V \frac{q\delta q}{Gt^2} dA dx = \int_0^L [\sum_{i=1}^{\text{No. walls}} \int_0^{s^{(i)}} \frac{q^{(i)}\delta q^{(i)}}{G^{(i)}t^{(i)2}}(t^{(i)}ds)]dx = L\sum_{i=1}^{\text{No. walls}} \int_0^{s^{(i)}} \frac{q^{(i)}\delta q^{(i)}}{G^{(i)}t^{(i)}} ds$$

where L is the length of the beam and $s^{(i)}$ is the length of the ith wall (measured along its midsurface). If, as is likely, the thickness and shear modulus of each wall are uniform over the length of the wall, then

$$\delta W^*_{\text{int, shear}} = L\sum_{i=1}^{\text{No. walls}} \frac{1}{G^{(i)}t^{(i)}} \int_0^{s^{(i)}} q^{(i)}\delta q^{(i)} ds \qquad \text{[8.4.4]}$$

The variable shear flow $q^{(i)}(s)$ in each wall is obtained from Equation 4.7.3.

To compute the complementary virtual work due to bending, recall that the flexure stress at a given section of the beam is according to Equation 4.6.6,

$$\sigma_x = K_y y + K_z z$$

where y and z are the coordinates of the point in the cross section relative to the centroid, and

$$K_y = -\frac{M_z I_y + M_y I_{yz}}{I_y I_z - I_{yz}^2} \qquad K_z = \frac{M_y I_z + M_z I_{yz}}{I_y I_z - I_{yz}^2} \qquad \text{[8.4.5]}$$

Thus,

$$\delta W^*_{\text{int, bending}} = \iiint_V \frac{1}{E}(K_y y + K_z z)(\delta K_y y + \delta K_z z)dV$$

Expanding the integrand, this becomes

$$\delta W^*_{\text{int, bending}} = \int_0^L \left\{ \iint_A \frac{1}{E} \left[K_y \delta K_y y^2 + (K_y \delta K_z + K_z \delta K_y) yz + K_z \delta K_z z^2 \right] dA \right\} dx$$

$$= \int_0^L \frac{1}{E} \left[K_y \delta K_y \left(\iint_A y^2 dA \right) + \left(K_y \delta K_z + K_z \delta K_y \right) \left(\iint_A yz\, dA \right) + K_z \delta K_z \left(\iint_A z^2 dA \right) \right] dx$$

The three integrals are the area moments and the product of inertia. Therefore,

$$\delta W^*_{\text{int,bending}} = \int_0^L \frac{1}{E} \left[K_y \delta K_y I_z + \left(K_y \delta K_z + K_z \delta K_y \right) I_{yz} + K_z \delta K_z I_y \right] dx \qquad \textbf{[8.4.6]}$$

Substituting Equation 8.4.5 into Equation 8.4.6, we finally obtain, after some simplification,

$$\delta W^*_{\text{int,bending}} = \int_0^L \frac{1}{E \left(I_y I_z - I_{yz}^2 \right)} \left\{ \left[M_y I_z + M_z I_{yz} \right] \delta M_y + \left[M_z I_y + M_y I_{yz} \right] \delta M_z \right\} dx \qquad \textbf{[8.4.7]}$$

Example 8.4.1   Find the displacement v_p in the direction of the 1000 lb load for the box beam in Figure 8.4.2. The elastic moduli are $E = 10 \times 10^6$ psi and $G = 4 \times 10^6$ psi.

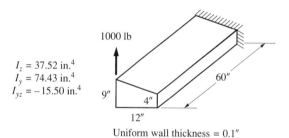

$I_z = 37.52$ in.^{4}
$I_y = 74.43$ in.^{4}
$I_{yz} = -15.50$ in.^{4}

1000 lb

60"

9"

4"

12"

Uniform wall thickness = 0.1"

Figure 8.4.2   Cantilever box beam. The section properties were calculated in Example 4.7.3.

The variable shear flows for this problem were calculated in Example 4.7.3, and they are reproduced in Figure 8.4.3a for convenience. Since the displacement in the direction of the applied load is required, the virtual load δP is applied in the direction of P on the same line of action, thereby producing virtual shear flows that differ from the true ones by the factor $\delta P / 1000$, as illustrated in Figure 8.4.3b.

Let us first use Equation 8.4.7 to obtain the internal complementary virtual work due to bending. Since $M_y = 0$ and $\delta M_y = 0$ for this problem, we have

$$\delta W^*_{\text{int,bending}} = \int_0^L \frac{I_y}{E \left(I_y I_z - I_{yz}^2 \right)} M_z \delta M_z dx \qquad \textbf{[a]}$$

Substituting the material and section properties (Figure 8.4.2) and the true and virtual bending moments $M_z = 1000x$ and $\delta M_z = \delta P x$, Equation a yields

$$\delta M_z = 9.720 \frac{L^3}{E} \delta P = 0.210 \delta P \qquad \textbf{[b]}$$

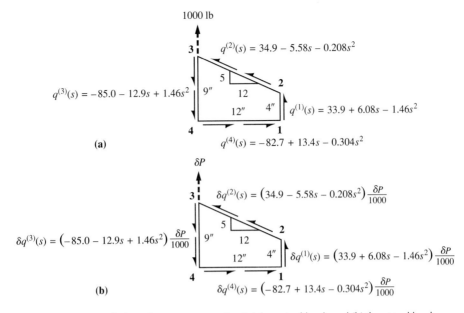

Figure 8.4.3   Wall shear flows accompanying (a) the actual load, and (b) the virtual load.

Since the four walls of the cross section have a common thickness and shear modulus, Equation 8.4.4 becomes

$$\delta W^*_{\text{int, shear}} = \frac{L}{Gt}\left(\int_0^4 q^{(1)}\delta q^{(1)}ds + \int_0^{13} q^{(2)}\delta q^{(2)}ds + \int_0^9 q^{(3)}\delta q^{(3)}ds + \int_0^{12} q^{(4)}\delta q^{(4)}ds \right) \qquad \text{[c]}$$

Substituting the true and virtual shear flows shown in Figure 8.4.3 and doing the integrations leads to

$$\delta W^*_{\text{int, shear}} = 135\frac{L}{Gt}\delta P = 0.0203\delta P \qquad \text{[d]}$$

The total internal complementary virtual work is therefore

$$\delta W^*_{\text{int}} = \delta W^*_{\text{int, bending}} + \delta W^*_{\text{int, shear}} = 0.230\delta P \qquad \text{[e]}$$

Setting this equal to the external complementary virtual work, we get $v_P \delta P = 0.230\delta P$, or

$$v_P = 0.230 \text{ in.} \qquad \text{[f]}$$

Observe that bending accounts for 91 percent of the displacement.

Example 8.4.2   Calculate the angle of twist per unit length for the beam in Figure 8.4.2.
Applying a virtual torque δT to the section produces the uniform shear flow

$$\delta q = \frac{\delta T}{2A} = \frac{\delta T}{2 \times \frac{1}{2}(9+4) \cdot 12} = \frac{\delta T}{156} \qquad \text{[a]}$$

as indicated in Figure 8.4.4.

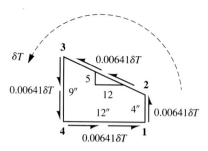

Figure 8.4.4 Constant shear flow due to a virtual torque δT.

Since the virtual bending moments are zero, it follows from Equation 8.4.7 that $\delta W^*_{int,bending} = 0$. Thus,

$$\delta W^*_{int} = \delta W^*_{int,shear} = \frac{L}{Gt} \left(\int_0^4 q^{(1)} ds + \int_0^{13} q^{(2)} ds + \int_0^9 q^{(3)} ds + \int_0^{12} q^{(4)} ds \right) \times \frac{\delta T}{156} \qquad \text{[b]}$$

where the constant virtual shear flow in Equation a is factored out of the right side of Equation 8.4.4. Substituting the shear flows from Figure 8.4.3a and integrating yields

$$\delta W^*_{int} = \frac{L}{Gt} (-7.39) \, \delta T = - \left(18.5 \times 10^{-6} \right) L \delta T \qquad \text{[c]}$$

Since $\delta W^*_{ext} = \theta_x \delta T$, the principle of complementary virtual work implies

$$\frac{\theta_x}{L} = -18.5 \times 10^{-6} \text{ rad/in.} = -0.00106 \text{ degree/in.}$$

The negative sign means that the angle of twist due to the actual load is clockwise, in the direction opposite to the virtual torque in Figure 8.4.4.

8.5   DEFLECTIONS IN IDEALIZED BEAMS

The complementary internal virtual work formula for idealized beams built up of stringers and shear panels combines Equation 8.4.7 for the stringers with the expressions obtained in section 8.2 for shear panels as follows:

$$\delta W^*_{int} = \int_0^L \frac{1}{E(I_y I_z - I_{yz}^2)} \{ [M_y I_z + M_z I_{yz}] \delta M_y + [M_z I_y + M_y I_{yz}] \delta M_z \} dx + \sum_{i=1}^{No.\ panels} \delta W^{*(i)}_{int} \qquad \text{[8.5.1]}$$

For each panel we substitute the appropriate virtual work expression, depending on whether it is a rectangle (Equation 8.2.6), parallelogram (Equation 8.2.8), or trapezoid (Equation 8.2.13).

Example 8.5.1   Calculate the horizontal (z) displacement of the centroid at the free end of the idealized, single-cell box beam depicted in Figure 8.5.1, given that a vertical shear of 1000 lb acts through the centroid. The location of the centroid and the values of the centroidal moments of inertia are shown.

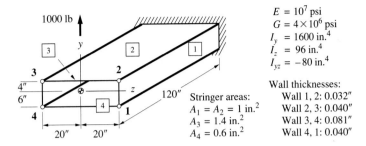

E = 10^7 psi
G = 4×10^6 psi
I_y = 1600 in.^4
I_z = 96 in.^4
I_yz = -80 in.^4

Wall thicknesses:
Wall 1, 2: 0.032"
Wall 2, 3: 0.040"
Wall 3, 4: 0.081"
Wall 4, 1: 0.040"

Stringer areas:
$A_1 = A_2 = 1$ in.²
$A_3 = 1.4$ in.²
$A_4 = 0.6$ in.²

Figure 8.5.1   Single-cell box beam subjected to a shear load. Material and geometric properties are given.

Figure 8.5.2a is a free-body diagram of a segment of the beam between the free end and station x. The internal reactive loads required for equilibrium at the cut are shown acting on the back side of the free body through the centroid G. Thus, at station x, $V_y = -1000$ lb, $V_z = 0$, $M_y = 0$, and $M_z = 1000x$ in.-lb. Proceeding in the usual fashion, starting with Equation 4.8.2, we first obtain the shear flows in each of the panels. Leaving the familiar details of their computation to the reader, we have sketched the results in Figure 8.5.3a.

To find the horizontal displacement of the centroid, we apply a horizontal virtual load δQ, as shown in Figure 8.5.2b. Also shown are the corresponding internal equilibrium reactions at the centroid of station x. They are: $\delta V_y = 0$, $\delta V_z = -\delta Q$,

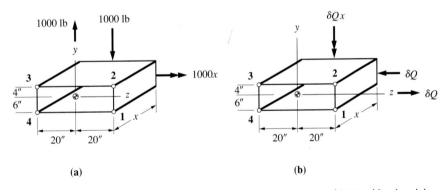

(a)   (b)

Figure 8.5.2   (a) Actual load and the internal reactions at station x. (b) Virtual load and the internal reactions at station x.

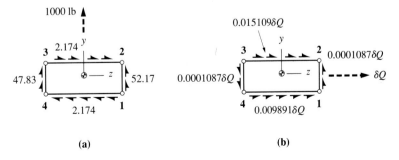

(a)   (b)

Figure 8.5.3   (a) Shear flows (lb/in.) accompanying the true 10,000 lb load.
(b) Virtual shear flows due to the virtual load.

$\delta M_y = -\delta Q\, x$, and $\delta M_z = 0$. We use these virtual loads, together with their true counterparts, to calculate the complementary virtual work of the structure.

Referring to Equation 8.5.1, let us first write down the internal complementary virtual work of the stringers, substituting the true and virtual quantities in their proper locations.

$$\delta W^*_{\text{int,stringers}} = \int_0^{120} \frac{1}{10^7 \left[1600 \cdot 96 - (-80)^2\right]} \left\{\left[0 \cdot 96 + (1000x)\,(-80)\right](-\delta Q\, x) + \left[1000x \cdot 1600 + 0 \cdot (-80)\right](0)\right\} dx$$

This reduces to

$$\delta W^*_{\text{int,stringers}} = 54.35 \times 10^{-9} \delta Q \int_0^{120} x^2 dx$$

and finally to

$$\delta W^*_{\text{int,stringers}} = 0.03130\delta Q$$

For the shear panels, we use Equation 8.2.6, starting with panel 1 and proceeding counterclockwise around the cell, as follows:

$$\delta W^*_{\text{int,panels}} = \sum_{\text{panels}} \frac{A}{Gt} q \delta q$$

$$= \frac{120 \cdot 10}{4(10^6)(0.032)}(52.17)(0.0001087\delta Q) + \frac{120 \cdot 40}{4(10^6)(0.04)}(-2.174)(-0.015109\delta Q)$$

$$+ \frac{120 \cdot 10}{4(10^6)(0.081)}(-47.83)(0.0001087\delta Q) + \frac{120 \cdot 40}{4(10^6)(0.04)}(-2.174)(0.009891\delta Q)$$

Completing the calculations leads to

$$\delta W^*_{\text{int,panels}} = 0.0003743\delta Q$$

Notice that we arbitrarily assumed that the shear flow in a web is positive if it is directed in the counterclockwise sense around the closed section. Whatever sign convention we choose, the product $q \delta q$ must be positive if the true and the virtual shear flows in a given web are in the same direction.

The total complementary virtual work for the beam is

$$\delta W^*_{\text{int}} = 0.03130\delta Q + 0.0003743\delta Q = 0.03168\delta Q$$

The complementary external virtual work is

$$\delta W^*_{\text{ext}} = w_G \times \delta Q$$

where w_G is the displacement of the centroid in the direction of the virtual load δQ. Equating the external and internal works yields

$$w_G = 0.03168 \text{ in.}$$

Thus, the 1000 lb load causes both an upward displacement of the centroid and a horizontal displacement component of 0.03168 inches to the right. The vertical component of displacement is found by repeating this analysis and using a virtual load δQ directed upward at the centroid instead of to the right. The reader is encouraged to do so and verify for comparison that the y component of deflection is 0.626 inches.

A common requirement is determining the angle of twist of a box beam with constant cross section. We will derive a general formula for this calculation. Figure 8.5.4 shows a portion of an idealized box beam of uniform cross section. The shear flows in part (a) of the figure correspond to the true loading of the structure. The true load is arbitrary and may consist of combined shear, bending, and torsion. The shear flows in part (b) of the figure arise from a virtual torque δT applied to the same structure. For our example, the webs of the box beam are numbered counterclockwise around the cross section. Let us denote the shear flow in web i as $q^{(i)}$. The length of the web measured in the plane of the cross section is $s^{(i)}$. All panels have a common length Δx, which is the length of the beam segment. A sufficient number of rigid ribs are spaced along the axis of the beam to maintain the shape of the cross section. Therefore, as the beam deforms, each section rotates as a rigid body around the twist axis (cf. section 6.8). The change in the angle of twist from one section to another can be found using Equation 8.5.1, together with the principle of complementary virtual work.

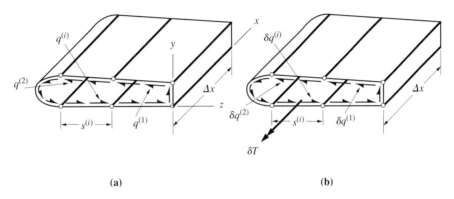

(a)   **(b)**

Figure 8.5.4   (a) Shear flows due to an arbitrary load on the box beam. (b) Virtual shear flows due to a virtual pure torque δT.

The virtual load is pure torsion. This means that the virtual bending moments are zero, so the stringers are not involved in the virtual work calculations. Since the beam is not tapered, the shear panels are either rectangular or cylindrical. For such panels, the internal complementary virtual work is given by Equation 8.2.6. Therefore, the internal complementary virtual work for panel i is

$$\delta W_{int}^{*(i)} = \frac{A^{(i)}}{G^{(i)}t^{(i)}} q^{(i)}\delta q^{(i)} = \frac{s^{(i)}\Delta x}{G^{(i)}t^{(i)}} q^{(i)}\delta q^{(i)}$$

Since the virtual load is pure torsion, the virtual shear flow is the same in every web and is

$$\delta q^{(1)} = \delta q^{(2)} = \ldots = \delta q^{(i)} = \ldots = \frac{\delta T}{2A}$$

where A is the area of the closed cross section, which is the region enclosed by the webs. Therefore, for each panel, we have

$$\delta W_{int}^{*(i)} = \frac{\delta T \, \Delta x}{2A} \frac{s^{(i)}}{G^{(i)}t^{(i)}} q^{(i)}$$

The external complementary virtual work is $\Delta\theta \, \delta T$, where $\Delta\theta$ is the true rotation of the cross section at $x + \Delta x$ relative to that at x. $\Delta\theta$ is positive if the rotation is clockwise around the positive x axis. In other words, a positive twist angle has the same direction as positive torque (cf. Figure 4.3.1). Setting the external complementary virtual work equal to the total internal complementary virtual work of all n panels comprising the cross section yields

$$\Delta\theta_x \, \delta T = \frac{\delta T \, \Delta x}{2A} \sum_{i=1}^{n} \frac{s^{(i)}}{G^{(i)} t^{(i)}} q^{(i)}$$

or

$$\frac{\Delta\theta_x}{\Delta x} = \frac{1}{2A} \sum_{i=1}^{n} \frac{s^{(i)}}{G^{(i)} t^{(i)}} q^{(i)}$$

In the limit as Δx approaches zero, we have

$$\frac{d\theta_x}{dx} = \frac{1}{2A} \sum_{i=1}^{n} \frac{s^{(i)}}{G^{(i)} t^{(i)}} q^{(i)} \qquad\qquad [8.5.2]$$

The shear flow, web length, and thickness are likely to vary from web to web, but the shear modulus G is usually the same all around the section. Therefore, we can move it outside the summation to get

$$\frac{d\theta_x}{dx} = \frac{1}{2AG} \sum_{i=1}^{n} q^{(i)} \frac{s^{(i)}}{t^{(i)}} \qquad \text{uniform G} \qquad [8.5.3]$$

If the actual load is pure torsion, then all of the true shear flows equal $T/2A$, and Equation 8.5.3 becomes

$$\frac{d\theta_x}{dx} = \frac{T}{4A^2 G} \sum_{i=1}^{n} \frac{s^{(i)}}{t^{(i)}} \qquad \text{pure torsion} \qquad [8.5.4]$$

According to Equation 4.4.14, the torsion constant J is given by $J = T/G\phi$, where $\phi = d\theta_x/dx$. For an idealized box beam, Equation 8.5.4 therefore yields

$$J = \frac{4A^2}{\displaystyle\sum_{i=1}^{n} \frac{s^{(i)}}{t^{(i)}}} \qquad\qquad [8.5.5]$$

Remember that Equations 8.5.2 through 8.5.5 are valid *only for beams of constant cross section*.

Example 8.5.2   Calculate the angle of twist of the free end of the cantilevered idealized box beam in Figure 8.5.5. The location of the centroid, as well as the values of the centroidal moments of inertia, are shown in the figure.

From Equation 4.8.2, we find the flange load gradients in stringers 1 through 4, as shown in parentheses in Figure 8.5.6. Starting with flange 1 and using the spanwise equilibrium condition at each stringer (shear flow out equals shear flow in plus P'), we find the shear flows $q^{(1)}$, $q^{(2)}$, and $q^{(3)}$ in terms of $q^{(4)}$, as follows:

$$\begin{aligned} q^{(1)} &= q^{(4)} + 40 \\ q^{(2)} &= q^{(1)} - 40 = \left(q^{(4)} + 40\right) - 40 = q^{(4)} \\ q^{(3)} &= q^{(2)} - 120 = q^{(4)} - 120 \end{aligned} \qquad [a]$$

Moment equivalence about flange 4 requires that

$$\left(10q^{(1)}\right) \times 40 + \left(40q^{(2)}\right) \times 10 = 0$$

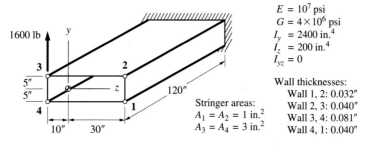

Figure 8.5.5   Single-cell box beam subjected to a shear load. The material and geometric properties are shown.

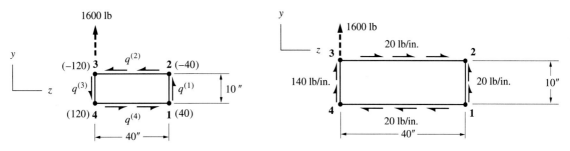

Figure 8.5.6   Flange load gradients (lb/in.) and the assumed shear flow directions.

Figure 8.5.7   Computed shear flows in the box beam.

or, using Equation a,

$$400\left[\left(q^{(4)} + 40\right) + q^{(4)}\right] = 0$$

Solving this for $q^{(4)}$ and substituting the result into Equation a yields the shear flows shown in Figure 8.5.7.

To find the angle of twist of the near end of the beam relative to the built-in end, we use Equation (8.5.3) and require that $q^{(i)}$ be considered positive if it acts in the counterclockwise sense around the cell. For our case, we get

$$\frac{d\theta_x}{dx} = \frac{1}{2(40 \times 10)(4 \times 10^6)}\left[(20)\frac{10}{0.032} + (-20)\frac{40}{0.04} + (-140)\frac{10}{0.081} + (-20)\frac{40}{0.04}\right]$$
$$= -15.9 \times 10^{-6} \text{ radians/in.}$$

Therefore,

$$\theta_x = \theta_x)_{x=0} - 15.9 \times 10^{-6}x$$

Since $\theta_x = 0$ at $x = 120$ in., we have

$$0 = \theta_x)_{x=0} - 15.9 \times 10^{-6}(120)$$

so that

$$\theta_x)_{x=0} = 0.00191 \text{ rad} = 0.109°$$

The positive sign means that the rotation of the section is clockwise.

The deflection analysis of tapered idealized box beams is complicated by the fact that the shear flow in the trapezoidal webs is not constant. For the shear panels, the problem reduces to calculating the average shear flow in each panel, since that is the load parameter that appears in the internal complementary virtual work expression, Equation 8.2.13. To calculate the average shear flows in each panel of a box beam, for both the true and the virtual loads, we use the procedure developed in Chapter 4. That is, we first calculate the flange loads in the stringers at each end of the tapered box segment, using Equation 4.8.1. We then require each stringer to be in equilibrium under the combined action of the end flange loads and the average shear flows exerted over the length of the stringers by the adjacent webs. Those equilibrium equations, together with moment equivalence about the beam axis, yield the average shear flows and hence the complementary virtual work for each panel.

In general, this procedure for tapered beams does not yield accurate results for the average shear flows if the procedure is applied over large spanwise distances. Accuracy can be improved by dividing the beam into several smaller spanwise segments, applying the method to each segment, and finally summing up the individual complementary virtual works to obtain $\delta W^*_{\text{int,panels}}$ for the complete beam.

To calculate the complementary virtual work of the stringers, which act in concert to carry the bending loads, we employ Equation 8.4.7, which accounts for the variation of both bending moments, and cross-sectional geometry with respect to span.

Example 8.5.3   For the idealized tapered box beam pictured in Figure 8.5.8, calculate: (a) the deflection in the direction of the applied load, and (b) the rotation of the free end. The figure shows the geometric and material property data for the structure.

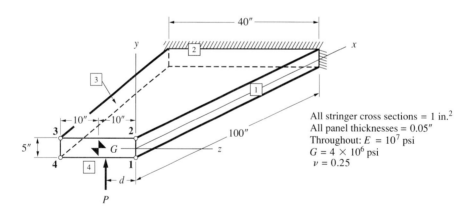

All stringer cross sections = 1 in.^{2}
All panel thicknesses = 0.05″
Throughout: $E = 10^7$ psi
$G = 4 \times 10^6$ psi
$\nu = 0.25$

Figure 8.5.8   Cantilevered, tapered, idealized box beam with a point load at the free end.

(a)

To solve this problem, we must calculate the internal complementary virtual work δW^*_{int} of the structure by adding together the complementary virtual work of the individual structural members, which are the rods and the shear panels. Since this idealized assembly is assumed to deform in bending according to beam theory, we can use Equation 8.4.7 to calculate the collective virtual work of the four stringers.

First, we note that the centroidal area moments of inertia are

$$I_{yz} = 0 \text{ and } I_z = 4 \times \left[1 \times \left(5/2\right)^2 \right] = 25 \text{ in.}^4 \qquad [a]$$

I_y is not required, since there is no bending about the y axis in this problem; that is,

$$M_y = 0 \qquad \text{[b]}$$

The bending moment M_z increases linearly with x, or

$$M_z = Px \qquad \text{[c]}$$

If we replace the true load P with the virtual load δP acting in the same direction, then the virtual bending moment is

$$\delta M_z = \delta P \, x$$

and, according to Equation 8.4.7, the internal complementary virtual work for the stringers is

$$\delta W^*_{\text{int,stringers}} = \int_0^{100} \frac{1}{EI_z} M_z \delta M_z dx = \int_0^{100} \frac{1}{10^7 \times 25} (Px)(\delta Px)\, dx = 0.001333 P\delta P$$

To calculate the internal complementary virtual work due to shear, we need to obtain the average shear flow $\bar{q}^{(i)}$ in each of the four panels. To do so, we first establish spanwise equilibrium of each of the stringers by means of Equation 4.9.9,

$$\sum_{\substack{\text{adjoining webs}}} \bar{q}_{\text{out}} = \frac{\overline{dP_x^{(i)}}}{dx} \qquad \text{[d]}$$

The average flange load gradient $\overline{dP_x^{(i)}/dx}$. is found by computing the flange load at each end of the beam and using Equation 4.9.8, as follows:

$$\frac{\overline{dP_x^{(i)}}}{dx} = \frac{P_x^{(i)}(L) - P_x^{(i)}(0)}{L} \qquad \text{[e]}$$

where L in this case is 100 inches. The flange loads in this expression are obtained from Equation 4.8.1, which, in view of Equations a through c reduces to

$$P_x^{(i)} = -\frac{Px_i y_i}{25} A_i \qquad \text{[f]}$$

All of the flange loads are zero at the free end of the beam ($x = 0$). However, at $x = 100$ in., Equation f yields

$$P_x^{(1)}(100) = 10P \qquad P_x^{(2)}(100) = -10P \qquad P_x^{(3)}(100) = -10P \qquad P_x^{(4)}(100) = 10P \qquad \text{[g]}$$

Substituting these into Equation e, we obtain the average flange load gradients

$$\frac{\overline{dP_x^{(1)}}}{dx} = \frac{P}{10} \qquad \frac{\overline{dP_x^{(2)}}}{dx} = -\frac{P}{10} \qquad \frac{\overline{dP_x^{(3)}}}{dx} = \frac{P}{10} \qquad \frac{\overline{dP_x^{(4)}}}{dx} = -\frac{P}{10} \qquad \text{[h]}$$

These are shown on a generic cross section of the beam in Figure 8.5.9, which also indicates the assumed directions of the average shear flows in the panels. By applying Equation e at flanges 2, 3, and 4, we can obtain the average shear flows $\bar{q}^{(2)}$, $\bar{q}^{(3)}$, and $\bar{q}^{(4)}$ in terms of $\bar{q}^{(1)}$, as follows:

$$\bar{q}^{(2)} = \bar{q}^{(1)} - \frac{P}{10} \qquad \bar{q}^{(3)} = \bar{q}^{(1)} - \frac{P}{5} \qquad \bar{q}^{(4)} = \bar{q}^{(1)} - \frac{P}{10} \qquad \text{[i]}$$

To obtain $\bar{q}^{(1)}$, we establish moment equivalence at the free end of the beam, where the flange loads are all zero. Setting the moments of the shear flows about flange 1 equal to the moment of the load P, we get (Figure 8.5.10)

$$20q^{(2)}(0) \times 5 + 5q^{(3)}(0) \times 20 = -P \times d \qquad \text{[j]}$$

Since the shear flow is constant throughout the rectangular side panels, we have $q^{(3)}(0) = \bar{q}^{(3)}$. In the top and bottom panels, the shear flow varies according to Equation 2.5.4, that is,

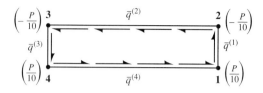

Figure 8.5.9   A generic cross section of the tapered beam, showing the average flange load gradients (in parentheses) and the assumed directions of the average shear flows.

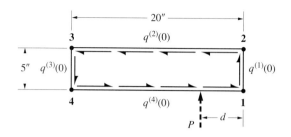

Figure 8.5.10   Shear flows at the free end of the tapered beam.

$$q(x) = \bar{q}\frac{h(0)h(100)}{h(x)^2}$$

where the variable width of these panels is given by

$$h(x) = 20(1 + 0.01x)$$

Thus, $q^{(2)}(0) = 2\bar{q}^{(2)}$. Equation j therefore becomes

$$200\bar{q}^{(2)} + 100\bar{q}^{(3)} = -P \times d$$

Substituting for $\bar{q}^{(2)}$ and $\bar{q}^{(3)}$ in terms of $\bar{q}^{(1)}$, using Equation i, yields

$$200\left(\bar{q}^{(1)} - \frac{P}{10}\right) + 100\left(\bar{q}^{(1)} - \frac{P}{5}\right) = -P \times d$$

We solve this for $\bar{q}^{(1)}$ and then use Equation i to find the remaining average shear flows in terms of the applied load P and its distance d from the rightmost vertical web, as follows:

$$\bar{q}^{(1)} = \frac{(40 - d)P}{300} \qquad \bar{q}^{(2)} = \frac{(10 - d)P}{300} \qquad \bar{q}^{(3)} = -\frac{(20 + d)P}{300} \qquad \bar{q}^{(4)} = \frac{(10 - d)P}{300} \qquad \text{[k]}$$

Since we are seeking the displacement at and in the direction of the applied load P, the virtual load is identical to the true load, replacing P with δP. Therefore, we use Equation k to find the virtual average shear flows:

$$\delta\bar{q}^{(1)} = \frac{(40 - d)\delta P}{300} \qquad \delta\bar{q}^{(2)} = \frac{(10 - d)\delta P}{300} \qquad \delta\bar{q}^{(3)} = -\frac{(20 + d)\delta P}{300} \qquad \delta\bar{q}^{(4)} = \frac{(10 - d)\delta P}{300} \qquad \text{[l]}$$

Substituting the true and virtual shear flows, along with the given geometric and material property data, into the virtual work expressions—Equation 8.2.6 for the rectangles—we get

$$\delta W_{int}^* = \frac{A}{Gt}q\delta q \qquad \text{[m]}$$

Using Equation 8.2.13 for the trapezoids, we get

$$\delta W_{int}^* = \left[1 + \frac{2}{3(1 + v)}\left(\cot^2\gamma - \cot\alpha\ \cot\gamma + \cot^2\alpha\right)\right]\frac{A}{Gt}\bar{q}\delta\bar{q} \qquad \text{[n]}$$

(Note that the included angles on the longer of the two parallel edges are 90° and 78.69°.) We therefore get

$$\delta W^*_{\text{int,panels}} = \overbrace{\frac{5 \cdot 100}{4(10^6) \cdot 0.05}\left[\frac{P(40-d)}{300}\right]\left[\frac{\delta P(40-d)}{300}\right]}^{\text{panel 1}} + \overbrace{\frac{5 \cdot \sqrt{100^2+20^2}}{4(10^6) \cdot 0.05}\left[-\frac{P(20+d)}{300}\right]\left[-\frac{\delta P(20+d)}{300}\right]}^{\text{panel 3}}$$

$$+ \overbrace{2 \times \left[1 + \frac{2}{3(1+0.25)}\cot^2 78.69°\right]\frac{\frac{1}{2}(20+40) \cdot 100}{4(10^6) \cdot 0.05}\left[\frac{P(10-d)}{300}\right]\left[\frac{\delta P(10-d)}{300}\right]}^{\text{panels 2 and 4}}$$

or

$$\delta W^*_{\text{int,panels}} = (3.966 \times 10^{-7}d^2 - 7.898 \times 10^{-6}d + 8.982 \times 10^{-5})P\delta P$$

The total internal complementary virtual work, $\delta W^*_{\text{int,stringers}} + \delta W^*_{\text{int,panels}}$, is therefore

$$\delta W^*_{\text{int}} = (3.966 \times 10^{-7}d^2 - 7.898 \times 10^{-6}d + 142.3 \times 10^{-5})P\delta P$$

Setting the right-hand side of this equation equal to the external complementary virtual work, $v\delta P$, yields the expression for v, the vertical component of displacement, in terms of d, the location of the applied load, as follows:

$$v = (3.966 \times 10^{-7}d^2 - 7.898 \times 10^{-6}d + 142.3 \times 10^{-5})P$$

For example, if we select $d = 5$ in. for the load application point, the vertical displacement of that point is

$$v = 0.001393P \text{ in.} \qquad \text{(where } P \text{ is measured in pounds)}$$

(b)

To compute the rotation of the free end of the beam, we must remove the true load P and apply a virtual torque δT to the section. Assuming the flange loads at both ends of the tapered box are zero, the pure torque δT produces an average virtual shear flow $\delta \bar{q}_T$, which is the same for every panel. The term $\delta \bar{q}_T$ is found by establishing moment equivalence at the free end of the beam. Just as for Equation j, we sum moments about flange 1:

$$\delta q^{(2)}(0) \times 20 \times 5 + \delta q^{(3)}(0) \times 5 \times 20 = \delta T$$

Again, since panel 3 is rectangular, $\delta q^{(3)}(0) = \delta \bar{q}_T$; however, according to Equation 2.5.4, $\delta q^{(2)}(0) = 2\delta \bar{q}_T$ because of the taper of panel 2. Thus,

$$2\delta \bar{q}_T \times 20 \times 5 + \delta \bar{q}_T \times 5 \times 20 = \delta T$$

so that

$$\delta \bar{q}_T = \frac{\delta T}{300} \qquad\qquad \text{[o]}$$

We substitute this virtual shear flow and the true shear flows (Equation k) into Equations m and n to get the total internal complementary virtual work. The stringers do not enter into the picture here, since there are no virtual bending moments associated with the applied virtual torque δT. That is, $\delta W^*_{\text{int}} = \delta W^*_{\text{int,panels}}$, so that

$$\delta W^*_{\text{int}} = \overbrace{\frac{5 \cdot 100}{4(10^6) \cdot 0.05}\left[\frac{P(40-d)}{300}\right]\left[\frac{\delta T}{300}\right]}^{\text{panel 1}} + \overbrace{\frac{5 \cdot \sqrt{100^2+20^2}}{4(10^6) \cdot 0.05}\left[-\frac{P(20+d)}{300}\right]\left[\frac{\delta T}{300}\right]}^{\text{panel 3}}$$

$$+ \overbrace{2 \times \left[1 + \frac{2}{3(1+0.25)}\cot^2 78.69°\right]\frac{\frac{1}{2}(20+40) \cdot 100}{4(10^6) \cdot 0.05}\left[\frac{P(10-d)}{300}\right]\left[\frac{\delta T}{300}\right]}^{\text{panels 2 and 4}}$$

Upon simplification, this becomes

$$\delta W_{int}^* = (3.949 \times 10^{-6} - 3.966 \times 10^{-7}d)P\delta T$$

Setting this expression equal to the external complementary virtual work, $\theta \, \delta T$, where θ is the rotation of the cross section, we conclude that

$$\theta = (3.949 \times 10^{-6} - 3.966 \times 10^{-7}d)P \quad \text{radians} \qquad \text{(where } P \text{ is measured in pounds)}$$

As expected, the twist angle depends on the location of the applied load P. That angle is zero if the load passes through the point $d = 9.96$ inches.

8.6 SHEAR CENTER OF CLOSED SECTIONS

The shear center of constant-cross-section beam is the point through which the shear load borne by a section must pass if the accompanying stresses in the cross section are those dictated by beam theory. For thin-walled sections, this means Equation 4.7.3 alone governs the shear stress distribution. Therefore, by definition, there are no torsional shear stresses if the shear load acts through the shear center. Since the beam carries no torque, it bends but does not twist.

The procedure for locating the shear center of open section beams is explained in section 4.6 and illustrated in Example 4.7.1. For a closed section, we find the shear center by applying a shear load of arbitrary magnitude and requiring that the angle of twist per unit length, as given by Equation 8.5.2, be zero. Instead of applying the shear load at an arbitrary point and calculating the angle of twist, we set the angle of twist equal to zero and find the required line of action of the load.

The following example illustrates the fact that, in general, the shear center and the centroid of a section do not coincide. However, if a section has an axis of symmetry, then the shear center, like the centroid, lies upon that axis. It follows that if a section is doubly symmetric, the shear center lies at the centroid.

Example 8.6.1 Find the shear center of the section shown in Figure 8.6.1.

First we calculate the location of the centroid G. Using the information in the figure, we find that

$$y_G = \frac{(8 \times 2) + (6 \times 0.5) + (4 \times 1)}{2 \times (2 + 0.5 + 1)} = 3.286 \text{ in.}$$

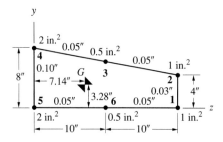

Figure 8.6.1   Idealized beam section.

$$z_G = \frac{2 \times (10 \times 0.5) + 2 \times (20 \times 1)}{2 \times (2 + 0.5 + 1)} = 7.143 \text{ in.}$$

The centroidal area moments of inertia are

$$I_{Gy} = 2 \times \left[2 \times (7.143)^2\right] + 2 \times \left[0.5 \times (10 - 7.143)^2\right] + 2 \times \left[1 \times (20 - 7.143)^2\right] = 542.9 \text{ in.}^4$$

$$\begin{aligned} I_{Gz} &= 2 \times \left[(8 - 3.286)^2 + (3.286)^2\right] + 0.5 \times \left[(6 - 3.286)^2 + (3.286)^2\right] \\ &\quad + 1 \times \left[(4 - 3.286)^2 + (3.286)^2\right] = 86.43 \text{ in.}^4 \end{aligned}$$

$$\begin{aligned} I_{Gyz} &= 2 \times [(8 - 3.286)(-7.143) + (-3.286)(-7.143)] \\ &\quad + 0.5 \times [(6 - 3.286)(10 - 7.143) + (-3.286)(10 - 7.143)] \\ &\quad + 1 \times [(4 - 3.286)(20 - 7.143) + (-3.286)(20 - 7.143)] = -54.29 \text{ in.}^4 \end{aligned}$$

From Equation 4.8.2, the flange load gradient at the ith flange is, therefore,

$$P_x'^{(i)} = 2.274 \times 10^{-5} \left\{ V_y [542.9(y_i - y_G) + 54.29(z_i - z_G)] + V_z [54.29(y_i - y_G) + 86.43(z_i - z_G)] \right\} A_i \qquad \text{[a]}$$

where y_i and z_i are the flange coordinates relative to the axes on Figure 8.6.1.

To find the z coordinate of the shear center, we apply an arbitrary shear force in the y direction and set $V_z = 0$. If $V_y = -1000$ lb, then Equation a gives the flange load gradients shown in parentheses by each flange in Figure 8.6.2. Starting, say, with web 6-1 on the lower right of the section, we calculate all of the web shear flows in terms of q_{6-1} using Equation 4.8.3. These are shown in Figure 8.6.2. We do not yet know where the 1000 lb load acts, but since it acts through the shear center we do know that the twist of the cross section is zero; that is, from Equation 8.5.3, we have

$$\theta' = \frac{1}{2AG} \sum_{i=1}^{\text{No. walls}} q^{(i)} \frac{s^{(i)}}{t^{(i)}} = 0$$

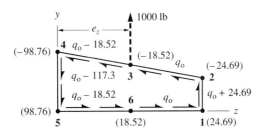

Figure 8.6.2   Flange load gradients (lb/in.) and wall shear flows.

Starting with the wall joining flanges 1 and 2, and moving counterclockwise around the cell, we get

$$\begin{aligned} \frac{1}{2AG} &\left[(q_o + 24.69)\left(\frac{4}{0.03}\right) + q_o\left(\frac{10.2}{0.05}\right) + (q_o - 18.52)\left(\frac{10.2}{0.05}\right) \right. \\ &\left. + (q_o - 117.3)\left(\frac{8}{0.1}\right) + (q_o - 18.52)\left(\frac{10}{0.05}\right) + q_o\left(\frac{10}{0.05}\right) \right] = 0 \end{aligned}$$

Collecting terms and solving for q_o yields

$$q_o = 13.29 \text{ lb}/\text{in.}$$

If we substitute this value into Figure 8.6.2, we get the shear flow pattern illustrated in Figure 8.6.3.

Finally, we locate the shear center by requiring that the moments of the shear flows corresponding to zero twist angle equal that of the 1000 lb load acting through the shear center. Summing the moments about the lower left corner of the section, we have

$$(2 \times 40 \times 37.98) + (2 \times 40 \times 13.29) - (2 \times 40 \times 5.229) = 1000 \times e_z$$

or

$$e_z = 3.683 \text{ in.}$$

Therefore, the shear center lies 3.68 inches to the right of the leftmost vertical wall of the cell.

To find the y coordinate of the shear center, we set $V_y = 0$, assign Vz an arbitrary value, and proceed as before, calculating the shear flows corresponding to a zero twist angle. For a 1000 lb load to the right, the results are shown in Figure 8.6.4.

As before, we locate the 1000 lb force by summing the moments around flange 5, as follows:

$$-(2 \times 40 \times 26.64) - (2 \times 40 \times 22.16) + (2 \times 40 \times 3.992) = -1000 \times e_y$$

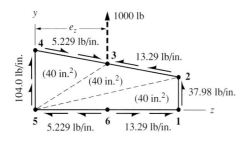

Figure 8.6.3   Shear flows required to prevent the section from rotating.

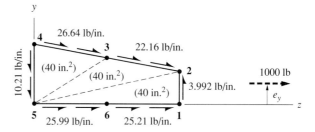

Figure 8.6.4   Shear flows for a 1000 lb load to the right and zero twist.

so that

$$e_y = 3.585 \text{ in.}$$

This places the shear center 3.58 inches above the bottom of the cell.

Compare the shear center coordinates with those of the centroid shown in Figure 8.6.1.

8.7   WARPING DEFLECTIONS

Figure 8.7.1a shows a thin-walled, idealized box beam in pure torsion. The dimensions of the rectangular cross section are h and w, where $w > h$. The cross section is shown a distance L from the rigid wall supporting the beam. As we know, an applied torque T produces the constant shear flow $q = T/2A$ around the cross section, the area of which is $A = wh$. The shear stress in a given wall (relative to its local xy axes as shown in Figure 8.7.1a) is $\tau_{xy} = q/t$, and it is not uniform around the section unless the wall thickness t is constant. In each shear panel, the shear stress τ_{xy} produces a shear strain $\gamma_{xy} = \tau_{xy}/G$, where G is the shear modulus.

The deformation at a given section of the torque box consists of two modes, shown in Figure 8.7.1b and c, respectively. The first is a rigid-body rotation of the cross section around the twist axis. This rotation increases lin-

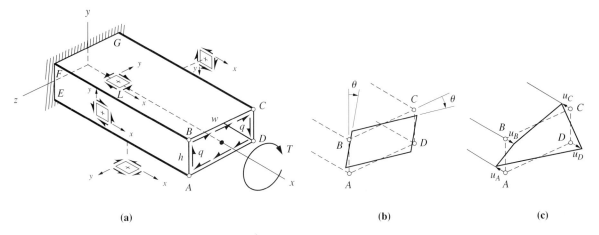

(a) (b) (c)

Figure 8.7.1 (a) Thin-walled box beam in pure torsion. Each panel shows a "local" xy axis relative to which the sign of stress and strain in that panel may be established. (b) Rigid-body rotation of the cross section. (c) Cross sectional warping, which is superimposed on the rigid rotation.

early with distance from the fixed end of the beam. The rectangular shape of the cross section is maintained by internal ribs. The rigid rotation is generally accompanied by the warping deformation illustrated in Figure 8.7.1c.

Warping occurs because of the torque-induced shear strain in the walls of the box beam. Consider any one of the panels, such as side panel $ABFE$. As illustrated in Figure 8.7.2a, rigid rotation of the beam cross section induces transverse components of corner displacement $v_A = v_B$ in the plane of the panel. These in turn produce the shear strain due solely to twist: $\gamma_{xy,\text{twist}} = \tau/G$. Since the displacement v_A depends on the distance between corner A and the axis of rotation, $\gamma_{xy,\text{twist}}$ is proportional to the distance of the panel from the twist axis and is greater in the narrow side panels than in the wide cover panels. Suppose for simplicity that all of the walls have the same thickness t, which means that the shear stress, and therefore the net shear strain γ_{xy}, is the same in every panel. Since $\gamma_{xy,\text{twist}}$ varies from panel to panel, there must be another component of the shear strain, $\gamma_{xy,\text{warp}}$, such that

$$\gamma_{xy} = \gamma_{xy,\text{twist}} + \gamma_{xy,\text{warp}}$$

Here, $\gamma_{xy,\text{warp}}$ compensates for the variation of $\gamma_{xy,\text{twist}}$ in going from one panel to the next, so that net shear strain γ_{xy} is constant. The only way $\gamma_{xy,\text{warp}}$ can develop is if axial components of displacement u_A and u_B arise,

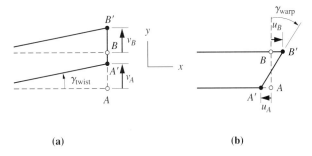

(a) (b)

Figure 8.7.2 Portion of the side panel $ABCD$ of Figure 8.7.1.
Shear strain due to (a) twist and (b) warping.

as shown in Figure 8.7.2b.[1] These axial displacements, normal to the beam cross section, warp the cross section in the manner illustrated in Figure 8.7.1c.

Note that if the cross section of the torque box is square $(h = w)$ and the wall thickness is uniform, then both $\gamma_{xy,\text{twist}}$ and γ_{xy} are identical in all four panels. No compensating warping strain is required and, indeed, does not occur. The same may be said of any uniform-thickness thin-walled cylindrical tube whose cross section is a polygon, all edges of which are tangent to an inscribed circle. In the limit as we increase the number of sides, a circular tube is obtained, and we showed by another means in Example 4.4.1 that the tube does not warp in pure torsion.

In our previous discussion of beam deflections, we ignored the out-of-plane displacements of a cross section that accompany the beam's rotation due to torsion. Let us assume that the tendency for each cross section to warp out of its original plane can occur freely and without restraint. In actuality, warping is restrained near rigid supports and transverse stiffeners, such as massive frames and bulkheads, which can resist loads normal to their plane. To suppress warping, local stresses not accounted for by pure torsion analysis must develop. It may be possible to estimate the stresses required to restrain warping, if we can first compute the warping displacements themselves. This will be explored further in section 8.9.

Consider the idealized box beam shown in Figure 8.7.3a. For purposes of illustration, let the four stringers all have the same area A and let the panels have a common thickness t and a common shear modulus. By symmetry, both the centroid and the shear center of the rectangular cross section lie at the geometric center G. The beam supports a vertical shear load P, offset a distance d from the left vertical web. The shear flows, computed by means of Equations 4.8.2 and 4.8.3, are illustrated in Figure 8.7.3b.

If the load P passes through the shear center at G, pure bending occurs and warping is zero. To calculate the warping deflections associated with an arbitrary value of d, we remove the true load, section the beam at a station $x = x_1$, and apply a self-equilibrating set of virtual loads to the stringers of the free body lying between $x = x_1$ and the support. This is shown in Figure 8.7.4a. Each virtual load is applied normal to the section and has a magnitude δP. The accompanying virtual shear flows $\delta q^{(1)}$, $\delta q^{(2)}$, $\delta q^{(3)}$, and $\delta q^{(4)}$ are found as usual in terms of δP by invoking spanwise equilibrium of the stringers between $x = x_1$ and $x = L$, in addition to requiring that their resultant virtual moment vanish. The results of these calculations appear in Figure 8.7.4b.

Assuming that the displacements of the stringers are in the positive x direction, the external complementary virtual work is

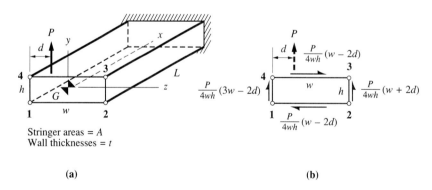

(a)   **(b)**

Figure 8.7.3   (a) Idealized box beam with a vertical shear load P. (b) The corresponding section shear flows.

---

[1]Recall the development of the strain–displacement relations in section 3.7. See, in particular, Equation 3.7.24 and Figure 3.7.4.

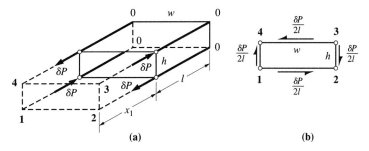

Figure 8.7.4   (a) Self-equilibrating set of virtual stringer loads applied at a given section of the box beam of the previous figure. (b) The corresponding virtual shear flows.

$$\delta W_{\text{ext}}^* = \delta P u_1 - \delta P u_2 + \delta P u_3 - \delta P u_4$$

The minus signs account for the opposite directions of virtual load and displacement at flanges 2 and 4. This expression can also be written as

$$\delta W_{\text{ext}}^* = \delta P \left[(u_1 - u_4) - (u_2 - u_3) \right]$$

From Figure 8.7.5, we see that $(u_1 - u_4)/h$ is the angle ϕ_{14} through which web 1-4 rotates, whereas web 2-3 rotates through the angle $\phi_{23} = (u_2 - u_3)/h$. If the warp angle ϕ is the difference between the rotations of the two webs, both measured positive counterclockwise around the z axis, the external complementary virtual work is

$$\delta W_{\text{ext}}^* = \delta P[\phi_{14} h - \phi_{23} h] = h \phi \delta P \qquad \text{[8.7.1]}$$

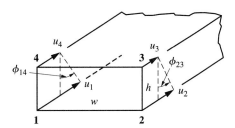

Figure 8.7.5   Relation between flange displacements and the rotations of the vertical sides of the cell.

Since the virtual loading in Figure 8.7.4 induces no virtual bending moments in the stringers, the internal complementary virtual work is just that of the panels, for which we use Equation 8.2.8. Thus,

$$\delta W_{\text{int}}^* = \sum_{\text{panels}} \frac{A}{Gt} q \delta q = \frac{1}{Gt} \left[wl q^{(1-2)} \delta q^{(1-2)} + hl q^{(2-3)} \delta q^{(2-3)} + wl q^{(3-4)} \delta q^{(3-4)} + hl q^{(4-1)} \delta q^{(4-1)} \right] \qquad \text{[8.7.2]}$$

Substituting the true and virtual shear flows from Figures 8.7.3 and 8.7.4, considering those directed counterclockwise around the cell as positive, we get

$$\delta W_{int}^* = \frac{1}{Gt}\left\{wl\left[-\frac{P}{4wh}(w-2d)\right]\left(\frac{\delta P}{2l}\right)+hl\left[\frac{P}{4wh}(w+2d)\right]\left(-\frac{\delta P}{2l}\right)\right.$$

$$\left.+wl\left[-\frac{P}{4wh}(w-2d)\right]\left(\frac{\delta P}{2l}\right)+hl\left[-\frac{P}{4wh}(3w-2d)\right]\left(-\frac{\delta P}{2l}\right)\right\}$$

Upon simplification, we obtain

$$\delta W_{int}^* = -\frac{P\delta P}{2Gtwh}(w-h)\left(\frac{w}{2}-d\right) \qquad \text{[8.7.3]}$$

Setting this equal to the external complementary virtual work in Equation 8.7.1 yields the warp angle for a section with uniform wall thickness, as follows:

$$\phi = -\frac{P}{2Gtwh^2}(w-h)\left(\frac{w}{2}-d\right) \qquad \text{[8.7.4]}$$

As expected, the warp angle is zero when the load passes through the shear center, that is, when $d = w/2$. Notice also that regardless of the line of action of the load, the warp angle is zero if $w = h$, that is, if the section is square.

The counterclockwise moment of the force P about the shear center G is $-P\left[(w/2)-d\right]$. The twisting effect of P is that of a pure torque with the same moment, that is,

$$T = -P\left(\frac{w}{2}-d\right)$$

Thus, according to Equation 8.7.4, if the beam is in pure torsion, the warp angle is

$$\phi = \frac{T}{2Gtwh^2}(w-h) \qquad \text{[8.7.5]}$$

Example 8.7.1   Calculate the warp angle for an idealized beam with constant cross section loaded in shear, as illustrated in Figure 8.7.6.

Since this beam is in bending, we must first locate the centroid in the usual fashion:

$$y_G = \frac{\sum\limits_{i=1}^{4} y_i A_i}{\sum\limits_{i=1}^{4} A_i} = \frac{4.050}{1.250} = 3.240 \text{ in.} \qquad z_G = \frac{\sum\limits_{i=1}^{4} z_i A_i}{\sum\limits_{i=1}^{4} A_i} = \frac{13.50}{1.250} = 10.80 \text{ in.} \qquad \text{[a]}$$

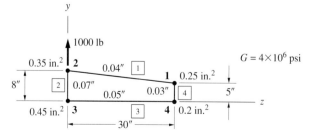

Figure 8.7.6   Section of a constant cross section box beam.

The centroidal area moments of inertia are then

$$I_{G_y} = \sum_{i=1}^{4} (z_i - z_G)^2 A_i = 259.2 \text{ in.}^4 \qquad I_{G_z} = \sum_{i=1}^{4} (y_i - y_G)^2 A_i = 15.53 \text{ in.}^4 \qquad I_{G_{yz}} = \sum_{i=1}^{4} (y_i - y_G)(z_i - z_G) A_i = -6.240 \text{ in.}^4 \quad \textbf{[b]}$$

Substituting $V_y = -1000$ lb and $V_z = 0$ into Equation 4.8.2, we get

$$P_x'^{(i)} = \frac{1}{I_{G_y} I_{G_z} - I_{G_{yz}}^2} [(I_{G_y} V_y - I_{G_{yz}} V_z)(y_i - y_G) + (I_{G_z} V_z - I_{G_{yz}} V_y)(z_i - z_G)] A_i$$

which yields the flange load gradients shown in parentheses in Figure 8.7.7, along with the assumed directions of the shear flows.

The stringer equilibrium relation, Equation 4.8.3, is used to express three of the shear flows in terms of the remaining one, say $q^{(4)}$. Thus, starting with flange 1, we have

$$q^{(1)} = q^{(4)} + P_x'^{(1)} = q^{(4)} - 36.13$$
$$q^{(2)} = q^{(4)} + P_x'^{(2)} = (q^{(4)} + P_x'^{(1)}) + P_x'^{(2)} = q^{(4)} - 138.6 \qquad \textbf{[c]}$$
$$q^{(3)} = q^{(2)} + P_x'^{(3)} = (q^{(4)} + P_x'^{(1)} + P_x'^{(2)}) + P_x'^{(3)} = q^{(4)} - 36.13$$

Invoking moment equivalence about flange 2, we have

$$(30q^{(3)}) \times 8 + (5q^{(4)}) \times 30 = 0 \qquad \textbf{[d]}$$

This can be written in terms of $q^{(4)}$ alone by substituting the third of Equation c:

$$240 \left(q^{(4)} - 36.13\right) + 150q^{(4)} = 0$$

Solving this for $q^{(4)}$ and then substituting into Equations c, we find all of the shear flows due to the 1000 lb shear load, as follows:

$$q^{(4)} = 22.23 \text{ lb/in.}$$
$$q^{(3)} = -13.90 \text{ lb/in.}$$
$$q^{(2)} = -116.3 \text{ lb/in.} \qquad \textbf{[e]}$$
$$q^{(1)} = -13.90 \text{ lb/in.}$$

To find the warp angle, we isolate a portion of the box adjacent to the section (Figure 8.7.8) and apply a self-equilibrating pair of virtual couples to the stringers, as was done in Figure 8.7.4. However, the vertical webs of this structure are unequal in

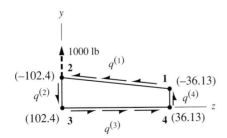

Figure 8.7.7   Flange load gradients (lb/in.) and assumed directions of shear flow.

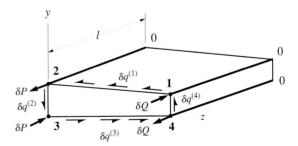

Figure 8.7.8   Self-equilibrating set of virtual loads applied to the stringers of an isolated free body of a portion of the box beam, together with the corresponding set of virtual shear flows.

height: $h_2 = 8$ in. and $h_4 = 5$ in. Therefore, if the net moment of the two pairs of virtual couples might be zero, we must require that $\delta P h_2 = \delta Q h_4$, that is,

$$\delta Q = \frac{h_2}{h_4} \delta P = 1.6 \delta P \tag{f}$$

Since the virtual flange loads are assumed to be zero at a distance l from the section, the equilibrium of stringers 2, 3 and 4, respectively, requires that

$$\delta q^{(2)} - \delta q^{(1)} = -\frac{\delta P}{l} \quad or \quad \delta q^{(2)} = \delta q^{(1)} - \frac{\delta P}{l}$$

$$\delta q^{(3)} - \delta q^{(2)} = \frac{\delta P}{l} \quad or \quad \delta q^{(3)} = \delta q^{(2)} \tag{g}$$

$$\delta q^{(4)} - \delta q^{(3)} = -\frac{\delta Q}{l} \quad or \quad \delta q^{(4)} = \delta q^{(3)} - \frac{\delta Q}{l}$$

The resultant moment of the shear flows in the plane of the section must vanish. Using flange 2 as the moment summation point, this means that

$$(30\delta q^{(3)}) \times 8 + (5\delta q^{(4)}) \times 30 = 0 \tag{h}$$

Substituting the second and third of Equation g into h yields

$$240\delta q^{(1)} + 150(\delta q^{(1)} - 1.6\frac{\delta P}{l}) = 0 \tag{i}$$

From this and Equations g, we find all of the virtual shear flows in terms of the virtual load δP, as follows:

$$\delta q^{(1)} = 0.6154\frac{\delta P}{l} \quad \delta q^{(2)} = -0.3846\frac{\delta P}{l} \quad \delta q^{(3)} = 0.6154\frac{\delta P}{l} \quad \delta q^{(4)} = -0.9846\frac{\delta P}{l} \tag{j}$$

With the true and virtual shear flows in hand, we can determine the internal virtual work, according to

$$\delta W_{int}^* = \sum_{panels} \frac{A}{Gt} q \delta q = \frac{l}{G} \left[\frac{h^{(1)}}{t^{(1)}} q^{(1)} \delta q^{(1)} + \frac{h^{(2)}}{t^{(2)}} q^{(2)} \delta q^{(2)} + \frac{h^{(3)}}{t^{(3)}} q^{(3)} \delta q^{(3)} + \frac{h^{(4)}}{t^{(4)}} q^{(4)} \delta q^{(4)} \right]$$

so that

$$\delta W_{int}^* = \frac{l}{4(10^6)} \left[\frac{\sqrt{30^2 + 3^2}}{0.04}(-13.90) \times 0.6154\frac{\delta P}{l} + \frac{8}{0.07}(-116.3) \times \left(-0.3846\frac{\delta P}{l}\right) \right.$$

$$\left. + \frac{30}{0.05}(-13.90) \times 0.6154\frac{\delta P}{l} + \frac{5}{0.03} 22.23 \times \left(-0.9846\frac{\delta P}{l}\right) \right]$$

or

$$\delta W_{int}^* = -0.0025288 \delta P \tag{k}$$

If the true displacements of the four stringers in the positive x direction at the section are designated u_1, u_2, u_3, and u_4, respectively, then the external complementary virtual work is

$$\delta W_{ext}^* = -\delta P u_2 + \delta P u_3 - \delta Q u_4 + \delta Q u_1 = (u_3 - u_2)\delta P - (u_4 - u_1)\delta Q \tag{l}$$

The counterclockwise rotation ϕ_2 about the z axis of web 2-3 in terms of the axial displacements of stringers 2 and 3, is

$$\phi_2 = \frac{u_3 - u_2}{h_2} \tag{m}$$

Likewise, for web 1-4, we have

$$\phi_4 = \frac{u_4 - u_1}{h_1} \tag{n}$$

Therefore, Equation l can be written

$$\delta W^*_{\text{ext}} = \phi_2 h_2 \delta P - \phi_4 h_4 \delta Q \tag{o}$$

Incorporating Equation f into o, we obtain the external complementary virtual work in terms of the warp angle $\phi = \phi_3 - \phi_1$,

$$\delta W^*_{\text{ext}} = \phi h_2 \delta P = 8\phi_2 \delta P \tag{p}$$

Equation k and p, together with the principle of complementary virtual work, imply that

$$\phi = -0.000316 \text{ radians} = -0.0181 \text{ degrees}$$

8.8   MULTICELL IDEALIZED BOX BEAMS

Let us now consider idealized box beams consisting of several closed cells, such as the four-cell, wing-like structure illustrated in Figure 8.8.1. Up to this point, our discussion of box beams has focused on those with a single cell. A single-cell box beam is perfectly capable of sustaining torsional as well as bending loads. The additional cells are not required for stability and are therefore redundant. The purpose of this section is to explore the means of computing the shear flows and deflections of this type of redundant structure. Initially we will focus on beams with constant cross section.

The analysis of idealized multicell sections begins with identifying the redundant members. In Figure 8.8.1, these are the three internal webs. The shear flows $q^{(10)}$, $q^{(11)}$, and $q^{(12)}$ (circled) are the redundant loads, in terms of which all the other shear flows will be expressed. The assumed directions of the shear flows are arbitrary. After locating the centroid of the cross section and computing the centroidal area moments of inertia, we calculate the flange load gradients using Equation 4.8.2. These are shown in parentheses alongside the respective stringers, as in Figure 8.8.1. Then, we impose stringer equilibrium, Equation 4.8.3, at each flange. For the section illustrated, that yields eight independent equations, by means of which we can express eight of the shear flows—say, $q^{(1)}$ through $q^{(8)}$—in terms of $q^{(9)}$ and the three redundants, $q^{(10)}$, $q^{(11)}$, and $q^{(12)}$. The procedure so far is identical to that for a single cell, except for the presence of the redundants. The next step is also the same as in a statically

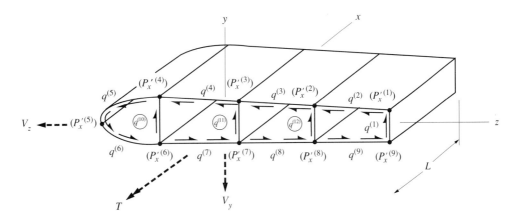

Figure 8.8.1   Multicell idealized box beam section.

determinate situation: we invoke moment equivalence, requiring the moment of the shear flows about an arbitrary point to be equal to the moment of the applied loads. With this additional, ninth equation we can reduce the number of independent unknown shear flows to just the three redundants. We now have the nine dependent shear flows as linear functions of the redundants and the applied loads, as follows:

$$q^{(i)} = a_i q^{(10)} + b_i q^{(11)} + c_i q^{(12)} + d_i V_z + e_i V_y + f_i T \qquad i = 1, \ldots, 9 \qquad \text{[8.8.1]}$$

The constants $a_i, \cdots, f_i$ depend on the geometry of the cross section.

We determine virtual shear flows $\delta q^{(1)}, \ldots, \delta q^{(9)}$ in terms of the virtual redundant shear flows by setting the true loads to zero and requiring that the 12 virtual shear flows form a self-equilibrating set. That is, we set the terms involving the applied loads in Equation (8.8.1) equal to zero and replace the shear flows by their virtual counterparts:

$$\delta q^{(i)} = a_i \delta q^{(10)} + b_i \delta q^{(11)} + c_i \delta q^{(12)} \qquad i = 1, \ldots, 9 \qquad \text{[8.8.2]}$$

The internal complementary virtual work is given by the familiar expression [cf. section 8.2]

$$\delta W_{\text{int}}^* = \sum_{i=1}^{12} \frac{k^{(i)} A^{(i)}}{G^{(i)} t^{(i)}} q^{(i)} \delta q^{(i)} \qquad \text{[8.8.3]}$$

where $A^{(i)}$ is the area of the ith shear panel, and $k^{(i)} = 1$ if the panel is rectangular or cylindrical. Since the redundants are internal to the structure, the external complementary virtual work is zero. The principle of complementary virtual work requires that the internal complementary virtual work also vanish. Equation (8.8.3) then implies that

$$\frac{L}{G} \sum_{i=1}^{12} \frac{s^{(i)}}{t^{(i)}} q^{(i)} \delta q^{(i)} = 0$$

or

$$\frac{s^{(i)}}{t^{(i)}} q^{(i)} \delta q^{(i)} + \frac{s^{(10)}}{t^{(10)}} q^{(10)} \delta q^{(10)} + \frac{s^{(11)}}{t^{(11)}} q^{(11)} \delta q^{(11)} + \frac{s^{(12)}}{t^{(12)}} q^{(12)} \delta q^{(12)} = 0 \qquad \text{[8.8.4]}$$

Substituting Equations (8.8.1) and (8.8.2) into the first nine terms of this equation and factoring out the independent virtual shear flows $\delta q^{(10)}$, $\delta q^{(11)}$ and $\delta q^{(12)}$, we obtain

$$\left(c_{1,1} q^{(10)} + c_{1,2} q^{(11)} + c_{1,3} q^{(12)} + b_{1,1} V_z + b_{1,2} V_y + b_{1,3} T \right) \delta q^{(10)}$$
$$+ \left(c_{2,1} q^{(10)} + c_{2,2} q^{(11)} + c_{2,3} q^{(12)} + b_{2,1} V_z + b_{2,2} V_y + b_{2,3} T \right) \delta q^{(11)}$$
$$+ \left(c_{3,1} q^{(10)} + c_{3,2} q^{(11)} + c_{3,3} q^{(12)} + b_{3,1} V_z + b_{3,2} V_y + b_{3,3} T \right) \delta q^{(12)} = 0$$

The coefficients $c_{i,j}$ and $b_{i,j}$ are numbers that depend only on the geometry and material properties of the particular cross section. Since this equation must remain valid for arbitrary values of the three virtual shear flows, their coefficients must vanish; we thereby obtain three equations for the three redundant shear flows as follows:

$$
\begin{aligned}
c_{1,1} q^{(10)} + c_{1,2} q^{(11)} + c_{1,3} q^{(12)} &= -\left(b_{1,1} V_z + b_{1,2} V_y + b_{1,3} T \right) \\
c_{2,1} q^{(10)} + c_{2,2} q^{(11)} + c_{2,3} q^{(12)} &= -\left(b_{2,1} V_z + b_{2,2} V_y + b_{2,3} T \right) \\
c_{3,1} q^{(10)} + c_{3,2} q^{(11)} + c_{3,3} q^{(12)} &= -\left(b_{3,1} V_z + b_{3,2} V_y + b_{3,3} T \right)
\end{aligned}
\qquad \text{[8.8.5]}
$$

The square array of coefficients on the left is a flexibility matrix (section 6.8); therefore, we know it will always be symmetric, so that $c_{1,2} = c_{2,1}$, $c_{1,3} = c_{3,1}$, and $c_{2,3} = c_{3,2}$. Furthermore, according to Equation 6.8.2, $c_{1,1}$, $c_{2,2}$, and $c_{3,3}$ must all be positive.

This procedure is also followed for tapered multicell beams. In that case the average shear flows are involved, as is the virtual work expression for tapered shear panels, Equation 8.2.13. The following examples illustrate the method described here.

Example 8.8.1   The section of a constant-cross-section beam in Figure 8.8.2 carries a pure torque of 50,000 in.-lb counterclockwise. Find the shear flows and rate of twist.

Let us choose the internal webs 9 and 10 as the redundants. Figure 8.8.3 shows the assumed directions for the shear flows, with the redundant shear flows q_9 and q_{10} circled. As usual, we first invoke longitudinal equilibrium of the stringers, Equation 4.8.3, taking into account that in the absence of bending, the flange load gradients are zero. Taking each stringer in turn, starting with number 1, we have

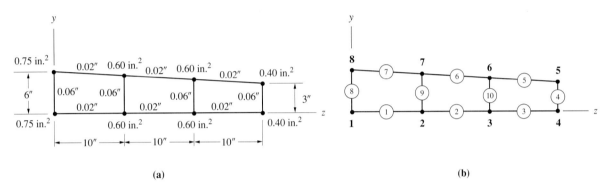

Figure 8.8.2   (a) Section size and properties. (b) Flange and wall numbering.

$$
\begin{aligned}
&1 \quad q^{(1)} - q^{(8)} = 0 \\
&2 \quad q^{(2)} - q^{(1)} - q^{(9)} = 0 \\
&3 \quad q^{(3)} - q^{(2)} - q^{(10)} = 0 \\
&4 \quad q^{(4)} - q^{(3)} = 0 \\
&5 \quad q^{(5)} - q^{(4)} = 0 \\
&6 \quad -q^{(5)} + q^{(6)} + q^{(10)} = 0 \\
&7 \quad -q^{(6)} + q^{(7)} + q^{(9)} = 0
\end{aligned}
\qquad\text{[a]}
$$

Solving for shear flows $q^{(1)}$ through $q^{(7)}$ in terms of $q^{(8)}$, $q^{(9)}$, and $q^{(10)}$, we get

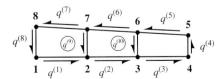

Figure 8.8.3   Assumed directions for the shear flows.

The circled shear flows are the chosen redundants.

$$q^{(1)} = q^{(8)}$$
$$q^{(2)} = q^{(8)} + q^{(9)}$$
$$q^{(3)} = q^{(8)} + q^{(9)} + q^{(10)}$$
$$q^{(4)} = q^{(8)} + q^{(9)} + q^{(10)}$$
$$q^{(5)} = q^{(8)} + q^{(9)} + q^{(10)}$$
$$q^{(6)} = q^{(8)} + q^{(9)}$$
$$q^{(7)} = q^{(8)}$$

[b]

For moment equivalence, we set the sum of the moments of the shear flows counterclockwise about flange 8 equal to the applied torque, as follows:

$$2\left(\frac{1}{2}10 \times 6\right)q^{(1)} + 2\left(\frac{1}{2}10 \times 6\right)q^{(2)} + 2\left(\frac{1}{2}10 \times 6\right)q^{(3)} + 2\left(\frac{1}{2}3 \times 30\right)q^{(4)} - 2\left(\frac{1}{2}5 \times 10\right)q^{(9)} - 2\left(\frac{1}{2}4 \times 20\right)q^{(10)} = 50,000 \quad \text{[c]}$$

Substituting Equations b and simplifying leads to an expression for q_8 in terms of the two redundant shear flows,

$$q^{(8)} = 185.2 - 0.5926q^{(9)} - 0.2593q^{(10)} \qquad \text{[d1]}$$

Therefore, from Equations b, we have

$$q^{(1)} = q^{(7)} = q^{(8)} = 185.2 - 0.5926q^{(9)} - 0.2593q^{(10)}$$
$$q^{(2)} = q^{(6)} = 185.2 + 0.4074q^{(9)} - 0.2593q^{(10)} \qquad \text{[d2]}$$
$$q^{(3)} = q^{(4)} = q^{(5)} = 185.2 + 0.4074q^{(9)} + 0.7407q^{(10)}$$

Equations d are relationships among the true shear flows due to the applied torque. As usual in the force method, we remove the true load and replace the redundants by virtual shear flows $\delta q_{(9)}$ and $\delta q_{(10)}$, which are considered to be external loads applied along the hypothetical lengthwise cuts. The relationships among the virtual shear flows are obtained from Equations d, with the effects of the true applied torque absent. Thus, we have

$$\delta q^{(1)} = \delta q^{(7)} = \delta q^{(8)} = -0.5926\delta q^{(9)} - 0.2593\delta q^{(10)}$$
$$\delta q^{(2)} = \delta q^{(6)} = 0.4074\delta q^{(9)} - 0.2593\delta q^{(10)} \qquad \text{[e]}$$
$$\delta q^{(3)} = \delta q^{(4)} = \delta q^{(5)} = 0.4074\delta q^{(9)} + 0.7407\delta q^{(10)}$$

Since there is neither a real nor a virtual bending load, the internal complementary virtual work is just that of the panels, or

$$\delta W_{\text{int}}^* = \sum_{i=1}^{10} \frac{A^{(i)}}{G^{(i)}t^{(i)}}q^{(i)}\delta q^{(i)} = \frac{L}{G}\sum_{i=1}^{10} \frac{s^{(i)}}{t^{(i)}}q^{(i)}\delta q^{(i)}$$

where L is the unspecified length of the box beam. From Equations d and e, it follows that

$$\delta W_{\text{int}}^* = \frac{L}{G}\left\{ \overbrace{\left(\frac{10}{0.02} + \frac{10.05}{0.02} + \frac{6}{0.06}\right)}^{\text{webs 1, 7, and 8}}\left(185.2 - 0.5926q^{(9)} - 0.2593q^{(10)}\right)\left(-0.5926\delta q^{(9)} - 0.2593\delta q^{(10)}\right) + \right.$$

$$+ \overbrace{\left(\frac{10}{0.02} + \frac{10.05}{0.02}\right)}^{\text{webs 2 and 6}}\left(185.2 + 0.4074q^{(9)} - 0.2593q^{(10)}\right)\left(0.4074\delta q^{(9)} - 0.2593\delta q^{(10)}\right)$$

$$\overbrace{+\left(\frac{10}{0.02}+\frac{3}{0.06}+\frac{10.05}{0.02}\right)\left(185.2+0.4074q^{(9)}+0.7407q^{(10)}\right)\left(0.4074\delta q^{(9)}+0.7407\delta q^{(10)}\right)}^{\text{webs 3, 4, and 5}}$$

$$\left.+\underbrace{\frac{5}{0.06}q^{(9)}\delta q^{(9)}}_{\text{web 9}}+\underbrace{\frac{4}{0.06}q^{(10)}\delta q^{(10)}}_{\text{web 10}}\right\}$$

Simplifying and combining terms leads to

$$\delta W_{\text{int}}^{*}=\frac{L}{G}\left[\left(811.6q^{(9)}+381.1q^{(10)}+34,050\right)\delta q^{(9)}+\left(381.1q^{(9)}+785.6q^{(10)}+43,310\right)\delta q^{(10)}\right]$$

The redundant shear flows $q^{(9)}$ and $q^{(10)}$ are internal to the structure; therefore we know $\delta W_{\text{ext}}^{*}=0$, which, according to the principle of complementary virtual work, means that

$$\frac{L}{G}\left[\left(811.6q^{(9)}+381.1q^{(10)}+34,050\right)\delta q^{(9)}+\left(381.1q^{(9)}+785.6q^{(10)}+43,310\right)\delta q^{(10)}\right]=0$$

for arbitrary, independent choices of $\delta q^{(9)}$ and $\delta q^{(10)}$. This results in a system of two equations for the two redundants,

$$811.6q^{(9)}+381.1q^{(10)}=-34,050$$
$$381.1q^{(9)}+785.6q^{(10)}=-43,313$$

[f]

The solution is $q^{(9)}=-20.81$ lb/in. and $q^{(10)}=-45.03$ lb/in. These two values are substituted into Equation d to get the remaining shear flows, which are illustrated in Figure 8.8.4.

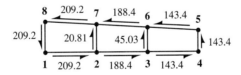

Figure 8.8.4   Shear flows (lb/in.) statically equivalent to a 50,000 in.-lb counter clockwise torque.

To find the twist of the section, we apply a virtual torque δT. Since the actual load is a pure torque of 50,000 in.-lb, the corresponding virtual shear flows will be those in Figure 8.8.4 multiplied by δT and divided by 50,000. Therefore, the internal complementary virtual work is

$$\delta W_{\text{int}}^{*}=\frac{L}{G}\sum_{i=1}^{10}\frac{s^{(i)}}{t^{(i)}}q^{(i)}\frac{q^{(i)}}{50,000}\delta T=\frac{L\delta T}{50,000G}\sum_{i=1}^{10}\frac{s^{(i)}}{t^{(i)}}q^{(i)2}$$

$$=\frac{L\delta T}{50,000G}\left[\frac{10}{0.02}(209.2)^{2}+\frac{10}{0.02}(188.4)^{2}+\frac{10}{0.02}(143.4)^{2}+\frac{3}{0.06}(143.4)^{2}+\frac{10.05}{0.02}(143.4)^{2}\right.$$

$$\left.+\frac{10.05}{0.02}(188.4)^{2}+\frac{10.05}{0.02}(209.2)^{2}+\frac{6}{0.06}(209.2)^{2}+\frac{5}{0.06}(-20.81)^{2}+\frac{4}{0.06}(-45.03)^{2}\right]$$

or

$$\delta W_{\text{int}}^{*}=2112.4\frac{L}{G}\delta T$$

[g]

Equating this to the complementary virtual work of the torque $\theta \delta T$, we find that the angle of twist (in radians) is

$$\theta = 2112.4 \frac{L}{G}$$

Since the section is redundant, we are free to choose any subsection to carry the applied virtual torque. Remember that the virtual loads need bear no resemblance to the true loads. They need not satisfy compatibility, but they must be self-equilibrating. Thus, instead of imagining that the entire cross section participates in the virtual loading, as we did thus far, let us assume that the virtual torque is taken up by, say, just the four webs of the rightmost cell, that is, by webs 3, 4, 5, and 10, as in Figure 8.8.5. The virtual shear flows are zero in the other webs. Since the virtual flange loads are zero, the virtual shear flows acting around the closed cell must all have the same value, denoted δq_T, so that

$$\delta q^{(10)} = \delta q^{(5)} = \delta q^{(4)} = \delta q^{(3)} = \delta q_T$$

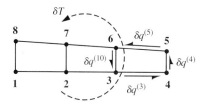

Figure 8.8.5   One alternative virtual shear load distribution representing the virtual torque.

For static equivalence, the moment of the constant shear flow δq_T must equal the virtual torque δT. That is, $2A_{\text{cell}} \delta q_T = \delta T$, or

$$\delta q_T = \frac{\delta T}{2 A_{\text{cell}}} = \frac{\delta T}{2 \times \left[\frac{1}{2} (3+4)(10) \right]} = 0.01428 \delta T$$

The internal complementary virtual work is

$$\delta W^*_{\text{int}} = \frac{L}{G} \left(\frac{s^{(3)}}{t^{(3)}} q^{(3)} \delta q^{(3)} + \frac{s^{(4)}}{t^{(4)}} q^{(4)} \delta q^{(4)} + \frac{s^{(5)}}{t^{(5)}} q^{(5)} \delta q^{(5)} + \frac{s^{(10)}}{t^{(10)}} q^{(10)} \delta q^{(10)} \right)$$

$$= 0.01428 \left(\frac{s^{(3)}}{t^{(3)}} q_3 + \frac{s^{(4)}}{t^{(4)}} q^{(4)} + \frac{s^{(5)}}{t^{(5)}} q^{(5)} + \frac{s^{(10)}}{t^{(10)}} q^{(10)} \right) \frac{L}{G} \delta T$$

where the common virtual shear flow is factored out. Substituting the web dimensions and the true shear flows previously computed, we get

$$\delta W^*_{\text{int}} = 0.01428 \left[\frac{10}{0.02} 143.4 + \frac{3}{0.06} 143.4 + \frac{\sqrt{10^2 + 1^2}}{0.02} 143.4 + \frac{4}{0.06}(-45.03) \right] \frac{L}{G} \delta T = 2112.4 \frac{L}{G} \delta T$$

which is identical to Equation g.

Example 8.8.2   Calculate the shear flows in the webs of the constant-cross-section idealized beam in the previous example if, instead of pure torsion, the beam is subjected to the shear loads shown in Figure 8.8.6b.

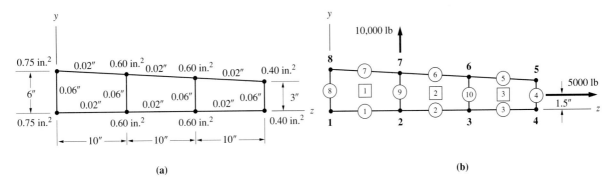

Figure 8.8.6   Three-cell section carrying shear loads.

In the previous example, this section was subjected to pure torsion, so there was no need to locate the centroid nor to compute the area moments of inertia. Here, we must compute the shear flows due to bending caused by the transverse loads on the beam. For that, we need the centroid and the area moments of inertia. The centroidal coordinates are

$$z_G = \frac{\sum\limits_{i=1}^{8} A_i z_i}{\sum\limits_{i=1}^{8} A_i} = \frac{60}{4.7} = 12.76 \text{ in.} \qquad y_G = \frac{\sum\limits_{i=1}^{8} A_i y_i}{\sum\limits_{i=1}^{8} A_i} = \frac{11.10}{4.7} = 2.362 \text{ in.} \qquad \text{[a]}$$

The area moments of inertia are

$$I_{G_y} = \sum_{i=1}^{8} (z_i - z_G)^2 A_i = 554.0 \text{ in.}^4 \qquad I_{G_z} = \sum_{i=1}^{8} (y_i - y_G)^2 A_i = 28.98 \text{ in.}^4 \qquad I_{G_{yz}} = \sum_{i=1}^{8} (y_i - y_G)(z_i - z_G) A_i = -27.70 \text{ in.}^4 \qquad \text{[b]}$$

Remember that in idealized beams, the webs themselves do not transmit loads due to bending moments, so their areas are not involved in these calculations.

To compute the flange load gradient at the ith flange, we use Equation 4.8.2, which is

$$P_x'^{(i)} = \frac{1}{I_{G_y} I_{G_z} - I_{G_{yz}}^2} [(I_{G_y} V_y - I_{G_{yz}} V_z)(y_i - y_G) + (I_{G_z} V_z - I_{G_{yz}} V_y)(z_i - z_G)] A_i \qquad \text{[c]}$$

Substituting the numbers from Equations a and b, this becomes

$$P_x'^{(i)} = [-371.4 y_i - 27.59 z_i + 1230] A_i \qquad \text{[d]}$$

Inserting the coordinates of the eight flanges along with their areas into this expression yields the flange load gradients shown in parentheses alongside each flange in Figure 8.8.7. Starting with flange 1, applying the equilibrium condition (shear flow out equals shear flow in plus flange load gradient) to each flange in turn, and moving clockwise around the section, we find that

$$q^{(1)} = q^{(8)} + 922.0$$
$$q^{(2)} = q^{(8)} + q^{(9)} + 1494$$
$$q^{(3)} = q^{(8)} + q^{(9)} + q^{(10)} + 1900$$
$$q^{(4)} = q^{(8)} + q^{(9)} + q^{(10)} + 2061 \qquad \text{[e]}$$
$$q^{(5)} = q^{(8)} + q^{(9)} + q^{(10)} + 1776$$
$$q^{(6)} = q^{(8)} + q^{(9)} + 1291$$
$$q^{(7)} = q^{(8)} + 749.2$$

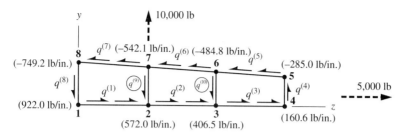

Figure 8.8.7 Flange load gradients (in parentheses) and the assumed directions for the shear flows, with $q^{(9)}$ and $q^{(10)}$ highlighted as the selected redundants.

We now invoke moment equivalence about flange 8 to get $q^{(8)}$ in terms of the redundants $q^{(9)}$ and $q^{(10)}$, or

$$2\left(\frac{1}{2}10 \times 6\right)q^{(1)} + 2\left(\frac{1}{2}10 \times 6\right)q^{(2)} + 2\left(\frac{1}{2}10 \times 6\right)q^{(3)} + 2\left(\frac{1}{2}3 \times 30\right)q^{(4)} - 2\left(\frac{1}{2}5 \times 10\right)q^{(9)} - 2\left(\frac{1}{2}4 \times 20\right)q^{(10)} =$$

$$10,000 \times 10 + 5000 \times 4.5$$

Using Equation e, we can reduce this to

$$270q^{(8)} + 160q^{(9)} + 70q^{(10)} + 4.445 \times 10^6 = 122,500$$

from which we obtain

$$q^{(8)} = -0.5926q^{(9)} - 0.2593q^{(10)} - 1,193 \qquad \text{[f1]}$$

With this result, we can use Equations e to express all of the dependent shear flows in terms of the redundants, as follows:

$$q^{(7)} = -0.5926q^{(9)} - 0.2593q^{(10)} - 1193$$
$$q^{(6)} = 0.4074q^{(9)} - 0.2593q^{(10)} + 98.70$$
$$q^{(5)} = 0.4074q^{(9)} + 0.7407q^{(10)} + 583.5$$
$$q^{(4)} = 0.4074q^{(9)} + 0.7407q^{(10)} + 868.6 \qquad \text{[f2]}$$
$$q^{(3)} = 0.4074q^{(9)} + 0.7407q^{(10)} + 707.9$$
$$q^{(2)} = 0.4074q^{(9)} - 0.2593q^{(10)} + 301.5$$
$$q^{(1)} = -0.5926q^{(9)} - 0.2593q^{(10)} - 270.6$$

Removing the true load and applying virtual shear flows in place of $q^{(9)}$ and $q^{(10)}$ causes virtual shear flows in the remaining webs. These virtual shear flows bear the same relationship to $\delta q^{(9)}$ and $\delta q^{(10)}$ as those expressed in Equation f if the terms arising from the externally applied loads are removed. That is,

$$\delta q^{(1)} = -0.5926\delta q^{(9)} - 0.2593\delta q^{(10)}$$
$$\delta q^{(2)} = 0.4074\delta q^{(9)} - 0.2593\delta q^{(10)}$$
$$\delta q^{(3)} = 0.4074\delta q^{(9)} + 0.7407\delta q^{(10)}$$
$$\delta q^{(4)} = 0.4074\delta q^{(9)} + 0.7407\delta q^{(10)}$$
$$\delta q^{(5)} = 0.4074\delta q^{(9)} + 0.7407\delta q^{(10)} \qquad \text{[g]}$$
$$\delta q^{(6)} = 0.4074\delta q^{(9)} - 0.2593\delta q^{(10)}$$
$$\delta q^{(7)} = -0.5926\delta q^{(9)} - 0.2593\delta q^{(10)}$$
$$\delta q^{(8)} = -0.5926\delta q^{(9)} - 0.2593\delta q^{(10)}$$

Substituting Equations f and g into the internal complementary virtual work expression, we get

$$\delta W^*_{int} = \sum_{i=1}^{10} \frac{L s^{(i)}}{G t^{(i)}} q^{(i)} \delta q^{(i)}$$

Setting the result equal to zero ($\delta W^*_{ext} = 0$ because the redundants are internal loads) leads to

$$\left[\left(811.6q^{(9)} + 381.1q^{(10)}\right) + 6.459 \times 10^5 \right] \delta q^{(9)} + \left[\left(381.1q^{(9)} + 785.6q^{(10)}\right) + 5.834 \times 10^5 \right] \delta q^{(10)} = 0$$

By the usual argument, this yields the following system of two equations:

$$811.6q^{(9)} + 381.1q^{(10)} = -645,900$$
$$381.1q^{(9)} + 785.6q^{(10)} = -583,400$$

[h]

As expected, the coefficient matrix on the left is identical to that of Equation g of the previous example. The solution of Equation h is $q^{(9)} = -579.0$ lb/in. and $q^{(10)} = -461.7$ lb/in. These, together with the remaining shear flows, obtained from Equation f, are sketched in Figure 8.8.8.

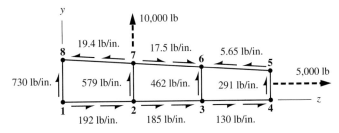

Figure 8.8.8   The shear flows, which are statically equivalent to the combination of an upward-directed 10,000 lb shear force through web 7–2 and a rightward-directed shear force whose line of action bisects web 4–5.

Example 8.8.3   Calculate the angle of twist per unit length for the beam in the previous example if $G = 4 \times 10^6$ lb/in.^{2}
     To find the twist rate, we apply a virtual torque δT to the section. Any stable, statically determinate portion of the multi-cell section may be imagined to take out the virtual torque. Let us arbitrarily choose the leftmost cell, the area of which is 55 in.^{2} The virtual shear flow due to δT is

$$\delta q_T = \frac{\delta T}{2 \times 55} = \frac{\delta T}{110}$$

Referring to Figures 8.8.6 and 8.8.8, the internal complementary virtual work is

$$\delta W^*_{int} = \frac{L}{G} \left(\frac{s^{(1)}}{t^{(1)}} q^{(1)} \delta q^{(1)} + \frac{s^{(9)}}{t^{(9)}} q^{(9)} \delta q^{(9)} + \frac{s^{(7)}}{t^{(7)}} q^{(7)} \delta q^{(7)} + \frac{s^{(8)}}{t^{(8)}} q^{(8)} \delta q^{(8)} \right)$$

$$= \frac{1}{110} \left[\frac{10}{0.02} 192 + \frac{5}{0.06} 579 + \frac{\sqrt{10^2 + 1^2}}{0.02} 19.4 + \frac{6}{0.06} (-730) \right] \frac{L}{G} \delta T$$

or

$$\delta W^*_{int} = 736.3 \frac{L}{G} \delta T$$

Noting that $\delta W^*_{ext} = \theta \delta T$ and setting $\delta W^*_{ext} = \delta W^*_{int}$, we find

$$\frac{\theta}{L} = \frac{736.3}{G} = 0.01055 \frac{\text{degrees}}{\text{in.}}$$

Example 8.8.4   Use the principle of complementary virtual work to calculate the maximum shear flow in the tapered, redundant cantilevered box beam illustrated in Figure 8.8.9. Loads in both transverse directions are applied to the free end, as shown. Figure 8.8.10 depicts the flange and web numbering and thickness information. Also, $E = 10 \times 10^6$ psi and $G = 4 \times 10^6$ psi ($\nu = 0.25$).

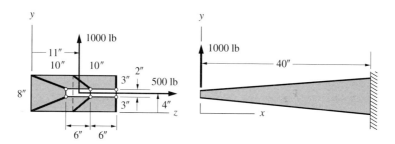

Figure 8.8.9   Tapered box beam with loads applied to the free end at $x=0$.

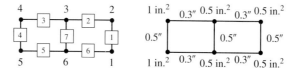

Figure 8.8.10   Flange and web numbering and, on the right, the flange areas and web thicknesses.

Since the center spar is redundant, this beam has one degree of static indeterminacy.

The initial step is to calculate the average flange load gradient for each of the six stringers. This requires locating the centroid and calculating the centroidal moments of inertia of the section at the wall. For the centroid, we find

$$z_G(40) = \frac{\sum\limits_{i=1}^{6} A_i z_i(40)}{\sum\limits_{i=1}^{6} A_i} = \frac{30}{4} = 7.5 \text{ in.} \qquad y_G(40) = \frac{\sum\limits_{i=1}^{6} A_i y_i(40)}{\sum\limits_{i=1}^{6} A_i} = \frac{16}{4} = 4.0 \text{ in.} \qquad \textbf{[a]}$$

and the area moments of inertia about the y and z axes through the centroid are

$$I_{G_y}(40) = \sum_{i=1}^{6} A_i [z_i(40) - z_G(40)]^2 = 275 \text{ in.}^4$$

$$I_{G_z}(40) = \sum_{i=1}^{6} A_i [y_i(40) - y_G(40)]^2 = 64.0 \text{ in.}^4 \qquad \textbf{[b]}$$

$$I_{G_{yz}}(40) = \sum_{i=1}^{6} A_i [y_i(40) - y_G(40)] [z_i(40) - z_G(40)] = 0$$

We then use Equation 4.8.1 to compute flange loads at $x = 40$ in.; that is,

$$P_x^{(i)}(40) = \left\{ \frac{A_i}{I_{G_y} I_{G_z} - I_{G_{yz}}^2} \left[-\left(M_z I_{G_y} + M_y I_{G_{yz}} \right)\left(y_i - y_G \right) + \left(M_y I_{G_z} + M_z I_{G_{yz}} \right)\left(z_i - z_G \right) \right] \right\}_{x=40''} \quad \text{[c]}$$

Therefore, the average flange load gradients are

$$\overline{P_x'^{(1)}} = \frac{P_x^{(1)}(40) - P_x^{(1)}(0)}{40} = \frac{795.4 - 0}{40} = 19.89 \text{ lb/in.}$$

$$\overline{P_x'^{(2)}} = \frac{P_x^{(2)}(40) - P_x^{(2)}(0)}{40} = \frac{-1704 - 0}{40} = -42.61 \text{ lb/in.}$$

$$\overline{P_x'^{(3)}} = \frac{P_x^{(3)}(40) - P_x^{(3)}(0)}{40} = \frac{-1341 - 0}{40} = -33.52 \text{ lb/in.} \quad \text{[d]}$$

$$\overline{P_x'^{(4)}} = \frac{P_x^{(4)}(40) - P_x^{(4)}(0)}{40} = \frac{-1954 - 0}{40} = -48.86 \text{ lb/in.}$$

$$\overline{P_x'^{(5)}} = \frac{P_x^{(5)}(40) - P_x^{(5)}(0)}{40} = \frac{3046 - 0}{40} = 76.14 \text{ lb/in.}$$

$$\overline{P_x'^{(6)}} = \frac{P_x^{(6)}(40) - P_x^{(6)}(0)}{40} = \frac{1159 - 0}{40} = 28.98 \text{ lb/in.}$$

These values are shown alongside the flanges in Figure 8.8.11. Starting arbitrarily with flange 2 and working our way counterclockwise around the section, we obtain

$$\bar{q}^{(2)} = \bar{q}^{(1)} + \overline{P_x'^{(2)}} = \bar{q}^{(1)} - 42.61$$

$$\bar{q}^{(3)} = \bar{q}^{(2)} - \bar{q}^{(7)} + \overline{P_x'^{(3)}} = \left(\bar{q}^{(1)} - 42.61 \right) - \bar{q}^{(7)} - 33.52 = \bar{q}^{(1)} - \bar{q}^{(7)} - 76.14$$

$$\bar{q}^{(4)} = \bar{q}^{(3)} + \overline{P_x'^{(4)}} = \left(\bar{q}^{(1)} - \bar{q}^{(7)} - 76.14 \right) - 48.86 = \bar{q}^{(1)} - \bar{q}^{(7)} - 125 \quad \text{[e]}$$

$$\bar{q}^{(5)} = \bar{q}^{(4)} + \overline{P_x'^{(5)}} = \left(\bar{q}^{(1)} - \bar{q}^{(7)} - 125 \right) + 76.14 = \bar{q}^{(1)} - \bar{q}^{(7)} - 48.86$$

$$\bar{q}^{(6)} = \bar{q}^{(5)} + \bar{q}^{(7)} + \overline{P_x'^{(6)}} = \left(\bar{q}^{(1)} - \bar{q}^{(7)} - 48.86 \right) + 28.98 = \bar{q}^{(1)} - \bar{q}^{(7)} - 19.89$$

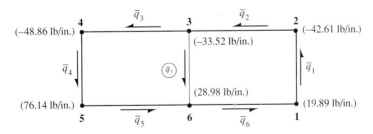

Figure 8.8.11 Computed average flange load gradients and assumed directions of the average shear flows, with $\bar{q}^{(7)}$ highlighted as the chosen redundant shear flow.

We now have all of the average shear flows in terms of $\bar{q}^{(1)}$ and the redundant shear flow $\bar{q}^{(7)}$. To eliminate $\bar{q}^{(1)}$, we invoke moment equivalence at the free end ($x = 0$) of the beam. Thus, summing moments about flange 5 we get

$$\left[q^{(1)}(0) \times 2 \right] \times 12 + \left[q^{(2)}(0) \times 6 \right] \times 2 + \left[q^{(3)}(0) \times 6 \right] \times 2 - \left[q^{(7)}(0) \times 2 \right] \times 6 = 1000 \times 3 - 500 \times 1 \quad \text{[f]}$$

The shear flows at $x = 0$ are related to the average shear flows by Equation 2.5.4,

$$q^{(i)}(0) = \bar{q}^{(i)} \frac{h^{(i)}(40)}{h^{(i)}(0)} \qquad i = 1, \cdots, 7 \qquad \text{[g]}$$

where $h^{(i)}(x)$ is the width of panel i at station x. We thus have

$$q^{(1)}(0) = \bar{q}^{(1)} \frac{8}{2} = 4\bar{q}^{(1)}$$

$$q^{(2)}(0) = \bar{q}^{(2)} \frac{10}{6} = 1.667\bar{q}^{(2)}$$

$$q^{(3)}(0) = \bar{q}^{(3)} \frac{10}{6} = 1.667\bar{q}^{(3)} \qquad \text{[h]}$$

$$q^{(4)}(0) = \bar{q}^{(7)} \frac{8}{2} = 4\bar{q}^{(7)}$$

Substituting Equations h into Equation f, we obtain the relationship between the average shear flows,

$$136.0\bar{q}^{(1)} - 68.0\bar{q}^{(7)} - 2375 = 2500$$

which yields $\bar{q}^{(1)}$ in terms of $\bar{q}^{(7)}$, as follows:

$$\bar{q}^{(1)} = 0.5\bar{q}^{(7)} + 35.85 \qquad \text{[i1]}$$

It follows from Equations e that

$$\bar{q}^{(2)} = 0.5\bar{q}^{(7)} - 6.768$$

$$\bar{q}^{(3)} = -0.5\bar{q}^{(7)} - 40.29$$

$$\bar{q}^{(4)} = -0.5\bar{q}^{(7)} - 89.15 \qquad \text{[i2]}$$

$$\bar{q}^{(5)} = -0.5\bar{q}^{(7)} - 13.02$$

$$\bar{q}^{(6)} = 0.5\bar{q}^{(7)} + 15.96$$

We have succeeded in expressing all of the average shear flows in terms of the single redundant shear flow $\bar{q}^{(7)}$. The next step is to remove the true loads from the structure, place an imaginary lengthwise cut in the web of the center spar, and apply an equal and opposite average virtual shear flow $\delta\bar{q}^{(7)}$ to each side of the cut, as indicated in Figure 8.8.12.

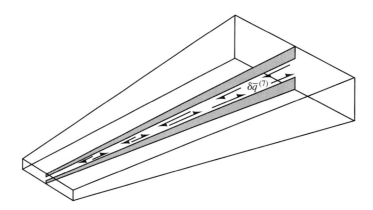

Figure 8.8.12   Virtual shear flow applied to the longitudinal cut in the redundant center spar.

The virtual average shear flows resulting from $\delta\bar{q}^{(7)}$ acting as the sole applied load are found from Equations i, as follows:

$$\delta\bar{q}^{(1)} = 0.5\delta\bar{q}^{(7)}$$
$$\delta\bar{q}^{(2)} = 0.5\delta\bar{q}^{(7)}$$
$$\delta\bar{q}^{(3)} = -0.5\delta\bar{q}^{(7)}$$
$$\delta\bar{q}^{(4)} = -0.5\delta\bar{q}^{(7)}$$
$$\delta\bar{q}^{(5)} = -0.5\delta\bar{q}^{(7)}$$
$$\delta\bar{q}^{(6)} = 0.5\delta\bar{q}^{(7)}$$

[j]

We are now in a position to calculate the complementary virtual work of each panel, using Equation 8.2.13

$$\delta W_{int}^{*(i)} = \frac{k^{(i)} A^{(i)}}{G^{(i)} t^{(i)}} \bar{q}^{(i)} \delta\bar{q}^{(i)}$$

$$k^{(i)} = 1 + \frac{2}{3\left(1 + \nu^{(i)}\right)} \left(\cot^2\alpha^{(i)} - \cot\alpha^{(i)}\cot\gamma^{(i)} + \cot^2\gamma^{(i)}\right)$$

[k]

From the geometry of the structure as presented in Figure 8.8.9, we can prepare Table 8.8.1.

Table 8.8.1   Shear panel geometry for the beam of Figure 8.8.9.

Panel	A (in.²)	α (degrees)	γ (degrees)	k
1	200.0	85.71	85.71	1.0030
2	320.9	90.00	84.30	1.0053
3	320.9	95.70	78.72	1.037
4	204.0	85.79	85.79	1.0029
5	320.9	78.72	95.70	1.0371
6	320.9	95.70	90.00	1.0053
7	201.0	85.73	85.73	1.0030

Substituting these data, together with the shear flows in Equations i and j, into the virtual work expression k for each panel and then summing, we find the total complementary internal virtual work of the beam, which is:

$$\delta W_{int}^* = \overbrace{0.0005015(0.5\bar{q}^{(7)} + 35.85)\delta\bar{q}^{(7)}}^{panel\ 1} + \overbrace{0.001344(0.5\bar{q}^{(7)} - 6.768)\delta\bar{q}^{(7)}}^{panel\ 2} + \overbrace{(-0.001387)(-0.5\bar{q}^{(7)} - 40.29)\delta\bar{q}^{(7)}}^{panel\ 3}$$

$$+ \overbrace{(-0.0005114)(-0.5\bar{q}^{(7)} - 89.15)\delta\bar{q}^{(7)}}^{panel\ 4} + \overbrace{(-0.001387)(-0.5\bar{q}^{(7)} - 13.02)\delta\bar{q}^{(7)}}^{panel\ 5} + \overbrace{0.001344(0.5\bar{q}^{(7)} + 15.96)\delta\bar{q}^{(7)}}^{panel\ 6}$$

$$+ \overbrace{0.001008\bar{q}^{(7)}\delta\bar{q}^{(7)}}^{panel\ 7}$$

Therefore,

$$\delta W_{int}^* = \left(0.004245\bar{q}^{(7)} + 0.1498\right)\delta\bar{q}^{(7)}$$

[l]

Since $\delta\bar{q}^{(7)}$ is internal to the structure, it induces no virtual bending moments in the stringers. This is why the internal virtual work in this case is associated exclusively with the panels. Also, since $\delta\bar{q}^{(7)}$ is an internal force quantity, then $\delta W_{ext}^* = 0$. It follows from Equation l and the principle of complementary virtual work that the redundant shear flow is

$$\bar{q}^{(7)} = -35.30 \text{ lb/in.}$$

By substituting this value into Equations i we obtain all of the remaining average shear flows, which are depicted in Figure 8.8.13. The greatest average shear flow exists in panel 4. Using Equation g, we can calculate the shear flow anywhere within the panels, particularly at the two ends of the beam, as illustrated in Figure 8.8.14.

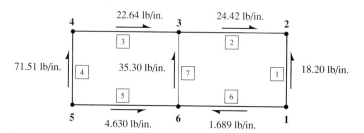

Figure 8.8.13   Average shear flows in the panels, displayed, for simplicity, on a generic cross section of the beam.

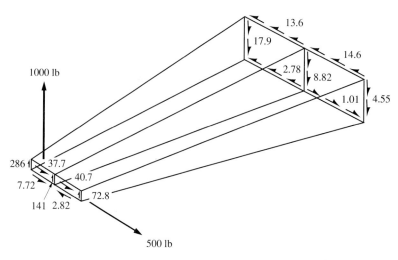

Figure 8.8.14   Shear flows (lb/in.) at each end of the box beam, in response to the loads at the left end.

Example 8.8.5   Calculate the angle of twist θ_x at the free end of the tapered beam in the previous example.

Apply a counterclockwise virtual torque δT at the free end of the beam. Only a statically determinate substructure is required to pick up this torque, so let us arbitrarily remove the center web—panel number 7—to produce a single cell. Since the pure virtual torque induces no virtual stringer loads, the average shear flow $\delta \bar{q}_T$ is the same in all of the walls. A generic section of the reduced structure is shown in Figure 8.8.15.

Moment equivalence about flange 5 at the free end of the beam requires that

$$\left[\delta q^{(1)}(0) \times 2\right] \times 12 + \left[\delta q^{(2)}(0) \times 6\right] \times 2 + \left[\delta q^{(3)}(0) \times 6\right] \times 2 = \delta T$$

Using Equation 2.5.4 to relate each of the shear flows in this equation to the average shear flow $\delta \bar{q}_T$ yields

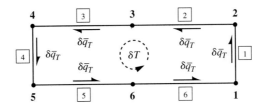

Figure 8.8.15   Generic section of the beam in Figure 8.8.9, without the center web, carrying pure virtual torsion.

$$[4\delta\bar{q}_T \times 2] \times 12 + \left[\frac{5}{3}\delta\bar{q}_T \times 6\right] \times 2 + \left[\frac{5}{3}\delta\bar{q}_T \times 6\right] \times 2 = \delta T$$

From this we obtain

$$\delta\bar{q}_T = \frac{\delta T}{136} \qquad \text{[a]}$$

Since the virtual shear flow $\delta\bar{q}_T$ is common to all of the panels except panel 7, which is left out $\left(\delta\bar{q}^{(7)} = 0\right)$, the internal complementary virtual work corresponding to the virtual torque is

$$\delta W_{int}^* = \sum_{i=1}^{6} \frac{k^{(i)} A^{(i)}}{G^{(i)} t^{(i)}} \bar{q}^{(i)} \delta\bar{q}^{(i)} = \frac{\delta\bar{q}_T}{G} \sum_{i=1}^{6} \frac{k^{(i)} A^{(i)}}{t^{(i)}} \bar{q}^{(i)} \qquad \text{[b]}$$

Substituting Equation a, along with panel data and the true shear flows from Example 8.8.4, leads to

$$\delta W_{int}^{*(i)} = -0.001287\delta T$$

Since $\delta W_{ext}^* = \theta_x \delta T$, the principle of complementary virtual work yields the following result for the angle of twist:

$$\theta_x = -0.001287 \text{ radians} = -0.07373 \text{ degrees}$$

The minus sign means that the rotation is clockwise looking inboard, towards the wall.

For consistency, our analysis of multicell box beams by the force method and complementary virtual work principle uses the same procedure we applied to all statically determinate problems up to now. An alternative approach for obtaining the equations needed to solve for multicell shear flows calls for all of the cells of the cross section to undergo the same rate of twist. In other words, it enforces the compatibility requirement that adjacent cells remain contiguous. We know that the principle of complementary virtual work enforces compatibility; therefore, the two procedures for dealing with multicell beams must be equivalent. It is instructive to demonstrate that they are.

Consider the two-cell section illustrated in Figure 8.8.16. Let us arbitrarily choose the interior web to be the redundant. The internal complementary virtual work for each wall is

$$\delta W_{int}^{*(i)} = \frac{L s^{(i)}}{G t^{(i)}} q^{(i)} \delta q^{(i)} \qquad i = 1, 2, 3$$

where L is the lengthwise dimension (perpendicular to the page). In Figure 8.8.16, $q^{(1)}$, $q^{(2)}$ and $q^{(3)}$ are the true shear flows due to the actual load (not shown) applied to the structure, and the virtual shear flows $\delta q^{(1)}$ and $\delta q^{(2)}$

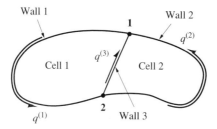

Figure 8.8.16   Two-cell idealized beam section where $q^{(3)}$ is redundant.

arise from the application of $\delta q^{(3)}$ in the redundant web. Since the redundant is an internal load, the external complementary virtual work is zero; therefore, so is the internal virtual work, so that

$$\sum_{i=1}^{3} \delta W_{\text{int}}^{*(i)} = 0$$

or

$$\frac{L}{G}\left(\frac{s^{(1)}}{t^{(1)}}q^{(1)}\delta q^{(1)} + \frac{s^{(2)}}{t^{(2)}}q^{(2)}\delta q^{(2)} + \frac{s^{(3)}}{t^{(3)}}q^{(3)}\delta q^{(3)}\right) = 0 \qquad \text{[8.8.6]}$$

With no virtual shear loads, the virtual flange load gradients vanish. For equilibrium, then, we have

$$\delta q^{(1)} = \delta q^{(2)} + \delta q^{(3)} \qquad \text{[8.8.7]}$$

The self-equilibrating virtual shear flows must be statically equivalent to zero applied virtual torsional moment. Summing the moments about flange 1 we therefore have

$$2A_1\delta q^{(1)} + 2A_2\delta q^{(2)} = 0 \qquad \text{[8.8.8]}$$

where A_1 and A_2 are the areas enclosed by cells 1 and 2, respectively. We can use Equations 8.8.7 and 8.8.8 to obtain $\delta q^{(1)}$ and $\delta q^{(2)}$ in terms of $\delta q^{(3)}$. Thus,

$$\delta q^{(1)} = \frac{A_2}{A_1 + A_2}\delta q^{(3)} \qquad \delta q^{(2)} = -\frac{A_2}{A_1 + A_2}\delta q^{(3)}$$

Substituting these expressions into Equation 8.8.6 yields

$$\frac{A_2}{A_1 + A_2}\frac{s^{(1)}}{t^{(1)}}q^{(1)} - \frac{A_2}{A_1 + A_2}\frac{s^{(2)}}{t^{(2)}}q^{(2)} + \frac{s^{(3)}}{t^{(3)}}q^{(3)} = 0$$

Then, multiplying by $(A_1 + A_2)/2$ and rearranging terms leads to

$$\frac{1}{2A_1}\left(q^{(1)}\frac{s^{(1)}}{t^{(1)}} + q^{(3)}\frac{s^{(3)}}{t^{(3)}}\right) = \frac{1}{2A_2}\left(q^{(1)}\frac{s^{(2)}}{t^{(2)}} - q^{(3)}\frac{s^{(3)}}{t^{(3)}}\right) \qquad \text{[8.8.9]}$$

According to Equation 8.5.3, this can be interpreted as

$$\left.\frac{d\theta_x}{dx}\right)_{\text{cell 1}} = \left.\frac{d\theta_x}{dx}\right)_{\text{cell 2}} \qquad \text{[8.8.10]}$$

That is, the angle of twist of each cell of the cross section is the same. The cross section rotates as a rigid body.

If the cross section has n cells, we therefore have

$$\left.\frac{d\theta}{dx}\right)_{\text{cell } 1} = \left.\frac{d\theta}{dx}\right)_{\text{cell } 2} = \cdots = \left.\frac{d\theta}{dx}\right)_{\text{cell } n} \tag{8.8.11}$$

which are $n - 1$ equations. Since the shear flow in the web separating adjacent cells is redundant, there are $n - 1$ redundant shear flows in a box beam cross section with n cells. Since the number of statically indeterminate shear flows always equals the number of additional governing equations, the shear flows are obtainable regardless of the number of cells.

8.9   RESTRAINT EFFECTS IN IDEALIZED BOX BEAMS

In section 8.7 we calculated the warp angle for a single-cell box beam of uniform rectangular cross section. The warp angle is a measure of the stringer displacement in the direction normal to the initially plane cross section. The twist angle is the rigid rotation of the cross section measured in the plane perpendicular to the beam's elastic axis. Previously, we assumed that warping deformations were free to occur without restraint. Let us now assess the effect that warping restraints have on the shear flow and flange loads.

Consider the box beam of uniform rectangular cross section shown in Figure 8.9.1. The flange areas are all A, which means that both the centroid and shear center are at the geometric center of the rectangle $(z = w/2$ and $y = h/2)$. For simplicity, let the four panels have a common thickness t. To calculate the warp angle, we apply a pair of self-equilibrating virtual couples of magnitude δPh to the cross section, acting in the direction in which the vertical webs rotate due to the true applied load P. As shown in section 8.7, the external complementary virtual work associated with these couples is

$$\delta W_{\text{ext}}^* = h\phi\delta P \tag{8.9.1}$$

where ϕ is the warp angle. Since warping is to be restrained, $\delta W_{\text{ext}}^* = 0$.

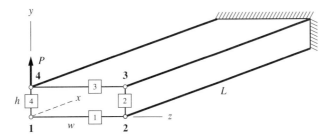

Figure 8.9.1   Idealized box beam of constant cross section. The webs have the same thickness t, and the cross-sectional area of each stringer is A.

The net applied virtual bending moment is zero, which means that the complementary internal virtual work due to bending is zero (cf. Equation 8.4.7). Therefore, the complementary internal virtual work is that of the panels alone and is given by (cf. Equation 8.2.6)

$$\delta W_{int}^* = \sum_{panels} \frac{A}{Gt} q \delta q = \frac{L}{Gt} \left[wq^{(1)} \delta q^{(1)} + hq^{(2)} \delta q^{(2)} + wq^{(3)} \delta q^{(3)} + hq^{(4)} \delta q^{(4)} \right] \qquad \text{[8.9.2]}$$

The true shear flows are assumed to be directed counterclockwise around the section, as indicated in Figure 8.9.2a, and the virtual shear flows corresponding to the applied virtual couples were shown in section 8.7 to be as drawn in Figure 8.9.2b.

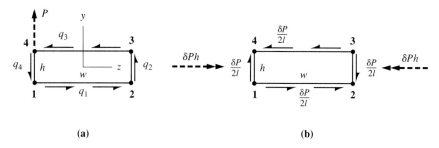

(a)   (b)

Figure 8.9.2   (a) Assumed directions for the shear flows due to P, with warping restrained.
(b) Virtual shear flows shown in section 8.7 to accompany the virtual couples applied in the planes of the vertical webs.

Because of the warping restraint, we cannot calculate the true shear flows using the beam formulas. However, we can impose the restriction that they be statically equivalent to the applied shear load P. Thus,

$$\sum F_y : \quad q^{(2)}h - q^{(4)}h = P$$
$$\sum F_z : \quad q^{(1)}w - q^{(3)}w = 0$$
$$\sum M_1 : \quad q^{(2)}hw + q^{(3)}wh = 0$$

These imply that

$$q^{(2)} = -q^{(1)} \quad q^{(3)} = q^{(1)} \quad q^{(4)} = -q^{(1)} - \frac{P}{h} \qquad \text{[8.9.3]}$$

Since the warping restraint means that $\delta W_{ext}^* = 0$, it follows from the principle of complementary virtual work that $\delta W_{int}^* = 0$. Substituting Equation 8.9.3 and the virtual shear flows into Equation 8.9.2, we therefore obtain

$$\frac{L}{Gt} \left[wq^{(1)} \frac{\delta P}{2L} + h \left(-q^{(1)} \right) \left(-\frac{\delta P}{2L} \right) + wq^{(1)} \frac{\delta P}{2L} + h \left(-q^{(1)} - \frac{P}{h} \right) \left(-\frac{\delta P}{2L} \right) \right] = 0$$

which simplifies to

$$\frac{\delta P}{2Gt} \left[2(w+h) q^{(1)} + P \right] = 0$$

Solving this for $q^{(1)}$ and substituting the result into Equation 8.9.3, we obtain all of the shear flows accompanying the warping restraint:

$$q^{(1)} = -\frac{P}{2(w+h)} \quad q^{(2)} = \frac{P}{2(w+h)} \quad q^{(3)} = -\frac{P}{2(w+h)} \quad q^{(4)} = -\frac{h+2w}{2(w+h)h} P \qquad \text{[8.9.4]}$$

These shear flows are sketched in Figure 8.9.3, along with the shear flows for the same load but with no warping restraint (cf. Figure 8.9.1b with $d = 0$). For the case $w = 4h$, the shear flows are as illustrated in Figure

8.9.4. As we can see, the warping restraint causes the shear flow in three of the panels to decrease 60 percent from the unrestrained case. A 20 percent increase occurs in the leftmost vertical web. According to Saint-Venant's principle, as we move away from the source of the warping restraint (e.g., a massive frame or rib), the effect on the shear flows becomes negligible within a distance roughly equal to w, the larger cross-sectional dimension.

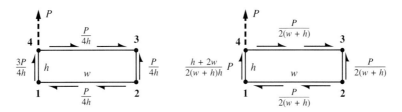

Figure 8.9.3   (a) Shear flows if the section of Figure 8.9.1 is free to warp.
(b) Shear flows if warping is prohibited.

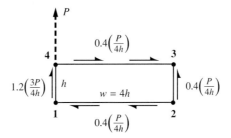

Figure 8.9.4   Shear flows in Figure 8.9.36
if $w = 4h$.

Example 8.9.1   If warping is restrained, calculate the shear flows in a beam with the cross section illustrated in Figure 8.9.5.

In Example 8.7.1, the unrestrained warp angle of this same section was calculated for the identical loading. To do so, we applied virtual couples to both of the vertical webs, as indicated in Figure 8.9.6a. Part (b) of the figure shows the corresponding virtual shear flows, the calculations of which are detailed in Example 8.7.1. The external complementary virtual work of the virtual load is proportional to the warp angle, which in this case we require to be zero. Therefore, $\delta W_{\text{ext}}^* = 0$ and, from the principle of complementary virtual work, $\delta W_{\text{int}}^* = 0$.

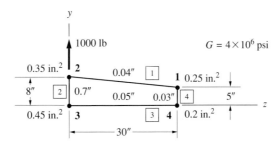

Figure 8.9.5   Load on a section of a box beam in
which warping is restrained.

Let us assume that the directions of the true shear flows are as sketched in Figure 8.9.7. The shear flows must be statically equivalent to the 1000 lb shear load directed upward through web 2–3. Therefore, the following three conditions apply:

$$\sum F_x = 0 \qquad \Rightarrow 30(q^{(1)} - q^{(3)}) = 0$$

$$\sum F_y = 1000 \qquad \Rightarrow (8 - 5)q^{(1)} - 8q^{(2)} + 5q^{(4)} = 1000$$

$$\sum M_{\text{flange } 2} = 0 \qquad \Rightarrow (30q^{(3)}) \times 8 + (5q^{(4)}) \times 30 = 0$$

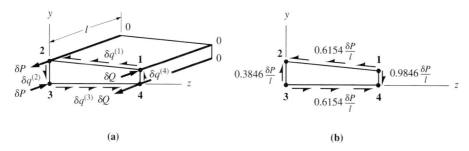

(a)   (b)

Figure 8.9.6   (a) Virtual couples applied to the vertical webs. (b) The corresponding virtual shear flows, calculated in Example 8.7.1.

Using these to express $q^{(2)}$, $q^{(3)}$, and $q^{(4)}$ in terms of $q^{(1)}$ we have

$$q^{(2)} = -\frac{5}{8}q^{(1)} - 125 \quad q^{(3)} = q^{(1)} \quad q^{(4)} = -\frac{8}{5}q^{(1)} \qquad \text{[a]}$$

The internal complementary virtual work is just that of the shear panels, which is

$$\delta W_{\text{int}}^{*} = \sum_{\text{panels}} \frac{A}{Gt} q \delta q = \frac{l}{G}\left(\frac{s^{(1)}}{t^{(1)}}q^{(1)}\delta q^{(1)} + \frac{s^{(2)}}{t^{(2)}}q^{(2)}\delta q^{(2)} + \frac{s^{(3)}}{t^{(3)}}q^{(3)}\delta q^{(3)} + \frac{s^{(4)}}{t^{(4)}}q^{(4)}\delta q^{(4)} \right)$$

The required geometric data appear in Figure 8.9.5, and the true and virtual shear flows are found in Equation a and Figure 8.9.6b, respectively. Thus,

$$\delta W_{\text{int}}^{*} = \frac{l}{G}\left[\frac{\sqrt{30^2 + 3^2}}{0.04}q^{(1)}\left(0.6153\frac{\delta P}{l}\right) + \frac{8}{0.07}\left(-\frac{5}{8}q^{(1)} - 125\right)\left(-0.3846\frac{\delta P}{l}\right) \right.$$

$$\left. + \frac{30}{0.05}q^{(1)}\left(0.6154\frac{\delta P}{l}\right) + \frac{5}{0.03}\left(-\frac{8}{5}q^{(1)}\right)\left(-0.9846\frac{\delta P}{l}\right) \right]$$

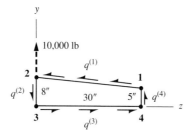

Figure 8.9.7   Assumed directions of the true shear flows.

or

$$\delta W^*_{int} = \frac{\delta P}{G} \left(1123 q^{(1)} + 5494\right) \qquad \text{[b]}$$

Since $\delta W^*_{int} = 0$, Equation b implies that $q^{(1)} = -4.892$ lb/in., and Equations a yield the rest of the shear flows, all of which are illustrated in Figure 8.9.8b, alongside the shear flows computed in Example 8.7.1, in which warping was unrestrained.

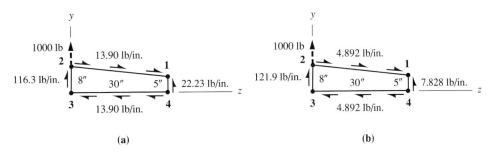

(a)   (b)

Figure 8.9.8   (a) Shear flows when warping is free to occur (cf. Example 8.7.1). (b) Shear flows when warping is prevented (the present case).

8.10   SHEAR LAG

Consider Figure 8.10.1, which shows a shear web bonded to two rigid walls and attached to a stiffener by means of which a point load P_0 is applied to the system. At the opposite end of the stiffener, the load is zero. The load is transferred to the walls by the shearing action of the panels. Previously, we have considered that based on equilibrium considerations alone, the stiffener load diffuses into the webs at a constant rate equal to the assumed constant shear flow q in the rectangular webs, such that $2qL = P_0$. To analyze stiffened web structures in section 8.3, we invoked the complementary virtual work principle. This required us to account for the elasticity of the webs and stiffeners. Nevertheless, we have continued to assume that the shear flow in a rectangular panel is constant.

Now, let us instead make an assumption about the nature of the web deformation and infer the shear flow from that assumption. In particular, assume that the x displacement (positive to the right) in the panel varies from zero at the wall to that of the stiffener, as shown in Figure 8.10.2a. It follows that the shear strain in the web is

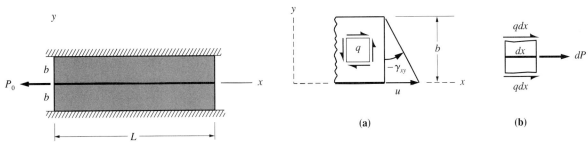

Figure 8.10.1   Load transfer to a shear web by means of a stiffener.

(a)   (b)

Figure 8.10.2   (a) Relationship between stiffener displacement and web shear in the upper panel. (b) Free-body diagram of a differential length of the stiffener.

$$\gamma_{xy} = -\frac{u}{b} \tag{8.10.1}$$

The minus sign reflects the fact that the initial right angle between the vertical edge of the panel and the horizontal stiffener increases.

The shear strain is related to the shear stress by Hooke's law, $\tau_{xy} = G\gamma_{xy}$. Since the shear flow equals the shear stress times the panel thickness t, we have

$$q = -\frac{Gt}{b}u \tag{8.10.2}$$

The term Gt/b is the *shear rigidity* of the panel; the larger this ratio, the larger the shear flow accompanying a given displacement of the edge of the panel attached to the longitudinal stiffener. Decreasing the panel thickness or increasing its width results in a more flexible web.

For the shear flow and flange load to be in equilibrium, we see from Figure 8.10.2b that $dP + 2qdx = 0$, or

$$q = -\frac{1}{2}\frac{dP}{dx} \tag{8.10.3}$$

The normal stress σ_x in the stiffener is related to the normal strain ε_x by Hooke's law, $\sigma_x = E\varepsilon_x$. The normal stress is just P/A, where A is the cross-sectional area, and the normal strain (cf. Chapter 3) is du/dx. Thus,

$$\frac{du}{dx} = \frac{P}{AE} \tag{8.10.4}$$

where AE is the axial rigidity of the stiffener. Differentiating Equation 8.10.2 with respect to x and substituting Equation 8.10.3 and 8.10.4 into the result yields a second-order differential equation involving just P,

$$\frac{1}{2}\frac{d^2P}{dx^2} = \frac{Gt}{b}\frac{P}{AE}$$

which we can write as

$$\frac{d^2P}{dx^2} - k^2P = 0 \qquad \left(k^2 = \frac{2Gt}{AEb}\right) \tag{8.10.5}$$

k is called the *shear lag parameter*, and it has dimensions of $[length]^{-1}$. The term k is a measure of the relative web stiffness, Gt/b, to that of the stiffener, AE. The solution of the homogeneous differential equation, 8.10.5, is found in terms of the hyperbolic sine and hyperbolic cosine, as follows:

$$P = C_1 \sinh kx + C_2 \cosh kx \tag{8.10.6}$$

where

$$\sinh kx = \frac{e^{kx} - e^{-kx}}{2} \qquad \cosh kx = \frac{e^{kx} + e^{-kx}}{2} \tag{8.10.7}$$

We determine the integration constants C_1 and C_2 by satisfying the boundary conditions on P. At $x = 0$, $P = P_0$. From Equation 8.10.6, we have

$$P_0 = C_1 \sinh k(0) + C_2 \cosh k(0)$$

whereas from Equation 8.10.7, $\sinh k(0) = 0$ and $\cosh k(0) = 1$, so that $C_2 = P_0$. At $x = L$, P vanishes. With $C_2 = P_0$, Equation 8.10.6 therefore yields

$$0 = C_1 \sinh kL + P_0 \cosh kL$$

which means that $C_1 = -P_0\left(\cosh kL / \sinh kL\right)$. Therefore, the stiffener load P as a function of x is

$$P = -P_0 \frac{\cosh kL}{\sinh kL} \sinh kx + P_0 \cosh kx = P_0\left(\frac{\sinh kL \cosh kx - \cosh kL \sinh kx}{\sinh kL}\right)$$

Using the definitions of hyperbolic sine and cosine given in Equation 8.10.7, we can show that

$$\sinh(kL - kx) = \sinh kL \cosh kx - \cosh kL \sinh kx$$

Therefore, the expression for the stiffener force can be written more compactly as

$$P = P_0 \frac{\sinh k(L - x)}{\sinh kL} = P_0 \frac{\sinh kL(1 - x/L)}{\sinh kL} \qquad [8.10.8]$$

Substituting this into Equation 8.10.3, we get the shear flow as a function of x, which is

$$q = \frac{P_0 k}{2} \frac{\cosh k(L - x)}{\sinh kL} \qquad [8.10.9]$$

The stiffener load variation in Equation 8.10.8 is plotted in Figure 8.10.3. As we can see, for a given stiffener length L, small values of the shear lag parameter k cause the stiffener load distribution to approach the linear distribution associated with constant shear flow panels. That is, if the shear panel is very flexible compared to the stiffener, the presence of the applied load is felt throughout the bulk of the structure. On the other hand, increasing the shear rigidity of the web magnifies the shear lag effect and the stiffener load falls off exponentially toward

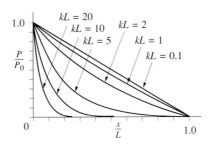

Figure 8.10.3   Stiffener load as a function of position for the stiffened web of Figure 8.10.1, according to Equation 8.10.8.

zero. The larger the value of k, the more rapid the decrease, so that the effect of the applied load is concentrated near its point of application. We can adjust the magnitude of the shear lag parameter by modifying the material or section properties, in terms of which it is defined in Equation 8.10.5. Quadrupling the web thickness t doubles the shear lag parameter; quadrupling the width b of the web reduces k by a factor of two.

Using Equation 8.10.7, we can write the expression for stiffener force, Equation 8.10.8, in an alternative form:

$$P = P_0 \frac{e^{kL} e^{-kx} - e^{-kL} e^{kx}}{e^{kL} - e^{-kL}} = P_0 \frac{e^{-kx} - e^{-2kL} e^{kx}}{1 - e^{-2kL}}$$

From this, we see that if x remains finite while kL increases without bound, then, in the limit,

$$P = P_0 e^{-kx} \qquad\qquad\qquad\qquad \textbf{[8.10.10]}$$

It follows from Equation 8.10.3 that

$$q = \frac{P_0 k}{2} e^{-kx} \qquad\qquad\qquad\qquad \textbf{[8.10.11]}$$

These expressions demonstrate the exponential nature of the decay of stiffener load and shear flow in the vicinity of the applied load.

Example 8.10.1   Using the shear lag approach, find the formulas for the stiffener loads and panel shear flows in the plane, stiffened web structure shown in Figure 8.10.4. The top and bottom stiffeners have the same cross-sectional area, and all other properties are uniform throughout. Plot the results for the special case $A_1 = A_2 = 0.5$ in.^{2}, $L = 40$ in., $t = 0.1$ in., $b = 2$ in., $Q = 1000$ lb, and $G = 0.4E$.

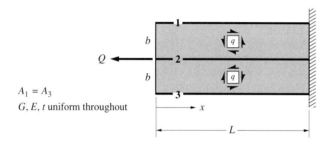

Figure 8.10.4   Stiffened panel with axial load applied to center stiffener.

First, we use equilibrium to establish a relationship betweeen the stiffener loads. From the free-body diagram in Figure 8.10.5a, we find that

$$2P_1 + P_2 = Q \qquad\qquad\qquad\qquad \textbf{[a]}$$

To relate the flange loads to the axially varying shear flow, we consider the equilibrium of a differential segment of the top stiffener, as illustrated in Figure 8.10.5b. We see that $qdx = dP_1$, or

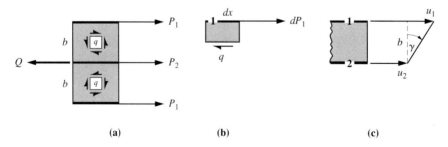

(a)   (b)   (c)

Figure 8.10.5   (a) Free-body diagram of a portion of the stiffened panels, revealing the flange loads. The upper and lower flange loads are equal by symmetry. (b) Free-body diagram of a differential length of the upper stiffener. (c) Relationship of the stiffener displacements to the web shear strain.

$$q = \frac{dP_1}{dx} \qquad \text{[b]}$$

Finally, as in Figure 8.10.5c, we establish the relationship between the axial displacements of the stiffeners and the shear strain in the web between them. The decrease in right angle γ, in radians, is the shear strain, which is very small; it is evident from the figure that $\gamma = (u_1 - u_2)/b$. The shear stress τ is q/t, where t is the thickness of the web. Therefore, from Hooke's law, $\tau = G\gamma$, we obtain

$$q = \frac{Gt}{b}(u_1 - u_2) \qquad \text{[c]}$$

Differentiating both sides of Equation c with respect to x yields

$$\frac{dq}{dx} = \frac{Gt}{b}\left(\frac{du_1}{dx} - \frac{du_2}{dx}\right) \qquad \text{[d]}$$

But du/dx is the normal strain ε, which is related to the normal stress by Hooke's law, $\sigma = E\varepsilon$. Furthermore, since the normal stress equals the axial load P divided by the cross-sectional area A, we can write Equation d as

$$\frac{dq}{dx} = \frac{Gt}{b}\left(\frac{P_1}{A_1 E} - \frac{P_2}{A_2 E}\right) = \frac{Gt}{Eb}\left(\frac{P_1}{A_1} - \frac{P_2}{A_2}\right) \qquad \text{[e]}$$

Substituting Equations a and b into this expression, we have

$$\frac{d^2 P_1}{dx^2} = \frac{Gt}{Eb}\left[\frac{P_1}{A_1} - \frac{(Q-2P_1)}{A_2}\right] = \frac{Gt}{Eb}\left(\frac{2A_1 + A_2}{A_1 A_2}\right)P_1 - \frac{Gt}{Eb}\frac{Q}{A_2} \qquad \text{[f]}$$

which can be written more compactly as

$$\frac{d^2 P_1}{dx^2} - k^2 P_1 = -\frac{Gt}{Eb}\frac{Q}{A_2} \qquad \text{[g]}$$

where

$$k^2 = \frac{Gt}{Eb}\left(\frac{2A_1 + A_2}{A_1 A_2}\right) \qquad \text{[h]}$$

The general solution of the second-order differential equation, Equation g, is

$$P_1 = \overbrace{C_1 \sinh kx + C_2 \cosh kx}^{\text{complementary solution}} + \overbrace{\frac{1}{k^2}\frac{Gt}{Eb}\frac{Q}{A_2}}^{\text{particular solution}} \qquad \text{[i]}$$

Substituting the shear lag parameter k from Equation h into the last term of Equation i, we obtain

$$P_1 = C_1 \sinh kx + C_2 \cosh kx + \frac{QA_1}{2A_1 + A_2} \qquad \text{[j]}$$

Since, according to Equation b, the shear flow is the derivative of P_1 with respect to x, Equation j implies that

$$q = kC_1 \cosh kx + kC_2 \sinh kx \qquad \text{[k]}$$

The integration constants C_1 and C_2, are found by applying the boundary conditions. At $x = 0$, we know that $P_1 = 0$. Setting P_1 and x equal to zero in Equation j, we see that

$$C_2 = -\frac{QA_1}{2A_1 + A_2} \qquad \text{[l]}$$

At $x = L$, $u_1 = u_2 = 0$, so Equation c implies that $q = 0$ at $x = L$. Therefore, from Equations k and l, we have

$$C_1 = \frac{QA_1}{2A_1 + A_2} \frac{\sinh kL}{\cosh kL} \qquad \text{[m]}$$

Substituting the integration constants, Equations l and m, back into Equation j, we get the expression for the flange load P_1,

$$P_1 = \frac{QA_1}{2A_1 + A_2} \frac{\sinh kL}{\cosh kL} \sinh kx - \frac{QA_1}{2A_1 + A_2} \cosh kx + \frac{QA_1}{2A_1 + A_2}$$

$$= \frac{QA_1}{2A_1 + A_2} \frac{1}{\cosh kL} [(\sinh kL \sinh kx - \cosh kL \cosh kx) + \cosh kL] \qquad \text{[n]}$$

Using the definition of the hyperbolic sine and cosine given in Equation 8.10.7, we can show that

$$\cosh kL \cosh kx - \sinh kL \sinh kx = \cosh k (L - x)$$

Therefore, Equation n can be written more compactly as

$$P_1 = \frac{QA_1}{2A_1 + A_2} \left[1 - \frac{\cosh k (L - x)}{\cosh kL} \right] \qquad \text{[o]}$$

Next, we obtain the formula for P_2 by substituting Equation o into Equation a, which leads to

$$P_2 = \frac{2QA_1}{2A_1 + A_2} \left[\frac{A_2}{2A_1} + \frac{\cosh k (L - x)}{\cosh kL} \right] \qquad \text{[p]}$$

Finally, from Equation b, we obtain the shear flow in the webs of the structure, as follows:

$$q = \frac{QA_1 k}{2A_1 + A_2} \frac{\sinh k(L - x)}{\cosh kL} \qquad \text{[q]}$$

The elementary solution of this problem would be to assume that the axial stress distribution is uniform over the cross section of total area $2A_1 + A_2$ (neglecting the webs). If that were so, the normal stress in each of the flanges would be $\sigma = Q/(2A_1 + A_2)$. Accordingly,

$$P_1 = \sigma A_1 = \frac{QA_1}{2A_1 + A_2} \qquad P_2 = \sigma A_2 = \frac{QA_2}{2A_1 + A_2} \qquad q = 0 \qquad \text{elementary solution} \qquad \text{[r]}$$

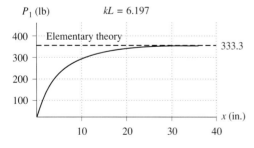

Figure 8.10.6   Axial load distribution in the upper and lower stiffeners of the structure in Figure 8.10.4, for the specified data.

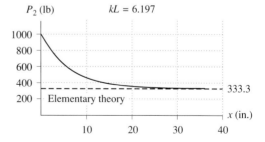

Figure 8.10.7   Axial load distribution in the center stiffener.

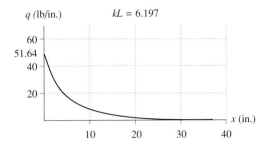

Figure 8.10.8 Shear flow in the panels of the structure in Figure 8.10.12.

The flange loads and shear flow given by Equations o, p, and q—the shear lag solution—are plotted in Figures 8.10.6, 7 and 8, using the numerical data supplied at the beginning of this example. Observe that as the location of the section moves farther away from the applied load, the stiffener loads and shear flow tend toward the elementary theory values.

Example 8.10.2 Using the shear lag approach, obtain an expression for the displacement of the left end of the center stiffener of the previous example.

From Equation p of Example 8.10.1, we know that the load in the center stiffener as a function of x (the distance from the applied load) is

$$P_2 = \frac{2QA_1}{2A_1 + A_2}\left[\frac{A_2}{2A_1} + \frac{\cosh k\,(L - x)}{\cosh kL}\right] \tag{a}$$

According to Equation 8.10.4, the axial displacement and the axial load in the stiffener are related through Hooke's law, as follows:

$$\frac{du_2}{dx} = \frac{P_2}{A_2 E} \tag{b}$$

Therefore,

$$\frac{du_2}{dx} = \frac{1}{A_2 E}\frac{2QA_1}{2A_1 + A_2}\left[\frac{A_2}{2A_1} + \frac{\cosh k\,(L - x)}{\cosh kL}\right] \tag{c}$$

Integrating this equation with respect to x, we get

$$u_2 = \frac{1}{A_2 E}\frac{2QA_1}{2A_1 + A_2}\left[\frac{A_2}{2A_1}x - \frac{1}{k}\frac{\sinh k\,(L - x)}{\cosh kL}\right] + C \tag{d}$$

The constant of integration, C, is found by applying the boundary condition $u_2(L) = 0$. Setting $x = L$ and $u_2 = 0$ in Equation d, we see that

$$C = -\frac{QL}{(2A_1 + A_2)E} \tag{e}$$

Substituting this into Equation d, we find that the displacement of any point on the center stiffener is

$$u_2 = -\frac{Q}{(2A_1 + A_2)\,E}\left[L - x + \frac{2A_1}{A_2}\frac{1}{k}\frac{\sinh k\,(L - x)}{\cosh kL}\right] \tag{f}$$

Evaluating this expression at $x = 0$ yields the displacement at the left end, which is

$$u_{2\text{left end}} = -\frac{Q}{(2A_1 + A_2)E}\left[L + \frac{2A_1}{A_2}\frac{1}{k}\frac{\sinh kL}{\cosh kL}\right]$$ [g]

The minus sign means that the displacement is to the left, in the direction of the applied load Q.

Consider the idealized box beam shown in Figure 8.10.9. The centroid G of the symmetric cross section is located at the geometric center of the rectangle, as is the shear center. Therefore, the vertical shear force Q on the free end, whose resultant passes through G, exerts no twist. To find the flange loads using beam theory, we use Equation 4.6.8, which in this case reduces to

$$\sigma_x = -\frac{M_z y}{I_z}$$

in which the area moment of inertia I_z is that of the six concentrated flange areas relative to G, namely,

$$I_z = 2\left[2A_1\left(\frac{a}{2}\right)^2 + A_2\left(\frac{a}{2}\right)^2\right] = \frac{a^2}{2}(2A_1 + A_2)$$

The bending moment as a function of spanwise coordinate x is $M_z = Qx$. Therefore, the axial stress in the top stringers, at $y = a/2$, is

$$\sigma_x = -\frac{(Qx)(a/2)}{(a^2/2)(2A_1 + A_2)} = -\frac{Q}{2A_1 + A_2}\frac{x}{a}$$

Multiplying the stress by the flange area gives the stringer load. Therefore, for stringers 1 and 2 on the top of the beam,

$$P_1 = -\frac{QA_1}{2A_1 + A_2}\frac{x}{a} \qquad P_2 = -\frac{QA_2}{2A_1 + A_2}\frac{x}{a}$$ [8.10.12]

The stringer loads on the bottom are the same except the sign is reversed. If the load is applied upwards as shown, the top stringers are obviously in compression and the bottom ones are in tension.

By symmetry, the shear flows in the vertical webs are both equal to q_v. From Figure 8.10.10 we see that

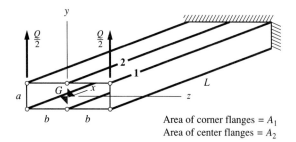

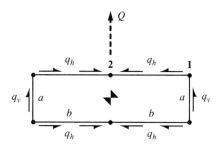

Figure 8.10.9 Idealized, cantilevered box beam of uniform, symmetrical cross section.

Figure 8.10.10 Shear flows in the cross section of the box beam.

$$q_v = \frac{Q}{2a}$$

[8.10.13]

At flange 1, $q_{\text{out}} = q_{\text{in}} + P_1'$, so that $q_h = q_v + P_1'$, where q_h is the shear flow in the top panels. Using Equations 8.10.12 and 8.10.13, we therefore have

$$q_h = \frac{Q}{2a} + \left(-\frac{QA_1}{2A_1 + A_2}\frac{1}{a}\right)$$

or

$$q_h = \frac{Q}{2a}\frac{A_2}{2A_1 + A_2}$$

[8.10.14]

The terms q_h and q_v are the constant shear flows in the top and side panels, according to beam theory. For the shear flow q_h to be as calculated, the top (and bottom) webs must be able to deform as shown in Figure 8.10.11 in

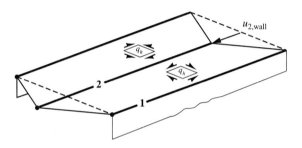

Figure 8.10.11   Free distortion of top panels due to a state of constant shear strain.

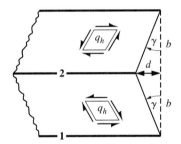

Figure 8.10.12   Relationship between the shear strain γ in the top panels and the displacement of the center stringer at the wall.

order to accommodate the shear *strain* accompanying the constant shear *stress* q_h/t, where t is the thickness of the panel. If the edges of the top panels at the wall rotate through an angle γ due to shear, then, as shown in Figure 8.10.12, the distance through which the right end of the center stringer moves is $d = \gamma \times b$. But γ is the engineering shear strain in the panel, that is, the decrease in right angle between lines in the x and z directions. The term γ is related to the shear stress by Hooke's law, $\gamma = \frac{(q_h/t)}{G}$, which means that

$$d = \frac{q_h b}{Gt}$$

[8.10.15]

If the stringers are attached to the wall, then the axial displacement of the central stringer must be zero at that end. This means that the edges of the panels at the wall remain at right angles to the stringers, thereby forcing the shear strain in that region to be zero. This in turn requires the shear stress, and hence the shear flow, to vanish at the wall. Clearly, the results of simple beam theory must be modified to account for this local effect, which is due to shear lag.

We need to determine the force that must be applied to the right end of the central stringer in Figure 8.10.12 in order to force it through the distance d, back against the wall. Once we have calculated that force and the corresponding shear flows in the panels, we can superimpose the results onto the beam solution to obtain a more accurate description of the loads throughout the structure.

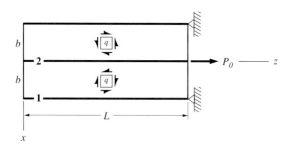

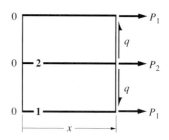

Figure 8.10.14 Free-body diagram of a portion of the top panel.

Figure 8.10.13 Top panel of the box beam in Figure 8.10.2.

Consider the top panel of the box beam, as illustrated in Figure 8.10.13. Apply a load P_0 at the right end of the center stringer, and keep the left end of all three stringers load free. The load is transferred through the webs to the outer stringers and through them to the wall, where the reactive loads are $P_0/2$. Figure 8.10.14 is a free-body diagram of the panel up to station x, in which the flange loads have been assumed to be tensile. The relationship between the central stringer load P_1 and the corner stringer load P_2 is

$$2P_1 + P_2 = 0 \qquad [8.10.16]$$

At $x = L$, $P_1 = -P_0/2$ and $P_2 = P_0$.

Figure 8.10.15a shows a differential length of stringer 2, from which we infer

$$\frac{dP_2}{dx} = -2q \qquad [8.10.17]$$

Part b of Figure 8.10.15 shows the axial displacements u_1 and u_2 of stringers 1 and 2 at any station x. Assuming that $u_1 > u_2$, the shear strain is

$$\gamma_{xz} = \frac{(u_1 - u_2)}{b}$$

The corresponding shear stress τ_{xz} is the shear flow q divided by the thickness of the panel. Thus, from Hooke's law, we have

$$q = \frac{Gt}{b}(u_1 - u_2) \qquad [8.10.18]$$

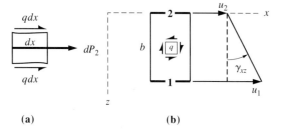

(a)   (b)

Figure 8.10.15 (a) Free-body diagram of a differential length of the center stringer. (b) Displacements of the center and corner stringers at station x.

For stringers 1 and 2 Hooke's law also yields

$$\frac{du_1}{dx} = \frac{P_1}{A_1 E} \qquad \frac{du_2}{dx} = \frac{P_2}{A_2 E} \qquad\qquad \text{[8.10.19]}$$

assuming that the modulus of elasticity is the same for all of the stringers.

Differentiating Equation 8.10.18 with respect to x and then substituting Equation 8.10.19, for the displacement derivatives we get

$$\frac{dq}{dx} = \frac{Gt}{b} \left(\frac{P_1}{A_1 E} - \frac{P_2}{A_2 E} \right) \qquad\qquad \text{[8.10. 20]}$$

Substituting the shear flow $q = -\frac{1}{2} dP_2/dx$ from Equation 8.10.17 into the left side of this equation, and the stringer load $P_2 = -P_1/2$ from Equation 8.10.16 into the right side, we again get a second-order differential equation, in this case involving just P_2

$$\frac{d^2 P_2}{dx^2} - k^2 P_2 = 0 \qquad\qquad \text{[8.10. 21a]}$$

where

$$k^2 = \frac{Gt}{Eb} \left(\frac{1}{A_1} + \frac{2}{A_2} \right) \qquad\qquad \text{[8.10.21b]}$$

k is the shear lag parameter. The solution of this homogeneous differential equation is

$$P_2 = C_1 \sinh kx + C_2 \cosh kx \qquad\qquad \text{[8.10. 22]}$$

We determine the integration constants C_1 and C_2 by satisfying the boundary conditions on P_2. At $x = 0$, the unsupported end of the panel, $P_2 = 0$. From Equation 8.10.22, we see that

$$0 = C_1 \sinh k(0) + C_2 \cosh k(0)$$

whereas from Equation 8.10.7, $\sinh k(0) = 0$ and $\cosh k(0) = 1$, so that $C_2 = 0$. At $x = L$, P_2 equals the applied load P_0. Therefore, Equation 8.10.22 with $C_2 = 0$ yields

$$P_0 = C_1 \sinh kL$$

so that $C_1 = P_0/\sinh kL$. Therefore, the stringer load P_2 as a function of x is

$$P_2 = P_0 \frac{\sinh kx}{\sinh kL} \qquad\qquad \text{[8.10. 23]}$$

With this we can obtain P_1 from Equation 8.10.16,

$$P_1 = -\frac{P_2}{2} = -\frac{P_0}{2} \frac{\sinh kx}{\sinh kL} \qquad\qquad \text{[8.10. 24]}$$

and the shear flow q from Equation 8.10.17,

$$q = -\frac{1}{2} \frac{dP_2}{dx} = -\frac{P_0 k}{2} \frac{\cosh kx}{\sinh kL} \qquad\qquad \text{[8.10. 25]}$$

where we used the fact that $d \sinh kx/dx = k \cosh kx$.

Evaluating Equation 8.10.18 at $x = L$, we obtain the shear flow at the wall,

$$q(L) = \frac{Gt}{b} [u_1(L) - u_2(L)]$$

But $u_1(L) = 0$ and $u_2(L)$ must be the displacement required to close the gap d between the right end of the center stringer and the wall, given by Equation 8.10.15. Thus, with the aid of Equation 8.10.25, we have

$$-\frac{P_0 k}{2} \frac{\cosh kL}{\sinh kL} = \frac{Gt}{b} \left[0 - \frac{q_h b}{Gt} \right]$$

which yields the force P_0 required to keep the center stringer attached to the wall, as follows:

$$P_0 = \frac{2q_h}{k} \frac{\sinh kL}{\cosh kL} \qquad\qquad [8.10.26]$$

Substituting this back into the above expressions for P_2, P_1, and q, we get

$$P_2 = \frac{2q_h}{k} \frac{\sinh kx}{\cosh kL} \qquad\qquad [8.10.27]$$

$$P_1 = -\frac{q_h}{k} \frac{\sinh kx}{\cosh kL} \qquad\qquad [8.10.28]$$

$$q = -q_h \frac{\cosh kx}{\cosh kL} \qquad\qquad [8.10.29]$$

These loads must be superimposed on those obtained from elementary beam theory, Equation 8.10.12 and 8.10.14, to enforce the condition of zero displacements at the wall. Thus,

$$P_1 = \overbrace{\left[-\frac{QA_1}{2A_1 + A_2} \frac{x}{a} \right]}^{\text{pure bending}} + \overbrace{\left[-\left(\frac{Q}{2a} \frac{A_2}{2A_1 + A_2} \right) \left(\frac{1}{k} \right) \frac{\sinh kx}{\cosh kL} \right]}^{\text{shear lag}}$$

$$P_2 = \overbrace{\left[-\frac{QA_2}{2A_1 + A_2} \frac{x}{a} \right]}^{\text{pure bending}} + \overbrace{\left[2\left(\frac{Q}{2a} \frac{A_2}{2A_1 + A_2} \right) \left(\frac{1}{k} \right) \frac{\sinh kx}{\cosh kL} \right]}^{\text{shear lag}}$$

$$q = \overbrace{\left[\frac{Q}{2a} \frac{A_1}{2A_1 + A_2} \right]}^{\text{pure bending}} + \overbrace{\left[-\left(\frac{Q}{2a} \frac{A_2}{2A_1 + A_2} \right) \frac{\cosh kx}{\cosh kL} \right]}^{\text{shear lag}}$$

or

$$P_1 = -Q \frac{A_1}{2A_1 + A_2} \left(\frac{x}{a} + \frac{A_2}{A_1} \frac{1}{2ka} \frac{\sinh kx}{\cosh kL} \right) \qquad \text{top corner stringers} \qquad [8.10.30]$$

$$P_2 = -Q \frac{A_2}{2A_1 + A_2} \left(\frac{x}{a} - \frac{1}{ka} \frac{\sinh kx}{\cosh kL} \right) \qquad \text{top center stringer} \qquad [8.10.31]$$

$$q = \frac{Q}{2a} \frac{A_2}{2A_1 + A_2} \left(1 - \frac{\cosh kx}{\cosh kL} \right) \qquad \text{horizontal webs} \qquad [8.10.32]$$

Example 8.10.3   Let the following numerical data apply to the box beam in Figure 8.10.9: $G = 0.4E$, $A_1 = A_2 = 1$ in.^{2}, $t = 0.1$ in., $L = 40$ in., $a = 2$ in., $b = 4$ in., and $Q = 1000$ lb. Plot the flange loads and web shear flow versus span for the shear lag solution just obtained and compare them with elementary beam theory.

For the given data, the shear lag parameter, Equation 8.10.21b, is,

$$k^2 = \frac{Gt}{Eb}\left(\frac{1}{A_1} + \frac{2}{A_2}\right) = 0.4 \times 0.025\,(1 + 2) = 0.03$$

so that $k = 0.1732$ in.$^{-1}$. Substituting the numbers into Equation 8.10.30, 8.10.31, and 8.10.32, we get

$$P_1 = -166.7\,[x + 0.005657\sinh(0.1732x)]$$
$$P_2 = -166.7\,[x\text{–}0.01131\sinh(0.1732x)]$$
$$q = 83.33\,[1\text{–}0.001960\cosh(0.1732x)]$$

These are plotted in Figures 8.10.16, 17, and 18.

Beam theory predicts that the flange loads in the center and corner stringers will be equal and will have the same linear variation from end to end. As we can see from Figures 8.10.16 and 8.10.17, that is the case over most of the span. However, near the wall, where the displacement constraints are imposed, discrepancies occur. The compressive load in the top corner stringers diverges to a value a bit over 7 percent greater than that obtained by beam theory. In the approach to the wall, the compressive flange load in the center stringer lags ever more significantly behind the corner flange's, ending up at the wall with a load that is only 86 percent of that predicted by beam theory. Figure 8.10.18 shows that throughout most of the first half of the beam, the shear flow in the panels has the constant value predicted by beam theory. But at the mid span the shear flow begins its rapid divergence toward zero at the wall. These various differences would be larger, as well as apparent over a larger portion of the span, if we decreased the shear lag parameter k. Increasing k would reduce the discrepancies and concentrate them even closer to the wall.

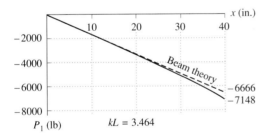

Figure 8.10.16   Corner stringer load distribution.

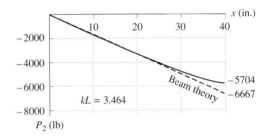

Figure 8.10.17   Center stringer load distribution.

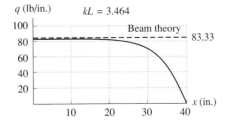

Figure 8.10.18   Shear flow in the horizontal webs.

EXERCISES

Use the force method to solve the following problems.

8.1   Calculate the horizontal displacement in the direction of the 1000 lb load on the stiffened web.

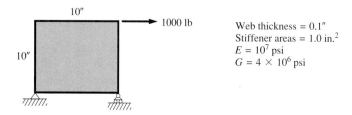

Exercise 8.1

8.2   The stiffened web has a horizontal load P applied as shown to point 3. The displacement of point 3 in the direction of P is 0.01 in. Calculate the value of the load.

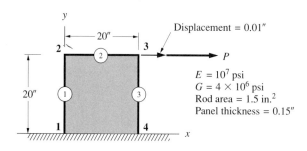

Exercise 8.2

8.3   The shear panel is surrounded by three stiffeners. Throughout, $E = 10^7$ psi and $G = 4 \times 10^6$ psi. The area of the stiffeners is 0.5 in.^{2}, and the panel thickness is 0.1 in. Find the vertical displacement at node 2 due to a horizontal 1000 lb load at node 1.

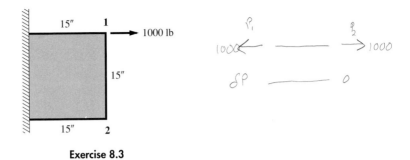

Exercise 8.3

8.4   The plane panel–stringer structure has a downward load P applied at node 4. Find the horizontal displacement of node 1 due to this loading.

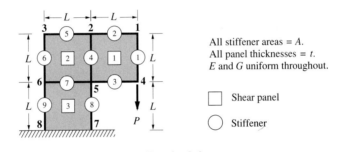

All stiffener areas = A.
All panel thicknesses = t.
E and G uniform throughout.

☐ Shear panel

◯ Stiffener

Exercise 8.4

8.5   Calculate the shear flow in each of the two panels of the stiffened web structure shown. The area of the three stiffeners is 1 in.^{2}, and the panel thicknesses are 0.1 in.^{2} Throughout, $E = 10^7$ psi and $G = 4 \times 10^6$ psi.

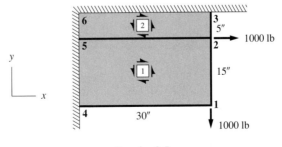

Exercise 8.5

8.6   Find the shear flow in all four of the webs. Panel thicknesses are 0.05 in., and stiffener areas are 0.5 in.^{2} Throughout, $E = 10^7$ psi and $G = 4 \times 10^6$ psi.

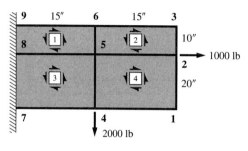

Exercise 8.6

8.7   Calculate the horizontal displacement of node 4. All stiffener areas are 0.5 in.^{2} All panel thicknesses are 0.1 in. Throughout, $E = 10^7$ psi and $\nu = 0.25$.

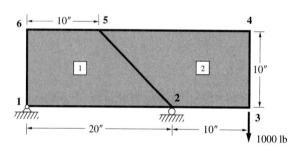

Exercise 8.7

8.8   Calculate the maximum shear flow q that can be applied to the periphery of the plane, stiffened panel structure if the maximum allowable shear stress in the panels is 10,000 psi. Choose panel 5 as the redundant member, and assume $E = 10^7$ psi, $G = 4 \times 10^6$ psi, $\nu = 0.25$. All 12 stiffener areas are 1.0 in.^{2}, and all five panel thicknesses are 0.1 in. Take advantage of symmetry.

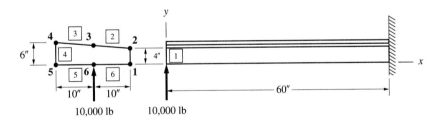

Exercise 8.8

8.9   Two views of a cantilever box beam are pictured. Throughout the structure, $E = 10^7$ psi and $G = 4 \times 10^7$ psi.

a. Calculate the z component of the displacement of the centroid of the free end of the beam.

b. Calculate the rotation of the free end of the beam about the x-axis.

Stringers: $A_1 = 0.5$ in.^{2} $A_2 = 0.6$ in.^{2} $A_3 = 0.7$ in.^{2}
$A_4 = 0.8$ in.^{2} $A_5 = 0.9$ in.^{2} $A_6 = 1.0$ in.^{2}

Panels: $t_1 = 0.040''$ $t_2 = 0.035''$ $t_3 = 0.030''$
$t_4 = 0.025''$ $t_5 = 0.020''$ $t_6 = 0.015''$

Exercise 8.9

8.10   Find the angle of twist per unit length for a box beam with the illustrated cross section if a 5000 lb load is applied as shown.

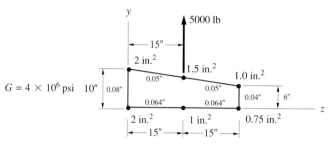

Exercise 8.10

8.11   Calculate the rate of twist for the single-cell box beam loaded as shown.

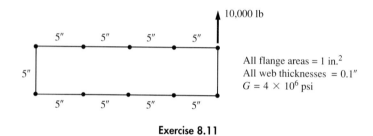

Exercise 8.11

8.12   For the plane, tapered cantilever idealized beam shown:

a.   Find the vertical displacement of the left (free) end of the beam.

b.   What percentage of the displacement is due to shear?

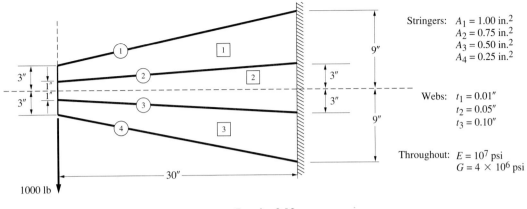

Stringers: $A_1 = 1.00$ in.^{2}
$A_2 = 0.75$ in.^{2}
$A_3 = 0.50$ in.^{2}
$A_4 = 0.25$ in.^{2}

Webs: $t_1 = 0.01''$
$t_2 = 0.05''$
$t_3 = 0.10''$

Throughout: $E = 10^7$ psi
$G = 4 \times 10^6$ psi

Exercise 8.12

8.13   Calculate the displacement in the direction of the load P. Throughout, $E = 10^7$ psi and $G = 4 \times 10^6$ psi. All panels are 0.04 in. thick, and the stringer areas are given in the figure.

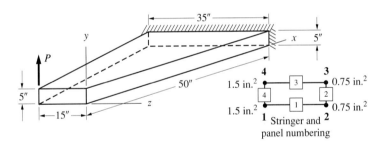

Exercise 8.13

8.14   Calculate the angle of twist at the free end of the box beam due to the pure torsional applied load. The front and side views are shown. All stringer areas are 1 in.^{2}, and all panel thicknesses are 0.1 in. Throughout, $E = 10^7$ psi and $G = 4 \times 10^6$ psi.

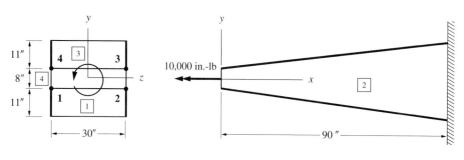

Exercise 8.14

8.15   Calculate the twist angle at the tip of the wing box in Exercise 4.24.

8.16   For the semicircular closed section shown, calculate the distance e of the shear center from the vertical.

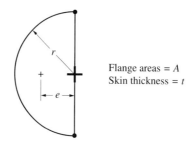

Flange areas = A
Skin thickness = t

Exercise 8.16

8.17   Find the *y* and *z* coordinates of the shear center of the section shown. The flange areas and web thicknesses are as given.

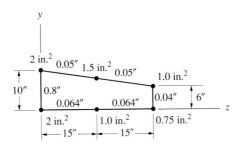

Exercise 8.17

8.18   Calculate the *z* coordinate of the shear center of the closed, triangular thin-walled section. All flanges have the same area *A* and all webs have the same thickness *t*.

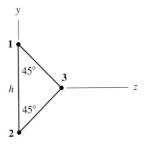

Exercise 8.18

8.19   The box beam of uniform cross section carries the loading shown. Calculate the shear flow $q^{(7)}$ in the interior wall.

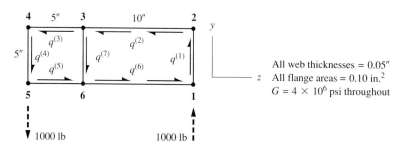

All web thicknesses = 0.05"
All flange areas = 0.10 in.^{2}
$G = 4 \times 10^6$ psi throughout

Exercise 8.19

8.20   The cantilever box beam has the four-cell cross section shown. The web thicknesses are shown in parentheses. For the 10,000 lb load applied as shown, the computed shear flows in lb/in. are shown for each web. Calculate the angle of twist of the free end of the beam.

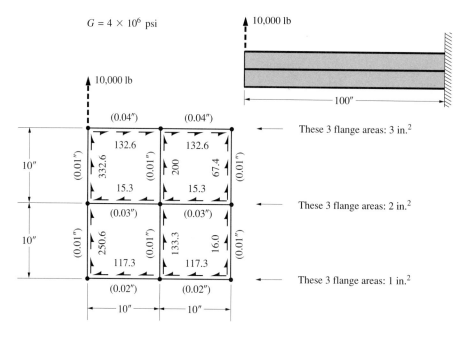

Exercise 8.20

8.21   The computed shear flows for the six-cell beam cross section, loaded as shown, are as given (lb/in.). Calculate the angle of twist per unit length.

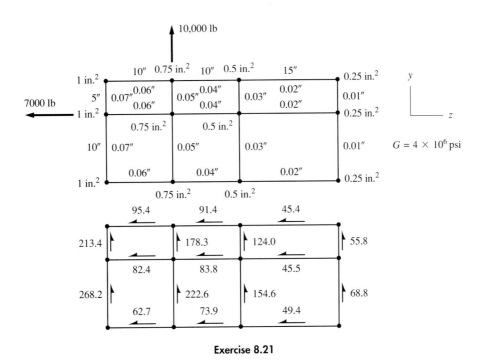

Exercise 8.21

8.22   Find the shear flows in the thin webs, if a pure torque of 10,000 in.-lb is applied to the section.

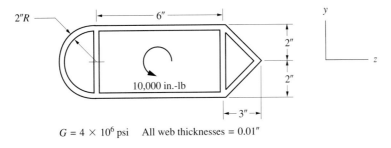

$G = 4 \times 10^6$ psi   All web thicknesses = 0.01″

Exercise 8.22

8.23 Find the shear flows in a box beam with the cross section shown if a 5000 lb shear load is applied as illustrated.

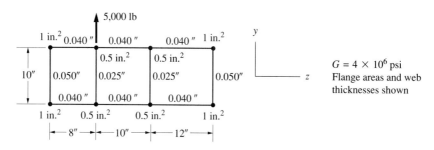

Exercise 8.23

8.24 Calculate the shear flow in each of the walls of the circular, doubly-redundant, multicell section. Flange areas are 1 in.^{2} and web thicknesses are 0.05 in. Throughout, $E = 10^7$ psi and $G = 4(10^6)$ psi.

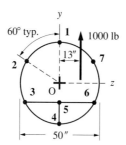

Exercise 8.24

8.25   If the shear modulus G and the panel thickness t are uniform throughout, calculate the average shear flow in each panel due to the pure torque T.

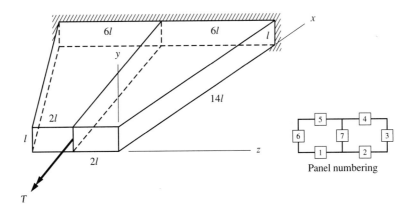

Exercise 8.25

8.26   The tapered, two-cell beam is acted on by a pure torque of 50,000 in.-lb counterclockwise. The length of the beam (perpendicular to the page) is 100 in. All flange areas are 1.0 in.^{2} and all wall thicknesses are 0.01 in. Also, $G = 4 \times 10^6$ psi. Calculate the shear flows at the wall (far end) and the angle of twist of the free (near) end.

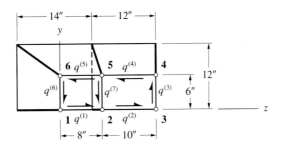

Exercise 8.26

8.27   The two-cell tapered wing box carries shear loads applied at the tip, as shown. Stringer areas and coordinate data are given in the table; panel data are in the figure. Find the shear flows on a section at each end of the beam.

Stringer	x (in.)	y (in.)	z (in.)	Area (in.²)
1	0	2	25	1.0
	40	0	0	1.0
2	0	2	28	0.75
	40	0	10	0.75
3	0	2	35	0.5
	40	0	20	0.5
4	0	4	35	0.5
	40	6	20	0.5
5	0	4	28	0.75
	40	6	10	0.75
6	0	4	25	1.0
	40	6	0	1.0

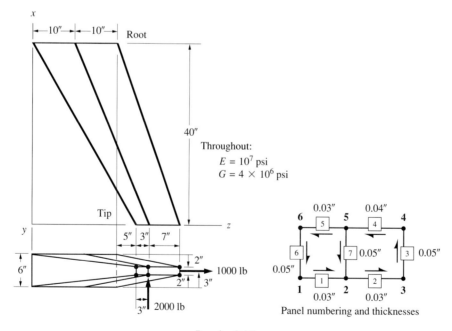

Throughout:
$$E = 10^7 \text{ psi}$$
$$G = 4 \times 10^6 \text{ psi}$$

Panel numbering and thicknesses

Exercise 8.27

8.28 Analyze the shear lag in the structure shown and find:

a. The load P_1 in the upper stringer at the wall.

b. The maximum shear flow in the web.

c. The displacement of the left (free) end of the top stringer.

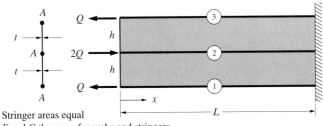

Stringer areas equal
E and G the same for webs and stringers

Exercise 8.28

8.29 Using shear lag analysis, calculate the maximum value of the load Q that can be applied as shown such that the maximum tensile stress anywhere in the box beam (including the webs) does not exceed 15,000 psi.

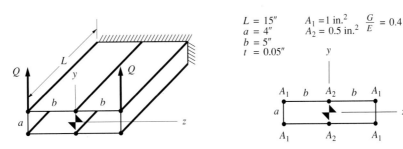

$$L = 15'' \qquad A_1 = 1 \text{ in.}^2 \qquad \frac{G}{E} = 0.4$$
$$a = 4'' \qquad A_2 = 0.5 \text{ in.}^2$$
$$b = 5''$$
$$t = 0.05''$$

Exercise 8.29

8.30 From a shear lag analysis, the shear flow in the structure shown is $q = -8.6 \sinh(2.064 - 0.1032x)$. Calculate the axial load in the top stringer at the wall.

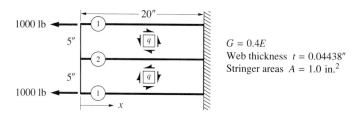

$G = 0.4E$
Web thickness $t = 0.04438''$
Stringer areas $A = 1.0$ in.^{2}

Exercise 8.30

The Matrix Displacement Method

CHAPTER OUTLINE

9.1   INTRODUCTION

Unlike the force method, the displacement method requires no special treatment of redundant members, if any are present. The procedure for solving statically determinate and indeterminate structures is the same, which makes the displacement method a straightforward, universal approach to the analysis of linearly elastic structures. Displacement variables are the primary unknowns. Since even modest structures possess many degrees of freedom, that is, many unknown displacements, the method is best implemented on a computer. That is one compelling reason to cast the subject in matrix notation.[1] Therefore, this chapter begins with an introduction to matrices and the essentials of matrix algebra.

Afterwards, the displacement method, in the form of the *direct stiffness method*, is introduced, in the context of linear elastic spring assemblies. These artificial "structures" embody the essence of all elastic structures and provide the simplest setting in which to establish the terminology and procedures of the displacement method. These procedures are subsequently formalized in matrix notation, setting the stage for their application to more practical structures, in Chapters 10 and 11.

9.2   VECTOR NOTATION

Vectors can be defined in a space of dimension N by listing the components in the same manner as in two or three dimensions. Two N dimensional vectors $\{\mathbf{x}\}$ and $\{\mathbf{y}\}$ are expressed in terms of their components as follows:

$$\{\mathbf{x}\} = \left\{ \begin{array}{c} x_1 \\ x_2 \\ \cdot \\ \cdot \\ x_N \end{array} \right\} \qquad \{\mathbf{y}\} = \left\{ \begin{array}{c} y_1 \\ y_2 \\ \cdot \\ \cdot \\ y_N \end{array} \right\}$$

Written in this fashion, the vectors are referred to as *column vectors*. The elements in row i of each vector are x_i and y_i, respectively. When the components of a vector are listed horizontally rather than vertically, the vector is called a *row vector*. The distinction between column vectors and row vectors is critical to the definition of matrix multiplication, which will be discussed shortly. When a vector is expressed as a row vector, we use the notation $\lfloor \mathbf{x} \rfloor$,

$$\lfloor \mathbf{x} \rfloor = \lfloor x_1 \quad x_2 \quad \cdot \quad \cdot \quad x_N \rfloor$$

9.3   MATRIX NOTATION

A matrix is a set of numbers arranged in rows and columns and represented symbolically by a letter enclosed in square brackets, as follows:

---

[1] The force method can also be cast in matrix formalism. See, for example, W. Mcguire, and R. H. Gallagher, *Matrix Structural Analysis*, New York, Wiley, 1979.

$$[\mathbf{A}] = \begin{bmatrix} A_{11} & A_{12} & \cdot & \cdot & A_{1N} \\ A_{21} & A_{22} & \cdot & \cdot & A_{2N} \\ \cdot & \cdot & \cdot & \cdot & \cdot \\ \cdot & \cdot & \cdot & \cdot & \cdot \\ A_{M1} & \cdot & \cdot & \cdot & A_{MN} \end{bmatrix}$$

A particular element of matrix $[\mathbf{A}]$ is denoted A_{ij}, where the first subscript indicates the row in which the element is located and the second subscript refers to the column. It is sometimes convenient to view an M by N matrix as a set of M row vectors, each having N components, or as a set of N column vectors with M terms in each of them.

A *square* matrix is one that has an equal number of rows and columns. The *principal diagonal* of a square matrix is composed of the elements on a line drawn from the upper left corner to the lower right corner of the square array. In other words, the row and column numbers of an element on the principal diagonal are equal.

9.4 BASIC MATRIX OPERATIONS

9.4.1 Vector Addition

When one vector is added to another, the respective components of each vector are summed, that is, the ith component the vector $\{\mathbf{z}\}$, where $\{\mathbf{z}\} = \{\mathbf{x}\} + \{\mathbf{y}\}$, is

$$z_i = x_i + y_i, \qquad i = 1, \cdots, N \qquad \text{[9.4.1]}$$

9.4.2 Scalar Multiplication of a Vector

Multiplication of a vector by a scalar b is accomplished by multiplying every component of vector $\{\mathbf{x}\}$ by that scalar. If $\{\mathbf{z}\} = b\{\mathbf{x}\}$, then

$$z_i = bx_i, \qquad i = 1, \cdots, N \qquad \text{[9.4.2]}$$

9.4.3 Dot Product

The dot product (or *inner* product) of two vectors is the sum of the products of the corresponding components of the two vectors. In two or three dimensions, the dot product represents the product of the *magnitude* of both vectors times the cosine of the angle between them. Written explicitly, we have

$$\{\mathbf{x}\} \cdot \{\mathbf{y}\} = \sum_{i=1}^{N} x_i y_i = x_1 y_1 + x_2 y_2 + \cdots + x_N y_N \qquad \text{[9.4.3]}$$

9.4.4 Matrix Addition

Suppose $[\mathbf{A}]$ and $[\mathbf{B}]$ are M by N matrices. Two matrices are added by summing the elements that have the same subscripts, that is, the same location within each array. Thus, if $[\mathbf{C}] = [\mathbf{A}] + [\mathbf{B}]$, then

$$C_{ij} = A_{ij} + B_{ij}, \qquad i = 1, \cdots, M \qquad j = 1, \cdots, N \qquad \textbf{[9.4.4]}$$

9.4.5   Scalar Multiplication of a Matrix

Multiplication of a matrix by a scalar is defined as multiplying every element of the matrix by the scalar. If $[\mathbf{C}] = b[\mathbf{A}]$, then

$$C_{ij} = bA_{ij} \qquad i = 1, \cdots, M \qquad j = 1, \cdots, N \qquad \textbf{[9.4.5]}$$

9.4.6   Matrix Multiplication

Multiplication of a matrix and a column vector is carried out by treating the rows of the matrix as individual row vectors. The result is a column vector whose components are found by taking the dot product (Equation 9.4.3) of each row vector with the column vector. For example, let

$$[\mathbf{A}] = \begin{bmatrix} A_{11} & A_{12} \\ A_{21} & A_{22} \end{bmatrix} \qquad \{\mathbf{x}\} = \begin{Bmatrix} x_1 \\ x_2 \end{Bmatrix}$$

Then, by definition,

$$[\mathbf{A}]\{\mathbf{x}\} = \begin{bmatrix} A_{11} & A_{12} \\ A_{21} & A_{22} \end{bmatrix} \begin{Bmatrix} x_1 \\ x_2 \end{Bmatrix} = \begin{Bmatrix} A_{11}x_1 + A_{12}x_2 \\ A_{21}x_1 + A_{22}x_2 \end{Bmatrix}$$

The matrix product in this case is a 2 by 1 column vector. We can use subscript notation to express row i of the product as follows:

$$\{[\mathbf{A}]\{\mathbf{x}\}\}_i = \sum_{j=1}^{N} A_{ij}x_j = A_{i1}x_1 + A_{i2}x_2 + \cdots + A_{iN}x_N \qquad \textbf{[9.4.6]}$$

Linear algebraic equations can be written compactly using matrix notation. For example, a system of two equations in two unknowns x and y is

$$ax + by = c$$
$$dx + ey = f$$

which can be expressed in matrix form as

$$\begin{bmatrix} a & b \\ d & e \end{bmatrix} \begin{Bmatrix} x \\ y \end{Bmatrix} = \begin{Bmatrix} c \\ f \end{Bmatrix}$$

The product of two matrices $[\mathbf{A}]$ and $[\mathbf{B}]$ is denoted $[\mathbf{A}][\mathbf{B}]$ and is found by taking the dot product of each row vector of $[\mathbf{A}]$ into each column vector of $[\mathbf{B}]$. The dot product of row i of $[\mathbf{A}]$ and column j of $[\mathbf{B}]$ is stored in element i, j of the product matrix. Thus, if $[\mathbf{C}] = [\mathbf{A}][\mathbf{B}]$, then component C_{ij} is obtained as follows:

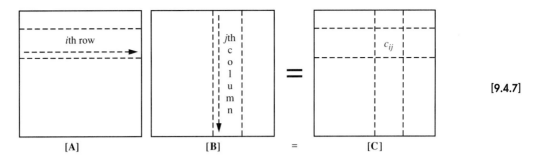

$$[9.4.7]$$

Clearly, if two matrices are to be multiplied, the number of columns in the first matrix [A] must equal the number of rows in the second matrix [B]. The product matrix [C] has the same number of rows as [A] and the same number of columns as [B], that is,

$$\underset{(M \times L)}{[\mathbf{A}]} \quad \underset{(L \times N)}{[\mathbf{B}]} \quad = \quad \underset{(M \times N)}{[\mathbf{C}]}$$

Representing the multiplication in subscript notation shows explicitly how to compute the value of any element of the product matrix.

$$C_{ij} = \sum_{k=1}^{L} A_{ik} B_{kj} = A_{i1} B_{1j} + A_{i2} B_{2j} + \cdots + A_{iL} B_{Lj} \qquad i = 1, \cdots, M \qquad j = 1, \cdots, N \qquad [9.4.8]$$

Note that the summation occurs over the *inner* subscripts, that is, the second subscript of the first matrix and the first subscript of the second matrix.

The order of matrix multiplication is important, because matrix multiplication is *not* commutative, that is, $[\mathbf{A}][\mathbf{B}] \neq [\mathbf{B}][\mathbf{A}]$.

The following example illustrates the rules we have introduced thus far.

Example 9.4.1   Let

$$[\mathbf{A}] = \begin{bmatrix} 2 & 4 \\ 1 & -1 \end{bmatrix} \qquad [\mathbf{B}] = \begin{bmatrix} 3 & 2 \\ 0 & 1 \end{bmatrix} \qquad b = 3$$

Calculate $b[\mathbf{A}]$, $[\mathbf{A}] + [\mathbf{B}]$, and $[\mathbf{A}][\mathbf{B}]$. Show that $[\mathbf{A}][\mathbf{B}] \neq [\mathbf{B}][\mathbf{A}]$.

Using the given information, we find that

$$b[\mathbf{A}] = 3 \begin{bmatrix} 2 & 4 \\ 1 & -1 \end{bmatrix} = \begin{bmatrix} 3 \cdot 2 & 3 \cdot 4 \\ 3 \cdot 1 & 3 \cdot (-1) \end{bmatrix} = \begin{bmatrix} 6 & 12 \\ 3 & -3 \end{bmatrix} \qquad [a]$$

$$[\mathbf{A}] + [\mathbf{B}] = \begin{bmatrix} 2 & 4 \\ 1 & -1 \end{bmatrix} + \begin{bmatrix} 3 & 2 \\ 0 & 1 \end{bmatrix} = \begin{bmatrix} 2+3 & 4+2 \\ 1+0 & -1+1 \end{bmatrix} = \begin{bmatrix} 5 & 6 \\ 1 & 0 \end{bmatrix} \qquad [b]$$

$$[\mathbf{A}][\mathbf{B}] = \begin{bmatrix} 2 & 4 \\ 1 & -1 \end{bmatrix} \begin{bmatrix} 3 & 2 \\ 0 & 1 \end{bmatrix} = \begin{bmatrix} \lfloor 2 \ 4 \rfloor \begin{Bmatrix} 3 \\ 0 \end{Bmatrix} & \lfloor 2 \ 4 \rfloor \begin{Bmatrix} 2 \\ 1 \end{Bmatrix} \\ \lfloor 1 \ -1 \rfloor \begin{Bmatrix} 3 \\ 0 \end{Bmatrix} & \lfloor 1 \ -1 \rfloor \begin{Bmatrix} 2 \\ 1 \end{Bmatrix} \end{bmatrix} = \begin{bmatrix} 6 & 8 \\ 3 & 1 \end{bmatrix} \qquad [c]$$

If we reverse the order of multiplication in Equation c, we obtain a different result:

$$[\mathbf{B}][\mathbf{A}] = \begin{bmatrix} 3 & 2 \\ 0 & 1 \end{bmatrix} \begin{bmatrix} 2 & 4 \\ 1 & -1 \end{bmatrix} = \begin{bmatrix} \lfloor 3 \;\; 2 \rfloor \begin{Bmatrix} 2 \\ 1 \end{Bmatrix} & \lfloor 3 \;\; 2 \rfloor \begin{Bmatrix} 4 \\ -1 \end{Bmatrix} \\ \lfloor 0 \;\; 1 \rfloor \begin{Bmatrix} 2 \\ 1 \end{Bmatrix} & \lfloor 0 \;\; 1 \rfloor \begin{Bmatrix} 4 \\ -1 \end{Bmatrix} \end{bmatrix} = \begin{bmatrix} 8 & 10 \\ 1 & -1 \end{bmatrix} \qquad \text{[d]}$$

Therefore, $[\mathbf{A}]\,[\mathbf{B}] \neq [\mathbf{B}]\,[\mathbf{A}]$.

9.4.7 Matrix Transpose

The transpose of a matrix, denoted by a capital superscript T, is found by interchanging its rows and columns. Taking the transpose of a matrix means simply interchanging the subscripts on a matrix element as follows:

$$[\mathbf{A}]^T = [A_{ij}^T] \quad \text{where} \quad A_{ij}^T = A_{ji} \qquad \text{[9.4.9]}$$

For example,

$$\begin{bmatrix} 1 & -2 & 4 \\ 6 & 9 & -3 \\ 5 & 7 & 8 \end{bmatrix}^T = \begin{bmatrix} 1 & 6 & 5 \\ -2 & 9 & 7 \\ 4 & -3 & 8 \end{bmatrix}$$

Obviously, for any matrix, the following is true:

$$\left([\mathbf{A}]^T\right)^T = [\mathbf{A}] \qquad \text{[9.4.10]}$$

The transpose of an N by 1 column vector $\{\mathbf{x}\}$ is the 1 by N row vector $\lfloor \mathbf{x} \rfloor$; conversely, the transpose of a 1 by N row vector $\lfloor \mathbf{x} \rfloor$ is a column vector:

$$\{\mathbf{x}\}^T = \lfloor \mathbf{x} \rfloor \qquad \lfloor \mathbf{x} \rfloor^T = \{\mathbf{x}\}$$

The transpose of the product of two matrices equals the product of the transposes of the individual matrices, in reverse order. This useful *reversal rule*, expressed in matrix notation, is

$$([\mathbf{A}][\mathbf{B}])^T = [\mathbf{B}]^T\,[\mathbf{A}]^T \qquad \text{[9.4.11]}$$

The validity of this rule can be established by using the definition of matrix multiplication given in Equation 9.4.8, according to which the i, j component of the matrix product $[\mathbf{A}][\mathbf{B}]$ is

$$[\mathbf{AB}]_{ij} = \sum_{k=1}^{N} A_{ik} B_{kj}$$

Then

$$\left[(\mathbf{AB})^T\right]_{ij} = [\mathbf{AB}]_{ji} = \sum_{k=1}^{N} A_{jk} B_{ki} = \sum_{k=1}^{N} B_{ki} A_{jk} = \sum_{k=1}^{N} B_{ik}^T A_{kj}^T = \left[\mathbf{B}^T \mathbf{A}^T\right]_{ij}$$

The following example illustrates the reversal rule for 2 by 2 matrices.

Example 9.4.2   Let

$$[\mathbf{A}] = \begin{bmatrix} A_{11} & A_{12} \\ A_{21} & A_{22} \end{bmatrix} \qquad [\mathbf{B}] = \begin{bmatrix} B_{11} & B_{12} \\ B_{21} & B_{22} \end{bmatrix}$$

Show that $([\mathbf{A}][\mathbf{B}])^T = [\mathbf{B}]^T [\mathbf{A}]^T$.

According to the rules of matrix multiplication,

$$[\mathbf{A}][\mathbf{B}] = \begin{bmatrix} A_{11} & A_{12} \\ A_{21} & A_{22} \end{bmatrix} \begin{bmatrix} B_{11} & B_{12} \\ B_{21} & B_{22} \end{bmatrix} = \begin{bmatrix} \lfloor A_{11} & A_{12} \rfloor \begin{Bmatrix} B_{11} \\ B_{21} \end{Bmatrix} & \lfloor A_{11} & A_{12} \rfloor \begin{Bmatrix} B_{12} \\ B_{22} \end{Bmatrix} \\ \lfloor A_{21} & A_{22} \rfloor \begin{Bmatrix} B_{11} \\ B_{21} \end{Bmatrix} & \lfloor A_{21} & A_{22} \rfloor \begin{Bmatrix} B_{12} \\ B_{22} \end{Bmatrix} \end{bmatrix} = \begin{bmatrix} A_{11}B_{11} + A_{12}B_{21} & A_{11}B_{12} + A_{12}B_{22} \\ A_{21}B_{11} + A_{22}B_{21} & A_{21}B_{12} + A_{22}B_{22} \end{bmatrix}$$

The transpose of the product matrix is therefore

$$([\mathbf{A}][\mathbf{B}])^T = \begin{bmatrix} A_{11}B_{11} + A_{12}B_{21} & A_{21}B_{11} + A_{22}B_{21} \\ A_{11}B_{12} + A_{12}B_{22} & A_{21}B_{12} + A_{22}B_{22} \end{bmatrix} \qquad \text{[a]}$$

On the other hand,

$$[\mathbf{B}]^T [\mathbf{A}]^T = \begin{bmatrix} B_{11} & B_{21} \\ B_{12} & B_{22} \end{bmatrix} \begin{bmatrix} A_{11} & A_{21} \\ A_{12} & A_{22} \end{bmatrix} = \begin{bmatrix} \lfloor B_{11} & B_{21} \rfloor \begin{Bmatrix} A_{11} \\ A_{12} \end{Bmatrix} & \lfloor B_{11} & B_{21} \rfloor \begin{Bmatrix} A_{21} \\ A_{22} \end{Bmatrix} \\ \lfloor B_{12} & B_{22} \rfloor \begin{Bmatrix} A_{11} \\ A_{12} \end{Bmatrix} & \lfloor B_{12} & B_{22} \rfloor \begin{Bmatrix} A_{21} \\ A_{22} \end{Bmatrix} \end{bmatrix} \qquad \text{[b]}$$

$$= \begin{bmatrix} B_{11}A_{11} + B_{21}A_{12} & B_{11}A_{21} + B_{21}A_{22} \\ B_{12}A_{11} + B_{22}A_{12} & B_{12}A_{21} + B_{22}A_{22} \end{bmatrix}$$

Comparing the right-hand sides of Equations a and b, we see that they are identical; therefore, $([\mathbf{A}][\mathbf{B}])^T = [\mathbf{B}]^T [\mathbf{A}]^T$.

9.4.8   Symmetric Matrices

A square matrix is said to be *symmetric* if the transpose operation does not alter the original matrix, that is, if $[\mathbf{A}] = [\mathbf{A}]^T$. The ith row vector of a symmetric matrix is identical to the ith column vector, so that interchanging the rows and columns yields the same matrix. The matrix

$$[\mathbf{B}] = \begin{bmatrix} 1 & 2 & 3 \\ 2 & 3 & 5 \\ 3 & 5 & 6 \end{bmatrix}$$

is symmetric, since row one equals column one, row two equals column two, and row three equals column three: $[\mathbf{B}]^T = [\mathbf{B}]$.

9.4.9   Identity Matrix

The identity matrix is a special form of square matrix that contains only 1s along its principal diagonal and zeroes everywhere else. The identity matrix is denoted by the special symbol $[\mathbf{I}]$.

$$[\mathbf{I}] = \begin{bmatrix} 1 & 0 & \cdot & \cdot & 0 \\ 0 & 1 & \cdot & \cdot & 0 \\ \cdot & & \cdot & & \cdot \\ \cdot & & & \cdot & \cdot \\ 0 & 0 & \cdot & \cdot & 1 \end{bmatrix} \qquad (N \text{ by } N \text{ matrix}) \qquad\qquad [9.4.12]$$

If an identity matrix is multiplied by another N by N matrix, the original matrix results. That is,

$$[\mathbf{I}][\mathbf{C}] = [\mathbf{C}] \qquad [\mathbf{C}][\mathbf{I}] = [\mathbf{C}] \qquad [\mathbf{I}]\{\mathbf{x}\} = \{\mathbf{x}\} \qquad \lfloor\mathbf{x}\rfloor[\mathbf{I}] = \lfloor\mathbf{x}\rfloor \qquad [9.4.13]$$

Example 9.4.3   Verify the property of the identity matrix given in Equation 9.4.13 for an arbitrary 3 by 3 matrix,

$$[\mathbf{A}] = \begin{bmatrix} A_{11} & A_{12} & A_{13} \\ A_{21} & A_{22} & A_{23} \\ A_{31} & A_{32} & A_{33} \end{bmatrix}$$

Carrying out the matrix multiplication, we get

$$[\mathbf{A}][\mathbf{I}] = \begin{bmatrix} A_{11} & A_{12} & A_{13} \\ A_{21} & A_{22} & A_{23} \\ A_{31} & A_{32} & A_{33} \end{bmatrix}\begin{bmatrix} 1 & 0 & 0 \\ 0 & 1 & 0 \\ 0 & 0 & 1 \end{bmatrix} = \begin{bmatrix} \lfloor A_{11} \ A_{12} \ A_{13} \rfloor \left\{ \begin{matrix}1\\0\\0\end{matrix}\right\} & \lfloor A_{11} \ A_{12} \ A_{13} \rfloor \left\{ \begin{matrix}0\\1\\0\end{matrix}\right\} & \lfloor A_{11} \ A_{12} \ A_{13} \rfloor \left\{ \begin{matrix}0\\0\\1\end{matrix}\right\} \\ \lfloor A_{21} \ A_{22} \ A_{23} \rfloor \left\{ \begin{matrix}1\\0\\0\end{matrix}\right\} & \lfloor A_{21} \ A_{22} \ A_{23} \rfloor \left\{ \begin{matrix}0\\1\\0\end{matrix}\right\} & \lfloor A_{21} \ A_{22} \ A_{23} \rfloor \left\{ \begin{matrix}0\\0\\1\end{matrix}\right\} \\ \lfloor A_{31} \ A_{32} \ A_{33} \rfloor \left\{ \begin{matrix}1\\0\\0\end{matrix}\right\} & \lfloor A_{31} \ A_{32} \ A_{33} \rfloor \left\{ \begin{matrix}0\\1\\0\end{matrix}\right\} & \lfloor A_{31} \ A_{32} \ A_{33} \rfloor \left\{ \begin{matrix}0\\0\\1\end{matrix}\right\} \end{bmatrix}$$

$$= \begin{bmatrix} A_{11} & A_{12} & A_{13} \\ A_{21} & A_{22} & A_{23} \\ A_{31} & A_{32} & A_{33} \end{bmatrix}$$

This shows that $[\mathbf{A}][\mathbf{I}] = [\mathbf{A}]$.

In matrix theory, the identity matrix $[\mathbf{I}]$ plays a role analogous to that of the number 1 in scalar algebra. Just as $1 \cdot a = a$, where a is any scalar quantity, so $[\mathbf{I}][\mathbf{A}] = [\mathbf{A}]$ for any matrix $[\mathbf{A}]$. Observe also that contrary to the general rule, multiplication by the identity matrix is commutative.

We close this section by proving a simple but important assertion. Let $[\mathbf{A}]$ be an M by N matrix that satisfies the equation

$$[\mathbf{A}]\{\mathbf{z}\} = \{\mathbf{0}\} \qquad\qquad [9.4.14]$$

regardless of the values assigned to the N components of vector $\{\mathbf{z}\}$. Then, all components of $[\mathbf{A}]$ must be zero. To prove this, set the first component of $\{\mathbf{z}\}$ equal to 1 and all others equal to 0. (Remember, $\{\mathbf{z}\}$ is arbitrary.) Carrying out the multiplication in Equation 9.4.14 for this choice of $\{\mathbf{z}\}$ yields

$$A_{i1} = 0 \qquad i = 1, 2, \cdots, M$$

that is, the first column of $[\mathbf{A}]$ must be filled with zeroes. Next, set the second component of $\{\mathbf{z}\}$ equal to 1 and all of the remaining components equal to zero. Carrying out the multiplication in Equation 9.4.14 this time shows that column 2 of $[\mathbf{A}]$ must be zero. Continuing in this fashion, using up the N possible locations of the nonzero entry in $\{\mathbf{z}\}$, we end up requiring that all N columns of $[\mathbf{A}]$ be filled with zeroes, in other words, that $[\mathbf{A}] = 0$. This result will be applied to matrix equations. In particular, it follows that

$$\boxed{\text{If } [\mathbf{B}]\{\mathbf{z}\} = [\mathbf{C}]\{\mathbf{z}\} \text{ for arbitrary } \{\mathbf{z}\}, \text{ then } [\mathbf{B}] = [\mathbf{C}].}$$

$$\text{[9.4.15]}$$

9.5 DETERMINANTS AND MATRIX INVERSE

The *minor* M_{ij} of element A_{ij} of an Nth order square matrix $[\mathbf{A}]$ is the *determinant* of the $N-1$ order matrix found by crossing out row i and column j of $[\mathbf{A}]$. The *cofactor* of A_{ij}, denoted $\mathrm{cof}(A_{ij})$, is the *signed* minor, which is

$$\mathrm{cof}(A_{ij}) = (-1)^{i+j} M_{ij}$$

$$\text{[9.5.1]}$$

When the sum of the row and column numbers of A_{ij} is even, a plus sign results; if the sum is odd, a negative sign is obtained. For example, let

$$[\mathbf{A}] = \begin{bmatrix} A_{11} & A_{12} & A_{13} \\ A_{21} & A_{22} & A_{23} \\ A_{31} & A_{32} & A_{33} \end{bmatrix}$$

Then the minors of A_{22} and A_{31}, respectively, are

$$M_{22} = \det \begin{bmatrix} A_{11} & A_{13} \\ A_{31} & A_{33} \end{bmatrix} \qquad M_{31} = \det \begin{bmatrix} A_{12} & A_{13} \\ A_{22} & A_{23} \end{bmatrix}$$

where det stands for determinant, which is computed as follows for 2 by 2 matrices:

$$\det \begin{bmatrix} A_{11} & A_{13} \\ A_{31} & A_{33} \end{bmatrix} = A_{11}A_{33} - A_{13}A_{31} \qquad \text{and} \qquad \det \begin{bmatrix} A_{12} & A_{13} \\ A_{22} & A_{23} \end{bmatrix} = A_{12}A_{23} - A_{13}A_{22}$$

The cofactors of A_{13} and A_{32}, respectively, are

$$\mathrm{cof}(A_{13}) = (+1) \det \begin{bmatrix} A_{21} & A_{22} \\ A_{31} & A_{32} \end{bmatrix} \qquad \mathrm{cof}(A_{32}) = (-1) \det \begin{bmatrix} A_{11} & A_{13} \\ A_{21} & A_{23} \end{bmatrix}$$

The determinant of an Nth order matrix $[\mathbf{A}]$ is found by the method of *expansion by cofactors*. This expansion can be carried out across any row or down any column of the matrix. Each element of the row (or column) is multiplied by its cofactor, and the results are summed over all the elements of the row (or column). The determinant expansion across row r can be written as

$$\det[\mathbf{A}] = A_{r1}\mathrm{cof}(A_{r1}) + A_{r2}\mathrm{cof}(A_{r2}) + \cdots + A_{rN}\mathrm{cof}(A_{rN})$$

$$= \sum_{i=1}^{N} A_{ri}\mathrm{cof}(A_{ri})$$

$$\text{[9.5.2]}$$

Similarly, the determinant expansion down column c can be expressed as

$$\det[\mathbf{A}] = \sum_{i=1}^{N} A_{ic}\mathrm{cof}(A_{ic}) \tag{9.5.3}$$

The determinant of any order matrix can be calculated by expanding by cofactors involving determinants of successively one order less, leading ultimately to determinants of 2 by 2 matrices. The determinant of a 2 by 2 matrix, as pointed out previously, is

$$\det \begin{bmatrix} A_{11} & A_{12} \\ A_{21} & A_{22} \end{bmatrix} = A_{11}A_{22} - A_{12}A_{21}$$

In terms of cofactor notation, this is

$$\det \begin{bmatrix} A_{11} & A_{12} \\ A_{21} & A_{22} \end{bmatrix} = A_{11}\mathrm{cof}(A_{11}) + A_{12}\mathrm{cof}(A_{12})$$

To compute the determinant of the 3 by 3 matrix

$$[\mathbf{A}] = \begin{bmatrix} A_{11} & A_{12} & A_{13} \\ A_{21} & A_{22} & A_{23} \\ A_{31} & A_{32} & A_{33} \end{bmatrix}$$

we can choose to expand by cofactors across, say, row 1, thereby obtaining

$$\det[\mathbf{A}] = A_{11}\mathrm{cof}(A_{11}) + A_{12}\mathrm{cof}(A_{12}) + A_{13}\mathrm{cof}(A_{13})$$

$$= A_{11}\det \begin{bmatrix} A_{22}A_{23} \\ A_{32}A_{33} \end{bmatrix} - A_{12}\det \begin{bmatrix} A_{21}A_{23} \\ A_{31}A_{33} \end{bmatrix} + A_{13}\det \begin{bmatrix} A_{21}A_{22} \\ A_{31}A_{32} \end{bmatrix}$$

After expanding the 2 by 2 determinants, this becomes

$$\det[\mathbf{A}] = A_{11}A_{22}A_{33} + A_{21}A_{32}A_{13} + A_{31}A_{12}A_{23} - A_{31}A_{22}A_{13} - A_{21}A_{12}A_{33} - A_{11}A_{32}A_{23} \tag{9.5.4}$$

The reader can readily verify that the same result is found for det[**A**] if the expansion is done along any other row or column.

The following properties are valid for matrices of any order.

• *If all elements of a matrix row or column are zeroes, then the determinant is zero.*

Expanding by cofactors along the row or column of zeroes will obviously yield a sum of zero.

• *The determinant of a matrix equals that of its transpose,*

$$\det[\mathbf{A}] = \det[\mathbf{A}]^{T}$$

This is readily shown by using Equations 9.5.2 and 9.5.3, together with the fact that $A_{ij}^{T} = A_{ji}$ as follows:

$$\det[\mathbf{A}]^{T} = \sum_{i=1}^{N} A_{ri}^{T}\mathrm{cof}\left(A_{ri}^{T}\right) \quad (\text{expansion by cofactors across row } r \text{ of } [\mathbf{A}]^{T})$$

$$= \sum_{i=1}^{N} A_{ir}\mathrm{cof}(A_{ir}) \quad (\text{expansion by cofactors down column } r \text{ of } [\mathbf{A}])$$

$$= \det[\mathbf{A}]$$

- *Interchanging any two rows or columns of a matrix changes the sign of its determinant.*

Let us interchange, say, rows 1 and 2 of the 3 by 3 matrix $[\mathbf{A}]$ and call the new matrix $[\mathbf{A}']$. Then, using Equation 9.5.4,

$$\det\left[\mathbf{A}'\right] = \det \begin{bmatrix} A_{21} & A_{22} & A_{23} \\ A_{11} & A_{12} & A_{13} \\ A_{31} & A_{32} & A_{33} \end{bmatrix} = A_{21}A_{12}A_{33} + A_{11}A_{32}A_{23} + A_{31}A_{22}A_{13} - A_{31}A_{12}A_{23} - A_{11}A_{22}A_{33} - A_{21}A_{32}A_{13}$$

$$= -A_{11}A_{22}A_{33} - A_{21}A_{32}A_{13} - A_{31}A_{12}A_{23} + A_{31}A_{22}A_{13} + A_{21}A_{12}A_{33} + A_{11}A_{32}A_{23}$$

$$= -\det[\mathbf{A}]$$

- *If any two rows or any two columns of a matrix are identical, then the determinant of the matrix is zero.*

Let $[\mathbf{A}]$ be a matrix having two identical rows (say), and let $[\mathbf{A}']$ be the matrix that results from interchanging those two rows. Then, $\det[\mathbf{A}] = -\det[\mathbf{A}']$. However, since the rows are identical, $[\mathbf{A}] = [\mathbf{A}']$. Therefore, $\det[\mathbf{A}] = -\det[\mathbf{A}]$, which means that $\det[\mathbf{A}] = 0$.

- *Addition of a scalar multiple of one row (or column) of a matrix to any other row (or column) of the matrix leaves the determinant of the matrix unchanged.*

To illustrate this, let us add b times row s of the matrix $[\mathbf{A}]$ to row r and call the new matrix $[\mathbf{A}']$. The determinant of $[\mathbf{A}']$, expanding by cofactors across row r, is

$$\det\left[\mathbf{A}'\right] = \sum_{i=1}^{N} (A_{ri} + bA_{si}) \operatorname{cof}(A_{ri}) = \sum_{i=1}^{N} A_{ri}\operatorname{cof}(A_{ri}) + b\sum_{i=1}^{N} A_{si}\operatorname{cof}(A_{ri})$$

The first summation on the far right is $\det[\mathbf{A}]$. The second summation is the expansion by cofactors across row r of $[\mathbf{A}]$, with the entries in row r replaced by those in row s. This is the determinant of a matrix with two identical rows, which we know is zero. Thus, $\det[\mathbf{A}'] = \det[\mathbf{A}]$.

Adding a scalar multiple of one row (or column) to another row (or column) is called an *elementary row (or column) operation*. Clearly, elementary row or column operations leave the determinant of a matrix unchanged.

- *If a row or a column of a matrix is multiplied by a scalar, the determinant of that matrix is multiplied by the same scalar.*

To verify this, multiply row r of the matrix $[\mathbf{A}]$ by the scalar b and call the new matrix $[\mathbf{A}']$. The determinant of $[\mathbf{A}']$, expanding by cofactors across row r, is

$$\det[\mathbf{A}'] = \sum_{i=1}^{N} bA_{ri}\operatorname{cof}(A_{ri}) = b\sum_{i=1}^{N} A_{ri}\operatorname{cof}(A_{ri}) = b\det[\mathbf{A}]$$

- *The determinant of the product of two square matrices equals the product of the determinants of each matrix, that is*

$$\det([\mathbf{A}][\mathbf{B}]) = (\det[\mathbf{A}])(\det[\mathbf{B}])$$

We can verify this important property of determinants using the following 2 by 2 matrices:

$$[\mathbf{A}] = \begin{bmatrix} A_{11} & A_{12} \\ A_{21} & A_{22} \end{bmatrix} \qquad [\mathbf{B}] = \begin{bmatrix} B_{11} & B_{12} \\ B_{21} & B_{22} \end{bmatrix}$$

After forming the product

$$[\mathbf{A}][\mathbf{B}] = \begin{bmatrix} A_{11}B_{11} + A_{12}B_{21} & A_{11}B_{12} + A_{12}B_{22} \\ A_{21}B_{11} + A_{22}B_{21} & A_{21}B_{12} + A_{22}B_{22} \end{bmatrix}$$

we calculate the determinant and find that

$$\det([A][B]) = (A_{11}B_{11} + A_{12}B_{21})(A_{21}B_{12} + A_{22}B_{22}) - (A_{11}B_{12} + A_{12}B_{22})(A_{21}B_{11} + A_{22}B_{21})$$
$$= (A_{11}A_{22} - A_{12}A_{21})(B_{11}B_{22} - B_{12}B_{21})$$
$$= (\det[A]\det[B])$$

According to Equation 9.5.2, the determinant of [A], calculated by cofactor expansion along row r, is

$$\det[A] = \sum_{i=1}^{N} A_{ri}\text{cof}(A_{ri})$$

Consider the following, similar expression:

$$S = \sum_{i=1}^{N} A_{pi}\text{cof}(A_{qi}) \qquad (p \neq q)$$

Since $p \neq q$, A_{pi} and A_{qi} belong to different rows of [A]; therefore, S is not a cofactor expansion of $\det[A]$. Rather, S may be interpreted as the determinant of the matrix obtained from [A] by setting the elements of row p equal to those in row q. But we observed previously that the determinant vanishes if a matrix has two identical rows. Thus,

$$\sum_{i=1}^{N} A_{pi}\text{cof}(A_{qi}) = 0 \qquad (p \neq q) \qquad\qquad [9.5.5]$$

If we replace $\text{cof}(A_{qi})$ with $\text{cof}(A_{iq}^T)$, Equations 9.5.2 and 9.5.5 can be combined and written as

$$\sum_{i=1}^{N} A_{pi}\text{cof}\left(A_{iq}^T\right) = \begin{cases} \det[A] & (p = q) \\ 0 & (p \neq q) \end{cases} \qquad\qquad [9.5.6]$$

Now, let [cof(A)] stand for the matrix of cofactors of [A], so that [cof(A^T)]is the matrix of cofactors of $[A]^T$, which is the transpose of the matrix of cofactors of [A]. The term [cof(A^T)] is called the *adjoint* of the matrix [A], or [*adj* (A)], which is

$$\left[\text{adj }(A)\right] \stackrel{\text{def}}{=} \left[\text{cof}\left(A^T\right)\right]$$

Referring to the matrix multiplication rule expressed in Equation 9.4.8, we see that the left side of Equation 9.5.6 represents the matrix product [A][adj (A)], which yields an N by N matrix, since [A] itself is N by N. The right side of Equation 9.5.6 states that all elements of this product matrix are zero, except those along the principal diagonal, which are all equal to $\det[A]$. Therefore, Equation 9.5.6 can be written in matrix notation as

$$[A][\text{adj }(A)] = (\det[A])[I] \qquad\qquad [9.5.7]$$

Expansion of $\det[A]$ by cofactors down column c can be written as follows, using subscript notation:

$$\det[A] = \sum_{i=1}^{N} A_{ic}\text{cof}(A_{ic}) = \sum_{i=1}^{N} \text{cof}\left(A_{ci}^T\right)A_{ci}$$

However, since the summation expresses the expansion by cofactors of the determinant of a matirx in which column i and column j are equal, we have

$$\sum_{i=1}^{N} A_{ic}\text{cof}(A_{id}) = \sum_{i=1}^{N} \text{cof}(A_{di}^T) A_{ic} = 0 \qquad (c \neq d)$$

These two results can be summarized in the following matrix equation:

$$[\text{adj}(\mathbf{A})][\mathbf{A}] = (\det[\mathbf{A}])[\mathbf{I}] \qquad\qquad \textbf{[9.5.8]}$$

Finally, we can rewrite and combine Equations 9.5.7 and 9.5.8 to obtain

$$\frac{[\mathbf{A}]\left[\text{adj}(\mathbf{A})\right]}{\det[\mathbf{A}]} = \frac{\left[\text{adj}(\mathbf{A})\right][\mathbf{A}]}{\det[\mathbf{A}]} = [\mathbf{I}] \qquad\qquad \textbf{[9.5.9]}$$

Given a square matrix $[\mathbf{A}]$, its *inverse* matrix, denoted $[\mathbf{A}]^{-1}$, is defined (if it exists) such that

$$[\mathbf{A}][\mathbf{A}]^{-1} = [\mathbf{A}]^{-1}[\mathbf{A}] = [\mathbf{I}] \qquad\qquad \textbf{[9.5.10]}$$

The inverse of a matrix multiplied by the matrix itself, in any order, yields the identity matrix. *Matrix inverse* is analogous to *scalar reciprocal*. Recall that if c is a scalar, its reciprocal, $1/c$, is written c^{-1}, and it satisfies the equation $cc^{-1} = 1$. The inverse matrix can be identified from Equation 9.5.9, as follows:

$$[\mathbf{A}]^{-1} = \frac{\left[\text{adj}(\mathbf{A})\right]}{\det[\mathbf{A}]} \qquad\qquad \textbf{[9.5.11]}$$

This formula for the inverse of a matrix requires that we first calculate the determinant, then transpose the matrix and form its matrix of cofactors (its adjoint matrix), and finally divide each element of that adjoint matrix by the determinant.

Example 9.5.1   Use Equation 9.5.11 to find the inverse of the matrix

$$[\mathbf{A}] = \begin{bmatrix} 1 & 2 & -2 \\ -1 & 1 & 3 \\ 2 & -1 & 2 \end{bmatrix}$$

First, we find the determinant of the matrix by expanding by cofactors across, say, row 1.

$$\det[\mathbf{A}] = (1)\det\begin{bmatrix} 1 & 3 \\ -1 & 2 \end{bmatrix} - (2)\det\begin{bmatrix} 1 & -2 \\ 2 & 2 \end{bmatrix} + (-2)\det\begin{bmatrix} -1 & 1 \\ 2 & -1 \end{bmatrix} = 23$$

Form the transpose matrix,

$$[\mathbf{A}]^T = \begin{bmatrix} 1 & -1 & 2 \\ 2 & 1 & -1 \\ -2 & 3 & 2 \end{bmatrix}$$

Next, we compute the matrix of cofactors of $[A]^T$, as follows:

$$[\text{cof}(\mathbf{A}^T)] = \begin{bmatrix} +\det\begin{bmatrix} 1 & -1 \\ 3 & 2 \end{bmatrix} & -\det\begin{bmatrix} 2 & -1 \\ -2 & 2 \end{bmatrix} & +\det\begin{bmatrix} 2 & 1 \\ -2 & 3 \end{bmatrix} \\ -\det\begin{bmatrix} -1 & 2 \\ 3 & 2 \end{bmatrix} & +\det\begin{bmatrix} 1 & 2 \\ -2 & 2 \end{bmatrix} & -\det\begin{bmatrix} 1 & -1 \\ -2 & 3 \end{bmatrix} \\ +\det\begin{bmatrix} -1 & 2 \\ 1 & -1 \end{bmatrix} & -\det\begin{bmatrix} 1 & 2 \\ 2 & -1 \end{bmatrix} & +\det\begin{bmatrix} 1 & -1 \\ 2 & 1 \end{bmatrix} \end{bmatrix} = \begin{bmatrix} 5 & -2 & 8 \\ 8 & 6 & -1 \\ -1 & 5 & 3 \end{bmatrix}$$

Dividing this matrix through by the determinant, we finally obtain

$$[\mathbf{A}]^{-1} = \frac{1}{23} \begin{bmatrix} 5 & -2 & 8 \\ 8 & 6 & -1 \\ -1 & 5 & 3 \end{bmatrix} = \begin{bmatrix} 0.2174 & -0.0870 & 0.3478 \\ 0.3478 & 0.2609 & -0.0435 \\ -0.0435 & 0.2174 & 0.1304 \end{bmatrix}$$

Linear algebraic equations can be solved by finding the inverse of the coefficient matrix. Consider, for example, the following 3 by 3 system of equations:

$$A_{11}x_1 + A_{12}x_2 + A_{13}x_3 = b_1$$
$$A_{21}x_1 + A_{22}x_2 + A_{23}x_3 = b_2$$
$$A_{31}x_1 + A_{32}x_2 + A_{33}x_3 = b_3$$

In matrix form, this system is written compactly as

$$[\mathbf{A}]\{\mathbf{x}\} = \{\mathbf{b}\} \qquad \text{[9.5.12]}$$

where

$$[\mathbf{A}] = \begin{bmatrix} A_{11} & A_{12} & A_{13} \\ A_{21} & A_{22} & A_{23} \\ A_{31} & A_{32} & A_{33} \end{bmatrix} \qquad \{\mathbf{x}\} = \begin{Bmatrix} x_1 \\ x_2 \\ x_3 \end{Bmatrix} \qquad \{\mathbf{b}\} = \begin{Bmatrix} b_1 \\ b_2 \\ b_3 \end{Bmatrix}$$

The solution of Equation 9.5.12 may be found by multiplying both sides of the equation by the inverse matrix $[\mathbf{A}]^{-1}$.

$$[\mathbf{A}]^{-1}[\mathbf{A}]\{\mathbf{x}\} = [\mathbf{A}]^{-1}\{\mathbf{b}\}$$
$$[\mathbf{I}]\{\mathbf{x}\} = [\mathbf{A}]^{-1}\{\mathbf{b}\} \qquad \left(\text{since } [\mathbf{A}]^{-1}[\mathbf{A}] = [\mathbf{I}]\right) \qquad \text{[9.5.13]}$$
$$\{\mathbf{x}\} = [\mathbf{A}]^{-1}\{\mathbf{b}\} \qquad ([\mathbf{I}]\{\mathbf{x}\} = \{\mathbf{x}\})$$

Equation 9.5.13 shows the solution for the column vector of unknowns $\{\mathbf{x}\}$ in terms of the inverse of the given coefficient matrix $[\mathbf{A}]$ and the given right-side vector $\{\mathbf{b}\}$.

Example 9.5.2   Solve the following linear system of equations for the three unknowns x_1, x_2, and x_3:

$$x_1 + 2x_2 - 2x_3 = -7$$
$$-x_1 + x_2 + 3x_3 = 3$$
$$2x_1 - x_2 + 2x_3 = 8$$

In matrix notation , $[\mathbf{A}]\{\mathbf{x}\} = \{\mathbf{b}\}$, where

$$[\mathbf{A}] = \begin{bmatrix} 1 & 2 & -2 \\ -1 & 1 & 3 \\ 2 & -1 & 2 \end{bmatrix} \quad \{\mathbf{x}\} = \begin{Bmatrix} x_1 \\ x_2 \\ x_3 \end{Bmatrix} \quad \{\mathbf{b}\} = \begin{Bmatrix} -7 \\ 3 \\ 8 \end{Bmatrix}$$

From Example 9.5.1, we have

$$[A]^{-1} = \frac{1}{23} \begin{bmatrix} 5 & -2 & 8 \\ 8 & 6 & -1 \\ -1 & 5 & 3 \end{bmatrix}$$

so that for the solution vector $\{x\}$, we obtain

$$\{x\} = [A]^{-1}\{b\} = \frac{1}{23} \begin{bmatrix} 5 & -2 & 8 \\ 8 & 6 & -1 \\ -1 & 5 & 3 \end{bmatrix} \begin{Bmatrix} -7 \\ 3 \\ 8 \end{Bmatrix} = \frac{1}{23} \begin{Bmatrix} 23 \\ -46 \\ 46 \end{Bmatrix} = \begin{Bmatrix} 1 \\ -2 \\ 2 \end{Bmatrix}$$

9.6 CRAMER'S RULE

The solution of linear equations can be cast in a form involving the ratio of determinants. This method is convenient for hand and calculator solutions of very small systems of equations (three or four unknowns at most), and for visualizing some characteristics of linear systems. Computing determinants is impractical for large problems, so other more efficient equation solvers are employed. Among them is the Gauss elimination procedure discussed in the next section.

Recall from Equation 9.5.11 and 9.5.13 that the solution of the system $[A]\{x\} = \{b\}$ is

$$\{x\} = [A]^{-1}\{b\} = \frac{1}{\det[A]} \left[\text{adj}(A) \right] \{b\} \qquad [9.6.1]$$

That is,

$$x_i = \frac{1}{\det[A]} \sum_{j=1}^{N} \text{cof}(A_{ij}^T) b_j = \frac{1}{\det[A]} \sum_{j=1}^{N} b_j \text{cof}(A_{ji}) \qquad [9.6.2]$$

Yet, according to Equation 9.5.3, we know that

$$\sum_{j=1}^{N} b_j \text{cof}(A_{ji}) = \det[A]_i$$

where $[A]_i$ is the matrix $[A]$ with the ith column replaced by the column vector $\{b\}$. Thus, we can write Equation 9.6.2 as

$$x_i = \frac{\det[A]_i}{\det[A]} \qquad i = 1, 2, \cdots, N \qquad [9.6.3]$$

Equation 9.6.3 expresses *Cramer's rule*, which states that the solution for each of the unknowns x_1, x_2, etc., is found from a ratio of two determinants. The determinant in the denominator is that of the coefficient matrix. The determinant in the numerator is that of the coefficient matrix, with the column corresponding to the solution variable replaced by the right-hand column vector.

A *nonhomogeneous* set of equations has a nonzero right side vector $\{b\}$, that is,

$$[A]\{x\} = \{b\} \qquad \{b\} \neq \{0\}$$

Cramer's rule allows us to categorize the solution of this system with respect to the determinant of the coefficient matrix. There are two cases: $\det[A] = 0$ and $\det[A] \neq 0$.

If $\det[\mathbf{A}] = 0$, Cramer's ratio becomes infinite and a solution to the nonhomogeneous system does not exist. In this case, both the system of equations and the coefficient matrix $[\mathbf{A}]$ are said to be *singular*. A zero determinant implies that the matrix has rows which are linear combinations of each other. In a system of N equations, if any equation is a linear combination of one or more others, it can be eliminated and the remaining $N - 1$ equations can be solved in terms of the eliminated unknown, whose value is unrestricted. Therefore, there are an infinite number of solutions, one for each arbitrary value of the eliminated unknown.

In the "regular case," $\det[\mathbf{A}] \neq 0$. The system is nonsingular and Cramer's rule gives the unique solution.

Example 9.6.1   Consider the following set of equations.

$$x_1 + 2x_2 - 2x_3 = -7 \qquad \text{[a]}$$

$$-x_1 + x_2 + 3x_3 = 3 \qquad \text{[b]}$$

$$3x_1 + 3x_2 - 7x_3 = -17 \qquad \text{[c]}$$

Let us try to solve this system by first expressing x_1 and x_2 in terms of x_3 using Equations a and b, then substituting these two expressions into Equation c, giving us a single equation that we can solve for x_3. Then, x_1 and x_2 will be found by back-substitution. Thus, from Equations a and b, we obtain

$$x_1 = -\frac{13}{3} + \frac{8}{3}x_3 \qquad x_2 = -\frac{4}{3} - \frac{1}{3}x_3 \qquad \text{[d]}$$

Substituting these into Equation c, we have

$$3\left(-\frac{13}{3} + \frac{8}{3}x_3\right) + 3\left(-\frac{4}{3} - \frac{1}{3}x_3\right) - 7x_3 = -17$$

$$8x_3 - 8x_3 - 17 = -17$$

that is,

$$0 = 0$$

an equation that is satisfied by any value of x_3. Since x_3 can have any of an infinite number of values, it follows from Equation d that x_1 and x_2 can also. Thus, the system composed of Equations a, b, and c has no unique solution, a fact we could have discovered at once by computing the determinant of the coefficients. Expanding by cofactors down, say, column 1, we find

$$\det \begin{bmatrix} 1 & 2 & -2 \\ -1 & 1 & 3 \\ 3 & 3 & -7 \end{bmatrix} = (1)\det\begin{bmatrix} 1 & 3 \\ 3 & -7 \end{bmatrix} - (-1)\det\begin{bmatrix} 2 & -2 \\ 3 & -7 \end{bmatrix} + (3)\det\begin{bmatrix} 2 & -2 \\ 1 & 3 \end{bmatrix}$$

$$= (1)(-16) - (-1)(-8) + (3)(8)$$

$$= 0$$

The determinant of the matrix of coefficients is zero. So, in fact are the determinants of this matrix, obtained by replacing each column, in turn, by the right side of Equation a through c. Thus, applying Cramer's rule yields

$$x_1 = \frac{\det\begin{bmatrix} -7 & 2 & -2 \\ 3 & 1 & 3 \\ -17 & 3 & -7 \end{bmatrix}}{\det\begin{bmatrix} 1 & 2 & -2 \\ -1 & 1 & 3 \\ 3 & 3 & -7 \end{bmatrix}} = \frac{0}{0} \qquad x_2 = \frac{\det\begin{bmatrix} 1 & -7 & -2 \\ -1 & 3 & 3 \\ 3 & -17 & -7 \end{bmatrix}}{\det\begin{bmatrix} 1 & 2 & -2 \\ -1 & 1 & 3 \\ 3 & 3 & -7 \end{bmatrix}} = \frac{0}{0} \qquad x_3 = \frac{\det\begin{bmatrix} 1 & 2 & -7 \\ -1 & 1 & 3 \\ 3 & 3 & -17 \end{bmatrix}}{\det\begin{bmatrix} 1 & 2 & -2 \\ -1 & 1 & 3 \\ 3 & 3 & -7 \end{bmatrix}} = \frac{0}{0}$$

The ratio $0/0$ is undefined. Therefore, there is no unique set of the quantities x_1, x_2, and x_3 that satisfies Equations a, b, and c. The system of equations is singular. The reason it is singular is that Equation c is a linear combination of Equations a and b. In fact, Equation c is found by multiplying Equation a throughout by 2 and subtracting Equation b from the result.

Example 9.6.2   Solve the following system of equations using Cramer's rule:

$$x_1 + 2x_2 - 2x_3 = -7$$
$$-x_1 + x_2 + 3x_3 = 3$$
$$2x_1 - x_2 + 2x_3 = 8$$

For the determinant of the coefficients, we have

$$\det[\mathbf{A}] = \det \begin{bmatrix} 1 & 2 & -2 \\ -1 & 1 & 3 \\ 2 & -1 & 2 \end{bmatrix} = 23$$

and Cramer's rule yields

$$x_1 = \frac{\det \begin{bmatrix} -7 & 2 & -2 \\ 3 & 1 & 3 \\ 8 & -1 & 2 \end{bmatrix}}{\det[\mathbf{A}]} = 1 \qquad x_2 = \frac{\det \begin{bmatrix} 1 & -7 & -2 \\ -1 & 3 & 3 \\ 2 & 8 & 2 \end{bmatrix}}{\det[\mathbf{A}]} = -2 \qquad x_3 = \frac{\det \begin{bmatrix} 1 & 2 & -7 \\ -1 & 1 & 3 \\ 2 & -1 & 8 \end{bmatrix}}{\det[\mathbf{A}]} = 2$$

9.7   GAUSS ELIMINATION

To solve the large systems of equations that result from the analysis of practical aerospace structures, an elimination or decomposition scheme is used. The easiest one to understand is the *Gauss elimination method*. This method does not require computing determinants, nor does it necessarily involve explicit computation of the inverse of a matrix. It is basically a sequence of elementary row operations. Each stage of the Gauss elimination procedure involves subtracting a scalar multiple of one equation from all those below it, to eliminate the leading term from those equations. This process is repeated until all equations have been treated and the system is reduced to *upper triangular form*. A matrix in upper triangular form has only zeroes below the principal diagonal.

For example, consider the following third-order system $[\mathbf{A}]\{\mathbf{x}\} = \{\mathbf{b}\}$, in which $A_{11} \neq 0$:

$$A_{11}x_1 + A_{12}x_2 + A_{13}x_3 = b_1$$
$$A_{21}x_1 + A_{22}x_2 + A_{23}x_3 = b_2 \qquad \text{[9.7.1]}$$
$$A_{31}x_1 + A_{32}x_2 + A_{33}x_3 = b_3$$

Multiply Equation $9.7.1_1$ by A_{21}/A_{11} and subtract the resulting equation from Equation $9.7.1_2$. Then, multiply the original Equation $9.7.1_1$ by A_{31}/A_{11} and subtract that result from Equation $9.7.1_3$. We thereby obtain an equivalent set of equations with x_1 eliminated from all but the first one, as follows:

$$A_{11}x_1 + A_{12}x_2 + A_{13}x_3 = b_1$$
$$A_{22}^{(1)}x_2 + A_{23}^{(1)}x_3 = b_2^{(1)} \qquad \text{[9.7.2a]}$$
$$A_{32}^{(1)}x_2 + A_{33}^{(1)}x_3 - b_3^{(1)}$$

where

$$A_{22}^{(1)} = A_{22} - \frac{A_{21}}{A_{11}}A_{12} \qquad A_{23}^{(1)} = A_{23} - \frac{A_{21}}{A_{11}}A_{13} \qquad b_2^{(1)} = b_2 - \frac{A_{21}}{A_{11}}b_1$$

$$A_{32}^{(1)} = A_{32} - \frac{A_{31}}{A_{11}}A_{12} \qquad A_{33}^{(1)} = A_{33} - \frac{A_{31}}{A_{11}}A_{13} \qquad b_3^{(1)} = b_3 - \frac{A_{31}}{A_{11}}b_1$$

[9.7.2b]

Next, if $A_{22}^{(1)} \neq 0$, Equation 9.7.2a$_2$ of this system is multiplied by $A_{32}^{(1)}/A_{22}^{(1)}$ and subtracted from Equation 9.7.2a$_3$. This eliminates x_2 from the third equation and completes the forward elimination process of converting the original system into an equivalent set of three equations in upper triangular form, as follows:

$$A_{11}x_1 + A_{12}x_2 + A_{13}x_3 = b_1$$
$$A_{22}^{(1)}x_2 + A_{23}^{(1)}x_3 = b_2^{(1)}$$
$$A_{33}^{(2)}x_3 = b_3^{(2)}$$

[9.7.3a]

where

$$A_{33}^{(2)} = A_{33}^{(1)} - \frac{A_{32}^{(1)}}{A_{22}^{(1)}}A_{23}^{(1)} \qquad b_3^{(2)} = b_3^{(1)} - \frac{A_{32}^{(1)}}{A_{22}^{(1)}}b_2^{(1)}$$

[9.7.3b]

With the equations in this form, it is easy to solve for the unknowns x_1, x_2, and x_3. Equation 9.7.3a$_3$ is solved for x_3, which is substituted into Equation 9.7.3a$_2$, whereupon x_2 is found. Finally, x_2 and x_3 are substituted into Equation 9.7.3a$_1$, which can then be solved for the remaining unknown, x_1. The operation of recovering the variables in this fashion is called *back substitution*.

Gauss elimination is therefore a direct two-stage process for solving a system of equations: forward elimination followed by back substitution. At the beginning of each step of the forward elimination procedure, in the lead equation the coefficient of the unknown to be eliminated from all remaining equations is called the *pivot*. In the illustration, A_{11} and $A_{22}^{(1)}$ were pivots. The pivot must be divided into each term of the first equation to effect the elimination of the corresponding unknown from the rest of the reduced system of equations. If the lead coefficient in the first equation is zero, it cannot be used as a pivot. Instead, the first equation must be interchanged with one of the remaining equations that has a nonzero leading coefficient for use as a pivot. Such a pivot will exist if the system is nonsingular. The *pivoting strategy* usually involves selecting the equation that will provide the pivot of largest magnitude.

The upper triangular system $[\mathbf{T}]\{\mathbf{x}\} = \{\mathbf{b}'\}$ remaining at the end of the forward elimination process is obtained from the original system $[\mathbf{A}]\{\mathbf{x}\} = \{\mathbf{b}\}$ by a series of elementary row operations, which we know leaves the determinant unchanged if no pivoting occurs. Pivoting—the switching of rows—simply changes the sign of the determinant. The determinant of an upper triangular matrix

$$[\mathbf{T}] = \begin{bmatrix} T_{11} & T_{12} & T_{13} & \cdot & T_{1N} \\ 0 & T_{22} & T_{23} & \cdot & \cdot \\ 0 & 0 & T_{33} & \cdot & \cdot \\ 0 & 0 & 0 & \cdot & \cdot \\ 0 & 0 & 0 & 0 & T_{NN} \end{bmatrix}$$

is just the product of its diagonal components, as we can see by expanding det$[\mathbf{T}]$ by cofactors, always down the first column. Therefore, the determinant of a matrix is a by-product of the forward elimination step in the Gauss elimination procedure.

Finally, note that if the matrix $[\mathbf{A}]$ is symmetric, then the symmetry is preserved after each step of the forward elimination procedure. Observe, for example, that $A_{32}^{(1)} = A_{23}^{(1)}$ in Equations 9.7.2b.

Example 9.7.1 Use the Gauss elimination method to solve the following system of three equations:

$$2x_1 + x_2 + x_3 = 1$$
$$4x_1 + x_2 \qquad = -2 \qquad \text{[a]}$$
$$-2x_1 + 2x_2 + x_3 = 7$$

These equations can be represented by the following shorthand notation, where the vertical line separates the coefficient matrix from the right side vector.

$$\begin{array}{ccc|c} 2 & 1 & 1 & 1 \\ 4 & 1 & 0 & -2 \\ -2 & 2 & 1 & 7 \end{array}$$

The goal is to zero out the columns below the principal diagonal. Step 1 is to multiply the first equation by 2 and subtract the result from the second equation, yielding

$$\begin{array}{ccc|c} 2 & 1 & 1 & 1 \\ 0 & -1 & -2 & -4 \\ -2 & 2 & 1 & 7 \end{array}$$

Notice that the right-hand side is changed along with the coefficient matrix. A zero appears in column 1, just below the principal diagonal. Step 2 is to zero out the last element in column 1. This can be achieved by multiplying the first equation by −1, then subtracting the result from the third equation to get

$$\begin{array}{ccc|c} 2 & 1 & 1 & 1 \\ 0 & -1 & -2 & -4 \\ 0 & 3 & 2 & 8 \end{array}$$

Having produced zeroes in all column 1 locations below the principal diagonal, we move down one row and in one column, focusing our attention on the remaining two equations,

$$\begin{array}{cc|c} -1 & -2 & -4 \\ 3 & 2 & 8 \end{array}$$

To produce a zero below the principal diagonal of this system, multiply the first equation by −3 and subtract the result from the last equation to obtain

$$\begin{array}{cc|c} -1 & -2 & -4 \\ 0 & -4 & -4 \end{array}$$

The complete reduced system is therefore

$$\begin{array}{ccc|c} 2 & 1 & 1 & 1 \\ 0 & -1 & -2 & -4 \\ 0 & 0 & -4 & -4 \end{array}$$

This of course represents the set of equations

$$2x_1 + x_2 + x_3 = 1$$
$$-x_2 - 2x_3 = -2 \qquad \text{[b]}$$
$$-4x_3 = -4$$

which is equivalent to the original set. We solve for the unknowns from the bottom equation upward. The last equation gives $x_3 = 1$. Substituting this into the second equation, we find $x_2 = 2$. Finally, with these two results, the first equation yields $x_1 = -1$.

Note that the determinant of the coefficient matrix in Equation b is the product of the diagonal components, $2 \times (-1) \times (-4) = 8$. This is the same as the determinant of the coefficients in Equation (a), since no pivoting was required.

The Gauss elimination method can be used in a slightly different form to find the inverse of the coefficient matrix, if it is needed. Since $[\mathbf{I}]\{\mathbf{b}\} = \{\mathbf{b}\}$, we can write the linear system $[\mathbf{A}]\{\mathbf{x}\} = \{\mathbf{b}\}$ in the form

$$[\mathbf{A}]\{\mathbf{x}\} = [\mathbf{I}]\{\mathbf{b}\} \qquad [9.7.4]$$

By going through the Gauss elimination procedure twice, first from the top downwards, and then from the bottom upwards, row combinations can be done in such a way that $[\mathbf{A}]$ is replaced on the left by the identity matrix $[\mathbf{I}]$, that is,

$$[\mathbf{I}]\{\mathbf{x}\} = [\mathbf{B}]\{\mathbf{b}\} \qquad [9.7.5]$$

where $[\mathbf{B}]$ is the matrix that results from performing the same sequence of operations on the identity matrix on the right side of Equation 9.7.4. However, the solution of Equation 9.7.4 is $\{\mathbf{x}\} = [\mathbf{A}]^{-1}\{\mathbf{b}\}$. Comparing this with Equation 9.7.5, we see that $[\mathbf{B}] = [\mathbf{A}]^{-1}$. That is, applying the same row operations to $[\mathbf{I}]$ that yield the identity matrix when applied to $[\mathbf{A}]$ will produce the inverse of $[\mathbf{A}]$.

Example 9.7.2   Find $[\mathbf{A}]^{-1}$ if

$$[\mathbf{A}] = \begin{bmatrix} 2 & 1 & 1 \\ 1 & 3 & 1 \\ 1 & 1 & 4 \end{bmatrix}$$

First, form the augmented matrix, placing the matrix $[\mathbf{A}]$ to the left and the identity matrix $[\mathbf{I}]$ to the right of the vertical line:

$$
\begin{array}{ccc|ccc}
2 & 1 & 1 & 1 & 0 & 0 \\
1 & 3 & 1 & 0 & 1 & 0 \\
1 & 1 & 4 & 0 & 0 & 1
\end{array}
$$

The goal is to combine the rows such that we end up with the identity matrix on the left, and therefore the inverse of $[\mathbf{A}]$ on the right-hand side. First, divide row 1 by 2, so that the leading element of row 1 becomes 1.

$$
\begin{array}{ccc|ccc}
1 & 0.5 & 0.5 & 0.5 & 0 & 0 \\
1 & 3 & 1 & 0 & 1 & 0 \\
1 & 1 & 4 & 0 & 0 & 1
\end{array}
$$

Next, subtract row 1 from row 2 and from row 3.

$$
\begin{array}{ccc|ccc}
1 & 0.5 & 0.5 & 0.5 & 0 & 0 \\
0 & 2.5 & 0.5 & -0.5 & 1 & 1 \\
0 & 0.5 & 3.5 & -0.5 & 0 & 1
\end{array}
$$

Now, divide row 2 by 2.5 (to place a 1 on the diagonal element), multiply the result by 0.5, and subtract that result from row 3.

$$
\begin{array}{ccc|ccc}
1 & 0.5 & 0.5 & 0.5 & 0 & 0 \\
0 & 1 & 0.2 & -0.2 & 0.4 & 0 \\
0 & 0 & 3.4 & -0.4 & -0.2 & 1
\end{array}
$$

Finally, divide the last equation by 3.4 to obtain

$$
\begin{array}{ccc|ccc}
1 & 0.5 & 0.5 & 0.5 & 0 & 0 \\
0 & 1 & 0.2 & -0.2 & 0.4 & 0 \\
0 & 0 & 1 & -0.1176 & -0.05882 & 0.2941
\end{array}
$$

This completes the first Gauss elimination, reducing [A] to the upper triangular form

$$
[T] = \begin{bmatrix} 1 & 0.5 & 0.5 \\ 0 & 1 & 0.2 \\ 0 & 0 & 1 \end{bmatrix}
$$

The identity matrix [I] on the right has been replaced by a *lower triangular* matrix.

As a note, at this point, the determinant of [A] can be computed as the determinant of [T] times the *reciprocal* of the product of the factors required to place 1's on the diagonal. That is,

$$
\det[T] = \frac{\det[A]}{3.4 \cdot 2.5 \cdot 2}
$$

so that

$$
\det[A] = \det[T] \times (3.4 \cdot 2.5 \cdot 2) = 1 \times 17 = 17
$$

To continue, we repeat the elimination procedure, this time from bottom to top, to place zeroes in all locations above the principal diagonal of the left-hand matrix. To begin, multiply row 3 by 0.2 and subtract the result from row 2; multiply row 3 by 0.5 and subtract that from row 1. These operations yield

$$
\begin{array}{ccc|ccc}
1 & 0.5 & 0 & 0.5588 & 0.0294 & -0.1470 \\
0 & 1 & 0 & -0.1764 & 0.4118 & -0.0588 \\
0 & 0 & 1 & -0.1176 & -0.0588 & 0.2941
\end{array}
$$

Finally, multiply row 2 by 0.5, then subtract from row one.

$$
\begin{array}{ccc|ccc}
1 & 0 & 0 & 0.6471 & -0.1765 & -0.1176 \\
0 & 1 & 0 & -0.1765 & 0.4118 & -0.0588 \\
0 & 0 & 1 & -0.1176 & -0.0588 & 0.2941
\end{array}
$$

This completes the second elimination procedure. The inverse matrix is thus identified as

$$
[A]^{-1} = \begin{bmatrix} 0.6471 & -0.1765 & -0.1176 \\ -0.1765 & 0.4118 & -0.0588 \\ -0.1176 & -0.0588 & 0.2941 \end{bmatrix}
$$

9.8   THE EIGENVALUE PROBLEM

If multiplying a vector $\{\mathbf{x}\}$ by a matrix $[\mathbf{A}]$ is equivalent simply to multiplying $\{\mathbf{x}\}$ by a scalar, then $\{\mathbf{x}\}$ is called an *eigenvector* of $[\mathbf{A}]$, and the scalar is the accompanying *eigenvalue*. To find the eigenvalues and eigenvectors of a matrix $[\mathbf{A}]$, we must find those scalars λ and vectors $\{\mathbf{x}\}$ for which

$$[\mathbf{A}]\{\mathbf{x}\} = \lambda\{\mathbf{x}\} \qquad\qquad \text{[9.8.1]}$$

An N by N matrix has N eigenvalues and N eigenvectors.

Let us rewrite Equation 9.8.1 as

$$[\mathbf{A} - \lambda\mathbf{I}]\{\mathbf{x}\} = \{\mathbf{0}\} \qquad\qquad \text{[9.8.2]}$$

where $\{\mathbf{0}\}$ is a column vector of zeroes and $[\mathbf{I}]$ is the identity matrix. The trivial solution of Equation 9.8.2 is $\{\mathbf{x}\} = \{\mathbf{0}\}$. We could obtain this zero solution from Equation 9.8.2 by multiplying through by the inverse of $[\mathbf{A} - \lambda\mathbf{I}]$, or

$$\{\mathbf{x}\} = [\mathbf{A} - \lambda\mathbf{I}]^{-1}\{\mathbf{0}\} = \{\mathbf{0}\} \qquad\qquad \text{[9.8.3]}$$

The eigenvalues and corresponding eigenvectors comprise the *nontrivial* solution of Equation 9.8.1. To obtain this nontrivial solution, the matrix $[\mathbf{A} - \lambda\mathbf{I}]^{-1}$ must not exist; otherwise, Equation 9.8.3 delivers the *trivial* solution $\{\mathbf{x}\} = \{\mathbf{0}\}$. The inverse of $[\mathbf{A} - \lambda\mathbf{I}]$ will not exist if we require that its determinant vanish, that is, if

$$\det[\mathbf{A} - \lambda\mathbf{I}] = 0 \qquad\qquad \text{[9.8.4]}$$

The matrix $[\mathbf{A} - \lambda\mathbf{I}]$ is the matrix $[\mathbf{A}]$ with the scalar λ subtracted from every diagonal element, as follows:

$$[\mathbf{A} - \lambda\mathbf{I}] = \begin{bmatrix} A_{11} - \lambda & A_{12} & \cdot & \cdot & A_{1N} \\ A_{21} & A_{22} - \lambda & \cdot & \cdot & \cdot \\ \cdot & \cdot & \cdot & \cdot & \cdot \\ \cdot & \cdot & \cdot & \cdot & \cdot \\ A_{N1} & \cdot & \cdot & \cdot & A_{NN} - \lambda \end{bmatrix}$$

If $[\mathbf{A}]$ is N by N, then $\det[\mathbf{A} - \lambda\mathbf{I}]$ is a polynomial of degree N in λ. The N roots of the *characteristic equation,* Equation 9.8.4, are the N eigenvalues of $[\mathbf{A}]$. Each eigenvalue, in turn, is substituted back into Equation 9.8.1 or 9.8.2, which is then solved for the corresponding eigenvector $\{\mathbf{x}\}$. A unique eigenvector cannot be found; it can be determined only to within an arbitrary scalar factor. To see this, multiply Equation 9.8.1 by any nonzero scalar a,

$$a[\mathbf{A}]\{\mathbf{x}\} = a\lambda\{\mathbf{x}\}$$

This can be written

$$[\mathbf{A}]\{a\mathbf{x}\} = \lambda\{a\mathbf{x}\}$$

which implies that $\{a\mathbf{x}\}$ is also an eigenvector. (This indeterminacy follows from the requirement of Equation 9.8.4 that $[\mathbf{A} - \lambda\mathbf{I}]$ be singular.)

If $[\mathbf{A}]$ is a symmetric matrix, then all of its eigenvalues, that is, all of the roots of Equation 9.8.4, are real. Otherwise, there may be complex eigenvalues and eigenvectors. Also, for symmetric matrices, the eigenvectors are orthogonal to one another. That is, if $\{\mathbf{x}^{(i)}\}$ and $\{\mathbf{x}^{(j)}\}$ are the eigenvectors for eigenvalues $\lambda^{(i)}$ and $\lambda^{(j)}$ $(\lambda^{(i)} \neq \lambda^{(j)})$, then $\{\mathbf{x}^{(i)}\}^T\{\mathbf{x}^{(j)}\} = 0$. As discussed in some detail in Chapter 3, the stress and strain tensors are symmetric matrices whose eigenvalues are the principal stresses and strains and whose orthogonal eigenvectors define the principal directions.

Example 9.8.1   Find the eigenvalues and eigenvectors of the matrix

$$[\mathbf{A}] = \begin{bmatrix} 16 & -24 & 18 \\ 3 & -2 & 0 \\ -9 & 18 & -17 \end{bmatrix}$$

Writing the problem in the form of Equation 9.8.2, we have

$$\begin{bmatrix} 16 - \lambda & -24 & 18 \\ 3 & -2 - \lambda & 0 \\ -9 & 18 & -17 - \lambda \end{bmatrix} \begin{Bmatrix} x_1 \\ x_2 \\ x_3 \end{Bmatrix} = \begin{Bmatrix} 0 \\ 0 \\ 0 \end{Bmatrix} \qquad [\text{a}]$$

The determinant of the coefficients must vanish; therefore,

$$\det \begin{bmatrix} 16 - \lambda & -24 & 18 \\ 3 & -2 - \lambda & 0 \\ -9 & 18 & -17 - \lambda \end{bmatrix} = 0$$

Expanding the determinant yields a cubic equation in λ, namely,

$$\lambda^3 + 3\lambda^2 - 36\lambda + 32 = 0$$

The roots of this characteristic equation are $\lambda^{(1)} = -8$, $\lambda^{(2)} = 1$, and $\lambda^{(3)} = 4$. They are all real in this case, even though the matrix $[\mathbf{A}]$ is not symmetric. We next substitute each eigenvalue in turn into Equation a to find the eigenvectors. For $\lambda^{(1)} = -8$, we get

$$24x_1 - 24x_2 + 18x_3 = 0$$
$$3x_1 + 6x_2 \qquad\quad = 0$$
$$-9x_1 + 18x_2 - 9x_3 = 0$$

The second of these equations yields $x_1 = -2x_2$. Substitute this into either the first or the third equation to find $x_3 = 4x_2$. Then, x_2 may be chosen arbitrarily. If we set it equal to 1 the eigenvector is

$$\{\mathbf{x}^{(1)}\} = \begin{Bmatrix} -2 \\ 1 \\ 4 \end{Bmatrix}$$

To find $\{\mathbf{x}^{(2)}\}$, substitute $\lambda^{(2)} = 1$ into Equation a:

$$15x_1 - 24x_2 + 18x_3 = 0$$
$$3x_1 - 3x_2 \qquad\quad = 0$$
$$-9x_1 + 18x_2 - 16x_3 = 0$$

The second equation, together with either the first or third equation, yields $x_1 = x_2$ and $x_3 = 0.5x_2$. If we arbitrarily set $x_2 = 2$, then the eigenvector can be written

$$\{\mathbf{x}^{(2)}\} = \begin{Bmatrix} 2 \\ 2 \\ 1 \end{Bmatrix}$$

Finally, for $\lambda^{(3)} = 4$, Equation a becomes

$$12x_1 - 24x_2 + 18x_3 = 0$$
$$3x_1 - 6x_2 \quad\quad = 0$$
$$-9x_1 + 18x_2 - 24x_3 = 0$$

The second equation implies $x_1 = 2x_2$, which, when substituted into either the first or third equation, yields $x_3 = 0$. If we then arbitrarily set $x_2 = 1$, the third eigenvector is

$$\{x^{(3)}\} = \begin{Bmatrix} 2 \\ 1 \\ 0 \end{Bmatrix}$$

The eigenvectors are not mutually orthogonal, since the matrix $[A]$ is not symmetric.

9.9   ORTHOGONAL COORDINATE TRANSFORMATIONS

The cartesian coordinates of a point depend on the choice of reference frame against which the coordinates are measured. The relationship between the coordinates of the same point in two different coordinate systems is called a *coordinate transformation*. The mathematical description of a cartesian coordinate transformation requires a particular form of matrix called an *orthogonal* matrix. An orthogonal matrix $[R]$, by definition, has the property that $[R]^T = [R]^{-1}$, that is

$$[R]^T[R] = [R][R]^T = [I] \qquad \text{(orthogonal matrix)} \qquad\qquad \textbf{[9.9.1]}$$

Recall that the determinant of a matrix product equals the product of the individual determinants, the determinant of a matrix and its transpose are identical, and $\det[I] = 1$. Therefore, Equation 9.9.1 implies that $(\det[R])^2 = 1$, so that $\det[R] = \pm 1$. If the determinant is positive, then $[R]$ is a *proper* orthogonal transformation; otherwise, it is *improper*. Proper orthogonal transformations relate right-hand cartesian systems to other right-hand cartesian frames.

$$\det[R] = +1 \qquad \text{(proper orthogonal matrix)} \qquad\qquad \textbf{[9.9.2]}$$

Consider two cartesian coordinate systems, the unbarred xyz system and the barred $\bar{x}\bar{y}\bar{z}$ system, which share a common origin, but are rotated with respect to each other, as shown in Figure 9.9.1. Let the unit vectors

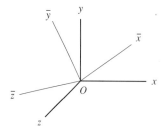

Figure 9.9.1   Two sets of cartesian coordinate axes.

along the $\bar{x}$, $\bar{y}$, and $\bar{z}$ axes be $\bar{\mathbf{i}}$, $\bar{\mathbf{j}}$, and $\bar{\mathbf{k}}$, respectively. Similarly, denote the xyz unit vectors as $\mathbf{i}$, $\mathbf{j}$, and $\mathbf{k}$. Let $\mathbf{v}$ be a vector, such as a force, couple, translational displacement, or small rotational displacement. Resolving $\mathbf{v}$ into components v_x, v_y, and v_z along the x, y, and z axes, we can write

$$\mathbf{v} = v_x \mathbf{i} + v_y \mathbf{j} + v_z \mathbf{k} \qquad [9.9.3]$$

The components of $\mathbf{v}$ along the $\bar{x}, \bar{y}$, and $\bar{z}$ axes are found by projecting $\mathbf{v}$ onto each axis by means of the vector dot product operation. Thus,

$$\bar{v}_x = \bar{\mathbf{i}} \cdot \mathbf{v} = (\bar{\mathbf{i}} \cdot \mathbf{i}) v_x + (\bar{\mathbf{i}} \cdot \mathbf{j}) v_y + (\bar{\mathbf{i}} \cdot \mathbf{k}) v_z$$

or

$$\bar{v}_x = \cos \phi_{\bar{x}x} v_x + \cos \phi_{\bar{x}y} v_y + \cos \phi_{\bar{x}z} v_z \qquad [9.9.4a]$$

where $\phi_{\bar{x}x}$ is the angle between the $\bar{x}$ and x axes, $\phi_{\bar{x}y}$ is the angle between the $\bar{x}$ and y axes, etc. Since $\bar{v}_y = \bar{\mathbf{j}} \cdot \mathbf{v}$ and $\bar{v}_z = \bar{\mathbf{k}} \cdot \mathbf{v}$, we find in similar fashion that

$$\bar{v}_y = \cos\phi_{\bar{y}x} v_x + \cos\phi_{\bar{y}y} v_y + \cos\phi_{\bar{y}z} v_z \qquad [9.9.4b]$$

$$\bar{v}_z = \cos\phi_{\bar{z}x} v_x + \cos\phi_{\bar{z}y} v_y + \cos\phi_{\bar{z}z} v_z \qquad [9.9.4c]$$

Equation 9.9.4 show how to compute the components of a vector in the barred system from the components given in the unbarred coordinates. The nine direction cosines of the barred system's three axes are required to describe the relative orientation of the two systems. Let us simplify the notation for the direction cosines as follows:

$$
\begin{array}{lll}
l_1 = \cos \phi_{\bar{x}x} & m_1 = \cos \phi_{\bar{x}y} & n_1 = \cos \phi_{\bar{x}z} \\
l_2 = \cos \phi_{\bar{y}x} & m_2 = \cos \phi_{\bar{y}y} & n_2 = \cos \phi_{\bar{y}z} \\
l_3 = \cos \phi_{\bar{z}x} & m_3 = \cos \phi_{\bar{z}y} & n_3 = \cos \phi_{\bar{z}z}
\end{array}
\qquad [9.9.5]
$$

Then, using matrix notation, we can write the coordinate transformation as

$$\{\bar{\mathbf{v}}\} = [\mathbf{R}]\{\mathbf{v}\} \qquad [9.9.6a]$$

where

$$
\{\bar{\mathbf{v}}\} = \begin{Bmatrix} \bar{v}_x \\ \bar{v}_y \\ \bar{v}_z \end{Bmatrix} \qquad
\{\mathbf{v}\} = \begin{Bmatrix} v_x \\ v_y \\ v_z \end{Bmatrix} \qquad
[\mathbf{R}] = \begin{bmatrix} l_1 & m_1 & n_1 \\ l_2 & m_2 & n_2 \\ l_3 & m_3 & n_3 \end{bmatrix}
\qquad [9.9.6b]
$$

Here, $[\mathbf{R}]$ is the coordinate transformation matrix, which we employed in sections 3.4 and 3.7 to obtain the stress and strain transformation relations. The rows of $[\mathbf{R}]$ are the unit vectors of the barred coordinate system, resolved into components along the unbarred axes; the columns of $[\mathbf{R}]$ are the unit vectors of the unbarred coordinate system, resolved into components along the barred axes. Since the rows of $[\mathbf{R}]$ are orthogonal unit vectors, the squared components of each row must sum to unity and the dot product of any two rows must vanish, as detailed in Equation 3.4.4.

Let us evaluate the matrix product $[\mathbf{R}][\mathbf{R}]^T$ which is

$$
[\mathbf{R}][\mathbf{R}]^T = \begin{bmatrix} l_1 & m_1 & n_1 \\ l_2 & m_2 & n_2 \\ l_3 & m_3 & n_3 \end{bmatrix} \begin{bmatrix} l_1 & l_2 & l_3 \\ m_1 & m_2 & m_3 \\ n_1 & n_2 & n_3 \end{bmatrix}
$$

Expanding, we have

$$[\mathbf{R}][\mathbf{R}]^T = \begin{bmatrix} (l_1)^2 + (m_1)^2 + (n_1)^2 & l_1l_2 + m_1m_2 + n_1n_2 & l_1l_3 + m_1m_3 + n_1n_3 \\ l_1l_2 + m_1m_2 + n_1n_2 & (l_2)^2 + (m_2)^2 + (n_2)^2 & l_2l_3 + m_2m_3 + n_2n_3 \\ l_1l_3 + m_1m_3 + n_1n_3 & l_2l_3 + m_2m_3 + n_2n_3 & (l_3)^2 + (m_3)^2 + (n_3)^2 \end{bmatrix}$$

From Equation 3.4.4, we see that the diagonal components are all equal to 1, while all of the off-diagonal terms are zero. That is,

$$[\mathbf{R}][\mathbf{R}]^T = \begin{bmatrix} 1 & 0 & 0 \\ 0 & 1 & 0 \\ 0 & 0 & 1 \end{bmatrix} = [\mathbf{I}]$$

Since the column vectors of $[\mathbf{R}]$ are also orthogonal unit vectors (cf. Equation 3.4.3), we can similarly deduce that $[\mathbf{R}]^T[\mathbf{R}] = [\mathbf{I}]$. In summary, then, we can write

$$[\mathbf{R}][\mathbf{R}]^T = [\mathbf{R}]^T[\mathbf{R}] = [\mathbf{I}] \qquad [9.9.7]$$

Comparing Equations 9.9.7 and 9.5.10, we see that $[\mathbf{R}]^T = [\mathbf{R}]^{-1}$, that is, the transpose of $[\mathbf{R}]$ is identical to its inverse. This means that $[\mathbf{R}]$ is an orthogonal matrix. To find its inverse, we only need to interchange its rows and columns, a simple operation compared to the lengthy calculations generally required to invert a matrix.

9.10   SOME GENERAL ASPECTS OF THE STIFFNESS METHOD

Consider the model of a linear spring, Figure 9.10.1, a simple structural element. The nodes at each end of the spring are labeled 1 and 2, respectively. Forces X_1 and X_2 are applied to each node of the spring in the direction of the spring axis. The nodes generally undergo axial displacements u_1 and u_2, as well.

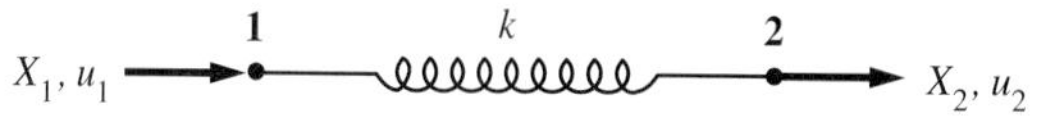

Figure 9.10.1   Spring element.

A spring exhibits linear elastic behavior if the force developed in it is directly proportional to the *relative* displacement of its two nodes, in which case

$$X_1 = k(u_1 - u_2)$$

The constant of proportionality k is called the *spring constant* or *spring rate*, and its value depends on the material with which the spring is made, as well as the details of the spring's geometry. For example, for a coiled helical spring, the spring constant is

$$k = \frac{d^4 G}{8D^3 N}$$

where G is the shear modulus of the coiled wire material, d is the wire diameter, D is the mean coil diameter, and N is the number of coils.

The magnitude of the spring deflection is limited. If the relative displacement of the nodes becomes too large, the spring will permanently deform due to plastic deformation, or may even break because of the excessive stresses developed in the spring material. In this text, structural displacements are limited to those which give rise only to linear elastic behavior throughout the structure.

The spring is a two-force member. For it to be in equilibrium, the forces at each end must be colinear, equal in magnitude, and opposite in direction:

$$X_2 = -X_1$$

Since $X_1 = k(u_1 - u_2)$, this implies that

$$X_2 = k(u_2 - u_1)$$

These simple results, summarized in matrix notation, yield

$$\left\{ \begin{array}{c} X_1 \\ X_2 \end{array} \right\} = \left[\begin{array}{cc} k & -k \\ -k & k \end{array} \right] \left\{ \begin{array}{c} u_1 \\ u_2 \end{array} \right\} \qquad \textbf{[9.10.1]}$$

More compactly, this is

$$\{Q\} = [k]\{q\} \qquad \textbf{[9.10.2]}$$

where

$$\{Q\} = \left\{ \begin{array}{c} X_1 \\ X_2 \end{array} \right\} \qquad [k] = \left[\begin{array}{cc} k & -k \\ -k & k \end{array} \right] \qquad \{q\} = \left\{ \begin{array}{c} u_1 \\ u_2 \end{array} \right\}$$

The vector $\{Q\}$ is the *element load vector*, $\{q\}$ is the *element displacement vector,* and $[k]$ is the *element stiffness matrix.* Notice that the stiffness matrix is singular, that is, the determinant of $[k]$ is zero. This reflects the fact that the spring has not been constrained against rigid-body translation along its axis. In rigid-body translation, both nodes have the same displacement. The reader can easily verify that the load vector on the left of Equation 9.10.1 is unaltered if an arbitrary additional displacement d is applied to both nodes of the unconstrained spring in Figure 9.10.1. Therefore, the relationship between $\{Q\}$ and $\{q\}$ is not unique for the unconstrained spring element, nor is it unique for any elastic structure that is free to move as a rigid body.

To establish the force–deflection, or *stiffness,* relationships for a structural element such as the spring, we focus on the element itself without regard to the surrounding structure. The stiffness properties of the spring element were completely defined in the natural, element-based, local frame of reference. An appropriate local frame of reference for the spring element in Figure 9.10.1 has its origin at node 1, the leftmost node, with the positive x axis directed to the right through node 2. The y axis is directed, say, upwards in the plane of the page, perpendicular to the x axis. The z axis is then perpendicular to the page. This makes the local frame of reference a standard right-hand cartesian coordinate system, which means that the cross product of the x axis unit vector into the y axis unit vector equals the z axis unit vector.

Our model of a spring is shown lying in a plane; an actual spring, like all material objects, resides in three-dimensional space. Therefore, material particles, that is, points within or upon a structure or structural element, generally have three translational degrees of freedom. Unless a given material particle is somehow physically constrained, it can be displaced along any of the three spatial dimensions. These three translational degrees of freedom will be called the x, y, and z freedoms relative to a chosen cartesian coordinate system. In addition, points of a structure may have rotational degrees of freedom about each of the coordinate axes. For example, on the neutral axis of a beam of symmetrical cross section loaded in a plane of symmetry, a material particle may displace vertically, while its immediate neighborhood of material particles may rotate in the plane through an angle defined by the slope of the elastic curve. In general, each material particle of a structure, taken together with its immediate material neighborhood, has six degrees of freedom: three translational freedoms and three rotational freedoms.

Structural nodes, such as those of the spring element, are material particles. They belong to and are carried along with the structure as it deforms in response to applied loads. A node may therefore have six degrees of freedom. However, the analytical model of a given structure might not require that all six degrees of freedom at its nodes be incorporated in order to account adequately for its mechanical behavior. Indeed, the goal of matrix structural analysis is to accurately model practical structures composed of an *infinite* number of material particles (and therefore possessing an infinite number of degrees of freedom) using assemblages of a *finite* number of appropriately selected elements joined together at a relatively few points (nodes), at which the essential structural properties are lumped in some rational fashion.

The spring element in Figure 9.10.1 is a two degree of freedom element, because incorporating only one (the axial) degree of freedom at each of its two nodes is sufficient to describe the essential structural properties of the element. These properties are expressed in Equations 9.10.1 and 9.10.2. The five ignored degrees of freedom at each node are deemed irrelevant to modeling the spring. Rotating either node point about any axis, by definition, induces no loads in the spring. Neither does displacing the nodes in a direction perpendicular to the spring axis. The details of the actual spring structure lying between nodes 1 and 2 are lumped completely and simply into a single quantity, the spring constant k. If the simplifying assumptions upon which the spring element is based fall short of modeling some aspects of its structure, then elements based on alternative or less restrictive assumptions must be chosen. Industrial computer programs have extensive libraries of finite elements from which the engineer can choose those needed to model real structures satisfactorily.

Let us now join several spring elements together to form a structure composed of parallel *colinear* springs. Assign each element an arbitrary, unique number, for identification purposes. If there are N elements, we usually number them from 1 to N. The points at which two or more springs are joined together are called *structural nodes*. A portion of such an array of spring elements is shown in Figure 9.10.2. Where more than two elements are attached to a node, that node point is shown as a rigid vertical spacer, for clarity only; the spacing between colinear springs is zero. Since all of the spring elements composing the assembly are parallel, the only relevant degree of freedom at each node is the axial or x freedom. If there are a total of M nodes in this entire structure, then it has a total of M degrees of freedom, one per node.

A major step in every matrix structural analysis problem is to infer the overall structural stiffness properties—the load versus deflection behavior—from that of the constituent elements. The relatively simple spring structure can be used to show how this is done, and the procedure extends to structures of more practical interest and complexity.

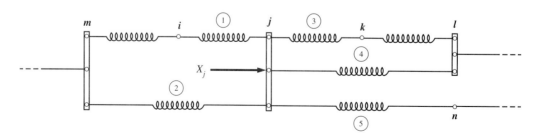

Figure 9.10.2   Portion of a linear spring structure.

The nodes of a structure are identified by a convenient, arbitrary numbering system that is independent of the node numbering scheme established for the individual elements comprising the structure. Each of a structure's M degrees of freedom has a unique number, which is determined by the structural node to which it belongs. Since there is but one degree of freedom per node in the colinear spring structure of Figure 9.10.2, the degree of freedom number at each node is set equal to the node number itself. Practical structures generally require the inclusion of more than one degree of freedom at a node. A three-dimensional truss structure, for example, must have at least three translational degrees of freedom per node. To ascertain the total number of degrees of freedom (d.o.f.) for a structure that is not supported or constrained, use the following formula:

$$\text{Total structural d.o.f.} = (\text{Number of d.o.f. per node}) \times (\text{Total number of nodes})$$

Quantities (loads, displacements, stiffness matrices, etc.) that pertain to an element of a structure are *local* quantities. We will use a superscript to denote the element to which a local quantity belongs. Quantities associated with the complete, assembled structure are *global* quantities, and they will appear without superscripts.

In Figure 9.10.2, an *external* (global) load X_j is shown applied to node j in the direction of freedom j. Spring elements 1, 2, 3, 4, and 5 are connected to node j, so each of these elements exerts a force on that node. For each element, this force is equal in magnitude but opposite in direction to the force exerted *by* node j *on* the element. Since the complete structure is required to be in equilibrium, any portion of it, including node j, must be in equilibrium as well. Figure 9.10.3 shows node j isolated as a free body in equilibrium under the action of the external force X_j and all of the *internal* forces coming from the elements attached to the node. The superscripts identify the elements to which the internal loads are applied. The subscripts and load directions for the local element quantities are consistent with the notation in Figure 9.10.1.

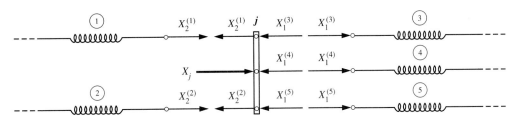

Figure 9.10.3   The external and internal forces action on node j of Figure 9.10.2.

For node j in Figure 9.10.3 to be in equilibrium, the sum of the forces in the direction of freedom j must be zero, or

$$X_j = X_2^{(1)} + X_2^{(2)} + X_1^{(3)} + X_1^{(4)} + X_1^{(5)} \qquad \text{[9.10.3]}$$

For element e, the load–deflection relationship, Equation 9.10.2, is valid, with superscripts added to single out individual elements from the group comprising the structure:

$$\{Q^{(e)}\} = [k^{(e)}]\{q^{(e)}\}$$

Thus, for elements 1 through 5,

$$X_2^{(1)} = k^{(1)}\left(u_2^{(1)} - u_1^{(1)}\right) = k^{(1)}\left(u_j - u_i\right)$$

$$X_2^{(2)} = k^{(2)}\left(u_2^{(2)} - u_1^{(2)}\right) = k^{(2)}\left(u_j - u_m\right)$$

$$X_1^{(3)} = k^{(3)}\left(u_1^{(3)} - u_2^{(3)}\right) = k^{(3)}\left(u_j - u_k\right) \qquad \text{[9.10.4]}$$

$$X_1^{(4)} = k^{(4)}\left(u_1^{(4)} - u_2^{(4)}\right) = k^{(4)}\left(u_j - u_l\right)$$

$$X_1^{(5)} = k^{(5)}\left(u_1^{(5)} - u_2^{(5)}\right) = k^{(5)}\left(u_j - u_n\right)$$

In writing these equations, we used Figure 9.10.2 to determine the structural nodes to which each element is attached. For example, node 1 of element 1 coincides with node i of the structure. Therefore, the displacement of node 1 of element 1 equals the displacement of node i of the structure, and the displacement of node 2 of element 1 equals the displacement of node j of the structure. That is,

$$u_1^{(1)} = u_i \qquad \text{and} \qquad u_2^{(1)} = u_j$$

In like manner,

$$u_1^{(2)} = u_m \qquad \text{and} \qquad u_2^{(2)} = u_j$$
$$u_1^{(3)} = u_j \qquad \text{and} \qquad u_2^{(3)} = u_k$$
$$u_1^{(4)} = u_j \qquad \text{and} \qquad u_2^{(4)} = u_l$$
$$u_1^{(5)} = u_j \qquad \text{and} \qquad u_2^{(5)} = u_n$$

Each local displacement component is therefore identified with a global counterpart. Implicit in setting up these relationships is the notion of *displacement compatibility*, which means that for all of the element nodes attached to a common structural node, the displacements are identical. This obvious requirement ensures that the structure does not come apart at its nodes. There are cases where displacement compatibility does not hold, such as the analysis of crack propagation, bearing stresses, and sliding surfaces. These are beyond the scope of this text.

Substituting the element forces from Equation 9.10.4 into the equilibrium equation, Equation 9.10.3, yields

$$X_j = \left[k^{(1)} + k^{(2)} + k^{(3)} + k^{(4)} + k^{(5)}\right]u_j - k^{(1)}u_i - k^{(2)}u_m - k^{(3)}u_k - k^{(4)}u_l - k^{(5)}u_n \qquad \text{[9.10.5]}$$

or

$$X_j = k_{ji}\,u_i + k_{jj}\,u_j + k_{jk}\,u_k + k_{jl}\,u_l + k_{jm}\,u_m + k_{jn}\,u_n \qquad \text{[9.10.6]}$$

where

$$k_{ji} = -k^{(1)}$$

$$k_{jj} = k^{(1)} + k^{(2)} + k^{(3)} + k^{(4)} + k^{(5)}$$

$$k_{jk} = -k^{(3)}$$

$$k_{jl} = -k^{(4)} \qquad \text{[9.10.7]}$$

$$k_{jm} = -k^{(2)}$$

$$k_{jn} = -k^{(5)}$$

Let $\{q\}$ be the vector whose components (some of which may be zero) form the set of all nodal displacements of the complete spring structure, only a portion of which appears in Figure 9.10.2. Likewise, let $\{Q\}$ be the vector whose components (some of which may be zero) form the set of all external loads applied to the nodes of that structure. Vector $\{q\}$ is the *global displacement vector*, and $\{Q\}$ is the *global load vector*. Since the structure is composed of linearly elastic spring elements, its overall behavior will be linearly elastic. Therefore, the global loads and global displacements are linearly related, that is,

$$\{Q\} = [k]\{q\} \qquad \text{[9.10.8]}$$

Matrix $[k]$ is the stiffness matrix of the structural assembly. It is called the *global* stiffness matrix, as distinguished from the *local* stiffness matrices of the individual elements. (Recall that in our notation, global quantities do not require superscripts.)

Without loss of generality, suppose for simplicity that nodes i, j, k, l, m, and n in Figure 9.10.2 are numbered consecutively in that same order, so that if node i were, say, 53, then nodes j, k, l, m, and n would be numbered 54, 55, 56, 57, and 58. Then, following the rules of matrix multiplication, we can find X_j, the jth component of the global load vector $\{Q\}$, from Equation 9.10.8. We do so by taking the dot product of row j of $[k]$ and the global displacement vector $\{q\}$, as follows:

$$X_j = \begin{bmatrix} \cdots & \cdots & \cdots & k_{ji} & k_{jj} & k_{jk} & k_{jl} & k_{jm} & k_{jn} & \cdots & \cdots & \cdots \end{bmatrix} \begin{Bmatrix} \cdots \\ \cdots \\ \cdots \\ u_i \\ u_j \\ u_k \\ u_l \\ u_m \\ u_n \\ \cdots \\ \cdots \\ \cdots \end{Bmatrix} \qquad \text{[9.10.9]}$$

where the explicit global stiffness matrix entries in row j are found in terms of the element stiffness matrices, as shown in Equation 9.10.7. All of the other entries, represented by the ellipses, are zeroes, since no other terms appear in Equation 9.10.6, the equation of equilibrium at node j. Thus, each term in row j of the global stiffness matrix $[k]$ can be found from the individual element stiffness matrices by writing the equilibrium equation in the direction of the degree of freedom j and then invoking displacement compatibility.

A nonnegative number will always appear in column j of row j, that is, on the principal diagonal of $[\mathbf{k}]$. A nonzero number will appear in any column i of row j only if degree of freedom i is joined to freedom j by an element that connects the nodes to which freedoms i and j belong. In the spring assembly of Figure 9.10.2, only freedoms i, m, k, l, and n are joined to freedom j, by means of elements 1, 2, 3, 4, and 5. Since the possibly many other nodes of the structure are not connected directly to node j, no entries are in the row j columns corresponding to their degrees of freedom. It is typical of all structures, no matter how complex, that any given node is only connected by structural elements to a few nodes in its immediate vicinity. Although a global stiffness matrix may reflect the structure's size and complexity by having many components, usually relatively few entries are distributed across its rows. Matrices whose components are mostly zeroes are called *sparse* matrices. Structures with many degrees of freedom have large but sparse stiffness matrices.

We arrived at Equation 9.10.9 by assuming that the six nodes i, j, k, l, m, and n in Figure 9.10.2 are numbered consecutively. This means that of this group, those nodes joined to node j have degree of freedom numbers that differ by at most four from that of node j. This in turn implies that there are no nonzero entries in row j of $[\mathbf{k}]$ beyond a distance of four columns from the diagonal element, just as in Equation 9.10.9. Since node numbering is arbitrary, suppose that instead of the consecutive numbering scheme selected, the nodes of the structure are numbered such that the numerical labels of node j and node i in Figure 9.10.2 differ by, say, 30. Then, in row j of $[\mathbf{k}]$, there would be a nonzero entry corresponding to freedom i *thirty* columns away from the diagonal, instead of just one column away. Obviously, in row j of $[\mathbf{k}]$, the greater the difference between the numerical label of degree of freedom j and those of the freedoms joined to it, the greater will be the distance between the diagonal element and the furthest nonzero entry from it.

Suppose we take in turn each row of the global stiffness matrix $[\mathbf{k}]$ and, starting with the diagonal element, count the number of columns to the last nonzero entry in that row. This number is called the *semibandwidth* of that row. The largest number thus obtained, after considering all of the rows, is called the semibandwidth of the matrix. In matrices with a small semibandwidth, the nonzero components are clustered within a band about the principal diagonal, in a fashion exemplified by Equation 9.10.9. The semibandwidth of a stiffness matrix of an unconstained structure may be computed as follows. Consider all pairs of nodes that have a structural element in common, in each case noting the difference between the global node numbers. Record the maximum node number difference for all such pairs. Then

$$N_b = N_{dof}(D_{\max} + 1) \qquad\qquad \text{[9.10.10]}$$

where N_b is the semibandwidth of the stiffness matrix, N_{dof} is the number of degrees of freedom per node, and $D_{\max}$ is the maximum node number difference. Different node numbering patterns for the same structure yield different arrangements of the rows and columns of the stiffness matrix and, consequently, different semibandwidths.

The *stiffness equations* $[\mathbf{k}]\{\mathbf{q}\} = \{\mathbf{Q}\}$ must eventually be solved for $\{\mathbf{q}\}$. The number of operations required by the Gauss elimination method for a symmetric matrix is proportional to N_b^2. Therefore, in the interest of computational efficiency, nodes should be numbered in such a way that the semibandwidth of $[\mathbf{k}]$ is as small as possible. This consideration limits the choice of node numbering strategies. For two- and three-dimensional structures with many nodes and an irregular geometry, determining the optimum node numbering scheme is best left to a computer, thus freeing the analyst to number the nodes in any convenient manner. Automatic optimal node renumbering is a feature of general purpose structural analysis codes.

It is instructive to consider an alternate method of generating the global stiffness matrix of a structure. Suppose the displacements are held to zero at every degree of freedom except the rth, at which the displacement is unity, $u_r = 1$. Then, the load-deflection relationship, Equation 9.10.8, can be used to find the set of external forces required to maintain static equilibrium, namely,

$$\{Q\} = [k] \begin{Bmatrix} 0 \\ 0 \\ \cdot \\ \cdot \\ \cdot \\ \cdot \\ 0 \\ 1 \\ 0 \\ \cdot \\ \cdot \\ \cdot \\ 0 \end{Bmatrix} \leftarrow r\text{th degree of freedom}$$

To carry out the matrix multiplication on the right, we take the dot product of each row of $[k]$ into the column vector, which has a 1 in the rth position and zeroes everywhere else. Clearly, the only entry in each row of $[k]$ that survives the dot product is the one in column r, so that

$$\{Q\} = \begin{Bmatrix} k_{1r} \\ k_{2r} \\ k_{3r} \\ \cdot \\ \cdot \\ k_{rr} \\ \cdot \\ \cdot \\ k_{Nr} \end{Bmatrix}$$

The right-hand side is column r of the stiffness matrix $[k]$. That is, *each term in column r of the global stiffness matrix is the force developed in the direction of that degree of freedom in response to a unit displacement in the direction of freedom r, with all other displacements held to zero.*

Returning to the spring assembly of Figure 9.10.2, let the displacement at node j be 1 unit, while all other nodes are held to zero displacement. Forces will only develop within those springs whose nodes are displaced, that is, only in those springs attached to node j. Referring to the figure, we see that this involves only elements 1 through 5. With node j displaced 1 unit to the right, we observe that

Force in spring 1 = $k^{(1)}$, tension

Force in spring 2 = $k^{(2)}$, tension

Force in spring 3 = $k^{(3)}$, compression

Force in spring 4 = $k^{(4)}$, compression

Force in spring 5 = $k^{(5)}$, compression

Therefore, spring 1 exerts a force to the *right* on node i, so that for equilibrium, an external force to the *left* is required in degree of freedom direction i. Spring 3 exerts a force to the *right* on node k, so that an external force to the *left* is required at degree of freedom k. All five springs exert forces to the *left* on node j; therefore, an external force equal to the sum of those forces and directed to the *right* must exist at degree of freedom j. The loads at nodes l, m, and n can be similarly deduced. All other nodes of the structure are free of external load, because no spring attached to those nodes is deformed. We conclude that the external nodal loads developed in response to the unit displacement at node j are

$$\{Q\} = \begin{Bmatrix} 0 \\ 0 \\ \cdot \\ \cdot \\ \cdot \\ -k^{(1)} \\ +k^{(1)} + k^{(2)} + k^{(3)} + k^{(4)} + k^{(5)} \\ -k^{(3)} \\ -k^{(4)} \\ -k^{(2)} \\ -k^{(5)} \\ \cdot \\ \cdot \\ 0 \\ 0 \end{Bmatrix} \begin{array}{l} \\ \\ \\ \\ \\ \longleftarrow \text{ degree of freedom } i \\ \longleftarrow \text{ degree of freedom } j \\ \longleftarrow \text{ degree of freedom } k \\ \longleftarrow \text{ degree of freedom } l \\ \longleftarrow \text{ degree of freedom } m \\ \longleftarrow \text{ degree of freedom } n \\ \\ \\ \\ \end{array}$$

As we just discussed, this column vector must be column j of the global stiffness matrix $[\mathbf{k}]$. Thus, all elements of column j of $[\mathbf{k}]$ are zero, except for the following:

$$\begin{aligned} k_{ij} &= -k^{(1)} \\ k_{jj} &= k^{(1)} + k^{(2)} + k^{(3)} + k^{(4)} + k^{(5)} \\ k_{kj} &= -k^{(3)} \\ k_{ij} &= -k^{(4)} \\ k_{mj} &= -k^{(2)} \\ k_{nj} &= -k^{(5)} \end{aligned}$$

[9.10.11]

Comparing Equations 9.10.11 and 9.10.7, we see that row j of $[\mathbf{k}]$ is identical to column j.

This analysis could be done at any node of the structure, with analogous conclusions. Any row r of the stiffness matrix can be found by considering equilibrium at degree of freedom r, while any column r can be found by considering equilibrium at all degrees of freedom due to a unit displacement at degree of freedom r. Row r of $[\mathbf{k}]$ will always be equal to column r, which means that $[\mathbf{k}]$ is symmetric. The global stiffness matrix is symmetric because the spring element stiffness matrices from which it is assembled (cf. Equation 9.10.2) are symmetric. In section 6.5, it was shown that stiffness matrices in general are symmetric.

Figure 9.10.4 depicts an unrestrained 11 degree of freedom colinear spring assembly composed of 15 spring elements, all having the same spring constant k. The 11 node structure is accompanied by its 11 by 11 global stiffness matrix $[\mathbf{k}]$, constructed in the manner just explained. (Additional examples to follow will illustrate the procedure.) Notice the symmetry of the matrix and the fact that all of its principal diagonal entries are *positive*.

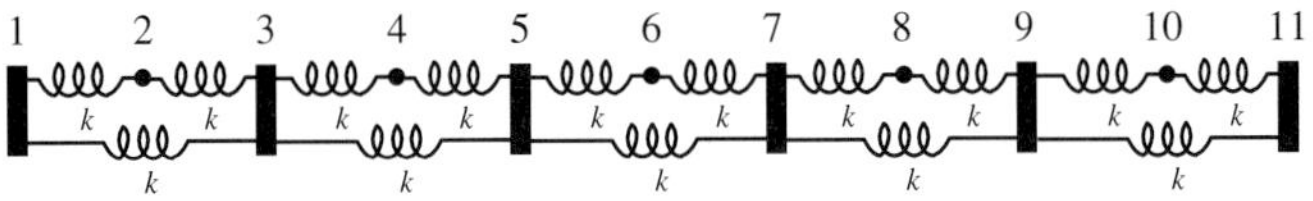

$N_b = 3$

$$
[\mathbf{k}] = \begin{array}{c} \text{dof} \\ 1 \\ 2 \\ 3 \\ 4 \\ 5 \\ 6 \\ 7 \\ 8 \\ 9 \\ 10 \\ 11 \end{array}
\begin{array}{c}
\begin{array}{ccccccccccc} 1 & 2 & 3 & 4 & 5 & 6 & 7 & 8 & 9 & 10 & 11 \end{array} \\
\left[\begin{array}{ccccccccccc}
2k & -k & -k & 0 & 0 & 0 & 0 & 0 & 0 & 0 & 0 \\
-k & 2k & -k & 0 & 0 & 0 & 0 & 0 & 0 & 0 & 0 \\
-k & -k & 4k & -k & -k & 0 & 0 & 0 & 0 & 0 & 0 \\
0 & 0 & -k & 2k & -k & 0 & 0 & 0 & 0 & 0 & 0 \\
0 & 0 & -k & -k & 4k & -k & -k & 0 & 0 & 0 & 0 \\
0 & 0 & 0 & 0 & -k & 2k & -k & 0 & 0 & 0 & 0 \\
0 & 0 & 0 & 0 & -k & -k & 4k & -k & -k & 0 & 0 \\
0 & 0 & 0 & 0 & 0 & 0 & -k & 2k & -k & 0 & 0 \\
0 & 0 & 0 & 0 & 0 & 0 & -k & -k & 4k & -k & -k \\
0 & 0 & 0 & 0 & 0 & 0 & 0 & 0 & -k & 2k & -k \\
0 & 0 & 0 & 0 & 0 & 0 & 0 & 0 & -k & -k & 2k
\end{array}\right]
\end{array}
$$

Figure 9.10.4 Colinear spring assembly and its banded global stiffness matrix.

The nodes of the structure are numbered such that the maximum node number difference between any two nodes joined by a single spring element is two. This yields a semibandwidth N_b of three, as can be inferred from Equation 9.10.10. No other node numbering scheme for this structure will produce a lower value of N_b.

Figure 9.10.4 clearly reveals the banded nature of the stiffness matrix. Observe that the sum of the components of any column (or row) of $[\mathbf{k}]$ is zero, as expected, since the columns of $[\mathbf{k}]$ form a set of external nodal loads in equilibrium. In Figure 9.10.5, the same structure is shown with an alternative node numbering pattern. The resulting stiffness matrix is symmetric, with positive numbers along the principal diagonal and columns that sum to zero, but it is no longer banded. In fact, this version of $[\mathbf{k}]$ has a semibandwidth equal to the entire width of the matrix. Of the two node numberings, that of Figure 9.10.4 is preferred because the smaller resulting semibandwidth will reduce the number of computations required to solve the system $[\mathbf{k}]\{\mathbf{q}\} = \{\mathbf{Q}\}$.

A structure in equilibrium must somehow be restrained by supports. There must be a sufficient number of constraints on its degrees of freedom at least to prevent rigid-body motion. For example, at least one nodal degree of freedom of the spring structure in Figures 9.10.4 and 9.10.5 must be fixed, to ensure that the assembly cannot move in the direction of its axis without deforming. Constraints of this kind are called *displacement boundary conditions.* In Equation 9.10.8 it is convenient to partition the components of the load vector $\{\mathbf{Q}\}$ and displacement vector $\{\mathbf{q}\}$ into two groups: the *unconstrained* degrees of freedom, and the *constrained* degrees of freedom. The components of $\{\mathbf{Q}\}$ and $\{\mathbf{q}\}$ that are unrestrained will be listed first, followed by the constrained components. This partitioning will generally require a reshuffling of the components of $\{\mathbf{Q}\}$ and $\{\mathbf{q}\}$, and therefore a rearrangement of the rows and columns of the global stiffness matrix $[\mathbf{k}]$.

The partitioned stiffness equations $\{\mathbf{Q}\} = [\mathbf{k}]\{\mathbf{q}\}$ have the form

$$
\left\{\begin{array}{c} \mathbf{Q}_f \\ \hdashline \mathbf{Q}_s \end{array}\right\} = \left[\begin{array}{c:c} \mathbf{k}_{ff} & \mathbf{k}_{fs} \\ \hdashline \mathbf{k}_{sf} & \mathbf{k}_{ss} \end{array}\right] \left\{\begin{array}{c} \mathbf{q}_f \\ \hdashline \mathbf{q}_s \end{array}\right\} \qquad [9.10.12]
$$

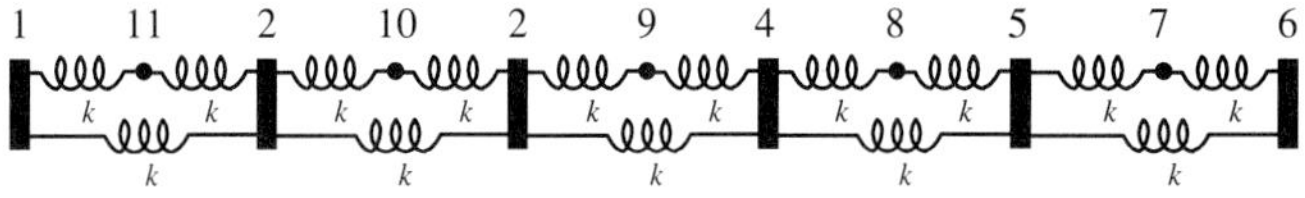

$$[\mathbf{k}] = \begin{array}{c} \\ \text{dof} \end{array} \begin{array}{ccccccccccc} 1 & 2 & 3 & 4 & 5 & 6 & 7 & 8 & 9 & 10 & 11 \end{array}$$

dof	1	2	3	4	5	6	7	8	9	10	11
1	$2k$	$-k$	0	0	0	0	0	0	0	0	$-k$
2	$-k$	$4k$	$-k$	0	0	0	0	0	0	$-k$	$-k$
3	0	$-k$	$4k$	$-k$	0	0	0	0	$-k$	$-k$	0
4	0	0	$-k$	$4k$	$-k$	0	0	$-k$	$-k$	0	0
5	0	0	0	$-k$	$4k$	$-k$	$-k$	$-k$	0	0	0
6	0	0	0	0	$-k$	$2k$	$-k$	0	0	0	0
7	0	0	0	0	$-k$	$-k$	$2k$	0	0	0	0
8	0	0	0	$-k$	$-k$	0	0	$2k$	0	0	0
9	0	0	$-k$	$-k$	0	0	0	0	$2k$	0	0
10	0	$-k$	$-k$	0	0	0	0	0	0	$2k$	0
11	$-k$	$-k$	0	0	0	0	0	0	0	0	$2k$

Figure 9.10.5 The same structure as Figure 9.10.4, with a different node numbering.

where the subscript f stands for *free* (unconstrained) degrees of freedom, and the subscript s stands for *supported* (constrained) degrees of freedom. An N degree of freedom structure has N_f unconstrained and N_s constrained degrees of freedom, and

$$N_f + N_s = N$$

Observe that

$$[\mathbf{k}_{ff}] \text{ is an } N_f \text{ by } N_f \text{ square matrix}$$
$$[\mathbf{k}_{ss}] \text{ is an } N_s \text{ by } N_s \text{ square matrix}$$
$$[\mathbf{k}_{fs}] \text{ is an } N_f \text{ by } N_s \text{ rectangular matrix}$$
$$[\mathbf{k}_{sf}] \text{ is an } N_s \text{ by } N_f \text{ rectangular matrix}$$

Also, since $[\mathbf{k}]$ is symmetric ($[\mathbf{k}] = [\mathbf{k}]^T$), we have

$$\begin{bmatrix} \mathbf{k}_{ff} & \mathbf{k}_{fs} \\ \hline \mathbf{k}_{sf} & \mathbf{k}_{ss} \end{bmatrix} = \begin{bmatrix} \mathbf{k}_{ff} & \mathbf{k}_{fs} \\ \hline \mathbf{k}_{sf} & \mathbf{k}_{ss} \end{bmatrix}^T = \begin{bmatrix} \mathbf{k}_{ff}^T & \mathbf{k}_{fs}^T \\ \hline \mathbf{k}_{sf}^T & \mathbf{k}_{ss}^T \end{bmatrix}$$

which means that $[\mathbf{k}_{ff}]$ and $[\mathbf{k}_{ss}]$ are symmetric and

$$[\mathbf{k}_{sf}] = [\mathbf{k}_{fs}]^T \qquad ([\mathbf{k}_{fs}] = [\mathbf{k}_{sf}]^T)$$

Writing out the partitioned stiffness equations in Equation 9.10.12, we obtain

$$\{Q_f\} = [k_{ff}]\{q_f\} + [k_{fs}]\{q_s\}$$
$$\{Q_s\} = [k_{sf}]\{q_f\} + [k_{ss}]\{q_s\}$$

or

$$[k_{ff}]\{q_f\} = \{Q_f\} - [k_{fs}]\{q_s\} \qquad\qquad\qquad \text{[9.10.13a]}$$

$$\{Q_s\} = [k_{sf}]\{q_f\} + [k_{ss}]\{q_s\} \qquad\qquad\qquad \text{[9.10.13b]}$$

At every degree of freedom of a structure, either the load or the displacement (not both) is known from the outset. The vector $\{Q_f\}$ is the set of all known or *prescribed* external loads, many or all of which may be zero. The displacements $\{q_f\}$ in the direction of the prescribed loads are the unknowns of the problem, and they are found by solving Equation 9.10.13a. The vector $\{q_s\}$ is the set of all prescribed displacements, some or all of which may be zero. The external loads $\{Q_s\}$ act at the freedoms whose displacements are prescribed, so they are initially unknown but can eventually be found from Equation 9.10.13b.

Equation 9.10.13a represents N_f equations in N_f unknowns, namely, the N_f components of the free displacements $\{q_f\}$. All other quantities in Equation 9.10.13a are known. The rows and columns of $[k_{ff}]$ are those that remain after eliminating all of the rows and columns of $[k]$ that correspond to the prescribed displacement degrees of freedom. The vector $[k_{fs}]\{q_s\}$ is an N_f by 1 load vector, which is found by multiplying each of the N_s columns of $[k]$ corresponding to a prescribed displacement degree of freedom by that displacement and summing the N_s results. This vector is subtracted from the applied load vector $\{Q_f\}$ to form the right-hand side of the system of Equation 9.10.13a, which is then solved for $\{q_f\}$. Finally, $\{q_f\}$ and $\{q_s\}$ are substituted into the right side of Equation 9.10.13b to find $\{Q_s\}$, the external loads (reactions) acting at the restrained degrees of freedom. If all of the prescribed displacements are zero, which is often the case, then Equation 9.10.13 take the simpler form

$$[k_{ff}]\{q_f\} = \{Q_f\} \qquad\qquad \text{Only if } \{q_s\} = \{0\} \qquad\qquad \text{[9.10.14]}$$
$$\{Q_s\} = [k_{sf}]\{q_f\}$$

Example 9.10.1   Use the direct stiffness method to produce the global stiffness equations of the linear spring assembly in Figure 9.10.6. There are a total of three nodes, and the spring stiffnesses are k_1 and k_2, respectively.

The numbering of the structure's degrees of freedom will depend on the way in which the nodes are numbered. Degree of freedom 1 goes with node 1, which arbitrarily but conveniently happens to be the leftmost node. Degree of freedom 2 goes with node 2 (wherever it may be located in the structure), and so on. The correspondence between local and global degrees of freedom is revealed in a nodal connectivity table (Table 9.10.1), showing the structural nodes to which each element node is attached.

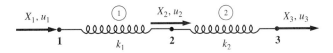

Figure 9.10.6   Two-spring assembly.

Table 9.10.1   Element connectivities for Figure 9.10.6

Element	Node 1	Node 2
1	1	2
2	2	3

The load–deflection relationships for the two springs are found using Equation 9.10.1:
Spring 1:

$$X_1^{(1)} = k_1 u_1^{(1)} - k_1 u_2^{(1)} \tag{a}$$

$$X_2^{(1)} = -k_1 u_1^{(1)} + k_1 u_2^{(1)} \tag{b}$$

Spring 2:

$$X_1^{(2)} = k_2 u_1^{(2)} - k_2 u_2^{(2)} \tag{c}$$

$$X_2^{(2)} = -k_2 u_1^{(2)} + k_2 u_2^{(2)} \tag{d}$$

Looking at Figure 9.10.6, we see that $X_1^{(1)} = X_1$ and $X_2^{(2)} = X_3$, while displacement compatibility requires that $u_1^{(1)} = u_1$, $u_2^{(1)} = u_1^{(2)} = u_2$, and $u_2^{(2)} = u_3$. Therefore, Equations a and d, respectively, can be rewritten in terms of global loads and displacements, as follows:

$$X_1 = k_1 u_1 - k_1 u_2 \tag{e}$$

$$X_3 = -k_2 u_2 + k_2 u_3 \tag{f}$$

For equilibrium at node 2, we must have

$$X_2 = X_2^{(1)} + X_1^{(2)}$$

which can be written in terms of the element displacements, using Equations b and c:

$$X_2 = [-k_1 u_1^{(1)} + k_1 u_2^{(1)}] + [k_2 u_1^{(2)} - k_2 u_2^{(2)}]$$

Using the aforementioned displacement compatibility relations, we can reduce this to an equation involving just global loads and displacements,

$$X_2 = -k_1 u_1 + (k_1 + k_2) u_2 - k_3 u_3 \tag{g}$$

Equations e, f, and g comprise the global stiffness equations, which, when written in the form $\{Q\} = [k]\{q\}$, become

$$\begin{Bmatrix} X_1 \\ X_2 \\ X_3 \end{Bmatrix} = \begin{bmatrix} k_1 & -k_1 & 0 \\ -k_1 & k_1 + k_2 & -k_2 \\ 0 & -k_2 & k_2 \end{bmatrix} \begin{Bmatrix} u_1 \\ u_2 \\ u_3 \end{Bmatrix} \tag{h}$$

We found the assembled global stiffness matrix in this above example by writing the equations of equilibrium and the displacement compatibility at each node. However, that most fundamental implementation of the direct stiffness method can become tedious for structures with many nodes. Alternatively, the global stiffness matrix can be found from the individual spring element stiffnesses by means of the *assembly rule,* a procedure that automatically accounts for nodal equilibrium and compatibility. This method, which applies to *any* structure, will be used throughout this chapter and is illustrated by the following example.

Example 9.10.2   Use the assembly rule to find the stiffness matrix of the spring array in Figure 9.10.6.

First, we write the element stiffness matrices for springs 1 and 2:

$$
\left[\mathbf{k}^{(1)}\right] = \begin{array}{c} 1 \\ 2 \end{array} \begin{array}{cc} 1 & 2 \\ \left[\begin{array}{cc} k_1 & -k_1 \\ -k_1 & k_1 \end{array}\right] \end{array}
\qquad
\left[\mathbf{k}^{(2)}\right] = \begin{array}{c} 2 \\ 3 \end{array} \begin{array}{cc} 2 & 3 \\ \left[\begin{array}{cc} k_2 & -k_2 \\ -k_2 & k_2 \end{array}\right] \end{array}
$$

From the nodal connectivity table in the previous example, observe that spring 1's local freedom 1 coincides with global freedom 1. Therefore, we label row 1 and column 1 of spring 1's stiffness matrix with 1s to designate that correspondence. Similarly, since local freedom 2 of spring 1 coincides with global freedom 2, we label row 2 and column 2 of $[\mathbf{k}^{(1)}]$ with 2s. Spring 2's local freedom 1 coincides with global freedom 2, whereas its local freedom 2 coincides with global freedom 3. Therefore, row 1 and column 1 of spring 2's stiffness matrix are labeled with 2s, while row 2 and column 2 are labeled with 3s.

Now, set aside a 3 by 3 global stiffness matrix and a 3 by 1 global load vector, with all elements of both initialized to zero. Their rows and columns are labeled with the three global degree of freedom numbers, in order.

$$
[\mathbf{k}] = \begin{array}{c} 1 \\ 2 \\ 3 \end{array} \begin{array}{ccc} 1 & 2 & 3 \\ \left[\begin{array}{ccc} 0 & 0 & 0 \\ 0 & 0 & 0 \\ 0 & 0 & 0 \end{array}\right] \end{array}
\qquad
\{\mathbf{Q}\} = \begin{array}{c} 1 \\ 2 \\ 3 \end{array} \left\{\begin{array}{c} 0 \\ 0 \\ 0 \end{array}\right\}
$$

Add each component of $[\mathbf{k}^{(1)}]$ into the row and column of $[\mathbf{k}]$ indicated by the global row and column labels on the element stiffness matrix. This yields

$$
[\mathbf{k}] = \begin{array}{c} 1 \\ 2 \\ 3 \end{array} \begin{array}{ccc} 1 & 2 & 3 \\ \left[\begin{array}{ccc} k_1 & -k_1 & 0 \\ -k_1 & k_1 & 0 \\ 0 & 0 & 0 \end{array}\right] \end{array}
$$

Next, add the components of $[\mathbf{k}^{(2)}]$ into the locations of $[\mathbf{k}]$ indicated by the global freedom labels on that element stiffness matrix, to obtain

$$
[\mathbf{k}] = \begin{array}{c} 1 \\ 2 \\ 3 \end{array} \begin{array}{ccc} 1 & 2 & 3 \\ \left[\begin{array}{ccc} k_1 & -k_1 & 0 \\ -k_1 & k_1 + k_2 & -k_2 \\ 0 & -k_2 & k_2 \end{array}\right] \end{array}
\qquad\qquad \text{[a]}
$$

This agrees with the stiffness matrix in Equation h of Example 9.10.1. The load vector $\{\mathbf{Q}\}$ is assembled by simply adding the external load at each global freedom i into row i of $\{\mathbf{Q}\}$. Guided by Figure 9.10.6, we find

$$
\{\mathbf{Q}\} = \begin{array}{c} 1 \\ 2 \\ 3 \end{array} \left\{\begin{array}{c} X_1 \\ X_2 \\ X_3 \end{array}\right\}
\qquad\qquad \text{[b]}
$$

which is in agreement with Equation h of the previous example. The semibandwidth for this structure is 2, as is obvious from Equation a. Alternatively, using Equation 9.10.10 and noting that the maximum node number difference across any element is 1, we get $N_b = 1 \times (1 + 1) = 2$.

Example 9.10.3   Use the assembly rule to build the stiffness matrix for the spring structure in Figure 9.10.7. Except for the different node numbering scheme, this structure is identical to that in Figure 9.10.6.

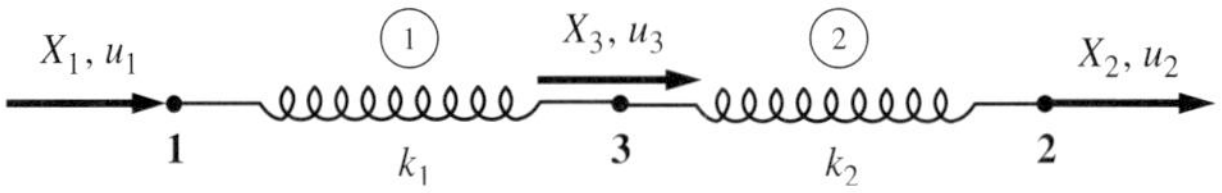

Figure 9.10.7   Spring assembly of Figure 9.10.6, with alternative node numbering.

In this case, the nodal connectivities are as shown in the following table:

Table 9.10.2   Element connectivities in Figure 9.10.7.

Element	Node	Node
1	1	3
2	3	2

As before, we write the stiffness matrices of each spring element, and we use the table to label their rows and columns with the global freedom numbers to which the corresponding local freedoms are bound.

$$
[\mathbf{k}^{(1)}] = \begin{array}{c} 1 \\ 3 \end{array} \begin{bmatrix} \overset{1}{k_1} & \overset{3}{-k_1} \\ -k_1 & k_1 \end{bmatrix} \qquad [\mathbf{k}^{(2)}] = \begin{array}{c} 3 \\ 2 \end{array} \begin{bmatrix} \overset{3}{k_2} & \overset{2}{-k_2} \\ -k_2 & k_2 \end{bmatrix}
$$

The global degrees of freedom are numbered in one-to-one correspondence with the node numbers. Therefore, assembling element 1, then element 2, in the same fashion as before, yields

$$
[\mathbf{k}] = \begin{array}{c} 1 \\ 2 \\ 3 \end{array} \begin{bmatrix} \overset{1}{k_1} & \overset{2}{0} & \overset{3}{-k_1} \\ 0 & k_2 & -k_2 \\ -k_1 & -k_2 & k_1 + k_2 \end{bmatrix} \tag{a}
$$

For the load vector, we obtain

$$
\{\mathbf{Q}\} = \begin{array}{c} 1 \\ 2 \\ 3 \end{array} \begin{Bmatrix} X_1 \\ X_2 \\ X_3 \end{Bmatrix} \tag{b}
$$

Since the second node-numbering sequence results in a maximum node-number difference across both elements of 2, the semibandwidth calculated from Equation 9.10.10 is 3, which is also evident from Equation a.

The straightforward assembly rule used in the previous examples to assemble global stiffness matrices and load vectors is valid in general, and we can use it for structures that are more complex than the simple spring arrangements we have studied so far. The reader should take time at this point to verify the stiffness matrices in Figures 9.10.4 and 9.10.5, using this assembly procedure.

It must be emphasized that this method ensures equilibrium and displacement compatibility at the nodes. It may seem from time to time in the sequence that equilibrium and compatibility considerations are being ignored; yet they are actually fully accounted for whenever the assembly rule is implemented.

In order for the spring assembly in Figure 9.10.6 to support an axial load and remain in equilibrium, the capability for pure translational rigid-body displacement along its axis must be eliminated. That is, at least one of the nodal degrees of freedom must be restrained.

Example 9.10.4   Use the direct stiffness method to find the force in each of the springs in Figure 9.10.8.

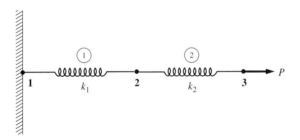

Figure 9.10.8   Constrained, axially-loaded two-spring structure.

The stiffness equations for this two-spring system are in Equation h of Example 9.10.1. Identifying X_1 as the reaction R at the wall and X_3 as the applied load P at node 3, and noting that the external load at node 2 is zero, we see that these equations become

$$\begin{bmatrix} k_1 & -k_1 & 0 \\ -k_1 & k_1 + k_2 & -k_2 \\ 0 & -k_2 & k_2 \end{bmatrix} \begin{Bmatrix} u_1 \\ u_2 \\ u_3 \end{Bmatrix} = \begin{Bmatrix} R \\ 0 \\ P \end{Bmatrix} \qquad [a]$$

The only prescribed displacement is that of global degree of freedom 1, at node 1, which is zero because of the rigid support. Since all of the prescribed displacements are zero, we can write Equation a in the form of Equations 9.10.14. All we need to do to form $[\mathbf{k}_{ff}]$ is simply to eliminate row 1 and column 1 from the stiffness matrix in Equation a. For this problem, then, Equation 9.10.14$_1$ becomes

$$\begin{bmatrix} k_1 + k_2 & -k_2 \\ -k_2 & k_2 \end{bmatrix} \begin{Bmatrix} u_2 \\ u_3 \end{Bmatrix} = \begin{Bmatrix} 0 \\ P \end{Bmatrix}$$

The solution of this system of two equations is

$$u_2 = \frac{P}{k_1} \qquad u_3 = \frac{P}{k_1} + \frac{P}{k_2}$$

Then, from the first line of Equation a, the reaction at the wall is as follows (in analogy to Equation $9.10.14_2$):

$$R = \lfloor k_1 \quad -k_2 \quad 0 \rfloor \begin{Bmatrix} u_1 \\ u_2 \\ u_3 \end{Bmatrix} = (k_1)(0) + (-k_1)\left(\frac{P}{k_1}\right) + 0\left(\frac{P}{k_1} + \frac{P}{k_2}\right) = -P$$

Since the reaction is equal in magnitude but opposite in direction to the applied load P, the structure is indeed in equilibrium.

To complete the analysis by the direct stiffness procedure, we must find the force in each spring element. Rather than using statics, we use the stiffness equations for each element, Equations a through d of Example 9.10.1. For spring 1 we have

$$\text{Force in spring 1} \; = \; X_2^{(1)} = k_1(u_2^{(1)} - u_1^{(1)}) = k_1(u_2 - u_1) = k_1\left(\frac{P}{k_1} - 0\right) = P$$

For spring 2,

$$\text{Force in spring 2} \; = \; X_2^{(2)} = k_2\left(u_2^{(2)} - u_1^{(2)}\right) = k_2(u_3 - u_2) = k_2\left[\left(\frac{P}{k_1} + \frac{P}{k_2}\right) - \frac{P}{k_1}\right] = P$$

The load P applied at the free end of the spring assembly is obviously transmitted through each spring directly to the wall.

The steps followed in solving the simple two-spring system in Figure 9.10.8 are common to the solution of any structures problem by the *displacement method*. They are:

1. Write each element stiffness matrix, identifying the local degrees of freedom with the global degrees of freedom.

2. Assemble the element stiffness matrices into the global stiffness matrix.

3. Assemble the global load vector.

4. Eliminate components of the global stiffness matrix and load vector corresponding to zero support displacements, and modify the load vector to account for nonzero prescribed displacements, if any.

5. Solve the reduced stiffness equations for the unknown displacements.

6. Substitute the solved-for displacements into the element stiffness equations, to find each element's internal loads.

7. Substitute the solved-for displacements into the equations eliminated in Step 4, to find the support reactions, if needed.

The final example of this section illustrates the matrix displacement method for a structure that is somewhat more complex than those of the previous two examples.

Example 9.10.5   Calculate the nodal displacements, element loads, and support reactions for the parallel/axial spring assembly shown in Figure 9.10.9. The springs are attached to the rigid spacer bars such that all points of the system move parallel to the colinear springs. All springs have the same stiffness k, and an external load P is applied to the right at node 5. First, we establish the table of nodal connectivities (Table 9.10.3), keeping in mind that the leftmost node of each spring element is its first node (node 1).

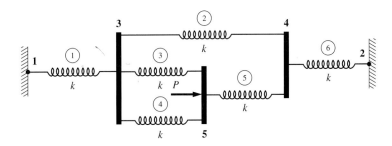

Figure 9.10.9   Statically indeterminate, colinear spring structure.

Table 9.10.3   Element connectivities in Figure 9.10.9.

Element	Node 1	Node 2
1	1	3
2	3	4
3	3	5
4	3	5
5	5	4
6	4	2

Next, guided by the connectivity table, we write each element's stiffness matrix and label its rows and columns with the global degree of freedom numbers with which the element freedoms coincide.

$$
[\mathbf{k}^{(1)}] = \begin{array}{c} 1 \\ 3 \end{array} \begin{bmatrix} \overset{1}{k} & \overset{3}{-k} \\ -k & k \end{bmatrix} \qquad
[\mathbf{k}^{(2)}] = \begin{array}{c} 3 \\ 4 \end{array} \begin{bmatrix} \overset{3}{k} & \overset{4}{-k} \\ -k & k \end{bmatrix} \qquad
[\mathbf{k}^{(3)}] = \begin{array}{c} 3 \\ 5 \end{array} \begin{bmatrix} \overset{3}{k} & \overset{5}{-k} \\ -k & k \end{bmatrix}
$$

$$
[\mathbf{k}^{(4)}] = \begin{array}{c} 3 \\ 5 \end{array} \begin{bmatrix} \overset{3}{k} & \overset{5}{-k} \\ -k & k \end{bmatrix} \qquad
[\mathbf{k}^{(5)}] = \begin{array}{c} 5 \\ 4 \end{array} \begin{bmatrix} \overset{5}{k} & \overset{4}{-k} \\ -k & k \end{bmatrix} \qquad
[\mathbf{k}^{(6)}] = \begin{array}{c} 4 \\ 2 \end{array} \begin{bmatrix} \overset{4}{k} & \overset{2}{-k} \\ -k & k \end{bmatrix}
$$

Then, we initialize a 5 by 5 array to zero, and we label each row and column in order with the global degree of freedom numbers 1 through 5.

$$
\begin{array}{c} \\ 1 \\ 2 \\ 3 \\ 4 \\ 5 \end{array}
\begin{array}{c} \begin{array}{ccccc} 1 & 2 & 3 & 4 & 5 \end{array} \\
\begin{bmatrix} 0 & 0 & 0 & 0 & 0 \\ 0 & 0 & 0 & 0 & 0 \\ 0 & 0 & 0 & 0 & 0 \\ 0 & 0 & 0 & 0 & 0 \\ 0 & 0 & 0 & 0 & 0 \end{bmatrix} \end{array}
$$

Using the assembly rule and starting with spring element 1, we add the components of its stiffness matrix into the applicable locations of the global array. We then have

$$
\begin{array}{c@{\quad}ccccc}
 & 1 & 2 & 3 & 4 & 5 \\
\begin{array}{c} 1 \\ 2 \\ 3 \\ 4 \\ 5 \end{array} &
\left[\begin{array}{ccccc}
k & 0 & -k & 0 & 0 \\
0 & 0 & 0 & 0 & 0 \\
-k & 0 & k & 0 & 0 \\
0 & 0 & 0 & 0 & 0 \\
0 & 0 & 0 & 0 & 0
\end{array}\right]
\end{array}
$$

Adding the stiffness matrix of element 2 into this array yields

$$
\begin{array}{c@{\quad}ccccc}
 & 1 & 2 & 3 & 4 & 5 \\
\begin{array}{c} 1 \\ 2 \\ 3 \\ 4 \\ 5 \end{array} &
\left[\begin{array}{ccccc}
k & 0 & -k & 0 & 0 \\
0 & 0 & 0 & 0 & 0 \\
-k & 0 & 2k & -k & 0 \\
0 & 0 & -k & k & 0 \\
0 & 0 & 0 & 0 & 0
\end{array}\right]
\end{array}
$$

Continuing in this fashion, applying the assembly rule to elements 3 through 6, we finally obtain the global stiffness matrix of the structure in Figure 9.10.9, as follows:

$$
[\mathbf{k}] =
\begin{array}{c@{\quad}ccccc}
 & 1 & 2 & 3 & 4 & 5 \\
\begin{array}{c} 1 \\ 2 \\ 3 \\ 4 \\ 5 \end{array} &
\left[\begin{array}{ccccc}
k & 0 & -k & 0 & 0 \\
0 & k & 0 & -k & 0 \\
-k & 0 & 4k & -k & -2k \\
0 & -k & -k & 3k & -k \\
0 & 0 & -2k & -k & 3k
\end{array}\right]
\end{array}
$$

Observe that this matrix is symmetric, all of the principal diagonal entries are positive, and the rows and columns sum to zero.

In the next step, we apply the boundary conditions, that is, the prescribed loads and displacements. First, recall Equations 9.10.13,

$$
[\mathbf{k}_{ff}]\{\mathbf{q}_f\} = \{\mathbf{Q}_f\} - [\mathbf{k}_{fs}]\{\mathbf{q}_s\} \tag{a}
$$

$$
\{\mathbf{Q}_s\} = [\mathbf{k}_{sf}]\{\mathbf{q}_f\} + [\mathbf{k}_{ss}]\{\mathbf{q}_s\} \tag{b}
$$

Nodes 1 and 2 of the structure are constrained; nodes 3, 4, and 5 are free. Thus, for the set of free (f) and restrained (s) degrees of freedom, we have

$$
f: \quad 3, 4, 5
$$

$$
s: \quad 1, 2
$$

It follows that the partitioned stiffness matrix, the load vector, and the displacement vector in Equations a and b are as follows:

$$
[\mathbf{k}_{ff}] =
\begin{array}{c@{\ }c}
\begin{array}{c} 3\ \ 4\ \ 5 \end{array} & \\
\begin{array}{c} 3 \\ 4 \\ 5 \end{array} &
\left[\begin{array}{c} \mathbf{k} \end{array}\right]
\end{array}
= \left[\begin{array}{ccc}
4k & -k & -2k \\
-k & 3k & -k \\
-2k & -k & 3k
\end{array}\right]
\qquad
[\mathbf{k}_{fs}] =
\begin{array}{c@{\ }c}
\begin{array}{c} 1\ \ \ 2 \end{array} & \\
\begin{array}{c} 3 \\ 4 \\ 5 \end{array} &
\left[\begin{array}{c} \mathbf{k} \end{array}\right]
\end{array}
= \left[\begin{array}{cc}
-k & 0 \\
0 & -k \\
0 & 0
\end{array}\right]
$$

$$
[\mathbf{k}_{sf}] =
\begin{array}{c@{\ }c}
\begin{array}{c} 3\ \ 4\ \ 5 \end{array} & \\
\begin{array}{c} 1 \\ 2 \end{array} &
\left[\begin{array}{c} \mathbf{k} \end{array}\right]
\end{array}
= \left[\begin{array}{ccc}
-k & 0 & 0 \\
0 & -k & 0
\end{array}\right]
\qquad
[\mathbf{k}_{ss}] =
\begin{array}{c@{\ }c}
\begin{array}{c} 1\ \ \ 2 \end{array} & \\
\begin{array}{c} 1 \\ 2 \end{array} &
\left[\begin{array}{c} \mathbf{k} \end{array}\right]
\end{array}
= \left[\begin{array}{cc}
k & 0 \\
0 & k
\end{array}\right]
$$

$$\{Q_f\} = \begin{array}{c} 3 \\ 4 \\ 5 \end{array} \left\{ Q \right\} = \left\{ \begin{array}{c} 0 \\ 0 \\ P \end{array} \right\} \qquad \{Q_s\} = \begin{array}{c} 1 \\ 2 \end{array} \left\{ Q \right\} = \left\{ \begin{array}{c} X_1 \\ X_2 \end{array} \right\}$$

$$\{q_f\} = \begin{array}{c} 3 \\ 4 \\ 5 \end{array} \left\{ q \right\} = \left\{ \begin{array}{c} u_3 \\ u_4 \\ u_5 \end{array} \right\} \qquad \{q_s\} = \begin{array}{c} 1 \\ 2 \end{array} \left\{ q \right\} = \left\{ \begin{array}{c} 0 \\ 0 \end{array} \right\}$$

The vector $\{Q_f\}$ is the set of prescribed loads at the free nodes 3, 4, and 5. The displacements $\{q_f\}$ at those nodes are the unknowns of the problem. Vector $\{q_s\}$ is the set of prescribed displacements, which are zero at the supported nodes 1 and 2, while $\{Q_s\}$ is the set of loads at those supports, that is, the reactions. Substituting $[k_{ff}]$, $[k_{fs}]$, $\{q_f\}$, $\{q_s\}$, and $\{Q_f\}$ into Equation a, we get

$$\begin{bmatrix} 4k & -k & -2k \\ -k & 3k & -k \\ -2k & -k & 3k \end{bmatrix} \begin{Bmatrix} u_3 \\ u_4 \\ u_5 \end{Bmatrix} = \begin{Bmatrix} 0 \\ 0 \\ P \end{Bmatrix}$$

It is easy to verify that the solution of this system is

$$\{q_f\} = \frac{1}{13}\frac{P}{k} \begin{Bmatrix} 7 \\ 6 \\ 11 \end{Bmatrix}$$

To obtain the reactions, we substitute $\{q_f\}$, together with $[k_{sf}]$, $[k_{ss}]$, and $\{q_s\}$, into Equation b:

$$\begin{Bmatrix} X_1 \\ X_2 \end{Bmatrix} = \begin{bmatrix} -k & 0 & 0 \\ 0 & -k & 0 \end{bmatrix} \begin{Bmatrix} 7 \\ 6 \\ 11 \end{Bmatrix} \frac{1}{13}\frac{P}{k} + \begin{bmatrix} k & 0 \\ 0 & k \end{bmatrix} \begin{Bmatrix} 0 \\ 0 \end{Bmatrix}$$

which yields

$$\{Q_s\} = \left\{ \begin{array}{c} -\dfrac{7P}{13} \\ -\dfrac{6P}{13} \end{array} \right\}$$

To find the element loads, we substitute the appropriate global displacement components into each of the spring element stiffness equations, as follows:

$$\{Q^{(e)}\} = [k^{(e)}]\{q^{(e)}\} \qquad e = 1, \cdots, 6$$

Element 1:

$$\begin{Bmatrix} X_1^{(1)} \\ X_2^{(1)} \end{Bmatrix} = \begin{bmatrix} k & -k \\ -k & k \end{bmatrix} \begin{Bmatrix} 0 \\ u_3 \end{Bmatrix} = \begin{bmatrix} k & -k \\ -k & k \end{bmatrix} \begin{Bmatrix} 0 \\ \frac{7}{13}\frac{P}{k} \end{Bmatrix} = \left\{ \begin{array}{c} -\frac{7}{13}\frac{P}{k} \\ \frac{7}{13}\frac{P}{k} \end{array} \right\}$$

Element 2:

$$\begin{Bmatrix} X_1^{(2)} \\ X_2^{(2)} \end{Bmatrix} = \begin{bmatrix} k & -k \\ -k & k \end{bmatrix} \begin{Bmatrix} u_3 \\ u_4 \end{Bmatrix} = \begin{bmatrix} k & -k \\ -k & k \end{bmatrix} \begin{Bmatrix} \frac{7}{13}\frac{P}{k} \\ \frac{6}{13}\frac{P}{k} \end{Bmatrix} = \left\{ \begin{array}{c} \frac{1}{13}\frac{P}{k} \\ -\frac{1}{13}\frac{P}{k} \end{array} \right\}$$

Element 3:

$$\left\{ \begin{array}{c} X_1^{(3)} \\ X_2^{(3)} \end{array} \right\} = \left[\begin{array}{cc} k & -k \\ -k & k \end{array} \right] \left\{ \begin{array}{c} u_3 \\ u_5 \end{array} \right\} = \left[\begin{array}{cc} k & -k \\ -k & k \end{array} \right] \left\{ \begin{array}{c} \frac{7}{13}\frac{P}{k} \\ \frac{11}{13}\frac{P}{k} \end{array} \right\} = \left\{ \begin{array}{c} -\frac{4}{13}\frac{P}{k} \\ \frac{4}{13}\frac{P}{k} \end{array} \right\}$$

Element 4:

$$\left\{ \begin{array}{c} X_1^{(4)} \\ X_2^{(4)} \end{array} \right\} = \left[\begin{array}{cc} k & -k \\ -k & k \end{array} \right] \left\{ \begin{array}{c} u_3 \\ u_5 \end{array} \right\} = \left[\begin{array}{cc} k & -k \\ -k & k \end{array} \right] \left\{ \begin{array}{c} \frac{7}{13}\frac{P}{k} \\ \frac{11}{13}\frac{P}{k} \end{array} \right\} = \left\{ \begin{array}{c} -\frac{4}{13}\frac{P}{k} \\ \frac{4}{13}\frac{P}{k} \end{array} \right\}$$

Element 5:

$$\left\{ \begin{array}{c} X_1^{(5)} \\ X_2^{(5)} \end{array} \right\} = \left[\begin{array}{cc} k & -k \\ -k & k \end{array} \right] \left\{ \begin{array}{c} u_5 \\ u_4 \end{array} \right\} = \left[\begin{array}{cc} k & -k \\ -k & k \end{array} \right] \left\{ \begin{array}{c} \frac{11}{13}\frac{P}{k} \\ \frac{6}{13}\frac{P}{k} \end{array} \right\} = \left\{ \begin{array}{c} \frac{5}{13}\frac{P}{k} \\ -\frac{5}{13}\frac{P}{k} \end{array} \right\}$$

Element 6:

$$\left\{ \begin{array}{c} X_1^{(6)} \\ X_2^{(6)} \end{array} \right\} = \left[\begin{array}{cc} k & -k \\ -k & k \end{array} \right] \left\{ \begin{array}{c} u_4 \\ u_0 \end{array} \right\} = \left[\begin{array}{cc} k & -k \\ -k & k \end{array} \right] \left\{ \begin{array}{c} \frac{6}{13}\frac{P}{k} \\ 0 \end{array} \right\} = \left\{ \begin{array}{c} \frac{6}{13}\frac{P}{k} \\ -\frac{6}{13}\frac{P}{k} \end{array} \right\}$$

The results thus obtained are summarized in Figure 9.10.10, which shows the nodes and elements of the structure in Figure 9.10.9 as individual free bodies, together with the forces acting on them. Note that all of the forces are in equilibrium. Also, note that it did not matter that the structure was statically indeterminate.

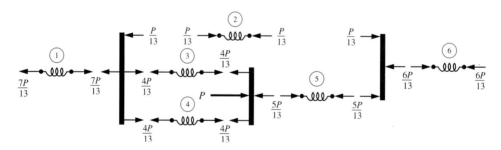

Figure 9.10.10 Computed forces within the colinear spring structure of Figure 9.10.9.

9.11 CONSTITUTIVE EQUATIONS IN MATRIX FORM

Up to this point, we have viewed stress and strain as tensors, expressed as 3 by 3 matrices in the manner of Equation 3.7.27. In matrix formalism, it is convenient to represent stress and engineering strain alternatively as column vectors $\{\sigma\}$ and $\{\varepsilon\}$, respectively. In three dimensions, the components of these vectors are the six components of stress and strain, listed in the order shown:

$$\{\boldsymbol{\sigma}\} = \begin{Bmatrix} \sigma_x \\ \sigma_y \\ \sigma_z \\ \tau_{xy} \\ \tau_{xz} \\ \tau_{yz} \end{Bmatrix} \qquad \{\boldsymbol{\varepsilon}\} = \begin{Bmatrix} \varepsilon_x \\ \varepsilon_y \\ \varepsilon_z \\ \gamma_{xy} \\ \gamma_{xz} \\ \gamma_{yz} \end{Bmatrix}$$

[9.11.1]

The relationships between stress and strain—the *constitutive equations*—for linearly elastic materials then have the compact form

$$\{\boldsymbol{\varepsilon}\} = [\mathbf{D}]\{\boldsymbol{\sigma}\} + \{\boldsymbol{\varepsilon}_T\}$$

[9.11.2]

and

$$\{\boldsymbol{\sigma}\} = [\mathbf{E}](\{\boldsymbol{\varepsilon}\} - \{\boldsymbol{\varepsilon}_T\})$$

[9.11.3]

The matrix $[\mathbf{D}]$ is the symmetric material flexibility (or *compliance*) matrix, and $[\mathbf{E}]$ is the symmetric material stiffness matrix. They are the inverse of one another, or

$$[\mathbf{E}] = [\mathbf{D}]^{-1}$$

[9.11.4]

Equations 3.11.4 and 3.11.5 show that for *isotropic* materials, we have

$$[\mathbf{D}] = \begin{bmatrix} \frac{1}{E} & -\frac{\nu}{E} & -\frac{\nu}{E} & 0 & 0 & 0 \\ -\frac{\nu}{E} & \frac{1}{E} & -\frac{\nu}{E} & 0 & 0 & 0 \\ -\frac{\nu}{E} & -\frac{\nu}{E} & \frac{1}{E} & 0 & 0 & 0 \\ 0 & 0 & 0 & \frac{2(1+\nu)}{E} & 0 & 0 \\ 0 & 0 & 0 & 0 & \frac{2(1+\nu)}{E} & 0 \\ 0 & 0 & 0 & 0 & 0 & \frac{2(1+\nu)}{E} \end{bmatrix}$$

[9.11.5]

and

$$[\mathbf{E}] = \begin{bmatrix} \frac{(1-\nu)E}{(1+\nu)(1-2\nu)} & \frac{\nu E}{(1+\nu)(1-2\nu)} & \frac{\nu E}{(1+\nu)(1-2\nu)} & 0 & 0 & 0 \\ \frac{\nu E}{(1+\nu)(1-2\nu)} & \frac{(1-\nu)E}{(1+\nu)(1-2\nu)} & \frac{\nu E}{(1+\nu)(1-2\nu)} & 0 & 0 & 0 \\ \frac{\nu E}{(1+\nu)(1-2\nu)} & \frac{\nu E}{(1+\nu)(1-2\nu)} & \frac{(1-\nu)E}{(1+\nu)(1-2\nu)} & 0 & 0 & 0 \\ 0 & 0 & 0 & \frac{E}{2(1+\nu)} & 0 & 0 \\ 0 & 0 & 0 & 0 & \frac{E}{2(1+\nu)} & 0 \\ 0 & 0 & 0 & 0 & 0 & \frac{E}{2(1+\nu)} \end{bmatrix}$$

[9.11.6]

The strain components in Equation 9.11.1$_2$ are related to the displacements by Equations 3.7.20. The components of $\{\boldsymbol{\varepsilon}\}_T$, the *thermal* strains, are proportional to the temperature change T relative to the reference state of zero strain. For isotropic materials, the thermal strain vector is obtained from the following constitutive relationship:

$$\{\boldsymbol{\varepsilon}_T\} = \lfloor \alpha T \quad \alpha T \quad \alpha T \quad 0 \quad 0 \quad 0 \rfloor^T$$

[9.11.7]

where α, the coefficient of thermal expansion, is a material property.

In plane stress, we have three independent nonzero stress and strain components, namely,

$$\{\boldsymbol{\sigma}\} = \begin{Bmatrix} \sigma_x \\ \sigma_y \\ \tau_{xy} \end{Bmatrix} \quad \{\boldsymbol{\varepsilon}\} = \begin{Bmatrix} \varepsilon_x \\ \varepsilon_y \\ \gamma_{xy} \end{Bmatrix} \tag{9.11.8}$$

For isotropic materials, we have

$$[\mathbf{D}] = \begin{bmatrix} \frac{1}{E} & -\frac{\nu}{E} & 0 \\ -\frac{\nu}{E} & \frac{1}{E} & 0 \\ 0 & 0 & \frac{2(1+\nu)}{E} \end{bmatrix} \quad [\mathbf{E}] = \begin{bmatrix} \frac{E}{1-\nu^2} & \frac{\nu E}{1-\nu^2} & 0 \\ \frac{\nu E}{1-\nu^2} & \frac{E}{1-\nu^2} & 0 \\ 0 & 0 & \frac{E}{2(1+\nu)} \end{bmatrix} \quad \{\boldsymbol{\varepsilon}_T\} = \begin{Bmatrix} \alpha T \\ \alpha T \\ 0 \end{Bmatrix} \tag{9.11.9}$$

The [**D**] and [**E**] matrices for anisotropic materials in plane stress may be inferred from section 3.17.

9.12 GENERAL FORM OF THE STIFFNESS EQUATIONS USING VIRTUAL WORK

In Chapter 6, we showed that the principle of virtual work is an exact, alternative statement of equilibrium. It can therefore be used to establish the stiffness equations of a structural element, instead of invoking equilibrium at the nodes, as was done for the spring element in the previous section. In fact, for two- and three-dimensional structural elements with several nodes, such as flat plates or curved shells, the proportionality constants relating the nodal loads and displacements cannot be readily inferred, because we have no simple, closed-form, exact solution for the stresses and strains within these arbitrarily shaped elements. We are forced to make reasonable simplifying assumptions about the internal stress or strain distribution, and then turn to virtual work methods (or their derived forms, the energy methods) to find the best element stiffness matrix consistent with our assumptions.

For simplicity, but without loss of generality, let us develop this procedure for an arbitrary two-dimensional, plane stress element with a total of N degrees of freedom (two per node), such as the one shown in Figure 9.12.1.

Let vector $\{\mathbf{q}\}$ represent the set of displacements in the degree of freedom directions, and let $\{\mathbf{Q}_P\}$ be the vector of external point loads acting at each degree of freedom. Then,

$$\{\mathbf{q}\} = \lfloor q_1 \quad q_2 \quad q_3 \quad \cdots \quad q_N \rfloor^T \qquad \{\mathbf{Q}_P\} = \lfloor Q_1 \quad Q_2 \quad Q_3 \quad \cdots \quad Q_N \rfloor^T$$

The displacements $\{\mathbf{u}\}$ *within* the element are assumed to be continuous functions of position throughout the element. The nature of this continuous variation in the element is expressed by the choice of *shape functions*,

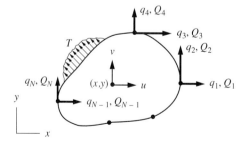

Figure 9.12.1　Planar element with N degrees of freedom, with surface traction and nodal point loads.

Body forces and a temperature field may also be present.

which relate the interior displacements to those at the nodes. In particular, for our two-dimensional, N degree of freedom element,

$$\{\mathbf{u}\} = [\mathbf{N}]\{\mathbf{q}\} \qquad [9.12.1]$$

where

$$\{\mathbf{u}\} = \begin{Bmatrix} u\,(x,\,y) \\ v\,(x,\,y) \end{Bmatrix}$$

which is the internal vector displacement field, and the 2 by N *shape function matrix* has the components

$$[\mathbf{N}] = \begin{bmatrix} N_{11}\,(x,\,y) & N_{12}\,(x,\,y) & N_{13}\,(x,\,y) & \cdot & \cdot & \cdot & N_{1N}\,(x,\,y) \\ N_{21}\,(x,\,y) & N_{22}\,(x,\,y) & N_{23}\,(x,\,y) & \cdot & \cdot & \cdot & N_{2N}\,(x,\,y) \end{bmatrix} \qquad [9.12.2]$$

Having expressed the displacements as functions of x and y in Equation 9.12.2, we can find the strains from the strain–displacement relations, Equations 3.7.20. Thus,

$$\varepsilon_x = \frac{\partial u}{\partial x} = \left\lfloor \frac{\partial N_{11}}{\partial x} \quad \frac{\partial N_{12}}{\partial x} \quad \frac{\partial N_{13}}{\partial x} \quad \cdots \quad \frac{\partial N_{1N}}{\partial x} \right\rfloor \{\mathbf{q}\}$$

$$\varepsilon_y = \frac{\partial v}{\partial y} = \left\lfloor \frac{\partial N_{21}}{\partial y} \quad \frac{\partial N_{22}}{\partial y} \quad \frac{\partial N_{23}}{\partial y} \quad \cdots \quad \frac{\partial N_{2N}}{\partial y} \right\rfloor \{\mathbf{q}\}$$

$$\gamma_{xy} = \frac{\partial u}{\partial y} + \frac{\partial v}{\partial x} = \left\lfloor \frac{\partial N_{11}}{\partial y} \quad \frac{\partial N_{12}}{\partial y} \quad \frac{\partial N_{13}}{\partial y} \quad \cdots \quad \frac{\partial N_{1N}}{\partial y} \right\rfloor \{\mathbf{q}\} + \left\lfloor \frac{\partial N_{21}}{\partial x} \quad \frac{\partial N_{22}}{\partial x} \quad \frac{\partial N_{23}}{\partial x} \quad \cdots \quad \frac{\partial N_{2N}}{\partial x} \right\rfloor \{\mathbf{q}\}$$

$$= \left\lfloor \left(\frac{\partial N_{11}}{\partial y} + \frac{\partial N_{21}}{\partial x} \right) \quad \left(\frac{\partial N_{12}}{\partial y} + \frac{\partial N_{22}}{\partial x} \right) \quad \cdots \quad \left(\frac{\partial N_{1N}}{\partial y} + \frac{\partial N_{2N}}{\partial x} \right) \right\rfloor \{\mathbf{q}\}$$

In terms of the 3 by N *strain–displacement matrix* $[\mathbf{B}]$, this is

$$\{\boldsymbol{\varepsilon}\} = [\mathbf{B}]\{\mathbf{q}\} \qquad [9.12.3]$$

where

$$\{\boldsymbol{\varepsilon}\} = \left\lfloor \varepsilon_x \quad \varepsilon_y \quad \gamma_{xy} \right\rfloor^T$$

and the components of $[\mathbf{B}]$ are

$$[\mathbf{B}] = \begin{bmatrix} \dfrac{\partial N_{11}}{\partial x} & \dfrac{\partial N_{12}}{\partial x} & \cdots & \dfrac{\partial N_{1N}}{\partial x} \\[2mm] \dfrac{\partial N_{21}}{\partial y} & \dfrac{\partial N_{22}}{\partial y} & \cdots & \dfrac{\partial N_{2N}}{\partial y} \\[2mm] \dfrac{\partial N_{11}}{\partial y} + \dfrac{\partial N_{21}}{\partial x} & \dfrac{\partial N_{12}}{\partial y} + \dfrac{\partial N_{22}}{\partial x} & \cdots & \dfrac{\partial N_{1N}}{\partial y} + \dfrac{\partial N_{2N}}{\partial x} \end{bmatrix} \qquad [9.12.4]$$

To find the element stresses $\{\boldsymbol{\sigma}\} = \left\lfloor \sigma_x \quad \sigma_y \quad \tau_{xy} \right\rfloor^T$, we substitute Equation 9.12.3 into the stress–strain relations $\{\boldsymbol{\sigma}\} = [\mathbf{E}](\{\boldsymbol{\varepsilon}\} - \{\boldsymbol{\varepsilon}_T\})$ to obtain

$$\{\boldsymbol{\sigma}\} = [\mathbf{E}][\mathbf{B}]\{\mathbf{q}\} - [\mathbf{E}]\{\boldsymbol{\varepsilon}_T\} \qquad [9.12.5]$$

These are stresses that accompany the displacements assumed in Equation 9.12.1. The thermal strains $\{\boldsymbol{\varepsilon}_T\}$ arise from temperature variations throughout the solid.

Since we *assumed* a form for both the displacement field in Equation 9.12.1 and the thermal strain field $\{\boldsymbol{\varepsilon}_T\}$, it is extremely unlikely that the stress field we compute from Equation 9.12.5 will satisfy equilibrium

(Equations 3.3.4). If it does, we will have assumed the *exact* solution! However, exact solutions for stress fields in elastic solids generally require extensive mathematical analysis and are not available for arbitrary shapes and boundary conditions. We accept the fact that, in general, an assumed displacement field is only an approximation, although it is selected so as to adequately represent the deformation of the solid body being modelled. The approximate displacement field is employed in the principle of virtual work to produce the set of equations that govern equilibrium *at the nodes* between the internal forces (the forces exerted on the nodes by the elements) and the externally applied forces. We are thus assured that the structural model reacts to equilibrate the loads applied to its nodes. The form of the assumed displacement field determines how the element forces are apportioned to the nodes.

According to Equation 6.3.3, the internal virtual work for the element using matrix notation is

$$\delta W_{\text{int}} = \iiint_V \{\delta \boldsymbol{\varepsilon}\}^T \{\boldsymbol{\sigma}\} dV \qquad [9.12.6]$$

where V is the volume of the solid. From Equation 9.12.3, the virtual strains are related to the virtual displacements by $\{\delta \boldsymbol{\varepsilon}\} = [\mathbf{B}]\{\delta \mathbf{q}\}$. Therefore, by the reversal rule, we have

$$\{\delta \boldsymbol{\varepsilon}\}^T = \{\delta \mathbf{q}\}^T [\mathbf{B}]^T$$

Using this, together with Equation 9.12.5, we can write the internal virtual work as

$$\delta W_{\text{int}} = \iiint_V \{\delta \mathbf{q}\}^T ([\mathbf{B}]^T [\mathbf{E}][\mathbf{B}]\{\mathbf{q}\} - [\mathbf{B}]^T [\mathbf{E}]\{\boldsymbol{\varepsilon}_T\}) dV$$

The displacement vectors are constants, so they can be taken out from under the integral sign to obtain

$$\delta W_{\text{int}} = \{\delta \mathbf{q}\}^T \left(\iiint_V [\mathbf{B}]^T [\mathbf{E}][\mathbf{B}] dV \right) \{\mathbf{q}\} - \{\delta \mathbf{q}\}^T \iiint_V [\mathbf{B}]^T [\mathbf{E}]\{\boldsymbol{\varepsilon}_T\} dV \qquad [9.12.7]$$

For the external virtual work, we can write Equation 6.3.5 in matrix notation, as follows:

$$\delta W_{\text{ext}} = \{\delta \mathbf{q}\}^T \{\mathbf{Q}\}_P + \iint_S \{\delta \mathbf{u}\}^T \{\mathbf{T}^{(n)}\} dS + \iiint_V \{\delta \mathbf{u}\}^T \{\mathbf{b}\} dV$$

The integrals on the right represent the virtual work of the point loads $\{\mathbf{Q}_P\}$, the surface tractions

$$\{\mathbf{T}^{(n)}\} = \lfloor T_x^{(n)} \quad T_y^{(n)} \rfloor^T$$

and the body forces

$$\{\mathbf{b}\} = \lfloor b_x \quad b_y \rfloor^T$$

We use Equation 9.12.1 to express the virtual displacement field in terms of virtual nodal displacements; that is,

$$\{\delta \mathbf{u}\} = [\mathbf{N}]\{\delta \mathbf{q}\}$$

It follows that

$$\{\delta \mathbf{u}\}^T = \{\delta \mathbf{q}\}^T [\mathbf{N}]^T \qquad [9.12.8]$$

For the total external virtual work, we can therefore write

$$\delta W_{\text{ext}} = \{\delta\mathbf{q}\}^T \{\mathbf{Q}\}_P + \{\delta\mathbf{q}\}^T \iint_S [\mathbf{N}]^T \{\mathbf{T}^{(n)}\} dS + \{\delta\mathbf{q}\}^T \iiint_V [\mathbf{N}]^T \{\mathbf{b}\} dV \qquad [9.12.9]$$

Now, according to the principle of virtual work, $\delta W_{\text{int}} = \delta W_{\text{ext}}$. Substituting Equations 9.12.7 and 9.12.9 and rearranging terms yields

$$\{\delta\mathbf{q}\}^T \left(\iiint_V [\mathbf{B}]^T [\mathbf{E}][\mathbf{B}] dV \right) \{\mathbf{q}\} =$$

$$\{\delta\mathbf{q}\}^T \{\mathbf{Q}_P\} + \{\delta\mathbf{q}\}^T \iint_S [\mathbf{N}]^T \{\mathbf{T}^{(n)}\} dS + \{\delta\mathbf{q}\}^T \iiint_V [\mathbf{N}]^T \{\mathbf{b}\} dV + \{\delta\mathbf{q}\}^T \iiint_V [\mathbf{B}]^T [\mathbf{E}] \{\boldsymbol{\varepsilon}_T\} dV \qquad [9.12.10]$$

This equality must hold for any arbitrary choice of the virtual nodal displacement vector $\{\delta\mathbf{q}\}$. Since none of the other factors in any of the terms depend on $\{\delta\mathbf{q}\}$, the only way to maintain the equality is to have the following be true (recall Equation 9.4.15):

$$\left(\iiint_V [\mathbf{B}]^T [\mathbf{E}][\mathbf{B}] dV \right) \{\mathbf{q}\} = \{\mathbf{Q}_P\} + \iint_S [\mathbf{N}]^T \{\mathbf{T}^{(n)}\} dS + \iiint_V [\mathbf{N}]^T \{\mathbf{b}\} dV + \iiint_V [\mathbf{B}]^T [\mathbf{E}] \{\boldsymbol{\varepsilon}_T\} dV \qquad [9.12.11]$$

The element load vector $\{\mathbf{Q}\}$ comprises *all* of the terms on the right side of this equation,

$$\{\mathbf{Q}\} = \{\mathbf{Q}_P\} + \{\mathbf{Q}_{\text{eq}}\} \qquad [9.12.12]$$

where the *equivalent point load vector* $\{\mathbf{Q}_{\text{eq}}\}$ is given by

$$\{\mathbf{Q}_{\text{eq}}\} = \iint_S [\mathbf{N}]^T \{\mathbf{T}^{(n)}\} dS + \iiint_V [\mathbf{N}]^T \{\mathbf{b}\} dV + \iiint_V [\mathbf{B}]^T [\mathbf{E}]\{\boldsymbol{\varepsilon}_T\} dV \qquad [9.12.13]$$

Thus, Equation 9.12.11 may be written

$$\left(\iiint_V [\mathbf{B}]^T [\mathbf{E}][\mathbf{B}] dV \right) \{\mathbf{q}\} = \{\mathbf{Q}\} \qquad [9.12.14]$$

or

$$[\mathbf{k}]\{\mathbf{q}\} = \{\mathbf{Q}\} \qquad [9.12.15]$$

where $[\mathbf{k}]$ is the stiffness matrix of the element:

$$[\mathbf{k}] = \iiint_V [\mathbf{B}]^T [\mathbf{E}][\mathbf{B}] dV \qquad [9.12.16]$$

The volume integrand on the right is an N by N matrix:

$$\underbrace{\underbrace{[\mathbf{B}]^T}_{N \text{ by } 3} \underbrace{[\mathbf{E}]}_{3 \text{ by } 3} \underbrace{[\mathbf{B}]}_{3 \text{ by } N}}_{\underbrace{N \text{ by } 3}_{N \text{ by } N}}$$

By comparison, the surface and volume integrals in Equation 9.12.13 are all N by 1, since $[\mathbf{N}]^T$ is N by 2 and both the surface traction vector $\{\mathbf{T}\}$ and body force density vector $\{\mathbf{b}\}$ are 2 by 1 column vectors. Furthermore, $[\mathbf{B}]^T [\mathbf{E}]$ is N by 3 and $\{\boldsymbol{\varepsilon}_T\}$ is a 3 by 1 column vector.

The principle of virtual work thus provides formulas for computing element stiffness matrices, Equation 9.12.16, and load vectors, Equation 9.12.13.

The development here was for a two-dimensional, plane stress element. However, the results, cast in matrix form, remain valid for one, two, or three dimensions and for elements with arbitrary stress–strain fields. The nature of a particular element will determine the size and contents of its matrices $[\mathbf{N}]$, $[\mathbf{B}]$, and $[\mathbf{E}]$, as well as the form and complexity of the integrals. In any case, Equation 9.12.15, the element stiffness equation, may be obtained by the same general procedure.

Observe that since the material stiffness matrix $[\mathbf{E}]$ in Equation 9.12.16 is always symmetric, that is, $[\mathbf{E}]^T = [\mathbf{E}]$, it follows that the stiffness matrix is symmetric, or

$$[[\mathbf{B}]^T [\mathbf{E}][\mathbf{B}]]^T = [\mathbf{B}]^T [\mathbf{E}]^T [[\mathbf{B}]^T]^T = [\mathbf{B}]^T [\mathbf{E}][\mathbf{B}]$$

a fact that was demonstrated previously in Equation 6.5.6.

Structures are usually assemblages of a number, and perhaps a variety, of smaller elements. Consistent with our previous notation, a superscript (e) will be attached to quantities associated with the elements of a structure. Those quantities without the superscript (e) will pertain to the structure as a whole.

The virtual work of a structure is the sum of the individual virtual works of all of its elements. The principle of virtual work, $\delta W_{\text{int}} = \delta W_{\text{ext}}$, may therefore be written

$$\sum_{\text{elements}} \delta W_{\text{int}}^{(e)} = \sum_{\text{elements}} \delta W_{\text{ext}}^{(e)}$$

where $\delta W_{\text{int}}^{(e)}$ and $\delta W_{\text{ext}}^{(e)}$ are the internal and external virtual works of element e. According to Equations 9.12.10, 9.12.12, and 9.12.13, this can be expressed as

$$\sum_{\text{elements}} \left\{\delta \mathbf{q}^{(e)}\right\}^T [\mathbf{k}^{(e)}]\{\mathbf{q}^{(e)}\} = \sum_{\text{elements}} \left\{\delta \mathbf{q}^{(e)}\right\}^T \left\{\mathbf{Q}^{(e)}\right\} \qquad \textbf{[9.12.17]}$$

The collection of all nodal displacements of a structure is the global displacement vector $\{\mathbf{q}\}$, which has N components, where N is the total number of global degrees of freedom. Each element's displacement vector $\{\mathbf{q}^{(e)}\}$ contains a few of the components of $\{\mathbf{q}\}$, depending on the nodes to which the element is attached. For example, the linear spring assembly in Figure 9.12.2 has five nodes, with one degree of freedom per node. So the global displacement vector has a total of five components, each of which is shown alongside the respective node. One of the six two-degree-of-freedom spring elements is pictured by itself in the figure, to show its own two local displacements, $q_1^{(4)}$ and $q_2^{(4)}$. Clearly, for displacement compatibility, $q_1^{(4)} = q_3$ and $q_2^{(4)} = q_5$.

In general, the displacement compatibility condition for each element of any structure can be written

$$\{\mathbf{q}^{(e)}\} = [\mathbf{\Omega}^{(e)}]\{\mathbf{q}\} \qquad e = 1, \cdots, \text{no. of elements} \qquad \textbf{[9.12.18]}$$

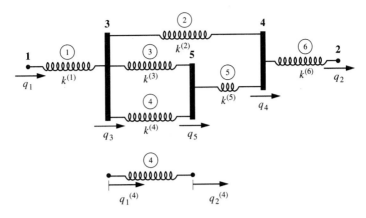

Figure 9.12.2   Linear spring assembly.

where $\left[\mathbf{\Omega}^{(e)}\right]$ is the *element connectivity matrix*. The number of rows in $\left[\mathbf{\Omega}^{(e)}\right]$ equals the number of *element* degrees of freedom, whereas the number of columns equals the number of *global* degrees of freedom in the structure. Each component of $\left[\mathbf{\Omega}^{(e)}\right]$ is either 0 or 1, according to the following definition:

$$\Omega_{\alpha K}^{(e)} = \begin{cases} 1 & \text{if degree of freedom } \alpha \text{ of element } e \text{ coincides with global degree of freedom } K \\ 0 & \text{if otherwise} \end{cases} \qquad \text{[9.12.19]}$$

Example 9.12.1   Write the element connectivity matrices for each of the springs in Figure 9.12.2.

In our convention, the leftmost node of a linear spring element is always element node 1. Since element 1 is attached to nodes 1 and 3 of the assembly, $q_1^{(1)} = q_1$ and $q_2^{(1)} = q_3$. We can write this as

$$\left\{ \begin{matrix} q_1^{(1)} \\ q_2^{(1)} \end{matrix} \right\} = \begin{bmatrix} 1 & 0 & 0 & 0 & 0 \\ 0 & 0 & 1 & 0 & 0 \end{bmatrix} \left\{ \begin{matrix} q_1 \\ q_2 \\ q_3 \\ q_4 \\ q_5 \end{matrix} \right\}$$

so that

$$\left[\mathbf{\Omega}^{(1)}\right] = \begin{bmatrix} 1 & 0 & 0 & 0 & 0 \\ 0 & 0 & 1 & 0 & 0 \end{bmatrix}$$

In a similar fashion, the reader should easily verify that

$$\left[\mathbf{\Omega}^{(2)}\right] = \begin{bmatrix} 0 & 0 & 1 & 0 & 0 \\ 0 & 0 & 0 & 1 & 0 \end{bmatrix}$$

$$\left[\mathbf{\Omega}^{(3)}\right] = \left[\mathbf{\Omega}^{(4)}\right] = \begin{bmatrix} 0 & 0 & 1 & 0 & 0 \\ 0 & 0 & 0 & 0 & 1 \end{bmatrix}$$

$$[\mathbf{\Omega}^{(5)}] = \begin{bmatrix} 0 & 0 & 0 & 0 & 1 \\ 0 & 0 & 0 & 1 & 0 \end{bmatrix}$$

$$[\mathbf{\Omega}^{(6)}] = \begin{bmatrix} 0 & 0 & 0 & 1 & 0 \\ 0 & 1 & 0 & 0 & 0 \end{bmatrix}$$

Aided by connectivity matrices, we can show how the principle of virtual work leads to the assembly rule. Substitute the compatibility condition, Equation 9.12.18, into the global virtual work expression, Equation 9.12.17, to obtain

$$\sum_{\text{elements}} \left\{ [\mathbf{\Omega}^{(e)}] \{\delta\mathbf{q}\} \right\}^T [\mathbf{k}^{(e)}] \left\{ [\mathbf{\Omega}^{(e)}] \{\mathbf{q}\} \right\} = \sum_{\text{elements}} \left\{ [\mathbf{\Omega}^{(e)}] \{\delta\mathbf{q}\} \right\}^T \left\{ \mathbf{Q}^{(e)} \right\}$$

Noting that $\{[\mathbf{\Omega}^{(e)}]\{\delta\mathbf{q}\}\}^T = \{\delta\mathbf{q}\}^T [\mathbf{\Omega}^{(e)}]^T$, we can rewrite this equation as

$$\{\delta\mathbf{q}\}^T \left\{ \sum_{\text{elements}} [\mathbf{\Omega}^{(e)}]^T [\mathbf{k}^{(e)}][\mathbf{\Omega}^{(e)}]\{\mathbf{q}\} \right\} = \{\delta\mathbf{q}\}^T \left\{ \sum_{\text{elements}} [\mathbf{\Omega}^{(e)}]^T \{\mathbf{Q}^{(e)}\} \right\}$$

The global virtual displacement vector $\{\delta\mathbf{q}\}$ is arbitrary. In the sense of Equation 9.4.15, then, we can cancel it from both sides of this equation, to obtain

$$\sum_{\text{elements}} [\mathbf{\Omega}^{(e)}]^T [\mathbf{k}^{(e)}][\mathbf{\Omega}^{(e)}]\{\mathbf{q}\} = \sum_{\text{elements}} [\mathbf{\Omega}^{(e)}]^T \{\mathbf{Q}^{(e)}\}$$

or

$$[\mathbf{k}]\{\mathbf{q}\} = \{\mathbf{Q}\} \tag{9.12.20}$$

where $[\mathbf{k}]$ is the global stiffness matrix,

$$[\mathbf{k}] = \sum_{\text{elements}} [\mathbf{\Omega}^{(e)}]^T [\mathbf{k}^{(e)}][\mathbf{\Omega}^{(e)}] \tag{9.12.21}$$

and $\{\mathbf{Q}\}$ is the global load vector,

$$\{\mathbf{Q}\} = \sum_{\text{elements}} [\mathbf{\Omega}^{(e)}]^T \{\mathbf{Q}^{(e)}\}$$

According to Equation 9.12.12, $\{\mathbf{Q}^{(e)}\} = \left\{ \mathbf{Q}_P^{(e)} \right\} + \left\{ \mathbf{Q}_{\text{eq}}^{(e)} \right\}$. Therefore,

$$\{\mathbf{Q}\} = \sum_{\text{elements}} [\mathbf{\Omega}^{(e)}]^T \left\{ \mathbf{Q}_P^{(e)} \right\} + \sum_{\text{elements}} [\mathbf{\Omega}^{(e)}]^T \left\{ \mathbf{Q}_{\text{eq}}^{(e)} \right\}$$

However, $\sum_{\text{elements}} [\mathbf{\Omega}^{(e)}]^T \{\mathbf{Q}_P^{(e)}\}$ is just the vector $\{\mathbf{Q}_P\}$ of external point loads applied directly to the nodes of the structure. Therefore, the global load vector $\{\mathbf{Q}\}$ for the entire structure is the sum of the point load vector and the assembled equivalent load vector; that is,

$$\{\mathbf{Q}\} = \{\mathbf{Q}_P\} + \left\{ \mathbf{Q}_{\text{eq}} \right\} \tag{9.12.22}$$

where

$$\{\mathbf{Q}_{eq}\} = \sum_{\text{elements}} [\mathbf{\Omega}^{(e)}]^T \{\mathbf{Q}_{eq}^{(e)}\}$$

[9.12.23]

Equations 9.12.21 and 9.12.23 show explicitly how the global stiffness matrix and the equivalent load vector of a structure are assembled from the element counterparts. Thus, they concisely depict the assembly rule.

Example 9.12.2   Use Equation 9.12.21 to form the global stiffness matrix of the linear spring array in Figure 9.12.2. Since there are six springs in the assembly, we have

$$[\mathbf{k}] = \sum_{\text{elements}} [\mathbf{\Omega}^{(e)}]^T [\mathbf{k}^{(e)}][\mathbf{\Omega}^{(e)}]$$

$$= [\mathbf{\Omega}^{(1)}]^T [\mathbf{k}^{(1)}][\mathbf{\Omega}^{(1)}] + [\mathbf{\Omega}^{(2)}]^T [\mathbf{k}^{(2)}][\mathbf{\Omega}^{(2)}] + [\mathbf{\Omega}^{(3)}]^T [\mathbf{k}^{(3)}][\mathbf{\Omega}^{(3)}]$$

$$+ [\mathbf{\Omega}^{(4)}]^T [\mathbf{k}^{(4)}][\mathbf{\Omega}^{(4)}] + [\mathbf{\Omega}^{(5)}]^T [\mathbf{k}^{(5)}][\mathbf{\Omega}^{(5)}] + [\mathbf{\Omega}^{(6)}]^T [\mathbf{k}^{(6)}][\mathbf{\Omega}^{(6)}]$$

Substituting the connectivity matrices found in Example 9.12.1 yields

$$[\mathbf{k}] = \begin{bmatrix} 1 & 0 \\ 0 & 0 \\ 0 & 1 \\ 0 & 0 \\ 0 & 0 \end{bmatrix} \begin{bmatrix} k^{(1)} & -k^{(1)} \\ -k^{(1)} & k^{(1)} \end{bmatrix} \begin{bmatrix} 1 & 0 & 0 & 0 & 0 \\ 0 & 0 & 1 & 0 & 0 \end{bmatrix} + \begin{bmatrix} 0 & 0 \\ 0 & 0 \\ 1 & 0 \\ 0 & 1 \\ 0 & 0 \end{bmatrix} \begin{bmatrix} k^{(2)} & -k^{(2)} \\ -k^{(2)} & k^{(2)} \end{bmatrix} \begin{bmatrix} 0 & 0 & 1 & 0 & 0 \\ 0 & 0 & 0 & 1 & 0 \end{bmatrix}$$

$$+ \begin{bmatrix} 0 & 0 \\ 0 & 0 \\ 1 & 0 \\ 0 & 0 \\ 0 & 1 \end{bmatrix} \begin{bmatrix} k^{(3)} & -k^{(3)} \\ -k^{(3)} & k^{(3)} \end{bmatrix} \begin{bmatrix} 0 & 0 & 1 & 0 & 0 \\ 0 & 0 & 0 & 0 & 1 \end{bmatrix} + \begin{bmatrix} 0 & 0 \\ 0 & 0 \\ 1 & 0 \\ 0 & 0 \\ 0 & 1 \end{bmatrix} \begin{bmatrix} k^{(4)} & -k^{(4)} \\ -k^{(4)} & k^{(4)} \end{bmatrix} \begin{bmatrix} 0 & 0 & 1 & 0 & 0 \\ 0 & 0 & 0 & 0 & 1 \end{bmatrix}$$

$$+ \begin{bmatrix} 0 & 0 \\ 0 & 0 \\ 0 & 0 \\ 0 & 1 \\ 1 & 0 \end{bmatrix} \begin{bmatrix} k^{(5)} & -k^{(5)} \\ -k^{(5)} & k^{(5)} \end{bmatrix} \begin{bmatrix} 0 & 0 & 0 & 0 & 1 \\ 0 & 0 & 0 & 1 & 0 \end{bmatrix} + \begin{bmatrix} 0 & 0 \\ 0 & 1 \\ 0 & 0 \\ 1 & 0 \\ 0 & 0 \end{bmatrix} \begin{bmatrix} k^{(6)} & -k^{(6)} \\ -k^{(6)} & k^{(6)} \end{bmatrix} \begin{bmatrix} 0 & 0 & 0 & 1 & 0 \\ 0 & 1 & 0 & 0 & 0 \end{bmatrix}$$

Each of the matrix multiplications yields a 6 by 6 matrix, so that we obtain the sum

$$[\mathbf{k}] = \begin{bmatrix} k^{(1)} & 0 & -k^{(1)} & 0 & 0 \\ 0 & 0 & 0 & 0 & 0 \\ -k^{(1)} & 0 & k^{(1)} & 0 & 0 \\ 0 & 0 & 0 & 0 & 0 \\ 0 & 0 & 0 & 0 & 0 \end{bmatrix} + \begin{bmatrix} 0 & 0 & 0 & 0 & 0 \\ 0 & 0 & 0 & 0 & 0 \\ 0 & 0 & k^{(2)} & -k^{(2)} & 0 \\ 0 & 0 & -k^{(2)} & k^{(2)} & 0 \\ 0 & 0 & 0 & 0 & 0 \end{bmatrix} + \begin{bmatrix} 0 & 0 & 0 & 0 & 0 \\ 0 & 0 & 0 & 0 & 0 \\ 0 & 0 & k^{(3)} & 0 & -k^{(3)} \\ 0 & 0 & 0 & 0 & 0 \\ 0 & 0 & -k^{(3)} & 0 & k^{(3)} \end{bmatrix}$$

$$+ \begin{bmatrix} 0 & 0 & 0 & 0 & 0 \\ 0 & 0 & 0 & 0 & 0 \\ 0 & 0 & k^{(4)} & 0 & -k^{(4)} \\ 0 & 0 & 0 & 0 & 0 \\ 0 & 0 & -k^{(4)} & 0 & k^{(4)} \end{bmatrix} + \begin{bmatrix} 0 & 0 & 0 & 0 & 0 \\ 0 & 0 & 0 & 0 & 0 \\ 0 & 0 & 0 & 0 & 0 \\ 0 & 0 & 0 & k^{(5)} & -k^{(5)} \\ 0 & 0 & 0 & -k^{(5)} & k^{(5)} \end{bmatrix} + \begin{bmatrix} 0 & 0 & 0 & 0 & 0 \\ 0 & k^{(6)} & 0 & -k^{(6)} & 0 \\ 0 & 0 & 0 & 0 & 0 \\ 0 & -k^{(6)} & 0 & k^{(6)} & 0 \\ 0 & 0 & 0 & 0 & 0 \end{bmatrix}$$

It follows that

$$[\mathbf{k}] = \begin{bmatrix} k^{(1)} & 0 & -k^{(1)} & 0 & 0 \\ 0 & k^{(6)} & 0 & -k^{(6)} & 0 \\ -k^{(1)} & 0 & k^{(1)} + k^{(2)} + k^{(3)} & -k^{(2)} & -k^{(3)} \\ 0 & -k^{(6)} & -k^{(2)} & k^{(2)} + k^{(5)} + k^{(6)} & -k^{(5)} \\ 0 & 0 & -k^{(3)} & -k^{(5)} & k^{(3)} + k^{(5)} \end{bmatrix}$$

The reader should check that this same result is obtained, more quickly and conveniently, by rote implementation of the assembly procedure explained in section 9.10.

Although connectivity matrices are not needed for assembling stiffness matrices and load vectors, in actual practice they do provide a convenient means of mathematically expressing the assembly rule.

EXERCISES

Do each problem by hand, showing all intermediate steps.

9.1 Given

$$[\mathbf{A}] = \begin{bmatrix} 1 & -4 & 9 \\ 2 & 5 & 8 \\ 4 & -3 & 10 \end{bmatrix} \quad [\mathbf{B}] = \begin{bmatrix} 3 & 4 & 5 \\ 2 & -6 & 4 \\ 1 & 3 & 7 \end{bmatrix} \quad [\mathbf{C}] = \begin{bmatrix} -7 & 6 \\ 5 & 8 \\ 9 & -4 \end{bmatrix} \quad \{\mathbf{u}\} = \begin{Bmatrix} 1 \\ -2 \\ 3 \end{Bmatrix} \quad \{\mathbf{v}\} = \begin{Bmatrix} -1 \\ 4 \\ 2 \end{Bmatrix}$$

perform the following matrix operations:

a. $[\mathbf{A}] + [\mathbf{B}]$ b. $[\mathbf{B}][\mathbf{A}]$ c. $[\mathbf{A}]^T [\mathbf{B}]^T$ d. $[\mathbf{A}][\mathbf{C}]$ e. $\{\mathbf{v}\}^T \{\mathbf{u}\}$ f. $\{\mathbf{u}\}\{\mathbf{v}\}^T$

g. $\{\mathbf{u}\}^T [\mathbf{A}]\{\mathbf{u}\}$ h. $[\mathbf{C}]^T [\mathbf{C}]$ i. $[\mathbf{C}][\mathbf{C}]^T$ j. $\det[\mathbf{A}]$ k. $\det[\mathbf{B}]$ l. $\det([\mathbf{A}][\mathbf{B}])$

9.2 Solve the following systems using Gauss elimination. Check your work by substituting your solutions back into the original equations.

a.
$$x_1 + 3x_2 - 2x_3 = -10$$
$$2x_1 - 4x_2 + 5x_3 = 5$$
$$-3x_1 - 2x_2 - 4x_3 = -6$$

b.
$$5x - 6y + 4z = 26$$
$$-6x + 6y - 2z = 13$$
$$4x - 2y + z = 52$$

9.3 Calculate the determinant of the matrix shown.

$$\begin{bmatrix} -3 & 1 & 4 & 8 & 6 & 9 \\ 0 & 2 & 1 & -3 & 5 & 8 \\ 0 & 0 & 9 & -2 & 7 & -4 \\ 0 & 0 & 0 & 4 & 1 & 9 \\ 0 & 0 & 0 & 0 & 5 & 2 \\ 0 & 0 & 0 & 0 & 0 & 6 \end{bmatrix}$$

9.4   Calculate the determinant of the following matrix by:

a. Expanding by cofactors.

b. Reducing the matrix to upper triangular form.

$$\begin{bmatrix} 4 & 3 & 2 & 1 \\ 3 & 4 & 3 & 2 \\ 2 & 3 & 4 & 3 \\ 1 & 2 & 3 & 4 \end{bmatrix}$$

9.5   Invert the following matrix by dividing the adjoint by the determinant (Equation 9.5.11):

$$\begin{bmatrix} 2 & 2 & 1 \\ 1 & 6 & 3 \\ 0 & 2 & 5 \end{bmatrix}$$

9.6   Calculate the inverse of the following matrices by two different methods.

a. Divide the adjoint matrix by the determinant.

b. Use Gauss elimination.

$$\begin{bmatrix} -1 & 1 & 1 \\ 1 & 0 & 2 \\ 1 & 1 & 1 \end{bmatrix} \qquad \begin{bmatrix} 3 & 2 & 1 \\ 2 & 3 & 2 \\ 1 & 2 & 3 \end{bmatrix} \qquad \begin{bmatrix} 1 & 0 & 1 & 0 \\ 1 & 1 & 0 & 1 \\ 0 & 1 & 1 & 0 \\ 1 & 0 & 1 & 1 \end{bmatrix}$$

9.7   Solve the following system by hand calculation, using Cramer's rule.

$$x_1 + 3x_2 - 2x_3 = -10$$
$$2x_1 - 4x_2 + 5x_3 = 5$$
$$-3x_1 - 2x_2 - 4x_3 = -6$$

9.8   Given the matrices

$$[\mathbf{A}] = \begin{bmatrix} 1 & -1 & 1 \\ 2 & 1 & -1 \end{bmatrix} \quad [\mathbf{B}] = \begin{bmatrix} 1 & 3 & 1 \\ 3 & 2 & 3 \\ 1 & 3 & 2 \end{bmatrix} \quad [\mathbf{C}] = \begin{bmatrix} 4 & 2 \\ 1 & -2 \\ -1 & 3 \end{bmatrix} \quad \{\mathbf{d}\} = \begin{Bmatrix} 1 \\ 2 \end{Bmatrix}$$

calculate by hand the product $[\mathbf{A}][\mathbf{B}]^{-1}[\mathbf{C}]\{\mathbf{d}\}$, without calculating the inverse of $[\mathbf{B}]$.

9.9   Calculate the eigenvalues and eigenvectors of the matrix shown.

$$\begin{bmatrix} 10 & 2 & 1 \\ 2 & 10 & 1 \\ 2 & 1 & 10 \end{bmatrix}$$

9.10   Find the eigenvalues and eigenvectors of the following symmetric matrices. In each case, show that the eigenvectors are orthogonal.

$$[\mathbf{A}] = \begin{bmatrix} 1 & -6 & 4 \\ -6 & 2 & -2 \\ 4 & -2 & -3 \end{bmatrix} \quad [\mathbf{B}] = \begin{bmatrix} 3 & 1 & 5 \\ 1 & 3 & 5 \\ 5 & 5 & -1 \end{bmatrix}$$

9.11 Calculate the 6 by 6 stiffness matrix [**k**] of the colinear spring assembly.

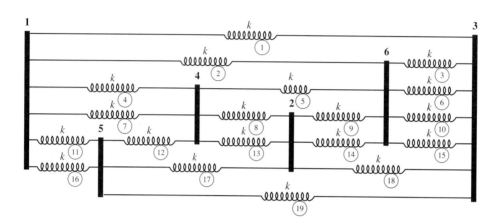

Exercise 9.11

9.12 If the structure in (a) is constrained as shown in (b), what is $\left[\mathbf{k}_{ff}\right]$?

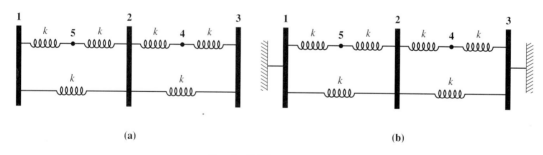

(a)   (b)

Exercise 9.12

9.13   Find the terms in row 3 of the stiffness matrix of the colinear spring structure shown. The number by each spring is its spring constant.

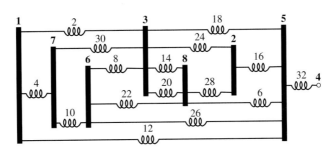

Exercise 9.13

9.14   For the one-dimensional spring assembly shown, find:

a. The stiffness matrix of the unconstrained structure.

b. The constrained stiffness matrix $\left[\mathbf{k}_{ff}\right]$ if node 2 is fixed.

c. The displacements of the nodes and the force in each spring if node 2 is fixed and the loading is as shown.

d. The displacements at the nodes and the force in each spring if the loading is as shown and node 2 is given a 5 mm displacement to the right.

e. The nodal displacements and the force in each spring if the loading is as shown and node 2 is fixed while node 1 is displaced 5 mm to the left.

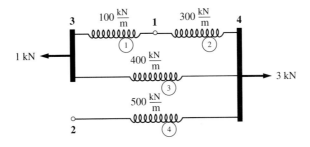

Exercise 9.14

9.15   The stiffnesses of the springs in the one-dimensional spring assembly shown are all multiples of k. An external load P is applied to the right at node 2. Using the given node numbering, find the force in each spring in terms of P. Check your answers by seeing that each node of the structure is in equilibrium.

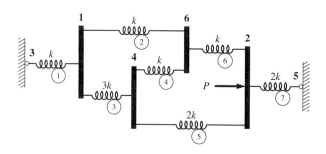

Exercise 9.15

9.16   Verify the stiffness matrix for the structure shown in Figure 9.10.4.

9.17   Verify the stiffness matrix for the structure shown in Figure 9.10.5.

9.18   Find the nodal displacements, element forces, and support reactions for the colinear spring assembly if node 4 is displaced d to the right. What external force is required to displace the node?

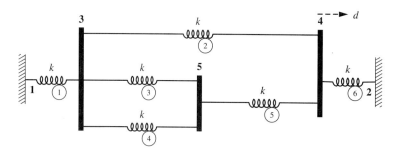

Exercise 9.18

9.19   A vector has components $\{v\}^T = \lfloor 1 \quad 2 \quad 3 \rfloor$ in the unbarred system. Find the components $\{\bar{v}\}$ of the vector in the barred system.

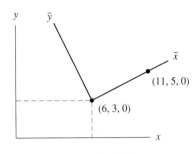

Exercise 9.19

9.20   The right-handed, barred $\bar{x}\,\bar{y}\,\bar{z}$ system is defined by the three points A, B, and C. The $\bar{x}\,\bar{y}$ plane is defined by the plane ABC. The $\bar{x}$ axis runs from A through B. The $\bar{z}$ axis is defined by the cross product of $\overrightarrow{AB}$ into $\overrightarrow{AC}$, so that the $+\bar{y}$ axis lies on the same side of the $\bar{x}$ axis as point C.

a. Find the orthogonal transformation matrix $[\mathbf{R}]$ relating the two coordinate systems.

b. If a vector in the barred system is $\{\bar{v}\}^T = \lfloor 2 \quad -1 \quad 3 \rfloor$, find the components of $\{v\}$ in the unbarred system.

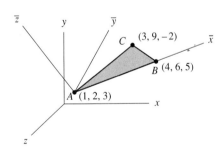

Exercise 9.20

Matrix Displacement Method:
Trusses, Beams, and Frames

CHAPTER OUTLINE

10.1   INTRODUCTION

In this chapter, the matrix displacement method formulated in Chapter 9 is applied to the analysis of the skeletal structures considered previously in the context of the force method (Chapter 7). We will use matrix notation and the formalized procedures of Chapter 9 to set up the stiffness matrix of a given structure, apply the point loads and distributed loads, enforce displacement constraints, and solve for the resulting displacements and internal loads. Detailed examples will be used to illustrate the procedure for problems that are large enough to be instructive, yet small enough to be solved using hand calculations. Both two- and three-dimensional structures will be considered, and the means of dealing with distributed loads, including thermal strain and inertia effects, will be explained.

Computer implementations of the displacement method abound, and many books contain applicable examples of algorithms and software, as well as references to the original literature.[1]

10.2   ROD ELEMENT STIFFNESS EQUATIONS

Let us apply the methods of section 9.11 to the constant-area, isotropic, homogeneous rod element of Figure 10.2.1. We wish to establish the stiffness equations required for the matrix displacement approach to truss analysis. First we will obtain the local form of the stiffness matrix $[\bar{k}]$, which relates the element load vector $\{\bar{Q}\}$ to the displacement vector $\{\bar{q}\}$, where

$$\{\bar{q}\} = \begin{Bmatrix} \bar{u}_1 \\ \bar{u}_2 \end{Bmatrix} \qquad \{\bar{Q}\} = \begin{Bmatrix} \bar{X}_1 \\ \bar{X}_2 \end{Bmatrix} \qquad\qquad [10.2.1]$$

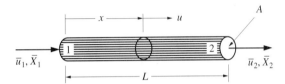

Figure 10.2.1   Elastic cylindrical rod element in its local coordinate system.

The components of $\{\bar{q}\}$ and $\{\bar{Q}\}$ lie along the axis of the rod. Since the rod is, by definition, a two-force member, there are no off-axis components of the load vector. On the other hand, there may be transverse components of displacement at either node. If so, however, they are components of rigid-body displacement and induce no axial strain. They are therefore not a factor in determining the stiffness properties of the rod.

Assume that the axial displacement u within the rod is a linear function of x, measured from local node 1 at the left end. That is,

$$u(x) = c_0 + c_1 x$$

This function must yield the endpoint displacements when $x = 0$ and $x = L$. Thus, $u(0) = \bar{u}_1$ implies that $c_0 = \bar{u}_1$. This, together with $u(L) = \bar{u}_2$, means that $c_1 = (\bar{u}_2 - \bar{u}_1)/L$. It follows that

$$u = \left(1 - \frac{x}{L}\right)\bar{u}_1 + \frac{x}{L}\bar{u}_2 \qquad\qquad [10.2.2]$$

or

$$u = \left\lfloor 1 - \frac{x}{L} \quad \frac{x}{L} \right\rfloor \begin{Bmatrix} \bar{u}_1 \\ \bar{u}_2 \end{Bmatrix}$$

Therefore, the shape function matrix $[\mathbf{N}]$ for the rod is the two-component row vector

[1] See, for example, T. Y. Yang, *Finite Element Structural Analysis*, Englewood Cliffs, NJ, Prentice-Hall, 1986; R. D. Cook, D. S. Malkus, and M. E. Plesha, *Concepts and Applications of Finite Element Analysis*, New York, Wiley, 1989; W. B. Bickford, *A First Course in the Finite Element Method*, Burr Ridge, IL, Irwin, 1994.

$$[\mathbf{N}] = \left\lfloor 1 - \frac{x}{L} \quad \frac{x}{L} \right\rfloor \qquad [10.2.3]$$

To find the normal strain in the rod, we differentiate the displacement field in Equation 10.2.2 with respect to x, as follows:

$$\varepsilon_x = \frac{\partial u}{\partial x} = \left\lfloor -\frac{1}{L} \quad \frac{1}{L} \right\rfloor \left\{ \begin{array}{c} \bar{u}_1 \\ \bar{u}_2 \end{array} \right\} \qquad [10.2.4]$$

Observe that the linear displacement assumption implies a state of uniform strain, and therefore a constant stress throughout the rod. From Equation 10.2.4, we identify the strain–displacement matrix $[\mathbf{B}]$ in this case to be the row vector

$$[\mathbf{B}] = \left\lfloor -\frac{1}{L} \quad \frac{1}{L} \right\rfloor$$

The stress–strain relationship for the isotropic rod, including thermal strain, is

$$\sigma_x = E(\varepsilon_x - \alpha T) \qquad [10.2.5]$$

so that the stress–strain matrix $[\mathbf{E}]$ is the scalar E, or

$$[\mathbf{E}] = E$$

The element stiffness matrix $[\bar{\mathbf{k}}]$ is found by substituting $\lfloor \mathbf{B} \rfloor$ and $[\mathbf{E}]$ into Equation 9.12.16:

$$[\bar{\mathbf{k}}] = \iiint_V [\mathbf{B}]^T E \lfloor \mathbf{B} \rfloor \, dV = \iiint_V \left\{ \begin{array}{c} -\frac{1}{L} \\ \frac{1}{L} \end{array} \right\} E \left\lfloor -\frac{1}{L} \quad \frac{1}{L} \right\rfloor dV = \iiint_V E \left[\begin{array}{cc} \frac{1}{L^2} & -\frac{1}{L^2} \\ -\frac{1}{L^2} & \frac{1}{L^2} \end{array} \right] dV = \frac{E}{L^2} \left[\begin{array}{cc} 1 & -1 \\ -1 & 1 \end{array} \right] \iiint_V dV \qquad [10.2.6]$$

However, for the cylindrical rod, we have $\iiint_V dV = AL$. Therefore, the rod element's local stiffness matrix is

$$[\bar{\mathbf{k}}] = \frac{AE}{L} \left[\begin{array}{cc} 1 & -1 \\ -1 & 1 \end{array} \right] \qquad [10.2.7]$$

Let us next determine the global stiffness matrix of the rod in two dimensions. Figure 10.2.2 shows a rod element in the plane of, and inclined with respect to, the global xy axes. From the perspective of the global two-dimensional system, the element has two degrees of freedom at each node, as illustrated in Figure 10.2.3.

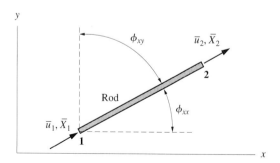

Figure 10.2.2 Planar rod element, with local degrees of freedom shown at each node.

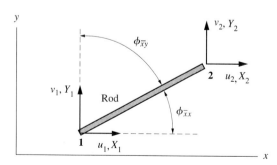

Figure 10.2.3 Rod element, with global degrees of freedom shown at each node.

To transform the displacement and load components from the element system to the global system, we need the orthogonal coordinate transformation matrix $[\mathbf{R}]$ presented in section 9.9:

$$[\mathbf{R}] = \begin{bmatrix} l_1 & m_1 & n_1 \\ l_2 & m_2 & n_2 \\ l_3 & m_3 & n_3 \end{bmatrix} \qquad\qquad \textbf{[10.2.8]}$$

Recall that the components of rows 1, 2, and 3 of $[\mathbf{R}]$ are the direction cosines of the element's $\bar{x}$, $\bar{y}$, and $\bar{z}$ axes, respectively, with respect to the global frame. Since we are focusing for now on planar systems, the xy axes of both the local and global systems are coplanar, so that their z axes are colinear. Therefore, $\phi_{\bar{x}z} = \phi_{\bar{y}z} = \phi_{\bar{z}x} = \phi_{\bar{z}y} = 90°$, which means that $l_3 = m_3 = n_1 = n_2 = 0$. Also, since $\phi_{\bar{z}z} = 0$, then $n_3 = 1$. It follows that in the planar case, $[\mathbf{R}]$ reduces to

$$[\mathbf{R}] = \begin{bmatrix} l_1 & m_1 & 0 \\ l_2 & m_2 & 0 \\ 0 & 0 & 1 \end{bmatrix} \qquad\qquad \textbf{[10.2.9]}$$

Recall from Equation 9.9.6 that the components of a vector $\{\bar{v}\}$ in the element system are related to those in the global system by

$$\{\bar{v}\} = [\mathbf{R}]\{v\} \qquad\qquad \textbf{[10.2.10a]}$$

or, alternatively,

$$\{v\} = [\mathbf{R}]^T \{\bar{v}\} \qquad\qquad \textbf{[10.2.10b]}$$

As long as we deal with planar systems with only translational degrees of freedom, the z component of all required vectors will be zero. Thus, in two dimensions, we can eliminate the third row and column of $[\mathbf{R}]$ in Equation 10.2.9, leaving a 2 by 2 orthogonal matrix for in-plane vector transformations, as follows:

$$[\mathbf{R}] = \begin{bmatrix} l_1 & m_1 \\ l_2 & m_2 \end{bmatrix} \qquad\qquad \textbf{[10.2.11]}$$

Also, Equations 10.2.10 remain valid for two-dimensional operations.

To obtain the components of nodal displacement in the element system, we simply project the global displacements at each node of the rod onto the element $\bar{x}$ axis, which has the direction cosines l_1 and m_1:

$$\bar{u}_1 = \lfloor l_1 \quad m_1 \rfloor \begin{Bmatrix} u_1 \\ v_1 \end{Bmatrix}$$

$$\bar{u}_2 = \lfloor l_1 \quad m_1 \rfloor \begin{Bmatrix} u_2 \\ v_2 \end{Bmatrix} \qquad\qquad \textbf{[10.2.12]}$$

We combine all of the nodal displacements into the single element displacement vector. The components of the vector are listed in order of increasing node number and, at each node, in order of degree of freedom, with the x freedom listed first, the y freedom listed second, and so on. Thus, we have the local rod element displacement vector, as given in Equation 10.2.1, and the global rod element displacement vector, as follows:

$$\{\mathbf{q}\} = \lfloor u_1 \quad v_1 \mathbin{\vdots} u_2 \quad v_2 \rfloor^T$$

Using this notation, we can write Equations 10.2.12 as

$$\left\{ \begin{array}{c} \bar{u}_1 \\ \bar{u}_2 \end{array} \right\} = \left[\begin{array}{cc:cc} l_1 & m_1 & 0 & 0 \\ 0 & 0 & l_1 & m_1 \end{array} \right] \left\{ \begin{array}{c} u_1 \\ v_1 \\ u_2 \\ v_2 \end{array} \right\} \tag{10.2.13}$$

or more compactly,

$$\{\bar{\mathbf{q}}\} = [\mathbf{\Lambda}]\{\mathbf{q}\} \tag{10.2.14}$$

where

$$[\mathbf{\Lambda}] = \left[\begin{array}{cc:cc} l_1 & m_1 & 0 & 0 \\ 0 & 0 & l_1 & m_1 \end{array} \right] \tag{10.2.15}$$

The resultant of the loads shown acting on each end of the rod in Figure 10.2.3 must lie completely on the axis of the rod, as illustrated in Figure 10.2.2. Therefore, we can calculate the global load components from those in the element system by applying Equations 10.2.10b and 10.2.11 to the load at node 1. The result is

$$\left\{ \begin{array}{c} X_1 \\ Y_1 \end{array} \right\} = \left[\begin{array}{cc} l_1 & l_2 \\ m_1 & m_2 \end{array} \right] \left\{ \begin{array}{c} \bar{X}_1 \\ 0 \end{array} \right\} = \bar{X}_1 \left\{ \begin{array}{c} l_1 \\ m_1 \end{array} \right\}$$

Likewise, at node 2, we have

$$\left\{ \begin{array}{c} X_2 \\ Y_2 \end{array} \right\} = \left[\begin{array}{cc} l_1 & l_2 \\ m_1 & m_2 \end{array} \right] \left\{ \begin{array}{c} \bar{X}_2 \\ 0 \end{array} \right\} = \bar{X}_2 \left\{ \begin{array}{c} l_1 \\ m_1 \end{array} \right\}$$

These can be written as the single equation

$$\{\mathbf{Q}\} = [\mathbf{\Lambda}]^T \{\bar{\mathbf{Q}}\} \tag{10.2.16}$$

where the local rod element load vector is found in Equation 10.2.1 and the global rod element load vector is

$$\{\mathbf{Q}\} = \lfloor X_1 \quad Y_1 \quad X_2 \quad Y_2 \rfloor^T$$

We are now in a position to find the global form of the rod element stiffness matrix. First, we substitute the local stiffness equations $\{\bar{\mathbf{Q}}\} = [\bar{\mathbf{k}}]\{\bar{\mathbf{q}}\}$ into the right side of Equation 10.2.16 to obtain

$$\{\mathbf{Q}\} = [\mathbf{\Lambda}]^T [\bar{\mathbf{k}}]\{\bar{\mathbf{q}}\}$$

Then, we substitute for $\{\bar{\mathbf{q}}\}$ on the right, using Equation 10.2.14:

$$\{\mathbf{Q}\} = [\mathbf{\Lambda}]^T [\bar{\mathbf{k}}][\mathbf{\Lambda}]\{\mathbf{q}\} \tag{10.2.17}$$

This is a linear relationship between the global load vector $\{\mathbf{Q}\}$ and the global displacement vector $\{\mathbf{q}\}$. The matrix of coefficients in such an expression is, by definition, the global stiffness matrix $[\mathbf{k}]$ of the element, that is,

$$\{\mathbf{Q}\} = [\mathbf{k}]\{\mathbf{q}\}$$

Comparing this with Equation 10.2.17, we see that

$$[\mathbf{k}] = [\mathbf{\Lambda}]^T [\bar{\mathbf{k}}][\mathbf{\Lambda}] \tag{10.2.18}$$

Equation 10.2.18 is the formula for transforming the local stiffness matrix in Equation 10.2.6 into its global coordinates, employing the rectangular transformation matrix in Equation 10.2.15. Although we derived the formula here for the special case of a rod element, as a matrix equation, Equation 10.2.18 is completely general and can be used to find the global stiffness matrix of any structural element, given its local form. The size and composition of the transformation matrix $[\Lambda]$ will vary, depending on the number of element nodes, as well as on the number and type of degrees of freedom at each node.

Before carrying out the matrix operations in Equation 10.2.18, observe that only the direction cosines, l_1 and m_1, of the rod element centerline appear in $[\Lambda]$. Therefore, we can drop the subscript 1 and proceed with the computations to find

$$[\mathbf{k}] = [\Lambda]^T \, [\bar{\mathbf{k}}] \, [\Lambda] = \begin{bmatrix} l & 0 \\ m & 0 \\ 0 & l \\ 0 & m \end{bmatrix} \frac{AE}{L} \begin{bmatrix} 1 & -1 \\ -1 & 1 \end{bmatrix} \begin{bmatrix} l & m & 0 & 0 \\ 0 & 0 & l & m \end{bmatrix} = \frac{AE}{L} \begin{bmatrix} l & -l \\ m & -m \\ -l & l \\ -m & m \end{bmatrix} \begin{bmatrix} l & m & 0 & 0 \\ 0 & 0 & l & m \end{bmatrix}$$

We thus obtain the global stiffness matrix of a rod element in two dimensions:

$$[\mathbf{k}] = \frac{AE}{L} \begin{array}{c} \begin{matrix} u_1 & \quad v_1 & \quad u_2 & \quad v_2 \end{matrix} \\ \begin{bmatrix} l^2 & lm & -l^2 & -lm \\ lm & m^2 & -lm & -m^2 \\ -l^2 & -lm & l^2 & lm \\ -lm & -m^2 & lm & m^2 \end{bmatrix} \end{array} \begin{matrix} X_1 \\ Y_1 \\ X_2 \\ Y_2 \end{matrix} \qquad \textbf{[10.2.19]}$$

where l and m are the global direction cosines of the axis of the rod, which is always understood to be directed *from* element node 1 *towards* element node 2.

Example 10.2.1   Find the nodal displacements, rod element loads, and support reactions for the plane truss structure loaded and constrained as shown in Figure 10.2.4.

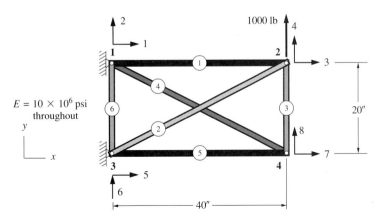

Figure 10.2.4   Statically indeterminate plane truss, with the global degrees of freedom shown alongside each node.

The truss is composed of rod elements, numbered from 1 to 6, and four nodes. The node numbering shown is one of the 24 possible schemes we could choose. Alongside each node are the two associated global degrees of freedom. Note that the degree of freedom numbering is *not* arbitrary; it is fixed, once the nodes have been numbered. The first two freedoms are always associated with node 1, the second two with node 2, and so forth, with the last two belonging to the highest-numbered node. In other words, the node i degrees of freedom are $2i - 1$ and $2i$, the first of which is in the x direction. The unconstrained truss, without the supports at nodes 3 and 4, is an eight degree of freedom structure, so its stiffness matrix is 8 by 8. With the supports, the degrees of freedom at nodes 1 and 3 are eliminated, leaving a 4 by 4 constrained stiffness matrix $[\mathbf{k}_{ff}]$.

The first step of the solution is to write the global stiffness matrix of each element, using Equation 10.2.19. Before doing so, we must specify the direction cosines of the rods. Either end of each rod may be chosen as its local node 1. Then, a table of connectivities is set up, listing the global node numbers of each element's first and second node. The orientation of each element is given by the angle ϕ measured positive counterclockwise around local node 1 from the positive global axis x to the positive element axis $\bar{x}$. The length of each element is obtained from the geometry of the structure. The cross-sectional area of each rod must be provided, as well as its modulus of elasticity. For our problem, the modulus of elasticity is common to all of the rods and is given as 10,000,000 psi. The areas are listed in Table 10.2.1 of element properties, which includes the connectivity data.

Table 10.2.1   Element properties for Figure 10.2.4.

Element Number	Area (in.²)	Length (in.)	AE/L (lb/in.)	Node 1	Node 2	ϕ (degrees)	l	m
1	1.000	40.00	250,000	1	2	0.00	1.000	0.000
2	0.500	44.72	111,800	3	2	26.56	0.8944	0.4472
3	1.000	20.00	500,000	4	2	90.00	0.000	1.000
4	0.500	44.72	111,800	1	4	−26.56	0.8944	−0.4472
5	1.000	40.00	250,000	3	4	0.00	1.000	0.000
6	1.000	20.00	500,000	1	3	−90.00	0.000	−1.000

The maximum node number difference equals 3 and occurs across element 4. Therefore, the semibandwidth of the unrestrained structure's stiffness matrix is $N_b = 2(3 + 1) = 8$.

The global stiffness matrix of each rod can now be computed by substituting the element data into Equation 10.2.19. The global degree of freedom numbers corresponding to each row and column are noted on the element matrices by referring to the connectivity information in Table 10.2.1.

$$
[\mathbf{k}^{(1)}] = 250,000 \begin{array}{c} \begin{array}{cccc} 1 & 2 & 3 & 4 \end{array} \\ \left[\begin{array}{cccc} 1 & 0 & -1 & 0 \\ 0 & 0 & 0 & 0 \\ -1 & 0 & 1 & 0 \\ 0 & 0 & 0 & 0 \end{array}\right] \begin{array}{c} 1 \\ 2 \\ 3 \\ 4 \end{array} \end{array}
$$

$$
[\mathbf{k}^{(2)}] = 250,000 \begin{array}{c} \begin{array}{cccc} 5 & 6 & 3 & 4 \end{array} \\ \left[\begin{array}{cccc} 0.3578 & 0.1789 & -0.3578 & -0.1789 \\ 0.1789 & 0.0894 & -0.1789 & -0.0894 \\ -0.3578 & -0.1789 & 0.3578 & 0.1789 \\ -0.1789 & -0.0894 & 0.1789 & 0.0894 \end{array}\right] \begin{array}{c} 5 \\ 6 \\ 3 \\ 4 \end{array} \end{array}
$$

$$
[\mathbf{k}^{(3)}] = 250,000 \begin{array}{c} \begin{array}{cccc} 7 & 8 & 3 & 4 \end{array} \\ \left[\begin{array}{cccc} 0 & 0 & 0 & 0 \\ 0 & 2 & 0 & -2 \\ 0 & 0 & 0 & 0 \\ 0 & -2 & 0 & 2 \end{array}\right] \begin{array}{c} 7 \\ 8 \\ 3 \\ 4 \end{array} \end{array}
$$

$$
[\mathbf{k}^{(4)}] = 250,000 \begin{array}{c} \begin{array}{cccc} 1 & 2 & 7 & 8 \end{array} \\ \left[\begin{array}{cccc} 0.3578 & -0.1789 & -0.3578 & 0.1789 \\ -0.1789 & 0.0894 & 0.1789 & -0.0894 \\ -0.3578 & 0.1789 & 0.3578 & -0.1789 \\ 0.1789 & -0.0894 & -0.1789 & 0.0894 \end{array}\right] \begin{array}{c} 1 \\ 2 \\ 7 \\ 8 \end{array} \end{array}
$$

$$
\left[\mathbf{k}^{(5)}\right] = 250{,}000
\begin{array}{c}
\begin{array}{cccc} 5 & 6 & 7 & 8 \end{array} \\
\left[\begin{array}{cc:cc}
1 & 0 & -1 & 0 \\
0 & 0 & 0 & 0 \\
\hdashline
-1 & 0 & 1 & 0 \\
0 & 0 & 0 & 0
\end{array}\right]
\begin{array}{c} 5 \\ 6 \\ 7 \\ 8 \end{array}
\end{array}
\qquad
\left[\mathbf{k}^{(6)}\right] = 250{,}000
\begin{array}{c}
\begin{array}{cccc} 1 & 2 & 5 & 6 \end{array} \\
\left[\begin{array}{cc:cc}
0 & 0 & 0 & 0 \\
0 & 2 & 0 & -2 \\
\hdashline
0 & 0 & 0 & 0 \\
0 & -2 & 0 & 2
\end{array}\right]
\begin{array}{c} 1 \\ 2 \\ 5 \\ 6 \end{array}
\end{array}
$$

Note that the element stiffness matrices all possess a common scalar multiplier. This facilitates assembling the global stiffness matrix by hand. To accomplish this, we developed an adjustment factor for each matrix in which $AE/L \neq 250{,}000$ lb/in., and that matrix's components were modified accordingly. For example, for rod number 2, the element properties table lists AE/L as 111,800. Therefore, all of the stiffness matrix terms for rod 2, computed from Equation 10.2.19, were multiplied by the factor $111{,}800/250{,}000 = 0.4472$ to yield $\left[\mathbf{k}^{(2)}\right]$ as shown.

Using the assembly rule, we take the element stiffness matrices one at a time and add its components into the appropriate row and column of the global stiffness matrix, which starts out as an 8 by 8 matrix of zeroes (blanks). The global freedom numbers listed along the tops and sides of the element matrices guide us in this effort, which yields

$$
[\mathbf{k}] = 250{,}000
\begin{array}{c}
\begin{array}{cccccccc} 1 & \quad 2 & \quad 3 & \quad 4 & \quad 5 & \quad 6 & \quad 7 & \quad 8 \end{array} \\
\left[\begin{array}{cc:cc:cc:cc}
1+0.3578 & -0.1789 & -1 & 0 & 0 & 0 & -0.3578 & 0.1789 \\
-0.1789 & 0.0894+2 & 0 & 0 & 0 & -2 & 0.1789 & -0.0894 \\
\hdashline
-1 & 0 & 1+0.3578 & 0.1789 & -0.3578 & -0.1789 & 0 & 0 \\
0 & 0 & 0.1789 & 0.0894+2 & -0.1789 & -0.0894 & 0 & -2 \\
\hdashline
0 & 0 & -0.3578 & -0.1789 & 0.3578+1 & 0.1789 & -1 & 0 \\
0 & -2 & -0.1789 & -0.0894 & 0.1789 & 0.0894+2 & 0 & 0 \\
\hdashline
-0.3578 & 0.1789 & 0 & 0 & -1 & 0 & 0.3578+1 & -0.1789 \\
0.1789 & -0.0894 & 0 & -2 & 0 & 0 & -0.1789 & 2+0.0894
\end{array}\right]
\begin{array}{c} 1 \\ 2 \\ 3 \\ 4 \\ 5 \\ 6 \\ 7 \\ 8 \end{array}
\end{array}
$$

Summing the terms, we have

$$
[\mathbf{k}] = 250{,}000
\begin{array}{c}
\begin{array}{cccccccc} 1 & \quad 2 & \quad 3 & \quad 4 & \quad 5 & \quad 6 & \quad 7 & \quad 8 \end{array} \\
\left[\begin{array}{cc:cc:cc:cc}
1.358 & -0.1789 & -1 & 0 & 0 & 0 & -0.3578 & 0.1789 \\
-0.1789 & 2.089 & 0 & 0 & 0 & -2 & 0.1789 & -0.0894 \\
\hdashline
-1 & 0 & 1.358 & 0.1789 & -0.3578 & -0.1789 & 0 & 0 \\
0 & 0 & 0.1789 & 2.089 & -0.1789 & -0.0894 & 0 & -2 \\
\hdashline
0 & 0 & -0.3578 & -0.1789 & 1.358 & 0.1789 & -1 & 0 \\
0 & -2 & -0.1789 & -0.0894 & 0.1789 & 2.089 & 0 & 0 \\
\hdashline
-0.3578 & 0.1789 & 0 & 0 & -1 & 0 & 1.358 & -0.1789 \\
0.1789 & -0.0894 & 0 & -2 & 0 & 0 & -0.1789 & 2.089
\end{array}\right]
\begin{array}{c} 1 \\ 2 \\ 3 \\ 4 \\ 5 \\ 6 \\ 7 \\ 8 \end{array}
\end{array}
\qquad \text{[a]}
$$

Assembling a stiffness matrix by hand, as we have done here, is straightforward but somewhat tedious and error-prone, even for a relatively small stiffness matrix, such as this one. As a check on our work, we observe that $[\mathbf{k}]$ is symmetric and all of its diagonal components are positive, which are requirements established in section 4.2. Furthermore, since all of the degrees of freedom of a truss are translational, we know that every row and column of its stiffness matrix must sum to zero; such is the case for the matrix in Equation a.

The next step is to assemble the 8 by 1 global load vector $\{\mathbf{Q}\}$. Since in this problem there are no body forces or thermal strains, the equivalent load vector $\{\mathbf{Q}_{eq}^{(e)}\}$ is zero for each element. Therefore, the load vector (Equation 9.12.12) is just the point-load vector $\{\mathbf{Q}_P\}$, which is assembled by adding the external loads in each degree of freedom direction into $\{\mathbf{Q}\}$. Figure 10.2.4 shows that the loads at freedoms 3, 7, and 8 are zero and the load at freedom 4 is 1000 lb. The remaining load components are the unknown reactions applied by the fixed supports at nodes 1 and 3. The global load vector is therefore

$$\{Q\} = \begin{Bmatrix} X_1 \\ Y_1 \\ \hdashline 0 \\ 1000 \\ \hdashline X_3 \\ Y_3 \\ \hdashline 0 \\ 0 \end{Bmatrix} \begin{matrix} 1 \\ 2 \\ 3 \\ 4 \\ 5 \\ 6 \\ 7 \\ 8 \end{matrix}$$

We are now in a position to write the stiffness equations, partitioned with respect to the free (f) and supported (s) degrees of freedom (cf. Equations 9.10.14), as follows:

$$\{Q_f\} = \left[k_{ff}\right]\{q_f\} \qquad \text{[b]}$$

since $\{q_s\} = \{0\}$

$$\{Q_s\} = [k_{sf}]\{q_f\} \qquad \text{[c]}$$

Note for this problem that f is the degree of freedom subset 3, 4, 7, 8, and s is the degree of freedom subset 1, 2, 5, 6. Therefore,

$$\{Q_f\} = \{Q\}\begin{matrix}3\\4\\7\\8\end{matrix} = \begin{Bmatrix} 0 \\ 1000 \\ 0 \\ 0 \end{Bmatrix} \qquad \{q_f\} = \{q\}\begin{matrix}3\\4\\7\\8\end{matrix} = \begin{Bmatrix} u_2 \\ v_2 \\ u_4 \\ v_4 \end{Bmatrix} \qquad \text{[d]}$$

$$\{Q_s\} = \{Q\}\begin{matrix}1\\2\\5\\6\end{matrix} = \begin{Bmatrix} X_1 \\ Y_1 \\ X_3 \\ Y_3 \end{Bmatrix} \qquad \text{[e]}$$

$$\left[k_{ff}\right] = \begin{matrix}3\\4\\7\\8\end{matrix}\begin{bmatrix}3\ 4\ 7\ 8\\ k \end{bmatrix} = 250,000 \begin{bmatrix} 1.358 & 0.1789 & 0 & 0 \\ 0.1789 & 2.089 & 0 & -2 \\ 0 & 0 & 1.358 & -0.1789 \\ 0 & -2 & -0.1789 & 2.089 \end{bmatrix} \qquad \text{[f]}$$

$$\left[k_{sf}\right] = \begin{matrix}1\\2\\5\\6\end{matrix}\begin{bmatrix}1\ 2\ 5\ 6\\ k \end{bmatrix} = 250,000 \begin{bmatrix} -1 & 0 & -0.3578 & 0.1789 \\ 0 & 0 & 0.1789 & -0.0894 \\ -0.3578 & -0.1789 & -1 & 0 \\ -0.1789 & -0.0894 & 0 & 0 \end{bmatrix} \qquad \text{[g]}$$

Substituting Equations d and f into Equation b yields

$$250,000 \begin{bmatrix} 1.358 & 0.1789 & 0 & 0 \\ 0.1789 & 2.089 & 0 & -2 \\ 0 & 0 & 1.358 & -0.1789 \\ 0 & -2 & -0.1789 & 2.089 \end{bmatrix} \begin{Bmatrix} u_2 \\ v_2 \\ u_4 \\ v_4 \end{Bmatrix} = \begin{bmatrix} 0 \\ 1000 \\ 0 \\ 0 \end{bmatrix} \qquad \text{[h]}$$

The solution to this set of equations is

$$\{\mathbf{q}_f\} = \begin{Bmatrix} u_2 \\ v_2 \\ u_4 \\ v_4 \end{Bmatrix} = \begin{Bmatrix} -0.004065 \\ 0.03087 \\ 0.003938 \\ 0.02989 \end{Bmatrix} \text{ (in.)}$$ [i]

The fourth-order linear system of equations, h, can be found without first assembling the complete 8 by 8 stiffness matrix of the unconstrained structure, as we have done. Instead, while computing each element's stiffness matrix, we can note whether any of its freedoms are constrained to zero by virtue of the element being attached to a structural support. If so, the corresponding rows and columns of the element stiffness matrix can be eliminated at once, before assembly into the global array. Proceeding in this fashion through all of the elements yields the smaller matrix $[\mathbf{k}_{ff}]$ instead of the larger matrix $[\mathbf{k}]$. Applying constraints at the element level is recommended for hand calculations. Also, since the stiffness matrix $[\mathbf{k}_{ff}]$ is symmetric, we need only assemble the upper or lower triangular part, further reducing the computational effort.

Having found the structure's nodal displacements in Equation i, we can now solve for the stress in each of the rod elements. First, we determine which of the structure's displacements pertain to a given element e, and we form that element's global displacement vector $\{\mathbf{q}^{(e)}\}$. We then project those global displacements into that element's system using Equation 10.2.14, to find the local displacement vector $\{\bar{\mathbf{q}}^{(e)}\} = [\boldsymbol{\Lambda}^{(e)}]\{\mathbf{q}^{(e)}\}$. The transformation matrix $[\boldsymbol{\Lambda}^{(e)}]$ for each element is obtained by substituting its direction cosines from Table 10.2.1 into Equation 10.2.15. The local displacements are used to determine the strain, by means of Equation 10.2.2, $\varepsilon_x^{(e)} = \lfloor \mathbf{B}^{(e)} \rfloor \{\bar{\mathbf{q}}^{(e)}\}$, after which Hooke's law yields the stress, $\sigma_x^{(e)} = E^{(e)}\varepsilon_x^{(e)}$. These computations are done as follows for all six elements of the truss.

Element 1

$$\{\bar{\mathbf{q}}^{(1)}\} = [\boldsymbol{\Lambda}^{(1)}]\{\mathbf{q}^{(1)}\}$$

$$\begin{Bmatrix} \bar{u}^{(1)} \\ \bar{u}^{(2)} \end{Bmatrix} = [\boldsymbol{\Lambda}^{(1)}] \begin{Bmatrix} u_1^{(1)} \\ v_1^{(1)} \\ u_2^{(1)} \\ v_2^{(1)} \end{Bmatrix} = \begin{bmatrix} 1 & 0 & 0 & 0 \\ 0 & 0 & 1 & 0 \end{bmatrix} \begin{Bmatrix} 0 \\ 0 \\ -0.004065 \\ 0.03085 \end{Bmatrix} = \begin{Bmatrix} 0 \\ -0.004065 \end{Bmatrix} \text{ (in.)}$$

$$\sigma_x^{(1)} = E^{(1)} \left\lfloor -\frac{1}{L^{(1)}} \quad \frac{1}{L^{(1)}} \right\rfloor \begin{Bmatrix} \bar{u}_1^{(1)} \\ \bar{u}_2^{(1)} \end{Bmatrix} = 10^7 \frac{-(0) + (-0.004065)}{40} = -1016 \text{ psi}$$

Element 2

$$\{\bar{\mathbf{q}}^{(2)}\} = [\boldsymbol{\Lambda}^{(2)}]\{\mathbf{q}^{(2)}\}$$

$$\begin{Bmatrix} \bar{u}_1^{(2)} \\ \bar{u}_2^{(2)} \end{Bmatrix} = [\boldsymbol{\Lambda}^{(2)}] \begin{Bmatrix} u_3^{(2)} \\ v_3^{(2)} \\ u_2^{(2)} \\ v_2^{(2)} \end{Bmatrix} = \begin{bmatrix} 0.8944 & 0.4472 & 0 & 0 \\ 0 & 0 & 0.8944 & 0.4472 \end{bmatrix} \begin{Bmatrix} 0 \\ 0 \\ -0.004065 \\ 0.03085 \end{Bmatrix} = \begin{Bmatrix} 0 \\ 0.01016 \end{Bmatrix} \text{ (in.)}$$

$$\sigma_x^{(2)} = E^{(2)} \left\lfloor -\frac{1}{L^{(2)}} \quad \frac{1}{L^{(2)}} \right\rfloor \begin{Bmatrix} \bar{u}_1^{(2)} \\ \bar{u}_2^{(2)} \end{Bmatrix} = 10^7 \frac{-(0) + (0.01016)}{44.72} = +2272 \text{ psi}$$

Element 3

$$\{\bar{\mathbf{q}}^{(3)}\} = [\mathbf{\Lambda}^{(3)}]\{\mathbf{q}^{(3)}\}$$

$$\left\{\begin{array}{c} \bar{u}_1^{(3)} \\ \bar{u}_2^{(3)} \end{array}\right\} = [\mathbf{\Lambda}^{(3)}]\left\{\begin{array}{c} u_4^{(3)} \\ v_4^{(3)} \\ u_2^{(3)} \\ v_2^{(3)} \end{array}\right\} = \begin{bmatrix} 0 & 1 & 0 & 0 \\ 0 & 0 & 0 & 1 \end{bmatrix}\left\{\begin{array}{c} 0.003935 \\ 0.02987 \\ -0.004065 \\ 0.03085 \end{array}\right\} = \left\{\begin{array}{c} 0.02987 \\ 0.03085 \end{array}\right\} \quad \text{(in.)}$$

$$\sigma_x^{(3)} = E^{(3)}\left\lfloor -\frac{1}{L^{(3)}} \quad \frac{1}{L^{(3)}} \right\rfloor \left\{\begin{array}{c} \bar{u}_1^{(3)} \\ \bar{u}_2^{(3)} \end{array}\right\} = 10^7 \frac{-(0.02987) + (0.03085)}{20} = +490.0 \ \text{psi}$$

Element 4

$$\{\bar{\mathbf{q}}^{(4)}\} = [\mathbf{\Lambda}^{(4)}]\{\mathbf{q}^{(4)}\}$$

$$\left\{\begin{array}{c} \bar{u}_1^{(4)} \\ \bar{u}_2^{(4)} \end{array}\right\} = [\mathbf{\Lambda}^{(4)}]\left\{\begin{array}{c} u_1^{(4)} \\ v_1^{(4)} \\ u_4^{(4)} \\ v_4^{(4)} \end{array}\right\} = \begin{bmatrix} 0.8944 & -0.4472 & 0 & 0 \\ 0 & 0 & 0.8944 & -0.4472 \end{bmatrix}\left\{\begin{array}{c} 0 \\ 0 \\ 0.003935 \\ 0.02987 \end{array}\right\} = \left\{\begin{array}{c} 0 \\ -0.009838 \end{array}\right\} \quad \text{(in.)}$$

$$\sigma_x^{(4)} = E^{(4)}\left\lfloor -\frac{1}{L^{(4)}} \quad \frac{1}{L^{(4)}} \right\rfloor \left\{\begin{array}{c} \bar{u}_1^{(4)} \\ \bar{u}_2^{(4)} \end{array}\right\} = 10^7 \frac{-(0) + (-0.009838)}{44.72} = -2200 \ \text{psi}$$

Element 5

$$\{\bar{\mathbf{q}}^{(5)}\} = [\mathbf{\Lambda}^{(5)}]\{\mathbf{q}^{(5)}\}$$

$$\left\{\begin{array}{c} \bar{u}_1^{(5)} \\ \bar{u}_2^{(5)} \end{array}\right\} = [\mathbf{\Lambda}^{(5)}]\left\{\begin{array}{c} u_3^{(5)} \\ v_3^{(5)} \\ u_4^{(5)} \\ v_4^{(5)} \end{array}\right\} = \begin{bmatrix} 1 & 0 & 0 & 0 \\ 0 & 0 & 1 & 0 \end{bmatrix}\left\{\begin{array}{c} 0 \\ 0 \\ 0.003935 \\ 0.02987 \end{array}\right\} = \left\{\begin{array}{c} 0 \\ 0.003935 \end{array}\right\} \quad \text{(in.)}$$

$$\sigma_x^{(5)} = E^{(5)}\left\lfloor -\frac{1}{L^{(5)}} \quad \frac{1}{L^{(5)}} \right\rfloor \left\{\begin{array}{c} \bar{u}_1^{(5)} \\ \bar{u}_2^{(5)} \end{array}\right\} = 10^7 \frac{-(0) + (0.003935)}{40} = 983.8 \ \text{psi}$$

Element 6

$$\{\bar{\mathbf{q}}^{(6)}\} = [\mathbf{\Lambda}^{(6)}]\{\mathbf{q}^{(6)}\}$$

$$\left\{\begin{array}{c} \bar{u}_1^{(6)} \\ \bar{u}_2^{(6)} \end{array}\right\} = [\mathbf{\Lambda}^{(6)}]\left\{\begin{array}{c} u_1^{(6)} \\ v_1^{(6)} \\ u_3^{(6)} \\ v_3^{(6)} \end{array}\right\} = \begin{bmatrix} 0 & -1 & 0 & 0 \\ 0 & 0 & 0 & -1 \end{bmatrix}\left\{\begin{array}{c} 0 \\ 0 \\ 0 \\ 0 \end{array}\right\} = \left\{\begin{array}{c} 0 \\ 0 \end{array}\right\} \quad \text{(in.)}$$

$$\sigma_x^{(6)} = E^{(6)} \left\lfloor -\frac{1}{L^{(6)}} \quad \frac{1}{L^{(6)}} \right\rfloor \left\{ \begin{array}{c} \bar{u}_1^{(6)} \\ \bar{u}_2^{(6)} \end{array} \right\} = 10^7 \frac{-(0)+(0)}{20} = 0 \text{ psi}$$

Multiplying each of these stresses by the element's cross-sectional area yields the axial force data summarized in Table 10.2.2.

Table 10.2.2   Element axial loads in Figure 10.2.4

Element	Axial Load (lb)	
1	1016	Compression
2	1136	Tension
3	490	Tension
4	1100	Compression
5	984	Compression
6	0	

To calculate the reactions at the supports (nodes 3 and 4), we use Equation c. Substituting $[\mathbf{k}_{sf}]$ and $[\mathbf{q}_{sf}]$ from Equations g and i, respectively, we find

$$\left\{ \begin{array}{c} X_1 \\ Y_1 \\ X_2 \\ Y_2 \end{array} \right\} = 250,000 \begin{bmatrix} -1.000 & 0 & -0.3578 & 0.1789 \\ 0 & 0 & 0.1789 & -0.0894 \\ -0.3578 & -0.1789 & -1 & 0 \\ -0.1789 & -0.0894 & 0 & 0 \end{bmatrix} \left\{ \begin{array}{c} -0.004065 \\ 0.03087 \\ 0.003938 \\ 0.02989 \end{array} \right\} = \left\{ \begin{array}{c} 2000 \\ -492 \\ -2000 \\ -508 \end{array} \right\} \text{ (lb)}$$

These reactions, along with the internal element loads and the external load applied at node 2, are shown in Figure 10.2.5. The reader should verify that all of the nodes are in equilibrium.

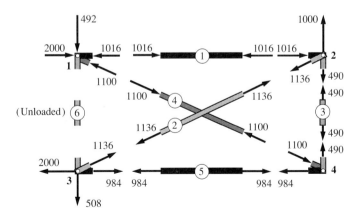

Figure 10.2.5   Computed loads (lb) for the structure of Figure 10.2.4.

To solve a space truss problem by the stiffness method, we need the three-dimensional form of the rod element stiffness matrix. The matrix $[\mathbf{k}]$ in three-dimensional space is calculated in the same manner as in two dimensions, with the z component included in all of the global vector quantities. The dot product operation projects the global displacement at node 1 onto the element axis $\bar{x}$, yielding the local displacement $\bar{u}_1$ (cf. Figure 10.2.1), just as it yields $\bar{u}_2$ at node 2:

$$\bar{u}_1 = \lfloor l \quad m \quad n \rfloor \begin{Bmatrix} u_1 \\ v_1 \\ w_1 \end{Bmatrix} \qquad \bar{u}_2 = \lfloor l \quad m \quad n \rfloor \begin{Bmatrix} u_2 \\ v_2 \\ w_2 \end{Bmatrix}$$

where l, m, and n are the three direction cosines of the rod element's axis. The local displacement vector $\{\bar{q}\}$ is obtained from the global displacement vector $\{q\}$ by

$$\{\bar{q}\} = [\mathbf{\Lambda}]\{q\} \tag{10.2.20a}$$

where

$$\{\bar{q}\} = \begin{Bmatrix} \bar{u}_1 \\ \bar{u}_2 \end{Bmatrix} \qquad [\mathbf{\Lambda}] = \begin{bmatrix} l & m & n & 0 & 0 & 0 \\ 0 & 0 & 0 & l & m & n \end{bmatrix} \qquad \{q\} = \begin{Bmatrix} u_1 \\ v_1 \\ w_1 \\ \hline u_2 \\ v_2 \\ w_2 \end{Bmatrix} \tag{10.2.20b}$$

The three global load components X_1, Y_1, and Z_1 at node 1 are the projections of the axially directed force $\bar{X}_1$ onto each of the three global xyz axes, or

$$X_1 = l\bar{X}_1 \qquad Y_1 = m\bar{X}_1 \qquad Z_1 = n\bar{X}_1$$

Likewise, at node 2, we have

$$X_2 = l\bar{X}_2 \qquad Y_2 = m\bar{X}_2 \qquad Z_2 = n\bar{X}_2$$

The global load vector $\{Q\}$ is obtained from the local load vector $\{\bar{Q}\}$ by the linear transformation

$$\{\bar{Q}\} = [\mathbf{\Lambda}]^T \{Q\} \tag{12.2.21a}$$

where $[\mathbf{\Lambda}]$ is given in Equation 10.2.20b and

$$\{\bar{Q}\} = \lfloor \bar{X}_1 \quad \bar{X}_2 \rfloor^T \qquad \{Q\} = \lfloor X_1 \quad Y_1 \quad Z_1 \; \vdots \; X_2 \quad Y_2 \quad Z_2 \rfloor^T \tag{10.2.21b}$$

Following the same reasoning as in the two-dimensional case, we are led to

$$[\mathbf{k}] = [\mathbf{\Lambda}]^T [\bar{\mathbf{k}}][\mathbf{\Lambda}] \tag{10.2.22}$$

Substituting $[\bar{\mathbf{k}}]$ and $[\mathbf{\Lambda}]$ from Equations 10.2.6 and 10.2.20b yields

$$[\mathbf{k}] = \begin{bmatrix} l & 0 \\ m & 0 \\ n & 0 \\ 0 & l \\ 0 & m \\ 0 & n \end{bmatrix} \frac{AE}{L} \begin{bmatrix} 1 & -1 \\ -1 & 1 \end{bmatrix} \begin{bmatrix} l & m & n & 0 & 0 & 0 \\ 0 & 0 & 0 & l & m & n \end{bmatrix} = \frac{AE}{L} \begin{bmatrix} l & -l \\ m & -m \\ n & -n \\ -l & l \\ -m & m \\ -n & n \end{bmatrix} \begin{bmatrix} l & m & n & 0 & 0 & 0 \\ 0 & 0 & 0 & l & m & n \end{bmatrix}$$

It follows that the global stiffness matrix of a rod element in three dimensions is

$$[\mathbf{k}] = \frac{AE}{L} \begin{bmatrix} & \overset{u_1}{} & \overset{v_1}{} & \overset{w_1}{} & \overset{u_2}{} & \overset{v_2}{} & \overset{w_2}{} \\ l^2 & ml & nl & -l^2 & -ml & -nl \\ lm & m^2 & nm & -lm & -m^2 & -nm \\ ln & mn & n^2 & -ln & -mn & -n^2 \\ -l^2 & -ml & -nl & l^2 & ml & nl \\ -lm & -m^2 & -nm & lm & m^2 & nm \\ -ln & -mn & -n^2 & ln & mn & n^2 \end{bmatrix} \begin{matrix} X_1 \\ Y_1 \\ Z_1 \\ X_2 \\ Y_2 \\ Z_2 \end{matrix}$$ [10.2.23]

The direction cosines l, m, and n are calculated using the global coordinates of the endpoints of the element. Referring to Figure 10.2.6, we see that the length of the rod is

$$L = \sqrt{(x_2 - x_1)^2 + (y_2 - y_1)^2 + (z_2 - z_1)^2}$$

Therefore, the direction cosines are

$$l = \frac{x_2 - x_1}{L} \qquad m = \frac{y_2 - y_1}{L} \qquad n = \frac{z_2 - z_1}{L}$$

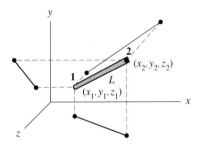

Figure 10.2.6   Global coordinates of rod element nodes in three dimensions.

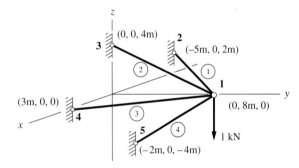

All rods: $E = 70$ GPa, $A = 4$ cm^2

Figure 10.2.7   Space truss composed of aluminum rods with a common cross section.

Example 10.2.2   Find the forces in the rods of the statically indeterminate space truss shown in Figure 10.2.7.

The cross-sectional area of every rod is 4 cm², and each rod has the same modulus of elasticity, 70 GPa (70×10^9 N/m²). Therefore, the axial rigidity AE equals 28×10^6 N and is uniform throughout the truss. Four of the structure's five nodes are rigidly supported. Only node 1 is unconstrained, and a downward-directed point load of 1 kN is applied to it. Let the first node of each element coincide with node 1 of the structure.

For element 1, the vector drawn from its first node to its far node is

$$\mathbf{r}_{12} = -5\mathbf{i} - 8\mathbf{j} + 2\mathbf{k}$$

so that the length of the element is

$$L^{(1)} = \sqrt{(-5)^2 + (-8)^2 + (2)^2} = 9.644 \text{ m}$$

Dividing each component of $\mathbf{r}_{12}$ by $L^{(1)}$ yields the three direction cosines of the element:

$$l^{(1)} = \frac{-5}{9.644} = -0.5184 \qquad m^{(1)} = \frac{-8}{9.644} = -0.8295 \qquad n^{(1)} = \frac{2}{9.644} = 0.2074$$

Proceeding in this fashion for the other three elements, we can summarize the element data in Table 10.2.3.

Table 10.2.3   Element properties for Figure 10.2.7.

Element Number	AE (N)	Length (m)	AE/L (N/m)	Node 1	Node 2	l	m	n
1	28,000,000	9.644	2,903,000	1	2	−0.5184	−0.8295	0.2074
2	28,000,000	8.944	3,131,000	1	3	0.0000	−0.8944	0.4472
3	28,000,000	8.544	3,277,000	1	4	0.3511	−0.9363	0.0000
4	28,000,000	9.165	3,055,000	1	5	−0.2182	−0.8729	−0.4364

This table contains the information required to calculate each element's global stiffness matrix, using Equation 10.2.23.

The three-dimensional structure in Figure 10.2.7 has five nodes, which means its stiffness matrix is 15 by 15 before the displacement boundary conditions are applied. To avoid unnecessary computations, we will apply the constraints at the element level. The resulting 3 by 3 constrained global stiffness matrix $[\mathbf{k}_{ff}]$ is therefore assembled directly. Since the second node of each element e is completely restrained, $[\mathbf{k}_{ff}^{(e)}]$ is found from Equation 10.2.23 by retaining only the degrees of freedom associated with the first node, that is, rows and columns 1, 2, and 3.

$$[\mathbf{k}_{ff}^{(1)}] = 10^6 \begin{bmatrix} 0.7801 & 1.248 & -0.3122 \\ 1.248 & 1.998 & -0.4992 \\ -0.3122 & -0.4992 & 0.1249 \end{bmatrix} \begin{matrix} 1 \\ 2 \\ 3 \end{matrix} \qquad [\mathbf{k}_{ff}^{(2)}] = 10^6 \begin{bmatrix} 0.000 & 0.000 & 0.000 \\ 0.000 & 2.504 & -1.252 \\ 0.000 & -1.252 & 0.6261 \end{bmatrix} \begin{matrix} 1 \\ 2 \\ 3 \end{matrix}$$

$$[\mathbf{k}_{ff}^{(3)}] = 10^6 \begin{bmatrix} 0.4040 & -1.077 & 0.000 \\ -1.077 & 1.998 & 0.000 \\ 0.000 & 0.000 & 0.000 \end{bmatrix} \begin{matrix} 1 \\ 2 \\ 3 \end{matrix} \qquad [\mathbf{k}_{ff}^{(4)}] = 10^6 \begin{bmatrix} 0.1455 & 0.5818 & 0.2909 \\ 0.5818 & 0.2909 & 1.1640 \\ 0.2909 & 1.1640 & 0.5816 \end{bmatrix} \begin{matrix} 1 \\ 2 \\ 3 \end{matrix}$$

The stiffness matrix assembly rule yields the global stiffness matrix of the constrained truss, as follows

$$[\mathbf{k}_{ff}] = 10^6 \begin{bmatrix} 1.329 & 0.7528 & -0.02128 \\ 0.7528 & 9.703 & -0.5872 \\ -0.02128 & -0.5872 & 1.333 \end{bmatrix}$$

and the stiffness equations $[\mathbf{k}_{ff}]\{\mathbf{q}_f\} = \{\mathbf{Q}_f\}$ are therefore

$$10^6 \begin{bmatrix} 1.329 & 0.7528 & -0.02128 \\ 0.7528 & 9.703 & -0.5872 \\ -0.02128 & -0.5872 & 1.333 \end{bmatrix} \begin{Bmatrix} u_1 \\ v_1 \\ w_1 \end{Bmatrix} = \begin{Bmatrix} 0 \\ 0 \\ -1000 \end{Bmatrix}$$

The solution for this system is

$$\left\{ \begin{array}{c} u_1 \\ v_1 \\ w_1 \end{array} \right\} = \left\{ \begin{array}{c} 1.473 \\ -4.780 \\ -77.09 \end{array} \right\} (10^{-5}) \;\; \text{meters}$$

As in the previous example, we project these displacements onto each rod to find the axial stress.

Element 1

$$\left\{ \bar{\mathbf{q}}^{(1)} \right\} = \left[\mathbf{\Lambda}^{(1)} \right] \left\{ \mathbf{q}^{(1)} \right\}$$

$$\left\{ \begin{array}{c} \bar{u}_1^{(1)} \\ \bar{u}_2^{(1)} \end{array} \right\} = \left[\begin{array}{cccccc} -0.5184 & -0.8295 & 0.2074 & 0.0000 & 0.0000 & 0.0000 \\ 0.0000 & 0.0000 & 0.0000 & -0.5184 & -0.8295 & 0.2074 \end{array} \right] \left\{ \begin{array}{c} 1.473 \\ -4.780 \\ -77.09 \\ \hline 0.000 \\ 0.000 \\ 0.000 \end{array} \right\} (10^{-5}) = \left\{ \begin{array}{c} -0.1279 \\ 0 \end{array} \right\} (10^{-3}) \; \text{m}$$

$$\sigma_x^{(1)} = E^{(1)} \left[-\frac{1}{L^{(1)}} \quad \frac{1}{L^{(1)}} \right] \left\{ \begin{array}{c} \bar{u}_1^{(1)} \\ \bar{u}_2^{(1)} \end{array} \right\} = 70 \times 10^9 \frac{-(-0.1279 \times 10^{-3}) + (0)}{9.644} = 928.3 \;\; \text{kPa}$$

Element 2

$$\left\{ \bar{\mathbf{q}}^{(2)} \right\} = \left[\mathbf{\Lambda}^{(2)} \right] \left\{ \mathbf{q}^{(2)} \right\}$$

$$\left\{ \begin{array}{c} \bar{u}_1^{(2)} \\ \bar{u}_2^{(2)} \end{array} \right\} = \left[\begin{array}{cccccc} 0.0000 & -0.8944 & 0.4472 & 0.0000 & 0.0000 & 0.0000 \\ 0.0000 & 0.0000 & 0.0000 & 0.0000 & -0.8944 & 0.4472 \end{array} \right] \left\{ \begin{array}{c} 1.473 \\ -4.780 \\ -77.09 \\ \hline 0.000 \\ 0.000 \\ 0.000 \end{array} \right\} (10^{-5}) = \left\{ \begin{array}{c} -0.3020 \\ 0 \end{array} \right\} (10^{-3}) \; \text{m}$$

$$\sigma_x^{(2)} = E^{(2)} \left[-\frac{1}{L^{(2)}} \quad \frac{1}{L^{(2)}} \right] \left\{ \begin{array}{c} \bar{u}_1^{(2)} \\ \bar{u}_2^{(2)} \end{array} \right\} = 70 \times 10^9 \frac{-(-0.3020 \times 10^{-3}) + (0)}{8.944} = 2.364 \;\; \textbf{MPa}$$

Element 3

$$\left\{ \bar{\mathbf{q}}^{(3)} \right\} = \left[\mathbf{\Lambda}^{(3)} \right] \left\{ \mathbf{q}^{(3)} \right\}$$

$$\left\{ \begin{array}{c} \bar{u}_1^{(3)} \\ \bar{u}_2^{(3)} \end{array} \right\} = \left[\begin{array}{cccccc} 0.3511 & -0.9363 & 0.0000 & 0.0000 & 0.0000 & 0.0000 \\ 0.0000 & 0.0000 & 0.0000 & 0.3511 & -0.9363 & 0.0000 \end{array} \right] \left\{ \begin{array}{c} 1.473 \\ -4.780 \\ -77.09 \\ \hline 0.000 \\ 0.000 \\ 0.000 \end{array} \right\} (10^{-5}) = \left\{ \begin{array}{c} 0.04993 \\ 0 \end{array} \right\} (10^{-3}) \; \text{m}$$

$$\sigma_x^{(3)} = E^{(3)} \left\lfloor -\frac{1}{L^{(3)}} \quad \frac{1}{L^{(3)}} \right\rfloor \left\{ \begin{array}{c} \bar{u}_1^{(3)} \\ \bar{u}_2^{(3)} \end{array} \right\} = 70 \times 10^9 \frac{-(0.04993 \times 10^{-3}) + (0)}{8.544} = -409.1 \text{ kPa}$$

Element 4

$$\{\bar{\mathbf{q}}^{(4)}\} = [\mathbf{\Lambda}^{(4)}]\{\mathbf{q}^{(4)}\}$$

$$\left\{ \begin{array}{c} \bar{u}_1^{(4)} \\ \bar{u}_2^{(4)} \end{array} \right\} = \left[\begin{array}{cccccc} -0.2182 & -0.8729 & -0.4364 & 0.0000 & 0.0000 & 0.0000 \\ 0.0000 & 0.0000 & 0.0000 & -0.2182 & -0.8729 & -0.4364 \end{array} \right] \left\{ \begin{array}{c} 1.473 \\ -4.780 \\ -77.09 \\ \hline 0.000 \\ 0.000 \\ 0.000 \end{array} \right\} (10^{-5}) = \left\{ \begin{array}{c} 0.3749 \\ 0 \end{array} \right\} (10^{-3}) \text{ m}$$

$$\sigma_x^{(4)} = E^{(4)} \left\lfloor -\frac{1}{L^{(4)}} \quad \frac{1}{L^{(4)}} \right\rfloor \left\{ \begin{array}{c} \bar{u}_1^{(4)} \\ \bar{u}_2^{(4)} \end{array} \right\} = 70 \times 10^9 \frac{-(0.3749 \times 10^{-3}) + (0)}{9.165} = -2.863 \text{ MPa}$$

The axial loads associated with each of these stresses are listed in Table 10.2.4.

Table 10.2.4   Axial loads in Figure 10.2.7.

Element	Axial Load (N)	
1	371	Tension
2	946	Tension
3	164	Compression
4	1140	Compression

If a temperature change occurs within an initially unstressed redundant truss structure, the accompanying thermal expansion or contraction of the affected members is manifested by a push or pull on the nodes to which they are attached. The thermally induced equivalent nodal point loads, in the local coordinates of a rod element, are found using Equation 9.12.13,

$$\{\bar{\mathbf{Q}}_{eq}\} = \iiint_V [\mathbf{B}]^T [\mathbf{E}] \{\boldsymbol{\varepsilon}_T\} dV = \int_0^L \left\{ \begin{array}{c} -\frac{1}{L} \\ \frac{1}{L} \end{array} \right\} E (\alpha T) A ds \qquad \textbf{[10.2.24]}$$

where T is the temperature change, and s is the distance along the rod. Let us include the possibility of a temperature change T that varies at most linearly from one end of the rod to the other,[2] or

---

[2] In the absence of body forces, the equilibrium equation for the cylindrical rod is $\partial \sigma_x / \partial = 0$. Since $\sigma_x = E(\partial u / \partial x - \alpha T)$, equilibrium therefore requires $\partial^2 u / \partial x^2 = \alpha \partial T / \partial x$. For a displacement field linear in x, $\partial^2 u / \partial x^2 = 0$. That means that T should be constant. Even though the assumed-displacement procedure relaxes the requirement for equilibrium within the element, more complex assumptions about the temperature field should be accompanied by upgrades of the displacement field, if we are to maintain confidence in the accuracy of our results.

$$T = T_1 + \frac{T_2 - T_1}{L}s \qquad \text{[10.2.25]}$$

where T_1 and T_2 are the temperature changes at node 1 and node 2, respectively. Substituting this expression into Equation 10.2.24 and doing the integrations yields the local equivalent load vector

$$\{\bar{Q}_{eq}\} = \left\{ \begin{array}{c} -AE\alpha T_{avg} \\ AE\alpha T_{avg} \end{array} \right\} \qquad \text{[10.2.26]}$$

where $T_{avg} = (T_1 + T_2)/2$, the average temperature change in the rod. As shown in Figure 10.2.8, the components of $\{\bar{Q}_{eq}\}$ are the forces exerted on the nodes by the element due to its thermal expansion.

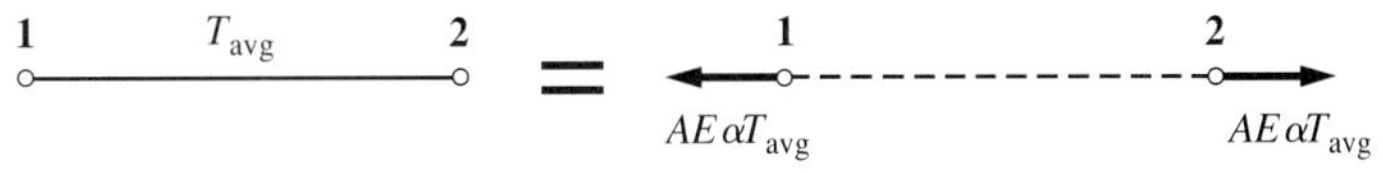

Figure 10.2.8   Nodal loads applied to the truss by the rod element due to an average temperature rise T_{avg}.

To obtain the global form of this equivalent load vector, we use Equation 10.2.16b

$$\{Q_{eq}\} = [\Lambda]^T \{\bar{Q}_{eq}\}$$

For the three-dimensional case, $[\Lambda]$ is given in Equation 10.2.20b, so that the global equivalent point-load vector for thermal strain is

$$\{Q_{eq}\} = \begin{bmatrix} l & 0 \\ m & 0 \\ n & 0 \\ 0 & l \\ 0 & m \\ 0 & n \end{bmatrix} \left\{ \begin{array}{c} -AE\alpha T_{avg} \\ AE\alpha T_{avg} \end{array} \right\} = AE\alpha T_{avg} \left\{ \begin{array}{c} -l \\ -m \\ -n \\ l \\ m \\ n \end{array} \right\} \qquad \text{[10.2.27a]}$$

In two-dimensional problems,

$$\{Q_{eq}\} = \begin{bmatrix} l & 0 \\ m & 0 \\ 0 & l \\ 0 & m \end{bmatrix} \left\{ \begin{array}{c} -AE\alpha T_{avg} \\ AE\alpha T_{avg} \end{array} \right\} = AE\alpha T_{avg} \left\{ \begin{array}{c} -l \\ -m \\ l \\ m \end{array} \right\} \qquad \text{[10.2.27b]}$$

If a truss is subjected to g loading, we can use the virtual work formulas of section 9.11 to calculate the equivalent point loads applied to the structural nodes due to the inertia of the attached elements. Let $\{a\}$ be the global acceleration vector at a point of a rod element. In three-dimensional space, $\{a\}$ is given by its three components along the global axes, as follows:

$$\{a\} = \lfloor a_x \quad a_y \quad a_z \rfloor^T \qquad \text{[10.2.28]}$$

If the mass density of the rod element material is ρ, then the body force per unit mass $\{b\}$ due to the acceleration is

$$\{b\} = -\rho\{a\} \qquad \text{[10.2.29]}$$

We are seeking the rod acceleration effects on the nodes to which the rod is attached. The net force of the nodes on the rod must be in the direction of the rod's acceleration vector. By the action–reaction principle, the force of the rod on the nodes is in the opposite direction. This is the reason for the minus sign in Equation 10.2.29.

Recall from Equation 9.12.13 that the equivalent load vector for body forces is

$$\{Q_{eq}\} = \int_0^L [N]^T \{b\} \, A ds \qquad\qquad [10.2.30]$$

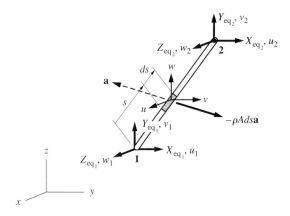

Figure 10.2.9   Rod element inertia loading and the equivalent nodal loads.

The global components u, v, w of the displacement of points on the rod's axis are presumed to be linear functions of position s; that is,

$$\left\{\begin{array}{c} u \\ v \\ w \end{array}\right\} = \begin{bmatrix} 1 - \frac{s}{L} & 0 & 0 & \frac{s}{L} & 0 & 0 \\ 0 & 1 - \frac{s}{L} & 0 & 0 & \frac{s}{L} & 0 \\ 0 & 0 & 1 - \frac{s}{L} & 0 & 0 & \frac{s}{L} \end{bmatrix} \left\{\begin{array}{c} u_1 \\ v_1 \\ w_1 \\ u_2 \\ v_2 \\ w_2 \end{array}\right\}$$

This ensures that the element remains straight (does not bend like a beam) while undergoing small rigid-body displacement and axial strain. Therefore, the global form of the shape function matrix $[N]$ is

$$[N] = \begin{bmatrix} 1 - \frac{s}{L} & 0 & 0 & \frac{s}{L} & 0 & 0 \\ 0 & 1 - \frac{s}{L} & 0 & 0 & \frac{s}{L} & 0 \\ 0 & 0 & 1 - \frac{s}{L} & 0 & 0 & \frac{s}{L} \end{bmatrix} \qquad\qquad [10.2.31]$$

Substituting Equations 10.2.28, 10.2.29, and 10.2.31 into Equation 10.2.30 yields

$$\{Q_{eq}\} = \int_0^L \begin{bmatrix} 1-\frac{s}{L} & 0 & 0 \\ 0 & 1-\frac{s}{L} & 0 \\ 0 & 0 & 1-\frac{s}{L} \\ \frac{s}{L} & 0 & 0 \\ 0 & \frac{s}{L} & 0 \\ 0 & 0 & \frac{s}{L} \end{bmatrix} \left(-\rho A \begin{Bmatrix} a_x \\ a_y \\ a_z \end{Bmatrix} \right) ds = -\rho A \int_0^L \begin{Bmatrix} \left(1-\frac{s}{L}\right)a_x \\ \left(1-\frac{s}{L}\right)a_y \\ \left(1-\frac{s}{L}\right)a_z \\ \frac{s}{L}a_x \\ \frac{s}{L}a_y \\ \frac{s}{L}a_z \end{Bmatrix} ds$$

If the acceleration is uniform over the element, then

$$\{Q_{eq}\} = -\frac{\rho A L}{2} \lfloor a_x \quad a_y \quad a_z \mid a_x \quad a_y \quad a_z \rfloor^T \qquad \text{[10.2.32a]}$$

In a two-dimensional context, this becomes

$$\{Q_{eq}\} = -\frac{\rho A L}{2} \lfloor a_x \quad a_y \mid a_x \quad a_y \rfloor \qquad \text{[10.2.32b]}$$

The global equivalent load vectors are computed for each element of a truss and then added, along with the actual point loads, into the global load vector, according to Equation 9.12.23. This application of the assembly rule is illustrated in the next example.

Example 10.2.3   Calculate the reactions at the supports and the stresses in the members of the truss in Figure 10.2.10. All of the elements have the same cross-sectional geometry and the same modulus of elasticity E, mass density ρ, and thermal expansion coefficient α. In addition to the downward load P at node 1, there is a uniform upward acceleration $a_y = 0.05P/\rho AL$. Elements 1 and 4 also undergo a uniform temperature rise $T = 0.1P/AE\alpha$.

The element properties of this structure, which is very similar to that of Example 10.2.1, are summarized in Table 10.2.5.

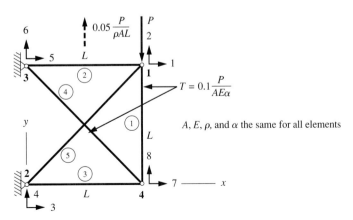

Figure 10.2.10   Plane truss subjected to point loading, inertia load, and thermal expansion, with the global degrees of freedom shown by each node.

Table 10.2.5 Element properties in Figure 10.2.10.

Element Number	Area	Length	ϕ (degrees)	Direction Cosines		Degrees of Freedom			
				l	m	1	2	3	4
1	A	L	$+90°$	0	1	7	8	1	2
2	A	L	0	1	0	5	6	1	2
3	A	L	0	1	0	3	4	7	8
4	A	$\sqrt{2}L$	$-45°$	$\sqrt{2}/2$	$-\sqrt{2}/2$	5	6	7	8
5	A	$\sqrt{2}L$	$-135°$	$-\sqrt{2}/2$	$-\sqrt{2}/2$	1	2	3	4

Based on the tabulated data, we compute the element stiffness matrices, using Equation 10.2.19. The rows and columns of each stiffness matrix are labeled with the corresponding global degree of freedom numbers.

$$[\mathbf{k}^{(1)}] = \frac{AE}{L} \begin{bmatrix} 0 & 0 & 0 & 0 \\ 0 & 1 & 0 & -1 \\ 0 & 0 & 0 & 0 \\ 0 & -1 & 0 & 0 \end{bmatrix} \begin{matrix} 7 \\ 8 \\ 1 \\ 2 \end{matrix}$$

$$[\mathbf{k}^{(2)}] = \frac{AE}{L} \begin{bmatrix} 1 & 0 & -1 & 0 \\ 0 & 0 & 0 & 0 \\ -1 & 0 & 1 & 0 \\ 0 & 0 & 0 & 0 \end{bmatrix} \begin{matrix} 5 \\ 6 \\ 1 \\ 2 \end{matrix}$$

$$[\mathbf{k}^{(3)}] = \frac{AE}{L} \begin{bmatrix} 1 & 0 & -1 & 0 \\ 0 & 0 & 0 & 0 \\ -1 & 0 & 1 & 0 \\ 0 & 0 & 0 & 0 \end{bmatrix} \begin{matrix} 3 \\ 4 \\ 7 \\ 8 \end{matrix}$$

$$[\mathbf{k}^{(4)}] = \frac{AE}{\sqrt{2}L} \begin{bmatrix} 0.5 & -0.5 & -0.5 & 0.5 \\ -0.5 & 0.5 & 0.5 & -0.5 \\ -0.5 & 0.5 & 0.5 & -0.5 \\ 0.5 & -0.5 & -0.5 & 0.5 \end{bmatrix} \begin{matrix} 5 \\ 6 \\ 7 \\ 8 \end{matrix}$$

$$[\mathbf{k}^{(5)}] = \frac{AE}{\sqrt{2}L} \begin{bmatrix} 0.5 & 0.5 & -0.5 & -0.5 \\ 0.5 & 0.5 & -0.5 & -0.5 \\ -0.5 & -0.5 & 0.5 & 0.5 \\ -0.5 & -0.5 & 0.5 & 0.5 \end{bmatrix} \begin{matrix} 1 \\ 2 \\ 3 \\ 4 \end{matrix}$$

The assembly rule yields the global stiffness matrix of the unconstrained truss:

$$[\mathbf{k}] = \frac{AE}{L} \begin{bmatrix} 1.3536 & 0.3536 & -0.3536 & -0.3536 & -1 & 0 & 0 & 0 \\ 0.3536 & 1.3536 & -0.3536 & -0.3536 & 0 & 0 & 0 & -1 \\ -0.3536 & -0.3536 & 1.3536 & 0.3536 & 0 & 0 & -1 & 0 \\ -0.3536 & -0.3536 & 0.3536 & 0.3536 & 0 & 0 & 0 & 0 \\ -1 & 0 & 0 & 0 & 1.3536 & -0.3536 & -0.3536 & 0.3536 \\ 0 & 0 & 0 & 0 & -0.3536 & 0.3536 & 0.3536 & -0.3536 \\ 0 & 0 & -1 & 0 & -0.3536 & 0.3536 & 1.3536 & -0.3536 \\ 0 & -1 & 0 & 0 & 0.3536 & -0.3536 & -0.3536 & 1.3536 \end{bmatrix} \begin{matrix} 1 \\ 2 \\ 3 \\ 4 \\ 5 \\ 6 \\ 7 \\ 8 \end{matrix}$$ [a]

Unlike the previous two examples, here we must assemble both the global stiffness matrix and the global equivalent load vector $\{\mathbf{Q}_{eq}\}$. Taking each rod in turn, we use Equations 10.2.27b and 10.2.32b, respectively, to calculate the equivalent load vectors due to thermal strain and the g load. The four components of the net equivalent load vector are labeled with the corresponding element degrees of freedom.

Element 1 $(l = 0,\ m = 1)$

$$
\{Q_{eq}^{(1)}\} = \overbrace{AE\alpha\left(0.1\frac{P}{AE\alpha}\right)\begin{Bmatrix} 0 \\ -1 \\ 0 \\ 1 \end{Bmatrix}}^{\text{thermal}} + \overbrace{\frac{\rho AL}{2}\begin{Bmatrix} 0 \\ -0.05\dfrac{P}{\rho AL} \\ 0 \\ -0.05\dfrac{P}{\rho AL} \end{Bmatrix}}^{\text{acceleration}} = \begin{Bmatrix} 0 \\ -0.125P \\ 0 \\ 0.075P \end{Bmatrix} \begin{matrix} 7 \\ 8 \\ 1 \\ 2 \end{matrix}
$$

Element 2 $(l = 1,\ m = 0)$

$$
\{Q_{eq}^{(2)}\} = \overbrace{\begin{Bmatrix} 0 \\ 0 \\ 0 \\ 0 \end{Bmatrix}}^{\text{thermal}} + \overbrace{\frac{\rho AL}{2}\begin{Bmatrix} 0 \\ -0.05\dfrac{P}{\rho AL} \\ 0 \\ -0.05\dfrac{P}{\rho AL} \end{Bmatrix}}^{\text{acceleration}} = \begin{Bmatrix} 0 \\ -0.025P \\ 0 \\ -0.025P \end{Bmatrix} \begin{matrix} 5 \\ 6 \\ 1 \\ 2 \end{matrix}
$$

Element 3 $(l = 1,\ m = 0)$

$$
\{Q_{eq}^{(3)}\} = \overbrace{\begin{Bmatrix} 0 \\ 0 \\ 0 \\ 0 \end{Bmatrix}}^{\text{thermal}} + \overbrace{\frac{\rho AL}{2}\begin{Bmatrix} 0 \\ -0.05\dfrac{P}{\rho AL} \\ 0 \\ -0.05\dfrac{P}{\rho AL} \end{Bmatrix}}^{\text{acceleration}} = \begin{Bmatrix} 0 \\ -0.025P \\ 0 \\ -0.025P \end{Bmatrix} \begin{matrix} 3 \\ 4 \\ 7 \\ 8 \end{matrix}
$$

Element 4 $(l = 1/\sqrt{2},\ m = -1/\sqrt{2})$

$$
\{Q_{eq}^{(4)}\} = \overbrace{AE\alpha\left(0.1\frac{P}{AE\alpha}\right)\begin{Bmatrix} -0.7071 \\ 0.7071 \\ 0.7071 \\ -0.7071 \end{Bmatrix}}^{\text{thermal}} + \overbrace{\frac{\rho A\sqrt{2}L}{2}\begin{Bmatrix} 0 \\ -0.05\dfrac{P}{\rho AL} \\ 0 \\ -0.05\dfrac{P}{\rho AL} \end{Bmatrix}}^{\text{acceleration}} = \begin{Bmatrix} -0.07071P \\ 0.03535P \\ 0.07071P \\ -0.1061\ P \end{Bmatrix} \begin{matrix} 5 \\ 6 \\ 7 \\ 8 \end{matrix}
$$

Element 5 $(l = -1/\sqrt{2},\ m = -1/\sqrt{2})$

$$
\{Q_{eq}^{(5)}\} = \overbrace{\begin{Bmatrix} 0 \\ 0 \\ 0 \\ 0 \end{Bmatrix}}^{\text{thermal}} + \overbrace{\frac{\rho A\sqrt{2}L}{2}\begin{Bmatrix} 0 \\ -0.05\dfrac{P}{\rho AL} \\ 0 \\ -0.05\dfrac{P}{\rho AL} \end{Bmatrix}}^{\text{acceleration}} = \begin{Bmatrix} 0 \\ -0.03535P \\ 0 \\ -0.03535P \end{Bmatrix} \begin{matrix} 1 \\ 2 \\ 3 \\ 4 \end{matrix}
$$

The complete eight-component global equivalent load vector $\{Q_{eq}\}$ of the truss is assembled from those of the elements according to Equation 9.12.23. The rule is the same as that used for assembling the stiffness matrices; that is, add the components of each element's equivalent load vector into its labeled global degree of freedom location in $\{Q_{eq}\}$. Thus,

$$\{Q_{eq}\} = \begin{Bmatrix} 0 \\ 0.075P \\ 0 \\ 0 \\ 0 \\ 0 \\ 0 \\ -0.125P \end{Bmatrix}\begin{matrix}1\\2\\3\\4\\5\\6\\7\\8\end{matrix} + \begin{Bmatrix} 0 \\ -0.025P \\ 0 \\ 0 \\ 0 \\ -0.025P \\ 0 \\ 0 \end{Bmatrix}\begin{matrix}1\\2\\3\\4\\5\\6\\7\\8\end{matrix} + \begin{Bmatrix} 0 \\ 0 \\ 0 \\ -0.025P \\ 0 \\ 0 \\ 0 \\ -0.025P \end{Bmatrix}\begin{matrix}1\\2\\3\\4\\5\\6\\7\\8\end{matrix} + \begin{Bmatrix} 0 \\ 0 \\ 0 \\ 0 \\ -0.07071P \\ 0.03535P \\ 0.07071P \\ -0.1061P \end{Bmatrix}\begin{matrix}1\\2\\3\\4\\5\\6\\7\\8\end{matrix} + \begin{Bmatrix} 0 \\ -0.03535P \\ 0 \\ -0.03535P \\ 0 \\ 0 \\ 0 \\ 0 \end{Bmatrix}\begin{matrix}1\\2\\3\\4\\5\\6\\7\\8\end{matrix} = \begin{Bmatrix} 0 \\ 0.01464P \\ 0 \\ -0.06036P \\ -0.07071P \\ 0.01035P \\ 0.07071P \\ -0.2561\ P \end{Bmatrix}\begin{matrix}1\\2\\3\\4\\5\\6\\7\\8\end{matrix}$$

El. 1   El. 2   El. 3   El. 4   El. 5

Finally, we add the equivalent load vector to the point-load vector, thereby obtaining the global load vector

$$\{Q\} = \{Q_P\} + \{Q_{eq}\} \tag{b}$$

Specifically,

$$\{Q\} = \begin{Bmatrix} 0 \\ -P \\ 0 \\ 0 \\ 0 \\ 0 \\ 0 \\ 0 \end{Bmatrix} + \begin{Bmatrix} 0 \\ 0.01464P \\ 0 \\ -0.06036P \\ -0.07071P \\ 0.01035P \\ 0.07071P \\ -0.2561P \end{Bmatrix} = \begin{Bmatrix} 0 \\ -0.9854P \\ 0 \\ -0.06036P \\ -0.07071P \\ 0.01035P \\ 0.07071P \\ -0.2561\ P \end{Bmatrix}\begin{matrix}1\\2\\3\\4\\5\\6\\7\\8\end{matrix} \tag{c}$$

As in the previous example, we appeal to Equations 9.10.13 for the partitioned stiffness equations:

$$[k_{ff}]\{q_f\} = \{Q_f\} - [k_{fs}]\{q_s\} \tag{d}$$

$$\{Q_s\} = [k_{sf}]\{q_f\} + [k_{ss}]\{q_s\} \tag{e}$$

For this structure, the free f and supported s degrees of freedom are f: 1, 2, 7, 8, and s: 3, 4, 5, 6. At the supports, all of the displacements are zero; therefore,

$$\{q_s\} = \{0\} \tag{f}$$

Substituting this, together with the appropriate components of Equations a and c into Equation d, guided by our labeling scheme, we get

$$\frac{AE}{L}\begin{matrix}1\\2\\7\\8\end{matrix}\begin{bmatrix} 1.3536 & 0.3536 & 0 & 0 \\ 0.3536 & 1.3536 & 0 & -1 \\ 0 & 0 & 1.3536 & -0.3536 \\ 0 & -1 & -0.3536 & 1.3536 \end{bmatrix}\begin{Bmatrix} u_1 \\ v_1 \\ u_4 \\ v_4 \end{Bmatrix} = \begin{Bmatrix} 0 \\ -0.9854P \\ 0.07071P \\ -0.2561P \end{Bmatrix}\begin{matrix}1\\2\\7\\8\end{matrix}$$

with column headers $1 \quad 2 \quad 7 \quad 8$.

The solution for this system is

$$\{q_f\} = \begin{Bmatrix} u_1 \\ v_1 \\ u_4 \\ v_4 \end{Bmatrix} = \begin{Bmatrix} 0.6547 \\ -2.506 \\ -0.5160 \\ -2.176 \end{Bmatrix}\frac{PL}{AE} \tag{g}$$

and the displacement vector of the truss is

$$[\mathbf{q}] = \frac{PL}{AE} \lfloor \overset{1}{0.6547} \quad \overset{2}{-2.506} \;\vdots\; \overset{3}{0} \quad \overset{4}{0} \;\vdots\; \overset{5}{0} \quad \overset{6}{0} \;\vdots\; \overset{7}{-0.5160} \quad \overset{8}{-2.176} \rfloor^T \qquad \text{[h]}$$

We find the reactions $\{\mathbf{Q}_{P_s}\}$ using Equations b and e, as follows:

$$\{\mathbf{Q}_{P_s}\} = [\mathbf{k}_{sf}]\{\mathbf{q}_f\} + [\mathbf{k}_{ss}]\{\mathbf{q}_s\} - \{\mathbf{Q}_{\text{eq}_s}\}$$

$$\begin{Bmatrix} X_2 \\ Y_2 \\ X_3 \\ Y_3 \end{Bmatrix} = \frac{AE}{L} \begin{matrix} 3 \\ 4 \\ 5 \\ 6 \end{matrix} \overset{\begin{matrix} 1 & 2 & 7 & 8 \end{matrix}}{\begin{bmatrix} -0.3536 & -0.3536 & -1 & 0 \\ -0.3536 & -0.3536 & 0 & 0 \\ -1 & 0 & -0.3536 & 0.3536 \\ 0 & 0 & 0.3536 & -0.3536 \end{bmatrix}} \begin{matrix} 1 \\ 2 \\ 7 \\ 8 \end{matrix} \begin{Bmatrix} 0.6547 \\ -2.506 \\ -0.5160 \\ -2.176 \end{Bmatrix} \frac{PL}{AE}$$

$$+ \frac{AE}{L} \begin{matrix} 3 \\ 4 \\ 5 \\ 6 \end{matrix} \overset{\begin{matrix} 3 & 4 & 5 & 6 \end{matrix}}{\begin{bmatrix} 1.3536 & 0.3536 & 0 & 0 \\ 0.3536 & 0.3536 & 0 & 0 \\ 0 & 0 & 1.3536 & -0.3536 \\ 0 & 0 & -0.3536 & 0.3536 \end{bmatrix}} \begin{matrix} 3 \\ 4 \\ 5 \\ 6 \end{matrix} \begin{Bmatrix} 0 \\ 0 \\ 0 \\ 0 \end{Bmatrix} - \begin{matrix} 3 \\ 4 \\ 5 \\ 6 \end{matrix} \begin{Bmatrix} 0 \\ -0.06036P \\ -0.07071P \\ 0.01035P \end{Bmatrix}$$

or

$$\begin{Bmatrix} X_2 \\ Y_2 \\ X_3 \\ Y_3 \end{Bmatrix} = \begin{Bmatrix} 1.171\ P \\ 0.6546P \\ -1.242\ P \\ 0.5869P \end{Bmatrix} - \begin{Bmatrix} 0 \\ -0.06036P \\ -0.07071P \\ 0.01035P \end{Bmatrix} = \begin{Bmatrix} 1.171\ P \\ 0.7150P \\ -1.171\ P \\ 0.5764P \end{Bmatrix}$$

The reactions are shown on Figure 10.2.11. Because of the upward acceleration, the reactions are not in equilibrium with the applied load. However, the reader should verify that the equations of motion, which for this case are the same as Equations 2.1.3, are indeed satisfied.

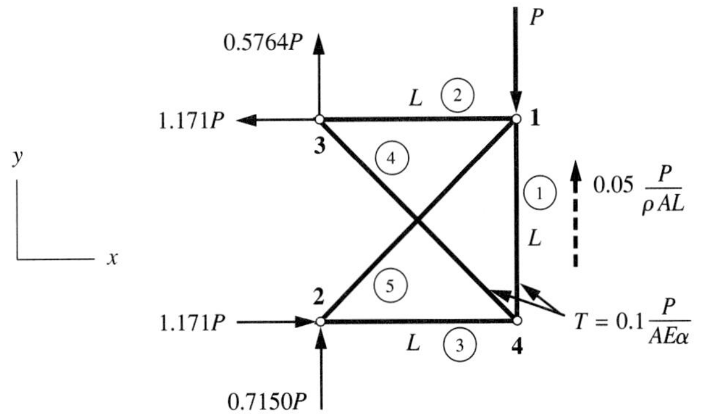

Figure 10.2.11 Reactions at the supports of the truss in Figure 10.2.10.

The stress in each element is calculated just as in Examples 10.2.1 and 10.2.2. We extract each element's global displacement vector from that of the truss, Equation h; project the result into the element coordinate system, using Equations 10.2.14; compute the strain from Equation 10.2.2; and obtain the stress, using Equation 10.2.5.

Element 1 $(l = 0, \ m = 1)$

$$\{q^{(1)}\} = \frac{PL}{AE} \left\lfloor \overset{7}{-0.5161} \quad \overset{8}{-2.176} \ \vdots \ \overset{1}{0.6547} \quad \overset{2}{-2.506} \right\rfloor^T$$

$$\{\bar{q}^{(1)}\} = [\Lambda^{(1)}]\{q^{(1)}\} = \begin{bmatrix} 0 & 1 & 0 & 0 \\ 0 & 0 & 0 & 1 \end{bmatrix} \begin{Bmatrix} -0.5160 \\ -2.176 \\ 0.6547 \\ -2.506 \end{Bmatrix} \frac{PL}{AE} = \begin{Bmatrix} -2.176 \\ -2.506 \end{Bmatrix} \frac{PL}{AE}$$

$$\varepsilon^{(1)} = [B^{(1)}]\{\bar{q}^{(1)}\} = \left\lfloor -\frac{1}{L} \quad \frac{1}{L} \right\rfloor \begin{Bmatrix} -2.176 \\ -2.506 \end{Bmatrix} \frac{PL}{AE} = -0.3307 \frac{P}{AE}$$

$$\sigma^{(1)} = E^{(1)} \left(\varepsilon^{(1)} - \alpha^{(1)} T_{\text{avg}}^{(1)} \right) = E \left[-0.3307 \frac{P}{AE} - \alpha \left(0.1 \frac{P}{AE\alpha} \right) \right] = -0.4307 \frac{P}{A}$$

Element 2 $(l = 1, \ m = 0)$

$$\{q^{(2)}\} = \frac{PL}{AE} \left\lfloor \overset{7}{0} \quad \overset{8}{0} \ \vdots \ \overset{1}{0.6547} \quad \overset{2}{-2.506} \right\rfloor^T$$

$$\{\bar{q}^{(2)}\} = [\Lambda^{(2)}]\{q^{(2)}\} = \begin{bmatrix} 1 & 0 & 0 & 0 \\ 0 & 0 & 1 & 0 \end{bmatrix} \begin{Bmatrix} 0 \\ 0 \\ 0.6547 \\ -2.506 \end{Bmatrix} \frac{PL}{AE} = \begin{Bmatrix} 0 \\ 0.6547 \end{Bmatrix} \frac{PL}{AE}$$

$$\varepsilon^{(2)} = [B^{(2)}]\{\bar{q}^{(2)}\} = \left\lfloor -\frac{1}{L} \quad \frac{1}{L} \right\rfloor \begin{Bmatrix} 0 \\ 0.6547 \end{Bmatrix} \frac{PL}{AE} = 0.6547 \frac{P}{AE}$$

$$\sigma^{(2)} = E^{(2)} \left(\varepsilon^{(2)} - \alpha^{(2)} T_{\text{avg}}^{(2)} \right) = E \left[0.6547 \frac{P}{AE} - \alpha (0) \right] = 0.6547 \frac{P}{A}$$

Element 3 $(l = 1, \ m = 0)$

$$\{q^{(3)}\} = \frac{PL}{AE} \left\lfloor \overset{3}{0} \quad \overset{4}{0} \ \vdots \ \overset{7}{-0.5160} \quad \overset{8}{-2.176} \right\rfloor^T$$

$$\{\bar{q}^{(3)}\} = [\Lambda^{(3)}]\{q^{(3)}\} = \begin{bmatrix} 1 & 0 & 0 & 0 \\ 0 & 0 & 1 & 0 \end{bmatrix} \begin{Bmatrix} 0 \\ 0 \\ -0.5160 \\ -2.176 \end{Bmatrix} \frac{PL}{AE} = \begin{Bmatrix} 0 \\ -0.5160 \end{Bmatrix} \frac{PL}{AE}$$

$$\varepsilon^{(3)} = \left[\mathbf{B}^{(3)}\right]\{\bar{\mathbf{q}}^{(3)}\} = \left\lfloor -\frac{1}{L} \quad \frac{1}{L} \right\rfloor \begin{Bmatrix} 0 \\ \overline{-0.5160} \end{Bmatrix} \frac{PL}{AE} = -0.5160\frac{P}{AE}$$

$$\sigma^{(3)} = E^{(3)}\left(\varepsilon^{(3)} - \alpha^{(3)}T_{\text{avg}}^{(3)}\right) = E\left[-0.5160\frac{P}{AE} - \alpha\,(0)\right] = -0.5160\frac{P}{A}$$

Element 4 ($l = 1/\sqrt{2}$, $m = -1/\sqrt{2}$)

$$\{\mathbf{q}^{(4)}\} = \frac{PL}{AE}\left\lfloor \overset{5}{0} \quad \overset{6}{0} \quad \overset{7}{-0.5160} \quad \overset{8}{-2.176} \right\rfloor^{T}$$

$$\{\bar{\mathbf{q}}^{(4)}\} = \left[\mathbf{\Lambda}^{(4)}\right]\{\mathbf{q}^{(4)}\} = \begin{bmatrix} \frac{1}{\sqrt{2}} & -\frac{1}{\sqrt{2}} & 0 & 0 \\ 0 & 0 & \frac{1}{\sqrt{2}} & -\frac{1}{\sqrt{2}} \end{bmatrix} \begin{Bmatrix} 0 \\ 0 \\ -0.5160 \\ -2.176 \end{Bmatrix}\frac{PL}{AE} = \begin{Bmatrix} 0 \\ 1.174 \end{Bmatrix}\frac{PL}{AE}$$

$$\varepsilon^{(4)} = \left[\mathbf{B}^{(4)}\right]\{\bar{\mathbf{q}}^{(4)}\} = \left\lfloor -\frac{1}{\sqrt{2}L} \quad \frac{1}{\sqrt{2}L} \right\rfloor \begin{Bmatrix} 0 \\ 1.174 \end{Bmatrix}\frac{PL}{AE} = 0.8298\frac{P}{AE}$$

$$\sigma^{(4)} = E^{(4)}\left(\varepsilon^{(4)} - \alpha^{(4)}T_{\text{avg}}^{(4)}\right) = E\left[0.8298\frac{P}{AE} - \alpha\left(0.1\frac{P}{AE\alpha}\right)\right] = 0.7298\frac{P}{A}$$

Element 5 ($l = -1/\sqrt{2}$, $m = -1/\sqrt{2}$)

$$\{\mathbf{q}^{(5)}\} = \frac{PL}{AE}\left\lfloor \overset{1}{0.6547} \quad \overset{2}{-2.506} \quad \overset{3}{0} \quad \overset{4}{0} \right\rfloor^{T}$$

$$\{\bar{\mathbf{q}}^{(5)}\} = \left[\mathbf{\Lambda}^{(5)}\right]\{\mathbf{q}^{(5)}\} = \begin{bmatrix} -\frac{1}{\sqrt{2}} & -\frac{1}{\sqrt{2}} & 0 & 0 \\ 0 & 0 & -\frac{1}{\sqrt{2}} & -\frac{1}{\sqrt{2}} \end{bmatrix} \begin{Bmatrix} 0.6547 \\ -2.506 \\ 0 \\ 0 \end{Bmatrix}\frac{PL}{AE} = \begin{Bmatrix} 1.309 \\ 0 \end{Bmatrix}\frac{PL}{AE}$$

$$\varepsilon^{(5)} = \left[\mathbf{B}^{(5)}\right]\{\bar{\mathbf{q}}^{(5)}\} = \left\lfloor -\frac{1}{\sqrt{2}L} \quad \frac{1}{\sqrt{2}L} \right\rfloor \begin{Bmatrix} 1.309 \\ 0 \end{Bmatrix}\frac{PL}{AE} = -0.9258\frac{P}{AE}$$

$$\sigma^{(5)} = E^{(5)}\left(\varepsilon^{(5)} - \alpha^{(5)}T_{\text{avg}}^{(5)}\right) = E\left[-0.9256\frac{P}{AE} - \alpha\,(0)\right] = -0.9258\frac{P}{A}$$

The stress analysis is summarized in Table 10.2.6. Note that the stress in each rod is constant throughout its length, because the temperature change T was assumed to be uniform. Therefore, the loads applied to the nodes at each end are identical.

The reader should verify that if each node of the truss is assigned one-half of the total mass of the rods attached to it, the equations of motion ($\sum F_x = ma_x$, $\sum F_y = ma_y$) are satisfied at each node.

Table 10.2.6   Axial loads in Figure 10.2.10.

Element	Axial Stress
1	0.4307 P/A Compression
2	0.6547 P/A Tension
3	0.5160 P/A Compression
4	0.7298 P/A Tension
5	0.9258 P/A Compression

The linear displacement field assumed at the outset in Equation 10.2.2 is the *exact* displacement field for a cylindrical bar in equilibrium under centroidal axial load. Since the stiffness matrix, Equation 10.2.7, is based on stress and strain fields that satisfy the equations of elasticity, it is the *exact* stiffness matrix for the rod element. For elements more complex than the two-node cylindrical bar, the assumed displacement field used to derive the stiffness matrix is likely to violate equilibrium within the element, even though the virtual work principle ensures that the element's nodal loads will be self-equilibrating.

To illustrate, consider the linearly tapered circular rod element in Figure 10.2.12. The radii of the two ends are r_1 and r_2, respectively, and the radius of an intermediate section is

$$r = r_1 + (r_2 - r_1)x/L$$

The taper is assumed to be slight, so that the *taper ratio* $\zeta = (r_2 - r_1)/r_1$ is small. In terms of ζ,

$$r = r_1[1 + \zeta(x/L)]$$

so that the area of a cross section at station x is

$$A = \pi r^2 = A_1 \left[1 + \zeta \frac{x}{L}\right]^2 \tag{10.2.33}$$

If we assume a linear displacement, just as we did for the straight rod, then we are once again led to Equation 10.2.6. However, for the tapered rod, we have

$$\iiint_V dV = \int_0^L A \, dx = \int_0^L A_1 \left[1 + \zeta \frac{x}{L}\right]^2 = A_1 L \left(1 + \zeta + \frac{1}{3}\zeta^2\right)$$

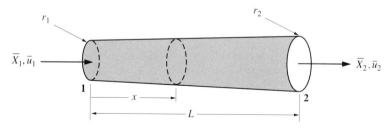

Figure 10.2.12   Tapered rod element.

Therefore, we obtain the following for the stiffness matrix of the tapered rod:

$$[\mathbf{k}] = \left(1 + \zeta + \frac{1}{3}\zeta^2\right) \frac{A_1 E}{L} \begin{bmatrix} 1 & -1 \\ -1 & 1 \end{bmatrix}$$ [10.2.34]

When the taper ratio ζ is zero, this reduces to that of a cylindrical rod, Equation 10.2.7. For a given displacement vector $\{\bar{\mathbf{q}}\} = \lfloor \bar{u}_1 \quad \bar{u}_2 \rfloor^T$, the stiffness equations $\{\bar{\mathbf{Q}}\} = [\bar{\mathbf{k}}]\{\bar{\mathbf{q}}\}$ imply that

$$\bar{X}_1 = -\left(1 + \zeta + \frac{1}{3}\zeta^2\right) \frac{A_1 E}{L}(\bar{u}_2 - \bar{u}_1)$$ [10.2.35]

and

$$\bar{X}_2 = -\bar{X}_1$$ [10.2.36]

For equilibrium of the free body shown in Figure 10.2.13, the axial load N at any point x within the rod must be related to $\bar{X}_1$ by the equation $N = -\bar{X}_1$; that is,

$$N = \left(1 + \zeta + \frac{1}{3}\zeta^2\right) \frac{A_1 E}{L}(\bar{u}_2 - \bar{u}_1)$$ [10.2.37]

On the other hand, N can be found independently in terms of the axial stress, as follows:

$$N = \sigma_x A = EA\varepsilon_x$$ [10.2.38]

Following our linear displacement assumption for the element, ε_x is obtained from Equation 10.2.4, which, with the area A given by Equation 10.2.33, implies that

$$N = \left(1 + 2\frac{x}{L}\zeta + \frac{x^2}{L^2}\zeta^2\right) \frac{A_1 E}{L}(\bar{u}_2 - \bar{u}_1)$$ [10.2.39]

This equation clearly does not agree with the equilibrium equation (Equation 10.2.37) unless $\zeta = 0$, that is, unless the rod has no taper. Therefore, requiring the displacement field in the tapered rod to be linear in x imposes a nonequilibrium stress distribution, even though the rod as a whole *is* in equilibrium, according to Equation 10.2.36. The absence of internal equilibrium reflects the fact that the displacement assumed in Equation 10.2.1 is not correct for a tapered rod. Therefore, the stiffness matrix of Equation 10.2.34 cannot be the rod's "exact" stiffness matrix.

We can find the exact stiffness matrix if we can come up with the right displacement field. To do so, we must require the rod to be in equilibrium and solve for the corresponding displacement function. Starting with the free body in Figure 10.2.13, the equilibrium condition is

$$N = -\bar{X}_1$$

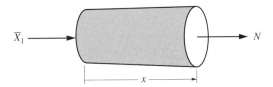

Figure 10.2.13   Internal axial load N at any point x in the rod of Figure 10.2.12.

Substituting Equation 10.2.38, the strain–displacement relation $\varepsilon_x = du/dx$, and the area expression, Equation 10.2.33, we get

$$\frac{du}{dx} = -\frac{\bar{X}_1}{AE} = -\frac{\bar{X}_1}{A_1 E} \frac{1}{(1 + \zeta x/L)^2}$$

Performing the indicated quadrature leads to

$$u = \frac{1 - x/L}{1 + \zeta x/L}\bar{u}_1 + \frac{(1 + \zeta)x/L}{1 + \zeta x/L}\bar{u}_2 \qquad \text{[10.2.40]}$$

In terms of the nodal displacements, the normal strain is therefore

$$\varepsilon_x = \frac{du}{dx} = \frac{1 + \zeta}{(1 + \zeta x/L)^2} \lfloor -\frac{1}{L} \quad \frac{1}{L} \rfloor \begin{Bmatrix} \bar{u}_1 \\ \bar{u}_2 \end{Bmatrix}$$

which implies that the strain–displacement matrix is

$$[\mathbf{B}] = \frac{1 + \zeta}{(1 + \zeta x/L)^2} \lfloor -\frac{1}{L} \quad \frac{1}{L} \rfloor \qquad \text{[10.2.41]}$$

Substituting $[\mathbf{B}]$ and $[\mathbf{E}] = E$ into Equation 9.12.16 yields

$$[\mathbf{k}] = \int_0^L E\,[\mathbf{B}]^T\,[\mathbf{B}]\,A\,dx = E \left\{ \int_0^L \left[\frac{1 + \zeta}{(1 + \zeta x/L)^2} \right]^2 A\,dx \right\} \begin{bmatrix} \frac{1}{L^2} & -\frac{1}{L^2} \\ -\frac{1}{L^2} & \frac{1}{L^2} \end{bmatrix}$$

This simplifies considerably to become

$$[\mathbf{k}] = (1 + \zeta)\frac{A_1 E}{L} \begin{bmatrix} 1 & -1 \\ -1 & 1 \end{bmatrix} \qquad \text{[10.2.42]}$$

We now have two stiffness matrices, Equations 10.2.34 and 10.2.42, for the slightly tapered rod element. Equation 10.2.42 is the more accurate of the two, since the displacement field (Equation 10.2.40) upon which it is based represents the exact stress distribution. A comparison of the two expressions for $[\mathbf{k}]$ shows that modeling the tapered rod with a linear displacement field yields an element that is too stiff. This is a general characteristic of assumed-displacement elements.

10.3   BEAM ELEMENT STIFFNESS EQUATIONS

Recall that a beam is a slender structural member that can transmit both bending and axial loads. Because we have adopted the small-deflection assumption of Bernoulli-Euler beam theory (section 4.5), the flexural stiffness of a beam is not coupled to its axial stiffness behavior,[3] which is the same as that of the rod element. Therefore, we may temporarily ignore a beam's axial load-bearing capability and focus on the shear and bending loads. Furthermore, as long as we deal with colinear assemblies of beam elements, there is no need to distinguish between global and local coordinates.

| [3]No axial load will arise if a transverse load is applied to a beam, regardless of how it is restrained at each end.

Consider a simple beam of symmetric cross section subjected to bending loads in the xy plane, as illustrated in Figure 10.3.1. Due to the loading, the neutral axis of the beam bends into a shape described by the elastic curve $v(x)$. This elastic curve is very shallow, so that $\tan^{-1}(dv/dx) = dv/dx$, that is, the slope of the curve at a point equals the angle (in radians) through which a differential line segment at that point rotates. A general load, such as that shown in part (a) of the figure, induces both shear force and bending moment throughout the beam. The shape of the deformed beam is influenced almost entirely by bending. Shear plays a negligible role in slender beams, as we showed in Chapter 7.

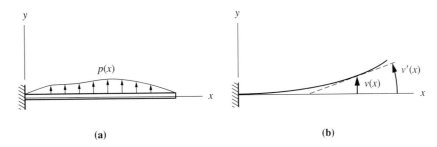

Figure 10.3.1   (a) Simple beam loaded arbitrarily in the xy plane. (b) Elastic curve.

Suppose we assume that the transverse deflection $v(x)$ at any point on a beam is a cubic polynomial in x, measured from the left end, as shown in Figure 10.3.2. Before discussing the validity of this assumption, let us draw the conclusions based on

$$v = a_o + a_1 x + a_2 x^2 + a_3 x^3 \qquad \text{[10.3.1]}$$

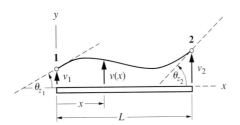

Figure 10.3.2   Deformed simple beam element and its nodal displacements.

If this is true, then the slope of the deflection curve, which may be referred to as the *rotation*,[4] is given by

$$\theta_z = \frac{dv}{dx} = a_1 + 2a_2 x + 3a_3 x^2 \qquad \text{[10.3.2]}$$

These expressions for the deflection and the slope must yield v_1 and θ_{z_1} at node 1 and v_2 and θ_{z_2} at node 2, that is,

| [4]The rotation of a line element on the neutral surface about the z axis.

$$v(0) = v_1 = a_o$$

$$\theta_z(0) = \theta_{z_1} = a_1$$

$$v(L) = v_2 = a_o + a_1 L + a_2 L^2 + a_3 L^3$$

$$\theta_z(L) = \theta_{z_2} = a_1 + 2a_2 L + 3a_3 L^2$$

In matrix form, these four equations are

$$
\begin{bmatrix}
1 & 0 & 0 & 0 \\
0 & 1 & 0 & 0 \\
1 & 1 & 1 & 1 \\
0 & 1 & 2 & 3
\end{bmatrix}
\begin{Bmatrix}
a_o \\
La_1 \\
L^2 a_2 \\
L^3 a_3
\end{Bmatrix}
=
\begin{Bmatrix}
v_1 \\
L\theta_{z_1} \\
v_2 \\
L\theta_{z_2}
\end{Bmatrix}
\qquad [10.3.3]
$$

which has the solution

$$a_o = v_1$$

$$a_1 = \theta_1$$

$$a_2 = -\frac{3}{L^2} v_1 - \frac{2}{L}\theta_{z_1} + \frac{3}{L^2} v_2 - \frac{1}{L}\theta_{z_2} \qquad [10.3.4]$$

$$a_3 = \frac{2}{L^3} v_1 + \frac{1}{L^2}\theta_{z_1} - \frac{2}{L^3} v_2 + \frac{1}{L^2}\theta_{z_2}$$

Substituting these values of the polynomial coefficients into Equation 10.3.1 yields the following formula for $v(x)$ in terms of the nodal displacements:

$$v = \lfloor N \rfloor \{q\} \qquad [10.3.5]$$

where $\lfloor N \rfloor$ is the row vector of four cubic-polynomial shape functions,

$$\lfloor N \rfloor = \left\lfloor 2\frac{x^3}{L^3} - 3\frac{x^2}{L^2} + 1 \quad \frac{x^3}{L^2} - 2\frac{x^2}{L} + x \quad -2\frac{x^3}{L^3} + 3\frac{x^2}{L^2} \quad \frac{x^3}{L^2} - \frac{x^2}{L} \right\rfloor \qquad [10.3.6]$$

and $\{q\}$ is the four-component element displacement vector,

$$\{q\} = \lfloor v_1 \quad \theta_{z_1} \quad v_2 \quad \theta_{z_2} \rfloor^T \qquad [10.3.7]$$

Since we are considering bending in only one plane (the xy plane), the axial mechanical strain, obtained from Equations 4.5.6a, is

$$\varepsilon_x = -y\frac{d^2 v}{dx^2} \qquad [10.3.8]$$

Differentiating Equation 10.3.5, and therefore each of the four polynomials in Equation 10.3.6, twice yields

$$v'' = \lfloor B \rfloor \{q\} \qquad [10.3.9]$$

where $\lfloor B \rfloor$ is the row matrix whose components are the second derivatives of the components of $\lfloor N \rfloor$:

$$\lfloor B \rfloor = \left\lfloor \frac{12x}{L^3} - \frac{6}{L^2} \quad \frac{6x}{L^2} - \frac{4}{L} \quad -\frac{12x}{L^3} + \frac{6}{L^2} \quad \frac{6x}{L^2} - \frac{2}{L} \right\rfloor \qquad [10.3.10]$$

Thus,

$$\varepsilon_x = -y\lfloor B \rfloor \{q\}$$

which means that the strain–displacement matrix for the beam element is

$$\lfloor \mathbf{B} \rfloor = -y \lfloor \boldsymbol{\mathcal{B}} \rfloor \qquad \text{[10.3.11]}$$

We may refer to $\lfloor \boldsymbol{\mathcal{B}} \rfloor$ as the generalized strain–displacement matrix, because it relates the beam's generalized strain—the curvature v''—to the displacement vector $\{\mathbf{q}\}$.

We are now in a position to calculate the beam element stiffness matrix. Since ε_x is the only nonzero strain (cf. Equations 4.5.6), the material stiffness matrix $[\mathbf{E}]$ reduces to $[\mathbf{E}] = E$, the modulus of elasticity, just as it did for the rod element. Substituting this and the strain–displacement matrix $[\mathbf{B}]$ into the formula for the stiffness matrix, Equation 9.12.16, yields

$$[\mathbf{k}] = \iiint_V [\mathbf{B}][\mathbf{E}][\mathbf{B}]^T \, dV = \iiint_V Ey^2 \lfloor \boldsymbol{\mathcal{B}} \rfloor \lfloor \boldsymbol{\mathcal{B}} \rfloor^T \, dV = \int_0^L \left(\iint_A Ey^2 dA \right) \lfloor \boldsymbol{\mathcal{B}} \rfloor \lfloor \boldsymbol{\mathcal{B}} \rfloor^T \, dx \qquad \text{[10.3.12]}$$

Assuming that the modulus of elasticity does not vary over the cross section A of the beam, we have

$$\iint_A Ey^2 dA = E \iint_A y^2 dA = EI_z$$

where I_z is the moment of inertia of the cross section. Substituting Equation 10.3.10 for $\lfloor \boldsymbol{\mathcal{B}} \rfloor$, Equation 10.3.12 becomes

$$[\mathbf{k}] = \int_0^L EI_z \begin{Bmatrix} \frac{12x}{L^3} - \frac{6}{L^2} \\ \frac{6x}{L^2} - \frac{4}{L} \\ -\frac{12x}{L^3} + \frac{6}{L^2} \\ \frac{6x}{L^2} - \frac{2}{L} \end{Bmatrix} \lfloor \frac{12x}{L^3} - \frac{6}{L^2} \quad \frac{6x}{L^2} - \frac{4}{L} \quad -\frac{12x}{L^3} + \frac{6}{L^2} \quad \frac{6x}{L^2} - \frac{2}{L} \rfloor dx$$

$$= \int_0^L EI_z \begin{bmatrix} \left(\frac{12x}{L^3} - \frac{6}{L^2}\right)^2 & \left(\frac{12x}{L^3} - \frac{6}{L^2}\right)\left(\frac{6x}{L^2} - \frac{4}{L}\right) & \left(\frac{12x}{L^3} - \frac{6}{L^2}\right)\left(-\frac{12x}{L^3} + \frac{6}{L^2}\right) & \left(\frac{12x}{L^3} - \frac{6}{L^2}\right)\left(\frac{6x}{L^2} - \frac{2}{L}\right) \\ \left(\frac{6x}{L^2} - \frac{4}{L}\right)\left(\frac{12x}{L^3} - \frac{6}{L^2}\right) & \left(\frac{6x}{L^2} - \frac{4}{L}\right)^2 & \left(\frac{6x}{L^2} - \frac{4}{L}\right)\left(-\frac{12x}{L^3} + \frac{6}{L^2}\right) & \left(\frac{6x}{L^2} - \frac{4}{L}\right)\left(\frac{6x}{L^2} - \frac{2}{L}\right) \\ \left(-\frac{12x}{L^3} + \frac{6}{L^2}\right)\left(\frac{12x}{L^3} - \frac{6}{L^2}\right) & \left(-\frac{12x}{L^3} + \frac{6}{L^2}\right)\left(\frac{6x}{L^2} - \frac{4}{L}\right) & \left(-\frac{12x}{L^3} + \frac{6}{L^2}\right)^2 & \left(-\frac{12x}{L^3} + \frac{6}{L^2}\right)\left(\frac{6x}{L^2} - \frac{2}{L}\right) \\ \left(\frac{6x}{L^2} - \frac{2}{L}\right)\left(\frac{12x}{L^3} - \frac{6}{L^2}\right) & \left(\frac{6x}{L^2} - \frac{2}{L}\right)\left(\frac{6x}{L^2} - \frac{4}{L}\right) & \left(\frac{6x}{L^2} - \frac{2}{L}\right)\left(-\frac{12x}{L^3} + \frac{6}{L^2}\right) & \left(\frac{6x}{L^2} - \frac{2}{L}\right)^2 \end{bmatrix} dx$$

With the additional assumption that the flexural rigidity EI_z is constant over the length of the beam, the 10 independent integrations can be carried out to obtain the local flexural stiffness matrix of a simple cubic-displacement beam element:

$$[\mathbf{k}] = \frac{EI_z}{L^3} \begin{array}{cccc} v_1 & \theta_{z_1} & v_2 & \theta_{z_2} \\ \begin{bmatrix} 12 & 6L & -12 & 6L \\ 6L & 4L^2 & -6L & 2L^2 \\ -12 & -6L & 12 & -6L \\ 6L & 2L^2 & -6L & 4L^2 \end{bmatrix} & \begin{matrix} Y_1 \\ M_{z_1} \\ Y_2 \\ M_{z_2} \end{matrix} \end{array} \qquad \text{[10.3.13]}$$

The four degree of freedom beam element is illustrated again in Figure 10.3.3, showing the transverse force Y and point couple M_z accompanying the displacements at each node. The local element load vector $\{\mathbf{Q}\}$ corresponding to the displacement vector $\{\mathbf{q}\}$ in Equation 10.3.7 is

$$\{\mathbf{Q}\} = \lfloor Y_1 \quad M_{z_1} \quad Y_2 \quad M_{z_2} \rfloor^T \qquad \text{[10.3.14]}$$

As in the rod element, the uniaxial stress at a given point in the isotropic beam is found from the constitutive equation

$$\sigma_x = E(\varepsilon_x - \alpha T)$$

The moment resultant of this stress distribution around the neutral axis is the bending moment M_z at station x (cf. section 4.6), which is

$$M_z = -\iint_A y\sigma_x dA$$

Substituting the stress–strain relationship and Equation 10.3.8 into the integral yields

$$M_z = \iint_A Ev''y^2 dA + \iint_A E\alpha Ty dA \qquad\qquad \text{[10.3.15]}$$

If the beam is homogeneous, then E and α do not depend on y, and, since $I_z = \iint_A y^2 dA$, we obtain

$$M_z = EI_z v'' + E\alpha \iint_A Ty dA \qquad\qquad \text{[10.3.16]}$$

In the absence of a transverse thermal gradient, the bending moment M_z within the beam is simply

$$M_z = EI_z v'' \qquad\qquad \text{[10.3.17]}$$

We have seen this moment–curvature relationship before (e.g., Equations 4.6.3). Substituting Equation 10.3.10 yields

$$M_z = EI_z \lfloor \boldsymbol{\mathcal{B}} \rfloor \{q\} \qquad\qquad \text{[10.3.18]}$$

The bending moment may be considered a generalized stress and the flexural rigidity EI_z a generalized elastic modulus.

Finding the bending moment from Equation 10.3.18, using just the computed displacements and without invoking statics, is consistent with the finite element displacement method of structural analysis. The finite element in this case is the simple beam.

We can now examine the validity of the assumption on which these findings depend, namely, that the variation of the transverse deflection of a beam is that of a cubic polynomial. Equation 4.6.7 gives the axial stress in a beam in terms of the bending moment. If $M_y = 0$, which is true for the case at hand, we have

$$\sigma_x = -\frac{M_z y}{I_z}$$

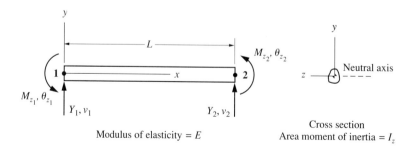

Modulus of elasticity = E

Cross section
Area moment of inertia = I_z

Figure 10.3.3   Flexural loads and displacements at the nodes of a two-dimensional beam element in the element reference system.

Combining this with Equation 4.6.1 (with $w = 0$) yields

$$\frac{d^2v}{dx^2} = \frac{M_z}{EI_z}$$

[10.3.19]

For v to be a cubic polynomial in x, its second derivative d^2v/dx^2 must be linear in x. This is so only if the bending moment itself is linear in x, according to Equation 10.3.19. The bending moment is the same linear function of x across the entire span of the beam only if no loads are applied between its endpoints. Thus, for Equation 10.3.6 to be exact, the only loads acting on the beam must be transverse shear forces and couples at each end, as shown in Figure 10.3.4. If intermediate point loads are applied to the beam, as in Figure 10.3.5, then the bending moment remains a linear function of x, $M(x) = c_0 + c_1 x$; however, the coefficients c_o and c_1 are different for regions of the beam between different pairs of point loads. The internal bending moment is piecewise linear. Therefore, according to Equation 10.3.19, the deflection curve $v(x)$ is a piecewise cubic polynomial. The only way to represent a piecewise cubic displacement field precisely, using the beam element of Figures 10.3.2 and 10.3.4, is to join several of these elements end to end, as illustrated in Figure 10.3.6. In this way, a single beam is regarded as an assembly of smaller beam elements.

Beams commonly support distributed loads, such as that illustrated in Figure 10.3.1a. The simplest of all such loads, the uniformly distributed load, produces a bending moment that, from statics, is quadratic in x, so that Equation 10.3.19 requires the displacement to be a fourth-degree polynomial. We cannot precisely represent a fourth-degree polynomial by a third-degree polynomial. However, by considering a beam with a distributed load to be composed of a number of smaller, cubic-displacement beam elements, we can approximate the actual displacement field as a patchwork of cubic displacement fields. The more elements we use, the better will be the approximation.

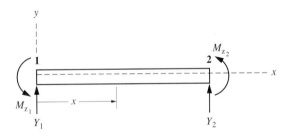

Figure 10.3.4   Simple beam to which no intermediate loads are applied.

Its transverse deflection is a cubic polynomial in x.

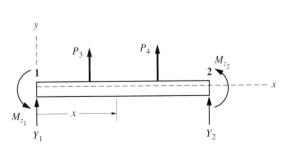

Figure 10.3.5   Simple beam in which the deflection is a piecewise cubic function of x.

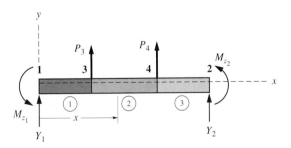

Figure 10.3.6   The simple beam of Figure 10.3.5 modeled as three cubic-displacement elements like that of Figure 10.3.4, joined end to end.

Instead of using several small beam elements to represent a single larger beam, we might seek an alternative to the cubic element of Figure 10.3.2, one with more built-in complexity that yields a more accurate display of the effects of intermediate loading. Such investigations, however, are beyond the scope of this text.

Example 10.3.1   Use the finite element method to find the bending moment distribution in the statically indeterminate beam in Figure 10.3.7.

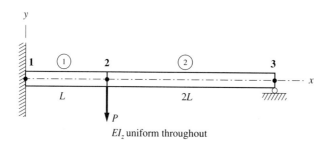

Figure 10.3.7   Loaded and constrained two-element beam assembly.

For greatest accuracy, a node is required where the point load P is applied. Therefore, we regard this beam as an assembly of two beam elements. The unconstrained structure has six degrees of freedom (three nodes, two degrees of freedom per node).

First, we employ Equation 10.3.13 to compute the element stiffness matrices, whose rows and columns will be labeled with the global degree of freedom numbers spanned by the element degrees of freedom. Keep in mind that the individual beam elements are aligned as shown in Figure 10.3.3, each with its first node to the left and last node to the right. Since the local and global coordinate directions coincide, the local and global forms of the element stiffness matrices are identical. Element 1 of length L has a flexural rigidity EI_z; therefore, its stiffness matrix is identical to that in Equation 10.3.13, or

$$
\left[\mathbf{k}^{(1)}\right] = \frac{EI_z}{L^3}
\begin{array}{c}
\begin{array}{cccc} 1 & 2 & 3 & 4 \end{array} \\
\left[
\begin{array}{cc:cc}
12 & 6L & -12 & 6L \\
6L & 4L^2 & -6L & 2L^2 \\
\hdashline
-12 & -6L & 12 & -6L \\
6L & 2L^2 & -6L & 4L^2
\end{array}
\right]
\begin{array}{c} 1 \\ 2 \\ 3 \\ 4 \end{array}
\end{array}
\qquad \text{[a]}
$$

Beam 2 has the same flexural rigidity EI_z as beam 1, but its length is $2L$. Replacing L by $2L$ everywhere in Equation 10.3.13 furnishes the applicable stiffness matrix

$$
\left[\mathbf{k}^{(2)}\right] = \frac{EI_z}{L^3}
\begin{array}{c}
\begin{array}{cccc} 3 & 4 & 5 & 6 \end{array} \\
\left[
\begin{array}{cc:cc}
\frac{12}{8} & \frac{12}{8}L & -\frac{12}{8} & \frac{12}{8}L \\
\frac{12}{8}L & \frac{16}{8}L^2 & -\frac{12}{8}L & \frac{8}{8}L^2 \\
\hdashline
-\frac{12}{8} & -\frac{12}{8}L & \frac{12}{8} & -\frac{12}{8}L \\
\frac{12}{8}L & \frac{8}{8}L^2 & -\frac{12}{8}L & 4L^2
\end{array}
\right]
\begin{array}{c} 3 \\ 4 \\ 5 \\ 6 \end{array}
\end{array}
\qquad \text{[b]}
$$

Next, we assemble these matrices into the 6 by 6 global array of the unconstrained structure, to obtain

$$
[\mathbf{k}] = \frac{EI_z}{L^3}
\begin{array}{c}
\begin{array}{cccccc} 1 & 2 & 3 & 4 & 5 & 6 \end{array} \\
\left[
\begin{array}{cccccc}
12 & 6L & -12 & 6L & 0 & 0 \\
6L & 4L^2 & -6L & 2L^2 & 0 & 0 \\
-12 & -6L & \frac{27}{2} & -\frac{9}{2}L & -\frac{3}{2} & \frac{3}{2}L \\
6L & 2L^2 & -\frac{9}{2}L & 6L^2 & -\frac{3}{2}L & L^2 \\
0 & 0 & -\frac{3}{2} & -\frac{3}{2}L & \frac{3}{2} & -\frac{3}{2}L \\
0 & 0 & \frac{3}{2}L & L^2 & -\frac{3}{2}L & 2L^2
\end{array}
\right]
\begin{array}{c} 1 \\ 2 \\ 3 \\ 4 \\ 5 \\ 6 \end{array}
\end{array}
\qquad [\mathbf{c}]
$$

The global load vector and displacement vector are

$$
\{\mathbf{Q}\} = \lfloor Y_1 \quad M_{z_1} \quad -P \quad 0 \quad Y_3 \quad 0 \rfloor^T \qquad \{\mathbf{q}\} = \lfloor 0 \quad 0 \quad v_2 \quad \theta_{z_2} \quad 0 \quad \theta_{z_3} \rfloor^T \qquad [\mathbf{d}]
$$

where Y_1, M_{z_1}, and Y_3 are the unknown support reactions, and v_2, θ_{z_2}, and θ_{z_3} are the unknown nodal displacements.

Global degrees of freedom 1, 2, and 5 are constrained by the supports to be zero. The set of constrained freedoms s is therefore: 1, 2, 5. The remaining three degrees of freedom belong to the set f of free degrees of freedom: 3, 4, 6. Thus, in terms of the partitioning scheme of Equation 9.10.12, we have

$$
[\mathbf{k}_{ff}] = \begin{array}{c} \begin{array}{ccc} 3 & 4 & 6 \end{array} \\ \begin{array}{c} 3 \\ 4 \\ 6 \end{array} \left[\quad \mathbf{k} \quad \right] \end{array}
\quad
[\mathbf{k}_{fs}] = \begin{array}{c} \begin{array}{ccc} 1 & 2 & 5 \end{array} \\ \begin{array}{c} 3 \\ 4 \\ 6 \end{array} \left[\quad \mathbf{k} \quad \right] \end{array}
\quad
[\mathbf{k}_{ss}] = \begin{array}{c} \begin{array}{ccc} 1 & 2 & 5 \end{array} \\ \begin{array}{c} 1 \\ 2 \\ 5 \end{array} \left[\quad \mathbf{k} \quad \right] \end{array}
\quad
[\mathbf{k}_{sf}] = \begin{array}{c} \begin{array}{ccc} 3 & 4 & 6 \end{array} \\ \begin{array}{c} 1 \\ 2 \\ 5 \end{array} \left[\quad \mathbf{k} \quad \right] \end{array}
\qquad [\mathbf{e}]
$$

where $[\mathbf{k}]$ is given by Equation c, and

$$
\{\mathbf{q}_f\} = \begin{Bmatrix} v_2 \\ \theta_{z_2} \\ \theta_{z_3} \end{Bmatrix}
\quad
\{\mathbf{q}_s\} = \begin{Bmatrix} 0 \\ 0 \\ 0 \end{Bmatrix}
\quad
\{\mathbf{Q}_f\} = \begin{Bmatrix} -P \\ 0 \\ 0 \end{Bmatrix}
\quad
\{\mathbf{Q}_s\} = \begin{Bmatrix} Y_1 \\ M_{z_1} \\ Y_3 \end{Bmatrix}
\qquad [\mathbf{f}]
$$

It follows that for this problem, the system of Equations 9.10.14 for the unknown displacements $\{\mathbf{q}_f\}$, $[\mathbf{k}_{ff}]\{\mathbf{q}_f\} = \{\mathbf{Q}_f\}$, is

$$
\frac{EI_z}{L^3}
\begin{bmatrix}
\frac{27}{2} & -\frac{9}{2}L & \frac{3}{2}L \\
-\frac{9}{2}L & 6L^2 & L^2 \\
\frac{3}{2}L & L^2 & 2L^2
\end{bmatrix}
\begin{Bmatrix} v_2 \\ \theta_{z_2} \\ \theta_{z_3} \end{Bmatrix}
=
\begin{Bmatrix} -P \\ 0 \\ 0 \end{Bmatrix}
\qquad [\mathbf{g}]
$$

Solving for the displacement yields

$$
\{\mathbf{q}_f\} = \begin{Bmatrix} v_2 \\ \theta_{z_2} \\ \theta_{z_3} \end{Bmatrix} = \begin{Bmatrix} -0.1358PL^3/EI_z \\ -0.1296PL^2/EI_z \\ 0.1666PL^2/EI_z \end{Bmatrix}
\qquad [\mathbf{h}]
$$

The deflected beam is shown in Figure 10.3.8. The curve is a blend of the two curves obtained separately for each element, using Equation 10.3.5.

Having found the displacements, we use Equations 10.3.11 and 10.3.18 to obtain the bending moments. For beam element 1, we have

$$
M_z^{(1)} = EI_z \lfloor \boldsymbol{\mathcal{B}}^{(1)} \rfloor \{\mathbf{q}^{(1)}\} = EI_z \left\lfloor \frac{12x}{L^3} - \frac{6}{L^2} \quad \frac{6x}{L^2} - \frac{4}{L} \quad -\frac{12x}{L^3} + \frac{6}{L^2} \quad \frac{6x}{L^2} - \frac{2}{L} \right\rfloor \begin{Bmatrix} 0 \\ 0 \\ -0.1358PL^3/EI_z \\ -0.1296PL^2/EI_z \end{Bmatrix}
\qquad [\mathbf{j}]
$$

which simplifies to

$$
M_z^{(1)} = -0.5556Px + 0.8520PL \qquad (0 \le x < L)
\qquad [\mathbf{k}]
$$

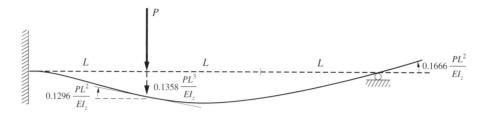

Figure 10.3.8   Elastic curve of the beam in Figure 10.3.2.

The maximum deflection is $0.1666PL^3 / EI_z$, at midspan.

For element 2, we have

$$M_z^{(2)} = EI_z \lfloor \boldsymbol{\mathcal{B}}^{(2)} \rfloor \{q^{(2)}\} = EI_z \lfloor \frac{12(x-L)}{(2L)^3} - \frac{6}{(2L)^2} \quad \frac{6(x-L)}{(2L)^2} - \frac{4}{2L} \quad \vdots \quad -\frac{12(x-L)}{(2L)^3} + \frac{6}{(2L)^2} \quad \frac{6(x-L)}{(2L)^2} - \frac{2}{2L} \rfloor \left\{ \begin{array}{c} -0.1358PL^3/EI_z \\ -0.1296PL^2/EI_z \\ \hdashline 0 \\ 0.1666PL^2/EI_z \end{array} \right\}$$ [l]

where x continues to be measured from the origin of the global system, which is at the left end of the beam. From this we obtain

$$M_z^{(2)} = -0.1482Px + 0.4445PL \qquad (L \leq x < 3L)$$ [m]

Equations k and m are plotted in Figure 10.3.9. For this problem, these bending moment functions, computed by the finite element method, are the exact solution. Why? Also, why will increasing the number of beam elements in the model of Figure 10.3.7 not improve on the solution for the deflection or bending moment?

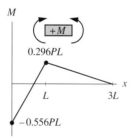

Figure 10.3.9   Bending moment distribution for the beam of Figure 10.3.7.

Let us now consider distributed loads on beam elements. In the matrix formulation, the equivalent loads are found using Equation 9.12.13,

$$\{Q_{\text{eq}}\} = \iint_S [\mathbf{N}]^T \{\mathbf{T}\} dS$$

For the beam element, we replace $\{\mathbf{T}\}dS$ with $p(x)dx$, where $p(x)$ is the distributed load per unit length:

$$\{\mathbf{Q}_{eq}\} = \int_0^L [\mathbf{N}]^T p\,dx \qquad \text{[10.3.20]}$$

Also,

$$[\mathbf{N}] = \lfloor N_1(x) \quad N_2(x) \quad N_3(x) \quad N_4(x) \rfloor^T$$

where the components of $[\mathbf{N}]$ are the cubic shape functions given in Equation 10.3.6. Thus,

$$\{\mathbf{Q}_{eq}\} = \left\lfloor \int_0^L \left(2\frac{x^3}{L^3} - 3\frac{x^2}{L^2} + 1\right) p(x)dx \quad \int_0^L \left(\frac{x^3}{L^2} - 2\frac{x^2}{L} + x\right) p(x)dx \right.$$

$$\left. \int_0^L \left(-2\frac{x^3}{L^3} + 3\frac{x^2}{L^2}\right) p(x)dx \quad \int_0^L \left(\frac{x^3}{L^2} - \frac{x^2}{L}\right) p(x)dx \right\rfloor^T \qquad \text{[10.3.21]}$$

If, for example, the distributed load varies linearly across the beam span, then

$$p(x) = p_1 + (p_2 - p_1)\frac{x}{L}$$

Substituting this expression into Equation 10.3.20 and carrying out the integrations yields

$$\{\mathbf{Q}_{eq}\} = \lfloor \left(\tfrac{7}{20}p_1 + \tfrac{3}{20}p_2\right)L \quad \left(\tfrac{1}{20}p_1 + \tfrac{1}{30}p_2\right)L^2 \quad \left(\tfrac{3}{20}p_1 + \tfrac{7}{20}p_2\right)L \quad \left(-\tfrac{1}{30}p_1 - \tfrac{1}{20}p_2\right)L^2 \rfloor^T \quad \text{linearly varying load} \quad \text{[10.3.22]}$$

These work-equivalent point loads and couples are shown alongside the linearly varying distributed load, in Figure 10.3.10.

Suppose a load P is uniformly distributed over a small subspan l of a beam, as illustrated in Figure 10.3.11. Then $p(x) = P/l$, and we have

$$\int_0^L p(x)N_i(x)dx = \int_a^{a+l} \frac{P}{l}N_i(x)dx = P\bar{N}_i$$

where $N_i(x)$ is any one of the shape functions, and $\bar{N}_i$ is its average value over l. In the limit as $l \to 0$, we arrive at a concentrated load P acting at the point $x = a$. Thus,

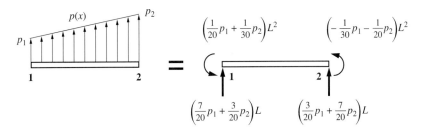

Figure 10.3.10 Linearly distributed load and its work-equivalent point loads.

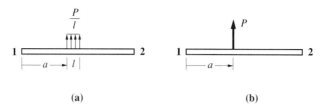

Figure 10.3.11   Concentrated load (b) viewed as a finite uniform load on a small subspan l (a).

$$\int_0^L p(x)N_i(x)dx = PN_i(a)$$

and computing the equivalent load vector from Equation 10.3.21, for a concentrated load P, yields

$$\{Q_{eq}\} = \left\lfloor \left(2\frac{a^3}{L^3} - 3\frac{a^2}{L^2} + 1\right)P \quad \left(\frac{a^3}{L^2} - 2\frac{a^2}{L} + a\right)P \quad \left(-2\frac{a^3}{L^3} + 3\frac{a^2}{L^2}\right)P \quad \left(\frac{a^3}{L^2} - \frac{a^2}{L}\right)P \right\rfloor \quad \text{point load at } x = a \quad \textbf{[10.3.23]}$$

Example 10.3.2   Use the finite element method to find the bending moment in the two-element propped cantilever beam in Figure 10.3.12. Compare the results with the exact solution.

This is the same beam as illustrated in Figure 10.3.7, but with a different loading. The downward point load at node 2 has been replaced by a downward uniform load p applied from $x = 0$ to $x = L$ (i.e., applied to beam element 1) plus a downward point load P of magnitude pL applied at $x = 2L$ (i.e., applied at the midpoint of beam element 2). Since the structure is identical to that of Example 10.3.1, the stiffness matrices computed there can be used for this problem. The same constraints also apply, which means that $[k_{ff}]$ will be the same as that for the beam in Figure 10.3.7. The difference between this problem and the previous one lies in the computation of the load vectors. In the previous example, all loads were applied at the nodes. In Figure 10.3.12, the external loading occurs between the nodes. Therefore, the point-load equivalents of these intermediate loads must be calculated and then added into the global load vector.

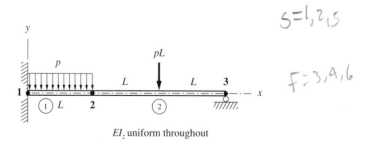

Figure 10.3.12   Loaded and constrained two-element beam assembly.

We consider the elements one at a time, forming each element's equivalent load vector $\{Q_{eq}^{(e)}\}$ and assembling it into the global equivalent load vector $\{Q_{eq}\}$, which is formed separately from the global point-load vector $\{Q_P\}$.

Element 1

Element 1 is spanned by the uniformly distributed load p, the point-load equivalents of which can be found from Equation 10.3.22. Noting that the length of the element is L and that the load is directed downward, so that $p_1 = p_2 = -p$, we find

$$\left\{Q_{eq}^{(1)}\right\} = \left\{\begin{array}{l} -pL/2 \\ -pL^2/12 \\ -pL/2 \\ pL^2/12 \end{array}\right\} \begin{array}{l} 1 \\ 2 \\ 3 \\ 4 \end{array} \qquad \textbf{[a]}$$

where the global degree of freedom correspondences are noted alongside each load vector component.

Element 2

Element 2 has a point load pL applied downward midway between its nodes. Observing that the length of the element is $2L$, we therefore have, from Equation 10.3.23,

$$\left\{Q_{eq}^{(1)}\right\} = \left\{\begin{array}{l} (-pL)/2 \\ (-pL)(2L)/8 \\ (-pL)/2 \\ -(-pL)(2L)/8 \end{array}\right\} = \left\{\begin{array}{l} -pL/2 \\ -pL^2/4 \\ -pL/2 \\ pL^2/4 \end{array}\right\} \begin{array}{l} 3 \\ 4 \\ 5 \\ 6 \end{array} \qquad \textbf{[b]}$$

Again, the corresponding global degrees of freedom are listed.

Having formed each element's equivalent load vector, the next step is to assemble them into the (initially zero) global equivalent load vector. Thus,

$$\left\{Q_{eq}\right\} = \left\{\begin{array}{l} (-pL/2) \\ (-pL^2/12) \\ (-pL/2) \quad + \quad (-pL/2) \\ (pL^2/12) \quad + \quad (-pL^2/4) \\ (-pL/2) \\ (pL^2/4) \end{array}\right\} = \left\{\begin{array}{l} -pL/2 \\ -pL^2/12 \\ -pL \\ -pL^2/6 \\ -pL/2 \\ pL^2/4 \end{array}\right\} \begin{array}{l} 1 \\ 2 \\ 3 \\ 4 \\ 5 \\ 6 \end{array} \qquad \textbf{[c]}$$

The vector of point loads applied directly to the nodes consists only of the reactions at the supports. Specifically,

$$\{Q_P\} = \left\{\begin{array}{l} Y_1 \\ M_1 \\ 0 \\ 0 \\ Y_3 \\ 0 \end{array}\right\} \begin{array}{l} 1 \\ 2 \\ 3 \\ 4 \\ 5 \\ 6 \end{array} \qquad \textbf{[d]}$$

The complete global load vector is the sum of $\{Q_P\}$ and $\{Q_{eq}\}$,

$$\{Q\} = \{Q_P\} + \{Q_{eq}\} \qquad \textbf{[e]}$$

Since the free degrees of freedom are 3, 4, and 6, we have

$$\{Q_f\} = \left\{Q_P\right\} \begin{array}{l} 3 \\ 4 \\ 6 \end{array} + \left\{Q_{eq}\right\} \begin{array}{l} 3 \\ 4 \\ 6 \end{array} = \left\{\begin{array}{l} 0 \\ 0 \\ 0 \end{array}\right\} + \left\{\begin{array}{l} -pL \\ -pL^2/6 \\ pL^2/4 \end{array}\right\} \qquad \textbf{[f]}$$

This load vector replaces that on the right of Equation g of Example 10.3.1, so that the system of equations for the free displacements v_2, θ_2, and θ_3 becomes

$$\frac{EI_z}{L^3} \begin{bmatrix} \frac{27}{2} & -\frac{9}{2}L & \frac{3}{2}L \\ -\frac{9}{2}L & 6L^2 & L^2 \\ \frac{3}{2}L & L^2 & 2L^2 \end{bmatrix} \left\{\begin{array}{l} v_2 \\ \theta_{z2} \\ \theta_{z3} \end{array}\right\} = \left\{\begin{array}{l} -pL \\ -\frac{pL^2}{6} \\ \frac{pL^2}{4} \end{array}\right\}$$

Solving for the displacements yields

$$\{\mathbf{q}_f\} = \begin{Bmatrix} v_2 \\ \theta_{z2} \\ \theta_{z3} \end{Bmatrix} = \begin{Bmatrix} -0.1991\,pL^4/EI_z \\ -0.2431\,pL^3/EI_z \\ 0.3958\,pL^3/EI_z \end{Bmatrix} \qquad \textbf{[g]}$$

The deflected shape of the beam is shown in Figure 10.3.13.

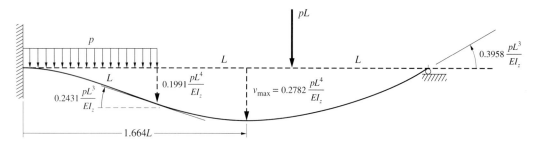

Figure 10.3.13   Elastic curve of the beam in Figure 10.3.12.

The reactions at the supports are obtained using Equation e,

$$\{\mathbf{Q}_{P_s}\} = \{\mathbf{Q}_s\} - \{\mathbf{Q}_{eq_s}\}$$

and Equation 9.10.13b of the partitioned stiffness equations,

$$\{\mathbf{Q}_s\} = \left[\mathbf{k}_{sf}\right]\{\mathbf{q}_f\} + [\mathbf{k}_{ss}]\{\mathbf{q}_s\}$$

Thus,

$$\{\mathbf{Q}_{P_s}\} = \left[\mathbf{k}_{sf}\right]\{\mathbf{q}_f\} + [\mathbf{k}_{ss}]\{\mathbf{q}_s\} - \{\mathbf{Q}_{eq_s}\} \qquad \textbf{[h]}$$

that is,

$$\begin{Bmatrix} \\ \mathbf{Q} \\ \\ \end{Bmatrix}\begin{matrix}1\\2\\5\end{matrix} = \begin{matrix}1\\2\\5\end{matrix}\overset{3\quad4\quad6}{\begin{bmatrix} & & \\ & \mathbf{k} & \\ & & \end{bmatrix}}\begin{Bmatrix} \\ \mathbf{q} \\ \\ \end{Bmatrix}\begin{matrix}3\\4+2\\6\end{matrix} + \begin{matrix}1\\2\\5\end{matrix}\overset{1\quad2\quad5}{\begin{bmatrix} & & \\ & \mathbf{k} & \\ & & \end{bmatrix}}\begin{Bmatrix} \\ \mathbf{q} \\ \\ \end{Bmatrix}\begin{matrix}1\\2\\5\end{matrix} - \begin{Bmatrix} \\ \mathbf{Q}_{eq} \\ \\ \end{Bmatrix}\begin{matrix}1\\2\\5\end{matrix}$$

Substituting the indicated components of $[\mathbf{k}]$, $\{\mathbf{q}\}$, $\{\mathbf{Q}_P\}$, and $\{\mathbf{Q}_{eq}\}$, we find

$$\begin{Bmatrix} Y_1 \\ M_{z1} \\ Y_3 \end{Bmatrix} = \frac{EI_z}{L^3}\begin{bmatrix} -12 & 6L & 0 \\ -6L & 2L^2 & 0 \\ -\frac{3}{2} & -\frac{3}{2}L & -\frac{3}{2}L \end{bmatrix}\begin{Bmatrix} -0.1991\,pL^4/EI_z \\ -0.2431\,pL^3/EI_z \\ 0.3958\,pL^3/EI_z \end{Bmatrix} + \frac{EI_z}{L^3}\begin{bmatrix} 12 & 6L & 0 \\ 6L & 4L^2 & 0 \\ 0 & 0 & \frac{3}{2} \end{bmatrix}\begin{Bmatrix} 0 \\ 0 \\ 0 \end{Bmatrix} - \begin{Bmatrix} -pL/2 \\ -pL^2/12 \\ -pL/2 \end{Bmatrix} = \begin{Bmatrix} 1.431\,pL \\ 0.7917\,pL^2 \\ 0.5696\,pL \end{Bmatrix}$$

The computed reactions are illustrated on the free-body diagram in Figure 10.3.14. (The reader should verify that the beam is in equilibrium under the set of loads shown.) The bending moment distribution can be found from this free-body diagram, using ordinary statics. It is plotted as the continuous curve labeled M_{exact} in Figure 10.3.15. To abide strictly by the finite element method, we calculate the bending moment within each of the two beam elements, using Equations 10.3.11 and 10.3.18, as in the previous example. By that approach, we obtain

$$M_z^{(1)} = EI_z \lfloor \boldsymbol{\mathcal{B}}^{(1)} \rfloor \{\mathbf{q}^{(1)}\}$$

$$= EI_z \lfloor \tfrac{12x}{L^3} - \tfrac{6}{L^2} \quad \tfrac{6x}{L^2} - \tfrac{4}{L} \quad\vdots\quad -\tfrac{12x}{L^3} + \tfrac{6}{L^2} \quad \tfrac{6x}{L^2} - \tfrac{2}{L} \rfloor \begin{Bmatrix} 0 \\ 0 \\ \hdashline -0.1991PL^3/EI_z \\ -0.2431PL^2/EI_z \end{Bmatrix}$$

$$= 0.9306pLx - 0.7084pL^2$$

$$M_z^{(2)} = EI_z \lfloor \boldsymbol{\mathcal{B}}^{(2)} \rfloor \{\mathbf{q}^{(2)}\}$$

$$= EI_z \lfloor \tfrac{12(x-L)}{(2L)^3} - \tfrac{6}{(2L)^2} \quad \tfrac{6(x-L)}{(2L)^2} - \tfrac{4}{2L} \quad\vdots\quad -\tfrac{12(x-L)}{(2L)^3} + \tfrac{6}{(2L)^2} \quad \tfrac{6(x-L)}{(2L)^2} - \tfrac{2}{2L} \rfloor \begin{Bmatrix} -0.1991PL^3/EI_z \\ -0.2431PL^2/EI_z \\ \hdashline 0 \\ 0.3958PL^2/EI_z \end{Bmatrix}$$

$$= 0.0696pLx + 0.4586pL$$

These two linear functions are shown, along with M_{exact}, in Figure 10.3.15. As pointed out previously in this text, the only way to reduce the obvious discrepancy, while retaining the cubic-displacement beam element, is to increase the number of nodes (degrees of freedom) in the model. For example, placing a node at $x = 2L$ will yield the exact solution in the rightmost two-thirds of the beam. Why? How many nodes must be added between nodes 1 and 2 to reduce the error in that region to zero?

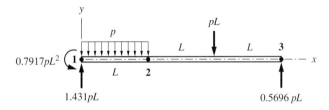

Figure 10.3.14   Computed reactions for the beam in Figure 10.3.13.

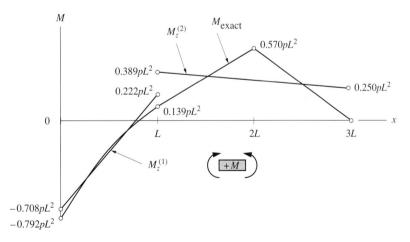

Figure 10.3.15   The exact (equilibrium) and the approximate (finite element) bending moment distribution for the beam in Figure 10.3.13.

Consistent equivalent loads for the simple beam element subjected to transverse g loading or transverse thermal gradients are found using Equation 9.12.13. For a body force density $\{\mathbf{b}\}$,

$$\{\mathbf{Q}_{eq}\} = \iiint\limits_{V} [\mathbf{N}]^T \{\mathbf{b}\} \, dV$$

For a lateral acceleration a_y, the body force density vector is simply $\{\mathbf{b}\} = -\rho a_y$, so that

$$\{\mathbf{Q}_{eq}\} = \int\limits_0^L [\mathbf{N}]^T \left(-\rho A a_y\right) dx$$

If we set $p(x) = -\rho A a_y(x)$, this equation is identical to Equation 10.3.20 for a distributed load per unit length. Therefore, equivalent nodal point loads for transverse acceleration may be found from Equation 10.3.22 or Figure 10.3.10.

For thermal strain in an isotropic simple beam, recalling Equation 10.3.11, we have

$$\{\mathbf{Q}_{eq}\} = \iiint\limits_{V} [\mathbf{B}]^T [\mathbf{E}]\{\boldsymbol{\varepsilon}_T\} dV = -\iiint\limits_{V} E\alpha[\boldsymbol{\mathcal{B}}]^T T y dV$$

where T is the temperature change at a point. For a cylindrical beam of uniform properties, neglecting temperature dependence and setting $dV = dA\,dx$, where dA is an element of the beam cross section, this becomes

$$\{\mathbf{Q}_{eq}\} = -E\alpha \iint\limits_{A} \int\limits_0^L [\boldsymbol{\mathcal{B}}]^T T y \, dA \, dx \qquad\qquad \text{[10.3.24]}$$

Consider the case of a bilinear temperature distribution in the beam, in which case the temperature change is

$$T = C_o + C_x x + C_y y + C_{xy} xy \qquad\qquad \text{[10.3.25]}$$

where C_o, C_x, C_y, and C_{xy} are constants. Substituting this expression into Equation 10.3.24 and collecting terms yields

$$\{\mathbf{Q}_{eq}\} = -E\alpha \left[C_o \int\limits_0^L [\boldsymbol{\mathcal{B}}]^T dx + C_x \int\limits_0^L [\boldsymbol{\mathcal{B}}]^T x dx \right] \iint\limits_A y\,dA - E\alpha \left[C_y \int\limits_0^L [\boldsymbol{\mathcal{B}}]^T dx + C_{xy} \int\limits_0^L [\boldsymbol{\mathcal{B}}]^T x dx \right] \iint\limits_A y^2 dA$$

Noting that $\iint\limits_A y\,dA = 0$ (since y is measured from the neutral axis) and $\iint\limits_A y^2 dA = I_z$, this reduces to

$$\{\mathbf{Q}_{eq}\} = -E\alpha I_z \left[C_y \int\limits_0^L [\boldsymbol{\mathcal{B}}]^T dx + C_{xy} \int\limits_0^L [\boldsymbol{\mathcal{B}}]^T x dx \right] \qquad\qquad \text{[10.3.26]}$$

Substituting $\lfloor \boldsymbol{\mathcal{B}} \rfloor$ from Equation 10.3.10 into the integrals, it is easy to show that

$$\int\limits_0^L [\boldsymbol{\mathcal{B}}]^T dx = \lfloor 0 \quad -1 \quad 0 \quad 1 \rfloor^T \qquad\qquad \int\limits_0^L [\boldsymbol{\mathcal{B}}]^T x dx = \lfloor 1 \quad 0 \quad -1 \quad L \rfloor^T$$

which means that Equation 10.3.26 can be written

$$\{\mathbf{Q}_{eq}\} = E\alpha I_z \lfloor -C_{xy} \quad C_y \ \vdots \ C_{xy} \quad -C_y - C_{xy}L \rfloor^T$$

Let $\Delta T(x) = T_u(x) - T_l(x)$ be the difference between the temperature change on the upper (u) and lower (l) surfaces of the beam, as a function of span x. Then, if h is the depth of the beam, we can show that the coefficients C_y and C_{xy} are given by

$$C_y = \frac{\Delta T(0)}{h} \qquad C_{xy} = \frac{\Delta T(L)/h - \Delta T(0)/h}{L} \qquad \text{[10.3.27]}$$

Therefore, we finally obtain the work-equivalent load vector for a bilinear temperature change in the beam, as follows:

$$\{\mathbf{Q}_{eq}\} = E\alpha I_z \lfloor -\tfrac{\Delta T(L)/h - \Delta T(0)/h}{L} \quad \tfrac{\Delta T(0)}{h} \ \vdots \ \tfrac{\Delta T(L)/h - \Delta T(0)/h}{L} \quad -\tfrac{\Delta T(L)}{h} \rfloor^T \qquad \text{[10.3.28]}$$

The equivalent loads are sketched in Figure 10.3.16.

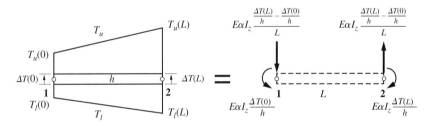

Figure 10.3.16   Bilinear temperature variation and the work-equivalent nodal loads for a simple beam.

The bending moment within the beam is given by Equations 10.3.20 and 10.3.22,

$$M_z = E I_z \lfloor \boldsymbol{\mathcal{B}} \rfloor \{\mathbf{q}\} + E\alpha \iint_A T y \, dA$$

Substituting Equations 10.3.26 and 10.3.27 into the integral and simplifying leads to

$$M_z = E I_z \lfloor \boldsymbol{\mathcal{B}} \rfloor \{\mathbf{q}\} + E I_z \alpha \left[\frac{\Delta T(0)}{h} \left(1 - \frac{x}{L}\right) + \frac{\Delta T(L)}{h} \frac{x}{L} \right] \qquad \text{[10.3.29]}$$

Example 10.3.3   Use the finite element method to calculate the deflection and internal bending moment distribution in the beam in Figure 10.3.17. On the top surface, the temperature rise T_u above ambient conditions is shown as a piecewise linear function of x. The temperature rise T_l on the bottom surface is zero. Assume that the temperature variation across the beam's depth is linear.

The only difference between this problem and the previous example is the nature of the equivalent load vectors, which are found using Equation 10.3.28. Thus,

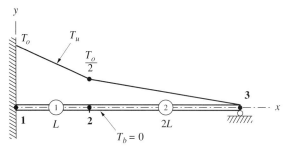

Figure 10.3.17   Beam loaded only by a bilinear temperature rise field.

$$\{\mathbf{Q}_{\text{eq}}^{(1)}\} = E\alpha I_z \left\{ \begin{array}{c} -\dfrac{\frac{T_o/2}{h} - \frac{T_o}{h}}{L} \\[2mm] \hdashline \dfrac{T_o}{h} \\[2mm] \hdashline \dfrac{\frac{T_o/2}{h} - \frac{T_o}{h}}{L} \\[2mm] \hdashline -\dfrac{T_o/2}{h} \end{array} \right\} = E\alpha I_z \left\{ \begin{array}{cc} \dfrac{T_o}{2hL} & 1 \\[2mm] \dfrac{T_o}{h} & 2 \\[2mm] -\dfrac{T_o}{2hL} & 3 \\[2mm] -\dfrac{T_o}{2h} & 4 \end{array} \right\}$$

$$\{\mathbf{Q}_{\text{eq}}^{(2)}\} = E\alpha I_z \left\{ \begin{array}{c} -\dfrac{0 - \frac{T_o/2}{h}}{2L} \\[2mm] \hdashline \dfrac{T_o/2}{h} \\[2mm] \hdashline \dfrac{0 - \frac{T_o/2}{h}}{2L} \\[2mm] \hdashline 0 \end{array} \right\} = E\alpha I_z \left\{ \begin{array}{cc} \dfrac{T_o}{4hL} & 3 \\[2mm] \dfrac{T_o}{2h} & 4 \\[2mm] -\dfrac{T_o}{4hL} & 5 \\[2mm] 0 & 6 \end{array} \right\} \qquad \text{[a]}$$

The global equivalent load vector is computed from these by means of the assembly rule:

$$\{\mathbf{Q}_{\text{eq}}\} = E\alpha I_z \left\{ \begin{array}{c} \dfrac{T_o}{2hL} \\[2mm] \dfrac{T_o}{h} \\[2mm] \hdashline -\dfrac{T_o}{2hL} + \dfrac{T_o}{4hL} \\[2mm] -\dfrac{T_o}{2h} + \dfrac{T_o}{2h} \\[2mm] \hdashline -\dfrac{T_o}{4hL} \\[2mm] 0 \end{array} \right\} = E\alpha I_z \left\{ \begin{array}{cc} \dfrac{T_o}{2hL} & 1 \\[2mm] \dfrac{T_o}{h} & 2 \\[2mm] -\dfrac{T_o}{4hL} & 3 \\[2mm] 0 & 4 \\[2mm] -\dfrac{T_o}{4hL} & 5 \\[2mm] 0 & 6 \end{array} \right\} \qquad \text{[b]}$$

The stiffness equations are those of the previous example,

$$[\mathbf{k}_{ff}]\{\mathbf{q}_f\} = \{\mathbf{Q}_{P_f}\} + \{\mathbf{Q}_{\text{eq}_f}\}$$

except for the far right-hand side,

$$\begin{array}{c} 3 \\ 4 \\ 6 \end{array} \begin{array}{c} 3 \quad 4 \quad 6 \\ \left[\quad \mathbf{k} \quad \right] \end{array} \{\mathbf{q}\} = \{\mathbf{Q}_P\} \begin{array}{c} 3 \\ 4 \\ 6 \end{array} + \{\mathbf{Q}_{\text{eq}}\} \begin{array}{c} 3 \\ 4 \\ 6 \end{array} = \left\{ \begin{array}{c} 0 \\ 0 \\ 0 \end{array} \right\} + \left\{ \begin{array}{c} -T_o/4hL \\ 0 \\ 0 \end{array} \right\}$$

That is,

$$\frac{EI_z}{L^3} \begin{bmatrix} \frac{27}{2} & -\frac{9}{2}L & \frac{3}{2}L \\[2mm] -\frac{9}{2}L & 6L^2 & L^2 \\[2mm] \frac{3}{2}L & L^2 & 2L^2 \end{bmatrix} \left\{ \begin{array}{c} v_2 \\ \theta_{z2} \\ \theta_{z3} \end{array} \right\} = \left\{ \begin{array}{c} -\frac{T_o}{4hL} \\ 0 \\ 0 \end{array} \right\}$$

Solving for the displacements, we obtain

$$\{\mathbf{q}_f\} = \begin{Bmatrix} v_2 \\ \theta_{z2} \\ \theta_{z3} \end{Bmatrix} = \frac{\alpha T_o L}{h} \begin{Bmatrix} -0.03395L \\ -0.03241 \\ 0.04167 \end{Bmatrix} \tag{a}$$

From these and the beam shape functions, Equations 10.3.6, we can plot the piecewise cubic displacement shown in Figure 10.3.18. We calculate the bending moment in each element using Equation 10.3.29. For element 1, we have

$$M_z^{(1)} = EI_z \lfloor \boldsymbol{\mathcal{B}}^{(1)} \rfloor \{\mathbf{q}^{(1)}\} + EI_z \alpha \left[\frac{\Delta T^{(1)}(0)}{h} \left(1 - \frac{x}{L^{(1)}}\right) + \frac{\Delta T^{(1)}\left(L^{(1)}\right)}{h} \frac{x}{L^{(1)}} \right]$$

$$= EI_z \lfloor \frac{12x}{L^3} - \frac{6}{L^2} \quad \frac{6x}{L^2} - \frac{4}{L} \quad \vdots \quad -\frac{12x}{L^3} + \frac{6}{L^2} \quad \frac{6x}{L^2} - \frac{2}{L} \rfloor \begin{Bmatrix} 0 \\ 0 \\ \cdots \\ -0.03395\alpha T_o L^2/h \\ -0.03241\alpha T_o L/h \end{Bmatrix} + EI_z \alpha \left[\frac{T_o}{h}\left(1 - \frac{x}{L}\right) + \frac{T_o}{2h}\frac{x}{L} \right]$$

$$= \frac{EI_z \alpha T_o}{hL}(-0.2870x + 0.8611L) \qquad (0 \le x \le L)$$

Continuing to measure x from the left end of the structure, we obtain for beam 2,

$$M_z^{(2)} = EI_z \lfloor \boldsymbol{\mathcal{B}}^{(2)} \rfloor \{\mathbf{q}^{(2)}\}$$

$$= EI_z \lfloor \frac{12(x-L)}{(2L)^3} - \frac{6}{(2L)^2} \quad \frac{6(x-L)}{(2L)^2} - \frac{4}{2L} \quad \vdots \quad -\frac{12(x-L)}{(2L)^3} + \frac{6}{(2L)^2} \quad \frac{6(x-L)}{(2L)^2} - \frac{2}{2L} \rfloor \begin{Bmatrix} -0.03395\alpha T_o L^2/h \\ -0.03241\alpha T_o L/h \\ \cdots \\ 0 \\ 0.04167\alpha T_o L/h \end{Bmatrix}$$

$$+ EI_z \alpha \left[\frac{T_o}{2h}\left(1 - \frac{x-L}{L}\right) + 0 \right]$$

$$= \frac{EI_z \alpha T_o}{hL}(-0.2870x + 0.8611L) \qquad (L \le x \le 3L)$$

Figure 10.3.18   Elastic curve of the beam in Figure 10.3.17.

The maximum deflection is $0.04167\,\alpha T_o L^2/h$ at the midspan $(x = 1.5L)$.

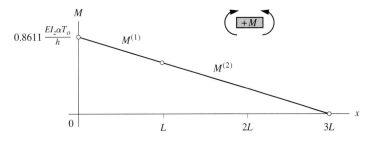

Figure 10.3.19   Bending moment in the beam of Figure 10.3.17.

Obviously, the bending moment has the same linear variation throughout the entire beam, as we should expect for a cantilever beam with a shear load opposite the clamped end. The bending moment is sketched in Figure 10.3.19.

10.4   TWO-DIMENSIONAL FRAMES

Recall that a two-dimensional frame is a plane, skeletal structure composed of slender elements that transmit in-plane bending loads, as well as axial loads. The frame element is a combination of the rod element of section 10.2 and the beam element of section 10.3. Figure 10.4.1 depicts a frame element in its local coordinate system. At each node, the element has three degrees of freedom, which are ordered as shown in Table 10.4.1 and in Figure 10.4.1. The plane frame element has a total of six degrees of freedom, so its stiffness matrix is 6 by 6. The matrix is found by assembling the rod element stiffness matrix and the beam element stiffness matrix into a 6 by 6 array.

Table 10.4.1   Plane frame element degrees of freedom.

Node	Degree of Freedom Number	Displacement, Force	
1	1	u_1, X_1:	Axial displacement, axial load
	2	v_1, Y_1:	Transverse displacement, shear load
	3	θ_1, M_{z_1}:	Rotation, moment about z axis
2	4	u_2, X_2:	Axial displacement, axial load
	5	v_2, Y_2:	Transverse displacement, shear load
	6	θ_2, M_{z_2}:	Rotation, moment about z axis

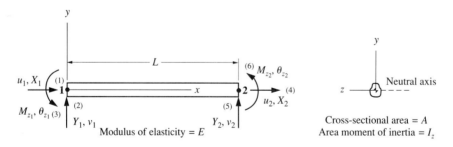

Figure 10.4.1   Two dimensional, six degree of freedom frame element.

From Equation 10.2.7, the rod element stiffness matrix is

$$
[\bar{\mathbf{k}}]_{\text{rod}} = \begin{array}{c} 1 4 \\ \left[\begin{array}{cc} AE/L & -AE/L \\ -AE/L & AE/L \end{array} \right] \begin{array}{c} 1 \\ 4 \end{array} \end{array}
$$

where the rows and columns are labeled with the corresponding degree of freedom numbers of the frame element. From Equation 10.3.13, the beam element stiffness matrix, labeled with the frame element degree of freedom numbers, is

$$[\bar{k}]_{beam} = \begin{array}{cccc} \quad2\quad & \quad3\quad & \quad5\quad & \quad6\quad \\ \begin{bmatrix} 12EI_z/L^3 & 6EI_z/L^2 & -12EI_z/L^3 & 6EI_z/L^2 \\ 6EI_z/L^2 & 4EI_z/L & -6EI_z/L^2 & 2EI_z/L \\ -12EI_z/L^3 & -6EI_z/L^2 & 12EI_z/L^3 & -6EI_z/L^2 \\ 6EI_z/L^2 & 2EI_z/L & -6EI_z/L^2 & 4EI_z/L \end{bmatrix} & \begin{matrix} 2 \\ 3 \\ 5 \\ 6 \end{matrix} \end{array}$$

Assembling these two matrices into the 6 by 6 frame element array yields the local form of the stiffness matrix for the plane frame element, as follows:

$$[\bar{k}] = \begin{array}{cccccc} u_1 & v_1 & \theta_1 & u_2 & v_2 & \theta_2 \\ \begin{bmatrix} \frac{AE}{L} & 0 & 0 & \frac{AE}{L} & 0 & 0 \\ 0 & \frac{12EI_z}{L^3} & \frac{6EI_z}{L^2} & 0 & -\frac{12EI_z}{L^3} & \frac{6EI_z}{L^2} \\ 0 & \frac{6EI_z}{L^2} & \frac{4EI_z}{L} & 0 & -\frac{6EI_z}{L^2} & \frac{2EI_z}{L} \\ -\frac{AE}{L} & 0 & 0 & \frac{AE}{L} & 0 & 0 \\ 0 & -\frac{12EI_z}{L^3} & -\frac{6EI_z}{L^2} & 0 & \frac{12EI_z}{L^3} & -\frac{6EI_z}{L^2} \\ 0 & \frac{6EI_z}{L^2} & \frac{2EI_z}{L} & 0 & -\frac{6EI_z}{L^2} & \frac{4EI_z}{L} \end{bmatrix} & \begin{matrix} X_1 \\ Y_1 \\ M_{z_1} \\ X_2 \\ Y_2 \\ M_{z_2} \end{matrix} \end{array} \qquad [10.4.1]$$

Figure 10.4.2 shows a frame element inclined relative to the global *xy* frame of reference. In part (a) of the figure, the element degrees of freedom relative to the local frame of reference are shown with bars, to distinguish these quantities from their counterparts in the unbarred global reference frame, illustrated in part (b) of the figure.

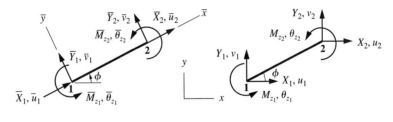

(a) Element degrees of freedom   (b) Global degrees of freedom

Figure 10.4.2   Inclined plane frame element.

The global stiffness matrix [**k**] is related to the element stiffness matrix [**k̄**] in Equation 10.4.1 by the transformation originally presented in Equation 10.2.18, namely,

$$[\mathbf{k}] = [\mathbf{\Lambda}]^T[\bar{\mathbf{k}}][\mathbf{\Lambda}] \qquad [10.4.2]$$

Recall from section 10.2 that [**Λ**] relates the local element displacement vector {**q̄**} to the global element displacement vector {**q**}:

$$\{\bar{\mathbf{q}}\} = [\mathbf{\Lambda}]\{\mathbf{q}\} \qquad [10.4.3a]$$

For the case at hand,

$$\{\bar{q}\} = \begin{Bmatrix} \bar{u}_1 \\ \bar{v}_1 \\ \bar{\theta}_{z_1} \\ \bar{u}_2 \\ \bar{v}_2 \\ \bar{\theta}_{z_2} \end{Bmatrix} \quad \text{and} \quad \{q\} = \begin{Bmatrix} u_1 \\ v_1 \\ \theta_{z_1} \\ u_2 \\ v_2 \\ \theta_{z_2} \end{Bmatrix} \qquad [10.4.3b]$$

To find $[\Lambda]$, we first note that the local displacements, $\bar{u}$ and $\bar{v}$, at either node are obtained simply by projecting the displacement vector, resolved into components along the global xy axes, onto the local $\bar{x}$ and $\bar{y}$ axes, respectively. The direction cosines of the element axis $\bar{x}$ are l_1 and m_1, while those for the $\bar{y}$ axis are l_2 and m_2. Using matrix notation,

$$\begin{Bmatrix} \bar{u} \\ \bar{v} \end{Bmatrix} = \begin{bmatrix} l_1 & m_1 \\ l_2 & m_2 \end{bmatrix} \begin{Bmatrix} u \\ v \end{Bmatrix} \qquad [10.4.4]$$

where, since the element axis $\bar{x}$ makes an angle ϕ with the global x axis,

$$\begin{aligned} l_1 &= \cos\phi_{\bar{x}x} = \cos\phi & m_1 &= \cos\phi_{\bar{x}y} = \sin\phi \\ l_2 &= \cos\phi_{\bar{y}x} = -\sin\phi & m_2 &= \cos\phi_{\bar{y}y} = \cos\phi \end{aligned} \qquad [10.4.5]$$

The rotational displacement at either node is unaffected by the coordinate transformation, since the z and $\bar{z}$ axes remain parallel to each other and perpendicular to the plane in which the element is constrained to move. Therefore,

$$\bar{\theta}_z = \theta_z \qquad [10.4.6]$$

Combining Equations 10.4.4 and 10.4.6, we have

$$\begin{Bmatrix} \bar{u} \\ \bar{v} \\ \bar{\theta}_z \end{Bmatrix} = \begin{bmatrix} l_1 & m_1 & 0 \\ l_2 & m_2 & 0 \\ 0 & 0 & 1 \end{bmatrix} \begin{Bmatrix} u \\ v \\ \theta_z \end{Bmatrix} \qquad [10.4.7]$$

Since this transformation of vector components holds at either node, the transformation for both sets of displacements can be written in a single matrix equation, as follows:

$$\begin{Bmatrix} \bar{u}_1 \\ \bar{v}_1 \\ \bar{\theta}_{z_1} \\ \bar{u}_2 \\ \bar{v}_2 \\ \bar{\theta}_{z_2} \end{Bmatrix} = \begin{bmatrix} l_1 & m_1 & 0 & 0 & 0 & 0 \\ l_2 & m_2 & 0 & 0 & 0 & 0 \\ 0 & 0 & 1 & 0 & 0 & 0 \\ 0 & 0 & 0 & l_1 & m_1 & 0 \\ 0 & 0 & 0 & l_2 & m_2 & 0 \\ 0 & 0 & 0 & 0 & 0 & 1 \end{bmatrix} \begin{Bmatrix} u_1 \\ v_1 \\ \theta_{z_1} \\ u_2 \\ v_2 \\ \theta_{z_2} \end{Bmatrix} \qquad [10.4.8]$$

This expression has the same form as Equation 10.4.3, so that we can identify $[\Lambda]$ as the 6 by 6 matrix on the right. Matrix $[\Lambda]$ can be written explicitly in terms of the angle of inclination ϕ by substituting the direction cosines from Equation 10.4.5,

$$[\Lambda] = \begin{bmatrix} \cos\phi & \sin\phi & 0 & 0 & 0 & 0 \\ -\sin\phi & \cos\phi & 0 & 0 & 0 & 0 \\ 0 & 0 & 1 & 0 & 0 & 0 \\ 0 & 0 & 0 & \cos\phi & \sin\phi & 0 \\ 0 & 0 & 0 & -\sin\phi & \cos\phi & 0 \\ 0 & 0 & 0 & 0 & 0 & 1 \end{bmatrix} \qquad [10.4.9]$$

Substituting the transformation matrix $[\Lambda]$ from Equation 10.4.9 and the local frame element stiffness matrix $[\bar{k}]$ from Equation 10.4.1 into Equation 10.4.2 and carrying out the matrix multiplications, we obtain the explicit form of the planar frame element global stiffness matrix:

$$[\mathbf{k}] = \frac{EI_z}{L^3}
\begin{array}{c}
\begin{array}{cccccc} u_1 & v_1 & \theta_{z_1} & u_2 & v_2 & \theta_{z_2} \end{array} \\
\left[
\begin{array}{cccccc}
\bar{A}l^2 + 12m^2 & (\bar{A}-12)\,lm & -6Lm & -\bar{A}l^2-12m^2 & (-\bar{A}+12)\,lm & -6Lm \\
(\bar{A}-12)\,lm & \bar{A}m^2 + 12l^2 & 6Ll & (-\bar{A}+12)\,lm & -\bar{A}m^2-12l^2 & 6Ll \\
-6Lm & 6Ll & 4L^2 & 6Lm & -6Ll & 2L^2 \\
-\bar{A}l^2-12m^2 & (-\bar{A}+12)\,lm & 6Lm & \bar{A}l^2 + 12m^2 & (\bar{A}-12)\,lm & 6Lm \\
(-\bar{A}+12)\,lm & -\bar{A}m^2-12l^2 & -6Ll & (\bar{A}-12)\,lm & \bar{A}m^2 + 12l^2 & -6Ll \\
-6Lm & 6Ll & 2L^2 & 6Lm & -6Ll & 4L^2
\end{array}
\right]
\begin{array}{c} X_1 \\ Y_1 \\ M_{z_1} \\ X_2 \\ Y_2 \\ M_{z_2} \end{array}
\end{array}$$

[10.4.10]

where $\bar{A} = AL^2/I_z$, and l and m are the direction cosines of the $\bar{x}$ axis: $l = \cos\phi$, $m = \sin\phi$. Note that, since $I_z = A\rho_z^2$, where ρ_z is the radius of gyration, $\bar{A} = (L/\rho_z)^2$; that is, $\bar{A}$ is the square of the slenderness ratio of the beam element.

It should be emphasized that the angle of inclination ϕ must be measured counterclockwise around node 1 *from* the global x axis *to* the element $\bar{x}$ axis, that is, to the beam centerline.

Example 10.4.1   Calculate the shear and bending moment throughout the two-element frame in Figure 10.4.3. The cross-sectional properties are uniform throughout and are given in the figure, along with geometric and load information. Also shown are the three global degrees of freedom associated with each node of the unconstrained structure.

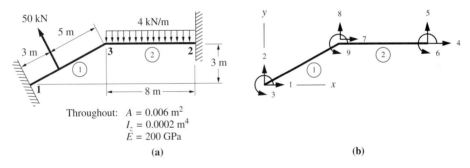

Throughout: $A = 0.006$ m^2
$I_z = 0.0002$ m^4
$E = 200$ GPa

(a)   (b)

Figure 10.4.3   Two-element frame. (a) Properties, loading and constraints (b) Global degrees of freedom.

First, we will compute each element's global stiffness matrix $[\mathbf{k}^{(e)}]$ ($e = 1,2$) and assemble them into the global array $[\mathbf{k}]$. Then, we will find the element equivalent load vectors $\{\bar{\mathbf{Q}}^{(e)}\}$, transformed into the global system and assembled into the global load vector $\{\mathbf{Q}\}$. Having assembled the global stiffness matrix and load vector, we will apply the support constraints to obtain the global stiffness equations $[\mathbf{k}_{ff}]\{\mathbf{q}_f\} = \{\mathbf{Q}_f\}$, which will be solved for the free displacements $\{\mathbf{q}_f\}$ at node 3. We will then project the global displacements into each element's local frame, to obtain the local displacement vectors $\{\bar{\mathbf{q}}^{(e)}\}$. Finally, we will find the element point-load vectors $\{\bar{\mathbf{Q}}_P^{(e)}\}$ from the individual element stiffness equations, $\{\bar{\mathbf{Q}}_P^{(e)}\} = [\bar{\mathbf{k}}^{(e)}]\{\bar{\mathbf{q}}^{(e)}\} - \{\bar{\mathbf{Q}}_{eq}^{(e)}\}$. With that information, we can use statics to obtain the shear and moment in each element. Because both elements have intermediate loadings, we know that Equation 10.3.18, $M_z = EI_z\lfloor\mathcal{B}\rfloor\{\mathbf{q}\}$, will yield inaccurate results for the bending moment in each part of the frame, unless we divide the frame into more beam elements. However, that would introduce more degrees of freedom, which could make a hand computation impractical.

Using the information in Figure 10.4.3, we can establish Table 10.4.2.

Table 10.4.2 Element properties for Figure 10.4.3.

Element	Node 1	Node 2	$\phi\ (°)$	l	m	E (N/m²)	L (m)	I_z (m⁴)	EI_z/L^3 (N/m)	A (m²)	AL^2/I_z
1	1	3	22.02	0.3750	0.9270	200×10^9	8	0.0002	78,125	0.006	1,920
2	3	2	0	0	1	200×10^9	8	0.0002	78,125	0.006	1,920

Substituting these data into Equation 10.4.10 yields the global stiffness matrices for the two elements:

$$\left[\mathbf{k}^{(1)}\right] = 10^6 \begin{array}{c} \\ \\ \\ \\ \\ \\ \end{array} \begin{bmatrix} \begin{array}{cccccc} 1 & 2 & 3 & 7 & 8 & 9 \\ 129.0 & 51.82 & -1.406 & -129.0 & -51.82 & -1.406 \\ 51.82 & 21.90 & 3.477 & -51.92 & -21.90 & 3.477 \\ -1.406 & 3.477 & 20.00 & 1.406 & -3.477 & 10.00 \\ -129.0 & -51.82 & 1.406 & 129.0 & 51.82 & 1.406 \\ -51.82 & -21.90 & -3.477 & 51.82 & 21.90 & -3.477 \\ -1.406 & 3.477 & 10.00 & 1.406 & -3.477 & 20.00 \end{array} \end{bmatrix} \begin{array}{c} 1 \\ 2 \\ 3 \\ 7 \\ 8 \\ 9 \end{array}$$ [a]

$$\left[\mathbf{k}^{(2)}\right] = 10^6 \begin{bmatrix} \begin{array}{cccccc} 7 & 8 & 9 & 4 & 5 & 6 \\ 150.0 & 0.000 & 0.000 & -150.0 & 0.000 & 0.000 \\ 0.000 & 0.9375 & 3.750 & 0.000 & -0.9375 & 3.750 \\ 0.000 & 3.750 & 20.00 & 0.000 & -3.750 & 10.00 \\ -150.0 & 0.000 & 0.000 & 150.0 & 0.000 & 0.000 \\ 0.000 & -0.9375 & -3.750 & 0.000 & 0.9375 & -3.750 \\ 0.000 & 3.750 & 10.00 & 0.000 & -3.750 & 20.00 \end{array} \end{bmatrix} \begin{array}{c} 7 \\ 8 \\ 9 \\ 4 \\ 5 \\ 6 \end{array}$$ [b]

As usual, the rows and columns are labeled with the corresponding global degree of freedom numbers. We use these labels as a guide to assemble $\left[\mathbf{k}^{(1)}\right]$ and $\left[\mathbf{k}^{(2)}\right]$ into the following 9 by 9 global stiffness matrix of the unconstrained nine degree of freedom structure:

$$[\mathbf{k}] = 10^6 \begin{bmatrix} \begin{array}{ccccccccc} 1 & 2 & 3 & 4 & 5 & 6 & 7 & 8 & 9 \\ 129.0 & 51.82 & -1.406 & 0.000 & 0.000 & 0.000 & -129.0 & -51.82 & -1.406 \\ 51.82 & 21.90 & 3.476 & 0.000 & 0.000 & 0.000 & -51.82 & -21.90 & 3.476 \\ -1.406 & 3.476 & 20.00 & 0.000 & 0.000 & 0.000 & 1.406 & -3.476 & 10.00 \\ 0.000 & 0.000 & 0.000 & 150.0 & 0.000 & 0.000 & -150.0 & 0.000 & 0.000 \\ 0.000 & 0.000 & 0.000 & 0.000 & 0.9375 & -3.750 & 0.000 & -0.9375 & -3.750 \\ 0.000 & 0.000 & 0.000 & 0.000 & -3.750 & 20.00 & 0.000 & 3.750 & 10.00 \\ -129.0 & -51.82 & 1.406 & -150.0 & 0.000 & 0.000 & 279.0 & 51.82 & 1.406 \\ -51.82 & -21.90 & -3.476 & 0.000 & -0.9375 & 3.750 & 51.82 & 22.84 & 0.2735 \\ -1.406 & 3.476 & 10.00 & 0.000 & -3.750 & 10.00 & 1.406 & 0.2735 & 40.00 \end{array} \end{bmatrix} \begin{array}{c} 1 \\ 2 \\ 3 \\ 4 \\ 5 \\ 6 \\ 7 \\ 8 \\ 9 \end{array}$$ [c]

Next, we compute the element equivalent load vectors. Element 1 has an intermediate 50 kN point load. The local components of the equivalent nodal point loads are found by setting $a = 3$ m, $L = 8$ m, and $P = 50$ kN in Equation 10.3.23:

$$\bar{Y}_1^{(1)}\big)_{eq} = \left[2\left(\frac{3}{8}\right)^3 - 3\left(\frac{3}{8}\right)^2 + 1\right](50) = 34.18 \text{ kN}$$

$$\bar{M}_{z_1}^{(1)}\big)_{eq} = \left[\frac{3^3}{8^2} - 2\left(\frac{3^2}{8}\right) + 3\right](50) = 58.59 \text{ kN-m}$$

$$\bar{Y}_2^{(1)}\big)_{eq} = \left[-2\left(\frac{3}{8}\right)^3 + 3\left(\frac{3}{8}\right)^2\right](50) = 15.82 \text{ kN}$$

$$\bar{M}_{z_2}^{(1)}\big)_{eq} = \left[\frac{3^3}{8^2} - \frac{3^2}{8}\right](50) = -35.16 \text{ kN-m}$$

Here, $\bar{X}_1^{(1)})_{eq} = \bar{X}_2^{(1)})_{eq} = 0$, since the 50 kN load has no component along the element axis. (Axial intermediate load components, if present, can be accounted for by using Equation 9.12.13 and the shape functions for axial deformation given in Equation 10.2.3.) Therefore, the local element equivalent load vector is

$$\left\{\bar{Q}_{eq}^{(1)}\right\} = 10^3 \begin{Bmatrix} 0 \\ 34.18 \\ 58.59 \\ \hline 0 \\ 15.82 \\ -35.16 \end{Bmatrix} \text{ (N)} \qquad \qquad \textbf{[d]}$$

The global equivalent load vector for element 1 is found by projecting the vector components in Equation d onto the global axes. This involves applying the transformation

$$\left\{Q_{eq}^{(1)}\right\} = [\Lambda^{(1)}]^T \left\{\bar{Q}_{eq}^{(1)}\right\}$$

The term $[\Lambda^{(1)}]$ is found from Equation 10.4.9, into which we substitute $l = 0.375$ and $m = 0.9270$ and take the transpose to obtain

$$\left\{Q_{eq}^{(1)}\right\} = \begin{bmatrix} 0.9270 & -0.3750 & 0 & 0 & 0 & 0 \\ 0.3750 & 0.9270 & 0 & 0 & 0 & 0 \\ 0 & 0 & 1 & 0 & 0 & 0 \\ 0 & 0 & 0 & 0.9270 & -0.3750 & 0 \\ 0 & 0 & 0 & 0.3750 & 0.9270 & 0 \\ 0 & 0 & 0 & 0 & 0 & 1 \end{bmatrix} \begin{Bmatrix} 0 \\ 34.18 \\ 58.59 \\ 0 \\ 15.82 \\ -35.16 \end{Bmatrix} (10^3) \qquad \textbf{[e]}$$

This yields the following equivalent load vector for element 1, resolved along the global reference axes:

$$\left\{Q_{eq}^{(1)}\right\} = (10^3) \begin{Bmatrix} -12.82 \\ 31.68 \\ 58.59 \\ \hline -5.932 \\ 14.66 \\ -35.15 \end{Bmatrix} \begin{matrix} 1 \\ 2 \\ 3 \\ 7 \\ 8 \\ 9 \end{matrix} \qquad \qquad \textbf{[f]}$$

Figure 10.4.4 illustrates the procedure for equating the intermediate point load to equivalent nodal load components, first in the local reference frame and then in the global reference frame.

Element 2 of the structure has a uniformly distributed load over its entire length. We use Equations 10.3.22, with $p_1 = p_2 = -4 \text{ kN/m}$ and $L = 8 \text{ m}$, to find the local element equivalent load vector components:

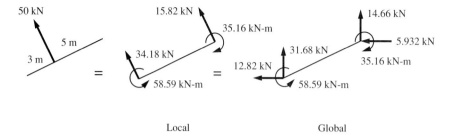

Figure 10.4.4 Equivalent load components for element 1 of Figure 10.4.3.

$$\bar{Y}_1^{(2)}\Big)_{eq} = \left[\frac{7}{20} + \frac{3}{20}\right](-4)\,(8) = -16 \text{ kN}$$

$$\bar{M}_{z_1}^{(2)}\Big)_{eq} = \left[\frac{1}{20} + \frac{1}{30}\right](-4)\,(8)^2 = -21.33 \text{ kN-m}$$

$$\bar{Y}_2^{(2)}\Big)_{eq} = \left[\frac{3}{20} + \frac{7}{20}\right](-4)\,(8) = -16 \text{ kN}$$

$$\bar{M}_{z_2}^{(2)}\Big)_{eq} = \left[-\frac{1}{30} - \frac{1}{20}\right](-4)\,(8)^2 = 21.33 \text{ kN-m}$$

The distributed load has no axial component, that is, $\bar{X}_1^{(2)})_{eq} = \bar{X}_2^{(2)})_{eq} = 0$. The global equivalent load vector is identical to the local one, because the coordinate system of element 2 is not inclined to the global frame. Therefore,

$$\left\{Q_{eq}^{(2)}\right\} = \left\{\bar{Q}_{eq}^{(2)}\right\} = (10^3) \begin{Bmatrix} 0.000 \\ -16.00 \\ -21.33 \\ \hdashline 0.000 \\ -16.00 \\ 21.33 \end{Bmatrix} \begin{matrix} 7 \\ 8 \\ 9 \\ 4 \\ 5 \\ 6 \end{matrix} \qquad \text{[g]}$$

The equivalent load system is illustrated in Figure 10.4.5.

The element global equivalent load vectors $\{Q_{eq}^{(1)}\}$ and $\{Q_{eq}^{(2)}\}$ are assembled into the nine-component global equivalent load vector $\{Q_{eq}\}$ of the structure, using the global degree of freedom number labels shown in Equations f and g as a guide. This procedure yields

$$\{Q_{eq}\} = (10^3) \begin{Bmatrix} -12.82 \\ 31.68 \\ 58.59 \\ \hdashline 0.000 \\ -16.00 \\ 21.33 \\ \hdashline -5.932 \\ -1.340 \\ -56.48 \end{Bmatrix} \begin{matrix} 1 \\ 2 \\ 3 \\ 4 \\ 5 \ (N) \\ 6 \\ 7 \\ 8 \\ 9 \end{matrix} \qquad \text{[h]}$$

The global point-load vector $\{Q_P\}$ consists of the six support reactions and the zero point load applied to node 3 of the structure:

Figure 10.4.5   Equivalent load components for element 2 of Figure 10.4.3.

$$\{Q_P\} = \begin{Bmatrix} X_1 \\ Y_1 \\ M_{z_1} \\ \hdashline X_2 \\ Y_2 \\ M_{z_2} \\ \hdashline 0 \\ 0 \\ 0 \end{Bmatrix} \begin{matrix} 1 \\ 2 \\ 3 \\ 4 \\ 5 \\ 6 \\ 7 \\ 8 \\ 9 \end{matrix}$$

[i]

To find the unknown displacements, we first apply the constraints by partitioning the global stiffness matrix [**k**], displacement vector {**q**}, and load vector {**Q**} in the manner of Equation 9.10.14. Degrees of freedom 7, 8, and 9 at node 3 are free; degrees of freedom 1 through 6 are constrained to zero by the fixed supports. The sets of free and constrained freedoms are f: 7, 8, 9 and s: 1, 2, 3, 4, 5, 6. Therefore,

$$[\mathbf{k}_{ff}] = \begin{matrix} & 7 & 8 & 9 \\ 7 \\ 8 \\ 9 \end{matrix} \begin{bmatrix} & \mathbf{k} & \end{bmatrix} = (10^6) \begin{bmatrix} 279.0 & 51.82 & 1.406 \\ 51.82 & 22.84 & 0.2735 \\ 1.406 & 0.2735 & 40.00 \end{bmatrix}$$

$$[\mathbf{k}_{fs}] = \begin{matrix} & 1 & 2 & 3 & 4 & 5 & 6 \\ 7 \\ 8 \\ 9 \end{matrix} \begin{bmatrix} & & \mathbf{k} & & \end{bmatrix} = (10^6) \begin{bmatrix} -120.0 & -51.82 & 1.406 & -150.0 & 0.000 & 0.000 \\ 51.82 & -21.90 & -3.476 & 0.000 & 0.000 & 0.000 \\ -1.406 & 3.476 & 10.00 & 0.000 & -3.750 & 10.00 \end{bmatrix}$$

$$\{\mathbf{Q}_{eq_f}\} = \begin{matrix} 7 \\ 8 \\ 9 \end{matrix} \begin{Bmatrix} \mathbf{Q}_{eq} \end{Bmatrix} = (10^3) \begin{Bmatrix} -5.932 \\ -1.340 \\ -56.48 \end{Bmatrix} \qquad \{\mathbf{Q}_{P_f}\} = \begin{matrix} 7 \\ 8 \\ 9 \end{matrix} \begin{Bmatrix} \mathbf{Q}_P \end{Bmatrix} = \begin{Bmatrix} 0 \\ 0 \\ 0 \end{Bmatrix}$$

$$\{\mathbf{q}_f\} = \begin{matrix} 7 \\ 8 \\ 9 \end{matrix} \begin{Bmatrix} \mathbf{q} \end{Bmatrix} = \begin{Bmatrix} u_3 \\ v_3 \\ \theta_{z3} \end{Bmatrix} \qquad \{\mathbf{q}_s\} = \begin{matrix} 1 \\ 2 \\ 3 \\ 4 \\ 5 \\ 6 \end{matrix} \begin{Bmatrix} \mathbf{q} \end{Bmatrix} = \begin{Bmatrix} 0 \\ 0 \\ 0 \\ 0 \\ 0 \\ 0 \end{Bmatrix}$$

It follows that the equations for the free displacements,

$$[\mathbf{k}_{ff}]\{\mathbf{q}_f\} = \{\mathbf{Q}_{P_f}\} + \{\mathbf{Q}_{\mathrm{eq}_f}\} - [\mathbf{k}_{fs}]\{\mathbf{q}_s\}$$

in this case are

$$(10^6)\begin{bmatrix} 279.0 & 51.82 & 1.406 \\ 51.82 & 22.84 & 0.2735 \\ 1.406 & 0.2735 & 40.00 \end{bmatrix}\begin{Bmatrix} u_3 \\ v_3 \\ \theta_{z_3} \end{Bmatrix} = (10^3)\begin{Bmatrix} -5.932 \\ -1.340 \\ -56.48 \end{Bmatrix} - \begin{Bmatrix} 0 \\ 0 \\ 0 \end{Bmatrix}$$

Solving this system of equations yields

$$\{\mathbf{q}_f\} = \begin{Bmatrix} u_3 \\ v_3 \\ \theta_{z_3} \end{Bmatrix} = \begin{Bmatrix} -0.01111 \text{ mm} \\ -0.01631 \text{ mm} \\ -0.001412 \text{ rad} \end{Bmatrix}$$ [j]

The shape of the structure subject to these displacements is shown in Figure 10.4.6.

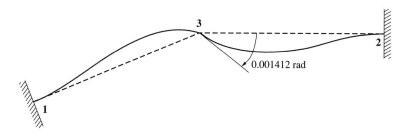

Figure 10.4.6   Deformed shape of structure loaded as shown in Figure 10.4.3.

Vectors $\{\mathbf{q}_f\}$, together with $\{\mathbf{q}_s\}$, comprise the now completely determined, nine-component global displacement vector of the structure. The element loads are found by applying the appropriate global displacements to each element in turn. For element 1, the correspondence between the element and structural displacements in the global system is

$$\{\mathbf{q}^{(1)}\} = \begin{Bmatrix} u_1^{(1)} \\ v_1^{(1)} \\ \theta_{z_1}^{(1)} \\ \hdashline u_2^{(1)} \\ v_2^{(1)} \\ \theta_{z_2}^{(1)} \end{Bmatrix} = \begin{Bmatrix} u_1 \\ v_1 \\ \theta_{z_1} \\ \hdashline u_3 \\ v_3 \\ \theta_{z_3} \end{Bmatrix} = (10^{-6})\begin{Bmatrix} 0.000 \\ 0.000 \\ 0.000 \\ \hdashline -11.11 \text{ m} \\ -16.31 \text{ m} \\ -1412 \text{ rad} \end{Bmatrix}$$

To find the loads in the element, we first project the components of $\{\mathbf{q}^{(1)}\}$ onto the local coordinate directions, using the transformation

$$\{\bar{\mathbf{q}}^{(1)}\} = [\mathbf{\Lambda}^{(1)}]\{\mathbf{q}^{(1)}\}$$

As in Equation e, the transformation matrix $[\mathbf{\Lambda}^{(1)}]$ is obtained from Equation 10.4.9, so that

$$\{\bar{q}^{(1)}\} = \begin{bmatrix} 0.9270 & 0.3750 & 0 & \vdots & 0 & 0 & 0 \\ -0.3750 & 0.9270 & 0 & \vdots & 0 & 0 & 0 \\ 0 & 0 & 1 & \vdots & 0 & 0 & 0 \\ \cdots & \cdots & \cdots & \vdots & \cdots & \cdots & \cdots \\ 0 & 0 & 0 & \vdots & 0.9270 & 0.3750 & 0 \\ 0 & 0 & 0 & \vdots & -0.3750 & 0.9270 & 0 \\ 0 & 0 & 0 & \vdots & 0 & 0 & 1 \end{bmatrix} \begin{Bmatrix} 0 \\ 0 \\ 0 \\ \cdots \\ -11.11 \\ -16.31 \\ -1412 \end{Bmatrix} (10^{-6}) = \begin{Bmatrix} 0 \\ 0 \\ 0 \\ \cdots \\ -16.42 \\ -10.95 \\ -1412 \end{Bmatrix} (10^{-6})$$

To calculate the point loads $\{\bar{Q}_P^{(1)}\}$ applied at the nodes of element 1, we substitute the local displacement vector $\{\bar{q}^{(1)}\}$, the local equivalent load vector $\{\bar{Q}_{eq}^{(1)}\}$ from Equation d, and the local element stiffness matrix $[\bar{k}^{(1)}]$ into the element stiffness equations

$$\{\bar{Q}_P^{(1)}\} = [\bar{k}^{(1)}]\{\bar{q}^{(1)}\} - \{\bar{Q}_{eq}^{(1)}\} \tag{k}$$

Ordinarily, at this point, we would have to compute $[\bar{k}^{(1)}]$ by substituting the properties of element 1 into Equation 10.4.1. However, in this example, element 1 is identical to element 2, except for being inclined. Therefore, although the global stiffness matrices (Equations a and b) of the two elements differ, their local stiffness matrices are identical: $[\bar{k}^{(1)}] = [\bar{k}^{(2)}]$. Furthermore, since element 2 has a zero angle of inclination to the global reference axes, its global and local stiffness matrices are the same: $[\bar{k}^{(2)}] = [k^{(2)}]$. This means that for $[\bar{k}^{(1)}]$, we can simply substitute the right side of Equation b into Equation k, which then becomes

$$\{\bar{Q}_P^{(1)}\} = 10^6 \begin{bmatrix} 150.0 & 0.000 & 0.000 & \vdots & -150.0 & 0.000 & 0.000 \\ 0.000 & 0.9375 & 3.750 & \vdots & 0.000 & -0.9375 & 3.750 \\ 0.000 & 3.750 & 20.00 & \vdots & 0.000 & -3.750 & 10.00 \\ \cdots & \cdots & \cdots & \vdots & \cdots & \cdots & \cdots \\ -150.0 & 0.000 & 0.000 & \vdots & 150.0 & 0.000 & 0.000 \\ 0.000 & -0.9375 & -3.750 & \vdots & 0.000 & 0.9375 & -3.750 \\ 0.000 & 3.750 & 10.00 & \vdots & 0.000 & -3.750 & 20.00 \end{bmatrix} \begin{Bmatrix} 0 \\ 0 \\ 0 \\ \cdots \\ -16.42 \\ -10.95 \\ -1412 \end{Bmatrix} (10^{-6}) - \begin{Bmatrix} 0 \\ 34,180 \\ 58,590 \\ \cdots \\ 0 \\ 15,820 \\ -35,160 \end{Bmatrix}$$

or

$$\{\bar{Q}_P^{(1)}\} = \begin{Bmatrix} 2463 \\ -5285 \\ -14,080 \\ -2463 \\ 5,285 \\ -28,200 \end{Bmatrix} - \begin{Bmatrix} 0 \\ 34,180 \\ 58,590 \\ 0 \\ 15,820 \\ -35,160 \end{Bmatrix} = \begin{Bmatrix} 2463 \text{ N} \\ -39,460 \text{ N} \\ -72,670 \text{ N-m} \\ -2463 \text{ N} \\ -10,540 \text{ N} \\ 6950 \text{ N-m} \end{Bmatrix} \tag{l}$$

The load vector components are sketched in Figure 10.4.7.

Moving to element 2, we see that the correspondence between the element and structural displacements is

$$\{\bar{q}^{(2)}\} = \{q^{(2)}\} = \begin{Bmatrix} u_1^{(2)} \\ v_1^{(2)} \\ \theta_{z1}^{(2)} \\ \cdots \\ u_2^{(2)} \\ v_2^{(2)} \\ \theta_{z2}^{(2)} \end{Bmatrix} = \begin{Bmatrix} u_3 \\ v_3 \\ \theta_{z3} \\ \cdots \\ u_2 \\ v_2 \\ \theta_{z2} \end{Bmatrix} = (10^{-6}) \begin{Bmatrix} -11.11 \text{ m} \\ -16.31 \text{ m} \\ -1412 \text{ rad} \\ 0.000 \\ 0.000 \\ 0.000 \end{Bmatrix}$$

The local and global displacement vectors of element 2 are identical, because the element coordinate axes were chosen to have the same orientation as the global system. Recall that this means $[\bar{k}^{(2)}] = [k^{(2)}]$, as well. Substituting $\{\bar{q}^{(2)}\}$, along with $[\bar{k}^{(2)}]$ and $\{Q_{eq}^{(2)}\}$ from Equations b and g, respectively, into the element stiffness equation

$$\{\bar{\mathbf{Q}}_P^{(2)}\} = [\bar{\mathbf{k}}^{(2)}]\{\bar{\mathbf{q}}^{(2)}\} - \{\bar{\mathbf{Q}}_{eq}^{(2)}\}$$ [m]

we get

$$\{\bar{\mathbf{Q}}_P^{(2)}\} = (10^6)\begin{bmatrix} 150.0 & 0.000 & 0.000 & -150.0 & 0.000 & 0.000 \\ 0.000 & 0.9375 & 3.750 & 0.000 & -0.9375 & 3.750 \\ 0.000 & 3.750 & 20.00 & 0.000 & -3.750 & 10.00 \\ -150.0 & 0.000 & 0.000 & 150.0 & 0.000 & 0.000 \\ 0.000 & -0.9375 & -3.750 & 0.000 & 0.9375 & -3.750 \\ 0.000 & 3.750 & 10.00 & 0.000 & -3.750 & 20.00 \end{bmatrix}\begin{Bmatrix} -11.11 \\ -16.31 \\ -1412 \\ 0 \\ 0 \\ 0 \end{Bmatrix}(10^{-6}) - \begin{Bmatrix} 0 \\ -16,000 \\ -21,330 \\ 0 \\ -16,000 \\ 21,330 \end{Bmatrix}$$

or

$$\{\bar{\mathbf{Q}}_P^{(2)}\} = \begin{Bmatrix} -1666 \\ -5310 \\ -28,300 \\ 1666 \\ 5310 \\ -14,180 \end{Bmatrix} - \begin{Bmatrix} 0 \\ -16,000 \\ -21,330 \\ 0 \\ -16,000 \\ 21,330 \end{Bmatrix} = \begin{Bmatrix} -1666\ \text{N} \\ 10,690\ \text{N} \\ -6,968\ \text{N-m} \\ 1666\ \text{N} \\ 23,310\ \text{N} \\ -35,510\ \text{N-m} \end{Bmatrix}$$ [n]

In Figure 10.4.7 the local nodal load components, together with the externally applied loads, are sketched for both elements in Figure 10.4.3. Using this information and statics, we can plot the shear and bending moment distributions over the compete frame, as in Figures 10.4.8 and 10.4.9.

At this point, a few comments are in order. We included all the detailed calculations for solving the planar frame problem in Figure 10.4.3 to fully illustrate the matrix displacement solution method. Some computations could have been avoided by applying the support conditions at the element level, computing only the 3 by 3 matrices $[\mathbf{k}_{ff}^{(1)}]$ and $[\mathbf{k}_{ff}^{(2)}]$, and then assembling

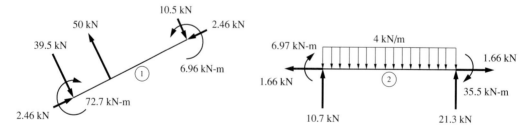

Figure 10.4.7   Element loads for the frame of Figure 10.4.3.

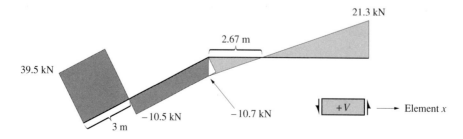

Figure 10.4.8   Internal shear force distribution in the frame of Figure 10.4.3.

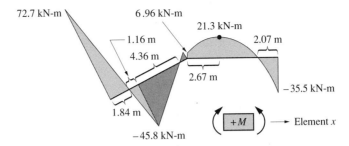

Figure 10.4.9   Internal bending moment distribution in the frame of Figure 10.4.3.

them to obtain the 3 by 3 constrained global stiffness matrix $[\mathbf{k}_{ff}]$ directly, rather than calculating the full 9 by 9 matrix $[\mathbf{k}]$ of Equation c. Furthermore, since stiffness matrices are symmetric, we really only need to compute the lower (or upper) triangular entries. If, as was done in this example, we initially ignore the support constraints and compute the complete, unconstrained structural stiffness matrix, then we can study the same structure subject to alternative support conditions, an option not readily available to us if we form the constrained stiffness matrix from the outset. However, if only the given set of support constraints are of interest, and a hand calculation is to be done, constraints should be applied at the element level to avoid unnecessary computations. Structures even slightly more complex than this example, that is, having just a few more elements and nodes (degrees of freedom), can become unwieldy if a displacement method solution is attempted by hand. The procedure can, and has widely been, programmed for digital computer implementation, relieving us of the computational burden, thereby allowing rapid and accurate analysis of complex planar frameworks, as well as other types of structures.

10.5   TORSIONAL ELEMENT

Consider a cylindrical shaft of arbitrary cross section subjected to pure torsion about the x axis, as illustrated in Figure 10.5.1. Due to the torsional loading, each cross section rotates through an angle θ_x. According to the Saint-Venant theory of (unrestrained) torsion (section 4.4), the angle of twist varies linearly from one end of the rod to the other, so that

$$\theta_x = \theta_{x_1}\left(1 - \frac{x}{L}\right) + \theta_{x_2}\frac{x}{L} \qquad \text{[10.5.1]}$$

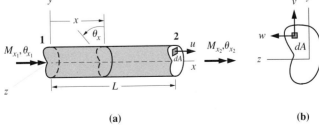

(a)   (b)

Figure 10.5.1   (a) Cylindrical bar of arbitrary cross section in pure torsion. (b) Displacements in a positively oriented plane of the cross section.

In general, the rotation is accompanied by a warping of the cross section, which is manifested by the axial displacement u of points on the initially plane section (cf. Figure 10.5.1a). The axial displacement u is assumed to be independent of x, proportional to the uniform rate of twist $d\theta_x/dx$, and otherwise specified by a warping function $\psi(y, z)$, as follows:

$$u = \frac{d\theta_x}{dx}\psi(y, z) = \frac{\theta_{x_2} - \theta_{x_1}}{L}\psi(y, z) \qquad [10.5.2]$$

The Saint-Venant theory assumes that no distortion occurs in the plane of the cross section. That is, viewed down the twist axis of the rod, each cross section appears to rotate as a rigid thin slice. Thus, the in-plane displacements v and w (Figure 10.5.1b) of points in the cross section are given by Equations 4.4.2,

$$v = -\theta_x z = -\left[\theta_{x_1}\left(1 - \frac{x}{L}\right) + \theta_{x_2}\frac{x}{L}\right]z \qquad w = \theta_x y = \left[\theta_{x_1}\left(1 - \frac{x}{L}\right) + \theta_{x_2}\frac{x}{L}\right]y$$

It is understood that the magnitude of the angle of twist θ_x is consistent with the small-displacement assumption of linear elasticity.

The only two nonvanishing strains, γ_{xy} and γ_{xz}, are given by Equations 4.7.7d and e. They are related to the two nodal displacements θ_{x_1} and θ_{x_2} by the strain–displacement expression

$$\left\{\begin{array}{c}\gamma_{xy}\\\gamma_{xz}\end{array}\right\} = \frac{1}{L}\left[\begin{array}{cc}-\left(\frac{\partial\psi}{\partial y} - z\right) & \frac{\partial\psi}{\partial y} - z\\-\left(\frac{\partial\psi}{\partial z} + y\right) & \frac{\partial\psi}{\partial z} + y\end{array}\right]\left\{\begin{array}{c}\theta_{x_1}\\\theta_{x_2}\end{array}\right\}$$

Therefore, in this case, the strain–displacement matrix $[\mathbf{B}]$ is the 2 by 2 array

$$[\mathbf{B}] = \frac{1}{L}\left[\begin{array}{c:c}-\left(\frac{\partial\psi}{\partial y} - z\right) & \frac{\partial\psi}{\partial y} - z\\\hdashline-\left(\frac{\partial\psi}{\partial z} + y\right) & \frac{\partial\psi}{\partial z} + y\end{array}\right] \qquad [10.5.3]$$

From Hooke's law for isotropic materials, we have

$$\left\{\begin{array}{c}\tau_{xy}\\\tau_{xz}\end{array}\right\} = \left[\begin{array}{cc}G & 0\\0 & G\end{array}\right]\left\{\begin{array}{c}\gamma_{xy}\\\gamma_{xz}\end{array}\right\}$$

so that the material stiffness matrix $[\mathbf{E}]$ for the torsion bar is

$$[\mathbf{E}] = \left[\begin{array}{cc}G & 0\\0 & G\end{array}\right] \qquad [10.5.4]$$

Substituting Equations 10.5.3 and 10.5.4 into the general formula for an element stiffness matrix, Equation 9.12.16, we get

$$[\mathbf{k}] = \int\!\!\int\!\!\int_V [\mathbf{B}]^T [\mathbf{E}][\mathbf{B}]\, dV$$

Noting that $dV = dA\,dx$, where dA is a differential element of cross-sectional area, we get, after some matrix multiplication,

$$[\bar{\mathbf{k}}] = \frac{G}{L^2} \int_0^L \iint_A \left[\begin{array}{c|c} \left(\frac{\partial\psi}{\partial y} - z\right)^2 + \left(\frac{\partial\psi}{\partial z} + y\right)^2 & -\left(\frac{\partial\psi}{\partial y} - z\right)^2 - \left(\frac{\partial\psi}{\partial z} + y\right)^2 \\ \hline -\left(\frac{\partial\psi}{\partial y} - z\right)^2 - \left(\frac{\partial\psi}{\partial z} + y\right)^2 & \left(\frac{\partial\psi}{\partial y} - z\right)^2 + \left(\frac{\partial\psi}{\partial z} + y\right)^2 \end{array} \right] dA\,dx$$

Since the integrand is independent of x, this expression simplifies to

$$[\bar{\mathbf{k}}] = \frac{GJ}{L} \left[\begin{array}{c|c} 1 & -1 \\ \hline -1 & 1 \end{array} \right]$$ [10.5.5]

where

$$J = \iint_A \left[\left(\frac{\partial\psi}{\partial y} - z\right)^2 + \left(\frac{\partial\psi}{\partial z} + y\right)^2 \right] dA$$

Here, J is the torsion constant introduced in Equation 4.4.14 and shown to have this equivalent expression in section 7.6. Observe the similarity between the local stiffness matrix $[\bar{\mathbf{k}}]$ of the rod element, Equation 10.2.7, and that of the torsion bar. Simply replace the axial rigidity AE in Equation 10.2.7 by the torsional rigidity GJ.

10.6   **THREE-DIMENSIONAL FRAMES**

A general, three-dimensional frame element transmits axial force, axial torque, shear in two planes, and bending in two planes. Each node has six degrees of freedom. The first three correspond to the three components of linear displacement and force at the node. The remaining three correspond to the three components of angular displacement and force (rotations and couples). Figure 10.6.1 shows a three-dimensional, 12 degree of freedom frame element, with the local nodal loads and displacements resolved into components along the three element coordinate axes. The correspondence between each load–displacement component and its degree of freedom number is also shown.

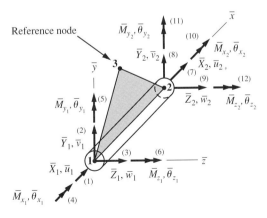

Figure 10.6.1   Three-dimensional frame element with 12 local degrees of freedom.

Node 3 lies in the $\bar{x}\bar{y}$ plane.

As in the case of the planar frame element, the $\bar{x}$ axis is directed along the axis of the beam, from node 1 to node 2. The element axes $\bar{y}$ and $\bar{z}$ in Figure 10.6.1 are defined to be the *principal axes of inertia* of the element cross section, with the $\bar{z}$ axis taken to be the *maximum* principal moment of inertia. The orientation of these two orthogonal principal axes relative to the global frame of reference is fixed by the assignment of the global coordinates of a third node (a reference node), located such that it and the $\bar{x}$ axis determine the element $\bar{x}\bar{y}$ plane. The positive $\bar{y}$ axis lies on the same side of the $\bar{x}$ axis as the reference node. The element $\bar{x}$, $\bar{y}$, and $\bar{z}$ axes are required to form a right-hand triad, thereby defining the positive direction of the axis, which is normal to the $\bar{x}\bar{y}$ plane.

With six degrees of freedom at both of its nodes, the general frame element has a total of 12 degrees of freedom and therefore a 12 by 12 stiffness matrix. The local form of this matrix can be formed by separately considering the axial stiffness (two degrees of freedom), the torsional stiffness (two degrees of freedom), and the bending stiffness (a total of eight degrees of freedom) in each of the two principal planes. Since these loading modes are not coupled, they can be superimposed, to assemble the complete stiffness matrix.

The axial stiffness matrix $[\mathbf{k}_A]$ is found in Equation 10.2.7,

$$[\mathbf{k}_A] = \frac{AE}{L} \begin{matrix} & 1 & 7 \\ & \begin{bmatrix} 1 & -1 \\ -1 & 1 \end{bmatrix} & \begin{matrix} 1 \\ 7 \end{matrix} \end{matrix}$$

[10.6.1]

Observe that the rows and columns are labeled with the degree of freedom numbers corresponding to pure axial loading in Figure 10.6.1.

According to Equation 10.5.5, the stiffness matrix $[\mathbf{k}_T]$ for pure axial torsion is

$$[\mathbf{k}_T] = \frac{GJ}{L} \begin{matrix} & 4 & 10 \\ & \begin{bmatrix} 1 & -1 \\ -1 & 1 \end{bmatrix} & \begin{matrix} 4 \\ 10 \end{matrix} \end{matrix}$$

[10.6.2]

where the torsional degree of freedom numbers for the element in Figure 10.6.1 are shown.

So far, we have accounted for four of the 12 degrees of freedom of the three-dimensional frame element. The remaining eight are incorporated by separately considering bending in the $\bar{x}\bar{y}$ and $\bar{x}\bar{z}$ planes. The bending loads in the $\bar{x}\bar{y}$ plane are sketched in Figure 10.6.2. The stiffness matrix for this load is found in Equation 10.3.13,

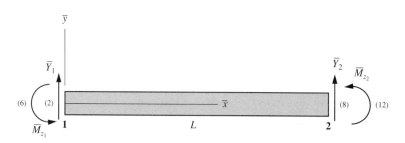

Figure 10.6.2   Bending in the $\bar{x}\bar{y}$ plane of Figure 10.6.1

$$[\mathbf{k}_{B_{\bar{x}\bar{y}}}] = \frac{EI_z}{L^3} \begin{array}{c} \begin{array}{cccc} 2 & 6 & 8 & 12 \end{array} \\ \begin{bmatrix} 12 & 6L & -12 & 6L \\ 6L & 4L^2 & -6L & 2L^2 \\ -12 & -6L & 12 & -6L \\ 6L & 2L^2 & -6L & 4L^2 \end{bmatrix} \begin{array}{c} 2 \\ 6 \\ 8 \\ 12 \end{array} \end{array} \qquad \textbf{[10.6.3]}$$

where the notation $[\mathbf{k}_{B_{\bar{x}\bar{y}}}]$ is used to designate the stiffness matrix for bending in the $\bar{x}\,\bar{y}$ plane. Notice that the degree of freedom labels agree with those in Figures 10.6.1 and 10.6.2.

To consider bending in the $\bar{x}\,\bar{z}$ plane, let us view the element in Figure 10.6.1 from the positive $\bar{y}$ axis towards the $\bar{x}\,\bar{z}$ plane. Figure 10.6.3 shows the in-plane bending loads observed from that direction.

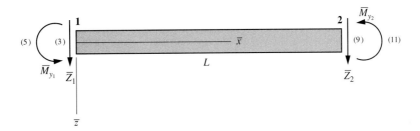

Figure 10.6.3   Bending in the $\bar{x}\bar{z}$ plane of Figure 10.6.1.

In Figure 10.6.3, the positive direction (downward) of the transverse loads and displacements is reversed from that of Figure 10.6.2 (upward). The positive direction of the nodal couples and rotations is the same. If we substitute the loading of Figure 10.6.3 into the stiffness equations for the beam shown in Figure 10.6.2, being careful to replace I_z by I_y and to use minus signs to account for the downward direction of the transverse quantities, we get

$$\begin{Bmatrix} -\bar{Z}_1 \\ \bar{M}_{y_1} \\ -\bar{Z}_2 \\ \bar{M}_{y_2} \end{Bmatrix} = \frac{EI_y}{L^3} \begin{bmatrix} 12 & 6L & -12 & 6L \\ 6L & 4L^2 & -6L & 2L^2 \\ -12 & -6L & 12 & -6L \\ 6L & 2L^2 & -6L & 4L^2 \end{bmatrix} \begin{Bmatrix} -\bar{w}_1 \\ \bar{\theta}_{y_1} \\ -\bar{w}_2 \\ \bar{\theta}_{y_2} \end{Bmatrix}$$

or

$$\begin{Bmatrix} \bar{Z}_1 \\ \bar{M}_{y_1} \\ \bar{Z}_2 \\ \bar{M}_{y_2} \end{Bmatrix} = \frac{EI_y}{L^3} \begin{bmatrix} 12 & -6L & -12 & -6L \\ -6L & 4L^2 & 6L & 2L^2 \\ -12 & 6L & 12 & 6L \\ -6L & 2L^2 & 6L & 4L^2 \end{bmatrix} \begin{Bmatrix} \bar{w}_1 \\ \bar{\theta}_{y_1} \\ \bar{w}_2 \\ \bar{\theta}_{y_2} \end{Bmatrix}$$

so that

$$[\mathbf{k}_{B\bar{x}\bar{z}}] = \frac{EI_y}{L^3} \begin{array}{c} \\ \\ \end{array} \begin{bmatrix} 12 & -6L & -12 & -6L \\ -6L & 4L^2 & 6L & 2L^2 \\ \hdashline -12 & 6L & 12 & 6L \\ -6L & 2L^2 & 6L & 4L^2 \end{bmatrix} \begin{array}{c} 3 \\ 5 \\ 9 \\ 11 \end{array}$$

[10.6.4]

Assembling the stiffness matrices in Equations 10.6.1 through 10.6.4 into the 12 by 12 array, and then observing that the result can be partitioned into four independent 3 by 3 submatrices, we finally obtain the complete local stiffness matrix for the three-dimensional frame element, as follows:

$$[\bar{\mathbf{k}}] = \begin{bmatrix} [\bar{\mathbf{k}}_1] & [\bar{\mathbf{k}}_2] & -[\bar{\mathbf{k}}_1] & [\bar{\mathbf{k}}_2] \\ [\bar{\mathbf{k}}_2]^T & [\bar{\mathbf{k}}_3] & -[\bar{\mathbf{k}}_2]^T & [\bar{\mathbf{k}}_4] \\ -[\bar{\mathbf{k}}_1] & -[\bar{\mathbf{k}}_2] & [\bar{\mathbf{k}}_1] & -[\bar{\mathbf{k}}_2] \\ [\bar{\mathbf{k}}_2]^T & [\bar{\mathbf{k}}_4] & -[\bar{\mathbf{k}}_2]^T & [\bar{\mathbf{k}}_3] \end{bmatrix}$$

[10.6.5a]

where the submatrices are given by

$$[\bar{\mathbf{k}}_1] = \begin{bmatrix} AE/L & 0 & 0 \\ 0 & 12EI_z/L^3 & 0 \\ 0 & 0 & 12EI_y/L^3 \end{bmatrix} \quad [\bar{\mathbf{k}}_2] = \begin{bmatrix} 0 & 0 & 0 \\ 0 & 0 & 6EI_z/L^2 \\ 0 & -6EI_y/L^2 & 0 \end{bmatrix}$$

$$[\bar{\mathbf{k}}_3] = \begin{bmatrix} GJ/L & 0 & 0 \\ 0 & 4EI_y/L & 0 \\ 0 & 0 & 4EI_z/L \end{bmatrix} \quad [\bar{\mathbf{k}}_4] = \begin{bmatrix} -GJ/L & 0 & 0 \\ 0 & 2EI_y/L & 0 \\ 0 & 0 & 2EI_z/L \end{bmatrix}$$

[10.6.5b]

As before, the overbar designates quantities in the local reference frame, rather than the global one.

The frame element stiffness matrix [**k**] in global coordinates is found by once again applying the transformation originally given in Equation 10.2.18,

$$[\mathbf{k}] = [\mathbf{\Lambda}]^T [\bar{\mathbf{k}}][\mathbf{\Lambda}]$$

[10.6.6]

where the matrix [**Λ**] relates the vector of 12 local element displacement components $\{\bar{\mathbf{q}}\}$ to its global counterpart $\{\mathbf{q}\}$, according to

$$\{\bar{\mathbf{q}}\} = [\mathbf{\Lambda}]\{\mathbf{q}\}$$

[10.6.7]

Both linear displacement and small rotational displacement are vector quantities, so their components transform according to Equation 9.9.6a; that is,

$$\begin{Bmatrix} \bar{u}_1 \\ \bar{v}_1 \\ \bar{w}_1 \end{Bmatrix} = \begin{bmatrix} l_1 & m_1 & n_1 \\ l_2 & m_2 & n_2 \\ l_3 & m_3 & n_3 \end{bmatrix} \begin{Bmatrix} u_1 \\ v_1 \\ w_1 \end{Bmatrix} \qquad \begin{Bmatrix} \bar{\theta}_{x_1} \\ \bar{\theta}_{y_1} \\ \bar{\theta}_{z_1} \end{Bmatrix} = \begin{bmatrix} l_1 & m_1 & n_1 \\ l_2 & m_2 & n_2 \\ l_3 & m_3 & n_3 \end{bmatrix} \begin{Bmatrix} \theta_{x_1} \\ \theta_{y_1} \\ \theta_{z_1} \end{Bmatrix}$$

with similar relations applying to the displacements at element node 2. (Recall that $\lfloor l_1 \quad m_1 \quad n_1 \rfloor$, $\lfloor l_2 \quad m_2 \quad n_2 \rfloor$, and $\lfloor l_3 \quad m_3 \quad n_3 \rfloor$—the rows of the transformation matrix [**Λ**]—are the unit vectors of the element $\bar{x}$, $\bar{y}$, and $\bar{z}$ axes, respectively, relative to the global coordinate system.) For the set of 12 element displacements, we can therefore write Equation 10.6.7 as

$$
\begin{Bmatrix} \bar{u}_1 \\ \bar{v}_1 \\ \bar{w}_1 \\ \bar{\theta}_{x_1} \\ \bar{\theta}_{y_1} \\ \bar{\theta}_{z_1} \\ \bar{u}_2 \\ \bar{v}_2 \\ \bar{w}_2 \\ \bar{\theta}_{x_2} \\ \bar{\theta}_{y_2} \\ \bar{\theta}_{z_2} \end{Bmatrix}
=
\begin{bmatrix}
[\boldsymbol{\lambda}] & [\mathbf{0}] & [\mathbf{0}] & [\mathbf{0}] \\
[\mathbf{0}] & [\boldsymbol{\lambda}] & [\mathbf{0}] & [\mathbf{0}] \\
[\mathbf{0}] & [\mathbf{0}] & [\boldsymbol{\lambda}] & [\mathbf{0}] \\
[\mathbf{0}] & [\mathbf{0}] & [\mathbf{0}] & [\boldsymbol{\lambda}]
\end{bmatrix}
\begin{Bmatrix} u_1 \\ v_1 \\ w_1 \\ \theta_{x_1} \\ \theta_{y_1} \\ \theta_{z_1} \\ u_2 \\ v_2 \\ w_2 \\ \theta_{x_2} \\ \theta_{y_2} \\ \theta_{z_2} \end{Bmatrix}
\qquad \textbf{[10.6.8]}
$$

where

$$
[\boldsymbol{\lambda}] = \begin{bmatrix} l_1 & m_1 & n_1 \\ l_2 & m_2 & n_2 \\ l_3 & m_3 & n_3 \end{bmatrix}
\qquad \text{and} \qquad
[\mathbf{0}] = \begin{bmatrix} 0 & 0 & 0 \\ 0 & 0 & 0 \\ 0 & 0 & 0 \end{bmatrix}
\qquad \textbf{[10.6.9]}
$$

compose the 12 by 12 transformation matrix $[\boldsymbol{\Lambda}]$ on the right of Equation 10.6.8.

To compute the direction cosines in Equation 10.6.9, we refer to Figure 10.6.4 and proceed as follows. First observe that the vector $\mathbf{r}_1$ drawn from element node 1 to node 2 defines the element axis $\bar{x}$. Using the global coordinates of the nodes, this vector is

$$
\mathbf{r}_1 = r_{1x}\mathbf{i} + r_{1y}\mathbf{j} + r_{1z}\mathbf{k} \qquad \textbf{[10.6.10a]}
$$

where

$$
r_{1x} = x_2 - x_1 \qquad r_{1y} = y_2 - y_1 \qquad r_{1z} = z_2 - z_1 \qquad \textbf{[10.6.10b]}
$$

Nodes 1 and 2, together with reference node 3, define the element $\bar{x}\,\bar{y}$ plane, which is shaded in Figure 10.6.4. The vector directed from node 1 to reference node 3 is

$$
\mathbf{r}_{1-3} = (x_3 - x_1)\mathbf{i} + (y_3 - y_1)\mathbf{j} + (z_3 - z_1)\mathbf{k}
$$

The cross product of $\mathbf{r}_1$ into $\mathbf{r}_{1-3}$ is normal to the $\bar{x}\,\bar{y}$ plane and defines the direction of the element $\bar{z}$ axis. Let $\mathbf{r}_3$ be the result of that cross product. Then,

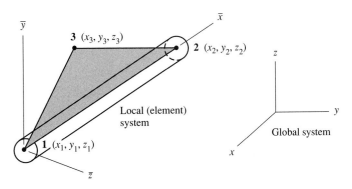

Figure 10.6.4   Global coordinates of an arbitrarily-oriented frame element.

$$\mathbf{r}_3 = \mathbf{r}_1 \times \mathbf{r}_{1-3} = r_{3x}\mathbf{i} + r_{3y}\mathbf{j} + r_{3z}\mathbf{k} \qquad [10.6.11a]$$

where

$$r_{3x} = r_{1y}(z_3 - z_1) - r_{1z}(y_3 - y_1) \qquad r_{3y} = r_{1z}(x_3 - x_1) - r_{1x}(z_3 - z_1) \qquad r_{3z} = r_{1x}(y_3 - y_1) - r_{1y}(x_3 - x_1) \qquad [10.6.11b]$$

Finally, we can define a vector $\mathbf{r}_2$ in the direction of the $\bar{y}$ axis by taking the cross product of $\mathbf{r}_3$ into $\mathbf{r}_1$, as follows:

$$\mathbf{r}_2 = \mathbf{r}_3 \times \mathbf{r}_1 = r_{2x}\mathbf{i} + r_{2y}\mathbf{j} + r_{2z}\mathbf{k} \qquad [10.6.12a]$$

where

$$r_{2x} = r_{3y}r_{1z} - r_{3z}r_{1y} \qquad r_{2y} = r_{3z}r_{1x} - r_{3x}r_{1z} \qquad r_{2z} = r_{3x}r_{1y} - r_{3y}r_{1x} \qquad [10.6.12b]$$

From Equations 10.6.10, 10.6.11, and 10.6.12, we have a triad of orthogonal vectors, $\mathbf{r}_1$, $\mathbf{r}_2$, and $\mathbf{r}_3$, directed along the element axes $\bar{x}$, $\bar{y}$, and $\bar{z}$, respectively. Dividing each of the vectors by its magnitude therefore yields the set of unit vectors $\bar{\mathbf{i}}$, $\bar{\mathbf{j}}$, and $\bar{\mathbf{k}}$ for the element system. Their components are the direction cosines required for the matrix $[\boldsymbol{\lambda}]$ in Equation 10.6.9:

$$l_1 = r_{1x}/|\mathbf{r}_1| \quad m_1 = r_{1y}/|\mathbf{r}_1| \quad n_1 = r_{1z}/|\mathbf{r}_1|$$
$$l_2 = r_{2x}/|\mathbf{r}_2| \quad m_2 = r_{2y}/|\mathbf{r}_2| \quad n_2 = r_{2z}/|\mathbf{r}_2| \qquad [10.6.13]$$
$$l_3 = r_{3x}/|\mathbf{r}_3| \quad m_2 = r_{3y}/|\mathbf{r}_3| \quad n_2 = r_{3z}/|\mathbf{r}_3|$$

Using the global coordinates of nodes 1, 2, and 3 of the frame element, we can find the direction cosines of the element coordinate axes and then form the matrix $[\boldsymbol{\lambda}]$, from which the matrix $[\boldsymbol{\Lambda}]$ in Equation 10.6.8 is constructed. We can then calculate the components of the frame element global stiffness matrix. We substitute Equation 10.6.5a and $[\boldsymbol{\Lambda}]$ from Equation 10.6.8 into Equation 10.6.6, which yields

$$[\mathbf{k}] = \begin{bmatrix} [\mathbf{k}_1] & [\mathbf{k}_2] & -[\mathbf{k}_1] & [\mathbf{k}_2] \\ [\mathbf{k}_2]^T & [\mathbf{k}_3] & -[\mathbf{k}_2]^T & [\mathbf{k}_4] \\ -[\mathbf{k}_1] & -[\mathbf{k}_2] & [\mathbf{k}_1] & -[\mathbf{k}_2] \\ [\mathbf{k}_2]^T & [\mathbf{k}_4] & -[\mathbf{k}_2]^T & [\mathbf{k}_3] \end{bmatrix} \qquad [10.6.14a]$$

where

$$[\mathbf{k}_1] = [\boldsymbol{\lambda}]^T[\bar{\mathbf{k}}_1][\boldsymbol{\lambda}] \qquad [\mathbf{k}_2] = [\boldsymbol{\lambda}]^T[\bar{\mathbf{k}}_2][\boldsymbol{\lambda}] \qquad [\mathbf{k}_3] = [\boldsymbol{\lambda}]^T[\bar{\mathbf{k}}_3][\boldsymbol{\lambda}] \qquad [\mathbf{k}_4] = [\boldsymbol{\lambda}]^T[\bar{\mathbf{k}}_4][\boldsymbol{\lambda}] \qquad [10.6.14b]$$

The explicit forms of these four submatrices are found by substituting Equations 10.6.5b for the overbarred submatrices and Equation 10.6.9, for $[\boldsymbol{\lambda}]$. The matrix multiplications in Equation 10.6.14b produce the following expressions for $[\mathbf{k}_1]$ through $[\mathbf{k}_4]$:

$$[\mathbf{k}_1] = \begin{bmatrix} \frac{AE}{L}l_1^2 + \frac{12E}{L^3}\left(I_z l_2^2 + I_y l_3^2\right) & \frac{AE}{L}l_1 m_1 + \frac{12E}{L^3}\left(I_z l_2 m_2 + I_y l_3 m_3\right) & \frac{AE}{L}l_1 n_1 + \frac{12E}{L^3}\left(I_z l_2 n_2 + I_y l_3 n_3\right) \\ \frac{AE}{L}l_1 m_1 + \frac{12E}{L^3}\left(I_z l_2 m_2 + I_y l_3 m_3\right) & \frac{AE}{L}m_1^2 + \frac{12E}{L^3}\left(I_z m_2^2 + I_y m_3^2\right) & \frac{AE}{L}m_1 n_1 + \frac{12E}{L^3}\left(I_z m_2 n_2 + I_y m_3 n_3\right) \\ \frac{AE}{L}l_1 n_1 + \frac{12E}{L^3}\left(I_z l_2 n_2 + I_y l_3 n_3\right) & \frac{AE}{L}m_1 n_1 + \frac{12E}{L^3}\left(I_z m_2 n_2 + I_y m_3 n_3\right) & \frac{AE}{L}n_1^2 + \frac{12E}{L^3}\left(I_z n_2^2 + I_y n_3^2\right) \end{bmatrix} \qquad [10.6.15a]$$

$$[\mathbf{k}_2] = \begin{bmatrix} \frac{6E}{L^2}\left(I_z - I_y\right)l_2 l_3 & \frac{6E}{L^2}\left(I_z l_2 m_3 - I_y l_3 m_2\right) & \frac{6E}{L^2}\left(I_z l_2 n_3 - I_y l_3 n_2\right) \\ \frac{6E}{L^2}\left(I_z l_3 m_2 - I_y l_2 m_3\right) & \frac{6E}{L^2}\left(I_z - I_y\right)m_2 m_3 & \frac{6E}{L^2}\left(I_z m_2 n_3 - I_y m_3 n_2\right) \\ \frac{6E}{L^2}\left(I_z l_3 n_2 - I_y l_2 n_3\right) & \frac{6E}{L^2}\left(I_z m_3 n_2 - I_y m_2 n_3\right) & \frac{6E}{L^2}\left(I_z - I_y\right)n_2 n_3 \end{bmatrix} \qquad [10.6.15b]$$

$$[k_3] = \begin{bmatrix} \frac{GJ}{L}l_1^2 + \frac{4E}{L}\left(I_yl_2^2 + I_zl_3^2\right) & \frac{GJ}{L}l_1m_1 + \frac{4E}{L}\left(I_yl_2m_2 + I_zl_3m_3\right) & \frac{GJ}{L}l_1n_1 + \frac{4E}{L}\left(I_yl_2n_2 + I_zl_3n_3\right) \\ \frac{GJ}{L}l_1m_1 + \frac{4E}{L}\left(I_yl_2m_2 + I_zl_3m_3\right) & \frac{GJ}{L}m_1^2 + \frac{4E}{L}\left(I_ym_2^2 + I_zm_3^2\right) & \frac{GJ}{L}m_1n_1 + \frac{4E}{L}\left(I_ym_2n_2 + I_zm_3n_3\right) \\ \frac{GJ}{L}l_1n_1 + \frac{4E}{L}\left(I_yl_2n_2 + I_zl_3n_3\right) & \frac{GJ}{L}m_1n_1 + \frac{4E}{L}\left(I_ym_2n_2 + I_zm_3n_3\right) & \frac{GJ}{L}n_1^2 + \frac{4E}{L}\left(I_yn_2^2 + I_zn_3^2\right) \end{bmatrix} \quad \textbf{[10.6.15c]}$$

$$[k_4] = \begin{bmatrix} -\frac{GJ}{L}l_1^2 + \frac{2E}{L}\left(I_yl_2^2 + I_zl_3^2\right) & -\frac{GJ}{L}l_1m_1 + \frac{2E}{L}\left(I_yl_2m_2 + I_zl_3m_3\right) & -\frac{GJ}{L}l_1n_1 + \frac{2E}{L}\left(I_yl_2n_2 + I_zl_3n_3\right) \\ -\frac{GJ}{L}l_1m_1 + \frac{2E}{L}\left(I_yl_2m_2 + I_zl_3m_3\right) & -\frac{GJ}{L}m_1^2 + \frac{2E}{L}\left(I_ym_2^2 + I_zm_3^2\right) & -\frac{GJ}{L}m_1n_1 + \frac{2E}{L}\left(I_ym_2n_2 + I_zm_3n_3\right) \\ -\frac{GJ}{L}l_1n_1 + \frac{2E}{L}\left(I_yl_2n_2 + I_zl_3n_3\right) & -\frac{GJ}{L}m_1n_1 + \frac{2E}{L}\left(I_ym_2n_2 + I_zm_3n_3\right) & -\frac{GJ}{L}n_1^2 + \frac{2E}{L}\left(I_yn_2^2 + I_zn_3^2\right) \end{bmatrix}$$

$$\textbf{[10.6.15d]}$$

Example 10.6.1 Use the matrix displacement method to find the location and value of the maximum bending moment in the frame in Figure 10.6.5. The two-element frame is built-in at nodes 2 and 3. Both elements have a solid circular cross section and are made of the same material. The geometric and material property data are given in the figure. A downward uniformly distributed load of 200 N/m is applied to the horizontal element.

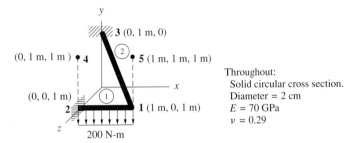

Figure 10.6.5 Two element frame.

From the figure, we can write the following properties, which are common to both elements:

Young's modulus $\qquad\qquad$ $E = 70$ GPa

Poisson's ratio $\qquad\qquad$ $\nu = 0.29$

Shear modulus $\qquad\qquad$ $G = \frac{E}{2(1+\nu)} = 27.13$ GPa

Area $\qquad\qquad$ $A = \pi r^2 = \pi(0.01 \text{ m})^2 = 314.2 \times 10^{-6} \text{ m}^2$

Area moments of inertia $\qquad$ $I_z = I_y = \frac{\pi r^4}{4} = \frac{\pi(0.01 \text{ m})^4}{4} = 7.854 \times 10^{-9} \text{ m}^4$

Polar moment of inertia $\qquad$ $J = I_y + I_z = 15.71 \times 10^{-9} \text{ m}^4$

Let us specify the orientation of the elements' coordinate axes by choosing the relationships between the local and global nodes, as indicated in Table 10.6.1.

Table 10.6.1 $\qquad\qquad$ **Element connectivities in Figure 10.6.5.**

Element	Node 1	Node 2	Node 3 (Reference Node)
1	2	1	4
2	1	3	5

Locating the reference nodes as we did positions the barred element systems as sketched in Figure 10.6.6. For both elements, the $\bar{x}\,\bar{y}$ plane, determined by the element axis and its reference node, is shown shaded in the figure, for clarity.

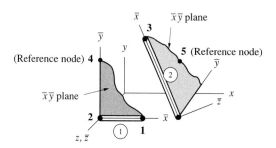

Figure 10.6.6   Orientation of the element coordinate systems for the frame of Figure 10.6.5.

The next step is to calculate the element stiffness matrices, starting with element 1. We must first find the direction cosines of each of the element's three reference axes. Ordinarily, this is done by proceeding through the calculations in Equations 10.6.10 through 10.6.13. However, the node numbering and reference node location for element 1 were chosen so as to align the element axes with the global axes, as Figure 10.6.6 shows. Since $\bar{\mathbf{i}} = \mathbf{i}$, $\bar{\mathbf{j}} = \mathbf{j}$, and $\bar{\mathbf{k}} = \mathbf{k}$, it follows that for element 1, we have simply

$$\left[\boldsymbol{\lambda}^{(1)}\right] = \begin{bmatrix} 1 & 0 & 0 \\ 0 & 1 & 0 \\ 0 & 0 & 1 \end{bmatrix}$$

The global stiffness matrix $[\mathbf{k}^{(1)}]$ of element 1 is given by Equation 10.6.14,

$$[\mathbf{k}^{(1)}] = \begin{bmatrix} [\mathbf{k}_1^{(1)}] & [\mathbf{k}_2^{(1)}] & -[\mathbf{k}_1^{(1)}] & [\mathbf{k}_2^{(1)}] \\ [\mathbf{k}_2^{(1)}]^T & [\mathbf{k}_3^{(1)}] & -[\mathbf{k}_1^{(1)}]^T & [\mathbf{k}_4^{(1)}] \\ -[\mathbf{k}_1^{(1)}] & -[\mathbf{k}_2^{(1)}] & [\mathbf{k}_1^{(1)}] & -[\mathbf{k}_2^{(1)}] \\ [\mathbf{k}_2^{(1)}]^T & [\mathbf{k}_4^{(1)}] & -[\mathbf{k}_2^{(1)}]^T & [\mathbf{k}_3^{(1)}] \end{bmatrix}$$

Global dof: 7 8 9   10 11 12   1 2 3   4 5 6
Local dof: 1 2 3   4 5 6   7 8 9   10 11 12

(row labels: 1 7 / 2 8 / 3 9 / 4 10 / 5 11 / 6 12 / 7 1 / 8 2 / 9 3 / 10 4 / 11 5 / 12 6)

[a]

Global degrees of freedom (dof) 7 through 12 are constrained to zero by the support at structural node 2. Therefore, we need only calculate $[\mathbf{k}_{ff}^{(1)}]$, the lower right 6 by 6 submatrix in Equation a:

$$[\mathbf{k}_{ff}^{(1)}] = \begin{bmatrix} [\mathbf{k}_1^{(1)}] & -[\mathbf{k}_2^{(1)}] \\ -[\mathbf{k}_2^{(1)}]^T & [\mathbf{k}_3^{(1)}] \end{bmatrix} \begin{matrix} 1 \\ 2 \\ 3 \\ 4 \\ 5 \\ 6 \end{matrix}$$

Global dof: 1 2 3   4 5 6

Substituting the material and cross-sectional properties, direction cosines, and element length (1 m) into Equations 10.6.15 yields

$$[\mathbf{k}_{ff}^{(1)}] = \begin{array}{cccccc} 1 & 2 & 3 & 4 & 5 & 6 \end{array} \begin{bmatrix} 21.99 \times 10^6 & 0 & 0 & 0 & 0 & 0 \\ 0 & 6597 & 0 & 0 & 0 & -3299 \\ 0 & 0 & 6597 & 0 & 3299 & 0 \\ 0 & 0 & 0 & 426.2 & 0 & 0 \\ 0 & 0 & 3299 & 0 & 2199 & 0 \\ 0 & -3299 & 0 & 0 & 0 & 2199 \end{bmatrix} \begin{array}{c} 1 \\ 2 \\ 3 \\ 4 \\ 5 \\ 6 \end{array}$$

Turning next to element 2, we first compute the direction cosines of its coordinate axes. From Equation 10.6.10b, the components of the vector $\mathbf{r}_1$, which is directed along the $\bar{x}$ axis, are

$$r_{1x} = 0 - 1 = 1 \qquad r_{1y} = 1 - 0 = 1 \qquad r_{1z} = 0 - 1 = 1 \tag{b}$$

Substituting these values into Equation 10.6.11b, we get the components of vector $\mathbf{r}_3$, which points in the $\bar{z}$ direction:

$$r_{3x} = 1(1 - 1) - (-1)(1 - 0) = 1 \qquad r_{3y} = (-1)(1 - 1) - (-1)(1 - 1) = 0 \qquad r_{3z} = (-1)(1 - 0) - (1)(1 - 0) = -1 \tag{c}$$

Employing Equations b and c in Equation 10.6.12b produces the components of $\mathbf{r}_2$, which defines the $\bar{y}$ direction:

$$r_{2x} = 0(-1) - (-1)(1) = 1 \qquad r_{2y} = (-1)(-1) - (1)(-1) = 2 \qquad r_{3z} = (1)(1) - (0)(-1) = 1 \tag{d}$$

The magnitudes of the three vectors $\mathbf{r}_1$, $\mathbf{r}_2$, and $\mathbf{r}_3$ are:

$$|\mathbf{r}_1| = 1.732 \qquad |\mathbf{r}_2| = 2.449 \qquad |\mathbf{r}_3| = 1.414 \tag{e}$$

We then divide the components of each vector by its magnitude, which establishes the three sets of direction cosines and thereby defines the rows of $[\boldsymbol{\lambda}^{(2)}]$:

$$[\boldsymbol{\lambda}^{(2)}] = \begin{bmatrix} -0.5774 & 0.5774 & -0.5774 \\ 0.4082 & 0.8165 & 0.4082 \\ 0.7071 & 0 & -0.7071 \end{bmatrix} \tag{f}$$

To ascertain the stiffness matrix $[\mathbf{k}^{(2)}]$ of element 2, we use Equation 10.6.14 to write

$$\begin{array}{cccc} \text{Global dof:} & \begin{array}{c} 1\ 2\ 3 \end{array} & \begin{array}{c} 4\ 5\ 6 \end{array} & \begin{array}{c} 13\ 14\ 15 \end{array} & \begin{array}{c} 16\ 17\ 18 \end{array} \\ \text{Local dof:} & \begin{array}{c} 1\ 2\ 3 \end{array} & \begin{array}{c} 4\ 5\ 6 \end{array} & \begin{array}{c} 7\ 8\ 9 \end{array} & \begin{array}{c} 10\ 11\ 12 \end{array} \end{array}$$

$$[\mathbf{k}^{(2)}] = \begin{bmatrix} [\mathbf{k}_1^{(2)}] & [\mathbf{k}_2^{(2)}] & -[\mathbf{k}_1^{(2)}] & [\mathbf{k}_2^{(2)}] \\ [\mathbf{k}_2^{(2)}]^T & [\mathbf{k}_3^{(2)}] & -[\mathbf{k}_2^{(2)}]^T & [\mathbf{k}_4^{(2)}] \\ \hdashline -[\mathbf{k}_1^{(2)}] & -[\mathbf{k}_2^{(2)}] & [\mathbf{k}_1^{(2)}] & -[\mathbf{k}_2^{(2)}] \\ [\mathbf{k}_2^{(2)}]^T & [\mathbf{k}_4^{(2)}] & -[\mathbf{k}_2^{(2)}]^T & [\mathbf{k}_3^{(2)}] \end{bmatrix} \begin{array}{cc} 1 & 1 \\ 2 & 2 \\ 3 & 3 \\ 4 & 4 \\ 5 & 5 \\ 6 & 6 \\ 7 & 13 \\ 8 & 14 \\ 9 & 15 \\ 10 & 16 \\ 11 & 17 \\ 12 & 18 \end{array} \tag{g}$$

Global freedoms 13 through 18 are constrained to be zero because of the rigid support at global node 3. Therefore, we need only compute $[\mathbf{k}_{ff}^{(2)}]$, which is the upper left 6 by 6 submatrix in Equation g:

$$\begin{array}{cc} \text{Global dof:} & \begin{array}{cc} 1\ 2\ 3 & 4\ 5\ 6 \end{array} \end{array}$$

$$[\mathbf{k}_{ff}^{(2)}] = \begin{bmatrix} [\mathbf{k}_1^{(2)}] & [\mathbf{k}_2^{(2)}] \\ [\mathbf{k}_2^{(2)}]^T & [\mathbf{k}_3^{(2)}] \end{bmatrix} \begin{array}{c} 1 \\ 2 \\ 3 \\ 4 \\ 5 \\ 6 \end{array}$$

Substituting the material and sectional data, the direction cosines, and element length (which equals $|\mathbf{r}_1|$) into Equations 10.6.15, we find

$$[\mathbf{k}_{ff}^{(2)}] = \begin{matrix} & 1 & 2 & 3 & 4 & 5 & 6 \\ \left[\begin{matrix} 4.234 \times 10^6 & -4.234 \times 10^6 & 4.234 \times 10^6 & 0 & -634.8 & -634.8 \\ -4.234 \times 10^6 & 4.234 \times 10^6 & -4.234 \times 10^6 & 634.8 & 0 & -634.8 \\ 4.234 \times 10^6 & -4.234 \times 10^6 & 4.234 \times 10^6 & 634.8 & 634.8 & 0 \\ 0 & 634.8 & 634.8 & 928.5 & 341.2 & -341.2 \\ -634.8 & 0 & 634.8 & 341.2 & 928.5 & 341.2 \\ -634.8 & -634.8 & 0 & -341.2 & 341.2 & 928.5 \end{matrix}\right] & \begin{matrix} 1 \\ 2 \\ 3 \\ 4 \\ 5 \\ 6 \end{matrix} \end{matrix}$$

[h]

The 6 by 6 global stiffness matrix $[\mathbf{k}_{ff}]$ of the constrained structure is found, as usual, by assembling $[\mathbf{k}_{ff}^{(1)}]$ and $[\mathbf{k}_{ff}^{(2)}]$:

$$[\mathbf{k}_{ff}] = \begin{bmatrix} 26.22 \times 10^6 & -4.232 \times 10^6 & 4.232 \times 10^6 & 0 & -634.8 & -634.8 \\ -4.232 \times 10^6 & 4.240 \times 10^6 & -4.232 \times 10^6 & 634.8 & 0 & -3934 \\ 4.232 \times 10^6 & -4.232 \times 10^6 & 4.240 \times 10^6 & 634.8 & 3934 & 0 \\ 0 & 634.8 & 634.8 & 1355 & 341.2 & -341.2 \\ -634.8 & 0 & 3934 & 341.2 & 3128 & 341.2 \\ -634.8 & -3934 & 0 & -341.2 & 341.2 & 3128 \end{bmatrix}$$

[i]

There are no point loads or point couples applied directly to node 1 of the frame; therefore, the components of the global point-load vector of the constrained structure are zero, or

$$\left\{\mathbf{Q}_{P_f}\right\} = \{\mathbf{0}\}$$

[j]

However, the distributed load on element 1 must be accounted for in the global equivalent load vector $\{\mathbf{Q}_{eq_f}\}$. The local components of the equivalent load vector $\{\bar{\mathbf{Q}}_{eq}^{(1)}\}$ of element 1 are given by Equation 10.3.22 and are sketched in Figure 10.6.7, from which we see that

$$\{\bar{\mathbf{Q}}_{eq}^{(1)}\} = \begin{Bmatrix} 0 \\ -100 \\ 0 \\ 0 \\ 0 \\ -16.67 \\ \hdashline 0 \\ -100 \\ 0 \\ 0 \\ 0 \\ 16.67 \end{Bmatrix} \begin{matrix} 1 \\ 2 \\ 3 \\ 4 \\ 5 \\ 6 \\ 7 \\ 8 \\ 9 \\ 10 \\ 11 \\ 12 \end{matrix} \quad \begin{matrix} 7 \\ 8 \\ 9 \\ 10 \\ 11 \\ 12 \\ 1 \\ 2 \\ 3 \\ 4 \\ 5 \\ 6 \end{matrix}$$

Local dof
Global dof

[k]

Since the local axes of element 1 have the same orientation as those of the global system, the local and global equivalent load vectors are identical: $\{\mathbf{Q}_{eq}^{(1)}\} = \{\bar{\mathbf{Q}}_{eq}^{(1)}\}$. Since element 2 has no intermediate loads, $\{\mathbf{Q}_{eq}^{(2)}\} = \{\mathbf{0}\}$. Therefore, the global load vector for the unconstrained degrees of freedom is

$$\{\mathbf{Q}_f\} = \left\{\mathbf{Q}_{eq_f}^{(1)}\right\} = \begin{Bmatrix} 0 \\ -100 \text{ N} \\ 0 \\ 0 \\ 0 \\ 16.67 \text{ N-m} \end{Bmatrix}$$

[l]

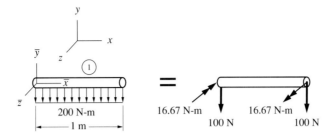

Figure 10.6.7   Equivalent loads on element 1 of the structure in Figure 10.6.5.

The system of equations for the free displacements of the structure is

$$[\mathbf{k}_{ff}]\{\mathbf{q}_f\} = \{\mathbf{Q}_{P_f}\} + \{\mathbf{Q}_{\text{eq}_f}\}$$

Substituting Equations i, j, and l and then solving for the six components of $\{\mathbf{q}_f\}$ yields

$$\{\mathbf{q}_f\} = \begin{Bmatrix} u_1 \\ v_1 \\ w_1 \\ \theta_{x1} \\ \theta_{y1} \\ \theta_{z1} \end{Bmatrix} = (10^{-3}) \begin{Bmatrix} -0.001712 \text{ m} \\ -16.89 \text{ m} \\ -16.88 \text{ m} \\ 5.674 \text{ rad} \\ 22.54 \text{ rad} \\ -17.75 \text{ rad} \end{Bmatrix} \qquad \textbf{[m]}$$

With the structure's displacements at hand, we can now determine the local nodal loads $\{\bar{\mathbf{Q}}_P^{(e)}\}$ for frame elements 1 and 2, using the element stiffness equations,

$$\{\bar{\mathbf{Q}}_P^{(e)}\} = [\bar{\mathbf{k}}^{(e)}]\{\bar{\mathbf{q}}^{(e)}\} - \{\bar{\mathbf{Q}}_{\text{eq}}^{(e)}\}$$

which we can partition as follows with respect to the free and supported degrees of freedom:

$$\{\bar{\mathbf{Q}}_{P_f}^{(e)}\} = [\bar{\mathbf{k}}_{ff}^{(e)}]\{\bar{\mathbf{q}}_f^{(e)}\} + [\bar{\mathbf{k}}_{fs}^{(e)}]\{\bar{\mathbf{q}}_s^{(e)}\} - \{\bar{\mathbf{Q}}_{\text{eq}_f}^{(e)}\}$$
$$\{\bar{\mathbf{Q}}_{P_s}^{(e)}\} = [\bar{\mathbf{k}}_{sf}^{(e)}]\{\bar{\mathbf{q}}_f^{(e)}\} + [\bar{\mathbf{k}}_{ss}^{(e)}]\{\bar{\mathbf{q}}_s^{(e)}\} - \{\bar{\mathbf{Q}}_{\text{eq}_s}^{(e)}\}$$
$$\qquad \textbf{[n]}$$

The appropriate components of the global displacement vector $\{\mathbf{q}\}$ are assigned to each element's global displacement vector $\{\mathbf{q}^{(e)}\}$, which is then transformed into the element reference frame by means of Equation 10.6.7, to obtain $\{\bar{\mathbf{q}}^{(e)}\}$.

Starting with element 1, we can find its 12-component global displacement vector in terms of the 18 components of $\{\mathbf{q}\}$ by setting

$$\{\mathbf{q}^{(1)}\} = \begin{Bmatrix} \mathbf{q} \end{Bmatrix} \begin{smallmatrix} 7 \\ 8 \\ 9 \\ 10 \\ 11 \\ 12 \\ 1 \\ 2 \\ 3 \\ 4 \\ 5 \\ 6 \end{smallmatrix}$$

Since its $\bar{x}\,\bar{y}\,\bar{z}$ axes coincide in direction with the global xyz system, $\{\bar{\mathbf{q}}^{(1)}\} = \{\mathbf{q}^{(1)}\}$. Therefore,

$$\left\{\bar{\mathbf{q}}_f^{(1)}\right\} = (10^{-3}) \begin{Bmatrix} -0.001712 \text{ m} \\ -16.89 \text{ m} \\ -16.88 \text{ m} \\ 5.674 \text{ rad} \\ 22.54 \text{ rad} \\ -17.75 \text{ rad} \end{Bmatrix} \qquad \left\{\bar{\mathbf{q}}_s^{(1)}\right\} = \{\mathbf{0}\}$$

[o]

Since $\{\bar{\mathbf{q}}_s^{(1)}\} = \{\mathbf{0}\}$, Equations n simplify to

$$\left\{\bar{\mathbf{Q}}_P^{(e)}\right\} = \left[\bar{\mathbf{k}}^{(e)}\right]\left\{\bar{\mathbf{q}}^{(e)}\right\} - \left\{\bar{\mathbf{Q}}_{\text{eq}}^{(e)}\right\}$$

$$\left\{\bar{\mathbf{Q}}_{P_f}^{(1)}\right\} = \left[\bar{\mathbf{k}}_{ff}^{(1)}\right]\left\{\bar{\mathbf{q}}_f^{(1)}\right\} - \left\{\bar{\mathbf{Q}}_{\text{eq}_f}^{(1)}\right\}$$

[p]

The components $\{\bar{\mathbf{Q}}_{\text{eq}_f}^{(1)}\}$ and $\{\bar{\mathbf{Q}}_{\text{eq}_s}^{(1)}\}$ of the local equivalent load vector are found in Equation k,

$$\left\{\bar{\mathbf{Q}}_{\text{eq}_f}^{(1)}\right\} = \begin{Bmatrix} 0 \\ -100 \\ 0 \\ 0 \\ 0 \\ 16.67 \end{Bmatrix} \text{(N)} \qquad \left\{\bar{\mathbf{Q}}_{\text{eq}_s}^{(1)}\right\} = \begin{Bmatrix} 0 \\ -100 \\ 0 \\ 0 \\ 0 \\ 16.67 \end{Bmatrix} \text{(N)}$$

[q]

The local stiffness matrix $[\bar{\mathbf{k}}^{(1)}]$ of element 1 is given by Equation 10.6.5, from which we obtain the two 6 by 6 submatrices $[\bar{\mathbf{k}}_{ff}^{(1)}]$ and $[\bar{\mathbf{k}}_{sf}^{(1)}]$:

$$\left[\bar{\mathbf{k}}_{ff}^{(1)}\right] = \begin{matrix} \scriptstyle 7\ 8\ 9\ 10\ 11\ 12 \\ \begin{matrix} \scriptstyle 7 \\ \scriptstyle 8 \\ \scriptstyle 9 \\ \scriptstyle 10 \\ \scriptstyle 11 \\ \scriptstyle 12 \end{matrix} \begin{bmatrix} & & \bar{\mathbf{k}}^{(1)} & & \end{bmatrix} \end{matrix} = \begin{bmatrix} [\bar{\mathbf{k}}_1] & -[\bar{\mathbf{k}}_2] \\ -[\bar{\mathbf{k}}_2] & [\bar{\mathbf{k}}_3] \end{bmatrix}$$

$$\left[\bar{\mathbf{k}}_{sf}^{(1)}\right] = \begin{matrix} \scriptstyle 7\ 8\ 9\ 10\ 11\ 12 \\ \begin{matrix} \scriptstyle 1 \\ \scriptstyle 2 \\ \scriptstyle 3 \\ \scriptstyle 4 \\ \scriptstyle 5 \\ \scriptstyle 6 \end{matrix} \begin{bmatrix} & & \bar{\mathbf{k}}^{(1)} & & \end{bmatrix} \end{matrix} = \begin{bmatrix} -[\bar{\mathbf{k}}_1] & [\bar{\mathbf{k}}_2] \\ -[\bar{\mathbf{k}}_2]^T & [\bar{\mathbf{k}}_4] \end{bmatrix}$$

Substituting the element properties into Equation 10.6.5b yields

$$\left[\bar{\mathbf{k}}_{ff}^{(1)}\right] = \begin{bmatrix} 21.99 \times 10^6 & 0 & 0 & 0 & 0 & 0 \\ 0 & 6597 & 0 & 0 & 0 & -3299 \\ 0 & 0 & 6597 & 0 & 3299 & 0 \\ 0 & 0 & 0 & 426.2 & 0 & 0 \\ 0 & 0 & 3299 & 0 & 2199 & 0 \\ 0 & -3299 & 0 & 0 & 0 & 2199 \end{bmatrix}$$

[r]

$$\left[\bar{\mathbf{k}}_{sf}^{(1)}\right] = \begin{bmatrix} -21.99 \times 10^6 & 0 & 0 & 0 & 0 & 0 \\ 0 & -6597 & 0 & 0 & 0 & 3299 \\ 0 & 0 & 6597 & 0 & -3299 & 0 \\ 0 & 0 & 0 & -426.2 & 0 & 0 \\ 0 & 0 & 3299 & 0 & 1100 & 0 \\ 0 & -3299 & 0 & 0 & 0 & 1100 \end{bmatrix}$$

Finally, substituting the displacements (Equations n), equivalent loads (Equations q), and stiffness submatrices (Equations r) into Equations p, we get the element 1 nodal loads,

$$
\{\bar{\mathbf{Q}}_f^{(1)}\} = \begin{Bmatrix} -37.66\ \text{N} \\ 47.14\ \text{N} \\ -36.98\ \text{N} \\ 2.418\ \text{N-m} \\ -6.097\ \text{N-m} \\ 0.00416\ \text{N-m} \end{Bmatrix} \qquad
\{\bar{\mathbf{Q}}_s^{(1)}\} = \begin{Bmatrix} 37.66\ \text{N} \\ 152.9\ \text{N} \\ 36.98\ \text{N} \\ -2.418\ \text{N-m} \\ -3.088\ \text{N-m} \\ 52.85\ \text{N-m} \end{Bmatrix} \qquad [\text{s}]
$$

Here, $\{\bar{\mathbf{Q}}_f^{(1)}\}$ is the load vector at node 2 of element 1, and $\{\bar{\mathbf{Q}}_s^{(1)}\}$ is the load vector at node 1, which is built-in at the wall. The components of these vectors are sketched in Figure 10.6.8.

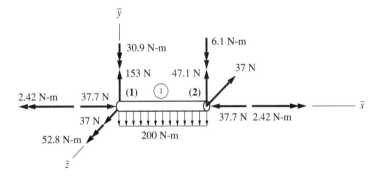

Figure 10.6.8   Nodal loads computed for element 1 of Figure 10.6.5.

The local node numbers are shown in parentheses.

Turning to element 2, we note that in the absence of any applied intermediate loads, the equivalent load vector $\{\bar{\mathbf{Q}}_{eq}^{(2)}\}$ is zero. Also, $\{\bar{\mathbf{q}}_s^{(2)}\} = \{\mathbf{0}\}$, as was the case for element 1. Therefore, the stiffness equations, Equations n, for element 2 are

$$
\{\bar{\mathbf{Q}}_f^{(2)}\} = [\bar{\mathbf{k}}_{ff}^{(2)}]\{\bar{\mathbf{q}}_f^{(2)}\}
$$
$$
\{\bar{\mathbf{Q}}_s^{(2)}\} = [\bar{\mathbf{k}}_{sf}^{(2)}]\{\bar{\mathbf{q}}_f^{(2)}\} \qquad [\text{t}]
$$

The local stiffness matrix $[\bar{\mathbf{k}}_{ff}^{(2)}]$ of element 2 is given by Equation 10.6.8, from which we obtain the two 6 by 6 submatrices $[\bar{\mathbf{k}}_{ff}^{(2)}]$ and $[\bar{\mathbf{k}}_{sf}^{(2)}]$:

$$
\begin{bmatrix} \bar{\mathbf{k}}_{ff}^{(2)} \end{bmatrix} = \begin{matrix} \\ 1 \\ 2 \\ 3 \\ 4 \\ 5 \\ 6 \end{matrix} \overset{\displaystyle 1\,2\,3\,4\,5\,6}{\begin{bmatrix} & & & & \\ & \bar{\mathbf{k}}^{(2)} & & \\ & & & & \end{bmatrix}} = \begin{bmatrix} [\bar{\mathbf{k}}_1] & [\bar{\mathbf{k}}_2] \\ [\bar{\mathbf{k}}_2]^T & [\bar{\mathbf{k}}_3] \end{bmatrix}
$$

$$
\begin{bmatrix} \bar{\mathbf{k}}_{sf}^{(2)} \end{bmatrix} = \begin{matrix} \\ 7 \\ 8 \\ 9 \\ 10 \\ 11 \\ 12 \end{matrix} \overset{\displaystyle 1\,2\,3\,4\,5\,6}{\begin{bmatrix} & & & & \\ & \bar{\mathbf{k}}^{(2)} & & \\ & & & & \end{bmatrix}} = \begin{bmatrix} -[\bar{\mathbf{k}}_1] & -[\bar{\mathbf{k}}_2] \\ [\bar{\mathbf{k}}_2]^T & [\bar{\mathbf{k}}_4] \end{bmatrix}
$$

These matrices are found from Equations 10.6.5b. Substituting the material and geometric properties of element 2, we get

$$
\left[\bar{\mathbf{k}}_{ff}^{(2)}\right] = \begin{bmatrix} 12.70 \times 10^6 & 0 & 0 & 0 & 0 & 0 \\ 0 & 1270 & 0 & 0 & 0 & 1100 \\ 0 & 0 & 1270 & 0 & -1100 & 0 \\ \hdashline 0 & 0 & 0 & 246.1 & 0 & 0 \\ 0 & 0 & -1100 & 0 & 1270 & 0 \\ 0 & 1100 & 0 & 0 & 0 & 1270 \end{bmatrix}
$$

[u]

$$
\left[\bar{\mathbf{k}}_{sf}^{(2)}\right] = \begin{bmatrix} -12.70 \times 10^6 & 0 & 0 & 0 & 0 & 0 \\ 0 & -1270 & 0 & 0 & 0 & -1100 \\ 0 & 0 & -1270 & 0 & 1100 & 0 \\ \hdashline 0 & 0 & 0 & -246.1 & 0 & 0 \\ 0 & 0 & -1100 & 0 & 634.8 & 0 \\ 0 & 1100 & 0 & 0 & 0 & 634.8 \end{bmatrix}
$$

The 12-component global displacement vector of element 2 is expressed in terms of the 18 components of the structure's global displacement vector $\{\mathbf{q}\}$, as follows:

$$
\{\mathbf{q}^{(2)}\} = \left\{ \mathbf{q} \begin{smallmatrix} 1 \\ 2 \\ 3 \\ 4 \\ 5 \\ 6 \\ 13 \\ 14 \\ 15 \\ 16 \\ 17 \\ 18 \end{smallmatrix} \right\} = (10^{-3}) \left\{ \begin{array}{c} -0.001712 \text{ m} \\ -16.89 \text{ m} \\ -16.88 \text{ m} \\ 5.674 \text{ rad} \\ 22.54 \text{ rad} \\ -17.75 \text{ rad} \\ 0 \\ 0 \\ 0 \\ 0 \\ 0 \\ 0 \end{array} \right\}
$$

We project the global displacement vector $\{\mathbf{q}^{(2)}\}$ onto the element axes by means of Equation 10.6.7,

$$
\{\bar{\mathbf{q}}^{(2)}\} = [\boldsymbol{\Lambda}^{(2)}]\{\mathbf{q}^{(2)}\}
$$

[v]

where, according to Equations 10.6.8 and 10.6.9, we have

$$
[\boldsymbol{\Lambda}^{(2)}] = \begin{bmatrix} [\boldsymbol{\lambda}^{(2)}] & [\mathbf{0}] & [\mathbf{0}] & [\mathbf{0}] \\ [\mathbf{0}] & [\boldsymbol{\lambda}^{(2)}] & [\mathbf{0}] & [\mathbf{0}] \\ [\mathbf{0}] & [\mathbf{0}] & [\boldsymbol{\lambda}^{(2)}] & [\mathbf{0}] \\ [\mathbf{0}] & [\mathbf{0}] & [\mathbf{0}] & [\boldsymbol{\lambda}^{(2)}] \end{bmatrix}
$$

and $[\boldsymbol{\lambda}^{(2)}]$ is found in Equation e. Therefore, Equation v yields

$$
\{\bar{\mathbf{q}}_s^{(2)}\} = \{\mathbf{0}\} \qquad \{\bar{\mathbf{q}}_f^{(2)}\} = (10^{-3}) \left\{ \begin{array}{c} -0.004785 \text{ m} \\ -20.68 \text{ m} \\ 11.93 \text{ m} \\ 19.99 \text{ rad} \\ 13.47 \text{ rad} \\ 16.56 \text{ rad} \end{array} \right\}
$$

[w]

Substituting these displacements and the stiffness submatrices, Equations u, into Equations t, we find the local nodal loads for element 2, as follows:

$$\{\bar{Q}_f^{(2)}\} = \begin{Bmatrix} -70.32 \text{ N} \\ -8.042 \text{ N} \\ 0.3328 \text{ N} \\ 4.918 \text{ N-m} \\ 3.989 \text{ N-m} \\ -1.707 \text{ N-m} \end{Bmatrix} \qquad \{\bar{Q}_s^{(2)}\} = \begin{Bmatrix} 70.32 \text{ N} \\ 8.042 \text{ N} \\ -0.3328 \text{ N} \\ -4.918 \text{ N-m} \\ -4.565 \text{ N-m} \\ -12.22 \text{ N-m} \end{Bmatrix}$$

Here, $\{\bar{Q}_f^{(2)}\}$ is the vector of local load components at node 1 of element 2, and $\{\bar{Q}_s^{(2)}\}$ is the set of loads at node 2 of element 2. These loads are sketched in Figure 10.6.9.

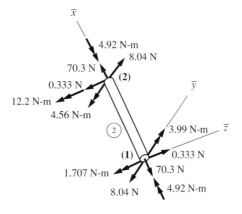

Figure 10.6.9   Nodal loads computed for element 2 of Figure 10.6.5.

The local node numbers are shown in parentheses.

The bending moment at any section is the vector sum of $\bar{M}_y$ and $\bar{M}_z$, which are the bending moments about the element $\bar{y}$ and $\bar{z}$ axes, respectively. From Figures 10.6.8 and 10.6.9, we see that the maximum bending moment occurs at node 1—the built-in end—of element 1. Since $\bar{M}_y = 30.9$ N-m and $\bar{M}_z = 52.8$ N-m at that point, the resultant bending moment is

$$M_{\max} = \sqrt{30.9^2 + 52.8^2} = 61.2 \text{ N-m}$$

The normal stress due to this bending moment is found from the bending stress formula $\sigma = Mc/I$. Using the value of $M_{\max}$, along with $c = 0.01$ m (the radius of the bar) and $I = 7.854(10^{-9})$ m^4, we find that the maximum bending stress is 77.9 MPa (11,300 psi).

With the rod and frame elements, we can mathematically model, for displacement method analysis, a variety of two- and three-dimensional skeletal structures typical of aircraft construction (e.g., fuselage trusses, spar and rib assemblies, landing gear, etc.). In Chapter 7, we looked at the deflection and load analysis of circular frame structures, using the force method. How do we apply the matrix displacement method to curved frames? We could seek a single, displacement-based, curved beam element,[5] but that would go beyond this introductory text. Instead, we can employ the finite element method, representing the curved structure as a series of straight beam elements. We will illustrate this procedure in the following example.

Example 10.6.2   Figure 10.6.10a shows a frame composed of a semicircular beam and a horizontal floor beam. The frame is in equilibrium under the action of a uniform shear flow on the curved portion and a point load applied at the middle of the floor beam. Use the finite element displacement method to find the maximum bending moment in the frame. Take advantage of symmetry.

This is the same problem solved in Example 7.8.9 by the classical force method. Again, we can reduce our work by observing the symmetry of both the structure and the load about the vertical line joining points A and O. We need only focus on one-half of the frame on either side of AO, applying one-half of the total load to each side.

To be symmetric, a horizontal displacement at either A or O must cause those points to move both to the left and the right by the same amount. That is impossible without opening up gaps or requiring material particles to pass through each other. Therefore, symmetry requires a zero horizontal displacement at both A and O. Likewise, there can be no rotation at A and O, for that would destroy the symmetry of the deformed shape of the frame about AO. The constraint shown at O in Figure 10.6.10b represents the enforcement of zero horizontal displacement and zero rotation, leaving motion in the vertical direction unrestrained. An identical support exists at A, with an additional restraint on vertical displacement, to prevent rigid-body motion.

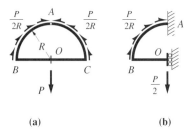

(a)   **(b)**

Figure 10.6.10   (a) Frame with self-equilibrating applied loads. (b) Constraints required to inforce symmetry and eliminate rigid-body motion.

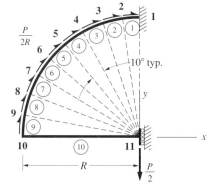

Figure 10.6.11   Finite element model of the quarter circular frame in Figure 10.6.10b.

Since the net force of the applied loads in the vertical direction is zero, no vertical force will develop at A; however, restraining A vertically will make the global stiffness matrix nonsingular. Obviously, the constraint on vertical motion could be applied to *any* point of the frame, except point O, where we have already specified the vertical load.

We divide the quarter circle into nine equal segments. We then model each segment, as well as the straight horizontal member at the bottom, as a plane frame element. As we can see in Figure 10.6.11, this yields a 10-element frame with 33

---

[5]See, for example: W. Weaver, Jr., and P. R. Johnston, *Finite Elements for Structural Analysis*, § 8.3, Englewood Cliffs, NJ, Prentice-Hall, 1984; W. McGuire and R. H. Gallagher, *Matrix Structural Analysis*, New York, Wiley, 1979, p. 318; J. J. Azar, *Matrix Structural Analysis*, § 3.11, New York, Pergamon Press, 1972.

degrees of freedom, before the constraints are applied at nodes 1 and 11. The 90 degree circular arc is approximated reasonably well by the nine straight-line segments. A finer subdivision would yield a more accurate representation, but would add more degrees of freedom to the structure, and therefore more unknowns.

To compute the global stiffness matrix of all ten elements, we use Equation 10.4.10, which contains the dimensionless term $\bar{A}$. Recalling that $\bar{A}$ is the square of the slenderness ratio, let us assume for this problem that $\bar{A} = 1000$. The assembly procedure

$$[\mathbf{k}] = \sum_{e=1}^{10} [\mathbf{\Omega}^{(e)}]^T [\mathbf{k}^{(e)}][\mathbf{\Omega}^{(e)}]$$

is carried out as illustrated in previous examples. However, in this problem, the number of calculations required makes a hand calculation impractical, and the procedure should be implemented on a digital computer.

The shear flow on elements 1 through 9 is converted into consistent equivalent nodal loads, as illustrated in Figure 10.6.12. Since the axial displacement in a beam element varies linearly from end to end, the resultant of the uniform load per unit length (Figure 10.6.12a) is apportioned equally to both nodes of that element (Figure 10.6.12b).[6] These axial loads are then projected into the global coordinate system (Figure 10.6.12c) and assembled into the global equivalent load vector $\{\mathbf{Q}_{eq}\}$. The point load vector $\{\mathbf{Q}_{P_f}\}$ has only one nonzero entry, the load $P/2$ at node 11.

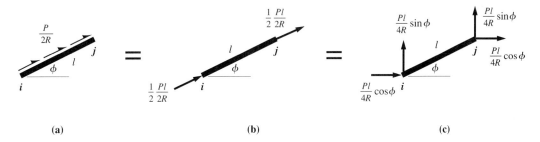

(a)   (b)   (c)

Figure 10.6.12   Work-equivalent loads for beam elements in the circular portion of the frame.

From the node numbering in Figure 10.6.11, we see that the free (f) and constrained (s) degrees of freedom are f: 4 through 30, plus 32 (28 total) and s: 1, 2, 3, 31, 33. Since $\{\mathbf{q}_s\} = \{\mathbf{0}\}$, the stiffness equations for this structure are

$$[\mathbf{k}_{ff}]\{\mathbf{q}_f\} = \{\mathbf{Q}_{P_f}\} + \{\mathbf{Q}_{eq_f}\}$$

the solution of which yields the 28-component displacement vector $\{\mathbf{q}_f\}$. For example, the computed vertical displacement at node 11 is

$$q_{32} = v_{11} = -0.0928\frac{PR^3}{EI_z} \qquad \text{[a]}$$

It is interesting to note that the force method yields $-0.0932PR^3/EI_z$ for this displacement (see Exercise 7.46).

The reactions at nodes 1 and 11 are obtained by substituting $\{\mathbf{q}_f\}$ into

$$\{\mathbf{Q}_{P_s}\} = [\mathbf{k}_{sf}]\{\mathbf{q}_f\} - \{\mathbf{Q}_{eq_s}\} \qquad \text{[b]}$$

---

[6] In an aircraft structure, the shear flow is offset from the beam centerline, acting along the outer surface of the beam where the skin is attached. This offset is accounted for by point couples at the nodes. Such a refinement is beyond the scope of this chapter.

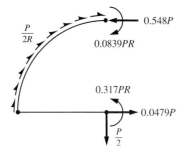

Figure 10.6.13 Reactions on the frame, as computed by the finite element displacement method.

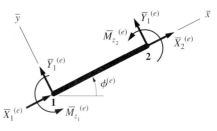

Figure 10.6.14 Element load vector components obtained from Equation c.

The computed results are shown in Figure 10.6.13. They agree very well with those found in Example 7.8.9, Figure 7.8.28. The bending moment diagrams are constructed using statics and, for all practical purposes, are identical.

To find the bending moment distribution using the matrix displacement method, we employ the local form of the element stiffness equations,

$$\{\bar{\mathbf{Q}}_P^{(e)}\} = [\bar{\mathbf{k}}^{(e)}]\{\bar{\mathbf{q}}^{(e)}\} - \{\bar{\mathbf{Q}}_{eq}^{(e)}\} \tag{c}$$

These equations produce the six components of the element load vector shown in Figure 10.6.14. The moments at each element node are listed in Table 10.6.2, along with the values computed from Equations c and f of Example 7.8.9. Again, the numbers for the two methods are very close.

Table 10.6.2 Computed nodal couples for each element of the frame, with the moments computed in Example 7.8.9 shown in parentheses.

Element	Node 1	Node 2	$\bar{M}_{z_1}/PR$	$\bar{M}_{z_2}/PR$
1	1	2	−0.0839 (−0.0828)	0.0755
2	2	3	−0.0755 (−0.0749)	0.0535
3	3	4	−0.0535 (−0.0531)	0.0209
4	4	5	−0.0209 (−0.0208)	−0.0184
5	5	6	0.0184 (0.0184)	−0.0608
6	6	7	0.0608 (0.0606)	−0.102
7	7	8	0.102 (0.102)	−0.139
8	8	9	0.139 (0.139)	−0.167
9	9	10	0.167 (0.166)	−0.183
10	10	11	0.183 (0.183)	0.317 (0.317)

EXERCISES

All of the problems are to be solved using the displacement method. Hand calculations should be employed, rather than canned software, unless otherwise directed.

10.1   Find the displacement of node 1 of the structure loaded and constrained as shown.

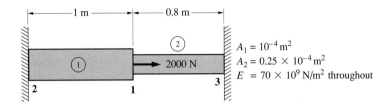

Exercise 10.1

10.2   Find the horizontal and vertical displacement components at node 1 of the truss. The cross-sectional area of the rods is 1 in.^{2}, and $E = 10^7$ psi throughout.

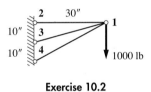

Exercise 10.2

10.3   Calculate the horizontal displacement of node 2 of the truss. All elements have the same AE. The displacement at node 3 is prescribed as $0.001L$ up the 30 degree incline, as shown.

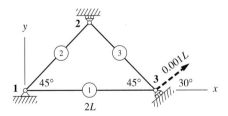

Exercise 10.3

10.4   All members of the truss have the same E and A. Find the displacements of nodes 1 and 2 due to a horizontal load at node 1.

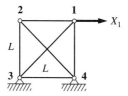

Exercise 10.4

10.5   Axial loads are applied to the stepped steel shaft as shown. The length and diameter of each section are shown. Calculate the axial load in each part of the shaft. The modulus of elasticity is 30×10^6 psi.

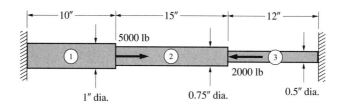

Exercise 10.5

10.6   Assemble the stiffness matrix of the pin-jointed truss shown, solve for the nodal displacements using Gauss elimination, and use the displacements to find the member forces. All rods have an area A of 1 in.^{2} and a modulus of elasticity E of 10^7 psi. In addition, $L = 20$ in., $P = 100$ lb (to the right), and $Q = 200$ lb (downward).

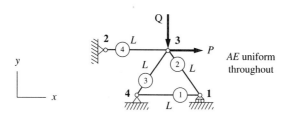

Exercise 10.6

10.7   Find the forces in the rods if they all have the same modulus of elasticity E and their cross-sectional areas are:

a. $A_1 = A$, $A_2 = 2A$,   and $A_3 = 3A$.

b. $A_1 = 3A$, $A_2 = 2A$,   and $A_3 = A$.

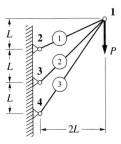

Exercise 10.7

10.8   Find the axial force in each member of the five-bar welded-steel tube truss. All rods have an outside diameter of 0.5 in. and a wall thickness of 0.01 in. In addition to the 100 lb load, each bar undergoes the indicated temperature rise. The modulus of elasticity is $E = 30 \times 10^6$ psi, and the coefficient of thermal expansion is $\alpha = 6 \times 10^{-6} \, °\text{F}^{-1}$.

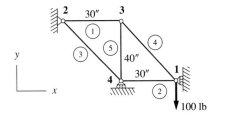

Element	Temperature rise
1	50 °F
2	80 °F
3	40 °F
4	70 °F
5	60 °F

Exercise 10.8

10.9 A load P is applied downward at node 1 of the truss, as shown. The displacement of node 1 in the direction of the load is 0.01 in. Find the value of load P. All rods have the same cross-sectional area A and modulus of elasticity E.

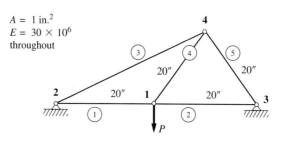

$A = 1$ in.^{2}
$E = 30 \times 10^6$
throughout

Exercise 10.9

10.10 The temperature of all members of the three-dimensional truss increases by 100 °C, and a load is applied as shown. The resulting displacements are listed in the table. Calculate the stress in the element joining nodes 2 and 4.

Nodal displacements (mm)

Node	x displ.	y displ.	z displ.
1	0	0	0
2	3.593	0.5057	−1.288
3	0	0	0
4	4.996	−5.957	−1.977
5	3.413	2.232	0.5957
6	0	0	0

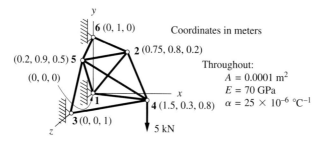

6 (0, 1, 0)   Coordinates in meters
2 (0.75, 0.8, 0.2)
(0.2, 0.9, 0.5) 5
(0, 0, 0)   Throughout:
                $A = 0.0001$ m^2
                $E = 70$ GPa
4 (1.5, 0.3, 0.8) $\alpha = 25 \times 10^{-6}$ °C^{-1}
3 (0, 0, 1)
5 kN

Exercise 10.10

10.11   Assemble the constrained stiffness matrix $[\mathbf{k}_{ff}]$ of the structure shown. Solve for the displacements of node 1, in terms of Pl/AE, and the load in each rod, in terms of P.

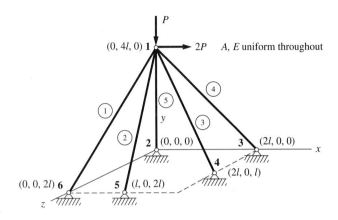

Exercise 10.11

10.12   Use the principle of virtual work to derive the stiffness matrix for the three-node rod element shown, using a quadratic to represent the axial displacement, that is, $u = a_0 + a_1 x + a_2 x^2$.

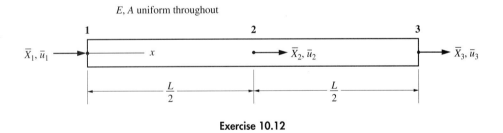

Exercise 10.12

10.13   Show that the three-dimensional global stiffness matrix of the rod in Exercise 10.12 is

$$
[\mathbf{k}] = \left(\frac{AE}{3L}\right)
\begin{bmatrix}
7l^2 & 7lm & 7ln & -8l^2 & -8lm & -8ln & l^2 & lm & ln \\
7lm & 7m^2 & 7mn & -8lm & -8m^2 & -8mn & lm & m^2 & mn \\
7ln & 7mn & 7n^2 & -8ln & -8mn & -8n^2 & ln & mn & n^2 \\
-8l^2 & -8lm & -8ln & 16l^2 & 16lm & 16ln & -8l^2 & -8lm & -8ln \\
-8lm & -8m^2 & -8mn & 16lm & 16m^2 & 16mn & -8lm & -8m^2 & -8mn \\
-8ln & -8mn & -8n^2 & 16ln & 16mn & 16n^2 & -8ln & -8mn & -8n^2 \\
l^2 & lm & ln & -8l^2 & -8lm & -8ln & 7l^2 & 7lm & 7ln \\
lm & m^2 & mn & -8lm & -8m^2 & -8mn & 7lm & 7m^2 & 7mn \\
ln & mn & n^2 & -8ln & -8mn & -8n^2 & 7ln & 7mn & 7n^2
\end{bmatrix}
$$

10.14   If a uniform load per unit length q acts along the entire length of the rod in Exercise 10.12, find the vector of equivalent nodal loads $\{Q_{eq}\}$.

Exercise 10.14

10.15   Find the semibandwidth of the stiffness matrix of the pin-jointed truss structure shown. Also, in terms of AE/L, find the following components of its stiffness matrix: $k_{15,18}$, $k_{31,42}$, and $k_{23,24}$. (For this problem assume that the structure is unloaded and unconstrained.)

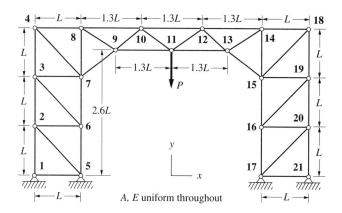

Exercises 10.15 and 10.16

10.16   Assume the rods of the truss are made of an aluminum alloy ($E = 70$ GPa) with a solid circular cross section, all of the same diameter, and let $L = 1$ m. The maximum allowable tensile or compressive stress in a member must not exceed 300 MPa. In addition, the compressive load must not exceed P_{cr}, the buckling load for a pinned-pinned column. ($P_{cr} = \pi^2 EI/L^2$, where I is the minimum area moment of inertia of the cross section, and L is the length of the member.)

a. If the rod diameter is 2 cm, find the maximum load P, in kN, that can be applied as shown to the structure.

b. If the load P is 20 kN, find the minimum allowable rod diameter (common to all rods).

Use available computer resources as an aid.

10.17 Two views of a three-dimensional truss are shown. The truss is made of aluminum rods ($E = 10^7$ psi), all of the same solid circular cross section. Nodes 9, 10, 11, and 12 at the base of the structure are completely supported. The structure is required to carry the loads $P = 1000$ lb and $Q = \pm 1000$ lb, applied as shown.

The material's ultimate tensile (and compressive) strength is 30,000 psi. The buckling load for a pinned bar is $P_{cr} = \pi^2 EI/L^2$, where I is the area moment of inertia, which can be written $I = A^2/4\pi$ for a solid circular section, where A is the cross-sectional area.

For either of the load combinations described, find the minimum rod area A for which the truss will not fail.

Let L (in inches) equal the last two digits of your student number, or 10, whichever is greater.

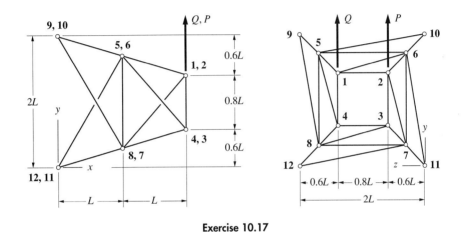

Exercise 10.17

10.18 For the cantilever beam of length L and flexural rigidity EI_z, find the displacements at node 1 due to the vertical load P.

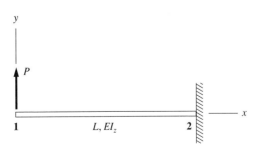

Exercise 10.18

10.19 Using two beam elements, calculate the rotation at node 3 of the beam, which is built-in at 1 and 2. The transverse displacement at node 3 is required to be $0.001l$ downwards, as shown. Assume EI_z is uniform across the span.

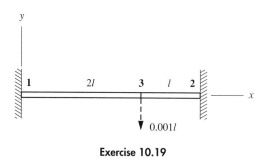

Exercise 10.19

10.20 Using two beam elements, find the vertical displacement of node 1 of the beam, which is clamped at both ends. Assume $I_z = 0.75\,\text{in.}^4$ and $E = 10^7\,\text{psi}$.

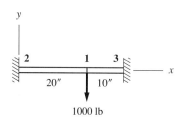

Exercise 10.20

10.21 Using two beam elements, calculate the vertical displacement of node 2 of the beam, which is clamped at each end, with a uniformly distributed load over half the span. Assume $I_z = 0.1\,\text{in.}^4$ and $E = 10^7\,\text{psi}$.

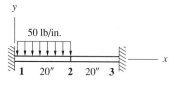

Exercise 10.21

10.22   The beam of length $2l$ has a uniform flexural rigidity EI_z. Find the rotations at nodes 1, 2, and 3 in terms of the couple M applied to node 2.

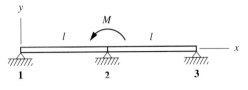

Exercise 10.22

10.23   Calculate the reaction at the left support of the beam supporting the uniform load p.

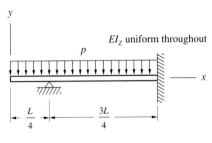

Exercise 10.23

10.24   Calculate the rotation at node 2 of the beam. Use a single beam element. Assume EI_z is uniform across the span.

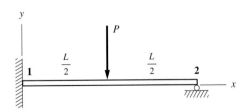

Exercise 10.24

10.25   Repeat Exercise 10.24 using two beam elements, as illustrated in the figure.

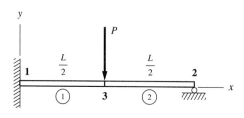

Exercise 10.25

10.26   If the displacements are as shown, calculate the bending moment at the middle of element 3 (section *a-a*).

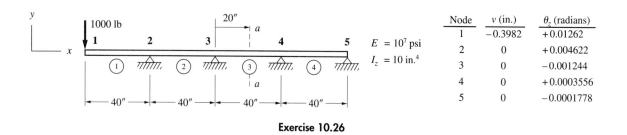

Node	v (in.)	θ_z (radians)
1	−0.3982	+0.01262
2	0	+0.004622
3	0	−0.001244
4	0	+0.0003556
5	0	−0.0001778

$E = 10^7$ psi
$I_z = 10$ in.^{4}

Exercise 10.26

10.27   The computed displacements of the beam are as shown. Draw the moment diagram for the entire beam, using positive for compression in the top fibers. Assume $A = 8$ in.^{2}, $I_z = 10$ in.^{4} and $E = 10^7$ psi.

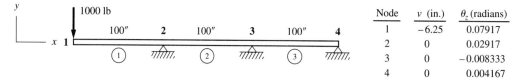

Node	v (in.)	θ_z (radians)
1	−6.25	0.07917
2	0	0.02917
3	0	−0.008333
4	0	0.004167

Exercise 10.27

10.28 Draw accurate shear and bending moment diagrams for the indeterminate beam shown.

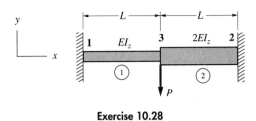

Exercise 10.28

10.29 Calculate the equivalent load vector for the given distributed load.

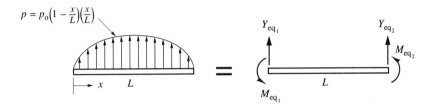

Exercise 10.29

10.30 Use available computer resources as an aid to determine the maximum bending stress and its location in the 11 degree of freedom beam loaded as shown.

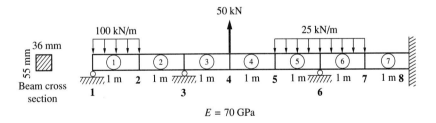

Exercise 10.30

10.31   A two-element beam has a varying, upward-directed distributed load, as shown. Calculate the exact value of the internal bending moment at node 2.

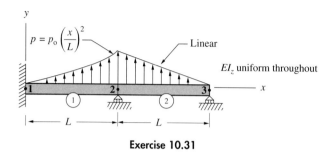

Exercise 10.31

10.32   A beam of length L and elastic modulus E is slightly tapered, and the area moment of inertia is a linear function of x

$$I_z = I_o[1 + \alpha(x/L)]$$

where $|\alpha| << 1$, and I_o is the area moment of inertia at node 1 ($x = 0$). Calculate the components of the stiffness matrix of the tapered beam, using the same cubic shape functions as for a beam element without taper.

10.33   Using the two beam elements shown, find:

a. the displacements at nodes 1 and 2.

b. the shear and bending moment at the wall.

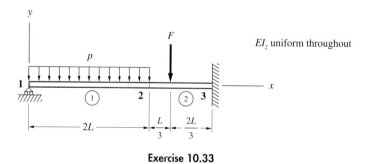

Exercise 10.33

10.34   The structure shown consists of two identical, solid, circular steel rods 1 cm in diameter and 10 cm long. Rod 1 is built into the wall at node 2. Rod 2 is pinned to rod 1 and to the support at node 3. Rod 1 is loaded at three points as shown. If $E = 210$ GPa, find the point on rod 1 at which the (exact) bending moment vanishes.

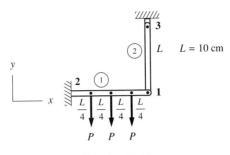

Exercise 10.34

10.35

a. Find the exact shear and bending moment at node 2.

b. Calculate the finite element values of the shear and bending moments at node 2.

c. Find the maximum value and location of the exact bending moment in the beam.

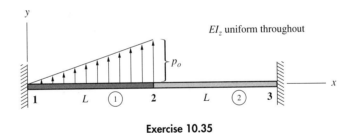

Exercise 10.35

10.36   Repeat Exercise 10.35, but model the structure as a single beam element of length $2L$ on which the triangular load acts over only one-half of its span.

10.37

a. Find the exact bending moment at the midspan ($L/2$) of the beam shown.

b. Calculate the finite element value of the bending moment at the same point.

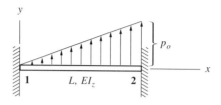

Exercise 10.37

10.38   Both elements of the plane frame have the same E, A, and L. For both, $I_z = 0.01AL^2$. Find the vertical displacement of node 1.

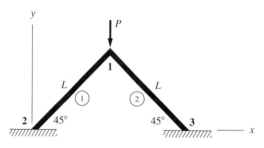

Exercise 10.38

10.39   Given the two-dimensional frame loaded and constrained as shown, find:

a. u_1, v_1, and θ_{z_1}.

b. Those points in each frame element where the (exact) bending moment is zero.

c. Those points in the frame where the finite element value of the bending moment is zero.

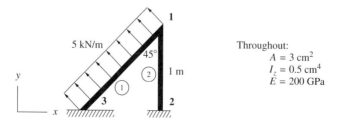

Throughout:
$A = 3\ \text{cm}^2$
$I_z = 0.5\ \text{cm}^4$
$E = 200\ \text{GPa}$

Exercise 10.39

10.40   Given the two-dimensional frame loaded and constrained as shown, find:

a. u_1, v_1, and θ_{z_1}.

b. Those points in each frame element where the (exact) bending moment is zero.

c. Those points in the frame where the finite element values of the bending moment are zero.

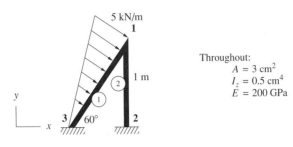

Throughout:
$$A = 3 \text{ cm}^2$$
$$I_z = 0.5 \text{ cm}^4$$
$$E = 200 \text{ GPa}$$

Exercise 10.40

10.41   Given the loading and the support conditions for the frame, find:

a. The displacements at node 1.

b. The location and exact value of the maximum bending moment in the structure.

c. The location of the maximum finite element value of the bending moment.

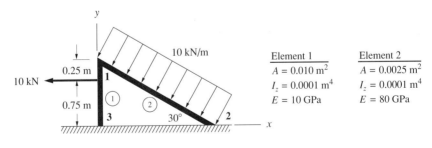

Element 1	Element 2
$A = 0.010 \text{ m}^2$	$A = 0.0025 \text{ m}^2$
$I_z = 0.0001 \text{ m}^4$	$I_z = 0.0001 \text{ m}^4$
$E = 10 \text{ GPa}$	$E = 80 \text{ GPa}$

Exercise 10.41

10.42   The displacements at node 2 of the frame, due to the applied loads, are $u_2 = 0.0593$ in., $v_2 = -0.109$ in., and $\theta_{z2} = 0.00131$ rad. Find:

a. The displacements at node 3.

b. The location and exact value of the maximum bending moment in the structure.

c. The locations of those points in the structure where the exact bending moment is zero.

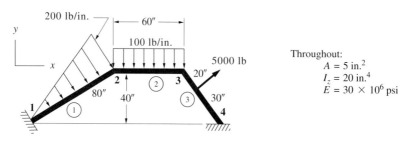

Exercise 10.42

10.43   For the frame loaded and constrained as shown, the displacements at node 1 are: $u_1 = -1.15$ mm, $v_1 = -1.77$ mm,   and $\theta_{z1} = -0.00464$ rad. Find:

a. The displacements at node 2.

b. The location and exact value of the maximum bending moment in the structure.

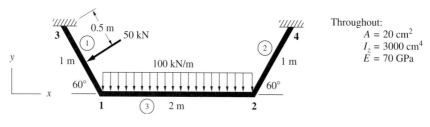

Exercise 10.43

10.44   A 200 lb load is applied as shown to the wooden portal frame, nodes 3 and 4 of which are built-in.

a. Find the displacements of nodes 1 and 2.

b. Locate those points in the frame where the bending moment is zero.

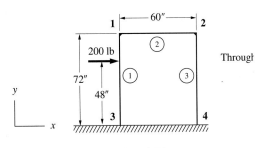

Exercise 10.44

10.45   All elements of the frame have the same A, E, and I_z. Find component $k_{10,18}$ of the 24 by 24 global stiffness matrix.

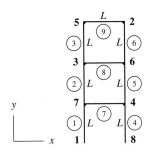

Exercise 10.45

10.46   Use available computer resources as an aid to find the exact maximum bending moment in the frame, which is loaded and supported as shown.

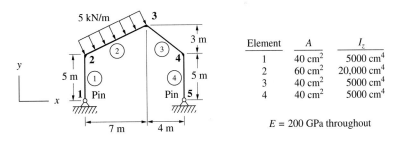

Element	A	I_z
1	40 cm^2	5000 cm^4
2	60 cm^2	20,000 cm^4
3	40 cm^2	5000 cm^4
4	40 cm^2	5000 cm^4

$E = 200$ GPa throughout

Exercise 10.46

10.47   The displacements of node 1 of the frame, under the given loads and constraints, are: $u_1 = 0$, $v_1 = -80.2\,\text{mm}$, $w_1 = 0$, $\theta_{x1} = 0.0177\,\text{rad}$, $\theta_{y1} = 0$, and $\theta_{z1} = -0.111\,\text{rad}$. Calculate the axial torque in element 2 of the structure, and verify that the load at node 1 is 200 N downward, as shown.

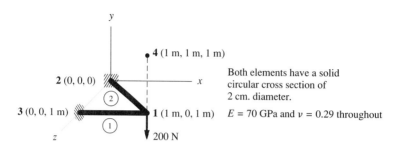

Exercise 10.47

10.48   Find the displacements at node 1 and the maximum stress due to bending.

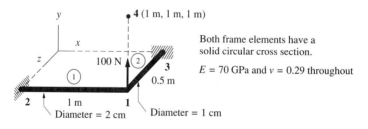

Exercise 10.48

10.49 For the three-dimensional frame loaded and constrained as shown, the calculated displacements at node 2 are: $u_1 = 0.0112$ mm, $v_2 = -0.00208$ mm, $w_2 = 35.2$ mm, $\theta_{x2} = -0.0214$ rad, $\theta_{y2} = -0.0380$ rad, and $\theta_{z2} = -0.0277$ rad. Find:

 a. The displacements at node 3.

 b. The location and exact value of the maximum bending moment in the structure.

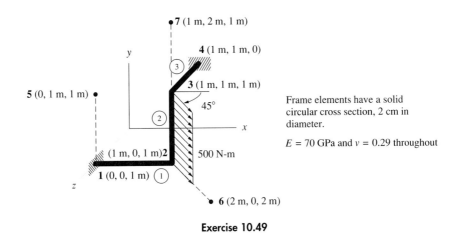

Frame elements have a solid circular cross section, 2 cm in diameter.

$E = 70$ GPa and $v = 0.29$ throughout

Exercise 10.49

10.50 Use available computer resources as an aid to find the maximum allowable value of the distributed load p (lb/in.) if the stress anywhere in the frame must not exceed 30 ksi, tension or compression. The length of each of the seven frame elements is L (in.), where L is the last two digits of your student number, or 10, whichever is greater.

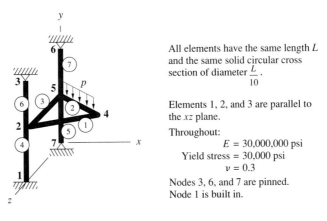

All elements have the same length L and the same solid circular cross section of diameter $\dfrac{L}{10}$.

Elements 1, 2, and 3 are parallel to the xz plane.

Throughout:
$$E = 30{,}000{,}000 \text{ psi}$$
$$\text{Yield stress} = 30{,}000 \text{ psi}$$
$$v = 0.3$$

Nodes 3, 6, and 7 are pinned.
Node 1 is built in.

Exercise 10.50

11

Matrix Displacement Method: Thin-Walled Structures

CHAPTER OUTLINE

11.1   INTRODUCTION

In Chapter 10, we used the principle of virtual work to solve truss and frame problems by the displacement method. Because the displacement method does not require special treatment of redundant members, it is a straightforward, universal approach to the equilibrium analysis of elastic structures. The principle of virtual work also leads to a universal procedure for deriving the stiffness matrices and load vectors of arbitrary finite subregions, that is, *finite elements*, of solid continua. Libraries of finite elements can be drawn upon to yield accurate computer models of the shape and structural behavior of the most complex aircraft components. In the interest of simplicity, our focus will be restricted to the simplest two-dimensional finite elements, the shear panel and the constant-strain triangle.

11.2   RECTANGULAR SHEAR PANEL

The rectangular shear panel illustrated in Figure 11.2.1 (cf. section 2.5) is a thin material sheet, every point of which is in a state of pure shear stress parallel to the element edges. The shear stress is required to be constant throughout the element.

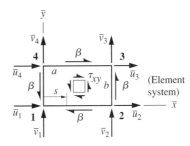

Figure 11.2.1   Rectangular shear panel of thickness t.

It follows that the element shear flow, which is the shear stress multiplied by the element thickness, is also constant everywhere, including along the element boundaries. The element is therefore referred to as the constant-shear flow panel.

Let us number the four corner nodes of the element consecutively, in either a clockwise or counterclockwise direction around the element. The origin of the element coordinate system is defined as node 1. The local x axis is directed from node 1 to node 2, and the local y axis is directed from node 1 to node 4. Although methods exist for handling warped elements,[1] we will assume that all four nodes lie in the same plane before deformation.

In the absence of body force and temperature fields the assumed state of uniform pure shear stress is

$$\tau_{xy} = \beta \text{ (constant)} \qquad \sigma_x = \sigma_y = 0 \qquad \text{[11.2.1]}$$

It clearly satisfies equilibrium, Equations 3.6.2, as well as compatibility, Equation 3.12.4.

The internal complementary virtual work δW^*_{int} of the shear panel (found from Equation 6.6.4) is

$$\delta W^*_{\text{int}} = \iiint_V \delta\tau_{xy}\gamma_{xy}dV = \iiint_V \delta\beta\frac{\beta}{G}dV = \frac{V}{G}\beta\delta\beta$$

where $V = abt$ is the volume of the panel, and t is its thickness. Thus,

$$\delta W^*_{\text{int}} = \frac{abt}{G}\beta\delta\beta \qquad \text{[11.2.2]}$$

The external complementary virtual work δW^*_{ext} is

$$\delta W^*_{\text{ext}} = \{\delta\bar{\mathbf{Q}}\}^T\{\bar{\mathbf{q}}\} \qquad \text{[11.2.3]}$$

where $\{\bar{\mathbf{q}}\}$ is the local element displacement vector, the components of which are illustrated in Figure 11.2.1 and are

$$\{\bar{\mathbf{q}}\} = \lfloor\,\bar{u}_1 \quad \bar{v}_1 \quad \bar{u}_2 \quad \bar{v}_2 \quad \bar{u}_3 \quad \bar{v}_3 \quad \bar{u}_4 \quad \bar{v}_4\,\rfloor^T \qquad \text{[11.2.4]}$$

Since point loads cannot be applied directly to the corners of the shear panel, $\{\bar{\mathbf{Q}}\}$ must be the vector of corner forces that are equivalent to the constant shear flow on each edge of the panel. Let us assume that the tangential displacement $\bar{u}$ on *and in the direction of* an edge of the panel is a linear function of position s along

| [1] J. Robinson, *Integrated Theory of Finite Elements*, New York, Wiley, 1973.

that edge, just as we assumed for the rod element. Then, along side 1-2, the principle of virtual work equivalency requires that

$$
\bar{X}_1 \delta \bar{u}_1 + \bar{X}_2 \delta \bar{u}_2 = - \int_0^a \delta \bar{u} \beta \, (t \, ds) = -\beta t \int_0^a \left[\delta \bar{u}_1 \left(1 - \frac{s}{a} \right) + \delta \bar{u}_2 \frac{s}{a} \right] ds
$$

Taking the integral and collecting terms to the left-hand side, yields

Edge 1-2: $\quad \left(\bar{X}_1 + \dfrac{\beta a t}{2} \right) \delta \bar{u}_1 + \left(\bar{X}_2 + \dfrac{\beta a t}{2} \right) \delta \bar{u}_2 = 0$ $\qquad$ [11.2.5a]

Proceeding in a similar fashion for the remaining three sides of the panel, we get

Edge 2-3: $\quad \left(\bar{Y}_2 - \dfrac{\beta b t}{2} \right) \delta \bar{v}_2 + \left(\bar{Y}_3 - \dfrac{\beta b t}{2} \right) \delta \bar{v}_3 = 0$ $\qquad$ [11.2.5b]

Edge 3-4: $\quad \left(\bar{X}_3 - \dfrac{\beta a t}{2} \right) \delta \bar{u}_3 + \left(\bar{X}_4 - \dfrac{\beta a t}{2} \right) \delta \bar{u}_4 = 0$ $\qquad$ [11.2.5c]

Edge 4-1: $\quad \left(\bar{Y}_4 + \dfrac{\beta b t}{2} \right) \delta \bar{v}_4 + \left(\bar{Y}_1 + \dfrac{\beta b t}{2} \right) \delta \bar{v}_1 = 0$ $\qquad$ [11.2.5d]

Since the virtual displacements are arbitrary, we can set their coefficients equal to zero in each of these equations, to obtain

$$
\{\bar{Q}\} = \{F\} \beta \qquad \text{[11.2.6a]}
$$

where

$$
\{\bar{Q}\} = \lfloor \bar{X}_1 \quad \bar{Y}_1 \vdots \bar{X}_2 \quad \bar{Y}_2 \vdots \bar{X}_3 \quad \bar{Y}_3 \vdots \bar{X}_4 \quad \bar{Y}_4 \rfloor^T \qquad \{F\} = \frac{t}{2} \lfloor -a \quad -b \vdots -a \quad b \vdots a \quad b \vdots a \quad -b \rfloor^T \qquad \text{[11.2.6b]}
$$

Then, from Equation 11.2.6a, we have

$$
\{\delta \bar{Q}\}^T = \delta \beta \{F\}^T
$$

Substituting this into Equation 11.2.3, yields the external complementary virtual work formula,

$$
\delta W_{\text{ext}}^* = \delta \beta \, \{F\}^T \, \{\bar{q}\}
$$

Substituting this expression and Equation 11.2.2 into the complementary virtual work equality, $\delta W_{\text{int}}^* = \delta W_{\text{ext}}^*$, and solving for β yields the element shear stress in terms of the local displacement vector $\{\bar{q}\}$,

$$
\beta = \tau_{xy} = \frac{G}{abt} \{F\}^T \{\bar{q}\} \qquad \text{[11.2.7]}
$$

Inserting this expression for the shear stress parameter β into Equation 11.2.6a, we obtain the element load vector $\{\bar{Q}\}$ in terms of $\{\bar{q}\}$, as follows:

$$
\{\bar{Q}\} = \frac{G}{abt} \{F\} \{F\}^T \{\bar{q}\} \qquad \text{[11.2.8]}
$$

With the aid of Equation 6.5.1, we can therefore specify the local stiffness matrix $[\bar{\mathbf{k}}]$ of the shear panel:

$$[\bar{\mathbf{k}}] = \frac{G}{abt}\{\mathbf{F}\}^T\{\mathbf{F}\} \qquad [11.2.9]$$

Recalling $\{\mathbf{F}\}$ from Equation 11.2.6 and carrying out the matrix multiplication yields

$$[\bar{\mathbf{k}}] = \frac{Gt}{4}
\begin{array}{c}
\begin{array}{cccccccc}
\bar{u}_1 & \bar{v}_1 & \bar{u}_2 & \bar{v}_2 & \bar{u}_3 & \bar{v}_3 & \bar{u}_4 & \bar{v}_4
\end{array} \\
\begin{bmatrix}
a/b & 1 & a/b & -1 & -a/b & -1 & -a/b & 1 \\
1 & b/a & 1 & -b/a & -1 & -b/a & -1 & b/a \\
a/b & 1 & a/b & -1 & -a/b & -1 & -a/b & 1 \\
-1 & -b/a & -1 & b/a & 1 & b/a & 1 & -b/a \\
-a/b & -1 & -a/b & 1 & a/b & 1 & a/b & -1 \\
-1 & -b/a & -1 & b/a & 1 & b/a & 1 & -b/a \\
-a/b & -1 & -a/b & 1 & a/b & 1 & a/b & -1 \\
1 & b/a & 1 & -b/a & -1 & -b/a & -1 & b/a
\end{bmatrix}
\begin{array}{c}
\bar{X}_1 \\ \bar{Y}_1 \\ \bar{X}_2 \\ \bar{Y}_2 \\ \bar{X}_3 \\ \bar{Y}_3 \\ \bar{X}_4 \\ \bar{Y}_4
\end{array}
\end{array} \qquad [11.2.10]$$

Let us expand the right-hand side of Equation 11.2.7 by substituting the components of $\{\mathbf{F}\}$ and $\{\bar{\mathbf{q}}\}$:

$$\tau_{xy} = \frac{G}{abt}\frac{t}{2}\lfloor -a \;\; -b \;\; -a \;\; b \;\; a \;\; b \;\; a \;\; -b \rfloor
\begin{Bmatrix}
\bar{u}_1 \\ \bar{v}_1 \\ \bar{u}_2 \\ \bar{v}_2 \\ \bar{u}_3 \\ \bar{v}_3 \\ \bar{u}_4 \\ \bar{v}_4
\end{Bmatrix}$$

Carrying out the multiplication operation and collecting terms, yields

$$\tau_{xy} = G\left(\frac{\phi_1 + \varphi_1}{2} + \frac{\phi_2 + \varphi_2}{2}\right) \qquad [11.2.11a]$$

where

$$\phi_1 = \frac{\bar{v}_2 - \bar{v}_1}{a} \qquad \phi_2 = \frac{\bar{v}_3 - \bar{v}_4}{a} \qquad \varphi_1 = \frac{\bar{u}_4 - \bar{u}_1}{b} \qquad \varphi_2 = \frac{\bar{u}_3 - \bar{u}_2}{b} \qquad [11.2.11b]$$

Studying Figure 11.2.2, we see that $\phi_1 + \varphi_1$ is the decrease in the right angle at node 1, and $\phi_2 + \varphi_2$ is the decrease in the right angle at node 3. Therefore, Equation 11.2.7 may be interpreted to mean that the shear stress is the shear modulus times the average shear strain in the element.

The global shear panel stiffness matrix is obtained, as usual, through the transformation

$$[\mathbf{k}] = [\mathbf{\Lambda}]^T[\bar{\mathbf{k}}][\mathbf{\Lambda}] \qquad [11.2.12]$$

where again, $[\mathbf{\Lambda}]$ is the transformation matrix that relates the components of the element displacement vector $\{\mathbf{q}\}$ along the global axes to the components along the local coordinate directions, $\{\bar{\mathbf{q}}\}$. Figure 11.2.3 illustrates a shear panel at an arbitrary orientation relative to the global system.

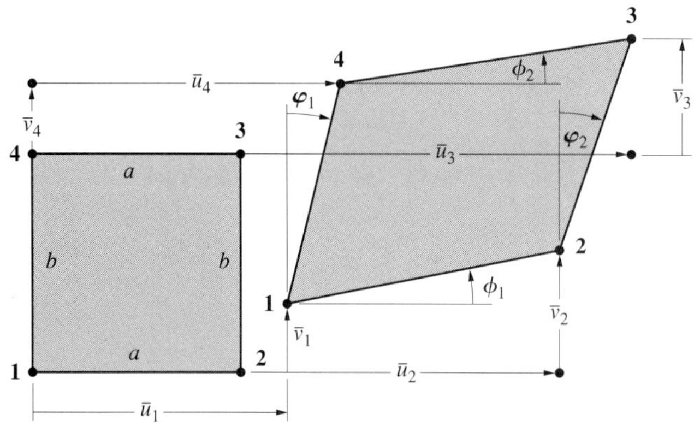

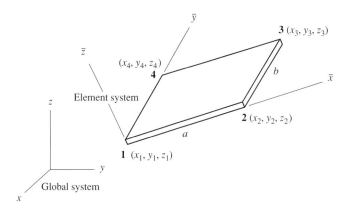

Figure 11.2.2 Deformation of the shear panel (element system).

Figure 11.2.3 Rectangular shear panel oriented in three dimensions.

The panel element stiffness matrix in Equation 11.2.10 was derived using just the two in-plane degrees of freedom at each node. Displacement components perpendicular to the plane of the element were ignored, since the shear panel is required to transmit only shear forces within its own plane. However, the in-plane loads and displacements will still have three global components, if the panel is positioned arbitrarily in space, as it is in Figure 11.2.3. The two local displacement components, $\bar{u}$ and $\bar{v}$, at a node are obtained by projecting the three-dimensional displacement vector, with global components u, v, and w, onto the element coordinate axes; that is,

$$\begin{Bmatrix} \bar{u} \\ \bar{v} \end{Bmatrix} = \begin{bmatrix} l_1 & m_1 & n_1 \\ l_2 & m_2 & n_2 \end{bmatrix} \begin{Bmatrix} u \\ v \\ w \end{Bmatrix}$$

where $\lfloor l_1 \ m_1 \ n_1 \rfloor$ and $\lfloor l_2 \ m_2 \ n_2 \rfloor$ are the direction cosines of the element axes $\bar{x}$ and $\bar{y}$, respectively, relative to the global system. Applying this transformation to all four sets of nodal displacements, we get

$$\{\bar{\mathbf{q}}\} = [\Lambda]\{\mathbf{q}\} \qquad [11.2.13]$$

where

$$\{\bar{\mathbf{q}}\} = \lfloor \bar{u}_1 \quad \bar{v}_1 \mid \bar{u}_2 \quad \bar{v}_2 \mid \bar{u}_3 \quad \bar{v}_3 \mid \bar{u}_4 \quad \bar{v}_4 \rfloor^T$$

$$\{\mathbf{q}\} = \lfloor u_1 \quad v_1 \quad w_1 \mid u_2 \quad v_2 \quad w_2 \mid u_3 \quad v_3 \quad w_3 \mid u_4 \quad v_4 \quad w_4 \rfloor^T$$

and

$$[\mathbf{\Lambda}] = \begin{bmatrix} [\boldsymbol{\lambda}] & [0] & [0] & [0] \\ [0] & [\boldsymbol{\lambda}] & [0] & [0] \\ [0] & [0] & [\boldsymbol{\lambda}] & [0] \\ [0] & [0] & [0] & [\boldsymbol{\lambda}] \end{bmatrix} \quad \text{(8 by 12)} \qquad [11.2.14a]$$

in which

$$[\boldsymbol{\lambda}] = \begin{bmatrix} l_1 & m_1 & n_1 \\ l_2 & m_2 & n_2 \end{bmatrix} \qquad [11.2.14b]$$

Substituting the local stiffness matrix $[\bar{\mathbf{k}}]$, Equation 11.2.10, and the transformation matrix $[\mathbf{\Lambda}]$ into the transformation, Equation 11.2.13, yields the 12 by 12 global shear panel stiffness matrix,

$$[\mathbf{k}] = \begin{bmatrix} [\mathbf{k}_1] & [\mathbf{k}_2] & -[\mathbf{k}_1] & -[\mathbf{k}_2] \\ [\mathbf{k}_2]^T & [\mathbf{k}_3] & -[\mathbf{k}_2] & -[\mathbf{k}_3] \\ -[\mathbf{k}_1] & -[\mathbf{k}_2] & [\mathbf{k}_1] & [\mathbf{k}_2] \\ -[\mathbf{k}_2]^T & -[\mathbf{k}_3] & [\mathbf{k}_2]^T & [\mathbf{k}_3] \end{bmatrix} \qquad [11.2.15a]$$

in which

$$[\mathbf{k}_1] = \frac{Gt}{4} \begin{bmatrix} \frac{a}{b}l_1^2 + \frac{b}{a}l_2^2 + 2l_1l_2 & \frac{a}{b}l_1m_1 + \frac{b}{a}l_2m_2 + l_1m_2 + l_2m_1 & \frac{a}{b}l_1n_1 + \frac{b}{a}l_2n_2 + l_1n_2 + l_2n_1 \\ \frac{a}{b}l_1m_1 + \frac{b}{a}l_2m_2 + l_1m_2 + l_2m_1 & \frac{a}{b}m_1^2 + \frac{b}{a}m_2^2 + 2m_1m_2 & \frac{a}{b}m_1n_1 + \frac{b}{a}m_2n_2 + m_1n_2 + m_2n_1 \\ \frac{a}{b}l_1n_1 + \frac{b}{a}l_2n_2 + l_1n_2 + l_2n_1 & \frac{a}{b}m_1n_1 + \frac{b}{a}m_2n_2 + m_1n_2 + m_2n_1 & \frac{a}{b}n_1^2 + \frac{b}{a}n_2^2 + 2n_1n_2 \end{bmatrix} \quad [11.2.15b]$$

$$[\mathbf{k}_2] = \frac{Gt}{4} \begin{bmatrix} \frac{a}{b}l_1^2 - \frac{b}{a}l_2^2 & \frac{a}{b}l_1m_1 - \frac{b}{a}l_2m_2 + l_1m_2 - l_2m_1 & \frac{a}{b}l_1n_1 - \frac{b}{a}l_2n_2 + l_1n_2 - l_2n_1 \\ \frac{a}{b}l_1m_1 - \frac{b}{a}l_2m_2 + l_1m_2 - l_2m_1 & \frac{a}{b}m_1^2 - \frac{b}{a}m_2^2 & \frac{a}{b}m_1n_1 - \frac{b}{a}m_2n_2 + m_1n_2 - m_2n_1 \\ \frac{a}{b}l_1n_1 - \frac{b}{a}l_2n_2 + l_1n_2 - l_2n_1 & \frac{a}{b}m_1n_1 - \frac{b}{a}m_2n_2 + m_1n_2 - m_2n_1 & \frac{a}{b}n_1^2 - \frac{b}{a}n_2^2 \end{bmatrix} \quad [11.2.15c]$$

$$[\mathbf{k}_3] = \frac{Gt}{4} \begin{bmatrix} \frac{a}{b}l_1^2 + \frac{b}{a}l_2^2 - 2l_1l_2 & \frac{a}{b}l_1m_1 + \frac{b}{a}l_2m_2 - l_1m_2 - l_2m_1 & \frac{a}{b}l_1n_1 + \frac{b}{a}l_2n_2 - l_1n_2 - l_2n_1 \\ \frac{a}{b}l_1m_1 + \frac{b}{a}l_2m_2 - l_1m_2 - l_2m_1 & \frac{a}{b}m_1^2 + \frac{b}{a}m_2^2 - 2m_1m_2 & \frac{a}{b}m_1n_1 + \frac{b}{a}m_2n_2 - m_1n_2 - m_2n_1 \\ \frac{a}{b}l_1n_1 + \frac{b}{a}l_2n_2 - l_1n_2 - l_2n_1 & \frac{a}{b}m_1n_1 + \frac{b}{a}m_2n_2 - m_1n_2 - m_2n_1 & \frac{a}{b}n_1^2 + \frac{b}{a}n_2^2 - 2n_1n_2 \end{bmatrix} \quad [11.2.15d]$$

Once we determine the global displacements, we use Equation 11.2.13 to obtain the local displacement vector, which, together with Equation 11.2.7, establishes the shear stress in the panel.

We compute the direction cosines for Equations 11.2.14 and 11.2.15 using the global coordinates of the nodes, as depicted in Figure 11.2.3. The vector directed from node 1 to node 2 along the element axis $\bar{x}$ is

$$\mathbf{r}_{12} = (x_2 - x_1)\mathbf{i} + (y_2 - y_1)\mathbf{j} + (z_2 - z_1)\mathbf{k}$$

Therefore, the direction cosines of the element axis $\bar{x}$ are

$$l_1 = \frac{x_2 - x_1}{|\mathbf{r}_{12}|} \qquad m_1 = \frac{y_2 - y_1}{|\mathbf{r}_{12}|} \qquad n_1 = \frac{z_2 - z_1}{|\mathbf{r}_{12}|}$$

where $|\mathbf{r}_{12}|$ is the length of $\mathbf{r}_{12}$.

Likewise, the direction cosines of the element axis $\bar{y}$, which is directed from node 1 to node 4, are

$$l_2 = \frac{x_4 - x_1}{|\mathbf{r}_{14}|} \qquad m_2 = \frac{y_4 - y_1}{|\mathbf{r}_{14}|} \qquad n_2 = \frac{z_4 - z_1}{|\mathbf{r}_{14}|}$$

Since the shear panel must be surrounded by stiffeners, we will use the linear-displacement rod element of section 10.2. The displacement varies linearly along an edge of the panel, just as it does along the length of the rod attached to it. Therefore, as illustrated in Figure 11.2.4, the displacement of adjacent points on the panel and the rod are the same everywhere along the shared boundary. This is what we would physically expect along the bond line between two elements. Recall that the stress is constant throughout the linear-displacement rod. No shear flow is applied between its nodes, so the stress does not change linearly from end to end, as was the case for the stiffener we used for the force method analysis of stiffened web structures (section 8.2). We could accommodate the linear-stress rod by adding one node to its midpoint and, for displacement compatibility, another node to the midpoint of each side of the shear panel, as well. With three nodes, we can fit a quadratic displacement field, $\bar{u} = C_0 + C_1 s + C_2 s^2$, within the rod and along the element edges. The corresponding strain $\varepsilon = \partial \bar{u} / \partial s$, therefore the stress, will vary linearly in the rod. Also, the additional nodes will increase the size of the stiffness matrices. With three nodes, the rod element's local stiffness matrix becomes 3 by 3; adding four nodes to the panel will increase its local stiffness matrix from 8 by 8 to 16 by 16. The additional complexity is neither necessary nor justified for treating the rectangular shear panel, with its simple, exact, uniform, pure shear stress field.

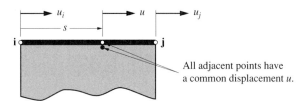

Figure 11.2.4   Displacement compatibility between the linear-displacement rod and the shear panel.

Example 11.2.1   Use the displacement method to calculate the shear flow and stiffener loads for the structure in Figure 11.2.5.

Using statics, we can find the shear flow in the panel immediately, by dividing the shear load by the height of the panel, recalling the sign convention for shear stress, to get

$$\text{Shear flow} = -\frac{1000\text{N}}{0.3\text{m}} = -3333 \frac{\text{N}}{\text{m}}$$

Let us first list the properties of the rod elements and the shear panel in Table 11.2.1.

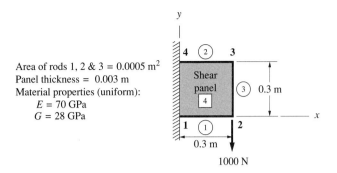

Area of rods 1, 2 & 3 = 0.0005 m²
Panel thickness = 0.003 m
Material properties (uniform):
E = 70 GPa
G = 28 GPa

Figure 11.2.5 Shear panel surrounded by rod elements.

Table 11.2.1 Element properties for Figure 11.2.5.

Element Number	Element Type	Area (m²)	Length (m)	$\frac{AE}{L}$ (N/m)	Node 1	Node 2	θ (degrees)	l	m
1	Rod	0.0005	0.3	116.7×10^6	1	2	0	1	0
2	Rod	0.0005	0.3	116.7×10^6	4	3	0	1	0
3	Rod	0.0005	0.3	116.7×10^6	2	3	90	0	1

		Thickness (m)	$\frac{Gt}{4}$ (N/m)	Node 1	Node 2	Node 3	Node 4	(l_1, m_1, n_1)	(l_2, m_2, n_2)
4	Shear panel	0.003	21×10^6	1	2	3	4	(1,0,0)	(0,1,0)

The two-dimensional form of the rod element stiffness matrix is Equation 10.2.18. Using the element data, applying the constraints to the elements, and labeling the global freedom numbers, we have, for the three rods,

$$
\left[k_{ff}^{(1)} \right] = (10^6) \begin{bmatrix} 116.7 & 0 \\ 0 & 0 \end{bmatrix} \begin{matrix} 3 \\ 4 \end{matrix}
\qquad
\left[k_{ff}^{(2)} \right] = (10^6) \begin{bmatrix} 116.7 & 0 \\ 0 & 0 \end{bmatrix} \begin{matrix} 5 \\ 6 \end{matrix}
\qquad
\left[k_{ff}^{(3)} \right] = (10^6) \begin{matrix} 3 & 4 & 5 & 6 \\ \begin{bmatrix} 0 & 0 & 0 & 0 \\ 0 & 116.7 & 0 & -116.7 \\ 0 & 0 & 0 & 0 \\ 0 & -116.7 & 0 & 116.7 \end{bmatrix} \end{matrix} \begin{matrix} 3 \\ 4 \\ 5 \\ 6 \end{matrix}
$$

Since the shear panel's local axes are aligned with the global system, it is convenient to use Equation 11.2.10 to calculate the panel's stiffness matrix. Applying the constraints and the global freedom numbering, we get

$$
\left[k_{ff}^{(4)} \right] = (10^6) \begin{matrix} 3 & 4 & 5 & 6 \\ \begin{bmatrix} 21 & -21 & -21 & -21 \\ -21 & 21 & 21 & 21 \\ -21 & 21 & 21 & 21 \\ -21 & 21 & 21 & 21 \end{bmatrix} \end{matrix} \begin{matrix} 3 \\ 4 \\ 5 \\ 6 \end{matrix}
$$

As usual, the 4 by 4 global stiffness matrix $[\mathbf{k}_{ff}]$ of the constrained structure is found by assembling these four element matrices, guided by their row and column labels, to yield

$$
[\mathbf{k}_{ff}] = (10^6)\begin{bmatrix} 137.7 & -21.00 & -21.00 & -21.00 \\ -21.00 & 137.7 & 21.00 & -95.67 \\ -21.00 & 21.00 & 137.7 & 21.00 \\ -21.00 & -95.67 & 21.00 & 137.7 \end{bmatrix}
$$

The stiffness equations, $[\mathbf{k}_{ff}]\{\mathbf{q}_f\} = \{\mathbf{Q}_f\}$, of the constrained structure are therefore

$$
(10^6)\begin{bmatrix} 137.7 & -21.00 & -21.00 & -21.00 \\ -21.00 & 137.7 & 21.00 & -95.67 \\ -21.00 & 21.00 & 137.7 & 21.00 \\ -21.00 & -95.67 & 21.00 & 137.7 \end{bmatrix}\begin{Bmatrix} u_2 \\ v_2 \\ u_3 \\ v_3 \end{Bmatrix} = \begin{Bmatrix} 0 \\ -1000 \\ 0 \\ 0 \end{Bmatrix}
$$

the solution of which is

$$
\{\mathbf{q}_f\} = (10^{-6})\lfloor -4.285 \quad -18.33 \quad 4.285 \quad -14.05 \rfloor^T \text{ (m)}
$$

We can now calculate the loads in the rod elements and the shear panel. As before, for each element e, we identify the global displacement components at each node, thereby forming the global element displacement vector $\{\mathbf{q}^{(e)}\}$. We then project this vector into the element frame of reference to form $\{\bar{\mathbf{q}}^{(e)}\}$, which we use to calculate the internal element loads.

Rod Element 1:
The global displacement vector for this element is

$$
\{\mathbf{q}^{(1)}\} = (10^{-6})\lfloor 0 \quad 0 \quad -4.285 \quad -18.33 \rfloor^T
$$

From Equation 10.2.15 and the direction cosines of the element, we have

$$
[\boldsymbol{\Lambda}^{(1)}] = \begin{bmatrix} 1 & 0 & 0 & 0 \\ 0 & 0 & 1 & 0 \end{bmatrix}
$$

Therefore, the projection operation $\{\bar{\mathbf{q}}^{(1)}\} = [\boldsymbol{\Lambda}^{(1)}]\{\mathbf{q}^{(1)}\}$ yields

$$
\{\bar{\mathbf{q}}^{(1)}\} = (10^{-6})\begin{Bmatrix} 0 \\ -4.285 \end{Bmatrix} \text{ (m)}
$$

Multiplying this displacement vector by the local element stiffness matrix $[\bar{\mathbf{k}}^{(1)}]$ yields the element load vector $\{\bar{\mathbf{Q}}^{(1)}\}$, as follows:

$$
\{\bar{\mathbf{Q}}^{(1)}\} = \begin{Bmatrix} \bar{X}_1^{(1)} \\ \bar{X}_2^{(1)} \end{Bmatrix} = \left(\frac{AE}{L}\right)^{(1)}\begin{bmatrix} 1 & -1 \\ -1 & 1 \end{bmatrix}\begin{Bmatrix} \bar{u}_1^{(1)} \\ \bar{u}_2^{(1)} \end{Bmatrix} = (116.7 \times 10^6)\begin{bmatrix} 1 & -1 \\ -1 & 1 \end{bmatrix}(10^{-6})\begin{Bmatrix} 0 \\ -4.285 \end{Bmatrix} = \begin{Bmatrix} 500 \\ -500 \end{Bmatrix} \text{ (N)}
$$

Rod Element 2
The global displacement vector is

$$
\{\mathbf{q}^{(2)}\} = (10^{-6})\lfloor 0 \quad 0 \quad 4.285 \quad -14.05 \rfloor^T
$$

The direction cosines are the same as for the previous element, so the projection operation yields

$$\{\bar{\mathbf{q}}^{(2)}\} = \left(10^{-6}\right) \left\{ \begin{array}{c} 0 \\ 4.285 \end{array} \right\} \text{ (m)}$$

Since it is clear that $\{\bar{\mathbf{q}}^{(2)}\} = -\{\bar{\mathbf{q}}^{(1)}\}$, and since the elements have identical local element stiffness matrices, it follows that $\{\bar{\mathbf{Q}}^{(2)}\} = -\{\bar{\mathbf{Q}}^{(1)}\}$, or

$$\left\{ \begin{array}{c} \bar{X}_1^{(2)} \\ \bar{X}_2^{(2)} \end{array} \right\} = \left\{ \begin{array}{c} -500 \\ 500 \end{array} \right\} \text{ (N)}$$

Rod Element 3

The global displacement vector for this vertical rod element is

$$\{\mathbf{q}^{(3)}\} = \left(10^{-6}\right) \lfloor -4.284 \quad -18.33 \quad 4.285 \quad -14.05 \rfloor^T$$

and its transformation matrix $[\mathbf{\Lambda}^{(3)}]$ is

$$[\mathbf{\Lambda}^{(3)}] = \begin{bmatrix} 0 & 1 & 0 & 0 \\ 0 & 0 & 0 & 1 \end{bmatrix}$$

The local displacement vector, $\{\bar{\mathbf{q}}^{(3)}\} = [\mathbf{\Lambda}^{(3)}]\{\mathbf{q}^{(3)}\}$, is therefore

$$\{\bar{\mathbf{q}}^{(3)}\} = \left(10^{-6}\right) \left\{ \begin{array}{c} -18.33 \\ -14.05 \end{array} \right\} \text{ (m)}$$

Since the local stiffness matrix of this rod is identical to that of the others, we have

$$\left\{ \begin{array}{c} \bar{X}_1^{(3)} \\ \bar{X}_2^{(3)} \end{array} \right\} = \left(116.7 \times 10^6\right) \begin{bmatrix} 1 & -1 \\ -1 & 1 \end{bmatrix} \left(10^{-6}\right) \left\{ \begin{array}{c} -18.33 \\ -14.05 \end{array} \right\} = \left\{ \begin{array}{c} -500 \\ 500 \end{array} \right\} \text{ (N)}$$

Shear Panel (Element 4)

The local and global element displacement vectors are identical, because the local shear panel axes coincide with the global system. Thus,

$$\{\bar{\mathbf{q}}^{(4)}\} = \{\mathbf{q}^{(4)}\} = \left(10^{-6}\right) \lfloor 0 \quad 0 \quad -4.285 \quad -18.33 \quad 4.285 \quad -14.05 \quad 0 \quad 0 \rfloor^T \text{ (m)}$$

The shear stress in the panel is found from Equation 11.2.7,

$$\tau_{xy} = \frac{G}{abt} [\mathbf{F}]^T \{\mathbf{q}\}$$

where $[\mathbf{F}]$ is given by Equation 11.2.6b, with $a = b = 0.3$ m and $t = 0.003$ m,

$$[\mathbf{F}]^T = \left(450 \times 10^6\right) \lfloor -1 \quad -1 \ \vdots \ -1 \quad 1 \ \vdots \ 1 \quad 1 \ \vdots \ 1 \quad -1 \rfloor$$

Therefore,

$$\tau_{xy} = \frac{28\,(10^9)}{(0.3)\,(0.3)\,(0.003)}\,(450 \times 10^6)\,\lfloor -1 \quad -1 \;\vdots\; -1 \quad 1 \;\vdots\; 1 \quad 1 \;\vdots\; 1 \quad -1 \rfloor \begin{Bmatrix} 0 \\ 0 \\ \hdashline -4.285 \times 10^{-6} \\ -18.33 \times 10^{-6} \\ \hdashline 4.285 \times 10^{-6} \\ -14.05 \times 10^{-6} \\ \hdashline 0 \\ 0 \end{Bmatrix} = -1.111\ \text{MPa}$$

The shear flow q equals $\tau_{xy}t$, or

$$q = -3.333\ \text{kN/m}$$

where $t = 0.003$ m.

The loads in each element of the stiffened shear panel assembly are shown in Figure 11.2.6. Observe that one-half of the shear force (1 kN) in the direction of each panel edge is distributed to the two corner nodes on that edge. Also, the sum of the element forces at each node of the structure equals the externally applied load at that node. At nodes 1 and 4, this summation gives the 1 kN support reactions; at node 2, it gives the applied downward 1 kN load.

It should be emphasized that the shear panel itself cannot transmit loads applied directly to its nodes. That capability is provided by the rod elements attached to the panel along each of its edges. Shear panels must be surrounded by rod elements, except along edges whose end nodes are totally constrained, as in Figure 11.2.5.

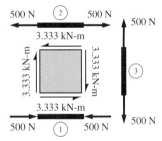

Figure 11.2.6   Element loads corresponding to the loading in Figure 11.2.5.

Example 11.2.2   Use the displacement method to calculate the axial load in each rod and the shear flow in each panel of the box-like structure shown in Figure 11.2.7. The box consists of four shear panels (front, top, bottom, and right end) and five stiffeners. Nodes 3 through 6 are rigidly constrained, and a 5 kN load is applied to node 1.

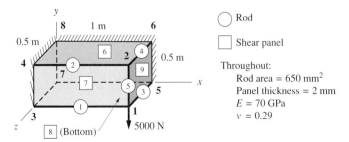

Figure 11.2.7   Rod and panel box structure.

The properties of the elements are listed in Tables 11.2.2 and 11.2.3. The sets of free (f) and constrained (s) global degrees of freedom are f: 1 through 6 and s: 7 through 24. The three-dimensional rod element stiffness matrices are found using Equation 10.2.22. Using the property table data and constraining at the element level, we find

$$[\mathbf{k}_{ff}^{(1)}] = (45.50 \times 10^6) \begin{array}{c} \begin{array}{ccc} 1 & 2 & 3 \end{array} \\ \begin{bmatrix} 1 & 0 & 0 \\ 0 & 0 & 0 \\ 0 & 0 & 0 \end{bmatrix} \begin{array}{c} 1 \\ 2 \\ 3 \end{array} \end{array} \qquad [\mathbf{k}_{ff}^{(2)}] = (45.50 \times 10^6) \begin{array}{c} \begin{array}{ccc} 4 & 5 & 6 \end{array} \\ \begin{bmatrix} 1 & 0 & 0 \\ 0 & 0 & 0 \\ 0 & 0 & 0 \end{bmatrix} \begin{array}{c} 4 \\ 5 \\ 6 \end{array} \end{array}$$

$$[\mathbf{k}_{ff}^{(3)}] = (91.00 \times 10^6) \begin{array}{c} \begin{array}{ccc} 1 & 2 & 3 \end{array} \\ \begin{bmatrix} 0 & 0 & 0 \\ 0 & 0 & 0 \\ 0 & 0 & 1 \end{bmatrix} \begin{array}{c} 1 \\ 2 \\ 3 \end{array} \end{array} \qquad [\mathbf{k}_{ff}^{(4)}] = (91.00 \times 10^6) \begin{array}{c} \begin{array}{ccc} 4 & 5 & 6 \end{array} \\ \begin{bmatrix} 0 & 0 & 0 \\ 0 & 0 & 0 \\ 0 & 0 & 1 \end{bmatrix} \begin{array}{c} 4 \\ 5 \\ 6 \end{array} \end{array}$$

$$[\mathbf{k}_{ff}^{(5)}] = (91.00 \times 10^6) \begin{array}{c} \begin{array}{cccccc} 1 & 2 & 3 & 4 & 5 & 6 \end{array} \\ \begin{bmatrix} 0 & 0 & 0 & 0 & 0 & 0 \\ 0 & 1 & 0 & 0 & -1 & 0 \\ 0 & 0 & 0 & 0 & 0 & 0 \\ 0 & 0 & 0 & 0 & 0 & 0 \\ 0 & -1 & 0 & 0 & 1 & 0 \\ 0 & 0 & 0 & 0 & 0 & 0 \end{bmatrix} \begin{array}{c} 1 \\ 2 \\ 3 \\ 4 \\ 5 \\ 6 \end{array} \end{array}$$

Table 11.2.2   Rod element properties for Figure 11.2.7.

Element No.	L (m)	$\frac{AE}{L}$ (N/m)	Node 1	Node 2	l	m	n
		For all rod elements: $E = 70$ GPa, $A = 650$ mm^2					
1	1	45.50×10^6	3	1	1	0	0
2	1	45.50×10^6	4	2	1	0	0
3	0.5	91.00×10^6	5	1	0	0	1
4	0.5	91.00×10^6	6	2	0	0	1
5	0.5	91.00×10^6	1	2	0	1	0

Table 11.2.3   Panel element properties for Figure 11.2.7.

Element No.	Node 1	Node 2	Node 3	Node 4	a (m)	b (m)	(l_1, m_1, n_1)	(l_2, m_2, n_2)
		For all panel elements: $G = 27.13$ GPa, $t = .002$ m; $Gt/4 = 13.5 \times 10^6$ N-m						
6	4	2	6	8	1	0.5	(1,0,0)	(0,0,−1)
7	3	1	2	4	1	0.5	(1,0,0)	(0,1,0)
8	3	1	5	7	1	0.5	(1,0,0)	(0,0,−1)
9	1	5	6	2	0.5	0.5	(0,0,−1)	(0,1,0)

For the shear panels, we turn to Equations 11.2.15 to obtain[2]

$$[\mathbf{k}_{ff}^{(6)}] = \begin{smallmatrix}4\\5\\6\end{smallmatrix} \begin{bmatrix}\mathbf{k}^{(6)}\end{bmatrix} = \begin{smallmatrix}4\\5\\6\end{smallmatrix} \begin{bmatrix}\mathbf{k}_3^{(6)}\end{bmatrix} = (10)^6 \begin{bmatrix} 27.13 & 0 & 13.56 \\ 0 & 0 & 0 \\ 13.56 & 0 & 6.783 \end{bmatrix} \begin{smallmatrix}4\\5\\6\end{smallmatrix}$$

Local dofs 4 5 6, Global dofs 4 5 6, Global dofs 4 5 6 / Global dofs.

$$[\mathbf{k}_{ff}^{(7)}] = \begin{smallmatrix}4\\5\\6\\7\\8\\9\end{smallmatrix} \begin{bmatrix}\mathbf{k}^{(7)}\end{bmatrix} = \begin{bmatrix} [\mathbf{k}_3^{(7)}] & [\mathbf{k}_2^{(7)}]^T \\ [\mathbf{k}_2^{(7)}] & [\mathbf{k}_1^{(7)}] \end{bmatrix} \begin{smallmatrix}1\\2\\3\\4\\5\\6\end{smallmatrix} = (10)^6 \begin{bmatrix} 27.13 & -13.56 & 0 & -27.13 & -13.56 & 0 \\ -13.56 & 6.783 & 0 & 13.56 & 6.783 & 0 \\ 0 & 0 & 0 & 0 & 0 & 0 \\ -27.13 & 13.56 & 0 & 27.13 & 13.56 & 0 \\ -13.56 & 6.783 & 0 & 13.56 & 6.783 & 0 \\ 0 & 0 & 0 & 0 & 0 & 0 \end{bmatrix} \begin{smallmatrix}1\\2\\3\\4\\5\\6\end{smallmatrix}$$

Local dofs 123 456, Global dofs 1 2 3 / 4 5 6. Global dofs 1 2 3 4 5 6.

$$[\mathbf{k}_{ff}^{(8)}] = \begin{smallmatrix}4\\5\\6\end{smallmatrix} \begin{bmatrix}\mathbf{k}^{(8)}\end{bmatrix} = \begin{smallmatrix}1\\2\\3\end{smallmatrix} \begin{bmatrix}\mathbf{k}_3^{(8)}\end{bmatrix} = (10)^6 \begin{bmatrix} 27.13 & 0 & 13.56 \\ 0 & 0 & 0 \\ 13.56 & 0 & 6.783 \end{bmatrix} \begin{smallmatrix}1\\2\\3\end{smallmatrix}$$

Local dofs 1 2 3, Global dofs 1 2 3. Global dofs 1 2 3.

$$[\mathbf{k}_{ff}^{(9)}] = \begin{smallmatrix}1\\2\\3\\10\\11\\12\end{smallmatrix} \begin{bmatrix}\mathbf{k}^{(9)}\end{bmatrix} = \begin{bmatrix} [\mathbf{k}_1^{(9)}] & -[\mathbf{k}_2^{(9)}]^T \\ -[\mathbf{k}_2^{(9)}] & [\mathbf{k}_3^{(9)}] \end{bmatrix} \begin{smallmatrix}1\\2\\3\\4\\5\\6\end{smallmatrix} = (10)^6 \begin{bmatrix} 0 & 0 & 0 & 0 & 0 & 0 \\ 0 & 13.56 & 13.56 & 0 & 13.56 & 13.56 \\ 0 & 13.56 & 13.56 & 0 & -13.56 & -13.56 \\ 0 & 0 & 0 & 0 & 0 & 0 \\ 0 & 13.56 & -13.56 & 0 & 13.56 & 13.56 \\ 0 & 13.56 & -13.56 & 0 & 13.56 & 13.56 \end{bmatrix} \begin{smallmatrix}1\\2\\3\\4\\5\\6\end{smallmatrix}$$

Local dofs 1 2 3 10 11 12, Global dofs 123 / 456. Global dofs 1 2 3 4 5 6.

The global stiffness matrix $[\mathbf{k}_{ff}]$ of the constrained structure is formed in the usual manner to obtain the stiffness equations $[\mathbf{k}_{ff}]\{\mathbf{q}_f\} = \{\mathbf{Q}_f\}$,

$$(10^6) \begin{bmatrix} 99.76 & -13.56 & 13.56 & -27.13 & -13.56 & 0 \\ -13.56 & 111.3 & -13.56 & 13.56 & -70.65 & 13.56 \\ 13.56 & -13.56 & 111.3 & 0 & -13.56 & -13.56 \\ -27.13 & 13.56 & 0 & 99.76 & 13.56 & 13.56 \\ -13.56 & -70.65 & -13.56 & 13.56 & 111.3 & 13.56 \\ 0 & 13.56 & -13.56 & 13.56 & 13.56 & 111.3 \end{bmatrix} \begin{Bmatrix} u_1 \\ v_1 \\ w_1 \\ u_2 \\ v_2 \\ w_2 \end{Bmatrix} = \begin{Bmatrix} 0 \\ -5000 \\ 0 \\ 0 \\ 0 \\ 0 \end{Bmatrix}$$

the solution of which is

$$\{\mathbf{q}_f\} = \lfloor -0.01599 \quad -0.09668 \quad -0.01628 \quad 0.01599 \quad -0.06921 \quad 0.01628 \rfloor^T \text{ (mm)}$$

[2]dof = degree of freedom

The global displacements are projected into the element coordinate systems to find the internal loads. For example, the global displacement vector of rod element 3 is

$$\{\mathbf{q}^{(3)}\} = \lfloor 0 \quad 0 \quad 0 \quad -0.01599 \quad -0.09668 \quad -0.01628 \rfloor^T \text{ (mm)}$$

From Equation 10.2.20b and the direction cosines of the element, we have

$$[\mathbf{\Lambda}^{(3)}] = \begin{bmatrix} 0 & 0 & 1 & 0 & 0 & 0 \\ 0 & 0 & 0 & 0 & 0 & 1 \end{bmatrix}$$

Therefore, the projection operation $\{\bar{\mathbf{q}}^{(3)}\} = [\mathbf{\Lambda}^{(3)}]\{\mathbf{q}^{(3)}\}$ yields

$$\{\bar{\mathbf{q}}^{(3)}\} = \begin{Bmatrix} 0 \\ -0.01628 \end{Bmatrix} \text{ (mm)}$$

Multiplying this displacement vector by the local element stiffness matrix $\{\bar{\mathbf{k}}^{(3)}\}$, yields the element load vector $\{\bar{\mathbf{Q}}^{(3)}\}$, which is

$$\begin{Bmatrix} \bar{X}_1^{(3)} \\ \bar{X}_2^{(3)} \end{Bmatrix} = (91 \times 10^6) \begin{bmatrix} 1 & -1 \\ -1 & 1 \end{bmatrix} (10^{-6}) \begin{Bmatrix} 0 \\ -16.28 \end{Bmatrix} = \begin{Bmatrix} 1481 \\ -1481 \end{Bmatrix} \text{ (N)}$$

The loads in the other stiffeners are found in the same manner.

To illustrate the calculation of the stress in the shear panels, we consider panel element 9, whose global displacement vector is

$$\{\mathbf{q}^{(9)}\} = \lfloor -0.01599 \quad -0.09668 \quad -0.01628 \quad 0 \quad 0 \quad 0 \quad 0 \quad 0 \quad 0 \quad 0.01599 \quad -0.06921 \quad 0.01628 \rfloor^T \text{ (mm)}$$

We find the local element displacement vector by the operation $\{\bar{\mathbf{q}}^{(9)}\} = [\mathbf{\Lambda}^{(9)}]\{\mathbf{q}^{(9)}\}$, where $[\mathbf{\Lambda}^{(9)}]$ is found in Equation 11.2.15a. Using the direction cosines of element 9 from Table 11.2.3, we get

$$\{\bar{\mathbf{q}}^{(9)}\} = \begin{bmatrix} \begin{bmatrix} 0 & 0 & -1 \\ 0 & 1 & 0 \end{bmatrix} & \begin{bmatrix} 0 & 0 & 0 \\ 0 & 0 & 0 \end{bmatrix} & \begin{bmatrix} 0 & 0 & 0 \\ 0 & 0 & 0 \end{bmatrix} & \begin{bmatrix} 0 & 0 & 0 \\ 0 & 0 & 0 \end{bmatrix} \\ \begin{bmatrix} 0 & 0 & 0 \\ 0 & 0 & 0 \end{bmatrix} & \begin{bmatrix} 0 & 0 & -1 \\ 0 & 1 & 0 \end{bmatrix} & \begin{bmatrix} 0 & 0 & 0 \\ 0 & 0 & 0 \end{bmatrix} & \begin{bmatrix} 0 & 0 & 0 \\ 0 & 0 & 0 \end{bmatrix} \\ \begin{bmatrix} 0 & 0 & 0 \\ 0 & 0 & 0 \end{bmatrix} & \begin{bmatrix} 0 & 0 & 0 \\ 0 & 0 & 0 \end{bmatrix} & \begin{bmatrix} 0 & 0 & -1 \\ 0 & 1 & 0 \end{bmatrix} & \begin{bmatrix} 0 & 0 & 0 \\ 0 & 0 & 0 \end{bmatrix} \\ \begin{bmatrix} 0 & 0 & 0 \\ 0 & 0 & 0 \end{bmatrix} & \begin{bmatrix} 0 & 0 & 0 \\ 0 & 0 & 0 \end{bmatrix} & \begin{bmatrix} 0 & 0 & 0 \\ 0 & 0 & 0 \end{bmatrix} & \begin{bmatrix} 0 & 0 & 0 \\ 0 & 0 & 0 \end{bmatrix} \end{bmatrix} \begin{Bmatrix} -0.01599 \\ -0.09668 \\ -0.01628 \\ 0 \\ 0 \\ 0 \\ 0 \\ 0 \\ 0 \\ 0.01599 \\ -0.06921 \\ 0.01628 \end{Bmatrix} = \begin{Bmatrix} 0.01628 \\ -0.09668 \\ 0 \\ 0 \\ 0 \\ 0 \\ -0.01628 \\ -0.06921 \end{Bmatrix} \text{ (mm)}$$

The shear stress in the panel is found from Equation 11.2.7,

$$\tau_{xy} = \frac{G}{abt}[\mathbf{F}]^T \{\bar{\mathbf{q}}\}$$

where $[\mathbf{F}]$ is given by Equation 11.2.6b, with $a = b = 0.5$ m and $t = 0.002$ m,

$$[\mathbf{F}]^T = (500 \times 10^{-6})\lfloor -1 \quad -1 \quad \vdots \quad -1 \quad 1 \quad \vdots \quad 1 \quad 1 \quad \vdots \quad 1 \quad -1 \rfloor$$

so that

$$\tau_{xy} = \frac{27.13 \times 10^9}{(0.5)(0.5)(0.002)}(500 \times 10^{-6})\lfloor -1 \quad -1 \quad \vdots \quad -1 \quad 1 \quad \vdots \quad 1 \quad 1 \quad \vdots \quad 1 \quad -1 \rfloor \begin{Bmatrix} 16.28 \times 10^{-6} \\ -96.68 \times 10^{-6} \\ \hdashline 0 \\ 0 \\ \hdashline 0 \\ 0 \\ \hdashline -16.28 \times 10^{-6} \\ -69.21 \times 10^{-6} \end{Bmatrix} = 3.617 \text{ MPa}$$

The shear flow q equals $\tau_{xy}t$, or

$$q = 7.236 \text{ kN/m}$$

since $t = 0.002$ m.

The other shear panels are treated in a similar fashion. Figure 11.2.8 shows the calculated internal loads for all of the rods and shear panels of the box structure.

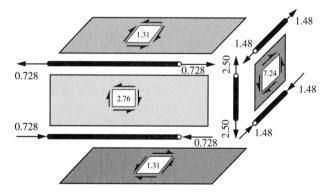

Figure 11.2.8   Axial loads (kN) and shear flows (kN/m) in the structure of Figure 11.2.7.

Example 11.2.3   Use the displacement method to calculate the shear flow in each of the panels of the two-cell, two-bay cantilevered box beam shown in Figure 11.2.9. The structure is modeled by 18 shear panels and 26 rod elements. A 1000 lb load is applied as shown.

This statically indeterminate structure has 36 unconstrained degrees of freedom. Therefore, a hand computation in the manner of the previous examples is impractical. Using a digital computer to carry out the calculations yields the shear flows summarized in Figure 11.2.10 and the stringer (rod) loads plotted in Figure 11.2.11.

It is interesting to note that engineering beam theory (section 8.8) yields the shear flows shown in Figure 11.2.12. The fairly good agreement with the forward bay shear flows in Figure 11.2.10 is due to the rigidity of the cap rib, which in turn is due to the surrounding 1 in.² rod elements. Also, engineering beam theory requires that there be no deformation in the plane of the cross section, and this implies the presence of ribs that are completely rigid in their plane. Engineering beam theory

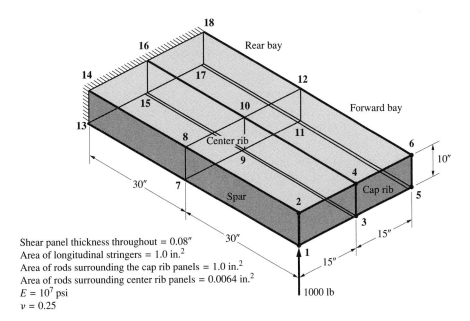

Shear panel thickness throughout = 0.08″
Area of longitudinal stringers = 1.0 in.^{2}
Area of rods surrounding the cap rib panels = 1.0 in.^{2}
Area of rods surrounding center rib panels = 0.0064 in.^{2}
$E = 10^7$ psi
$\nu = 0.25$

1000 lb

Figure 11.2.9   Cantilevered two-cell two-bay box beam.

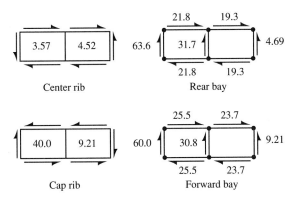

Figure 11.2.10   Shear flows (lb/in) in the ribs and around the bays of the box beam in Figure 11.2.9.

The views are inboard, toward the wall.

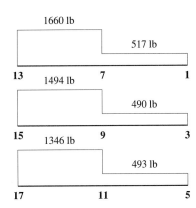

Figure 11.2.11   Computed tensile load distribution in the three bottom stringers.

Node numbers are those of Figure 11.2.9.

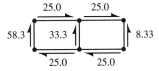

Figure 11.2.12   Shear flows computed by engineering beam theory.

accounts for the presence of ribs only where external shear loads are applied to the structure (section 5.2). Accordingly, the shear flows of Figure 11.2.12 extend throughout the entire structure. However, near the wall, these stresses are modified by the complete restraint against bending, twisting, and warping at that end of the box beam. The effect of these restraints appears in the difference between the forward and rear bay shear flows in Figure 11.2.10.

11.3 CONSTANT-STRAIN TRIANGLE

Figure 11.3.1 shows a plane, triangular, finite element used to model plane stress or plane strain problems. It has two degrees of freedom at each of its three corner nodes, labeled 1, 2, and 3, counterclockwise around the element. The two components of displacement $\{\mathbf{u}\} = \lfloor u \ \ v \rfloor^T$, at an interior point are interpolated from the nodal displacement vector $\{\mathbf{q}\}$ according to Equation 9.12.1,

$$\{\mathbf{u}\} = [\mathbf{N}]\{\mathbf{q}\}$$

where the element displacement vector is

$$\{\mathbf{q}\} = \lfloor u_1 \quad v_1 \mid u_2 \quad v_2 \mid u_3 \quad v_3 \rfloor^T$$

and the matrix of shape functions is

$$[\mathbf{N}] = \begin{bmatrix} N_1 & 0 & N_2 & 0 & N_3 & 0 \\ 0 & N_1 & 0 & N_2 & 0 & N_3 \end{bmatrix} \tag{11.3.1}$$

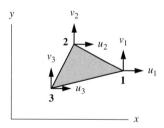

Figure 11.3.1 Constant-strain triangle.

Thus,

$$\begin{aligned} u &= N_1(x, y)u_1 + N_2(x, y)u_2 + N_3(x, y)u_3 \\ v &= N_1(x, y)v_1 + N_2(x, y)v_2 + N_3(x, y)v_3 \end{aligned} \tag{11.3.2}$$

The shape functions N_1, N_2, and N_3 are assumed to be linear functions of x and y, as follows;

$$\begin{aligned} N_1 &= a_1 + b_1 x + c_1 y \\ N_2 &= a_2 + b_2 x + c_2 y \\ N_3 &= a_3 + b_3 x + c_3 y \end{aligned} \tag{11.3.3}$$

As illustrated in Figure 11.3.2, if two elements m and n are attached at nodes i and j, they share a common boundary ij. Since the displacements u_i and v_i of node i are common to both elements, as are the displacements u_j and v_j at node j, and since the displacements in both elements vary linearly along ij, the displacements u and v of any point a on the common boundary are uniquely determined by Equation 11.3.2 and are common to both elements; they are compatible. That is, point a, which lies on the common boundary *before* deformation, does not end up in two different places a' and a'' *after* deformation, as suggested in Figure 11.3.3, where the displacements the two adjacent elements are incompatible on the common edge. In such a case, gaps, overlaps, or other kinds of physically (if not mathematically) unacceptable discontinuities can occur. Displacement compatibility between elements is generally enforced in finite element models.

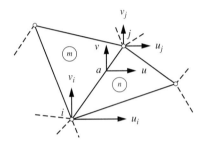

Figure 11.3.2   Displacement compatibility along a common boundary.

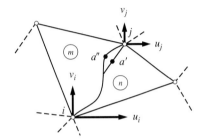

Figure 11.3.3   Displacement incompatibility between two connected elements.

For Equation 11.3.2 to yield $u = u_1$ and $v = v_1$ when $x = x_1$ and $y = y_1$, N_1 must equal 1 at node 1 and must vanish at the other two nodes. Using Equation $11.3.3_1$ to evaluate N_1 at the three nodes of the triangle, we get

$$a_1 + b_1 x_1 + c_1 y_1 = 1$$
$$a_1 + b_1 x_2 + c_1 y_2 = 0$$
$$a_1 + b_1 x_3 + c_1 y_3 = 0$$

or

$$\begin{bmatrix} 1 & x_1 & y_1 \\ 1 & x_2 & y_2 \\ 1 & x_3 & y_3 \end{bmatrix} \begin{Bmatrix} a_1 \\ b_1 \\ c_1 \end{Bmatrix} = \begin{Bmatrix} 1 \\ 0 \\ 0 \end{Bmatrix} \qquad [11.3.4]$$

Similarly, since $N_2 = 1$ at node 2 and vanishes at nodes 1 and 3, we have

$$\begin{bmatrix} 1 & x_1 & y_1 \\ 1 & x_2 & y_2 \\ 1 & x_3 & y_3 \end{bmatrix} \begin{Bmatrix} a_2 \\ b_2 \\ c_2 \end{Bmatrix} = \begin{Bmatrix} 0 \\ 1 \\ 0 \end{Bmatrix} \qquad [11.3.5]$$

and for N_3 we obtain

$$\begin{bmatrix} 1 & x_1 & y_1 \\ 1 & x_2 & y_2 \\ 1 & x_3 & y_3 \end{bmatrix} \begin{Bmatrix} a_3 \\ b_3 \\ c_3 \end{Bmatrix} = \begin{Bmatrix} 0 \\ 0 \\ 1 \end{Bmatrix} \qquad [11.3.6]$$

Solving these three equations for the unknown coefficients yields

$$\left\{\begin{array}{c} a_1 \\ b_1 \\ c_1 \end{array}\right\} = \left[\begin{array}{ccc} 1 & x_1 & y_1 \\ 1 & x_2 & y_2 \\ 1 & x_3 & y_3 \end{array}\right]^{-1} \left\{\begin{array}{c} 1 \\ 0 \\ 0 \end{array}\right\} \qquad \left\{\begin{array}{c} a_2 \\ b_2 \\ c_2 \end{array}\right\} = \left[\begin{array}{ccc} 1 & x_1 & y_1 \\ 1 & x_2 & y_2 \\ 1 & x_3 & y_3 \end{array}\right]^{-1} \left\{\begin{array}{c} 0 \\ 1 \\ 0 \end{array}\right\} \qquad \left\{\begin{array}{c} a_3 \\ b_3 \\ c_3 \end{array}\right\} = \left[\begin{array}{ccc} 1 & x_1 & y_1 \\ 1 & x_2 & y_2 \\ 1 & x_3 & y_3 \end{array}\right]^{-1} \left\{\begin{array}{c} 0 \\ 0 \\ 1 \end{array}\right\} \qquad \text{[11.3.7]}$$

Computing the matrix inverse using Equation 9.5.11 leads to

$$\left[\begin{array}{ccc} 1 & x_1 & y_1 \\ 1 & x_2 & y_2 \\ 1 & x_3 & y_3 \end{array}\right]^{-1} = \frac{1}{x_1 y_2 + x_2 y_3 + x_3 y_1 - x_1 y_3 - x_2 y_1 - x_3 y_2} \left[\begin{array}{ccc} x_2 y_3 - x_3 y_2 & x_3 y_1 - x_1 y_3 & x_1 y_2 - x_2 y_1 \\ y_2 - y_3 & y_3 - y_1 & y_1 - y_2 \\ -x_2 + x_3 & -x_3 + x_1 & -x_1 + x_2 \end{array}\right] \qquad \text{[11.3.8]}$$

Before proceeding, consider Figure 11.3.4, which shows two position vectors, $\mathbf{r}_{12}$ and $\mathbf{r}_{13}$, drawn from node 1 towards nodes 2 and 3, respectively. If A is the area of the triangle, then, by definition, the magnitude of the cross product of $\mathbf{r}_{12}$ and $\mathbf{r}_{13}$ is $2A$. Computing the cross product, we get

$$\mathbf{r}_{12} \times \mathbf{r}_{13} = [(x_2 - x_1)\mathbf{i} + (y_2 - y_1)\mathbf{j}] \times [(x_3 - x_1)\mathbf{i} + (y_3 - y_1)\mathbf{j}] = [(x_2 - x_1)(y_3 - y_1) - (y_2 - y_1)(x_3 - x_1)]\mathbf{k}$$

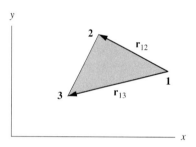

Figure 11.3.4 Position vectors used to compute the area of the triangle.

Thus,

$$2A = (x_2 - x_1)(y_3 - y_1) - (y_2 - y_1)(x_3 - x_1) = (x_2 y_3 - x_2 y_1 - x_1 y_3 + x_1 y_1) - (y_2 x_3 - y_2 x_1 - y_1 x_3 + y_1 x_1)$$

Canceling terms leaves

$$2A = x_1 y_2 + x_2 y_3 + x_3 y_1 - x_1 y_3 - x_2 y_1 - x_3 y_2 \qquad \text{[11.3.9]}$$

which means that Equation 11.3.8 can be written

$$\left[\begin{array}{ccc} 1 & x_1 & y_1 \\ 1 & x_2 & y_2 \\ 1 & x_3 & y_3 \end{array}\right]^{-1} = \frac{1}{2A} \left[\begin{array}{ccc} x_2 y_3 - x_3 y_2 & x_3 y_1 - x_1 y_3 & x_1 y_2 - x_2 y_1 \\ y_2 - y_3 & y_3 - y_1 & y_1 - y_2 \\ -x_2 + x_3 & -x_3 + x_1 & -x_1 + x_2 \end{array}\right]$$

Substituting this into Equation 11.3.7, we find

$$\begin{Bmatrix} a_1 \\ b_1 \\ c_1 \end{Bmatrix} = \frac{1}{2A} \begin{Bmatrix} x_2 y_3 - x_3 y_2 \\ y_2 - y_3 \\ -x_2 + x_3 \end{Bmatrix} \qquad \begin{Bmatrix} a_2 \\ b_2 \\ c_2 \end{Bmatrix} = \frac{1}{2A} \begin{Bmatrix} x_3 y_1 - x_1 y_3 \\ y_3 - y_1 \\ -x_3 + x_1 \end{Bmatrix} \qquad \begin{Bmatrix} a_3 \\ b_3 \\ c_3 \end{Bmatrix} = \frac{1}{2A} \begin{Bmatrix} x_1 y_2 - x_2 y_1 \\ y_1 - y_2 \\ -x_1 + x_2 \end{Bmatrix} \quad \text{[11.3.10]}$$

Therefore,

$$N_1 = \frac{1}{2A}(\alpha_1 + \beta_1 x + \gamma_1 y)$$

$$N_2 = \frac{1}{2A}(\alpha_2 + \beta_2 x + \gamma_2 y) \qquad \text{[11.3.11]}$$

$$N_3 = \frac{1}{2A}(\alpha_3 + \beta_3 x + \gamma_3 y)$$

where the coefficients of the shape functions N_i ($i = 1, 2, 3$) are given by

$$\begin{array}{lll} \alpha_1 = x_2 y_3 - x_3 y_2 & \beta_1 = y_2 - y_3 & \gamma_1 = -x_2 + x_3 \\ \alpha_2 = x_3 y_1 - x_1 y_3 & \beta_2 = y_3 - y_1 & \gamma_2 = -x_3 + x_1 \\ \alpha_3 = x_1 y_2 - x_2 y_1 & \beta_3 = y_1 - y_2 & \gamma_3 = -x_1 + x_2 \end{array} \qquad \text{[11.3.12]}$$

Equation 9.12.4 is used to compute the strain-displacement matrix $[\mathbf{B}]$, which is

$$[\mathbf{B}] = \begin{bmatrix} \dfrac{\partial N_1}{\partial x} & 0 & \dfrac{\partial N_2}{\partial x} & 0 & \dfrac{\partial N_3}{\partial x} & 0 \\ 0 & \dfrac{\partial N_2}{\partial y} & 0 & \dfrac{\partial N_2}{\partial y} & 0 & \dfrac{\partial N_2}{\partial y} \\ \dfrac{\partial N_1}{\partial y} & \dfrac{\partial N_1}{\partial x} & \dfrac{\partial N_2}{\partial y} & \dfrac{\partial N_2}{\partial x} & \dfrac{\partial N_3}{\partial y} & \dfrac{\partial N_3}{\partial x} \end{bmatrix}$$

Substituting Equation 11.3.11 yields

$$[\mathbf{B}] = \left(\frac{1}{2A}\right) \begin{bmatrix} \beta_1 & 0 & \beta_2 & 0 & \beta_3 & 0 \\ 0 & \gamma_1 & 0 & \gamma_2 & 0 & \gamma_3 \\ \gamma_1 & \beta_1 & \gamma_2 & \beta_2 & \gamma_3 & \beta_3 \end{bmatrix} \qquad \text{[11.3.13]}$$

Since this matrix does not depend on x and y, the strains ε_x, ε_y, and γ_{xy} are uniform throughout the element, which is why the element is called the *constant-strain* triangle.

The stiffness matrix of the triangle is found using Equation 9.12.16:

$$[\mathbf{k}] = \iiint_V [\mathbf{B}]^T [\mathbf{E}] [\mathbf{B}] \, dV$$

where $[\mathbf{E}]$ is the material stiffness matrix. For isotropic elastic materials in plane stress, $[\mathbf{E}]$ given by Equation 9.11.9$_2$,

$$[\mathbf{E}] = \begin{bmatrix} E' & vE' & 0 \\ vE' & E' & 0 \\ 0 & 0 & G \end{bmatrix} \qquad E' = \frac{E}{1 - v^2} \qquad G = \frac{E}{2(1 + v)} \qquad \text{[11.3.14]}$$

Since both $[\mathbf{B}]$ and $[\mathbf{E}]$ are constant, the integral simplifies to

$$[\mathbf{k}] = [\mathbf{B}]^T\, [\mathbf{E}]\, [\mathbf{B}] \left(\iiint_V dV \right)$$

The volume of the element is its area A times the thickness t, which we assume is uniform throughout the element. Therefore,

$$[\mathbf{k}] = (At) \left(\frac{1}{2A}\right) \begin{bmatrix} \beta_1 & 0 & \gamma_1 \\ 0 & \gamma_1 & \beta_1 \\ \beta_2 & 0 & \gamma_2 \\ 0 & \gamma_2 & \beta_2 \\ \beta_3 & 0 & \gamma_3 \\ 0 & \gamma_3 & \beta_3 \end{bmatrix} \begin{bmatrix} E' & vE' & 0 \\ vE' & E' & 0 \\ 0 & 0 & G \end{bmatrix} \left(\frac{1}{2A}\right) \begin{bmatrix} \beta_1 & 0 & \beta_2 & 0 & \beta_3 & 0 \\ 0 & \gamma_1 & 0 & \gamma_2 & 0 & \gamma_3 \\ \gamma_1 & \beta_1 & \gamma_2 & \beta_2 & \gamma_3 & \beta_3 \end{bmatrix}$$

so that in the global frame, the 6 by 6 stiffness matrix of the constant-strain triangle can be written

$$[\mathbf{k}] = \frac{t}{4A} \begin{bmatrix} E'\beta_1^2 + G\gamma_1^2 & vE'\beta_1\gamma_1 + G\beta_1\gamma_1 & E'\beta_1\beta_2 + G\gamma_1\gamma_2 & vE'\beta_1\gamma_2 + G\beta_2\gamma_1 & E'\beta_1\beta_3 + G\gamma_1\gamma_3 & vE'\beta_1\gamma_3 + G\beta_3\gamma_1 \\ vE'\beta_1\gamma_1 + G\beta_1\gamma_1 & E'\gamma_1^2 + G\beta_1^2 & vE'\beta_2\gamma_1 + G\beta_1\gamma_2 & E'\gamma_1\gamma_2 + G\beta_1\beta_2 & vE'\beta_3\gamma_1 + G\beta_1\gamma_3 & E'\gamma_1\gamma_3 + G\beta_1\beta_3 \\ E'\beta_1\beta_2 + G\gamma_1\gamma_2 & vE'\beta_2\gamma_1 + G\beta_1\gamma_2 & E'\beta_2^2 + G\gamma_2^2 & vE'\beta_2\gamma_2 + G\beta_2\gamma_2 & E'\beta_2\beta_3 + G\gamma_2\gamma_3 & vE'\beta_2\gamma_3 + G\beta_3\gamma_2 \\ vE'\beta_1\gamma_2 + G\beta_2\gamma_1 & E'\gamma_1\gamma_2 + G\beta_1\beta_2 & vE'\beta_2\gamma_2 + G\beta_2\gamma_2 & E'\gamma_2^2 + G\beta_2^2 & vE'\beta_3\gamma_2 + G\beta_2\gamma_3 & E'\gamma_2\gamma_3 + G\beta_2\beta_3 \\ E'\beta_1\beta_3 + G\gamma_1\gamma_3 & vE'\beta_3\gamma_1 + G\beta_1\gamma_3 & E'\beta_2\beta_3 + G\gamma_2\gamma_3 & vE'\beta_3\gamma_2 + G\beta_2\gamma_3 & E'\beta_3^2 + G\gamma_3^2 & vE'\beta_3\gamma_3 + G\beta_3\gamma_3 \\ vE'\beta_1\gamma_3 + G\beta_3\gamma_1 & E'\gamma_1\gamma_3 + G\beta_1\beta_3 & vE'\beta_2\gamma_3 + G\beta_3\gamma_2 & E'\gamma_2\gamma_3 + G\beta_2\beta_3 & vE'\beta_3\gamma_3 + G\beta_3\gamma_3 & E'\gamma_3^2 + G\beta_3^2 \end{bmatrix}$$

$$[11.3.15]$$

Figure 11.3.5 shows some elements along the edge of a plane stress region which is subject to a varying surface traction $\mathbf{T}^{(n)}$. To find the point-load vector $\{\mathbf{Q}_{eq}^{(e)}\}$ equivalent to the surface traction on element e, we refer to Equation 9.12.13 to get

$$\left\{\mathbf{Q}_{eq}^{(e)}\right\} = \iint_{S^{(e)}} [\mathbf{N}]^T \left\{\mathbf{T}^{(n)}\right\} dS$$

where $S^{(e)}$ is the surface of the element. Thus,

$$\left\{\mathbf{Q}_{eq}^{(e)}\right\} = \int_{edge\ ij} [\mathbf{N}]^T \left\{\mathbf{T}^{(n)}\right\} t\,ds + \int_{edge\ ik} [\mathbf{N}]^T \left\{\mathbf{T}^{(n)}\right\} t\,ds + \int_{edge\ kl} [\mathbf{N}]^T \left\{\mathbf{T}^{(n)}\right\} t\,ds$$

$$+ \iint_{front\ face} [\mathbf{N}]^T \left\{\mathbf{T}^{(n)}\right\} dx\,dy + \iint_{back\ face} [\mathbf{N}]^T \left\{\mathbf{T}^{(n)}\right\} dx\,dy$$

The front and back faces of the triangular element are free of traction, by the plane stress assumption; only edge ij is shown with a nonzero in-plane traction. Therefore, all of the integrals except the first one vanish, yielding

$$\left\{\mathbf{Q}_{eq}^{(e)}\right\} = \int_0^l \begin{bmatrix} N_i & 0 \\ 0 & N_i \\ N_j & 0 \\ 0 & N_j \\ N_k & 0 \\ 0 & N_k \end{bmatrix} \left\{\begin{matrix} T_x^{(n)} \\ T_y^{(n)} \end{matrix}\right\} t\,ds = \int_0^l \begin{bmatrix} N_i & 0 \\ 0 & N_i \\ N_j & 0 \\ 0 & N_j \\ 0 & 0 \\ 0 & 0 \end{bmatrix} \left\{\begin{matrix} T_x^{(n)} \\ T_y^{(n)} \end{matrix}\right\} t\,ds \qquad [11.3.16]$$

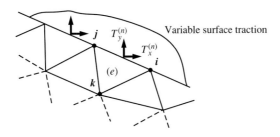

Figure 11.3.5   A boundary exposed to surface traction.

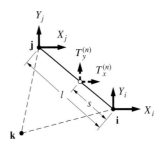

Figure 11.3.6   Point loads that are work-equivalent to the surface traction.

where l is the length of edge ij. Notice that $N_k = 0$ along edge ij, so that point loads are apportioned only to those nodes that lie on the edge to which the surface traction distribution is applied, as illustrated in Figure 11.3.6. For the equivalent point loads at nodes i and j, we get

$$\begin{Bmatrix} X_i \\ Y_i \\ X_j \\ Y_j \end{Bmatrix} = \int_0^l \begin{Bmatrix} N_i(s)T_x^{(n)}(s) \\ N_i(s)T_y^{(n)}(s) \\ N_j(s)T_x^{(n)}(s) \\ N_j(s)T_y^{(n)}(s) \end{Bmatrix} t\,ds \qquad [11.3.17]$$

Along edge ij,

$$x(s) = (1 - \xi)x_i + \xi x_j$$

and

$$y(s) = (1 - \xi)y_i + \xi y_j$$

where $\xi = s/l$. Therefore,

$$\begin{aligned} N_i(\xi) &= \frac{1}{2A}[\alpha_i + \beta_i x(\xi) + \gamma_i y(\xi)] \\ &= \frac{1}{2A}[x_j y_k - x_k y_j + (y_j - y_k)x(\xi) + (-x_j + x_k)y(\xi)] \\ &= \frac{1}{2A}\{x_j y_k - x_k y_j + (y_j - y_k)[(1 - \xi)x_i + \xi x_j] + (-x_j + x_k)[(1 - \xi)y_i + \xi y_j]\} \\ &= \frac{1}{2A}[x_i y_j + x_j y_k + x_k y_i - x_i y_k - x_j y_i - x_k y_j + (x_i y_j + x_j y_k + x_k y_i - x_i y_k - x_j y_i - x_k y)\xi] \\ &= \frac{1}{2A}[2A - 2A\xi] \end{aligned}$$

so that on edge ij, we have

$$N_i = 1 - \xi \qquad [11.3.18a]$$

Similarly,

$$
\begin{aligned}
N_j &= \frac{1}{2A}[\alpha_j + \beta_j x(\xi) + \gamma_j y(\xi)] \\
&= \frac{1}{2A}[x_k y_i - x_i y_k + (y_k - y_i)x(\xi) + (-x_k + x_i)y(\xi)] \\
&= \frac{1}{2A}\{x_k y_i - x_i y_k + (y_k - y_i)[(1-\xi)x_i + \xi x_j] + (-x_k + x_i)[(1-\xi)y_i + \xi y_j]\} \\
&= \frac{1}{2A}[(x_i y_j + x_j y_k + x_k y_i - x_i y_k - x_j y_i - x_k y_j)\xi] \\
&= \frac{1}{2A}[(2A)\xi]
\end{aligned}
$$

Therefore, along edge ij, we have

$$
N_j = \xi \tag{11.3.18b}
$$

Let us assume that the surface tractions vary linearly along the edge, that is, that

$$
\begin{aligned}
T_x^{(n)} &= (1-\xi)\,T_{xi}^{(n)} + \xi T_{xj}^{(n)} \\
T_y^{(n)} &= (1-\xi)\,T_{yi}^{(n)} + \xi T_{yj}^{(n)}
\end{aligned} \tag{11.3.19}
$$

Then, from Equations 11.3.17, 11.3.18, and 11.3.19, we get

$$
\begin{Bmatrix} X_i \\ Y_i \\ X_j \\ Y_j \end{Bmatrix} = \int_0^l \begin{Bmatrix} N_i(s)T_x^{(n)}(s) \\ N_i(s)T_y^{(n)}(s) \\ N_j(s)T_x^{(n)}(s) \\ N_j(s)T_y^{(n)}(s) \end{Bmatrix} t\,ds = tl \int_0^1 \begin{Bmatrix} N_i(\xi)T_x^{(n)}(\xi) \\ N_i(\xi)T_y^{(n)}(\xi) \\ N_j(\xi)T_x^{(n)}(\xi) \\ N_j(\xi)T_y^{(n)}(\xi) \end{Bmatrix} d\xi = tl \int_0^1 \begin{Bmatrix} (1-\xi)\left[(1-\xi)\,T_{xi}^{(n)} + \xi T_{xj}^{(n)}\right] \\ (1-\xi)\left[(1-\xi)\,T_{yi}^{(n)} + \xi T_{yj}^{(n)}\right] \\ \xi\left[(1-\xi)\,T_{xi}^{(n)} + \xi T_{xj}^{(n)}\right] \\ \xi\left[(1-\xi)\,T_{yi}^{(n)} + \xi T_{yj}^{(n)}\right] \end{Bmatrix} d\xi
$$

$$
= tl \int_0^1 \begin{Bmatrix} (1-\xi)^2\, T_{xi}^{(n)} + \left(\xi - \xi^2\right) T_{xj}^{(n)} \\ (1-\xi)^2\, T_{yi}^{(n)} + \left(\xi - \xi^2\right) T_{yj}^{(n)} \\ \left(\xi - \xi^2\right) T_{xi}^{(n)} + \xi^2 T_{xj}^{(n)} \\ \left(\xi - \xi^2\right) T_{yi}^{(n)} + \xi^2 T_{yj}^{(n)} \end{Bmatrix} d\xi
$$

Carrying out the integrals shows that

$$
\begin{Bmatrix} X_i \\ Y_i \\ X_j \\ Y_j \end{Bmatrix} = tl \begin{Bmatrix} \frac{1}{3}T_{xi}^{(n)} + \frac{1}{6}T_{xj}^{(n)} \\ \frac{1}{3}T_{yi}^{(n)} + \frac{1}{6}T_{yj}^{(n)} \\ \frac{1}{6}T_{xi}^{(n)} + \frac{1}{3}T_{xj}^{(n)} \\ \frac{1}{6}T_{yi}^{(n)} + \frac{1}{3}T_{yj}^{(n)} \end{Bmatrix} \qquad (\text{linearly} - \text{varying surface traction}) \tag{11.3.20}
$$

As expected, the equivalent loads depend only on the length and thickness of the boundary segment and are independent of the geometry of the element to which the traction is applied.

If body forces exist, then the element equivalent point loads are found by means of Equation 11.3.13,

$$
\left\{Q_{\text{eq}}^{(e)}\right\} = \iiint\limits_{V^{(e)}} [N]^T \{b\}\, dV
$$

where $V^{(e)}$ is the volume of the element. Assuming, as before, that the thickness of the element is uniform, we get

$$\{Q_{eq}^{(e)}\} = \iint\limits_{A} \begin{bmatrix} N_1 & 0 \\ 0 & N_1 \\ N_2 & 0 \\ 0 & N_2 \\ N_3 & 0 \\ 0 & N_3 \end{bmatrix} \begin{Bmatrix} b_x \\ b_y \end{Bmatrix} t\,dA = t \iint\limits_{A} \begin{Bmatrix} N_1 b_x \\ N_1 b_y \\ N_2 b_x \\ N_2 b_y \\ N_3 b_x \\ N_3 b_y \end{Bmatrix} dA \qquad \textbf{[11.3.21]}$$

If the body force vector is constant throughout the element, then

$$\{Q_{eq}^{(e)}\} = t \begin{Bmatrix} b_x \iint\limits_{A} N_1 dA \\ b_y \iint\limits_{A} N_1 dA \\ b_x \iint\limits_{A} N_2 dA \\ b_y \iint\limits_{A} N_2 dA \\ b_x \iint\limits_{A} N_3 dA \\ b_y \iint\limits_{A} N_3 dA \end{Bmatrix} \qquad \textbf{[11.3.22]}$$

Also,

$$\iint\limits_{A} N_1 dA = \iint\limits_{A} \frac{1}{2A}(\alpha_1 + \beta_1 x + \gamma_1 y)\,dA = \frac{\alpha_1}{2A}\iint\limits_{A} dA + \frac{\beta_1}{2A}\iint\limits_{A} x\,dA + \frac{\gamma_1}{2A}\iint\limits_{A} y\,dA = \frac{1}{2}(\alpha_1 + \beta_1 \bar{x} + \gamma_1 \bar{y})$$

where $\bar{x}$ and $\bar{y}$ are the coordinates of the triangle's centroid. Thus, we have

$$\iint\limits_{A} N_1 dA = \frac{1}{2}\left[x_2 y_3 - x_3 y_2 + (y_2 - y_3) \times \frac{1}{3}(x_1 + x_2 + x_3) + (-x_2 + x_3) \times \frac{1}{3}(y_1 + y_2 + y_3) \right]$$

$$= \frac{1}{6}(x_1 y_2 + x_2 y_3 + x_3 y_1 - x_1 y_3 - x_2 y_1 - x_3 y_2)$$

$$= \frac{1}{6}(2A)$$

$$= \frac{1}{3}A$$

The integrals of the other two shape functions yield the same result, which means that

$$\iint\limits_{A} N_1 dA = \iint\limits_{A} N_2 dA = \iint\limits_{A} N_3 dA = \frac{1}{3}A \qquad \textbf{[11.3.23]}$$

Therefore, Equation 11.3.22 becomes

$$\{Q_{eq}^{(e)}\} = \frac{At}{3}\begin{Bmatrix} b_x \\ b_y \\ b_x \\ b_y \\ b_x \\ b_y \end{Bmatrix} \qquad \text{(uniform body-force density)} \qquad \textbf{[11.3.24]}$$

Since the body force resultants in the x and y directions are $b_x \times At$ and $b_y \times At$, Equation 11.3.24 states that one-third of the body force resultant in each direction is apportioned to each of the element's nodes.

As before, the element equivalent point-load vector due to thermal strain is found from Equation 9.12.13,

$$\left\{Q_{eq}^{(e)}\right\} = \iiint_{V^{(e)}} [B]^T [E] \{\boldsymbol{\varepsilon}_T\} dV \tag{11.3.25}$$

For plane stress, an isotropic linearly elastic material, and a uniformly thick element, this becomes

$$\left\{Q_{eq}^{(e)}\right\} = \frac{t}{2A} \iint_A \begin{bmatrix} \beta_1 & 0 & \gamma_1 \\ 0 & \gamma_1 & \beta_1 \\ \beta_2 & 0 & \gamma_2 \\ 0 & \gamma_2 & \beta_2 \\ \beta_3 & 0 & \gamma_3 \\ 0 & \gamma_3 & \beta_3 \end{bmatrix} \begin{bmatrix} E' & \nu E' & 0 \\ \nu E' & E' & 0 \\ 0 & 0 & G \end{bmatrix} \left\{ \begin{array}{c} \alpha \Delta T \\ \alpha \Delta T \\ 0 \end{array} \right\} dA$$

$$= \frac{t}{2A} \iint_A \begin{bmatrix} \beta_1 & 0 & \gamma_1 \\ 0 & \gamma_1 & \beta_1 \\ \beta_2 & 0 & \gamma_2 \\ 0 & \gamma_2 & \beta_2 \\ \beta_3 & 0 & \gamma_3 \\ 0 & \gamma_3 & \beta_3 \end{bmatrix} \left\{ \begin{array}{c} E'(1+\nu)\alpha T \\ E'(1+\nu)\alpha T \\ 0 \end{array} \right\} dA$$

$$= \frac{E\alpha t}{2A(1-\nu)} \iint_A \begin{bmatrix} \beta_1 & 0 & \gamma_1 \\ 0 & \gamma_1 & \beta_1 \\ \beta_2 & 0 & \gamma_2 \\ 0 & \gamma_2 & \beta_2 \\ \beta_3 & 0 & \gamma_3 \\ 0 & \gamma_3 & \beta_3 \end{bmatrix} \left\{ \begin{array}{c} 1 \\ 1 \\ 0 \end{array} \right\} T dA$$

so that

$$\left\{Q_{eq}^{(e)}\right\} = \frac{E\alpha t}{2A(1-\nu)} \left\{ \begin{array}{c} \beta_1 \\ \gamma_1 \\ \beta_2 \\ \gamma_2 \\ \beta_3 \\ \gamma_3 \end{array} \right\} \iint_A T dA \tag{11.3.26}$$

Let us assume that the temperature change T is a linear function of position in the element, that is,

$$T = N_1 T_1 + N_2 T_2 + N_3 T_3$$

Then, we have

$$\iint_A T dA = T_1 \iint_A N_1 dA + T_2 \iint_A N_2 dA + T_3 \iint_A N_3 dA$$

From Equation 11.3.23, we obtain

$$\iint\limits_{A} T\,dA = A\left(\frac{T_1 + T_2 + T_3}{3}\right) = AT_{\text{avg}}$$

so that the element equivalent point-load vector due to thermal strain becomes

$$\left\{Q^{(e)}_{\text{eq}}\right\} = \frac{E\alpha T_{\text{avg}}t}{2(1-v)}\begin{Bmatrix}\beta_1\\\gamma_1\\\beta_2\\\gamma_2\\\beta_3\\\gamma_3\end{Bmatrix} \qquad \text{(linear temperature change over the element)} \qquad \textbf{[11.3.27]}$$

The thermal strain equivalent-load vector is strongly dependent on the element geometry.

Example 11.3.1[3] Calculate the stresses in the plane stress structure in Figure 11.3.7 for the two load cases shown.

We can take advantage of symmetry and model only one quadrant of the region, if we constrain it properly, as shown in Figure 11.3.8. For simplicity, we will use two constant-strain triangles, the minimum number required to grid the quadrant. Table 11.3.1 shows the correspondence among the element and global node numbers.

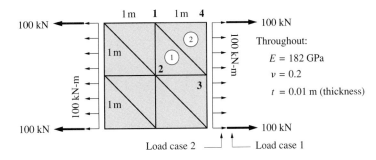

Figure 11.3.7   Coarse finite element grid of a square sheet with two sets of applied loads.

Table 11.3.1   Element connectivities in Figure 11.3.8.

Element	Node 1	Node 2	Node 3
1	1	2	3
2	1	3	4

The free and constrained degree of freedom groups are f: 2, 5, 7, 8 and s: 1, 3, 4, 6. For both elements, we know that

$$E' = \frac{E}{1-v^2} = 200\text{ GPa} \qquad G = \frac{E}{2(1+v)} = 70\text{ GPa} \qquad A = 0.5\text{ m}^2 \qquad \textbf{[a]}$$

From Equation 11.3.12, we find for element 1 that

---

[3]D.J. Dawe, *Matrix and Finite Element Displacement Analysis of Structures*, Section 8.3, Oxford, Clarendon Press, 1984.

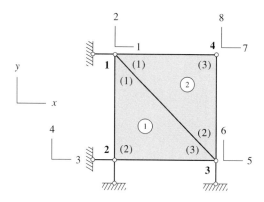

Figure 11.3.8   Upper right quadrant of the grid in Figure 11.3.7.

Local element node numbers are shown in parentheses.

$$\beta_1 = y_2 - y_3 = 0 \qquad \beta_2 = y_3 - y_1 = -1 \text{ m} \qquad \beta_3 = y_1 - y_2 = 1 \text{ m}$$

$$\gamma_1 = -x_2 + x_3 = 0 \qquad \gamma_2 = -x_3 + x_1 = -1 \text{ m} \qquad \gamma_3 = -x_1 + x_2 = 0$$

[b]

Applying the constraints at the element level, we extract the unconstrained components of the stiffness matrix, Equation 11.3.15,

$$[\mathbf{k}_{ff}^{(1)}] = \begin{matrix} & \text{2} & \text{5} & \leftarrow\downarrow \text{Local dof} \\ & & & \\ \left[\begin{matrix} & \mathbf{k}^{(1)} & \\ & & \end{matrix} \right] & \begin{matrix} 2 \\ 5 \end{matrix} \end{matrix} = \frac{t}{4A} \begin{bmatrix} E'\gamma_1^2 + G\beta_1^2 & \nu E'\beta_3\gamma_1 + G\beta_1\gamma_3 \\ \nu E'\beta_3\gamma_1 + G\beta_1\gamma_3 & E'\beta_3^2 + G\gamma_3^2 \end{bmatrix} = \left(\frac{0.01}{4 \times 0.5} \times 10^9 \right) \begin{matrix} & \text{2} & \text{5} & \leftarrow\downarrow \text{Global dof} \\ \begin{bmatrix} 200 & 60 \\ 60 & 200 \end{bmatrix} & \begin{matrix} 2 \\ 5 \end{matrix} \end{matrix}$$

[c]

$$= (10^6) \begin{bmatrix} \text{2} & \text{5} & \\ 1000 & 300 \\ 300 & 1000 \end{bmatrix} \begin{matrix} 2 \\ 5 \end{matrix} \text{ (N/m)}$$

For element 2,

$$\beta_1 = y_2 - y_3 = -1 \text{ m} \qquad \beta_2 = y_3 - y_1 = 0 \qquad \beta_3 = y_1 - y_2 = 1 \text{ m}$$

$$\gamma_1 = -x_2 + x_3 = 0 \qquad \gamma_2 = -x_3 + x_1 = -1 \text{ m} \qquad \gamma_3 = -x_1 + x_2 = 1 \text{ m}$$

[d]

$$[\mathbf{k}_{ff}^{(2)}] = \begin{matrix} & \text{2 3 5 6} & \leftarrow\downarrow \text{Local dof} \\ & & \\ \left[\begin{matrix} & \mathbf{k}^{(2)} & \\ & & \end{matrix} \right] & \begin{matrix} 2 \\ 3 \\ 5 \\ 6 \end{matrix} \end{matrix} = \frac{t}{4A} \begin{bmatrix} E'\gamma_1^2 + G\beta_1^2 & \nu E'\beta_2\gamma_1 + G\beta_1\gamma_2 & \nu E'\beta_3\gamma_1 + G\beta_1\gamma_3 & E'\gamma_1\gamma_3 + G\beta_1\beta_3 \\ \nu E'\beta_2\gamma_1 + G\beta_1\gamma_2 & E'\beta_2^2 + G\gamma_2^2 & E'\beta_2\beta_3 + G\gamma_2\gamma_3 & \nu E'\beta_2\gamma_3 + G\beta_3\gamma_2 \\ \nu E'\beta_3\gamma_1 + G\beta_1\gamma_3 & E'\beta_2\beta_3 + G\gamma_2\gamma_3 & E'\beta_3^2 + G\gamma_3^2 & \nu E'\beta_3\gamma_3 + G\beta_3\gamma_3 \\ E'\gamma_1\gamma_3 + G\beta_1\beta_3 & \nu E'\beta_2\gamma_3 + G\beta_3\gamma_2 & \nu E'\beta_3\gamma_3 + G\beta_3\gamma_3 & E'\gamma_3^2 + G\beta_3^2 \end{bmatrix}$$

$$= \left(\frac{0.01}{4 \times 0.5} \times 10^9 \right) \begin{matrix} & \text{2} & \text{5} & \text{7} & \text{8} & \leftarrow\downarrow \text{Global dof} \\ \begin{bmatrix} 70 & 70 & -70 & -70 \\ 70 & 70 & -70 & -70 \\ -70 & -70 & 270 & 130 \\ -70 & -70 & 130 & 270 \end{bmatrix} & \begin{matrix} 2 \\ 5 \\ 7 \\ 8 \end{matrix} \end{matrix}$$

$$= (10^6) \begin{array}{c} \begin{array}{cccc} 2 & 5 & 7 & 8 \end{array} \\ \begin{bmatrix} 350 & 350 & -350 & -350 \\ 350 & 350 & -350 & -350 \\ -350 & -350 & 1350 & 650 \\ -350 & -350 & 650 & 1350 \end{bmatrix} \begin{array}{c} 2 \\ 5 \\ 7 \\ 8 \end{array} \end{array} \text{(N/m)} \qquad \text{[e]}$$

Using the assembly rule, we form the global stiffness matrix $[\mathbf{k}_{ff}]$ for the quadrant, as follows:

$$[\mathbf{k}_{ff}] = (10^6) \begin{bmatrix} 1000 + 350 & 300 + 350 & -350 & -350 \\ 300 + 350 & 1000 + 350 & -350 & -350 \\ -350 & -350 & 1350 & 650 \\ -350 & -350 & 6530 & 1350 \end{bmatrix} = (10^6) \begin{bmatrix} 1350 & 650 & -350 & -350 \\ 650 & 1350 & -350 & -350 \\ -350 & -350 & 1350 & 650 \\ -350 & -350 & 650 & 1350 \end{bmatrix} \text{(N/m)} \qquad \text{[f]}$$

Load case 1

Load case 1 is a single 100 kN point load in the x direction at node 4. Thus, we have

$$\{\mathbf{Q}_f\} = \left\{ \mathbf{Q} \right\} \begin{array}{c} 2 \\ 5 \\ 7 \\ 8 \end{array} = \left\{ \begin{array}{c} Y_1 \\ X_3 \\ X_4 \\ Y_4 \end{array} \right\} = \left\{ \begin{array}{c} 0 \\ 0 \\ 100 \\ 0 \end{array} \right\} (10^3) \text{ (N)}$$

Therefore, the stiffness equations, $[\mathbf{k}_{ff}]\{\mathbf{q}_f\} = \{\mathbf{Q}_f\}$, are

$$(10^6) \begin{bmatrix} 1350 & 650 & -350 & -350 \\ 650 & 1350 & -350 & -350 \\ -350 & -350 & 1350 & 650 \\ -350 & -350 & 650 & 1350 \end{bmatrix} \left\{ \begin{array}{c} v_1 \\ u_3 \\ u_4 \\ v_4 \end{array} \right\} = \left\{ \begin{array}{c} 0 \\ 0 \\ 100 \\ 0 \end{array} \right\} (10^3) \qquad \text{[g]}$$

Solving for the displacements yields

$$\{\mathbf{q}_f\} = \left\{ \begin{array}{c} v_1 \\ u_3 \\ u_4 \\ v_4 \end{array} \right\} = \left\{ \begin{array}{c} 9.9715 \\ 9.9715 \\ 99.919 \\ -42.938 \end{array} \right\} (10^{-6}) \quad \text{(m)} \qquad \text{[h]}$$

We then use the displacements to calculate the stress in each element by means of Equation 9.12.5

$$\{\boldsymbol{\sigma}^{(e)}\} = [\mathbf{E}^{(e)}][\mathbf{B}^{(e)}]\{\mathbf{q}^{(e)}\} \qquad \text{[i]}$$

For both elements, the material stiffness matrix, Equation 11.3.14, is

$$[\mathbf{E}] = (10^9) \begin{bmatrix} 200 & 60 & 0 \\ 60 & 200 & 0 \\ 0 & 0 & 70 \end{bmatrix} \text{ (N/m)} \qquad \text{[j]}$$

The displacement vector of element 1 is

$$\{\mathbf{q}^{(1)}\} = \lfloor u_1 \quad v_1 \quad u_2 \quad v_2 \quad u_3 \quad v_3 \rfloor^T = (10^{-6})\lfloor 0 \quad 9.9715 \quad 0 \quad 0 \quad 9.9715 \quad 0 \rfloor^T$$

For the strain-displacement matrix, we use Equation 11.3.13 and Equations b, to get

$$\left[\mathbf{B}^{(1)}\right] = \left(\frac{1}{2A^{(1)}}\right)\begin{bmatrix} \beta_1 & 0 & \beta_2 & 0 & \beta_3 & 0 \\ 0 & \gamma_1 & 0 & \gamma_2 & 0 & \gamma_3 \\ \gamma_1 & \beta_1 & \gamma_2 & \beta_2 & \gamma_3 & \beta_3 \end{bmatrix}^{(1)} = \begin{bmatrix} 0 & 0 & -1 & 0 & 1 & 0 \\ 0 & 1 & 0 & -1 & 0 & 0 \\ 1 & 0 & -1 & -1 & 0 & 1 \end{bmatrix}$$

Therefore,

$$\{\boldsymbol{\sigma}^{(1)}\} = (10^9)\begin{bmatrix} 200 & 60 & 0 \\ 60 & 200 & 0 \\ 0 & 0 & 70 \end{bmatrix}\begin{bmatrix} 0 & 0 & -1 & 0 & 1 & 0 \\ 0 & 1 & 0 & -1 & 0 & 0 \\ 1 & 0 & -1 & -1 & 0 & 1 \end{bmatrix}\begin{Bmatrix} 0 \\ 9.9715 \\ 0 \\ 0 \\ 9.9715 \\ 0 \end{Bmatrix}(10^{-6}) \qquad \textbf{[k]}$$

which yields the stress in element 1,

$$\begin{Bmatrix} \sigma_x \\ \sigma_y \\ \tau_{xy} \end{Bmatrix}^{(1)} = \begin{Bmatrix} 2.593 \\ 2.593 \\ 0 \end{Bmatrix} \quad \text{MPa}$$

Following the identical procedure for element 2, we have

$$\{\mathbf{q}^{(2)}\} = \lfloor u_1 \quad v_1 \quad u_3 \quad v_3 \quad u_4 \quad v_4 \rfloor^T = 10^{-6} \times \lfloor 0 \quad 9.9715 \quad 9.9715 \quad 0 \quad 99.919 \quad -42.938 \rfloor^T$$

and

$$\left[\mathbf{B}^{(2)}\right] = \begin{bmatrix} -1 & 0 & 0 & 0 & 1 & 0 \\ 0 & 0 & 0 & -1 & 0 & 1 \\ 0 & -1 & -1 & 0 & 1 & 1 \end{bmatrix}$$

so that

$$\{\boldsymbol{\sigma}^{(2)}\} = (10^9)\begin{bmatrix} 200 & 60 & 0 \\ 60 & 200 & 0 \\ 0 & 0 & 70 \end{bmatrix}\begin{bmatrix} -1 & 0 & 0 & 0 & 1 & 0 \\ 0 & 0 & 0 & -1 & 0 & 1 \\ 0 & -1 & -1 & 0 & 1 & 1 \end{bmatrix}\begin{Bmatrix} 0 \\ 9.9715 \\ 9.9715 \\ 0 \\ 99.919 \\ -42.938 \end{Bmatrix}(10^{-6}) = \begin{Bmatrix} 17.407 \\ -2.5925 \\ 2.5926 \end{Bmatrix} \quad \text{MPa} \qquad \textbf{[l]}$$

The stresses, which are constant throughout each element, are shown in Figure 11.3.9. Across the boundary 1–3 between the two elements, the stresses, unlike the displacements, are clearly not continuous; therefore, the adjacent elements are not in equilibrium. This is typical of displacement-based finite elements.

Load Case 2

Load case 2 is a uniform surface traction of 100 kN/m along edge 3–4. To obtain the equivalent point loads at nodes 3 and 4 we employ Equation 11.3.20, recognizing that for this case, $T_{x3}^{(x)} = T_{x4}^{(x)} = 100$ kN/m and $T_{y3}^{(x)} = T_{y4}^{(x)} = 0$. Thus,

$$\begin{Bmatrix} X_3 \\ Y_3 \\ X_4 \\ Y_4 \end{Bmatrix} = tl\begin{Bmatrix} \frac{1}{3}T_{x3}^{(x)} + \frac{1}{6}T_{x4}^{(x)} \\ \frac{1}{3}T_{y3}^{(x)} + \frac{1}{6}T_{y4}^{(x)} \\ \frac{1}{6}T_{x3}^{(x)} + \frac{1}{3}T_{x4}^{(x)} \\ \frac{1}{6}T_{y3}^{(x)} + \frac{1}{3}T_{y4}^{(x)} \end{Bmatrix} = (0.01 \times 1)\begin{Bmatrix} \frac{1}{3}\left(100 \times 10^3\right) + \frac{1}{6}\left(100 \times 10^3\right) \\ 0 \\ \frac{1}{6}\left(100 \times 10^3\right) + \frac{1}{3}\left(100 \times 10^3\right) \\ 0 \end{Bmatrix} = \begin{Bmatrix} 500 \\ 0 \\ 500 \\ 0 \end{Bmatrix} \quad \text{(N)}$$

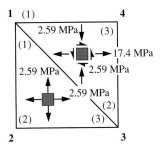

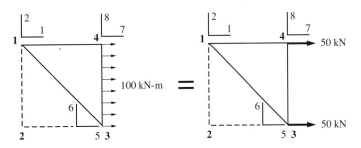

Figure 11.3.9   Stresses computed in the two triangular finite elements for load case 1.

Figure 11.3.10   Surface traction equivalent loads for load case 2.

These equivalent loads are illustrated in Figure 11.3.10. Since there are no other applied loads, the load vector of the constrained structure is

$$\{Q_f\} = \left\{ Q \begin{array}{c} 2 \\ 5 \\ 7 \\ 8 \end{array} \right\} = \left\{ \begin{array}{c} 0 \\ 50 \times 10^3 \\ 50 \times 10^3 \\ 0 \end{array} \right\}$$

Except for the right-hand side, the stiffness equations are identical to Equation g, which is

$$(10^6) \begin{bmatrix} 1350 & 650 & -350 & -350 \\ 650 & 1350 & -350 & -350 \\ -350 & -350 & 1350 & 650 \\ -350 & -350 & 650 & 1350 \end{bmatrix} \begin{Bmatrix} v_1 \\ u_3 \\ u_4 \\ v_4 \end{Bmatrix} = \begin{Bmatrix} 0 \\ 50 \\ 50 \\ 0 \end{Bmatrix} (10^3)$$

The solution of this system is therefore

$$\{q_f\} = \begin{Bmatrix} v_1 \\ u_3 \\ u_4 \\ v_4 \end{Bmatrix} = \begin{Bmatrix} -16.48 \\ 54.94 \\ 54.94 \\ -16.48 \end{Bmatrix} (10^{-6}) \quad m$$

The stresses, are found in the same manner as in Equations k and l for load case 1. In this case, they are the same for both elements, or

$$\begin{Bmatrix} \sigma_x \\ \sigma_y \\ \tau_{xy} \end{Bmatrix}^{(1)} = \begin{Bmatrix} \sigma_x \\ \sigma_y \\ \tau_{xy} \end{Bmatrix}^{(2)} = \begin{Bmatrix} 10 \\ 0 \\ 0 \end{Bmatrix} \quad MPa$$

as shown in Figure 11.3.11.

The exact solution to the problem of a uniform normal traction applied to opposite edges of a rectangular sheet is a state of uniform, uniaxial stress. The finite element solution yields the exact result, even with our very coarse mesh. Of course, the problem is not complex, so we expect the element to perform well, especially since we hope to use it to model more complex problems.

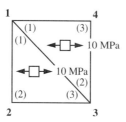

Figure 11.3.11　Stresses due to the uniform loading.

The constant-strain triangle was the first practical finite element. It was developed at the Boeing Aircraft Company in the early 1950s by engineers who coined the term "finite element method."[4,5] The triangle is the simplest two-dimensional finite element, which makes it a good choice for introducing the subject. Because the strain is constant throughout the element, a large number of elements are required to obtain acceptable results in problems where the stress distribution is even moderately complex. We can see this by revisiting Example 3.12.3, in which the plane stress problem shown in Figure 11.3.12a was solved using the theory of elasticity. A finite element model of the same problem is illustrated in part (b) of the figure. For $L/2h = 3$, Figure 11.3.13 compares the "exact" solution from the theory of elasticity with that obtained by the finite element method and the constant-strain triangle. The axial displacement u and stress σ_x at point P (on the top of the region at $L/2$) are compared using the ratio of the finite element results to those obtained from Equations m and u of Example 3.12.3.

As expected, for all three meshes, the finite element solution for the displacement is better than that for the stress, which is the average of stresses in the elements attached at P. The 48-element grid is too coarse to yield reliable values. Refining the mesh improves the solution, and with 432 elements, we are well within 10 percent of the exact solution for both displacement and stress. Further mesh refinement would bring us even closer. In the limit as the elements approach differential size (and the number of degrees of freedom approaches infinity), we attain the exact solution.

For a given number of elements, an infinite variety of grid patterns are obtainable by varying the locations of those nodes not required to define the boundary of the region. In Figure 11.3.13, we could just as easily have chosen the diagonals to run from lower left to upper right, rather than as shown. However, changing the grid

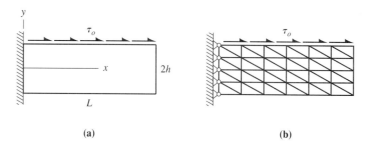

(a)　(b)

Figure 11.3.12　(a) The plane stress problem solved using the theory of elasticity in Example 3.12.3. (b) A finite element model.

[4]M. Turner, R. W. Clough, H. C. Martin, and L. J. Topp, "Stiffness and Deflection Analysis of Complex Structures," *J. Aeronautical Sci.* 23,9, pp. 805–823.

[5]R. W. Clough, "The Finite Element Method in Plane Stress Analysis," *Proc. 2nd ASCE Conference on Electronic Computation*, Pittsburgh, PA, September 1960, pp. 345–378.

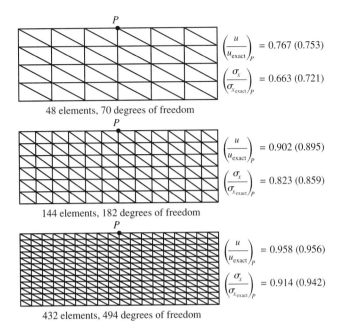

$$\left(\frac{u}{u_{\text{exact}}}\right)_P = 0.767\ (0.753)$$

$$\left(\frac{\sigma_x}{\sigma_{x_{\text{exact}}}}\right)_P = 0.663\ (0.721)$$

48 elements, 70 degrees of freedom

$$\left(\frac{u}{u_{\text{exact}}}\right)_P = 0.902\ (0.895)$$

$$\left(\frac{\sigma_x}{\sigma_{x_{\text{exact}}}}\right)_P = 0.823\ (0.859)$$

144 elements, 182 degrees of freedom

$$\left(\frac{u}{u_{\text{exact}}}\right)_P = 0.958\ (0.956)$$

$$\left(\frac{\sigma_x}{\sigma_{x_{\text{exact}}}}\right)_P = 0.914\ (0.942)$$

432 elements, 494 degrees of freedom

Figure 11.3.13   Comparison of the finite element (constant strain triangle) and the exact solutions of the problem in Example 3.12.3 for $L/2h = 3$.

Numbers in parentheses are for the reversed grid pattern (not shown), in which the diagonals run from lower left to upper right.

pattern can affect the results, as the numbers in parentheses in Figure 11.3.13 clearly reveal. The directionality of the constant-strain triangle is illustrated in the following example.

Example 11.3.2   Calculate the strain in each of the elements of the two grids in Figure 11.3.14, subjected to the given displacements.

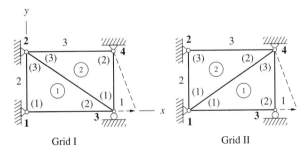

Grid I   Grid II

Figure 11.3.14   Rectangular region gridded with two constant-strain triangles.

Unit horizontal displacement occurs at global node 1. Local node numbers are in parentheses.

Grid I

The nodal coordinates of element 1 are: $x_1 = 0$, $y_1 = 0$; $x_2 = 3$, $y_1 = 0$; and $x_2 = 3$, $y_2 = 0$. Its area is 6; therefore, the strain-displacement matrix, from Equations 11.3.12 and 11.3.13, is

$$[\mathbf{B}^{(1)}] = \left(\frac{1}{6}\right) \begin{bmatrix} -2 & 0 & 2 & 0 & 0 & 0 \\ 0 & -3 & 0 & 0 & 0 & 3 \\ -3 & -2 & 0 & 2 & 3 & 0 \end{bmatrix}$$

For element 2: $x_1 = 3$, $y_1 = 0$; $x_2 = 3$, $y_2 = 2$; *and* $x_3 = 0$, $y_3 = 2$. It follows that

$$[\mathbf{B}^{(2)}] = \left(\frac{1}{6}\right) \begin{bmatrix} 0 & 0 & 2 & 0 & -2 & 0 \\ 0 & -3 & 0 & 3 & 0 & 0 \\ -3 & 0 & 3 & 2 & 0 & -2 \end{bmatrix}$$

The displacement vector of element 1 is

$$\{\mathbf{q}^{(1)}\} = \lfloor 0 \quad 0 \quad 1 \quad 0 \quad 0 \quad 0 \rfloor$$

while that of element 2 is

$$\{\mathbf{q}^{(2)}\} = \lfloor 1 \quad 0 \quad 0 \quad 0 \quad 0 \quad 0 \rfloor$$

Therefore, the strains in each of the elements of grid 1, using Equation 11.3.3, are

$$\begin{Bmatrix} \varepsilon_x^{(1)} \\ \varepsilon_y^{(1)} \\ \gamma_{xy}^{(1)} \end{Bmatrix} = \left(\frac{1}{6}\right) \begin{bmatrix} -2 & 0 & 2 & 0 & 0 & 0 \\ 0 & -3 & 0 & 0 & 0 & 3 \\ -3 & -2 & 0 & 2 & 3 & 0 \end{bmatrix} \begin{Bmatrix} 0 \\ 0 \\ 1 \\ 0 \\ 0 \\ 0 \end{Bmatrix} = \begin{Bmatrix} 1/3 \\ 0 \\ 0 \end{Bmatrix} ; \quad \begin{Bmatrix} \varepsilon_x^{(2)} \\ \varepsilon_y^{(2)} \\ \gamma_{xy}^{(2)} \end{Bmatrix} = \left(\frac{1}{6}\right) \begin{bmatrix} 0 & 0 & 2 & 0 & -2 & 0 \\ 0 & -3 & 0 & 3 & 0 & 0 \\ -3 & 0 & 3 & 2 & 0 & -2 \end{bmatrix} \begin{Bmatrix} 1 \\ 0 \\ 0 \\ 0 \\ 0 \\ 0 \end{Bmatrix} = \begin{Bmatrix} 0 \\ 0 \\ -1/2 \end{Bmatrix} \quad [a]$$

Grid 2

Observing the nodal coordinates of the elements in their new orientation, we find that

$$[\mathbf{B}^{(1)}] = \left(\frac{1}{6}\right) \begin{bmatrix} -2 & 0 & 2 & 0 & 0 & 0 \\ 0 & 0 & 0 & -3 & 0 & 3 \\ 0 & -2 & -3 & 2 & 3 & 0 \end{bmatrix} \quad \text{and} \quad [\mathbf{B}^{(2)}] = \left(\frac{1}{6}\right) \begin{bmatrix} 0 & 0 & 2 & 0 & -2 & 0 \\ 0 & -3 & 0 & 0 & 0 & 3 \\ -3 & 0 & 0 & 2 & 3 & -2 \end{bmatrix}$$

The element displacement vectors for this grid are

$$\{\mathbf{q}^{(1)}\} = \lfloor 0 \quad 0 \quad 1 \quad 0 \quad 0 \quad 0 \rfloor \quad \text{and} \quad \{\mathbf{q}^{(2)}\} = \lfloor 0 \quad 0 \quad 0 \quad 0 \quad 0 \quad 0 \rfloor$$

Therefore, the strains are

$$\begin{Bmatrix} \varepsilon_x^{(1)} \\ \varepsilon_y^{(1)} \\ \gamma_{xy}^{(1)} \end{Bmatrix} = \left(\frac{1}{6}\right) \begin{bmatrix} -2 & 0 & 2 & 0 & 0 & 0 \\ 0 & 0 & 0 & -3 & 0 & 3 \\ 0 & -2 & -3 & 2 & 3 & 0 \end{bmatrix} \begin{Bmatrix} 0 \\ 0 \\ 1 \\ 0 \\ 0 \\ 0 \end{Bmatrix} = \begin{Bmatrix} 1/3 \\ 0 \\ -1/2 \end{Bmatrix} ; \quad \begin{Bmatrix} \varepsilon_x^{(2)} \\ \varepsilon_y^{(2)} \\ \gamma_{xy}^{(2)} \end{Bmatrix} = \left(\frac{1}{6}\right) \begin{bmatrix} 0 & 0 & 2 & 0 & -2 & 0 \\ 0 & -3 & 0 & 0 & 0 & 3 \\ -3 & 0 & 0 & 2 & 3 & -2 \end{bmatrix} \begin{Bmatrix} 0 \\ 0 \\ 0 \\ 0 \\ 0 \\ 0 \end{Bmatrix} = \begin{Bmatrix} 0 \\ 0 \\ 0 \end{Bmatrix} \quad [b]$$

Equations a and b show how the same region, modeled by two different grids of constant-strain triangles, has two different strain distributions for the same set of displacements. Notice, however, that the average strain for the rectangle is the same for both grids, that is,

$$\{\varepsilon\}_{\text{ave., Grid 1}} = \{\varepsilon\}_{\text{ave., Grid 2}} = \frac{1}{2}\left(\{\varepsilon^{(1)}\} + \{\varepsilon^{(2)}\}\right) = \left\{\begin{array}{c} 1/6 \\ 0 \\ -1/4 \end{array}\right\}$$

For further insight into the behavior of the constant-strain triangle, consider a rectangular plane stress region of thickness t, subjected to pure bending, as illustrated in Figure 11.3.15. The exact solution for the displacements u and v throughout the region is (cf. Chapter 3, Exercise 3.17)

$$u = -Cxy \qquad v = C(x^2 + vy^2) \qquad\qquad\text{[11.3.28]}$$

where C is a constant that is proportional to the applied bending moment, and v is Poisson's ratio. According to Equations 3.7.20, the strains in the plane of the sheet are therefore

$$\varepsilon_x = -Cy \qquad \varepsilon_y = vCy \qquad \gamma_{xy} = 0 \qquad\qquad\text{[11.3.29]}$$

Let us model the region with a grid of constant-strain triangles. To the nodes of this mesh, we will apply the exact bending displacement field and then calculate the strains in a given element, such as element e in Figure 11.3.15. Elements e and f form a rectangle whose centroidal coordinates are denoted $\bar{x}^{(r)}$ and $\bar{y}^{(r)}$ (r stands for "rectangle"). Relative to the rectangle's centroid, the nodal coordinates of element e are

$$x_i = \bar{x}^{(r)} - \frac{a}{2} \qquad x_j = \bar{x}^{(r)} + \frac{a}{2} \qquad x_k = \bar{x}^{(r)} - \frac{a}{2}$$
$$y_i = \bar{y}^{(r)} - \frac{b}{2} \qquad y_j = \bar{y}^{(r)} - \frac{b}{2} \qquad y_k = \bar{y}^{(r)} + \frac{b}{2} \qquad\text{[11.3.30]}$$

Substituting these corner coordinates into Equations 11.3.12 and then those results into Equation 11.3.13 yields the strain–displacement matrix

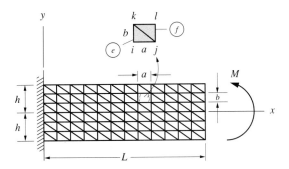

Figure 11.3.15   Rectangular sheet of thickness t subjected to pure bending.

A typical pair of adjacent constant-strain triangles are shown isolated from the mesh.

$$[\mathbf{B}^{(e)}] = \frac{1}{ab} \begin{bmatrix} -b & 0 & b & 0 & 0 & 0 \\ 0 & -a & 0 & 0 & 0 & a \\ -a & -b & 0 & b & a & 0 \end{bmatrix}$$ [11.3.31]

From this, we find the element strains in terms of the nodal displacements,

$$\{\boldsymbol{\varepsilon}^{(e)}\} = \frac{1}{ab} \begin{bmatrix} -b & 0 & b & 0 & 0 & 0 \\ 0 & -a & 0 & 0 & 0 & a \\ -a & -b & 0 & b & a & 0 \end{bmatrix} \begin{Bmatrix} u_i \\ v_i \\ u_j \\ v_j \\ u_k \\ v_k \end{Bmatrix} = \begin{Bmatrix} \dfrac{u_j - u_i}{a} \\ \dfrac{v_k - v_i}{b} \\ \dfrac{u_k - u_i}{b} + \dfrac{v_j - v_i}{a} \end{Bmatrix}$$

The nodal displacements u_i, v_i, $\cdots$, v_k in this expression are obtained from Equations 11.3.28 and the nodal coordinate data in Equation 11.3.30 finally yielding the element strains, as follows:

$$\varepsilon_x^{(e)} = -C \left(\bar{y}^{(r)} - \frac{b}{2} \right) \qquad \varepsilon_y^{(e)} = \nu C \bar{y}^{(r)} \qquad \gamma_{xy}^{(e)} = C \frac{a}{2}$$ [11.3.32a]

Observe that shear strain, which is independent of position in the grid, develops in the element, whereas, in pure bending, we know from Equations 11.3.29 that only normal strains occur. In a similar fashion, we find the strains in element f:

$$\varepsilon_x^{(f)} = -C \left(\bar{y}^{(r)} + \frac{b}{2} \right) \qquad \varepsilon_y^{(f)} = \nu C \bar{y}^{(r)} \qquad \gamma_{xy}^{(f)} = -C \frac{a}{2}$$ [11.3.32b]

Consider a virtual pure-bending displacement field

$$\delta u = -\delta C x y \qquad \delta v = \delta C \left(x^2 + \nu y^2 \right)$$

The corresponding virtual strains of elements e and f are:

$$\delta \varepsilon_x^{(e)} = -\delta C \left(\bar{y}^{(r)} - \frac{b}{2} \right) \qquad \delta \varepsilon_y^{(e)} = \nu \delta C \bar{y}^{(r)} \qquad \delta \gamma_{xy}^{(e)} = \delta C \frac{a}{2}$$

$$\delta \varepsilon_x^{(f)} = -\delta C \left(\bar{y}^{(r)} + \frac{b}{2} \right) \qquad \delta \varepsilon_y^{(f)} = \nu \delta C \bar{y}^{(r)} \qquad \delta \gamma_{xy}^{(f)} = -\delta C \frac{a}{2}$$ [11.3.33]

The virtual work associated with the virtual strains is found using Equation 9.12.6 and the stress-strain relationship, $\{\boldsymbol{\sigma}\} = [\mathbf{E}]\{\boldsymbol{\varepsilon}\}$,

$$\delta W_{\text{int}} = \int_V \{\delta \boldsymbol{\varepsilon}\}^T [\mathbf{E}] \{\boldsymbol{\varepsilon}\} dV$$

where, according to Equation 9.11.9, for plane stress (the case at hand), we have

$$[\mathbf{E}] = \begin{bmatrix} \dfrac{E}{1 - \nu^2} & \dfrac{\nu E}{1 - \nu^2} & 0 \\ \dfrac{\nu E}{1 - \nu^2} & \dfrac{E}{1 - \nu^2} & 0 \\ 0 & 0 & \dfrac{E}{2(1 + \nu)} \end{bmatrix}$$

Since the strains are constant in each triangle, the volume of which is $abt/2$, we have, for element e,

$$\delta W_{\text{int}}^{(e)} = \left\{\delta \boldsymbol{\varepsilon}^{(e)}\right\}^T [\mathbf{E}] \left\{\boldsymbol{\varepsilon}^{(e)}\right\} \frac{abt}{2} = \left\lfloor -\bar{y}^{(r)} + \frac{b}{2} \quad v\bar{y}^{(r)} \quad \frac{a}{2} \right\rfloor \begin{bmatrix} \dfrac{E}{1-v^2} & \dfrac{vE}{1-v^2} & 0 \\[2mm] \dfrac{vE}{1-v^2} & \dfrac{E}{1-v^2} & 0 \\[2mm] 0 & 0 & \dfrac{E}{2(1+v)} \end{bmatrix} \left\{ \begin{array}{c} -\bar{y}^{(r)} + \dfrac{b}{2} \\[2mm] v\bar{y}^{(r)} \\[2mm] \dfrac{a}{2} \end{array} \right\} (C\delta C) \left(\frac{abt}{2}\right)$$

and a similar expression for element f. For the rectangle r formed by triangles e and f, the internal virtual work is

$$\delta W_{\text{int}}^{(r)} = \delta W_{\text{int}}^{(e)} + \delta W_{\text{int}}^{(f)}$$

Doing the computations leads to the following expression:

$$\delta W_{\text{int}}^{(r)} = \left[\left(\bar{y}^{(r)}\right)^2 + \frac{1}{8} \frac{(1-v)a^2 + 2b^2}{1-v^2} \right] (abt) C\delta C \qquad [11.3.34]$$

Suppose the grid of Figure 11.3.15 has n_x elements along the x direction and n_y pairs of triangles (composite rectangles) in the y direction. Thus, $L = n_x \times a$ and $h = (n_y/2) \times b$. If we sum the expression in Equation 11.3.34 over all of the $n_x \times n_y$ composite rectangular elements of the grid, we arrive at the internal virtual work for the mesh, as follows:

$$\delta W_{\text{int}}^{(\text{mesh})} = \left[1 + \frac{1}{n_y^2} \frac{\frac{3}{2}\left(\frac{a}{b}\right)^2 + \frac{2+v^2}{1-v}}{1+v} \right] \left(\frac{2}{3} E t L h^3\right) (C\delta C) \qquad [11.3.35]$$

This is an approximation of the exact virtual work, which is given by

$$\delta W_{\text{int}} = \int_0^L \int_{-h}^h \sigma_x \delta\varepsilon_x t\, dy\, dx = t \int_0^L \int_{-h}^h E\varepsilon_x \delta\varepsilon_x\, dy\, dx = Et \int_0^L \int_{-h}^h (-Cy)(-\delta Cy)\, dy\, dx = \frac{2}{3} E t L h^3 C \delta C \qquad [11.3.36]$$

Using the principle of virtual work, $\delta W_{\text{ext}} = \delta W_{\text{int}}$, we can calculate the bending moment M required to produce the assumed displacement field (Equation 11.3.28). The rotation at the right end of the region is

$$\theta = \left.\frac{\partial v}{\partial x}\right)_{x=L} = CL$$

which means that the virtual rotation is

$$\delta\theta = \left.\frac{\partial \delta v}{\partial x}\right)_{x=L} = L\delta C$$

Therefore, $\delta W_{\text{ext}} = M \times L\delta C$, so that for the exact solution, we have

$$M_{\text{exact}} L\delta C = \frac{2}{3} E t L h^3 C \delta C$$

or

$$M_{\text{exact}} = \frac{2}{3} E t h^3 C \qquad [11.3.37]$$

On the other hand, for the finite element mesh, we have

$$ML\delta C = \left[1 + \frac{1}{n_y^2} \frac{\frac{3}{2}\left(\frac{a}{b}\right)^2 + \frac{2+v^2}{1-v}}{1+v} \right] \left(\frac{2}{3} E t L h^3\right) (C\delta C)$$

which yields

$$\frac{M}{M_{\text{exact}}} = \left[1 + \frac{1}{n_y^2} \frac{\frac{3}{2}\left(\frac{a}{b}\right)^2 + \frac{2 + \nu^2}{1 - \nu}}{1 + \nu} \right]$$ [11.3.38]

Since the second term in the brackets is positive, it is clear from this expression that, for the same pure bending displacement field, the mesh of constant-strain triangles always requires a larger bending moment than does the non-discretized region. That is, the finite element model is stiffer than the actual continuum. Let us not lose sight of the fundamental reason for this: The imposed displacement field (Equation 11.3.28) is quadratic (involving the polynomial terms x^2, xy, and y^2), whereas the displacement field in the constant-strain element is required to be linear in x and y. As we have seen, this induces a shear strain, which should not be present in pure bending, and inducing the additional shear requires additional applied load. Reducing the size of the elements' y dimension a decreases the magnitude of the shear strain (cf. Equations 11.3.32), thereby reducing the stiffness of the mesh (while increasing the number of triangles and, hence, the number of nodes).

Equation 11.3.38 reveals the dramatic influence of element *aspect ratio* a/b on the stiffness of the mesh. We can solve Equation 11.3.38 for n_y and make a plot like that shown in Figure 11.3.16. The plot shows the minimum number of composite rectangles, of a given aspect ratio, required per vertical zone of the grid such that the bending moment will be within 10, 5, 1, and 0.1 percent of the exact value. Remember that a number read off the chart must be multiplied by twice the number of vertical zones to arrive at the total number of constant-strain triangles needed for the entire grid. For example, if $L/2h = 3$ in Figure 11.3.15 and the element aspect ratio a/b is 1.5, then to come within *five percent* of the exact bending moment we need 20 vertical zones of 20 triangles, for a total of 400 elements and 231 nodes. For the same accuracy, increasing a/b to 10 requires 100 triangles per vertical zone and 15 zones, for a total of 1500 elements and 816 nodes. Clearly, high-aspect-ratio, or "skinny," elements are to be avoided.

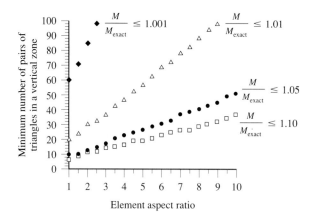

Figure 11.3.16   Influence of element aspect ratio on the minimum number of CST elements required to achieve a given accuracy for the pure bending problem. $\nu = 0.3$.

11.4   ANALYSIS BY SUBSTRUCTURES

Modeling a structure as an assemblage of finite elements, such as rods, beams, or shear panels, is actually a substructuring process. Viewed as such, the substructure is the finite element itself. When we are faced with the problem of analyzing a large structure, or modeling a relatively small structure in great detail, it is convenient—perhaps even necessary—to group finite elements together and handle these groups as separate substructures. Figure 11.4.1 suggests the manner in which an airplane might be substructured. The major substructures are identified, and each is modeled in appropriate detail by finite elements. These substructures are then eventually joined together at the nodes that lie on the substructure boundaries, such as the one dividing the crew module substructure A from the forward fuselage substructure B . Before the complete airplane structural model is assembled, the nodes interior to each substructure are eliminated, through a process called *static condensation*, which yields smaller (condensed) substructure stiffness matrices involving only the boundary degrees of freedom. Therefore, only those degrees of freedom associated with all of the substructure boundaries are required for the stiffness equations of the assembled airplane. After the boundary displacements have been computed for a given loading, the degrees of freedom that were previously eliminated are recovered and are used to calculate the substructure loads. Substructures can be viewed as large finite elements, or superelements, and may themselves be further subdivided into additional finite element groupings.

Through the technique of substructuring, the many major substructures of an airplane can be individually modeled and separately checked out before they are combined and then subjected to service loadings. If changes must be made within a substructure, only that substructure must be remodeled, and only its boundary stiffness matrix must be recalculated for assembly with other, unmodified substructure boundary matrices. This reduces the total solution cost for structures like aircraft, which have thousands of degrees of freedom and undergo numerous detailed design iterations.

As an example, consider the simple substructure illustrated in Figure 11.4.2. The global stiffness matrix $[\mathbf{k}]$ of this substructure is assembled as before from those of its finite element constituents. The stiffness equations of the substructure have the familiar form of

$$[\mathbf{k}]\{\mathbf{q}\} = \{\mathbf{Q}\} \qquad\qquad [11.4.1]$$

These equations incorporate all of the substructure's N nodal degrees of freedom. Once the boundary and interior nodes are identified, the stiffness equations can be partitioned with respect to the boundary and interior degrees of freedom. Let the subscript b refer to the N_b boundary degrees of freedom and the subscript i refer to the N_i

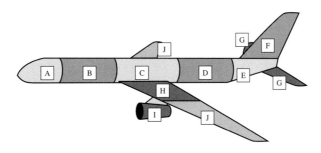

Figure 11.4.1   An aircraft partitioned into major substructures.

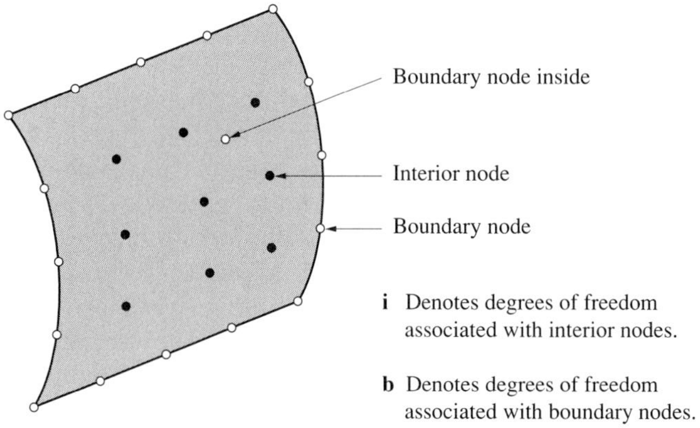

Figure 11.4.2 Interior and boundary nodes of a substructure.

interior degrees of freedom. Observe that $N_b + N_i = N$. Partitioned in this manner, the stiffness equations may be written as follows:

$$\begin{bmatrix} \mathbf{k}_{bb} & \mathbf{k}_{bi} \\ \mathbf{k}_{ib} & \mathbf{k}_{ii} \end{bmatrix} \begin{Bmatrix} \mathbf{q}_b \\ \mathbf{q}_i \end{Bmatrix} = \begin{Bmatrix} \mathbf{Q}_b \\ \mathbf{Q}_i \end{Bmatrix}$$ [11.4.2]

That is,

$$[\mathbf{k}_{bb}]\{\mathbf{q}_b\} + [\mathbf{k}_{bi}]\{\mathbf{q}_i\} = \{\mathbf{Q}_b\}$$ [11.4.3a]

$$[\mathbf{k}_{ib}]\{\mathbf{q}_b\} + [\mathbf{k}_{ii}]\{\mathbf{q}_i\} = \{\mathbf{Q}_i\}$$ [11.4.3b]

Equation 11.4.3b can be solved for the interior nodal displacements $[\mathbf{q}_i]$ to obtain

$$\{\mathbf{q}_i\} = [\mathbf{k}_{ii}]^{-1}\{\mathbf{Q}_i\} - [\mathbf{k}_{ii}]^{-1}[\mathbf{k}_{ib}]\{\mathbf{q}_b\}$$ [11.4.4]

Substituting this result into Equation 11.4.3a, we find that

$$\left([\mathbf{k}_{bb}] - [\mathbf{k}_{bi}][\mathbf{k}_{ii}]^{-1}[\mathbf{k}_{ib}]\right)\{\mathbf{q}_b\} = \{\mathbf{Q}_b\} - [\mathbf{k}_{bi}][\mathbf{k}_{ii}]^{-1}\{\mathbf{Q}_i\}$$ [11.4.5]

It is convenient at this point to let

$$[\mathbf{X}_{ib}] = [\mathbf{k}_{ii}]^{-1}[\mathbf{k}_{ib}]$$ [11.4.6]

and

$$\{\mathbf{Y}_i\} = [\mathbf{k}_{ii}]^{-1}\{\mathbf{Q}_i\}$$ [11.4.7]

Equation 11.4.5 can then be written more compactly as

$$[\mathbf{k}_b]\{\mathbf{q}_b\} = [\mathbf{R}_b]$$ [11.4.8]

where

$$[\mathbf{k}_b] = [\mathbf{k}_{bb}] - [\mathbf{k}_{bi}][\mathbf{X}_{ib}]$$ [11.4.9]

and

$$[\mathbf{R}_b] = \{\mathbf{Q}_b\} - [\mathbf{k}_{bi}]\{\mathbf{Y}_i\}$$ [11.4.10]

The matrix $[\mathbf{k}_b]$ is the *boundary stiffness matrix*, and $[\mathbf{R}_b]$ is the *boundary load vector*. The process of going from Equation 11.4.1 to Equation 11.4.8, and eliminating from the stiffness equations a subset of the unconstrained structure's degrees of freedom, is called *static condensation*. Notice that the eliminated, interior degrees of freedom have not in any way been constrained. This differs from the process of forming the constrained stiffness matrix $[\mathbf{k}_{ff}]$ (cf. Equations 9.10.13) by eliminating the degrees of freedom corresponding to prescribed displacements.

The procedure for implementing a structural analysis by substructures may be outlined as follows.

- Step 1

Assemble the substructure stiffness matrix $[\mathbf{k}]$ and load vector $\{\mathbf{Q}\}$ from the stiffness matrices and load vectors of the finite elements composing the substructure.

- Step 2

Partition $[\mathbf{k}]$ into $[\mathbf{k}_{bb}]$, $[\mathbf{k}_{bi}]$, $[\mathbf{k}_{ib}]$, and $[\mathbf{k}_{ii}]$. Partition $[\mathbf{Q}]$ into $[\mathbf{Q}_b]$ and $[\mathbf{Q}_i]$.

- Step 3

Use a single Gauss elimination procedure to solve the following augmented system for $[\mathbf{X}_{ib}]$ and $\{\mathbf{Y}_i\}$:

$$[\mathbf{k}_{ii}][\mathbf{X}_{ib} \mid \mathbf{Y}_i] = [\mathbf{k}_{ib} \mid \mathbf{Q}_i] \qquad [11.4.11]$$

Determine column 1 of $[\mathbf{X}_{ib}]$ by using column 1 of $[\mathbf{k}_{ib}]$ as the right-side vector in the Gauss elimination procedure. Determine column 2 of $[\mathbf{X}_{ib}]$ by using column 2 of $[\mathbf{k}_{ib}]$, and so on for all N_b columns of $[\mathbf{X}_{ib}]$. Find the vector $\{\mathbf{Y}_i\}$ by using $\{\mathbf{Q}_i\}$ as the right-side vector. In other words, during the single forward elimination process, all $N_b + 1$ columns of $[\mathbf{k}_{ib} \mid \mathbf{Q}_i]$ are operated on simultaneously, as if each were a single right-side vector. Similarly, the single back-substitution simultaneously yields the solution for all N_b columns of $[\mathbf{X}_{ib}]$ and for $\{\mathbf{Y}_i\}$. Thus, the inverse of $[\mathbf{k}_{ii}]$ in Equations 11.4.6 and 11.4.7 need not be computed.

- Step 4

Form the substructure's boundary stiffness matrix and load vector:

$$[\mathbf{k}_b] = [\mathbf{k}_{bb}] - [\mathbf{k}_{bi}][\mathbf{X}_{ib}]$$
$$[\mathbf{R}_b] = \{\mathbf{Q}_b\} - [\mathbf{k}_{bi}]\{\mathbf{Y}_i\}$$

- Step 5

Assemble the substructure boundary stiffness matrix and load vector into the respective global arrays for the complete structure.

Repeat steps 1 through 5 for all of the substructures comprising the complete structure. Then using Gauss elimination, solve the structural stiffness equations thus obtained, for the set of all boundary displacements.

- Step 6

Using the subset of boundary displacements, $\{\mathbf{q}_b\}$ pertaining to a given substructure, calculate its interior displacements, using Equation 11.4.4, which, in view of Equations 11.4.6 and 11.4.7, can be written

$$\{\mathbf{q}_i\} = \{\mathbf{Y}_i\} - [\mathbf{X}_{ib}]\{\mathbf{q}_b\} \qquad [11.4.12]$$

where $\{\mathbf{Y}_i\}$ and $[\mathbf{X}_{ib}]$ were found in step 3.

- Step 7

Apply the appropriate components of the complete set of substructure displacements $\{\mathbf{q}_b\}$ and $\{\mathbf{q}_i\}$ to the individual finite elements of the substructure to compute the loads in each element.

Let us illustrate the procedure by considering the plane truss substructures shown in Figure 11.4.3. Each substructure has its own node and element numbering scheme. However, as required, care is taken to insure that the number and types of degrees of freedom will match up along common substructure boundaries. For simplicity, the coordinate axes of both substructures are chosen to have the same orientation. Boundary and interior nodes are identified in the figure for each of the substructures. We will now compute the condensed boundary stiffness matrices for substructures 1 and 2, in turn.

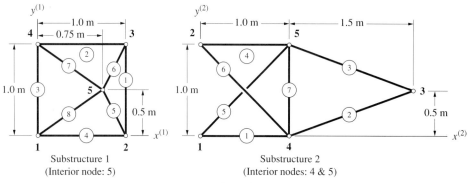

Figure 11.4.3 Two substructures composed of rod elements.

1. Substructure 1

The complete stiffness matrix of this 10 degrees of freedom truss structure is found by first assembling the stiffness matrices (cf. Equation 10.2.19) of its eight rod elements. The result is

$$[\mathbf{k}^{(1)}] = (10)^7$$

	1	2	3	4	5	6	7	8	9	10	
	3.713	1.075	−2.100	0.000	0.000	0.000	0.000	0.000	−1.613	−1.075	1
		2.817	0.000	0.000	0.000	0.000	0.000	−2.100	−1.075	−0.7168	2
			2.851	−1.503	0.000	0.000	0.000	0.000	−0.7513	1.503	3
				5.105	0.000	−2.100	0.000	0.000	1.503	−3.005	4
					2.851	1.503	−2.100	0.000	−0.7513	−1.503	5
						5.105	0.000	0.000	−1.503	−3.005	6
		(symmetric)					3.713	−1.075	−1.613	1.075	7
								2.817	1.075	−0.7168	8
									4.728	0.000	9
										7.444	10

Degrees of freedom 9 and 10 pertain to interior node 5. Thus,

$$[\mathbf{k}_{ii}^{(1)}] = \begin{matrix} 9 \\ 10 \end{matrix} \begin{matrix} 9 & 10 \\ [\ \mathbf{k}^{(1)}\ ] \end{matrix} = (10^7) \begin{bmatrix} 4.728 & 0.000 \\ 0.000 & 7.444 \end{bmatrix}$$

The boundary nodes are 1, 2, 3, and 4, which means that the boundary degree of freedom numbers are 1 through 8. Therefore,

$$[\mathbf{k}_{ib}^{(1)}] = \begin{smallmatrix} 9 \\ 10 \end{smallmatrix} \overset{1...8}{\left[\mathbf{k}^{(1)}\right]} = (10^7)\begin{bmatrix} -1.613 & -1.075 & -0.7513 & 1.503 & -0.7513 & -1.5026 & -0.1613 & 1.075 \\ -1.075 & -0.7168 & 1.503 & -3.005 & -1.503 & -3.005 & 1.075 & -0.7168 \end{bmatrix}$$

Node 5, the interior node, has no external applied load, so that $\{\mathbf{Q}_i^{(1)}\} = \{\mathbf{0}\}$. Using these values of the matrices $[\mathbf{k}_{ii}^{(1)}]$, $[\mathbf{k}_{ib}^{(1)}]$ and $\{\mathbf{Q}_i^{(1)}\}$, in Equation 11.4.11,

$$\left[\mathbf{k}_{ii}^{(1)}\right]\left[\mathbf{X}_{ib}^{(1)} \;\vdots\; \mathbf{Y}_i^{(1)}\right] = \left[\mathbf{k}_{ib}^{(1)} \;\vdots\; \mathbf{Q}_i^{(1)}\right]$$

yields the components of $[\mathbf{X}_{ib}^{(1)}]$ and $\{\mathbf{Y}_i^{(1)}\}$, as follows:

$$\left[\mathbf{X}_{ib}^{(1)}\right] = \begin{bmatrix} -0.3411 & -0.2274 & -0.1589 & 0.3178 & -0.1589 & -0.3178 & 0.3411 & 0.2274 \\ -0.1444 & -0.0963 & 0.2018 & -0.4037 & -0.2018 & -0.4037 & 0.1444 & -0.0963 \end{bmatrix}$$

$$\left\{\mathbf{Y}_i^{(1)}\right\} = \{\mathbf{0}\} \qquad \text{[11.4.13]}$$

The next step is to compute the matrix product $[\mathbf{k}_{bi}^{(1)}][\mathbf{X}_{ib}^{(1)}]$, recognizing that $[\mathbf{k}_{bi}^{(1)}] = [\mathbf{k}_{ib}^{(1)}]^T$, because the stiffness matrix is symmetric. The result of this matrix multiplication is

$$[\mathbf{k}_{bi}^{(1)}][\mathbf{X}_{ib}^{(1)}] = (10)^7 \begin{bmatrix} 0.7055 & 0.4703 & 0.3924 & -0.7847 & 0.4733 & 0.9466 & 0.3949 & -0.2632 \\ & 0.3135 & 0.2616 & -0.5231 & 0.3155 & 0.6311 & 0.2632 & -0.1755 \\ & & 0.4227 & -0.8454 & -0.1839 & -0.3679 & 0.4733 & -0.3155 \\ & & & 1.691 & 0.3679 & 0.7357 & -0.9466 & 0.6311 \\ & \text{(symmetric)} & & & 0.4227 & 0.8454 & 0.03924 & -0.2616 \\ & & & & & 1.691 & -0.7847 & 0.05231 \\ & & & & & & 0.7055 & -0.4703 \\ & & & & & & & 0.3135 \end{bmatrix}$$

Finally, we determine the boundary stiffness matrix $[\mathbf{k}_b^{(1)}]$ using Equation 11.4.9,

$$\left[\mathbf{k}_b^{(1)}\right] = \left[\mathbf{k}_{bb}^{(1)}\right] - \left[\mathbf{k}_{bi}^{(1)}\right]\left[\mathbf{X}_{ib}^{(1)}\right]$$

which shows that

$$[\mathbf{k}_b^{(1)}] = (10)^7 \begin{bmatrix} 3.007 & 0.6049 & -2.139 & 0.07847 & -0.4733 & -0.9466 & -0.3948 & 0.2632 \\ & 2.503 & -0.02616 & 0.05231 & -0.3155 & -0.6311 & -0.2632 & -1.924 \\ & & 2.429 & -0.6573 & 0.1839 & 0.3679 & -0.4733 & 0.3155 \\ & & & 3.414 & -0.3679 & -2.836 & 0.9466 & -0.6311 \\ & \text{(symmetric)} & & & 2.429 & 0.6573 & -2.139 & 0.02616 \\ & & & & & 3.414 & -0.07847 & 0.05231 \\ & & & & & & 3.007 & -0.6049 \\ & & & & & & & 2.503 \end{bmatrix}$$

Since $\{\mathbf{Y}_i^{(1)}\} = \{\mathbf{0}\}$, the computation of the boundary load vector $\{\mathbf{R}_b^{(1)}\}$, using Equation 11.4.10, yields simply

$$\{\mathbf{R}_b^{(1)}\} = \{\mathbf{Q}_b^{(1)}\}$$

2. Substructure 2

This 10 degree of freedom structure is composed of seven rod elements. Its stiffness matrix is

$$
[\mathbf{k}^{(2)}] = (10)^7
\begin{array}{c}
\begin{array}{cccccccccc}
1 & 2 & 3 & 4 & 5 & 6 & 7 & 8 & 9 & 10
\end{array} \\
\begin{bmatrix}
2.842 & 0.7425 & 0.000 & 0.000 & 0.000 & 0.000 & 0.000 & 0.000 & -0.7425 & -0.7425 \\
 & 0.7425 & 0.000 & 0.000 & 0.000 & 0.000 & 0.000 & 0.000 & -0.7425 & -0.7425 \\
 & & 2.851 & -0.7425 & 0.000 & 0.000 & -0.7425 & -0.7425 & -2.100 & 0.000 \\
 & & & 0.7425 & 0.000 & 0.000 & 0.7425 & -0.7425 & 0.000 & 0.000 \\
 & & & & 2.391 & 0.000 & -1.195 & -0.3984 & -1.195 & 0.3984 \\
 & & & & & 0.2656 & -0.3984 & -0.1328 & 0.3984 & -0.1382 \\
 & & \text{(symmetric)} & & & & 4.038 & -0.3440 & 0.000 & 0.000 \\
 & & & & & & & 2.975 & 0.000 & -2.100 \\
 & & & & & & & & 4.038 & 0.3440 \\
 & & & & & & & & & 2.975
\end{bmatrix}
\begin{array}{c}
1 \\ 2 \\ 3 \\ 4 \\ 5 \\ 6 \\ 7 \\ 8 \\ 9 \\ 10
\end{array}
\end{array}
$$

The interior node numbers are 7, 8, 9, and 10, which correspond to the interior nodes 4 and 5. Therefore,

$$
[\mathbf{k}_{ii}^{(2)}] = \begin{array}{c} 7 \\ 8 \\ 9 \\ 10 \end{array}
\begin{bmatrix} & & \mathbf{k}^{(2)} & \\ & & & \end{bmatrix}
= (10^7)
\begin{bmatrix}
4.038 & -0.3440 & 0.000 & 0.000 \\
-0.3440 & 2.975 & 0.000 & -2.100 \\
-2.100 & 0.000 & 4.038 & 0.3440 \\
0.000 & -2.100 & 0.3440 & 2.975
\end{bmatrix}
$$

The boundary degree of freedom numbers are 1 through 6, so that

$$
[\mathbf{k}_{ib}^{(2)}] = \begin{array}{c} 7 \\ 8 \\ 9 \\ 10 \end{array}
\begin{array}{c}
\begin{array}{cccccc} 1 & 2 & 3 & 4 & 5 & 6 \end{array} \\
\begin{bmatrix} & & & \mathbf{k}^{(2)} & & \end{bmatrix}
\end{array}
= (10^7)
\begin{bmatrix}
-2.100 & 0.000 & -0.7425 & 0.7425 & -1.195 & -0.3984 \\
0.000 & 0.000 & 0.7425 & -0.7425 & -0.3984 & -0.1328 \\
-0.7425 & -0.7425 & -2.100 & 0.000 & -1.195 & 0.3984 \\
-0.7425 & -0.7425 & 0.000 & 0.000 & 0.3984 & -0.1382
\end{bmatrix}
$$

As was the case for substructure 1, the internal nodes of substructure 2 are unloaded, that is,

$$
\{\mathbf{Q}_i^{(2)}\} = \{\mathbf{0}\}
$$

Substituting the matrices $[\mathbf{k}_{ii}^{(2)}]$, $[\mathbf{k}_{ib}^{(2)}]$, and $\{\mathbf{Q}_i^{(2)}\}$ into Equation 11.4.11, we get

$$
\left[\mathbf{k}_{ii}^{(2)} \right] \left[\mathbf{X}_{ib}^{(2)} \; \vdots \; \mathbf{Y}_i^{(2)} \right] = \left[\mathbf{k}_{ib}^{(2)} \; \vdots \; \mathbf{Q}_i^{(2)} \right]
$$

Solving for $[\mathbf{X}_{ib}^{(2)}]$, and $\{\mathbf{Y}_i^{(2)}\}$, using Gauss elimination, yields

$$
[\mathbf{X}_{ib}^{(2)}] =
\begin{bmatrix}
-0.5591 & -0.02847 & -0.1364 & 0.1439 & -0.3045 & -0.1155 \\
-0.4576 & -0.3341 & 0.5568 & -0.4687 & -0.09915 & -0.1971 \\
-0.1364 & -0.1439 & -0.5591 & 0.02847 & -0.3045 & 0.1155 \\
-0.5568 & -0.4687 & 0.4576 & -0.3341 & 0.09915 & -0.1971
\end{bmatrix}
\qquad \textbf{[11.4.14]}
$$

$$
\{\mathbf{Y}_i^{(2)}\} = \{\mathbf{0}\}
$$

The transpose of $[\mathbf{k}_{ib}^{(2)}]$ is $[\mathbf{k}_{bi}^{(2)}]$, which is multiplied by $[\mathbf{X}_{ib}^{(2)}]$ to produce the matrix

$$[\mathbf{k}_{bi}^{(2)}][\mathbf{X}_{bi}^{(2)}] = (10^7) \begin{bmatrix} 1.689 & 0.5147 & 0.3619 & -0.07533 & 0.7919 & 0.3031 \\ & 0.4549 & 0.07533 & 0.2269 & 0.1524 & 0.06062 \\ & & 1.689 & -0.5147 & 0.7919 & -0.3031 \\ & \text{(symmetric)} & & 0.4549 & -0.1524 & 0.06062 \\ & & & & 0.8069 & 0.000 \\ & & & & & 0.1444 \end{bmatrix}$$

According to Equation 11.4.9, we subtract this matrix from $[\mathbf{k}_{bb}^{(2)}]$ to obtain the boundary stiffness matrix $[\mathbf{k}_{b}^{(2)}]$:

$$[\mathbf{k}_{b}^{(2)}] = (10^7) \begin{bmatrix} 1.1537 & 0.2278 & -0.3619 & 0.0753 & -0.7919 & -0.3031 \\ & 0.2876 & -0.07533 & -0.2269 & -0.1524 & -0.0606 \\ & & 1.154 & -0.2278 & -0.7919 & -0.3031 \\ & \text{(symmetric)} & & 0.2876 & 0.1524 & -0.06062 \\ & & & & 1.584 & 0.000 \\ & & & & & 0.1212 \end{bmatrix}$$

Finally, according to Equation 11.4.10, since $\{\mathbf{Y}_i^{(2)}\} = \{\mathbf{0}\}$, the boundary load vector is

$$\{\mathbf{R}_b^{(2)}\} = \{\mathbf{Q}_b^{(2)}\}$$

Let us now join the two substructures in Figure 11.4.3, to form the loaded and constrained truss structure illustrated in Figure 11.4.4. The interior nodes and structural details are intentionally deleted in the figure, to emphasize that the stiffness properties of the substructures are conveyed entirely through the condensed degrees of freedom associated with the boundary nodes, which are the only nodes shown. The global node numbering, as always, is independent of the nodal numbering schemes used in the individual substructures.

To assemble the substructure boundary stiffness matrices, we first determine the global degree of freedom incidences, as usual. These are shown by labeling the rows and columns of $[\mathbf{k}_b^{(1)}]$ and $[\mathbf{k}_b^{(2)}]$ as follows, referring to Figures 11.4.3 and 11.4.4. (The numerical terms in these matrices are omitted, in the interest of brevity.)

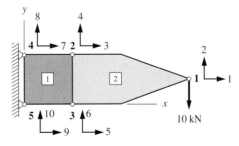

Figure 11.4.4   Structure composed of the two substructures of Figure 11.4.3.

$$[\mathbf{k}_b^{(1)}] = \begin{array}{c} \\ \\ \end{array} \begin{array}{cccccccc} 9 & 10 & 5 & 6 & 3 & 4 & 7 & 8 \end{array}$$

(8 by 8 matrix with row/column labels 9, 10, 5, 6, 3, 4, 7, 8)

$$[\mathbf{k}_b^{(2)}] = \begin{array}{c} \\ \\ \end{array} \begin{array}{cccccc} 5 & 6 & 3 & 4 & 1 & 2 \end{array}$$

(6 by 6 matrix with row/column labels 5, 6, 3, 4, 1, 2)

We assemble these two matrices in the usual fashion, to form the 10 by 10 stiffness matrix of the structure. Rows and columns 7, 8, 9, and 10 must then be eliminated to account for the rigid constraints at nodes 4 and 5. Therefore, the stiffness equations $[\mathbf{k}_{ff}]\{\mathbf{q}_f\} = \{\mathbf{Q}_f\}$, of the constrained structure become

$$(10^7) \begin{bmatrix} 1.584 & 00.000 & -0.7919 & 0.1524 & -0.7919 & -0.1524 \\ & 0.1212 & 0.3031 & 0.06062 & -0.3031 & -0.06062 \\ & & 3.582 & 0.4295 & -0.1779 & -0.4432 \\ & \text{(symmetric)} & & 3.702 & 0.4432 & -3.063 \\ & & & & 3.582 & -0.4295 \\ & & & & & 3.702 \end{bmatrix} \begin{Bmatrix} u_1 \\ v_1 \\ u_2 \\ v_2 \\ u_3 \\ v_3 \end{Bmatrix} = \begin{Bmatrix} 0 \\ -10000 \\ 0 \\ 0 \\ 0 \\ 0 \end{Bmatrix}$$

Solving this system for the six displacements, using Gauss elimination, yields

$$\{\mathbf{q}_f\} = \begin{Bmatrix} u_1 \\ v_1 \\ u_2 \\ v_2 \\ u_3 \\ v_3 \end{Bmatrix} = \begin{Bmatrix} 0.0000 \\ -16.31 \\ 1.309 \\ -1.519 \\ -1.309 \\ -1.518 \end{Bmatrix} \text{(mm)}$$

We can now complete the analysis of the structure by calculating the loads internal to each substructure. We start by ascertaining the displacements of the substructures' internal nodes. This is accomplished as indicated in Step 6 by applying Equation 11.4.12,

$$\{\mathbf{q}_i\} = \{\mathbf{Y}_i\} - \{\mathbf{X}_{ib}\}\{\mathbf{q}_b\}$$

to each substructure in turn.

1. Substructure 1
Equation 11.4.13 established that $\{Y_i^{(1)}\} = \{0\}$. Therefore, for substructure 1, we have

$$\left\{q_i^{(1)}\right\} = -\left[X_{ib}^{(1)}\right]\left\{q_b^{(1)}\right\}$$

$$\left\{q_i^{(1)}\right\} = -\begin{bmatrix} -0.3411 & -0.2274 & -0.1589 & 0.3178 & -0.1589 & -0.3178 & -0.3411 & 0.2274 \\ -0.1444 & -0.0963 & 0.2019 & -0.4037 & -0.2019 & -0.4037 & 0.1444 & -0.0963 \end{bmatrix} \begin{Bmatrix} 0 \\ 0 \\ -1.309 \\ -1.519 \\ 1.309 \\ -1.519 \\ 0 \\ 0 \end{Bmatrix} (10^{-3})$$

from which the displacements at its interior node 5 are found to be

$$\begin{Bmatrix} u_5^{(1)} \\ v_5^{(1)} \end{Bmatrix} = \begin{Bmatrix} -0.01057 \\ -0.7109 \end{Bmatrix} \quad (\text{mm})$$

2. Substructure 2

From Equation 11.4.14, we know that $\{Y_i^{(2)}\} = \{0\}$; thus, for substructure 2, we have

$$\left\{q_i^{(2)}\right\} = -\left[X_{ib}^{(2)}\right]\left\{q_b^{(2)}\right\}$$

$$\left\{q_i^{(2)}\right\} = -\begin{bmatrix} -0.5591 & -0.02847 & -0.1364 & 0.1439 & -0.3045 & -0.1155 \\ -0.4576 & -0.3341 & 0.5568 & -0.4687 & -0.09915 & -0.1971 \\ -0.1364 & -0.1439 & -0.5591 & 0.02847 & -0.3045 & 0.1154 \\ -0.5568 & -0.4687 & 0.4576 & -0.3341 & 0.09915 & -0.1971 \end{bmatrix} \begin{Bmatrix} -1.309 \\ -1.519 \\ 1.309 \\ -1.519 \\ 0 \\ -16.31 \end{Bmatrix} (10^{-3})$$

which yields the following displacements of its interior nodes 4 and 5

$$\begin{Bmatrix} u_4^{(2)} \\ v_4^{(2)} \\ u_5^{(2)} \\ v_5^{(2)} \end{Bmatrix} = \begin{Bmatrix} -2.262 \\ -5.763 \\ 2.262 \\ -5.763 \end{Bmatrix} \quad (\text{mm})$$

Upon applying the displacements of the substructures' boundary nodes and interior nodes to the rod elements within the substructures, the element stiffness equations will furnish all of the rod element loads. The results are summarized in Figure 11.4.5.

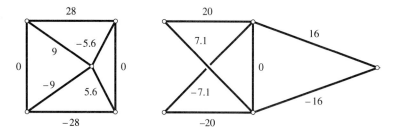

Figure 11.4.5   Axial loads (kN) in the elements of the substructures in Figure 11.4.4.

EXERCISES

Solve all of the problems using the displacement method. Use hand calculations, rather than canned software, unless otherwise directed.

11.1   Find the stiffness matrix $[\mathbf{k}_{ff}]$ of the structure composed of ten rods and three shear panels.

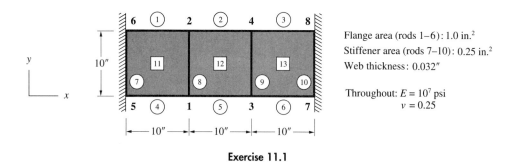

Flange area (rods 1–6): 1.0 in.²
Stiffener area (rods 7–10): 0.25 in.²
Web thickness: 0.032″

Throughout: $E = 10^7$ psi
$\nu = 0.25$

Exercise 11.1

11.2   For the structure of Exercise 11.1, find the maximum allowable rivet spacing d (to be uniform throughout the beam), based on an allowable shear of 375 lb per rivet, given that the structure must support a 10,000 lb upward load at node 2. Use available computer resources as an aid.

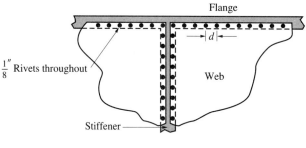

Exercise 11.2

11.3   Using the principle of complementary virtual work, show that the deflection of the free end of the cantilever beam is

$$v = \frac{2}{3}\frac{L^2}{h^2}\frac{PL}{AE} + 2(1+v)\frac{L}{h}\frac{P}{Et}$$

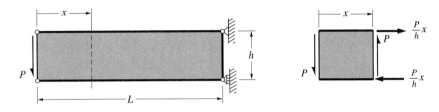

Area of rods = A. Thickness of shear web = t. E and v uniform throughout.

Exercise 11.3

11.4   Use the displacement method to calculate the vertical displacement at node 1 of the beam for the following values of h: 1 m, 0.75 m, 0.5 m, 0.25 m, and 0.1 m. Calculate the percentage difference between these five displacements and those calculated using the formula in Exercise 11.3. What do you conclude from your results?

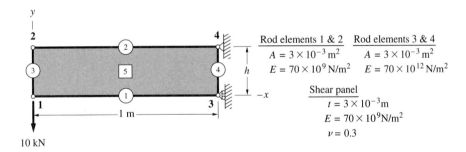

Exercise 11.4

11.5   The allowable stress in the shear panel is 120 MPa, and the allowable displacement in the direction of load P is 15 mm. What is the maximum load P that can be applied as shown to the structure?

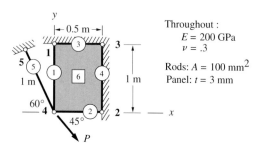

Throughout :
$E = 200$ GPa
$v = .3$

Rods: $A = 100$ mm^2
Panel: $t = 3$ mm

Exercise 11.5

11.6   Find the vertical displacement at node 1 of the three-rod, two-shear-panel assembly, loaded as shown.

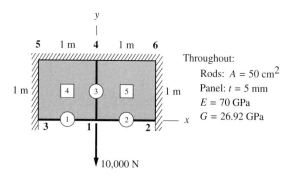

Throughout:
Rods: $A = 50$ cm^2
Panel: $t = 5$ mm
$E = 70$ GPa
$G = 26.92$ GPa

10,000 N

Exercise 11.6

11.7   Find the displacement in the direction of the 200 kN load.

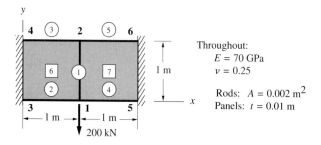

Throughout:
$E = 70$ GPa
$v = 0.25$

Rods: $A = 0.002$ m^2
Panels: $t = 0.01$ m

200 kN

Exercise 11.7

11.8   The global nodal displacements of the stiffened shear panel beam are given in the table. Calculate the maximum shear flow in the structure.

Nodal displacements

Node	u (in.)	v (in.)
1	0.3154	−0.7110
2	−0.1941	−0.2291
3	0.0000	0.0000
4	0.2405	−0.2188
5	−0.2691	−0.7860
6	0.1153	−0.2305
7	0.0000	0.0000
8	0.1153	−0.7193
9	0.0000	0.0000
10	−0.04628	−0.2173

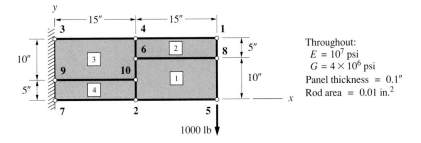

Throughout:
$E = 10^7$ psi
$G = 4 \times 10^6$ psi
Panel thickness = 0.1″
Rod area = 0.01 in.²

Exercise 11.8

11.9   The stiffened shear panel structure is loaded and constrained as shown. Find the shear flow in each panel.

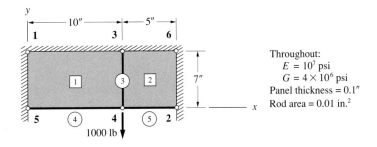

Throughout:
$E = 10^7$ psi
$G = 4 \times 10^6$ psi
Panel thickness = 0.1″
Rod area = 0.01 in.²

Exercise 11.9

11.10 Find the maximum tensile stress in the shear panel attached to the two rod elements. Note the externally applied, distributed load along rod element 1.

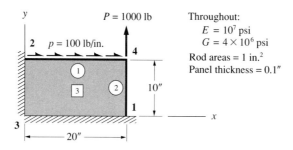

Exercise 11.10

11.11 The computed displacements of the stiffened panel structure, loaded as shown, are listed in the table. If the maximum allowable shear stress in any panel is 15,000 psi, what is the maximum downward load that can be applied at node 13?

Nodal displacements

Node	u (in. $\times 10^{-3}$)	v (in. $\times 10^{-3}$)
5	−0.1147	−0.3830
6	0.05465	−0.3724
7	−0.04506	−0.3741
8	0.1051	−0.3864
9	−0.1063	−1.158
10	0.07781	−1.158
11	−.04336	−1.139
12	0.07182	−1.120
13	0.06675	−1.911
14	0.01419	−1.512
15	−0.008577	−1.300
16	−0.07236	−1.238

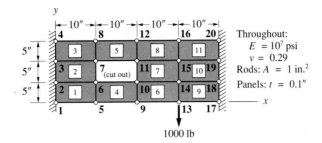

Exercise 11.11

11.12   The triangular box beam consists of nine rod elements and three shear panels. Loaded and constrained as shown, the calculated displacements at its free nodes are:

Node	u (in.)	v (in.)	w (in.)
1	0.0000	−0.4658	0.0000
2	0.2083	−0.3472	0.06555
3	−0.2083	−0.3472	−0.06555

Use these displacements to calculate the shear stress in panel element 1.

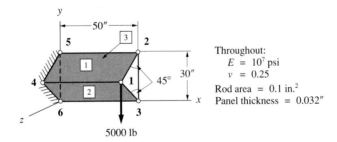

Exercise 11.12

11.13 The global displacements of the box beam are given in the table, for the loading shown. Calculate the shear stress in the top panel (panel 1).

Nodal displacements

Node	u (in.)	v (in.)	w (in.)
1	-0.6599×10^{-5}	0.2779×10^{-3}	-0.2297×10^{-3}
2	0.6599×10^{-5}	-0.3029×10^{-3}	-0.2297×10^{-3}
3	0.6599×10^{-5}	0.3029×10^{-3}	0.2297×10^{-3}
4	-0.6599×10^{-5}	-0.2779×10^{-3}	0.2297×10^{-3}
5	0	0	0
6	0	0	0
7	0	0	0
8	0	0	0

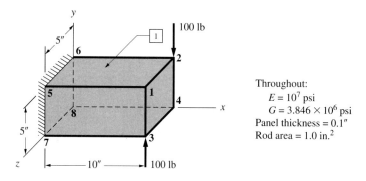

Throughout:
$E = 10^7$ psi
$G = 3.846 \times 10^6$ psi
Panel thickness $= 0.1''$
Rod area $= 1.0$ in.^{2}

Exercise 11.13

11.14 Two stiffened shear panels are joined as shown. Nodes 3, 4, 5, and 6 are rigidly attached to the wall. Node 2 is required to have the following displacement components: $u_2 = 2$ mm, $v_2 = -3$ mm, $w_2 = 1$ mm.

a. Calculate the displacements of node 1.
b. Calculate the shear flow in each panel.

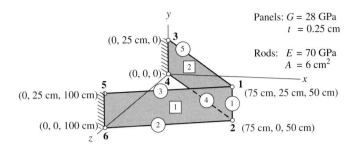

Panels: $G = 28$ GPa
         $t = 0.25$ cm

Rods: $E = 70$ GPa
        $A = 6$ cm^2

(0, 25 cm, 0) **3**

(0, 0, 0)

(0, 25 cm, 100 cm)   **5**

(0, 0, 100 cm)   **6**

1 (75 cm, 25 cm, 50 cm)

2 (75 cm, 0, 50 cm)

Exercise 11.14

11.15 The nine shear panels are riveted to the 13 rods (stiffeners and stringers). The rivet spacing is uniform throughout the structure. Find the maximum allowable rivet spacing, based on an allowable shear of 375 lb per rivet, for the loading shown. If the allowable panel shear stress is 22,500 psi, what is the overall margin of safety (MS) for the given loading? [MS = (allowable/calculated) − 1]. Use available computer resources as an aid.

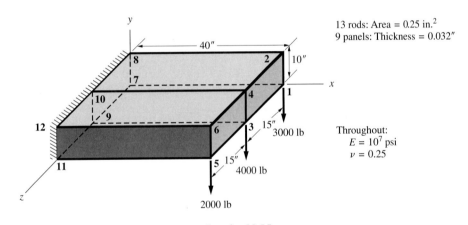

13 rods: Area = 0.25 in.^{2}
9 panels: Thickness = 0.032″

Throughout:
$E = 10^7$ psi
$\nu = 0.25$

Exercise 11.15

11.16 If the maximum allowable web shear stress is 10,000 psi, what is the maximum load P that can be applied as shown to the box beam? Note that the top left cover panel in the center bay is missing. Use available computer resources as an aid.

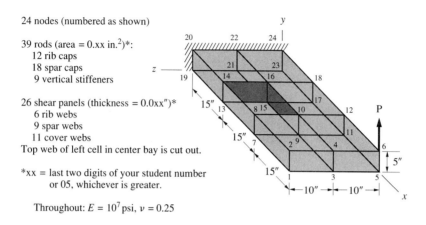

24 nodes (numbered as shown)

39 rods (area = 0.xx in.^{2})*:
  12 rib caps
  18 spar caps
  9 vertical stiffeners

26 shear panels (thickness = 0.0xx")*
  6 rib webs
  9 spar webs
  11 cover webs
Top web of left cell in center bay is cut out.

*xx = last two digits of your student number
  or 05, whichever is greater.

Throughout: $E = 10^7$ psi, $\nu = 0.25$

Exercise 11.16

11.17 The single constant-strain triangle element is loaded and constrained as shown. The nodal coordinates are given. Calculate the maximum principal stress in the element.

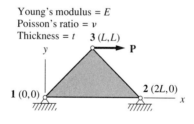

Young's modulus = E
Poisson's ratio = ν
Thickness = t

Exercise 11.17

11.18   Find components $k_{13,20}$ and $k_{11,21}$ of the 32 by 32 global stiffness matrix of the square sheet in plane stress, modeled as an assembly of 18 constant-strain triangles. Young's modulus is E, Poisson's ratio is v, and the thickness of the sheet is t. (Numbers in parentheses are the local node numbers for each element; global node numbers are boldface.)

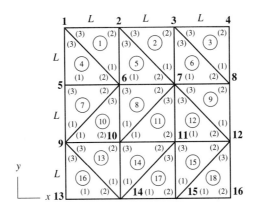

Exercise 11.18

11.19   The 0.1-inch thick plane region, modeled as an assembly of constant strain triangles, has a linearly-varying shear traction applied to the top edge, as shown. Calculate the equivalent point loads (in pounds) in the x direction at each of the seven nodes along the top edge of the region.

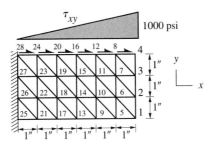

Exercise 11.19

11.20 The triangular bracket is modeled with eight constant-strain triangles. The computed nodal displacements are listed for the point load and constraints shown. Calculate the maximum shear stress in element 2.

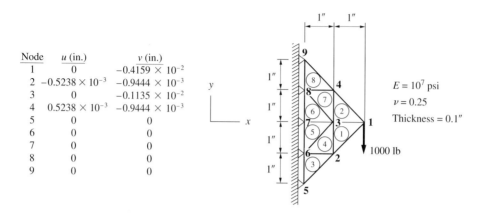

Node	u (in.)	v (in.)
1	0	-0.4159×10^{-2}
2	-0.5238×10^{-3}	-0.9444×10^{-3}
3	0	-0.1135×10^{-2}
4	0.5238×10^{-3}	-0.9444×10^{-3}
5	0	0
6	0	0
7	0	0
8	0	0
9	0	0

$E = 10^7$ psi

$\nu = 0.25$

Thickness = 0.1″

1000 lb

Exercise 11.20

11.21 Use the complementary virtual work principle to find the stiffness matrix for the slightly tapered beam shown. The element loads shown in the figure are required to be in equilibrium, and the bending moment varies linearly with x. The area moment of inertia I also varies linearly with x according to $I = I_0[1 + \alpha (x/L)]$, where $|\alpha| \ll 1$. Since $|\alpha| \ll 1$ (slight taper), you need only retain terms of order 1 in a. For example, $(1 + \alpha)^b = 1 + b\alpha$.

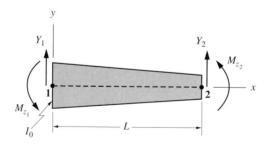

Exercise 11.21

11.22 Find the stiffness matrix for an axially-loaded rod (not a beam) element that has a uniform thickness b perpendicular to the page but is tapered in depth, as shown. The uniaxial stress field in the rod must be in equilibrium with the nodal loads.

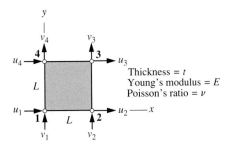

Exercise 11.22

11.23   In the square plate shown, assume plane stress. Also assume that the displacement field is bilinear in x and y, that is,

$$u = \left(1 - \frac{x}{L} - \frac{y}{L} + \frac{xy}{L^2}\right)u_1 + \left(\frac{x}{L} - \frac{xy}{L^2}\right)u_2 + \frac{xy}{L^2}u_3 + \left(\frac{y}{L} - \frac{xy}{L^2}\right)u_4$$

$$v = \left(1 - \frac{x}{L} - \frac{y}{L} + \frac{xy}{L^2}\right)v_1 + \left(\frac{x}{L} - \frac{xy}{L^2}\right)v_2 + \frac{xy}{L^2}v_3 + \left(\frac{y}{L} - \frac{xy}{L^2}\right)v_4$$

Find component k_{11} and k_{12} of the plate's 8 by 8 stiffness matrix.

Exercise 11.23

11.24   The stiffness matrix of the truss in part (a) of the figure is

$$[\mathbf{k}] = \frac{AE}{L}\begin{bmatrix} 2.549 & 1.183 & -1.000 & 0.000 & -.2500 & -.4330 & -1.299 & -.7500 \\ & 1.183 & 0.000 & 0.000 & -.4330 & -.7500 & -.7500 & -.4330 \\ & & 2.549 & -1.183 & -.2500 & .4330 & -1.299 & .7500 \\ & & & 1.183 & .4330 & -.7500 & .7500 & -.4330 \\ & & & & .5000 & 0.000 & 0.000 & 0.000 \\ \text{(symmetric)} & & & & & 3.232 & 0.000 & -1.732 \\ & & & & & & 2.598 & 0.000 \\ & & & & & & & 2.598 \end{bmatrix}$$

a. Treating node 4 as an interior node, compute the boundary stiffness matrix $[\mathbf{k}_b]$.

b. If the truss is loaded and constrained as shown in part (b) of the figure, calculate the displacements at nodes 2 and 3.

c. Use the results of (b) to calculate the displacements at node 4.

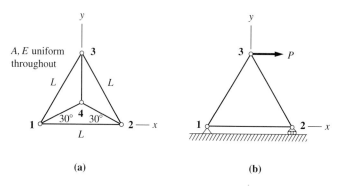

(a)   (b)

Exercise 11.24

11.25   Treating degree of freedom 4 as an internal degree of freedom, use static condensation to find the 3 by 3 stiffness matrix of a beam element with a pin (a *moment release*) at one end.

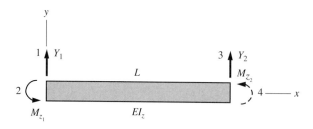

Exercise 11.25

11.26   Use the results of Exercise 11.25 to find the shear and bending moment distribution across the pinned beam structure.

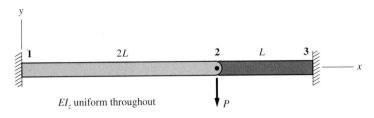

Exercise 11.26

11.27   Let node 1 be a boundary node and node 2 be an interior node of the stepped cantilever beam. Show that the boundary stiffness matrix is

$$[\mathbf{k}_b] = \frac{EI}{L^3} \begin{bmatrix} \frac{216}{99} & -\frac{180}{99}L \\ -\frac{180}{99}L & \frac{216}{99}L^2 \end{bmatrix}$$

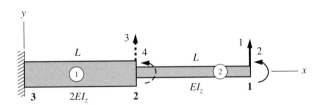

Exercise 11.27

11.28 Let node 1 be a boundary node and node 2 be an internal node of the constrained truss. Show that the boundary stiffness matrix is

$$[\mathbf{k}_b] = \frac{AE}{L} \begin{bmatrix} 2 & 0 \\ 0 & \sqrt{2} - 1 \end{bmatrix}$$

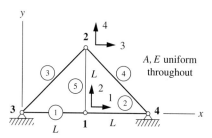

Exercise 11.28

11.29 The stiffness matrix of the constrained truss is

$$[\mathbf{k}] = \frac{AE}{L} \begin{bmatrix} 0.6078 & -0.1789 & -0.3578 & 0.1789 \\ -0.1789 & 0.08944 & 0.1789 & -0.08944 \\ -0.3578 & 0.1789 & 1.573 & -0.1789 \\ 0.1789 & -0.08944 & -0.1789 & 0.2683 \end{bmatrix}$$

Find the boundary stiffness matrix $[\mathbf{k}_b]$, if node 2 is designated an interior node.

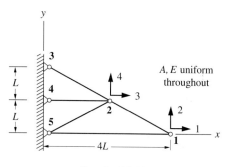

Exercise 11.29

11.30   If the center node of the three-node, quadratic-displacement rod element of Exercise 10.12 is taken to be the interior node, show that the 2 by 2 boundary stiffness matrix is that of a linear-displacement rod element, Equation 10.2.7.

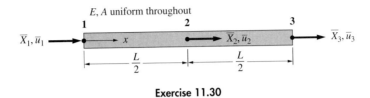

Exercise 11.30

11.31   The 8×8 stiffness matrix of the constrained structure is given in the following figure. Treating nodes 1 and 2 as interior nodes, find the 4 by 4 boundary stiffness matrix of the constrained structure. Throughout, $E = 10 \times 10^6$ psi, $G = 4 \times 10^6$ psi, panel thickness = 0.1 in., and rod area = 0.01 in.^{2}.

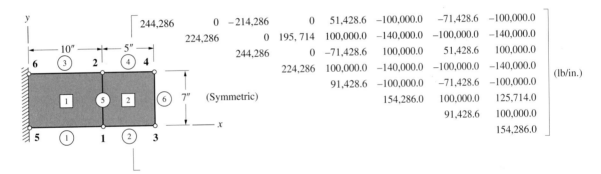

Exercise 11.31

c h a p t e r

12

Structural Stability

CHAPTER OUTLINE

12.1   INTRODUCTION

Buckling is the finite bowing, warping, wrinkling, or twisting deformation that accompanies the development of excessive compressive stresses throughout a structure or some portion thereof. *Primary* buckling deformation extends over the major dimensions of a structure; *secondary* buckling is confined to localized regions, such as the cross sections of individual members.[1] Primary buckling followed by sustained loading usually results in significant, if not total, loss of structural integrity. The large post-buckling displacements, if unrestrained, can lead to the permanent set or rupture of metal structures and the fracture of brittle material. Secondary buckling may cause some elements of the structure to fail, resulting in a redistribution of stress, which—depending on the design—may or may not degrade structural integrity.

To a large extent, modern aircraft structures consist of thin sheets attached to slender stiffeners. Buckling of these lightweight members can occur at stresses well below the elastic limit. Predicting the resistance of compression and shear members to elastic buckling is a critical part of the structural design process. This chapter is a brief introduction to elementary elastic-stability features that are relevant to the analysis and design of aircraft structures.

12.2 UNSTABLE BEHAVIOR

Let us first consider the behavior of a simplified system that characterizes some of the primary stability phenomena. Figure 12.2.1 shows a perfectly straight, rigid rod that is vertical when the linearly elastic spring is unstretched. The left end of the spring is free to move vertically in the smooth slot. An axial compressive load P is applied to the top of the rod. Neglecting gravity, let us write the equation of motion governing θ, the single coordinate required to specify the configuration of this system. Since the rod is constrained to rotate about the smooth pin at O, rigid-body dynamics states that[2]

$$M_O = I_O \ddot{\theta}$$

where I_O is the mass moment of inertia of the rod about O, and $\ddot{\theta}$ is the angular acceleration, measured positive clockwise, as is the net moment M_O about O. Summing the moments of P and the spring force in Figure 12.2.1b, we obtain

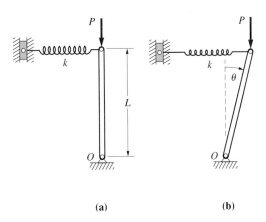

(a)   (b)

Figure 12.2.1 Rigid rod and linear spring assembly.
(a) Before deformation. (b) After deformation

[1] Primary and secondary do not refer to the relative importance of these types of buckling.
[2] See, for example, A. Bedford, and W. Fowler, *Engineering Mechanics: Dynamics*, Chapter 7, New York, Addison-Wesley, 1994. The dots over a symbol represent derivatives with respect to time: $\ddot{\theta} = d^2\theta/dt^2$.

$$P \times (L \sin\theta) - k(L \sin\theta) \times (L \cos\theta) = I_O \ddot{\theta} \qquad\qquad \textbf{[12.2.1]}$$

where k is the spring constant, which is the ratio of spring force to spring extension. Upon simplification, this becomes

$$\ddot{\theta} + \frac{L}{I_O}(kL \cos\theta - P)\sin\theta = 0 \qquad\qquad \textbf{[12.2.2]}$$

Let us assume that θ is small enough that the approximations $\sin\theta \cong \theta$ and $\cos\theta \cong 1$ are accurate. The differential equation can then be written

$$\ddot{\theta} + \frac{L}{I_O}(kL - P)\theta = 0 \qquad\qquad \textbf{[12.2.3]}$$

Let us also assume that at time $t = 0$, $\theta = \theta_0$, and $\dot{\theta} = 0$. That is, the rod is *perturbed* slightly from its vertical orientation by the amount of the small angular deflection θ_0. The form of the solution of Equation 12.2.3 depends on the sign of the coefficient of θ.

If $P < kL$, the coefficient of θ is positive and the solution is sinusoidal, as follows:

$$\theta = \theta_0 \cos\sqrt{\frac{L}{I_O}(kL - P)}\,t \qquad P < kL \qquad\qquad \textbf{[12.2.4a]}$$

The rod oscillates about the undeformed configuration, with the small amplitude of the disturbance θ_0. Eventually, ever-present friction, however small, will bring the rod to rest at equilibrium in the vertical, or nearly vertical, orientation. So, for $P < kL$, the system of Figure 12.2.1(a) is in *stable* equilibrium, much like a marble in the bottom of a teacup.

If $P > kL$, the solution is expressed in terms of the hyperbolic cosine:

$$\theta = \theta_0 \cosh\sqrt{\frac{L}{I_O}(P - kL)}\,t \qquad P > kL \qquad\qquad \textbf{[12.2.4b]}$$

Here, the rod deviation from the vertical does not remain small, but increases exponentially in time, rapidly exceeding the small perturbation θ_0 and never returning to the vertical. Thus, if $P > kL$, the system is in *unstable* equilibrium, like a marble perched on a basketball.

Finally, if $P = kL$, we have

$$\theta = \theta_0 \qquad P = kL \qquad\qquad \textbf{[12.2.4c]}$$

In this case, the system simply remains at rest in its slightly perturbed configuration, neither oscillating nor diverging, like a marble on a flat table top if nudged to a neighboring location. This is called *neutral* equilibrium. For an elastic system, the load corresponding to neutral equilibrium is called the *critical* load P_{cr}.

If the load in Figure 12.2.1a exceeds P_{cr}, the system becomes unstable and seeks another equilibrium configuration, if it exists. Observe that setting $\ddot{\theta} = 0$ in Equation 12.2.1 yields

$$P = kL \cos\theta_{eq} \qquad\qquad \textbf{[12.2.5]}$$

That is, the rod can be in equilibrium at the inclination $\theta > 0$ *if* the load is reduced to the value given in Equation 12.2.5. However, this "post-buckled" equilibrium state is unstable, since

$$\frac{dP}{d\theta_{eq}} = -kL \sin\theta_{eq} < 0$$

A small increase $\Delta\theta_{eq}$ in the angular orientation is accompanied by a *decrease* in the load, which in turn causes an additional increment of θ_{eq}, and so on. When this situation occurs, we say the structure has *negative stiffness*. We can also check the stability of the post-buckled configuration (Figure 12.2.1b) by holding θ_{eq} fixed and applying a small perturbation, replacing θ_{eq} by $\theta_{eq} + \delta\theta$ in Equation 12.2.2,

$$\delta\ddot{\theta} + \frac{L}{I_O}\left[kL\cos\left(\theta_{eq} + \delta\theta\right) - P\right]\sin\left(\theta_{eq} + \delta\theta\right) = 0 \qquad \text{[12.2.6]}$$

Since $\delta\theta << 1$, a Taylor series expansion to the first order[3] yields

$$\cos(\theta_{eq} + \delta\theta) = \cos\theta_{eq} - \sin\theta_{eq}\,\delta\theta$$

and

$$\sin(\theta_{eq} + \delta\theta) = \sin\theta_{eq} + \cos\theta_{eq}\,\delta\theta$$

Substituting these into Equation 12.2.6, applying Equation 12.2.5, and neglecting terms with $\delta\theta^2$, we get

$$\delta\ddot{\theta} - \frac{L}{I_O}\left[kL\sin^2\theta_{eq}\right]\delta\theta = 0 \qquad \text{[12.2.7]}$$

Since the coefficient of $\delta\theta$ is clearly negative, $\delta\theta$ increases exponentially with time, confirming that the post-buckled configuration is unstable.

We can summarize our results on a load–deflection diagram, such as that shown in Figure 12.2.2. As the load P increases from zero, the rod remains vertical and in stable equilibrium until the critical load P_{cr} is reached. At that point, the load path *bifurcates*, or divides into two branches. On one, the rod remains vertical, in unstable equilibrium as the load increases. On the other, the post-buckling path, the rod leans away from vertical, in unstable equilibrium, with diminished load-carrying capacity.

The post-buckled configuration *can* be stable. If we replace the linear spring in Figure 12.2.1 by a torsional spring at pin O, we have the situation illustrated in Figure 12.2.3.

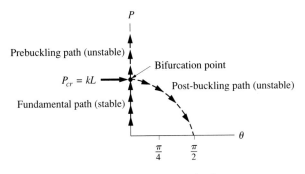

Figure 12.2.2   Load–deflection curve for the structure in Figure 12.2.1.

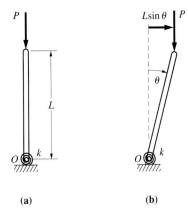

Figure 12.2.3   Rigid rod and torsional spring assembly. (a) Fundamental configuration. (b) Post-buckled configuration.

---

[3]Taylor series expansion of a function $f(x)$ in the neighborhood of x: $f(x + h) = f(x) + hf'(x) + \dfrac{h^2}{2!}f''(x) + \dfrac{h^3}{3!}f'''(x) + \cdots$.

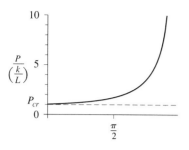

Figure 12.2.4 Load–deflection curve for the post-buckling path (Figure 12.2.3b).

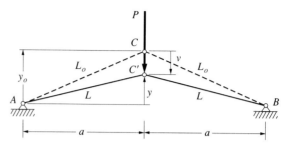

Figure 12.2.5 Shallow truss.

To investigate the stability of the vertically oriented rod, we proceed as before. Summing the moments about O leads to the following differential equation for *small* θ:

$$\ddot{\theta} + \frac{k - PL}{I_O}\theta = 0$$

where k is the torsional spring constant. It is clear that in this case,

$$P_{cr} = \frac{k}{L}$$

For equilibrium of the post-buckled configuration, $PL\sin\theta - k\theta = 0$, or

$$P = \frac{k}{L}\frac{\theta}{\sin\theta}$$

This load–deflection curve is plotted in Figure 12.2.4. Observe that in this case, $dP/d\theta$ is never negative, so the post-buckling path is stable. At the onset of buckling, the stiffness starts at zero and increases very slightly with increasing θ, until the rod approaches the horizontal orientation, and increases evermore rapidly thereafter. The small, positive, initial post-buckling stiffness means that for P even *slightly* greater than P_{cr}, a large angular deflection will occur after buckling and before the structure assumes its stable post-buckled orientation.

Another type of instability is illustrated by a shallow, symmetric truss, loaded in compression, as shown in Figure 12.2.5. During the application of load P, the point of application C moves through a displacement v from its initial vertical position y_o to point C' with coordinate y ($y < y_o$). Let us use the principle of virtual work to find the relationship between the load P and displacement v.

"Shallow" implies that $y_o/a \ll 1$. The initial length L_o and final length L of the two elastic rods are[4]

$$L_o = (a^2 + y_o^2)^{\frac{1}{2}} = a\left[1 + \left(\frac{y_o}{a}\right)^2\right]^{\frac{1}{2}} \cong a\left[1 + \frac{1}{2}\left(\frac{y_o}{a}\right)^2\right] \qquad \text{[12.2.8a]}$$

---

[4] $(1 + x)^{\frac{1}{2}} = 1 + \frac{x}{2} - \frac{x^2}{8} + \frac{x^3}{16} \cdots$.

$$L = (a^2 + y^2)^{\frac{1}{2}} = a\left[1 + \left(\frac{y}{a}\right)^2\right]^{\frac{1}{2}} \cong a\left[1 + \frac{1}{2}\left(\frac{y}{a}\right)^2\right] \qquad \text{[12.2.8b]}$$

where terms of higher order than $(y_o/a)^2$ are neglected. The axial strain in each rod is therefore

$$\varepsilon = \frac{L - L_o}{L_o} = \frac{L}{L_o} - 1 = \frac{\left[1 + \frac{1}{2}\left(\frac{y}{a}\right)^2\right]}{\left[1 + \frac{1}{2}\left(\frac{y_o}{a}\right)^2\right]} - 1$$

Again, using the fact that $(y_o/a)^2 \ll 1$, we get[5]

$$\varepsilon = \left[1 + \frac{1}{2}\left(\frac{y}{a}\right)^2\right]\left[1 - \frac{1}{2}\left(\frac{y_o}{a}\right)^2\right] - 1 = \frac{1}{2}\left[\left(\frac{y}{a}\right)^2 - \left(\frac{y_o}{a}\right)^2\right] \qquad \text{[12.2.9]}$$

From this expression for the true strain ε, we obtain the virtual strain $\delta\varepsilon$ by treating δ as the differential operator, to get

$$\delta\varepsilon = \frac{y}{a^2}\delta y \qquad \text{[12.2.10]}$$

According to Equation 6.3.4, the internal virtual work for a linearly-elastic rod in uniaxial stress is

$$\delta W_{\text{int}} = \iiint_V \sigma_x \delta\varepsilon_x dV = \int_0^L (E\varepsilon_x)\delta\varepsilon_x A dx$$

Therefore, for this truss, in which the axial rigidity AE and the axial strain ε are constant, we have

$$\delta W_{\text{int}} = 2 \times AEL\varepsilon\delta\varepsilon$$

which is valid as long as the stress in the rods remains below the elastic limit. Substituting Equations 12.2.8b, 12.2.9, and 12.2.10 yields

$$\delta W_{\text{int}} = 2AEa\left[1 + \frac{1}{2}\left(\frac{y}{a}\right)^2\right]\frac{1}{2}\left[\left(\frac{y}{a}\right)^2 - \left(\frac{y_o}{a}\right)^2\right]\frac{y}{a^2}\delta y$$

Neglecting the products of y_o/a and y/a that are higher than order 2 reduces this to

$$\delta W_{\text{int}} = \frac{AE}{a^3}\left(y^2 - y_o^2\right)y\delta y$$

which can be written in terms of the true and virtual displacements by observing that $y = y_o - v$ (and therefore $\delta y = -\delta v$), so that

$$\delta W_{\text{int}} = \frac{AE}{a^3}(2y_o - v)(y_o - v)v\delta v$$

---

[5] $(1+x)^{-1} = 1 - x + x^2 - 2x^3 + \cdots$.

The external virtual work is simply

$$\delta W_{ext} = P \delta v$$

Setting δW_{ext} equal to δW_{int}, we find that

$$P = \frac{AE}{a^3}(2y_o - v)(y_o - v)v$$

which is the load–deflection relationship we seek. Since the load is a cubic function of the displacement v, the equation is nonlinear. If we nondimensionalize both sides of the equation, it can be written

$$\bar{P} = (2 - \bar{v})(1 - \bar{v})\bar{v} \qquad\qquad [12.2.11]$$

where $\bar{P} = (P/AE)(a/y_o)^3$ and $\bar{v} = v/y_o$.

Figure 12.2.6 is a plot of this function. Observe that the structure is stable as the load is first applied; however, when the value of $\bar{P}$ reaches 0.385 at point A, the stiffness goes to zero and the unstable structure "snaps through," undergoing a five-fold increase in deflection, reaching point C on the second stable, positive-stiffness portion, which starts at B. Examples of elastic structures in which snap-through buckling can occur include slender shallow arches and thin-walled shallow domes, such as the bottom of an oilcan. Hence, the term "oilcanning" is commonly used to refer to the snap-through phenomenon.

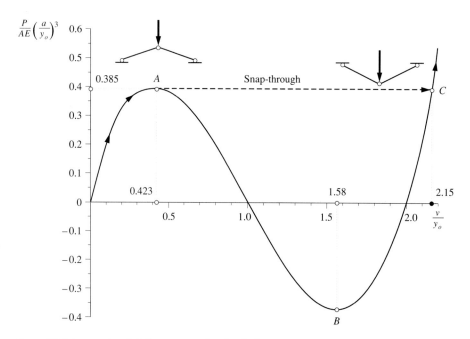

Figure 12.2.6   Load–deflection curve for the shallow truss in Figure 12.2.5. Inserts depict the configuration of the structure at A and C.

12.3   BEAM COLUMNS

A *column* is a straight bar subjected to compressive axial load. A *beam column* is a bar subjected to compressive axial load, as well as transverse load. Figure 12.3.1a shows a simply-supported, linearly-elastic, simple beam with a transverse load Q applied at its midspan and a compressive load P directed along its centroidal axis. The $x\,y$ plane in which bending occurs is a plane of symmetry of the cross section. Let us calculate the maximum lateral deflection of the beam, which symmetry dictates is at point C, where the load Q acts.

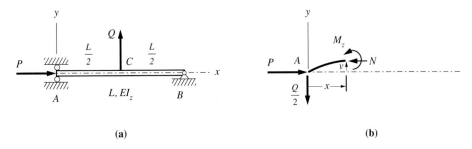

(a)   **(b)**

Figure 12.3.1   (a) Simply-supported beam column. (b) Free-body diagram of a portion of the deformed column.

Applying statics to the free body in Figure 12.3.1b and summing the moments around the neutral axis at the cut, we get

$$M_z + Pv + \frac{Q}{2}x = 0$$

where v is the lateral deflection, as illustrated in the figure. Maintaining the small strain/small curvature assumptions that we have used throughout our study of beams allows us to use Equation 10.3.19 to relate the bending moment M_z to the curvature d^2v/dx^2. The equilibrium equation can then be written

$$EI_z\frac{d^2v}{dx^2} + Pv + \frac{Q}{2}x = 0$$

Dividing by the flexural rigidity EI_z and rearranging terms, we get

$$\frac{d^2v}{dx^2} + \frac{P}{EI_z}v = -\frac{Q}{2EI_z}x \qquad \text{[12.3.1]}$$

It is easy to verify that the general solution of this differential equation is

$$v = \overbrace{A\sin(\sqrt{\frac{P}{EI_z}}x) + B\cos(\sqrt{\frac{P}{EI_z}}x)}^{\text{complementary solution}} + \overbrace{(-\frac{Q}{2P}x)}^{\text{particular solution}} \qquad \text{[12.3.2]}$$

To evaluate the constants of integration, A and B, we apply the boundary conditions. Requiring $v = 0$ at $x = 0$ implies $B = 0$, so we are left with

$$v = A \sin\left(\sqrt{\frac{P}{EI_z}}x\right) - \frac{Q}{2P}x \qquad \text{[12.3.3]}$$

Then, taking the first derivative yields the slope of the elastic curve, which is

$$\frac{dv}{dx} = A\sqrt{\frac{P}{EI_z}}\cos\left(\sqrt{\frac{P}{EI_z}}x\right) - \frac{Q}{2P}$$

By symmetry, the tangent to the elastic curve must be horizontal at the midspan. Setting $dv/dx = 0$ at $x = L/2$ requires that

$$A = \frac{1}{\sqrt{\frac{P}{EI_z}}\cos\left(\sqrt{\frac{P}{EI_z}}\frac{L}{2}\right)}\frac{Q}{2P}$$

Substituting this into Equation 12.3.3 and evaluating the resulting expression at $x = L/2$ yields

$$q = \frac{\frac{1}{2P}\tan\left(\sqrt{\frac{P}{EI_z}}\frac{L}{2}\right) - \frac{1}{2P}\sqrt{\frac{P}{EI_z}}\frac{L}{2}}{\sqrt{\frac{P}{EI_z}}}Q$$

where q is the displacement at the point of application of the load Q. Simplifying this equation, we get

$$q = \frac{\tan\xi - \xi}{\xi^3}\frac{QL^3}{16EI_z} \qquad \text{[12.3.4]}$$

where ξ is the dimensionless quantity

$$\xi = \sqrt{\frac{P}{EI_z}}\frac{L}{2} \qquad \text{[12.3.5]}$$

Solving Equation 12.3.4 for the load Q in terms of the displacement q, we get

$$Q = Kq \qquad \text{[12.3.6]}$$

where K is the *flexural stiffness coefficient* and is given by

$$K = \frac{\xi^3}{\tan\xi - \xi}\frac{16EI_z}{L^3} \qquad \text{[12.3.7]}$$

The term K is a measure of the resistance of the beam column to lateral displacement. For a given column, K is a function of the compressive axial load P. Figure 12.3.2 is a plot of K versus ξ. Observe that the flexural stiffness decreases with increasing axial load, finally going to zero at $\xi = \pi/2$ (and becoming negative thereafter). A nonpositive flexural stiffness means that the beam is unstable: it cannot resist even the slightest tendency to nudge it away from its straight, equilibrium configuration. A lateral perturbation, no matter how small, will precipitate buckling, a large lateral deflection of the column.

The magnitude of the buckled deflection cannot be calculated using small displacement theory. Nevertheless, we can predict the onset of buckling by noting when the flexural stiffness vanishes. The axial load at which this occurs is the critical load P_{cr}. Figure 12.3.2 reveals that for a simply-supported elastic column,

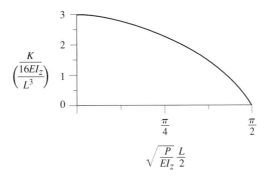

Figure 12.3.2   Bending stiffness of the simply-supported column as a function of compressive axial load.

$$P_{cr} = \frac{\pi^2 E I_z}{L^2} \qquad \text{[12.3.8]}$$

Equation 12.3.8 is known as the *Euler column formula* and P_{cr} is the *Euler buckling load.*

   To explore the notions of bending stiffness, elastic instability, and buckling load, we have chosen to apply the transverse load at the midspan of the beam column. However, it must be noted that the formula for the critical load of a simply-supported column is independent of the nature and location of the transverse disturbance.

12.4   SLENDER COLUMN BUCKLING

We have seen that a critical axial-load value exists, beyond which a column will buckle. To predict the onset of buckling using linear, engineering beam theory (neglecting shear), we presume that the initially perfectly straight column has deflected laterally (only slightly, and not beyond the limits of small deflection theory and linear elastic behavior) and is in equilibrium under the action of just the critical axial load and the reactions at the supports. Unlike beam theory, the equations of equilibrium for the deformed configuration are written so as to include the moment of the axial load about a section (cf. Figure 10.3.1b). We will apply this linear column theory approach to four common boundary conditions.

12.4.1   Pinned-Pinned Column

Let us start with the simply-supported, or "pinned-pinned," column of the previous section. Figure 12.4.1a shows the column, with no transverse load, but with an axial load directed precisely along the centroidal axis, at its critical value.

   As before, we treat the column as we would a simple beam in bending, so that the plane in which bending occurs contains a symmetry axis of the cross section. Applying statics to the deformed free body in Figure 12.4.1b, we are led again to Equation 12.3.1 with $Q = 0$, that is,

$$\frac{d^2v}{dx^2} + \frac{P_{cr}}{E I_z}v = 0 \qquad \text{[12.4.1]}$$

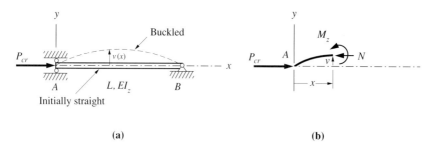

Figure 12.4.1   (a) Pinned-pinned column before and after buckling. (b) Free-body diagram of the elastically bent column.

It is customary to introduce the notation

$$\lambda^2 = \frac{P_{cr}}{EI_z}$$ [12.4.2]

so that

$$\frac{d^2v}{dx^2} + \lambda^2 v = 0$$ [12.4.3]

The solution of this homogeneous[6] differential equation is

$$v = A \sin \lambda x + B \cos \lambda x$$

To satisfy the boundary conditions, v must vanish at each end of the column. Setting $v = 0$ at $x = 0$ means that $B = 0$, leaving us with

$$v = A \sin \lambda x$$ [12.4.4]

Then, requiring $v = 0$ at $x = L$ implies that

$$A \sin \lambda L = 0$$

This equation can be satisfied by requiring $A = 0$. However, since we also have $B = 0$, that would mean $v = 0$ everywhere in the column. In other words, the column has not buckled. Since we are not interested in this "trivial" solution, we instead require that

$$\sin \lambda L = 0$$

This equality holds if

$$\lambda = \frac{n\pi}{L}$$

where n is a positive integer (but not zero, since that would again lead to the trivial solution). Substituting this expression for λ into Equation 12.4.2, we deduce

$$P_{cr} = n^2 \frac{\pi^2 EI_z}{L^2} \qquad n = 1, 2, 3, \cdots$$

| [6]The right-hand side is zero.

Apparently, there are a countably infinite number of buckling loads. For the nth such load, the deformed shape of the column, or the *mode shape* $v_n(x)$, is given by Equation 12.4.4:

$$v_n = A_n \sin \lambda_n x = A_n \sin \frac{n\pi x}{L}$$

The first three of these sine-wave modes is shown in Figure 12.4.2. In reality, if buckling occurs, it happens at the lowest possible mode. Thus, for the pinned-pinned condition, the buckling load is the Euler load found in the previous section,

$$P_{cr} = \frac{\pi^2 E I_z}{L^2} \qquad \text{Pinned-pinned column} \qquad \text{[12.4.5]}$$

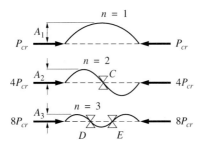

Figure 12.4.2   First three buckling modes of a pinned-pinned column.

The load P_{cr} is clearly a minimum for the smallest possible value of I_z. Thus, unless the column is restrained against it, buckling will occur in the plane normal to the minimum principal moment of inertia axis.

Observe that if the column is braced at the inflection points of the higher modes (point C, or points D and E in Figure 12.4.2), it is forced to buckle at loads significantly larger than the Euler load. The alternative to bracing is to raise the Euler load by increasing the cross-sectional moment of inertia.

Although we have obtained the buckled mode *shape*, its amplitude A_n remains undetermined. No matter what value we assign to the amplitude of a given mode, Equation 12.4.3 is satisfied. This indeterminacy is due to the small slope/small curvature assumption on which beam theory is based (cf. Section 4.5).

12.4.2   Fixed-Fixed Column

Both ends of a fixed-fixed column are restrained against rotation, as shown in Figure 12.4.3. The structure is clearly statically indeterminate. Using the free-body diagram of Figure 12.4.3b, we obtain the moment equilibrium equation,

$$M_z + P_{cr}v = M_A$$

Substituting the moment–curvature equation, $M_z = EI_z d^2v/dx^2$ and dividing by the flexural rigidity yields

$$\frac{d^2v}{dx^2} + \lambda^2 v = \frac{M_A}{EI_z} \qquad \text{[12.4.6]}$$

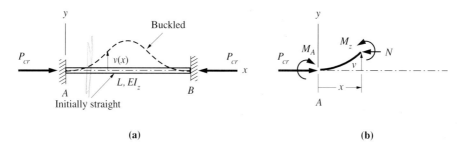

Figure 12.4.3   (a) Fixed-fixed column. (b) Free-body diagram of a portion of the deformed column.

where again

$$\lambda^2 = \frac{P_{cr}}{E I_z} \tag{12.4.7}$$

It is easy to verify that the solution to this differential equation is

$$v = \overbrace{A \sin \lambda x + B \cos \lambda x}^{\text{complementary solution}} + \overbrace{\frac{M_A}{\lambda^2 E I_z}}^{\text{particular solution}}$$

Taking the derivative of this expression defines the slope of the elastic curve,

$$\frac{dv}{dx} = \lambda A \cos \lambda x - \lambda B \sin \lambda x$$

There are four boundary conditions, two at each end of the column. At $x = 0$, we have $v = 0$ and $dv/dx = 0$, which requires that $A = 0$ and $B = -M_A/\lambda^2 E I_z$, so we are left with

$$v = \frac{M_A}{\lambda^2 E I_z}(1 - \cos \lambda x) \tag{12.4.8}$$

and

$$\frac{dv}{dx} = \frac{M_A}{\lambda E I_z} \sin \lambda x \tag{12.4.9}$$

For dv/dx to vanish at $x = L$, Equation 12.4.9 requires (for the nontrivial solution) that $\sin \lambda L = 0$. This in turn implies that

$$\lambda L = n\pi, \ n = 1, 2, 3, \cdots \tag{12.4.10}$$

Finally, Equation 12.4.8 yields $v = 0$ at $x = L$ only if $\cos \lambda L = 1$, and this is true if

$$\lambda L = n\pi, \ n = 2, 4, 6, \cdots \tag{12.4.11}$$

From Equations 12.4.10 and 12.4.11 we conclude that

$$\frac{P_{cr}}{E I_z} = \lambda^2 = \frac{n^2 \pi^2}{L^2}, \ n = 2, 4, 6, \cdots$$

Taking the lowest possible value of n yields the critical buckling load,

$$P_{cr} = 4\frac{\pi^2 E I_z}{L^2} \qquad \text{Fixed-fixed column} \qquad [12.4.12]$$

The mode shape is

$$v = \frac{M_A}{P_{cr}}\left(1 - \cos\frac{2\pi x}{L}\right)$$

Since we have no means of computing a value for M_A, the amplitude is indeterminate.

Observe that restraining the rotations at the supports increases the buckling load by a factor of four over that of the simply-supported beam. Another way of looking at this is that the buckling load of a fixed-fixed column equals that of a pinned-pinned column half as long, that is,

$$P_{cr} = \frac{\pi^2 E I_z}{(L/2)^2}$$

We say that the *effective length* L_e of a fixed-fixed column is $L/2$.

In general, for a long column of actual length L, we can express the critical load as

$$P_{cr} = \frac{\pi^2 E I}{L_e^2} = c\frac{\pi^2 E I}{L^2} \qquad [12.4.13]$$

where

$$L_e = \frac{L}{\sqrt{c}} \qquad [12.4.14]$$

The constant c is the *coefficient of constraint* or *end fixity factor*, which, as we have seen, depends on the manner in which the column is restrained at each end.

12.4.3   Pinned-Fixed Column

This type of column has a simple support at one end and is built in at the other. Figure 12.4.4b shows a free-body diagram of a deformed column portion lying between the simple support and station x. The diagram is identical to Figure 12.3.1 if we replace P with P_{cr} and $Q/2$ with $-Y_A$. Doing so in Equation 12.3.2 yields

$$v = \overbrace{A \sin \lambda x + B \cos \lambda x}^{\text{complementary solution}} + \overbrace{\frac{Y_A}{P_{cr}}x}^{\text{particular solution}}$$

The boundary condition at $x = 0$ is $v = 0$. Therefore, $B = 0$ and

$$v = A \sin \lambda x + \frac{Y_A}{P_{cr}}x \qquad [12.4.15]$$

and

$$\frac{dv}{dx} = \lambda A \cos \lambda x + \frac{Y_A}{P_{cr}}$$

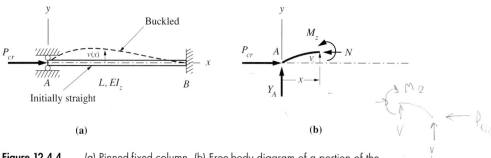

Figure 12.4.4   (a) Pinned-fixed column. (b) Free-body diagram of a portion of the deformed column, starting at the simple support.

Since $v = 0$ and $dv/dx = 0$ at $x = L$, we get

$$A \sin \lambda L = -\frac{Y_A}{P_{cr}} L \quad \text{and} \quad \lambda A \cos \lambda L = -\frac{Y_A}{P_{cr}}$$ [12.4.16]

Dividing the first of these two equations by the second yields

$$\frac{\sin \lambda L}{\lambda \cos \lambda L} = L$$

or

$$\tan \lambda L = \lambda L$$

The smallest root of this transcendental equation can be found by using an iterative algorithm, such as the Newton-Raphson method, or by simply plotting the function $f(x) = \tan x - x$, as in Figure 12.4.5, which shows that

$$\lambda L = 4.4934 = 1.4303\pi$$ [12.4.17]

Since $P_{cr} = EI_z \lambda^2$, the critical load for this set of constraints is

$$P_{cr} = 2.046 \frac{\pi^2 E I_z}{L^2} = \frac{\pi^2 E I_z}{(0.6992L)^2} \qquad \text{Pinned-fixed column}$$ [12.4.18]

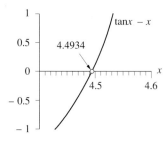

Figure 12.4.5   Graph of tan x − x versus x in the region of the first positive root, which is shown.

Clearly, $L_e = 0.6992L$ and $c = 2.046$. As expected, the buckling load for a column with the pinned-fixed restraint lies between that of the pinned-pinned ($c = 1$) and fixed-fixed ($c = 4$) end conditions.

According to Equations 12.4.13, 14, and 15, the buckled mode shape is

$$\frac{v}{L} = \frac{Y_A}{P_{cr}} \left(\frac{x}{L} - \frac{\sin \lambda x}{\sin \lambda L} \right) = \frac{Y_A}{P_{cr}} \left(\frac{x}{L} + 1.024 \sin 1.430\pi \frac{x}{L} \right)$$

The amplitude is undetermined, since the vertical pin reaction Y_A cannot be evaluated.

12.4.4   Free-Fixed Column

As indicated in Figure 12.4.6a, a *free-fixed* column has one end built in and the other end free of displacement constraints.

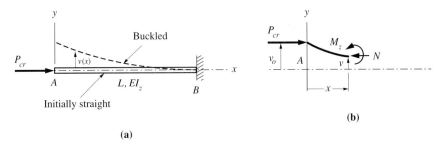

(a)

(b)

Figure 12.4.6   (a) Free-fixed column. (b) Free-body diagram of a portion of the deformed column, beginning with the free end.

Summing moments about the neutral axis at the right end of the free body in Figure 12.4.6b yields

$$M_z + P_{cr}v = P_{cr}v_o$$

where v_o is the deflection at the free end of the beam ($x = 0$). As before, we substitute the moment–curvature relation into this expression to obtain

$$\frac{d^2v}{dx^2} + \lambda^2 v = \lambda^2 v_o$$

in which

$$\lambda^2 = \frac{P_{cr}}{EI_z} \qquad\qquad \text{[12.4.19]}$$

The solution to the differential equation is

$$v = \overbrace{A \sin \lambda x + B \cos \lambda x}^{\text{complementary solution}} + \overbrace{v_o}^{\text{particular solution}}$$

Since $v = v_o$ at $x = 0$, it follows that $B = 0$; therefore,

$$v = A \sin \lambda x + v_o \qquad \text{[12.4.20]}$$

Accordingly,

$$\frac{dv}{dx} = \lambda A \cos \lambda x$$

Since the column is built in at $x = L$, we have $v = dv/dx = 0$ at that point, which means that

$$A \sin \lambda L + v_o = 0 \qquad \text{[12.4.21]}$$

and

$$\lambda A \cos \lambda L = 0 \qquad \text{[12.4.22]}$$

For a nontrivial solution, Equation 12.4.22 requires that $\cos \lambda L = 0$, the roots of which are

$$\lambda L = \frac{n\pi}{2}, \ n = 1, 3, 5, \cdots$$

As usual, we take only the smallest integer, or $\lambda L = \pi/2$. In this case, Equation 12.4.21 yields

$$A = -\frac{v_o}{\sin(\pi/2)} = -v_o \qquad \text{[12.4.23]}$$

From Equation 12.4.19, we find the critical load,

$$P_{cr} = \frac{1}{4}\frac{\pi^2 E I_z}{L^2} = \frac{\pi^2 E I_z}{(2L)^2} \qquad \text{Free-fixed column} \qquad \text{[12.4.24]}$$

With free-fixed constraints, a column buckles at *one-fourth* the critical load of a pinned-pinned column; therefore, $L_e = 2L$ and $c = 0.25$.

The shape of the fundamental mode is given by Equations 12.4.20 and 12.4.23:

$$v = v_o \left(1 - \sin\frac{\pi x}{2L}\right)$$

The amplitude v_o remains undetermined.

12.5   COLUMN IMPERFECTIONS AND LOAD MISALIGNMENT

In the previous section, we assumed that the column was initially straight and that the axial load was aligned perfectly with the centroidal axis. Let us investigate the effects of slight deviations from these conditions. We will first consider a column with initial imperfections.

Figure 12.5.1a shows a pinned-pinned column that is not straight, but is bent into an initial, unloaded shape. Unlike for the straight column, bending will occur immediately upon application of the axial load P, regardless of its magnitude, due to its offset from the slightly curved centerline of the bar. Figure 12.5.1b shows a free-body diagram of a portion of the column between the left end A and the cut at station x. Notice that the total deflection v_{tot} of the column at any point is the sum of its initial deviation v_o from a straight line and the additional deflection v due to the applied load P.

For moment equilibrium around the cut,

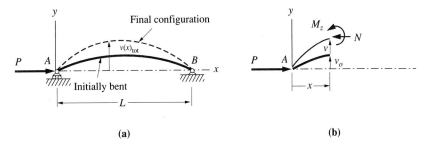

Figure 12.5.1   (a) Initially bent simply-supported column. (b) Free-body diagram of a portion of the column.

$$M_z + P v_{\text{tot}} = 0$$

However,

$$v_{\text{tot}} = v_o + v \qquad \text{[12.5.1]}$$

so that

$$M_z + Pv = -Pv_o$$

Let us assume that the initial shape of the bar is that of a sine function[7] with amplitude $a_o L$, where a_o, the *dimensionless* imperfection amplitude, is a very small number:

$$v_o = a_o L \sin \frac{\pi x}{L} \qquad \text{[12.5.2]}$$

Substituting this and the moment-curvature relation (Equation 10.3.19) into the equilibrium equation yields

$$\frac{d^2 v}{dx^2} + \lambda^2 v = -\lambda^2 a_o L \sin \frac{\pi x}{L} \qquad \text{[12.5.3]}$$

where

$$\lambda^2 = \frac{P}{E I_z} \qquad \text{[12.5.4]}$$

We can verify that the general solution to Equation 12.5.3 is

$$v = \overbrace{A \sin \lambda x + B \cos \lambda x}^{\text{complementary solution}} + \overbrace{\frac{\lambda^2 L^2}{\pi^2 - \lambda^2 L^2} a_o L \sin \frac{\pi x}{L}}^{\text{particular solution}}$$

At $x = 0$, v must vanish, which requires that $B = 0$. Therefore,

$$v = A \sin \lambda x + \frac{\lambda^2 L^2}{\pi^2 - \lambda^2 L^2} a_o L \sin \frac{\pi x}{L}$$

---

[7]Other functions (e.g., polynomials in x) could be chosen, but would only serve to complicate the analysis and would lead to the same conclusion reached by this choice.

The deflection v must also be zero at $x = L$, so that

$$A \sin \lambda L = 0$$

This can be true for *any* value of λ if $A = 0$, in which case

$$v = \frac{\lambda^2 L^2}{\pi^2 - \lambda^2 L^2} a_o L \sin \frac{\pi x}{L}$$

Substituting this expression and Equation 12.5.2 into Equation 12.5.1 yields the equation for the total deflection of the column:

$$v_{\text{tot}} = \left(a_o L \sin \frac{\pi x}{L} \right) + \left(\frac{\lambda^2 L^2}{\pi^2 - \lambda^2 L^2} a_o L \sin \frac{\pi x}{L} \right) = \frac{\pi^2}{\pi^2 - \lambda^2 L^2} a_o L \sin \frac{\pi x}{L}$$

At $x = L/2$, the lateral deflection takes on its maximum value,

$$\delta = \frac{\pi^2}{\pi^2 - \lambda^2 L^2} a_o L$$

Noting Equation 12.5.4, and recalling the Euler column formula, we can write this as

$$\delta = \frac{\pi^2}{\pi^2 - \frac{PL^2}{EI_z}} a_o L = \frac{\pi^2 EI_z}{\pi^2 EI_z - PL^2} a_o L = \frac{P_{cr} L^2}{P_{cr} L^2 - PL^2} a_o L$$

Setting $a = \delta/L$, we therefore obtain

$$a = \frac{1}{1 - P/P_{cr}} a_o \qquad \text{or} \qquad \frac{P}{P_{cr}} = 1 - \frac{a_o}{a} \qquad \text{[12.5.5]}$$

where a is the dimensionless midspan deflection.

The second of these two equations is plotted in Figure 12.5.2 for several values of a_o. Excessive lateral deflection of a slender column—$a \approx 0.1$ (ten percent of the column length)—occurs below the Euler buckling load, regardless of the size of the inevitable imperfection. However, the chart also shows that if a_o is sufficiently small, one can use the Euler buckling formula to predict the load capacity of a column, provided suitable safety factors are used (e.g., approximately 0.5 for steel).

Let us next consider a column that is geometrically perfect, but has an axial load applied off the centroidal axis, as illustrated in Figure 12.5.3a. The amount of the offset is called the *eccentricity*, e. To produce a symmetric deflection curve, thereby simplifying the analysis, the eccentricity must be the same at each end of the column. Using the free-body diagram in Figure 12.5.3b we obtain the following equilibrium equation:

$$M_z + P(e + v) = 0$$

Using the moment–curvature formula and Equation 12.5.4, we obtain

$$\frac{d^2 v}{dx^2} + \lambda^2 v = -\lambda^2 e$$

The general solution to this differential equation is

$$v = A \sin \lambda x + B \cos \lambda x - e$$

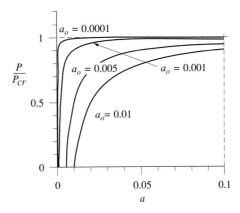

Figure 12.5.2   Axial load versus total (dimensionless) midspan deflection of a pinned-pinned column, for different amounts of initial imperfection.

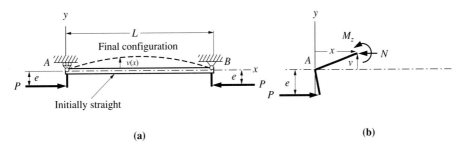

(a)   **(b)**

Figure 12.5.3   (a) Eccentrically loaded pinned-pinned column. (b) Free-body diagram of a portion of the column, beginning at the left support.

Setting the displacement equal to zero at $x = 0$, we find that $B = e$. Therefore,

$$v = A \sin \lambda x + e(\cos \lambda x - 1)$$

The displacement must also vanish at the right support, $x = L$. Therefore,

$$A = \frac{1 - \cos \lambda L}{\sin \lambda L} e$$

Simplifying, we finally obtain

$$v = \frac{\sin \lambda (L - x) + \sin \lambda x - \sin \lambda L}{\sin \lambda L} e$$

Since the lateral deflection is symmetric about the midspan of the column [that is, $v(x) = v(L - x)$], the maximum value δ occurs at $x = L/2$ and is given by

$$\delta = \frac{2 \sin \frac{\lambda L}{2} - \sin \lambda L}{\sin \lambda L} e$$

Using the trigonometric identity

$$\sin \lambda L = 2 \sin \frac{\lambda L}{2} \cos \frac{\lambda L}{2}$$

we reduce this to the following expression

$$a = \left(\frac{1}{\cos(\lambda L/2)} - 1 \right) a_e \qquad \text{[12.5.6]}$$

where $a = \delta/L$ and $a_e = e/L$ are the dimensionless midspan deflection and eccentricity, respectively. Writing Equation 12.5.4 in the form

$$\frac{\lambda L}{2} = \frac{L}{2} \sqrt{\frac{P}{EI_z}}$$

and the Euler column formula as

$$EI_z = \frac{L^2}{\pi^2} P_{cr}$$

we find that

$$\frac{\lambda L}{2} = \frac{\pi}{2} \sqrt{\frac{P}{P_{cr}}}$$

Substituting this expression into Equation 12.5.6, we get

$$\frac{a}{a_e} = \frac{1}{\cos \frac{\pi}{2} \sqrt{\frac{P}{P_{cr}}}} - 1$$

Solving this for P/P_{cr} yields

$$\frac{P}{P_{cr}} = \left[\frac{2}{\pi} \cos^{-1} \left(\frac{a_e}{a} \right) \right]^2 \qquad \text{[12.5.7]}$$

This load–deflection relation is plotted in Figure 12.5.4, which shows that if a column has even the smallest eccentricity, its load capacity is decreased. However, if the eccentricity is sufficiently small, the Euler buckling load can be used, given an appropriate safety factor, as previously discussed for a column with minor imperfections.

12.6   INELASTIC COLUMN BUCKLING

This section deals with plastic deformation effects on the buckling loads of columns made of ductile metal alloys. We will continue to assume that the columns are *torsionally stable*, that is, that the buckling mode is a lateral deflection only, and is not accompanied by a twist. In the context of buckling, it is convenient to set aside the usual notation and treat compressive stresses and strains as *positive* quantities.

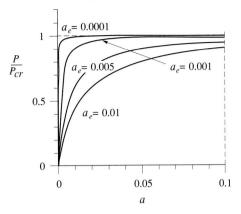

Figure 12.5.4   Load versus deflection for a pinned-pinned column with different eccentricities.

The Euler column formula, Equation 12.3.8, can be written generally as

$$P_{cr} = \frac{\pi^2 EI}{L_e^2} = c \frac{\pi^2 EI}{L^2}$$

[12.6.1]

where L is the length of the column, and L_e is the effective length, which is given by $L_e = L/\sqrt{c}$ (Section 12.4). Unless the column is constrained to buckle in a given plane, I is the minimum principal moment of inertia. Let σ_{cr} be the *critical stress* at which buckling occurs, so that

$$\sigma_{cr} = \frac{P_{cr}}{A}$$

where A is the cross-sectional area of the column. The moment of inertia is related to the area and the radius of gyration ρ by the formula $I = A\rho^2$. Therefore, Equation 12.6.1 becomes

$$\sigma_{cr} A = \frac{\pi^2 E\left(A\rho^2\right)}{L_e^2}$$

or

$$\sigma_{cr} = \frac{\pi^2 E}{(L_e/\rho)^2}$$

[12.6.2]

The term L_e/ρ is the *slenderness ratio* of the column.

Equation 12.6.2 is shown as the curve ABD in Figure 12.6.1a. In long columns, for which the slenderness ratio is large, σ_{cr} is below the proportional limit of the material. The figure shows that as the slenderness ratio is reduced, the critical stress increases, until we arrive at the proportional limit σ_P, which is followed more or less closely (depending on the material) by the yield point. As illustrated in Figure 12.6.1b, after σ_P, the stress-strain relation—assumed to be essentially the same for tension and compression—becomes nonlinear. Therefore, Equation 12.6.2 is no longer valid, and the curve in Figure 12.6.1a cannot be continued beyond point B upwards to the left towards D. Tests show that the reduced stiffness of the material near and after the onset of plastic deformation causes the curve to bend over and follow the trend BC. The curve CBA is the *column strength curve*.

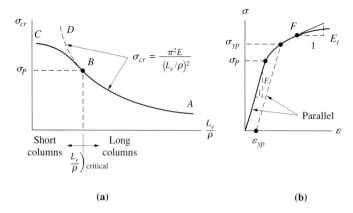

Figure 12.6.1   (a) Critical stress versus slenderness ratio. (b) Compressive stress–strain relation: σ_p is the proportional limit, and σ_{yp} is the yield point, defined in this case by the 0.2 percent offset method ($\varepsilon_{yp} = 0.002$ in. / in.).

The *critical slenderness ratio*, $L_e/\rho)_{\text{critical}}$, is the ratio at which the Euler column formula ceases to apply. It is therefore the demarcation between *long* columns and *short* columns.

To obtain a column strength curve that is in reasonable agreement with test data and is, in fact, on the conservative side, we can replace Young's modulus in Equation 12.6.2 by the *tangent modulus* E_t, which is the slope of the stress-strain curve, as follows:

$$E_t = \frac{d\sigma}{d\varepsilon}$$ [12.6.3]

As illustrated in Figure 12.6.1b, beyond the proportional limit, $E_t < E$ and E_t decreases with increasing stress. Thus, instead of Equation 12.6.2, we have

$$\sigma_{cr} = \frac{\pi^2 E_t}{(L_e/\rho)^2}$$ [12.6.4]

The idea behind the tangent modulus model is as follows. Suppose that when the compressive load in a straight column reaches P_{cr}, the state of uniformly-distributed compressive stress σ_x and strain ε_x on the cross section is beyond the proportional limit, say, at point F in Figure 12.6.1b. When buckling occurs, a small additional compressive strain $\Delta\varepsilon_x$ develops, composed of two parts,

$$\Delta\varepsilon_x = \Delta\varepsilon_a + \Delta\varepsilon_b$$

The term $\Delta\varepsilon_a$ is a uniformly-distributed compressive strain increment arising from a slight increase ΔP_{cr} in the critical load as the column changes from its straight to its slightly bent shape, and $\Delta\varepsilon_b = v''y \cdot$is the linearly-distributed flexure strain due to the bending (cf. Equation 4.5.6a). In the tangent modulus model, it is assumed that throughout the cross section, the stress increment $\Delta\sigma_x$ associated with the strain increment $\Delta\varepsilon_x$ is given by Equation 12.6.3, or

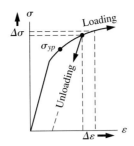

Figure 12.6.2   Stress–strain diagram showing the different paths followed when the stress is increased or decreased from a given state of plastic deformation.

$$\Delta\sigma_x = E_t \Delta\varepsilon_x \qquad\qquad \text{[12.6.5]}$$

This means that if the material is beyond the yield point when buckling occurs, there can be no *unloading* ($\Delta\varepsilon_x < 0$) anywhere in the cross section. Note that when unloading takes place beyond the yield point, E_t in Equation 12.6.5 must be replaced by Young's modulus E.

Figure 12.6.2 illustrates the difference between loading and unloading in the plastic range. Undoubtedly, some unloading actually does occur on the convex side of a bent column, where the flexure strain is tensile. Over the unloaded part of the cross section, E is the appropriate modulus and over the remainder it is E_t. The effective or *reduced* modulus E_R lies between the two, and its value depends on the shape of the cross section. It can be shown,[8] for example, that for a rectangular section,

$$\frac{E_R}{E} = \frac{4(E_t/E)}{\left(1 + \sqrt{E_t/E}\right)^2}$$

whereas for an idealized *I* section (area concentrated equally in the two flanges), we have

$$\frac{E_R}{E} = \frac{2(E_t/E)}{1 + (E_t/E)}$$

Tests of actual columns reveal that the critical load lies between those predicted by using E_t and then E_R in Equation 12.6.1, and is usually closest to the tangent modulus value.

Example 12.6.1   Using the stress–strain data for an aluminum alloy, as given in the first two columns of Table 12.6.1, plot the column strength curve, based on the tangent modulus model.

From the strain and stress data in the first two columns of the table, we can approximate the tangent modulus, using the central difference method:

| [8]A. Chajes, *Principles of Structural Stability Theory*, Section 1.14, Englewood Cliffs, NJ, Prentice-Hall, 1974.

Table 12.6.1   Raw stress–strain data for Al 2023-T3 (first two columns), and computed data required for the column strength curve.

[1] ε (in./in.)	[2] σ (psi)	[3] E_t (ksi)	[4] E_t/σ	[5] $L_e/\rho = \pi\sqrt{E_t/\sigma_{cr}}$
0	0			
0.0005	4500	9,750	2170	146.23
0.001	9750	10,300	1050	101.86
0.0015	14,750	9,750	661	80.77
0.002	19,500	9,500	487	69.34
0.0025	24,250	8,000	330	57.06
0.003	27,500	6,250	227	47.36
0.0035	30,500	5,250	172	41.22
0.004	32,750	4,000	122	34.72
0.0045	34,500	3,500	101	31.64
0.005	36,250	3,000	82.8	28.58
0.0055	37,500	2,250	60	24.33
0.006	38,500	1,500	39	19.61
0.0065	39,000	1,250	32.1	17.79
0.007	39,750	1,500	37.7	19.30
0.0075	40,500	1,000	24.7	15.61
0.008	40,750	750	18.4	13.48
0.0085	41,250	750	18.2	13.40
0.009	41,500	750	18.1	13.36
0.0095	42,000	600	14.3	11.87
0.01	42,100	545	13	11.31
0.015	45,000			

$$E_{t_i} = \frac{\Delta\sigma_i}{\Delta\varepsilon_i} = \frac{\sigma_{i+1} - \sigma_{i-1}}{\varepsilon_{i+1} - \varepsilon_{i-1}}$$ [a]

These are listed in column 3, while column 4 shows the computed ratio of the tangent modulus to the stress. We obtain the slenderness ratio from Equation 12.6.4,

$$\frac{L_e}{\rho} = \pi\sqrt{\frac{E_t}{\sigma_{cr}}}$$ [b]

where E_t/σ_{cr} is found in column 4. Finally, from columns 2 and 5, we plot the column strength curve, as shown in Figure 12.6.3. For comparison, the Euler column formula is also shown, for a modulus of elasticity $E = 9.85 \times 10^6$ psi, which is found from the linear portion of the stress–strain data (cf. Figure 12.6.4). Observe that the demarcation between short columns and long columns, in this case, occurs at $L_e/\rho \cong 75$.

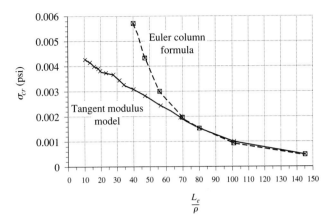

Figure 12.6.3   Column strength curve obtained from the stress–strain data in Table 12.6.1.

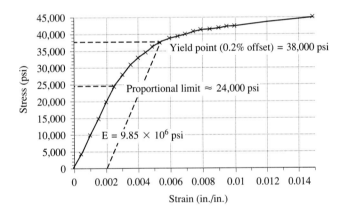

Figure 12.6.4   Plot of the stress–strain data in columns 1 and 2 of Table 12.6.1.

For short columns, column strength formulas have been proposed that adequately approximate the tangent modulus model, but do not require having a complete stress–strain curve available. They also tend to correct for errors that occur in the tangent modulus model if the column material has residual stress. The power formulas have the form

$$\frac{\sigma_{cr}}{E} = \frac{\sigma_{co}}{E} - K\left(\frac{L_e}{\rho}\right)^n \qquad \text{[12.6.6]}$$

where σ_{co} is the *column yield stress*. The column yield stress is usually greater than the compressive yield stress σ_{cy},

$$\sigma_{co} = f\sigma_{cy}$$

where f ranges from 1.0 to 1.2 or higher, depending on the specific alloy. Equation 12.6.6 is assumed to be valid below the critical slenderness ratio, whereas the Euler formula, Equation 12.6.2, is used for long columns,

$$\frac{\sigma_{cr}}{E} = \frac{\pi^2}{(L_e/\rho)^2}$$

[12.6.7]

The two equations are required to match in value and slope at $L_e/\rho)_{critical}$, as illustrated in Figure 12.6.5.

Equating the derivatives of Equations 12.6.6 and 12.6.7 at $L_e/\rho)_{critical}$ yields

$$K = \frac{2\pi^2 E}{n} \frac{1}{[L_e/\rho)_{critical}]^{n+2}}$$

[12.6.8]

Setting the right-hand sides of Equations 12.6.6 and 12.6.7 equal at $L_e/\rho)_{critical}$ and substituting Equation 12.6.8, we find

$$\left.\frac{L_e}{\rho}\right)_{critical} = \pi \sqrt{\left(1 + \frac{2}{n}\right) \frac{E}{\sigma_{co}}}$$

[12.6.9]

Finally, substituting Equations 12.6.8 and 12.6.9 into Equation 12.6.6 leads to

$$\frac{\sigma_{cr}}{E} = \frac{\sigma_{co}}{E}\left[1 - \frac{2}{n\pi^n}\frac{1}{\left(1 + \frac{2}{n}\right)^{1+\frac{n}{2}}}\left(\frac{\sigma_{co}}{E}\right)^{\frac{n}{2}}\left(\frac{L_e}{\rho}\right)^n\right]$$

[12.6.10]

The values commonly used for n are 1.0—primarily for aluminum alloys—and 2.0 for steel and other ductile metal alloys. In those cases we have the *straight-line formula*,

$$\frac{\sigma_{cr}}{E} = \frac{\sigma_{co}}{E}\left[1 - \frac{2\sqrt{3}}{9\pi}\left(\frac{\sigma_{co}}{E}\right)^{\frac{1}{2}}\left(\frac{L_e}{\rho}\right)\right] \qquad \left.\frac{L_e}{\rho}\right)_{critical} = \pi\sqrt{\frac{3}{\sigma_{co}/E}}$$

[12.6.11]

and the *Johnson 2.0 parabola formula,*

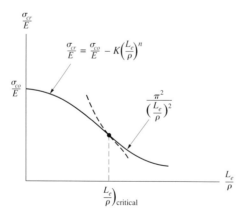

Figure 12.6.5 Power formula for short-column strength, dovetailing into the Euler equation for long columns.

$$\frac{\sigma_{cr}}{E} = \frac{\sigma_{co}}{E}\left[1 - \frac{1}{4\pi^2}\left(\frac{\sigma_{co}}{E}\right)\left(\frac{L_e}{\rho}\right)^2\right] \qquad \left.\frac{L_e}{\rho}\right)_{critical} = \pi\sqrt{\frac{2}{\sigma_{co}/E}} \qquad \textbf{[12.6.12]}$$

These are *typical* of the short-column design formulas in use. There are many specialized or alternate versions, reflecting such variables as material properties, manufacturing methods, type of service, and particular code requirements. Applicable design manuals must be consulted to select the correct column formula.

Example 12.6.2   Replot the column strength curve of Example 12.6.1 to include the straight-line and Johnson parabola short-column formulas. Also, assume that $\sigma_{co} = 41800$ psi.

The nondimensionalized plot appears below in Figure 12.6.6. As we can see, the straight-line formula is conservative compared to the tangent modulus model, which is cut off at the column yield stress. According to Equation (12.6.11), the critical slenderness ratio is 83.5, which is indicated on the figure.

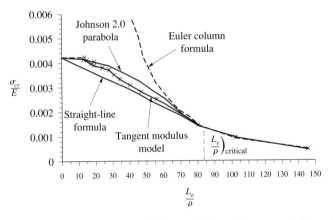

Figure 12.6.6   Comparison of the straight-line, Johnson parabola, and tangent modulus short-column equations for the stress–strain data of Figure 12.6.4, plus a specification that $\sigma_{co} = 41,800$ psi.

12.7   APPROXIMATE ANALYSIS METHODS

In this section, we will see how the principle of virtual work may be used to estimate the buckling load of columns by the *Rayleigh-Ritz* method. This method is approximate because it requires that we *assume* what the shape of the buckled column will be, rather than computing it precisely from a differential equation, as in Section 12.4. The accuracy of our buckling load estimate will depend largely on how close the assumed displacement matches the true one. In practical situations, an exact mathematical or tabulated solution may not be available, so an approximate solution may be the only choice. In that case, we must compare our results with those obtained by others for problems of similar geometry and boundary conditions, to judge their validity. Even in situations where a differential equation can be set up and solved, an approximate procedure may be justified, if it yields a satisfactory solution in less time and with less effort.

Figure 12.7.1 shows a column that was initially straight and had length L. As the axial load increases towards P_{cr}, some shortening of the column must occur, due to the compressive strain, but it is negligible compared to the displacement Δ that takes place when the shape of the column changes to that of the buckled configuration. Δ is due entirely to the lateral deflection caused by bending. From Figure 12.7.1, we see that

$$\Delta = L - L_x \tag{12.7.1}$$

where L is the length of the straight column and L_x is the projection of the deformed column on the x axis. The length of the deformed column is the same as it was before it buckled. That is,

$$L = \int_0^L ds \tag{12.7.2}$$

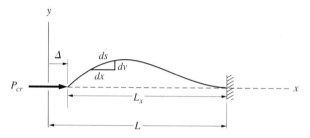

Figure 12.7.1   Column in neutral equilibrium after the onset of buckling.

However,

$$\overbrace{ds^2 = dx^2 + dv^2}^{\text{Pythagorean theorem}} = dx^2 + \left(\frac{dv}{dx}dx\right)^2 = \left[1 + \left(\frac{dv}{dx}\right)^2\right]dx^2$$

Therefore,

$$ds = \left[1 + \left(\frac{dv}{dx}\right)^2\right]^{\frac{1}{2}} dx \tag{12.7.3}$$

A Taylor series expansion of the bracketed expression yields

$$\left[1 + \left(\frac{dv}{dx}\right)^2\right]^{\frac{1}{2}} = 1 + \frac{1}{2}\left(\frac{dv}{dx}\right)^2 - \frac{1}{4}\left(\frac{dv}{dx}\right)^4 + \frac{3}{8}\left(\frac{dv}{dx}\right)^6 + \cdots$$

Enforcing the engineering beam theory assumption that the slopes are small requires that

$$\frac{dv}{dx} << 1 \qquad (ds \cong dx) \tag{12.7.4}$$

Therefore, we can neglect all but the first two terms of the series, so that the element of arc length is simply

$$ds = dx + \frac{1}{2}\left(\frac{dv}{dx}\right)^2 dx \tag{12.7.5}$$

Integrating the left side from 0 to L and the terms on the right from 0 to L_x, we get

$$L = L_x + \int_0^{L_x} \frac{1}{2}\left(\frac{dv}{dx}\right)^2 dx$$

Comparing this with Equation 12.7.1 leads to the conclusion that

$$\Delta = \int_0^{L_x} \frac{1}{2}\left(\frac{dv}{dx}\right)^2 dx \cong \int_0^L \frac{1}{2}(v')^2 dx \qquad [12.7.6]$$

We can now write an expression for the external virtual work,

$$\delta W_{\text{ext}} = P_{cr}\delta\Delta = P_{cr}\int_0^L v'\delta v' dx \qquad [12.7.7]$$

Since σ_x is the only nonzero component of stress, the internal virtual work is

$$\delta W_{\text{int}} = \iiint_V \sigma_x \delta\varepsilon_x dV$$

According to Equation 10.3.8, the true strain is

$$\varepsilon_x = -y\left(d^2v/dx^2\right) = -yv''$$

so that the virtual strain is

$$\delta\varepsilon_x = -y\left(d^2\delta v/dx^2\right) = -y\delta v''$$

It follows that

$$\delta W_{\text{int}} = \iiint_V E\varepsilon_x \delta\varepsilon_x dV = \int_0^L E\left(\iint_A y^2 dA\right)v''\delta v'' dx$$

or

$$\delta W_{\text{int}} = \int_0^L EI_z v''\delta v'' dx \qquad [12.7.8]$$

For the deformed column to be in a state of (neutral) equilibrium, the principle of virtual work ($\delta W_{\text{ext}} = \delta W_{\text{int}}$) requires

$$P_{cr}\int_0^L v'\delta v' dx = \int_0^L EI_z v''\delta v'' dx \qquad [12.7.9]$$

The assumed form of the true and virtual deflections—usually, sines, cosines, or polynomials—must satisfy the displacement boundary conditions (also referred to as *essential* or *kinematic* boundary conditions). Although not required by the Rayleigh-Ritz method, the approximate deflection may also be chosen to be consistent with the boundary conditions on the bending moment (the *nonessential* or *natural* boundary conditions). Since

$d^2v/dx^2 = M/EI_z$, we may, for example, require $d^2v/dx^2 = 0$, where M is known to vanish. Finally, the assumed deflection function should be capable of representing the buckled shape we expect based on physical grounds alone.

Example 12.7.1   Show that the assumed displacement $v = A \sin(\pi x/L)$ yields the Euler buckling load for a pinned-pinned column.

The form assumed for v satisfies the displacement and moment boundary conditions ($v = d^2v/dx^2 = 0$ at $x = 0$ and $x = L$. Since this is the first buckling mode of a pinned-pinned column, computed in Section 12.3 by solving a differential equation of equilibrium, we should expect it to yield the same (Euler) buckling load in conjunction with the principle of virtual work, which we know is an alternative statement of equilibrium.

We have

$$v' = A\frac{\pi}{L}\cos\frac{\pi x}{L} \qquad\qquad \delta v' = \delta A\frac{\pi}{L}\cos\frac{\pi x}{L}$$

$$v'' = -A\left(\frac{\pi}{L}\right)^2\sin\frac{\pi x}{L} \qquad \delta v'' = -\delta A\left(\frac{\pi}{L}\right)^2\sin\frac{\pi x}{L}$$

Thus, from Equation 12.7.7, we have

$$\delta W_{\text{ext}} = P_{cr}\int_0^L v'\delta v'\,dx = P_{cr}A\delta A\left(\frac{\pi}{L}\right)^2\int_0^L \cos^2\frac{\pi x}{L}\,dx = P_{cr}A\delta A\left(\frac{\pi}{L}\right)^2\left(\frac{L}{2}\right) = P_{cr}A\delta A\frac{\pi^2}{2L}$$

and from Equation 12.7.8, we have

$$\delta W_{\text{int}} = EI_z\int_0^L v''\delta v''\,dx = EI_z A\delta A\left(\frac{\pi}{L}\right)^4\int_0^L \sin^2\frac{\pi x}{L}\,dx = EI_z A\delta A\left(\frac{\pi}{L}\right)^4\left(\frac{L}{2}\right) = EI_z A\delta A\frac{\pi^4}{2L^3}$$

Thus,

$$P_{cr}A\delta A\frac{\pi^2}{2L} = EI_z A\delta A\frac{\pi^4}{2L^3}$$

or

$$P_{cr} = EI_z\frac{\pi^2}{L^2}$$

which is the Euler column formula.

Example 12.7.2   Find an approximate value for the buckling load of the pinned-pinned column in Figure 12.7.2, using the lowest possible degree polynomial.

The lowest-degree polynomial we may choose to approximate v is a quadratic

$$v = A + Bx + Cx^2 \tag{a}$$

To satisfy the displacement boundary conditions, v must be zero at both ends, $x = 0$ and $x = L$. This can be so only if $A = 0$ and $B = -CL$, so that

$v = 0 + (-CL)x + Cx^2$

$v = Cx^2 - CLx$

$v = Cx(x - L)$

$$v = Cx(x - L) \tag{b}$$

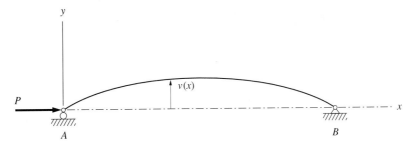

Figure 12.7.2   Pinned-pinned column.

Therefore,

$$v' = C(2x - L) \qquad \delta v' = \delta C(2x - L) \tag{c}$$

$$v'' = 2C \qquad \delta v'' = 2\delta C \tag{d}$$

Notice that although we know that the bending moment is zero at each end of the column, we cannot satisfy *that* boundary condition ($d^2v/dx^2 = 0$), since it would require $C = 0$. Then, according to Equation b, the column deflection would be zero. In other words, our assumed displacement field has too few degrees of freedom to satisfy both the essential displacement boundary conditions (which it must) *and* the non-essential moment boundary conditions (optional).

Substituting Equations c into Equation 12.7.7 yields

$$\delta W_{\text{ext}} = P_{cr} \int_0^L v' \delta v' \, dx = \frac{1}{3} P_{cr} L^3 C \delta C$$

and for the internal virtual work we find, from Equations d and 12.7.8,

$$\delta W_{\text{int}} = E I_z \int_0^L v'' \delta v'' \, dx = 4 E I_z C \delta C$$

Setting the internal and external virtual works equal and cancelling the term $C\delta C$ (the modal amplitude is indeterminate), we are left with

$$P_{cr} = 12 \frac{E I_z}{L^2} = 1.216 \frac{\pi^2 E I_z}{L^2} \tag{e}$$

This is 22 percent higher than the Euler buckling load and is therefore a poor approximation. A comparison of the true and approximate buckled mode shapes is presented in Figure 12.7.3.

Example 12.7.3   Find the buckling load of a pinned-pinned column using the minimum-degree polynomial that satisfies the moment boundary conditions, as well as those for displacement.

To satisfy all four boundary conditions, while not using up all the degrees of freedom, requires at least a fourth-degree polynomial approximation for the deflection, or

$$v = A + Bx + Cx^2 + Dx^3 + Fx^4 \tag{a}$$

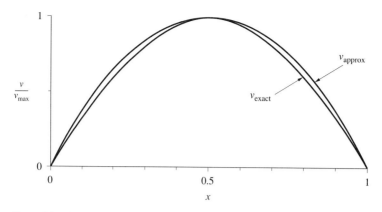

Figure 12.7.3   Exact (sine) and approximate (quadratic polynomial) buckling modes.

For v and v'' to vanish at $x = 0$ and $x = L$ requires that $A = C = 0$, $B = FL^2$, and $D = -2FL$, as the reader should verify. Therefore,

$$v = F(x^4 - 2Lx^3 + L^3x) \qquad \textbf{[b]}$$

From this, we obtain

$$v' = F(4x^3 - 6Lx^2 + L^3) \qquad \delta v' = \delta F(4x^3 - 6Lx^2 + L^3) \qquad \textbf{[c]}$$

$$v'' = F(12x^2 - 12Lx) \qquad \delta v'' = \delta F(12x^2 - 12Lx) \qquad \textbf{[d]}$$

Substituting these expressions into Equations 12.7.7 and 12.7.8, we get the following virtual work expressions:

$$\delta W_{\text{ext}} = \frac{17}{35} P_{cr} L^7 F \delta F \qquad \delta W_{\text{int}} = \frac{24}{5} EI_z L^5 F \delta F \qquad \textbf{[e]}$$

Setting $\delta W_{\text{ext}} = \delta W_{\text{int}}$ and solving for P_{cr} yields

$$P_{cr} = 1.0013 \frac{\pi^2 EI_z}{L^2} \qquad \textbf{[f]}$$

This is just 0.13 percent greater than the exact solution, an excellent approximation.

The last two examples suggest that polynomials can be used satisfactorily to approximate the buckled configuration, if we use enough terms. If we use a high-degree polynomial, the computational effort can become unwieldy. However, using too few terms can lead to poor results, as in Example 12.7.2. In any case, employing approximate displacement fields in the principle of virtual work will always yield a buckling load that exceeds the exact value. The reason is that replacing the true buckling mode by an approximate one is equivalent to placing constraints on the column: it is not allowed to buckle as it naturally would. Imposing constraints on the tendency of the column to displace laterally increases its resistance to buckling, which raises P_{cr} (see Figure 12.4.2.)

Polynomial approximations must be *complete*. That is, if a polynomial of degree n is to be used to represent the buckled mode shape, then *all* polynomial terms 1, x, x^2, $\cdots$ through x^n must be included in the expression for $v(x)$. Only boundary conditions and symmetry considerations, if any, justify the elimination of terms prior to applying the principle of virtual work.

Example 12.7.4   Use the principle of virtual work to compute the buckling load of the column shown in Figure 12.7.4.

This problem can be solved exactly by the method used in Sections 12.4 and 12.5. The procedure is straightforward, though somewhat lengthy, and leads to the solution

$$P_{cr})_{\text{exact}} = 3.729 \frac{\pi^2 E I_z}{L^2} \qquad \text{[a]}$$

Let us assume that the lateral deflection is given by a complete sixth-degree polynomial:

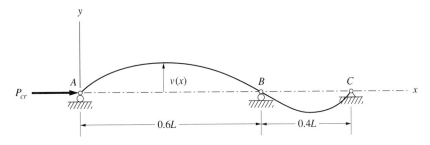

Figure 12.7.4   Column with unsymmetrical restraints.

$$v = A + Bx + Cx^2 + Dx^3 + Fx^4 + Gx^5 + Hx^6 \qquad \text{[b]}$$

It follows that

$$v' = B + 2Cx + 3Dx^2 + 4Fx^3 + 5Gx^4 + 6Hx^5 \qquad \text{[c]}$$

$$v'' = 2C + 6Dx + 12Fx^2 + 20Gx^3 + 30Hx^4 \qquad \text{[d]}$$

The three displacement boundary conditions are

$$v = 0 \quad \text{at} \quad x = 0, \quad x = 0.6L, \quad \text{and} \quad x = L \qquad \text{[e]}$$

The fact that the moment is zero at each end of the column implies

$$v'' = 0 \quad \text{at} \quad x = 0 \quad \text{and} \quad x = L \qquad \text{[f]}$$

Applying the boundary conditions in Equations e and f to v and v'' in Equations b and d imposes five relationships among the seven coefficients A through G, from which we obtain

$$
\begin{aligned}
A &= 0 \\
B &= -\frac{33}{155} L^4 G - \frac{459}{775} L^5 H \\
C &= 0 \\
D &= \frac{3243}{775} L^3 H + \frac{818}{465} L^3 G \\
F &= -\frac{3559}{625} L^2 H - \frac{1184}{465} LG
\end{aligned}
$$

Substituting these into Equations b yields

$$v = Gg(x) + Hh(x) \qquad \text{[g]}$$

where

$$g(x) = x^6 - \frac{3559}{775}L^2x^4 + \frac{3243}{775}L^3x^3 - \frac{459}{775}L^5x \qquad h(x) = x^5 - \frac{1184}{465}Lx^4 + \frac{818}{465}L^2x^3 - \frac{33}{155}L^4x \qquad \text{[h]}$$

Now that we have a displacement function that satisfies all of the boundary conditions, we are ready to proceed to the principle of virtual work. For the external virtual work, we appeal again to Equation 12.7.7, noting that $v' = Gg'(x) + Hh'(x)$ and $\delta v' = \delta Gg'(x) + \delta Hh'(x)$ to obtain

$$\delta W_{\text{ext}} = P_{cr}\int_0^L v'\delta v'\,dx = P_{cr}\int_0^L \left[\left(g'^2G + g'h'H\right)\delta G + \left(g'h'G + h'^2H\right)\delta H\right]dx \qquad \text{[i]}$$

where $g'(x)$ and $h'(x)$ are the first derivatives of the functions in Equation h. Since $v'' = Gg''(x) + Hh''(x)$ and $\delta v'' = \delta Gg''(x) + \delta Hh''(x)$, Equations 12.7.8 yields

$$\delta W_{\text{int}} = EI_z\int_0^L v''\delta v''\,dx = EI_z\int_0^L \left[\left(g''^2G + g''h''H\right)\delta G + \left(g''h''G + h''^2H\right)\delta H\right]dx \qquad \text{[j]}$$

In accordance with the principle of virtual work, we set $\delta W_{\text{int}} - \delta W_{\text{ext}} = 0$, and substitute Equations i and j to obtain

$$\left\{\left[\int_0^L \left(EI_zg''^2 - P_{cr}g'^2\right)dx\right]G + \left[\int_0^L \left(EI_zg''h'' - P_{cr}g'h'\right)dx\right]H\right\}\delta G$$
$$+ \left\{\left[\int_0^L \left(EI_zg''h'' - P_{cr}g'h'\right)dx\right]G + \left[\int_0^L \left(EI_zh''^2 - P_{cr}h'^2\right)dx\right]H\right\}\delta H = 0 \qquad \text{[k]}$$

This equality must be valid regardless of the values assigned to δG and δH. Therefore, the coefficients of δG and δH must vanish, giving us two equations in the two unknowns G and H. In matrix format, these two equations are

$$\begin{bmatrix} \int_0^L \left(EI_zg''^2 - P_{cr}g'^2\right)dx & \int_0^L \left(EI_zg''h'' - P_{cr}g'h'\right)dx \\ \int_0^L \left(EI_zg''h'' - P_{cr}g'h'\right)dx & \int_0^L \left(EI_zh''^2 - P_{cr}h'^2\right)dx \end{bmatrix}\begin{Bmatrix} G \\ H \end{Bmatrix} = \begin{Bmatrix} 0 \\ 0 \end{Bmatrix} \qquad \text{[l]}$$

The integrals required in Equation l are as follows:

$$\int_0^L g'^2\,dx = 0.01294L^9 \qquad \int_0^L g'h'\,dx = 0.03867L^{10} \qquad \int_0^L h'^2\,dx = 0.1161L^{11}$$
$$\int_0^L g''^2\,dx = 0.4864L^7 \qquad \int_0^L g''h''\,dx = 1.457L^8 \qquad \int_0^L h''^2\,dx = 4.415L^9$$

Substituting these into Equation l yields

$$\begin{bmatrix} 0.4864EI_zL^7 - 0.01294P_{cr}L^9 & 1.457EI_zL^8 - 0.03867P_{cr}L^{10} \\ 1.457EI_zL^8 - 0.03867P_{cr}L^{10} & 4.4153EI_zL^9 - 0.1161P_{cr}L^{11} \end{bmatrix}\begin{Bmatrix} G \\ H \end{Bmatrix} = \begin{Bmatrix} 0 \\ 0 \end{Bmatrix} \qquad \text{[m]}$$

A valid, but trivial, solution of this system of two linear, homogeneous equations is $G = H = 0$. If that is the case, the column deflection, Equation g, is zero; that is, buckling has not occurred. Since we are seeking the buckling load, we must reject the trivial solution and find a solution of Equation m such that G and H are not both zero. However, if the determinant of the coefficient matrix is not zero, then we can solve for G and H in terms of the right-hand side and get nothing *but* zero for both. Therefore, we must demand that the determinant vanish, that is, that the matrix be *singular*; it cannot be invertible. Calculating the determinant of the 2 by 2 matrix in Equation m and setting it equal to zero yields its characteristic equation:

$$7.444 \times 10^{-6} L^{20} P_{cr}^2 - 9.475 \times 10^{-4} E I_z L^{18} P_{cr} + 0.02508 \, (E I_z)^2 \, L^{16} = 0$$

or

$$\lambda^2 - 127.3\lambda + 3370 = 0 \qquad \text{[n]}$$

where

$$\lambda = \frac{P_{cr} L^2}{E I_z} \qquad \text{[o]}$$

There are two roots of this quadratic characteristic equation, $\lambda = 37.55$ and $\lambda = 89.74$. We are interested only in the smallest one, so that

$$P_{cr} = 37.55 \frac{E I_z}{L^2} = 3.8044 \frac{\pi^2 E I_z}{L^2} \qquad \text{[p]}$$

This is only 2 percent larger than the exact value given in Equation a.

To determine the buckled mode shape, we substitute P_{cr} into Equation m to get

$$\begin{bmatrix} -0.0004625 & -0.005106L \\ -0.005106L & -0.05638L^2 \end{bmatrix} \begin{Bmatrix} G \\ H \end{Bmatrix} = \begin{Bmatrix} 0 \\ 0 \end{Bmatrix}$$

Since the determinant of this coefficient matrix is zero, the two equations represented by this system are not independent. Only one of them can be used to find G and H. Choosing either one, we find $G = -11.04LH$. Thus, from Equations g and h, the approximate buckled mode shape is

$$v = H\left[-11.04Lg\,(x) + h\,(x)\right] = C\left(\frac{x^6}{L^6} - 11.04\frac{x^5}{L^5} + 25.52\frac{x^4}{L^4} - 15.24\frac{x^3}{L^3} + 1.758\frac{x}{L}\right) \qquad \text{[q]}$$

where C is the undetermined amplitude. This curve is plotted in Figure 12.7.5.

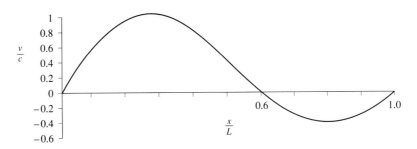

Figure 12.7.5   Sixth-degree polynomial approximation of the fundamental buckling mode shape for the column in Figure 12.7.6.

In Examples 12.7.1 through 12.7.4, we illustrated the Rayleigh-Ritz method, in which the assumed displacement field applies throughout the entire column. A variant of that method involves assuming separate displacement fields over several subregions of the column and patching them together to obtain the complete solution.

This is the *finite element* approach. From Equations 9.12.7 and 9.12.16 the internal virtual work, in the absence of thermal strain, is

$$\delta W_{int} = \{\delta \mathbf{q}\}^T [\mathbf{k}] \{\mathbf{q}\}$$

where the flexural stiffness matrix $[\mathbf{k}]$ for a simple, cubic-displacement beam element is given by Equation 10.3.13. We will continue to assume that beam deflections are adequately represented by cubic polynomials, so that Equation 10.3.13 remains applicable to columns subdivided into beam elements. It is convenient to develop a formula for the *external* virtual work, as well as for the internal virtual work.

The lateral deflection of a cubic-displacement beam element is found in Equation 10.3.5, which is

$$v(x) = \lfloor \mathbf{N} \rfloor \{\mathbf{q}\} = \lfloor N_1(x) \quad N_2(x) \quad N_3(x) \quad N_4(x) \rfloor \begin{Bmatrix} v_1 \\ \theta_{z_1} \\ v_2 \\ \theta_{z_2} \end{Bmatrix}$$

where the shape functions $N_1, \cdots, N_4$ are the cubic polynomials listed in Equation 10.3.6. Taking the first derivative of v yields

$$v'(x) = \lfloor \mathbf{N}' \rfloor \{\mathbf{q}\} = \lfloor N_1'(x) \quad N_2'(x) \quad N_3'(x) \quad N_4'(x) \rfloor \begin{Bmatrix} v_1 \\ \theta_{z_1} \\ v_2 \\ \theta_{z_2} \end{Bmatrix} \qquad \text{[12.7.10]}$$

The derivatives of the shape functions are readily found from Equation 10.3.6, as follows:

$$N_1' = 6\frac{x^2}{L^3} - 6\frac{x}{L^2} \qquad N_2' = 3\frac{x^2}{L^2} - 4\frac{x}{L} + 1 \qquad N_3' = -6\frac{x^2}{L^3} + 6\frac{x}{L^2} \qquad N_4' = 3\frac{x^2}{L^2} - 2\frac{x}{L} \qquad \text{[12.7.11]}$$

To obtain δW_{ext}, we substitute Equation 12.7.10 and $\delta v' = \lfloor \mathbf{N}' \rfloor \{\delta \mathbf{q}\}$ into Equation 12.7.7, to get

$$\delta W_{ext} = P_{cr} \int_o^L \delta v' v' dx = \{\delta \mathbf{q}\}^T \left(P_{cr} \int_0^L \lfloor \mathbf{N}' \rfloor^T \lfloor \mathbf{N}' \rfloor dx \right) \{\mathbf{q}\}$$

or

$$\delta W_{ext} = \{\delta \mathbf{q}\}^T [\mathbf{k}_\sigma] \{\mathbf{q}\}$$

where $[\mathbf{k}_\sigma]$ is the *geometric stiffness matrix*, and is given by

$$[\mathbf{k}_\sigma] = P_{cr} \int_0^L \lfloor \mathbf{N}' \rfloor^T \lfloor \mathbf{N}' \rfloor dx = P_{cr} \int_0^L \begin{bmatrix} (N_1')^2 & N_1'N_2' & N_1'N_3' & N_1'N_4' \\ N_2'N_1' & (N_2')^2 & N_2'N_3' & N_2'N_4' \\ N_3'N_1' & N_3'N_2' & (N_3')^2 & N_3'N_4' \\ N_4'N_1' & N_4'N_2' & N_4'N_3' & (N_4')^2 \end{bmatrix} dx$$

Substituting Equations 12.7.11 and computing the ten unique integrals yields the flexural geometric stiffness matrix for a beam element:

$$[\mathbf{k}_\sigma] = P_{cr} \begin{bmatrix} 6/(5L) & 1/10 & -6/(5L) & 1/10 \\ 1/10 & 2L/15 & -1/10 & -L/30 \\ -6/(5L) & -1/10 & 6/(5L) & -1/10 \\ 1/10 & -L/30 & -1/10 & 2L/15 \end{bmatrix}$$ [12.7.12]

For an individual beam element e, the principle of virtual work requires $\delta W_{\text{ext}}^{(e)} = \delta W_{\text{int}}^{(e)}$, or

$$\left\{\delta\mathbf{q}^{(e)}\right\}^T \left[\mathbf{k}_\sigma^{(e)}\right] \left\{\mathbf{q}^{(e)}\right\} = \left\{\delta\mathbf{q}^{(e)}\right\}^T \left[\mathbf{k}^{(e)}\right] \left\{\mathbf{q}^{(e)}\right\}$$

For an *assembly* of beam elements, the principle of virtual work requires

$$\sum_{\text{elements}} \delta W_{\text{ext}}^{(e)} = \sum_{\text{elements}} \delta W_{\text{int}}^{(e)}$$

or

$$\{\delta\mathbf{q}\}^T [\mathbf{k}_\sigma] \{\mathbf{q}\} = \{\delta\mathbf{q}\}^T [\mathbf{k}] \{\mathbf{q}\}$$ [12.7.13]

for arbitrary $\{\delta\mathbf{q}\}$. The term $\{\mathbf{q}\}$ is the global displacement vector, and $[\mathbf{k}]$ and $[\mathbf{k}_\sigma]$ are the global stiffness matrix and global geometric stiffness matrix, respectively. The matrix $[\mathbf{k}_\sigma]$ for the overall structure is obtained by using Equation 12.7.12 for each individual element and then using the assembly rule. Since $\{\delta\mathbf{q}\}$ is arbitrary, Equation 12.7.13, by the usual argument (Equation 9.4.15), leads to $[\mathbf{k}_\sigma] \{\mathbf{q}\} = [\mathbf{k}] \{\mathbf{q}\}$. Since the axial load P_{cr} is common to all members of a linear assembly of beam elements, we may rearrange terms and write this as

$$\left([\mathbf{k}] - \lambda \left[\mathbf{k}_\sigma'\right]\right) \{\mathbf{q}\} = \{0\}$$ [12.7.14]

where $\left[\mathbf{k}_\sigma'\right] = \frac{1}{\lambda} [\mathbf{k}_\sigma]$ and $\lambda = P_{cr}$ is an eigenvalue of this generalized eigenvalue problem. Recall from Section 9.8 that in the ordinary eigenvalue problem, the coefficient of λ is the identity matrix $[\mathbf{I}]$. Also bear in mind that in the context of a buckling problem, a positive value of P_{cr} denotes compression.

Example 12.7.5   Using the cubic-displacement element, calculate the critical buckling load for the stepped, pinned-pinned column in Figure 12.7.6a.

Observe that the flexural rigidity of the central portion of the column is five times that of the remainder. We should therefore expect the buckling load to lie in the range

$$0.2\frac{\pi^2 E I_z}{L^2} < P_{cr} < \frac{\pi^2 E I_z}{L^2}$$

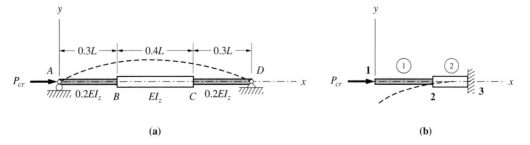

(a)   (b)

Figure 12.7.6   (a) Symmetric, stepped pinned-pinned column. (b) The equivalent free-fixed column modeled for solution.

A minimum of three finite elements and six degrees of freedom (four rotations—at A, B, C, and D—and two lateral displacements—at B and C) are needed to model the column in Figure 12.7.6a. To reduce the computational effort, let us instead take advantage of the structural symmetry and the loading, and treat the problem illustrated in Figure 12.7.6b. This reduces the degrees of freedom from six to four (displacement plus rotation at nodes 1 and 2, and 3 is built in).

Using Equation 10.3.13, we find the unconstrained stiffness matrices of elements 1 and 2 of the structure in Figure 12.7.6b, as follows:

$$[\mathbf{k}^{(1)}] = \frac{(EI_z)_1}{L_1^3} \begin{array}{cccc} 1 & 2 & 3 & 4 \\ \left[\begin{array}{cccc} 12 & 6L_1 & -12 & 6L_1 \\ 6L_1 & 4L_1^2 & -6L_1 & 2L_1^2 \\ -12 & -6L_1 & 12 & -6L_1 \\ 6L_1 & 2L_1^2 & -6L_1 & 4L_1^2 \end{array}\right] & \begin{array}{c} 1 \\ 2 \\ 3 \\ 4 \end{array} \end{array} = \frac{EI_z}{L^3} \begin{array}{cccc} 1 & 2 & 3 & 4 \\ \left[\begin{array}{cccc} 800/9 & 40L/3 & -800/9 & 40L/3 \\ 40L/3 & 8L^2/3 & -40L/3 & 4L^2/3 \\ -800/9 & -40L/3 & 800/9 & -40L/3 \\ 40L/3 & 4L^2/3 & -40L/3 & 8L^2/3 \end{array}\right] & \begin{array}{c} 1 \\ 2 \\ 3 \\ 4 \end{array} \end{array}$$

$$[\mathbf{k}^{(2)}] = \frac{(EI_z)_2}{L_2^3} \begin{array}{cccc} 3 & 4 & 5 & 6 \\ \left[\begin{array}{cccc} 12 & 6L_2 & -12 & 6L_2 \\ 6L_2 & 4L_2^2 & -6L_2 & 2L_2^2 \\ -12 & -6L_2 & 12 & -6L_2 \\ 6L_2 & 2L_2^2 & -6L_2 & 4L_2^2 \end{array}\right] & \begin{array}{c} 3 \\ 4 \\ 5 \\ 6 \end{array} \end{array} = \frac{EI_z}{L^3} \begin{array}{cccc} 3 & 4 & 5 & 6 \\ \left[\begin{array}{cccc} 1500 & 150L & -1500 & 150L \\ 150L & 20L^2 & -150L & 10L^2 \\ -1500 & -150L & 1500 & -150L \\ 150L & 10L^2 & -150L & 20L^2 \end{array}\right] & \begin{array}{c} 3 \\ 4 \\ 5 \\ 6 \end{array} \end{array}$$

Assembling these two matrices, guided by the global degree of freedom numbering along the rows and columns, we obtain the 6 by 6 global stiffness matrix $[\mathbf{k}]$. We then reduce it to 4 by 4 by eliminating rows and columns 5 and 6, because those degrees of freedom are constrained to zero. The result is

$$[\mathbf{k}_{ff}] = \frac{EI_z}{L^3} \begin{array}{cccc} 1 & 2 & 3 & 4 \\ \left[\begin{array}{cccc} 800/9 & 40L/3 & -800/9 & 40L/3 \\ 40L/3 & 8L^2/3 & -40L/3 & 4L^2/3 \\ -800/9 & -40L/3 & 14,300/9 & 410L/3 \\ 40L/3 & 4L^2/3 & 410L/3 & 68L^2/3 \end{array}\right] & \begin{array}{c} 1 \\ 2 \\ 3 \\ 4 \end{array} \end{array}$$ [a]

According to Equation 12.7.12, the geometric stiffness matrices are

$$[\mathbf{k}_\sigma^{(1)}] = P_{cr} \begin{array}{cccc} 1 & 2 & 3 & 4 \\ \left[\begin{array}{cccc} 6/(5L_1) & 1/10 & -6/(5L_1) & 1/10 \\ 1/10 & 2L_1/15 & -1/10 & -L_1/30 \\ -6/(5L_1) & -1/10 & 6/(5L_1) & -1/10 \\ 1/10 & -L_1/30 & -1/10 & 2L_1/15 \end{array}\right] & \begin{array}{c} 1 \\ 2 \\ 3 \\ 4 \end{array} \end{array} = P_{cr} \begin{array}{cccc} 1 & 2 & 3 & 4 \\ \left[\begin{array}{cccc} 4/L & 1/10 & -4/L & 1/10 \\ 1/10 & L/25 & -1/10 & -L/100 \\ -4/L & -1/10 & 4/L & -1/10 \\ 1/10 & -L/100 & -1/10 & L/25 \end{array}\right] & \begin{array}{c} 1 \\ 2 \\ 3 \\ 4 \end{array} \end{array}$$

$$[\mathbf{k}_\sigma^{(2)}] = P_{cr} \begin{array}{cccc} 3 & 4 & 5 & 6 \\ \left[\begin{array}{cccc} 6/(5L_2) & 1/10 & -6/(5L_2) & 1/10 \\ 1/10 & 2L_2/15 & -1/10 & -L_2/30 \\ -6/(5L_2) & -1/10 & 6/(5L_2) & -1/10 \\ 1/10 & -L_2/30 & -1/10 & 2L_2/15 \end{array}\right] & \begin{array}{c} 3 \\ 4 \\ 5 \\ 6 \end{array} \end{array} = P_{cr} \begin{array}{cccc} 3 & 4 & 5 & 6 \\ \left[\begin{array}{cccc} 6/L & 1/10 & -6/L & 1/10 \\ 1/10 & 2L/75 & -1/10 & -L/150 \\ -6/L & -1/10 & 6/L & -1/10 \\ 1/10 & -L/150 & -1/10 & 2L/25 \end{array}\right] & \begin{array}{c} 3 \\ 4 \\ 5 \\ 6 \end{array} \end{array}$$

Assembling these in the same fashion as the stiffness matrices, likewise eliminating degrees of freedom 5 and 6, we get

$$[\mathbf{k}_{\sigma_{ff}}] = P_{cr} \begin{array}{c} \begin{array}{cccc} \;\;1 & \;\;\;2 & \;\;\;\;3 & \;\;4 \end{array} \\ \begin{bmatrix} 4/L & 1/10 & -4/L & 1/10 \\ 1/10 & L/25 & -1/10 & -L/100 \\ \hdashline -4/L & -1/10 & 10/L & 0 \\ 1/10 & -L/100 & 0 & L/15 \end{bmatrix} \begin{array}{c} 1 \\ 2 \\ 3 \\ 4 \end{array} \end{array} \qquad \text{[b]}$$

For our example, then, Equation 12.7.14 becomes

$$\begin{bmatrix} 800/9 - 4\lambda & 40/3 - \lambda/10 & -800/9 + 4\lambda & 40/3 - \lambda/10 \\ 40/3 - \lambda/10 & 8/3 - \lambda/25 & -40/3 + \lambda/10 & 4/3 + \lambda/100 \\ \hdashline -800/9 + 4\lambda & -40/3 + \lambda/10 & 14300/9 - 10\lambda & 410/3 \\ 40/3 - \lambda/10 & 4/3 + \lambda/100 & 410/3 & 68/3 - \lambda/15 \end{bmatrix} \begin{Bmatrix} v_1/L \\ \theta_1 \\ v_2/L \\ \theta_2 \end{Bmatrix} = \begin{Bmatrix} 0 \\ 0 \\ 0 \\ 0 \end{Bmatrix} \qquad \text{[c]}$$

where

$$\lambda = \frac{P_{cr}L^2}{EI_z} \qquad \text{[d]}$$

Since we are not interested in the trivial solution ($v_1 = \theta_1 = v_2 = \theta_2 = 0$), the determinant of the coefficient matrix in Equation c must be set equal to zero. The determinant is found to be the following quartic polynomial in λ:

$$\det(\lambda) = \frac{21}{400}\lambda^4 - \frac{503}{15}\lambda^3 + \frac{35,485}{9}\lambda^2 - \frac{3,268,000}{27}\lambda + \frac{4,000,000}{9} \qquad \text{[e]}$$

Since λ is directly proportional to P_{cr}, only the smallest root of the equation $\det(\lambda) = 0$ is required. A relatively quick way to find that root is to plot the function $\det(\lambda)$, as shown in Figure 12.7.7. We can see that the smallest root lies at a point a bit over $\lambda = 4$. Trial and error with a hand calculator reveals that, more precisely, $\lambda = 4.235$. Therefore, referring to Equation d, we find that

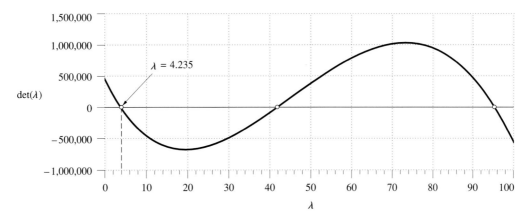

Figure 12.7.7   Plot of Equation f through the first three roots.

The fourth root lies at $\lambda = 497$.

$$P_{cr} = 4.235 \frac{EI_z}{L^2} = 0.4291 \frac{\pi^2 EI_z}{L^2}$$

This is only 0.36 percent greater than the exact[9] buckling load.

The last example suggests that the finite element approach to column analysis, using even the simple cubic-displacement beam element, will yield satisfactory results in practical applications. Even in that relatively simple problem, the lengthy calculations—especially the eigenvalue problem represented by Equation e of the example—border on the unwieldy if done by hand. However, the steps of the solution can and have been coded in software and are commonly available for digital computers.

12.8 THIN PLATE BUCKLING

A thin plate is a flat sheet of material, the thickness of which is very much smaller than its lateral dimensions. We will limit ourselves here to isotropic, elastic materials. Like a beam, of which the plate is a two-dimensional generalization, a plate is capable of supporting lateral or *bending* loads normal to its surface, as well as *membrane* forces that lie in the plane of the plate. Compressive in-plane loads may cause a thin plate to buckle at stresses well within the elastic limit.

Figure 12.8.1 shows an *a* by *b* buckled rectangular plate of uniform thickness *t*. The *middle surface* of the plate—the surface equidistant from the top and bottom ones—initially lies flat in the *xy* plane. Opposite ends of the plate, at $x = 0$ and $x = a$, are simply supported, or hinged. Like beams, simply-supported edges of plates cannot displace, and they are free of bending moment. A uniform compressive stress is shown acting

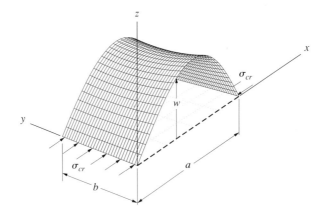

Figure 12.8.1   Buckled shape of a plate with unsupported sides and loaded in compression on its simply-supported ends.

The plate thickness is *t* and $w \ll t$.

| [9]S. P. Timoshenko and J. G. Gere, *Theory of Elastic Stability*, 2nd Edition, Section 2.14, New York, McGraw-Hill, 1961.

on the supported ends. Since its magnitude is σ_{cr}, buckling has occurred. The term $w(x, y)$ is the transverse displacement of points on the middle surface. As a function of both x and y, the displacement defines the shape of the deformed plate. In the figure, the deformation of the plate is highly magnified so its details can be seen. In thin-plate theory, w is on the order of the thickness of the plate itself. The edges of the plate parallel to the applied load are unsupported and free of load, as indicated.

Viewed edge-on, perpendicular to the xz plane, the plate looks like a column (cf. Figure 12.4.1) buckled into one-half of a sinusoid. We might therefore be tempted to treat this as a simply-supported wide column, concluding from Equation 12.4.5 that

$$\sigma_{cr} = \frac{P_{cr}}{A} = \frac{\pi^2 EI}{AL^2}$$

Since $A = bt$, $I = I_y = bt^3/12$, and $L = a$, we would find the critical stress to be given by

$$\sigma_{cr} = \frac{\pi^2 Et^2}{12a^2}$$

Although this is accurate enough for practical purposes, it is not strictly correct. The reason is that engineering beam theory deems the state of stress in a beam as uniaxial and ignores the Poisson effect (Section 3.11), essentially setting $v = 0$. A thin plate, however, is in a state of plane stress (Equations 3.12.2), which requires dividing E by $1 - v^2$. The exact formula for the critical compressive stress in the plate of Figure 12.8.1 is actually

$$\sigma_{cr} = C_1 \frac{\pi^2 E}{12(1 - v^2)} \left(\frac{t}{a}\right)^2 \qquad \text{loaded ends simply-supported; unloaded sides free} \qquad \textbf{[12.8.1]}$$

The term C_1 is a factor that depends on Poisson's ratio, as well as the plate aspect ratio a/b, where b is the width of the loaded sides. If the loaded ends of the plate are clamped instead of hinged, then, as we might expect from column theory, the expression in Equation 12.8.1 is multiplied by four:

$$\sigma_{cr} = C_2 \frac{4\pi^2 E}{12(1 - v^2)} \left(\frac{t}{a}\right)^2 \qquad \text{loaded ends clamped; unloaded sides free} \qquad \textbf{[12.8.2]}$$

The variation of C_1 and C_2 with aspect ratio, for a Poisson's ratio of 0.3, is shown in Figure 12.8.2. As v approaches zero, C_1 and C_2 approach 1 and the anticlastic curvature[10] of the buckled plate, evident in Figure 12.8.1, disappears. The horizontal asymptote of both curves is 0.910.

The differential equation for plate buckling, analogous to Equation 12.4.1 for columns, is[11]

$$\frac{\partial^4 w}{\partial x^4} + 2\frac{\partial^4 w}{\partial^2 x \partial y^2} + \frac{\partial^4 w}{\partial y^4} = \frac{12(1 - v^2)}{Et^4}\left(\sigma_x \frac{\partial^2 w}{\partial x^2} + \sigma_y \frac{\partial^2 w}{\partial y^2} + 2\tau_{xy}\frac{\partial^2 w}{\partial x \partial y}\right) \qquad \textbf{[12.8.2]}$$

where σ_x, σ_y, and τ_{xy} are the components of plane stress averaged across the thickness t of the plate. For a rectangular plate, the boundary conditions on w corresponding to the imposed constraints are as follows,[12] where ξ is the coordinate direction parallel to the edge in question and η is the coordinate direction normal to the edge.

---

[10] Having principal curvatures of opposite sign, like the surface of a saddle.

[11] See, for example, Timoshenko and Gere, *ibid.*, or A. C. Ugural, *Stresses in Plates in Shells*, New York, McGraw-Hill, 1981.

[12] Timoshenko and Gere, *ibid.*

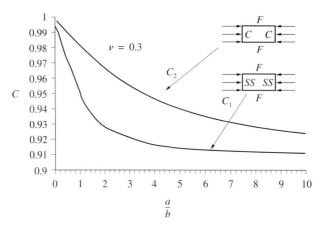

Figure 12.8.2 Variation of the coefficients in Equations 12.8.1 and 12.8.2: F = "free," C = "clamped," and SS = "simply-supported."

- Built-in edge (zero displacement and zero rotation)

$$w = 0 \quad \text{and} \quad \frac{\partial w}{\partial \eta} = 0 \qquad \text{at all points along the edge} \qquad \text{[12.8.3a]}$$

- Simply-supported edge (zero displacement and zero moment)

$$w = 0 \quad \text{and} \quad \frac{\partial^2 w}{\partial \eta^2} + \nu \frac{\partial^2 w}{\partial \xi^2} = 0 \qquad \text{at all points along the edge} \qquad \text{[12.8.3b]}$$

- Free (unsupported) edge (zero moment and zero transverse shear)

$$\frac{\partial^2 w}{\partial \eta^2} + \nu \frac{\partial^2 w}{\partial \xi^2} = 0 \qquad \text{and} \qquad \frac{\partial^3 w}{\partial \eta^3} + (2 - \nu) \frac{\partial^3}{\partial \eta \partial \xi^2} = 0 \qquad \text{at all points along the edge} \qquad \text{[12.8.3c]}$$

For rectangular plates in compression, with at least one of its unloaded edges constrained, the formula for critical stress is commonly presented in the form

$$\sigma_{cr} = C \frac{\pi^2 E}{12 \left(1 - \nu^2\right)} \left(\frac{t}{b}\right)^2 \qquad \text{[12.8.4]}$$

where C is a constraint coefficient that is a function of the boundary conditions, the aspect ratio, and—depending on the boundary conditions—Poisson's ratio ν (cf. Equation 12.8.3c). The term b is the width of the edges on which the uniform compressive stress acts. The presentation and illustration of the methods of obtaining exact and approximate solutions for C and the buckled mode shapes are beyond the scope of this text.

Some classical solutions for the constraint coefficient C are summarized in Figure 12.8.3. Observe that for aspect ratios greater than about 3, C is essentially constant, regardless of the manner in which the plate is supported. Thus, for a given length a of the unloaded sides of a plate, the buckling stress, according to Equation 12.8.4, is proportional to the *square* of the aspect ratio. For example, a plate whose length is five times its width has an elastic buckling strength 25 times that of a square plate of the same length, everything else being equal.

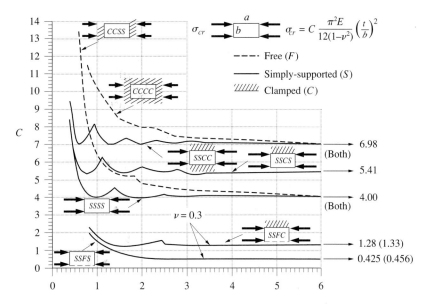

Figure 12.8.3   Coefficient C versus aspect ratio for several different boundary conditions. Only the two bottommost curves depend on Poisson's ratio, and their asymptotes for $\nu = 0.25$ are shown in parentheses.

The buckled mode shapes depend, of course, on the manner in which the plate is restrained. Consider, for example, a plate with all edges simply supported (the *SSSS* boundary conditions in Figure 12.8.3). Figure 12.8.4 shows a square plate, for which the buckled surface $w(x, y)$ is defined by a half-sine wave in the directions both parallel and perpendicular to the applied load,

$$w = A_1 \sin \frac{\pi x}{b} \sin \frac{\pi y}{b} \qquad 0 \le x \le b, \quad 0 \le y \le b$$

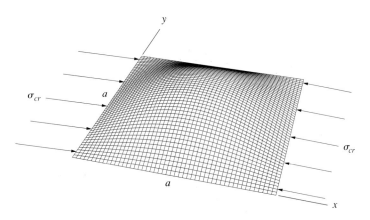

Figure 12.8.4   Buckled mode shape for a square plate, simply supported on all four sides.

For larger aspect ratios, the shape remains a half-sine wave across the width of the plate, and becomes an integer multiple of half-sine waves in the lengthwise direction:

$$w = A_m \sin \frac{m\pi x}{a} \sin \frac{\pi y}{b} = A_m \sin\left(\frac{m}{a/b}\frac{\pi x}{b}\right) \sin \frac{\pi y}{b} \qquad 0 \le x \le a, \ \ 0 \le y \le b \qquad \text{[12.8.5]}$$

The range of aspect ratios for which there are m half-sine waves is determined by the following expression:

$$\sqrt{m\,(m-1)} < \frac{a}{b} < \sqrt{m\,(m+1)} \qquad \text{[12.8.6]}$$

For example, $m = 1$ for aspect ratios up to $a/b = 1.414$, over which there is a significant variation in the buckling load, according to the *SSSS* curve in Figure 12.8.3. On the other hand, if $m = 5$, the aspect ratio of the plate must lie between 4.472 and 5.477. Clearly, if a/b is an integer, then $m = a/b$ and, according to Equation 12.8.5, the plate buckles into m half-sinusoids with a wavelength equal to the width of the plate. Figure 12.8.5 illustrates the buckled mode shape for a plate with an aspect ratio of 5. For large values of m, the left and right sides of the inequality (Equation 12.8.6) are nearly equal to m. Therefore, very long and narrow simply-supported plates buckle into a pattern of nearly square subregions.

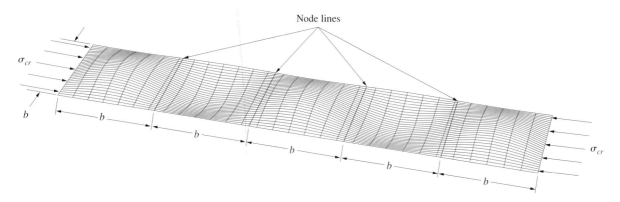

Figure 12.8.5   Mode shape of a simply-supported plate of aspect ratio $a/b = 5$.

In some cases, the buckled mode shape does not depend on the aspect ratio of the plate. A case in point is the *SSFS* condition (simply supported at the loaded ends and on one side, with the remaining side free) in Figure 12.8.3. Its mode shape, drawn in Figure 12.8.6 for $a/b = 5$, has the same appearance for any choice of that parameter.

12.9   CRIPPLING STRESS

In our discussion of columns in Sections 12.3–12.7, we were concerned with *primary*, or *general*, instability. The buckled mode shapes in that case have wavelengths on the order of the length of the column itself, and there is no distortion in the plane of the cross section. If the column is of thin-walled construction, then one has to be concerned as well about *secondary* or *local* instability—buckling of the individual thin-walled components of the column cross-section. Figure 12.9.1 shows a compression member with a thin-walled closed cross-section. Each

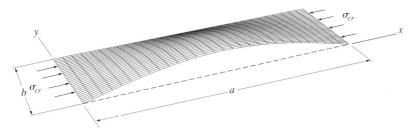

Figure 12.8.6   Buckled mode shape for a plate simply supported at the ends and on one side, with one side free.

The shape is independent of a/b.

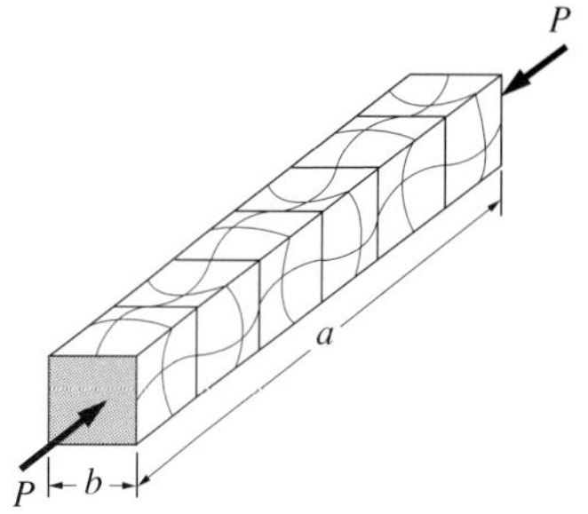

Figure 12.9.1   Secondary instability of a thin-walled column.

wall is a thin plate of large aspect ratio. If we conservatively consider each of the four narrow plates to be simply supported along their sides, then we might expect buckling of the type illustrated in the figure to occur (see Equation 12.8.5). Clearly, the cross-sections are distorted, and the wavelength of the buckled mode shapes is on the order of the dimensions of the cross-section, rather than the length of the column itself. Once the walls have buckled, the ability of the material in the bowed regions to carry additional load is greatly diminished. However, as long as the corners of the column remain straight, the material in those regions can continue to support the load until yielding occurs. The *average* compressive stress on the cross-section at the failure load is called the *crippling stress*, σ_{cc}.

To calculate the crippling stress of a column, we may use an empirical procedure[13] in which the n walls of the cross section are idealized as n rectangles, the ith one having thickness and length t_i and length b_i. These rectangles are considered the simply-supported, loaded ends of high-aspect-ratio thin plates whose common length a is that of the column. Plates with a common edge are conservatively considered to be simply supported along that side. Otherwise, the edge of a plate is free. The average crippling stress σ_{cc} in the ith wall (denoted $\sigma_{cc}^{(i)}$) is found by means of the empirical formula[14]

$$\frac{\sigma_{cc}}{\sigma_{cy}} = \alpha \left[\frac{\pi^2 C}{12 \left(1 - v^2\right) \lambda^2} \right]^{1-n} \tag{12.9.1}$$

where

$$\lambda = \sqrt{\frac{\sigma_{cy}}{E}} \frac{b}{t} \tag{12.9.2}$$

The term σ_{cy} is the compressive yield stress, C is the plate constraint coefficient for elastic buckling discussed in the previous section, and α and n are experimentally determined parameters to which we will conservatively assign the following values[15]

$$\alpha = 0.8 \qquad n = 0.6 \tag{12.9.3}$$

---

[13] The procedure for crippling stress analysis used here is adopted from Volume II, Part 2 of the *Astronautics Structures Manual*, NASA Marshall Space Flight Center, N76–76167, August 1975, available through the National Technical Information Service, U. S. Dept. of Commerce.
[14] R. M. Rivello, *Theory and Analysis of Flight Structures*, New York, McGraw-Hill, 1969.
[15] *ibid.*

We refer to Figure 12.8.3 to obtain the asymptotic values of C for the plate with neither side free ($SSSS$) and with one side free ($SSFS$, for $\nu = 0.3$)

$$\text{Neither side free :} \quad C = 4.0$$
$$\text{One side free :} \quad C = 0.425$$

[12.9.4]

Substituting Equations 12.9.2, 12.9.3, and 12.9.4 into Equation 12.9.1, we get

$$\text{Neither side free :} \quad \sigma_{cc}/\sigma_{cy} = 1.34\lambda^{-0.8}$$
$$\text{One side free :} \quad \sigma_{cc}/\sigma_{cy} = 0.546\lambda^{-0.8}$$

[12.9.5]

If the calculated crippling stress for a given wall exceeds the *cutoff stress* σ_{co} for the wall material, then the value of σ_{co} is taken to be the crippling stress. For many metal alloys, the cutoff stress for compressive buckling is just the compressive yield stress σ_{cy}. Some exceptions to that rule are shown in Table 12.9.1.

Table 12.9.1   Cutoff stresses for compressive buckling of unstiffened flat plates.

Material	Cutoff Stress
Aluminum 2014-T, 2024-T, and 6061-T alloys	$\sigma_{cy}\left(1 + \frac{\sigma_{cy}}{200,000}\right)$
Aluminum 7075-T	$1.075\sigma_{cy}$
Low carbon steel alloys	$1.1\sigma_{cy}$
Other materials	σ_{cy}

After calculating the crippling stress $\sigma_{cc}^{(i)}$ for each wall of the cross section, we find the average crippling stress for the column cross section, using the formula

$$\sigma_{cc} = \frac{\sum\limits_{i=1}^{n} \sigma_{cc}^{(i)} A_i}{\sum\limits_{i=1}^{n} A_i} \qquad A_i = b_i t_i$$

[12.9.6]

At failure, the crippling stress in the corner regions of the cross section is likely to be higher than in the walls, so the average crippling stress will be greater than that given by Equation 12.9.6. In an effort to account for this, we ignore corner material in computing each wall length b_i, as illustrated in Figure 12.9.2. Figure 12.9.3 shows one means of determining the effective wall lengths of stiffeners formed from sheet metal.

To find the crippling load P_{cc} on a section, we multiply the average crippling stress σ_{cc} by the total area A of the section,

$$P_{cc} = \sigma_{cc} A$$

where $A > \sum\limits_{i=1}^{n} b_i t_i$.

Once we have calculated σ_{cc} for the column, we can find its primary buckling stress by using the Johnson parabola column curve, Equation 12.6.12. The cutoff stress σ_{co} in that formula is taken to be either the crippling stress σ_{cc} or the compressive yield stress σ_{cy}, whichever is smallest.

t_4 $b_1 = b_3$ t_2
t_1
$b_2 = b_4$
t_3
b_1 t_2
t_1
b_2
$t_3 = t_4$ b_3 b_4
R_b
b_1 t_2
t_1 b_2
b_2
b_1 t_2

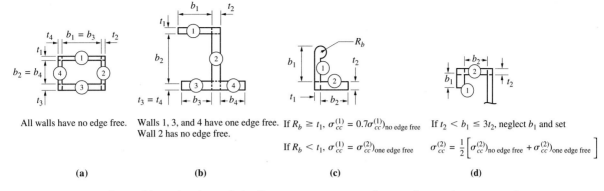

All walls have no edge free. | Walls 1, 3, and 4 have one edge free. Wall 2 has no edge free. | If $R_b \geq t_1$, $\sigma_{cc}^{(1)} = 0.7\sigma_{cc}^{(1)})_{\text{no edge free}}$ If $R_b < t_1$, $\sigma_{cc}^{(1)} = \sigma_{cc}^{(2)})_{\text{one edge free}}$ | If $t_2 < b_1 \leq 3t_2$, neglect b_1 and set $\sigma_{cc}^{(2)} = \frac{1}{2}\left[\sigma_{cc}^{(2)})_{\text{no edge free}} + \sigma_{cc}^{(2)})_{\text{one edge free}}\right]$

(a) **(b)** **(c)** **(d)**

Figure 12.9.2 Resolution of formed and extruded stiffener cross sections into elements, for crippling stress analysis.

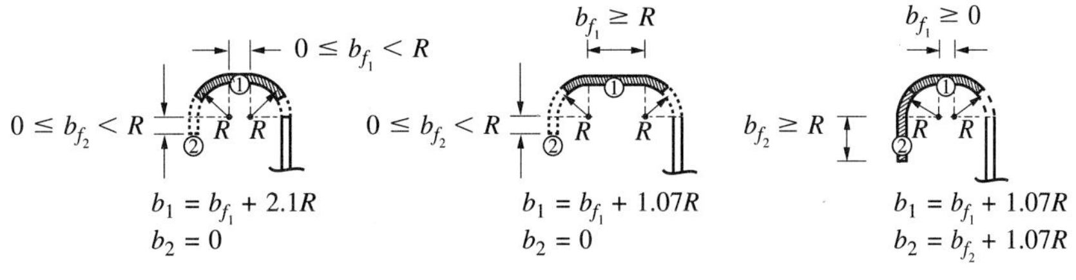

$0 \leq b_{f_1} < R$
$0 \leq b_{f_2} < R$
$b_1 = b_{f_1} + 2.1R$
$b_2 = 0$
$b_{f_1} \geq R$
$0 \leq b_{f_2} < R$
$b_1 = b_{f_1} + 1.07R$
$b_2 = 0$
$b_{f_1} \geq 0$
$b_{f_2} \geq R$
$b_1 = b_{f_1} + 1.07R$
$b_2 = b_{f_2} + 1.07R$

(a) **(b)** **(c)**

Figure 12.9.3 Resolution of formed stiffener cross sections with equal radii having centers on the same side of the sheet.

Example 12.9.1 Figure 12.9.4a shows the details of an extruded aluminum 7075-T6 section, for which $E = 10.5 \times 10^6$ psi and $\sigma_{cy} = 70,000$ psi. Calculate the crippling stress.

The section is resolved into five rectangles, as illustrated in Figure 12.9.4(b). The corner (and fillet) areas are omitted.

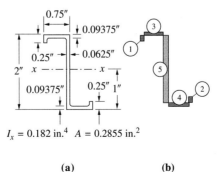

0.75″
0.09375″
0.25″
0.0625″
2″
x
x
0.09375″
0.25″
1″
$I_x = 0.182$ in.^{4} $A = 0.2855$ in.^{2}

(a) **(b)**

Figure 12.9.4 (a) 7075-T6 extruded section. (b) The same section discretized into rectangles.

Element 1

The length of this small lip is

$$b_1 = 0.25 - 0.09375 = 0.1562 \text{ in.}$$

Since b_1 is less than three times the thickness of the flange to which it is attached, this element is not wide enough to be treated as a flange itself and may therefore be neglected, according to Figure 12.9.2d. Its presence, however, does add some elastic restraint to the adjacent flange, and this is taken into account in computing the crippling stress of element 3.

Element 2

This element is neglected for the same reason as element 1.

Element 3

In this case, we have

$$b_3 = 0.75 - 0.09375 - 0.0625 = 0.594 \text{ in.} \qquad t_3 = 0.09375 \text{ in.} \qquad \text{[a]}$$

Thus,

$$\lambda_3 = \sqrt{\frac{70,000}{10.5 \times 10^6} \frac{0.594}{0.09375}} = 0.517$$

According to Figure 12.9.2d, the effect of the lip (element 1) is included by computing the crippling stress as the average of those for the "no edge free" and "one edge free" conditions. Thus,

$$\sigma_{cc}^{(3)} = 70,000 \left(\frac{1.34 + 0.546}{2} \right) \lambda^{-0.8} = 112,000 \text{ psi}$$

This far exceeds the cutoff stress of $1.075 \times 70,000 = 75,300$ psi (Table 12.9.1). So we set the crippling stress for this element equal to the cutoff stress, or

$$\sigma_{cc}^{(3)} = 75,300 \text{ psi} \qquad \text{[b]}$$

Element 4

This is identical to element 3, so we have

$$b_4 = 0.5625 \text{ in.} \qquad t_4 = 0.09375 \text{ in.} \qquad \text{[c]}$$

$$\sigma_{cc}^{(4)} = 75,300 \text{ psi} \qquad \text{[d]}$$

Element 5

The width and thickness of this rectangle are

$$b_5 = 2.00 - 2 \times 0.09375 = 1.812 \text{ in.} \qquad t_5 = 0.0625 \text{ in.} \qquad \text{[e]}$$

Therefore,

$$\lambda_5 = \sqrt{\frac{70,000}{10.5 \times 10^6} \frac{1.812}{0.0625}} = 2.37$$

Since this element has no edge free, we have

$$\sigma_{cc}^{(5)} = 70,000(1.34)\lambda^{-0.8} = 47,100 \text{ psi} \qquad \text{[f]}$$

Overall

Finally, Equation 12.9.6 yields the crippling stress for the overall cross section:

$$\sigma_{cc} = \frac{\sigma_{cc}^{(3)} b_3 t_3 + \sigma_{cc}^{(4)} b_4 t_4 + \sigma_{cc}^{(5)} b_5 t_5}{b_3 t_3 + b_4 t_4 + b_5 t_5} = \frac{2 \times (75,300 \times 0.5625 \times 0.09375) + 47,100 \times 1.812 \times 0.0625}{2 \times 0.5625 \times 0.09375 + 1.812 \times 0.0625} = 60,700 \text{ psi} \quad \text{[g]}$$

Recall that to find the crippling load P_{cc}, we multiply this average stress by the total area of the cross section, which is given in Figure 12.9.4:

$$P_{cc} = 60,700 \frac{\text{lb}}{\text{in.}^2} \times 0.2855 \text{ in.}^2 = 17,300 \text{ lb} \quad \text{[h]}$$

Example 12.9.2   Calculate the critical load for the primary buckling of two columns, both of which have the Z-section of the previous example and are constrained to buckle about the xx axis (cf. Figure 12.9.4a). One column is 30 inches long, and the other is 70 inches long. The end fixity factor in each case is $c = 1.5$.

We use the Johnson parabola column strength curve, Equation 12.6.12,

$$\sigma_{cr} = \sigma_{cc} \left[1 - \frac{1}{4\pi^2} \frac{\sigma_{cc}}{E} \left(\frac{L_e}{\rho} \right)^2 \right] \quad \text{[a]}$$

if the slenderness ratio is less than the critical value given by

$$\left. \frac{L_e}{\rho} \right)_{\text{critical}} = \pi \sqrt{\frac{2}{\sigma_{cc}/E}} \quad \text{[b]}$$

where σ_{cc} is the crippling stress for the column, which was found in the previous example. If the slenderness ratio exceeds the value given by Equation b, then the critical stress is found by means of the Euler column formula Equation 12.6.2,

$$\sigma_{cr} = \frac{\pi^2 E}{(L_e/\rho)^2} \quad \text{[c]}$$

We obtain the radius of gyration about the xx axis from the cross-sectional area A and the moment of inertia I_x, both of which are listed in Figure 12.9.4a:

$$\rho = \sqrt{\frac{I_x}{A}} = \sqrt{\frac{0.1820}{0.2855}} = 0.798 \text{ in.}$$

Therefore, according to Equation b, the critical slenderness ratio is

$$\left. \frac{L_e}{\rho} \right)_{\text{critical}} = \pi \sqrt{\frac{2}{60,700/10,500,000}} = 58.4$$

We now consider each column separately.

30-inch column

For this column, the effective length is

$$L_e = \frac{L}{\sqrt{c}} = \frac{30}{\sqrt{1.5}} = 24.5 \text{ in.}$$

and the slenderness ratio is

$$\frac{L_e}{\rho} = \frac{24.5}{0.798} = 30.7$$

This is less than the critical slenderness ratio, so we use Equation a to find the critical stress

$$\sigma_{cr} = 60,700 \left[1 - \frac{1}{4\pi^2} \frac{60,700}{10.5 \times 10^6} (30.7)^2 \right] = 52,300 \text{ psi}$$

The critical load for the 30-inch column is therefore

$$P_{cr} = \sigma_{cr} A = 52,300 \times 0.2855 = 14,900 \text{ lb}$$

70-inch column

The slenderness ratio for this length is

$$\frac{L_e}{\rho} = \frac{70 / \sqrt{1.5}}{0.798} = 71.6$$

Since this exceeds the critical value, we turn to Equation c for the primary buckling stress,

$$\sigma_{cr} = \frac{\pi^2 \left(10.5 \times 10^6 \right)}{71.6^2} = 20,200 \text{ psi}$$

This yields a buckling load of 5770 lb.

Figure 12.9.5 shows a portion of a thin panel to which stiffeners are bonded, with the stiffeners spaced a distance b apart. The stiffened panel is shown supporting a compressive load. The stiffeners act to divide the panel into narrow plates of width b and length a, where a is the distance between the end supports of the panel. Taken individually, these plates have a higher aspect ratio than the panel by itself, so the buckling stress is increased, as mentioned previously. The degree of edge restraint exerted by a stiffener on the skin, along a continuous attachment line, depends on the size and shape of the stiffener cross section. For our purposes, it is sufficient to assume that the stiffeners apply a hinged edge support to the panels. At a given panel section, the situation is somewhat like that illustrated in Figure 12.9.6, where the stringers rotate around the bond line with the buckling skin. This is conservative, and one would have to justify, by using empirical design curves and other data, whether or not, in individual cases, a larger resistance to buckling—approaching the clamped edge condition—may be assumed. For example, holding everything else fixed while increasing the spacing between the stiffeners would lower the

Figure 12.9.5   Stiffened panel in compression.

Figure 12.9.6   Local twist of stiffeners that apply simply-supported edge constraint to the skin subregions.

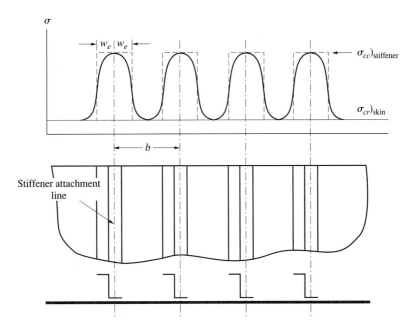

Figure 12.9.7   Actual versus equivalent stress distribution in stiffened panel structure after buckling of the skin.

buckling stress of the skin, perhaps to the point where the torsional rigidity of the stiffeners would be sufficient
to impose the clamped edge, zero rotation restraint.

At low load levels, both the skin and the stiffeners are active in distributing the stress. As the compressive
load increases, the skin buckles elastically, and any increase in load will be carried mainly by the stiffeners and
the adjacent plate material, as illustrated in Figure 12.9.7. The sinusoidal-like curve is only suggestive of the
complex nature of the post-buckled stress distribution in the skin and the stiffeners. That distribution is replaced
by the (pictured) uniform stress distribution extending across each stiffener and into the adjacent skins by a distance w_e called the *effective width* of the plate. The effective width is one-half the width of a plate in which the
buckling stress equals the stress in the stiffener. Thus, if $\sigma_{\text{stiffener}}$ is the stress in the stiffener and $\sigma_{cr})_{\text{skin}}$ is the
stress in the buckled skin, we set

$$\sigma_{cr})_{\text{skin}} = \sigma_{\text{stiffener}}$$

Substituting Equation 12.8.4 for the stress in the skin, replacing b with $2w_e$, yields

$$C\frac{\pi^2 E}{12\left(1 - v^2\right)}\left(\frac{t}{2w_e}\right)^2 = \sigma_{\text{stiffener}}$$

From this, we obtain

$$\frac{w_e}{t} = \frac{1}{2}\sqrt{\frac{C\pi^2}{12\left(1 - v^2\right)}}\sqrt{\frac{E}{\sigma_{\text{stiffener}}}}$$

This formula must be modified to fit test data, to allow for the possibility of stiffener and skin being made of different materials, and to account for the fact that crippling can occur at stresses above the proportional limit. Therefore, we will employ the following empirical expression to calculate the effective width w_e of a strip of material adjacent to a stringer:

$$\frac{w_e}{t} = K \frac{E_s)_{\text{skin}}}{\sqrt{E_s)_{\text{stiffener}}}} \sqrt{\frac{1}{\sigma_{\text{stiffener}}}} \qquad [12.9.7]$$

where $K = 0.85$ if the strip is simply supported (no edge free), and $K = 0.65$ if the strip has one edge free (see Figure 12.9.8). The term E_s is the value of the secant modulus at the level of the stiffener stress $\sigma_{\text{stiffener}}$. Below the proportional limit, $E_s = E$. If the skin and stiffener are the same alloy, then

$$\frac{w_e}{t} = K \sqrt{\frac{E_s}{\sigma_{\text{stiffener}}}} \qquad [12.9.8]$$

It must be pointed out that Equations 12.9.7 and 12.9.8 are not valid if the skin buckles between the fasteners (e.g., rivets or spot welds) that bond the skin to the stiffener, as illustrated in Figure 12.9.9. *Interfastener buckling* may occur if the fastener spacing is relatively large compared to the skin thickness. In such a case, the skin wrinkles between the fasteners, but the stiffener flange remains essentially straight. If the fastener pitch is small enough to prevent interfastener buckling, there is the possibility of *wrinkling failure*. In this longer-range buckling mode, illustrated in Figure 12.9.10, the skin does not bow away from the stiffener between the fasteners, but tends to force the stiffener flange to deform with it. This in turn induces stress in the stiffener web, possibly leading to local crippling of the stiffener. Clearly, tensile failure of the fasteners joining the skin and stiffener

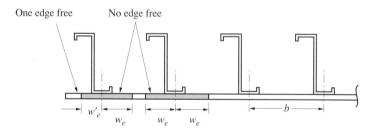

Figure 12.9.8   Cross section of a stiffened panel, the left end of which is free.

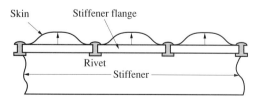

Figure 12.9.9   Interfastener (in this case inter*rivet*) buckling.

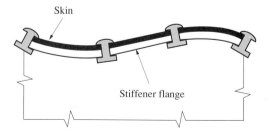

Figure 12.9.10   Wrinkling failure.

is a matter of concern when designing against this type of failure. For more detailed discussions of the design and stability of stiffened panels, see Bruhn,[16] Rivello,[17] and Niu.[18]

Example 12.9.3   Calculate the maximum compressive load of the stiffener-skin combination in Figure 12.9.11a if the section is stabilized against primary buckling. The stiffener is an aluminum 2014-T6 extrusion ($E = 10.7 \times 10^6$ psi, $\sigma_{cy} = 53,000$ psi) and the sheet is aluminum 7075-T6 ($E = 10.5 \times 10^6$ psi, $\sigma_{cy} = 67,000$ psi). Assume no interrivet buckling.

In the first step, we calculate the crippling stress of the stiffener by itself. To do so, we resolve the section into two rectangles, deleting corners, as indicated in Figure 12.9.11b.

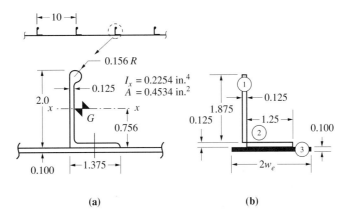

(a)   **(b)**

Figure 12.9.11   (a) Riveted stiffener and sheet unit. (b) Discretized stiffener and the effective skin. (All dimensions are inches.)

Element 1

$$b_1 = 1.875 \text{ in.} \qquad t_1 = 0.125 \text{ in.}$$

Therefore, from Equation 12.9.2,

$$\lambda = \sqrt{\frac{53,000}{10.7 \times 10^6} \frac{1.875}{0.125}} = 1.06$$

Substituting this into Equation 12.9.51, we get

$$\sigma_{cc}^{(1)} = 1.34 \times 1.06^{-0.8} \times 53,000 = 67,800 \text{ psi}$$

This just exceeds the cutoff stress for this alloy, which, according to Table 12.9.1, is

$$53,000 \times (1 + 53,000/200,000) = 67,000 \text{ psi}$$

---

[16]E. F. Bruhn, *Analysis and Design of Flight Vehicle Structures*, Indianapolis, S. R. Jacobs, 1973.
[17]*op. cit.*
[18]M. C. Y. Niu, *Airframe Structural Design*, Hong Kong, Conmilit Press, 1988.

so we must reduce our computed $\sigma_{cc}^{(1)}$ to this value. Also, since the free edge of this element has a bulb with a radius greater than the thickness, we are advised by Figure 12.9.2c to compute the crippling stress as 0.7 times that of the "no edge free" condition. This accounts for the restraint against buckling exerted by the bulb. Therefore, we finally obtain

$$\sigma_{cc}^{(1)} = 0.7 \times 67,000 = 46,900 \text{ psi}$$

Element 2

$$b_2 = 1.25 \text{ in.} \qquad t_2 = 0.125 \text{ in.}$$

Therefore,

$$\lambda = \sqrt{\frac{53,000}{10.7 \times 10^6} \frac{1.25}{0.125}} = 0.704$$

This segment has one edge free, so we use Equation 12.9.5$_2$ to find

$$\sigma_{cc}^{(2)} = 0.546 \times 0.704^{-.8} \times 53,000 = 38,300 \text{ psi}$$

According to Equation 12.9.6, the crippling stress of the stiffener is

$$\sigma_{cc} = \frac{1.875 \times 0.125 \times 46,900 + 1.25 \times 0.125 \times 38,300}{1.875 \times 0.125 + 1.25 \times 0.125} = 43,500 \text{ psi}$$

The next step is to find the effective width of the skin w_e. We must use Equation 12.9.7, because the extrusion and the sheet are made of different alloys. Since the crippling stress σ_{cc} is considerably less than the compressive yield stress, and therefore the proportional limit, of both the stiffener, and the skin material, we can replace the secant moduli by the moduli of elasticity. Therefore, since the "no edge free" condition applies to the sheet, we have

$$w_e = \left(0.85 \frac{10.5 \times 10^6}{\sqrt{10.7 \times 10^6}} \sqrt{\frac{1}{43,500}} \right) \times 0.1 = 1.31 \text{ in.}$$

To find the load-carrying capability of the stringer-sheet combination, we multiply the crippling stress σ_{cc} by the sum of the *actual* area of the stiffener and the *effective* area of the plate:

$$P_{cc} = 43,500 \times (0.4534 + 2 \times 1.31 \times 0.100) = 31,100 \text{ lb}$$

To estimate the primary buckling strength of stiffened compression panels, we can treat them as columns. Each stiffener, together with the effective width of skin, is viewed as a single column of effective length L_e that can bend about an axis parallel to the plane of the sheet. We first calculate the crippling stress σ_{cc} of the stiffener. That stress is then used to find the effective width w_e of the skin, as illustrated in the previous example. With that information, we can find the radius of gyration ρ of the stiffener *plus* the plate about the centroid of the composite region. At this point, the buckling stress is calculated using the Johnson parabola column strength formula, Equation 12.6.12,

$$\sigma_{cr} = \sigma_{cc} \left[1 - \frac{1}{4\pi^2} \frac{\sigma_{cc}}{E} \left(\frac{L_e}{\rho} \right)^2 \right]$$

We use this value of σ_{cr} to recompute an effective plate width, which leads to a revised radius of gyration and therefore a new value for σ_{cr}. The process is repeated several times until we obtain convergence with a reasonable precision.

Example 12.9.4   Calculate the compressive failing strength of the stiffened panel unit of Example 12.9.3 if the length of the panel is 42 inches and the end fixity coefficient is 2.

The stiffener and sheet unit are reproduced in Figure 12.9.12, with some additional notation to facilitate the calculations. The subscripts *st* and *sk* stand for "stiffener" and "skin," respectively. The centroid G of the composite area is located by the vertical coordinate y_G relative to the horizontal x axis through the centroid of the plate, as shown. A parallel axis through the centroid of the stiffener, with coordinate $y_{G_{st}} = 0.806$ in., is also shown.

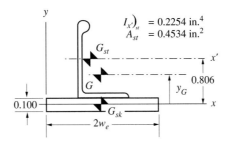

Figure 12.9.12   Bulb angle stiffener and effective sheet.
(All dimensions are inches.)

We found the stiffener crippling stress in the previous example. We can now use Equation 12.6.12 to determine the critical slenderness ratio of the stiffener and skin acting as a column:

$$\left.\frac{L_e}{\rho}\right)_{critical} = \pi\sqrt{\frac{2E}{\sigma_{cc}}} = \pi\sqrt{\frac{2 \times 10.6 \times 10^6}{43,500}} = 69.4 \tag{a}$$

where E is the average of the two slightly different elastic moduli for the stiffener and the sheet. The effective length of the column is

$$L_e = \frac{42}{\sqrt{2}} = 29.7 \text{ in.}$$

We know from the previous example that for $\sigma_{cc} = 43,500$ psi, the effective width w_e of the sheet is 1.31 in. Therefore, the centroid of the composite area relative to the x axis is

$$y_G = \frac{y_{G_{st}} A_{st} + y_{G_{sk}} A_{sk}}{A_{st} + A_{sk}} = \frac{0.806 \times 0.4534 + 0}{0.4534 + 2 \times 1.31 \times 0.100} = 0.5110 \text{ in.}$$

The moment of inertia of the composite area about a horizontal axis through its centroid is

$$\bar{I}_x = \overbrace{0.2254 + 0.4534 \times (0.805 - 0.511)^2}^{\text{stiffener}} + \overbrace{\frac{1}{12}(2 \times 1.31)(0.1)^3 + (2 \times 1.31)(0.1)(0 - 0.511)^2}^{\text{skin}} = 0.3334 \text{ in.}^4$$

From this, we obtain the centroidal radius of gyration of the composite area:

$$\bar{\rho}_x = \sqrt{\frac{\bar{I}_x}{A_{st} + A_{sk}}} = \sqrt{\frac{0.3334}{0.4534 + 0.262}} = 0.6828 \text{ in.}$$

We can now find the slenderness ratio of the column, using

$$\frac{L_e}{\bar{\rho}_x} = \frac{29.7}{0.6828} = 43.49$$

This is smaller than the critical slenderness ratio in Equation a, so we have a short column and the Johnson parabola column curve is applicable. Therefore, the critical stress is

$$\sigma_{cr} = \sigma_{cc}\left[1 - \frac{1}{4\pi^2}\left(\frac{\sigma_{cc}}{E}\right)\left(\frac{L_e}{\rho}\right)^2\right] = 43,500\left[1 - \frac{1}{4\pi^2}\frac{43,500}{10.6 \times 10^6}(43.49)^2\right] = 34,930\ \text{psi}$$

We use this as our stiffener stress to compute a new value of the effective plate width, obtaining

$$w_e = \left(0.85\frac{10.5 \times 10^6}{\sqrt{10.7 \times 10^6}}\sqrt{\frac{1}{34,930}}\right) \times 0.1 = 1.460\ \text{in.}$$

From this, we calculate, in the same way as before,

$$A_{sk} = 0.2920\ \text{in.} \Rightarrow y_G = 0.4903\ \text{in.} \Rightarrow \bar{I}_x = 0.3412\ \text{in.}^4 \Rightarrow \bar{\rho}_x = 0.6763\ \text{in.} \Rightarrow \frac{L_e}{\bar{\rho}_x} = 43.91 \Rightarrow \sigma_{cr} = 34,770\ \text{psi}$$

Using this stress, we find $w_e = 1.463$ in., which is the same to three significant figures as the previous step, so there is no need to repeat. The critical load is

$$P_{cr} = 34,770 \times (0.4534 + 2 \times 1.46 \times 0.1) = 25,900\ \text{lb}$$

12.10   SHEAR PANEL STABILITY

A thin plate, loaded in pure shear on its edges, will buckle when the shear stress reaches a critical value, given by

$$\tau_{cr} = C\frac{\pi^2 E}{12\left(1 - v^2\right)}\left(\frac{t}{b}\right)^2 \tag{12.10.1}$$

Just as in Equation 12.8.4 for compression buckling, C is a constraint coefficient that is a function of the boundary conditions and the aspect ratio a/b. Figure 12.10.1 illustrates the variation of C with aspect ratio for the two extreme cases, all edges simply supported and all edges clamped. Again, b is the length of the smaller side of the rectangle, and t is the wall thickness.

Using Mohr's circle, it is easy to show that a state of pure shear stress on orthogonal planes is accompanied by a state of principal stress on planes oriented 45 degrees thereto. One of the principal stresses is tensile and the other is compressive, and both are equal in magnitude to the shear stress, as shown in Figure 12.10.2. The compressive principal stress is responsible for the plate becoming unstable. It is not surprising that the buckled mode shape consists of wrinkles roughly transverse to the direction of the compressive principal stress. Figure 12.10.3 pictures the buckled mode of a simply-supported plate with $a/b = 2$.

For curved, rectangular, shear panels, such as that pictured in Figure 12.10.4, the theoretical buckling curves for all edges simply supported are presented in Figures 12.10.5 and 6. Figure 12.10.5 is for long panels, that is, those in which the small dimension b is in the circumferential direction. Figure 12.10.6 applies to short panels, in which the small dimension b is parallel to the axis of the cylinder. The variable on the abscissa is

$$z_b = \frac{b^2}{rt}\sqrt{1 - v^2}$$

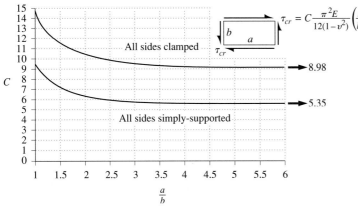

Figure 12.10.1 Theoretical buckling strength curves for a rectangular flat panel in pure shear.

Figure 12.10.2 Pure shear produces equal tensile and compressive principal stresses on planes at 45 degrees to the edges of the plate.

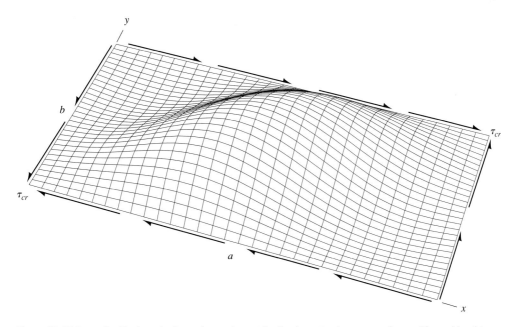

Figure 12.10.3 Buckled mode shape for a plate with all edges simply supported, $a = 2b$, and load in pure shear.

The term z_b is a measure of both the curvature and thinness of the panel. As z_b approaches zero, the radius of curvature r approaches infinity, which is true of a flat plate. Therefore, in Figures 12.10.5 and 6, the values of C for $z_b < 1$ are those for flat plates. Clearly, for a given aspect ratio a/b, a curved, rectangular panel buckles at a higher shear stress than its flat counterpart.

Curved shells are sensitive to initial imperfections, which are inevitable in real, fabricated panels. Tests and experience show that buckling may occur at stresses below those predicted by theory. Thus, the theoretical

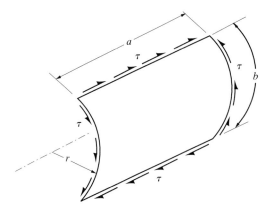

Figure 12.10.4 Curved (circular cylindrical) shear panel of thickness t.

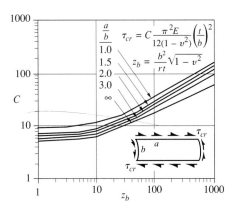

Figure 12.10.5 Theoretical shear buckling coefficient for long, simply-supported curved panels.

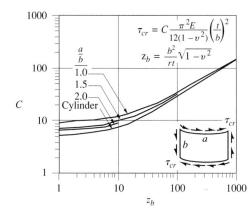

Figure 12.10.6 Theoretical shear buckling coefficient for short, simply-supported curved panels.

buckling curves are not conservative for design purposes. Additional empirical information may be required to accurately predict the buckling strength of a given design.

12.11 COMPLETE TENSION FIELD BEAMS

Figure 12.11.1 shows a cantilevered, stiffened web beam of depth h and length $4l$, with a transverse load P applied at the free end. Throughout this text, we have treated this type of structure as an idealized beam. Transverse sections of the web are in pure shear, the horizontal flanges transfer the axial loads due to the bending moment, and the vertical stiffeners pick up transverse applied loads. If the thickness of the web is t, then the shear flow q throughout the beam is the uniform shear $V_y = P$ divided by the height of the web, or

$$q = \frac{P}{h}$$

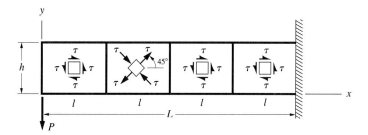

Figure 12.11.1   Idealized beam of panel thickness t, prior to buckling of the webs.

The shear stress τ on the transverse sections is

$$\tau = \frac{q}{t} = \frac{P}{ht} \qquad \textbf{[12.11.1]}$$

Associated with this state of pure shear are the principal stresses—one tensile, the other compressive—acting on perpendicular planes at 45 degrees to the vertical. Both principal stresses are equal in magnitude to the shear stress τ.

As P increases, so does τ, and eventually the webs buckle in shear, tending to wrinkle normal to the buckling direction, as suggested in Figure 12.11.2. At that point, if the web is quite thin, the compressive principal stress in the buckling direction vanishes, and a single tensile principal stress σ_p develops, acting diagonally at an angle α, which is less than 45 degrees, as illustrated in Figure 12.11.2. The marked change in the principal stresses is reflected in the fact that the state of stress on horizontal and transverse sections of the web is no longer pure shear. Normal stresses σ_x and σ_y appear on those planes, as shown. However, the shear stress must remain the same as before, given by Equation 12.11.1, if we maintain the assumption that the flanges carry none of the applied shear loading.

Since the shear stress τ is known, for a given angle α, the stresses σ_x, σ_y, and σ_p can be found using the stress transformation relations in Section 3.6. Let the $\bar{x}\,\bar{y}$ axes be aligned with the diagonal tension field, as shown in Figure 12.11.2. According to Equations 3.6.4, the transformation of the stresses from the barred system to the xy system is given by

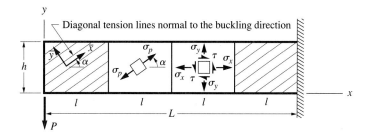

Figure 12.11.2   Postbuckling diagonal tension field and accompanying state of stress.

The shear stress is the same as in Figure 12.11.1.

$$\sigma_x = \bar{\sigma}_x \cos^2(-\alpha) + \bar{\sigma}_y \sin^2(-\alpha) + 2\bar{\tau}_{xy} \sin(-\alpha) \cos(-\alpha)$$
$$\sigma_y = \bar{\sigma}_x \sin^2(-\alpha) + \bar{\sigma}_y \cos^2(-\alpha) - 2\bar{\tau}_{xy} \sin(-\alpha) \cos(-\alpha) \qquad \text{[12.11.2]}$$
$$\tau_{xy} = (\bar{\sigma}_y - \bar{\sigma}_x) \sin(-\alpha) \cos(-\alpha) + \bar{\tau}_{xy}\left[\cos^2(-\alpha) - \sin^2(-\alpha)\right]$$

The minus sign on α is required because Equations 3.6.4 were derived for a counterclockwise rotation from xy to $\bar{x}\bar{y}$, but here we have a clockwise rotation from $\bar{x}\bar{y}$ to xy.

Since $\bar{\sigma}_x = \sigma_p$ and $\bar{\sigma}_y = \bar{\tau}_{xy} = 0$, we obtain

$$\sigma_x = \sigma_p \cos^2 \alpha \qquad \text{[12.11.3a]}$$

$$\sigma_y = \sigma_p \sin^2 \alpha \qquad \text{[12.11.3b]}$$

$$\tau_{xy} = \tau = \sigma_p \sin \alpha \cos \alpha \qquad \text{[12.11.3c]}$$

Solving Equation 12.11.3c for σ_p in terms of τ, we get

$$\sigma_p = \frac{\tau}{\sin \alpha \cos \alpha} = \frac{2P}{ht \sin 2\alpha} \qquad \text{[12.11.4]}$$

Substituting this into Equations 12.11.3a and b yields

$$\sigma_x = \frac{\tau}{\tan \alpha} = \frac{P}{ht \tan \alpha} \qquad \text{[12.11.5]}$$

and

$$\sigma_y = \tau \tan \alpha = \frac{P}{ht} \tan \alpha \qquad \text{[12.11.6]}$$

The postbuckled, diagonal tension field induces compressive stresses in the flanges and upright stiffeners over and above those that exist in the purely shear-resistant beam. Figure 12.11.3 is a free-body diagram of a portion of the beam to the left of a vertical section. In addition to the familiar tensile and compressive flange loads required to equilibrate the moment of the applied load P, compressive forces N_{DT_F} must act to balance the web normal stress σ_x accompanying the diagonal tension (DT) field. Force equilibrium in the x direction requires $\sigma_x ht - 2N_{DT_F} = 0$. Therefore, if σ_{UF} denotes the stress in the upper flange, using Equation 12.11.5 shows that

$$\sigma_{UF} = \frac{P}{A_F} \frac{x}{h} - \frac{P}{2A_F} \frac{1}{\tan \alpha} \qquad \text{[12.11.7]}$$

where A_F is the cross-sectional area of the flanges. Likewise, for the lower flange, we have

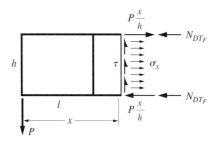

Figure 12.11.3   Free-body diagram of a portion of the tension field beam in Figure 12.11.2.

$$\sigma_{LF} = -\frac{P}{A_F}\frac{x}{h} - \frac{P}{2A_F}\frac{1}{\tan\alpha} \qquad \text{[12.11.8]}$$

If we isolate the entire upper flange as a free body, the forces are as shown in Figure 12.11.4. The stress, $\sigma_y = \tau\tan\alpha$, in the tension field web exerts a uniform transverse force per unit length, $\tau l\tan\alpha$, on the flange, which itself acts as a continuous beam clamped at each end and simply supported at each upright stiffener. Using the methods of Chapter 7, it is easy to show that the bending moment at the ends of the flange, due solely to the diagonal tension field, is

$$M_{DT} = \frac{1}{12}\tau t l^2 \tan\alpha = \frac{1}{12}\frac{Pl^2}{h}\tan\alpha \qquad \text{[12.11.9]}$$

The bending moment direction is indicated in Figure 12.11.4. This is a *secondary* bending moment, as opposed to the *primary* bending moment exerted by the top and bottom flanges acting together to resist the moment of the applied load (Figure 12.11.3). Figure 12.11.5 shows the bending moment distribution in the top flange, which is reflected in the bottom flange, as well.

We assume that the distributed load is shared equally by the two upright stiffeners on either side of a given panel. The compressive axial load N_{DT_S} in each of the interior stiffeners, due solely to the diagonal tension field, is

$$N_{DT_S} = \sigma_y t l = \frac{Pl}{h}\tan\alpha \qquad \text{[12.11.10]}$$

The load is one-half that value in the leftmost stiffener, which is adjacent to only one panel. The axial stress in an interior stiffener (*IS*) is

$$\sigma_{IS} = -\frac{N_{DT_S}}{A_{IS}} \qquad \text{[12.11.11]}$$

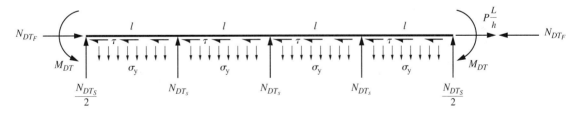

Figure 12.11.4   Free-body diagram of the upper flange.

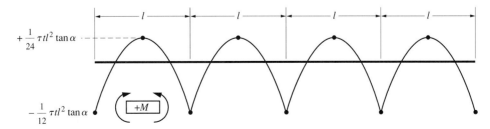

Figure 12.11.5   The secondary moment distribution on the top (and bottom) flange of the tension field beam in Figs. 12.11.1 and 2.

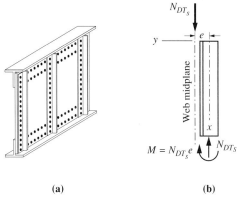

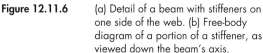

$M = N_{DT_S} e$ N_{DT_S}

(a) **(b)**

Figure 12.11.6 (a) Detail of a beam with stiffeners on one side of the web. (b) Free-body diagram of a portion of a stiffener, as viewed down the beam's axis.

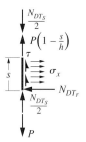

Figure 12.11.7 Free-body diagram of the stiffener at the left end of the beam in Figure 12.11.2.

If the stiffeners are attached to only one side of the web, as illustrated in Figure 12.11.6a, then N_{DT_S} is offset from the centroidal axis of the stiffener by an amount e. In Figure 12.11.6b, we see that this eccentric loading produces a bending moment $N_{DT_S} e$. In that case, the stress on the stiffener cross section is the superposition of flexural stress onto the otherwise uniform compressive stress due to N_{DT_S}:

$$\sigma_{IS} = -\frac{N_{DT_S}}{A_{IS}} - \frac{\left(N_{DT_S} e\right) y}{I_{IS}} = -\frac{N_{DT_S}}{A_{IS}}\left(1 + \frac{ey}{\rho_{IS}^2}\right)$$ [12.11.12]

The term ρ is the radius of gyration of the stiffener about its centroid, which means that $I_{IS} = A_{IS}\rho_{IS}^2$. Aided by the free-body diagram in Figure 12.11.7, we can also show that the axial stress σ_{ES} in the end stiffener on the left of the beam is

$$\sigma_{ES} = -\frac{N_{DT_S}/2 - P\left(1 - s/h\right)}{A_{ES}}\left(1 + \frac{ey}{\rho_{ES}^2}\right)$$ [12.11.13]

The compressive axial load N_{DT_S} in the stiffeners due to diagonal tension tends to promote buckling of the intermediate stiffeners out of the plane of the web. A stiffener may be treated as a column with an effective length L_e chosen according to the following empirical formulae:

$$L_e = \frac{h}{\sqrt{4-2l/h}} \qquad \text{if} \qquad \frac{l}{h} < 1.5$$
$$L_e = h \qquad \text{if} \qquad \frac{l}{h} > 1.5$$ [12.11.14]

For the web stresses to be uniformly distributed, as depicted in Figures 12.11.3 and 5 and as assumed throughout the previous discussion, the flanges and stiffeners must be infinitely rigid in the plane of the web: they must not bend, and the edges of the panels must remain straight. Otherwise, the transverse loading on the flanges will tend to pull them inward towards each other, as illustrated in Figure 12.11.8. That deformation will unload the webs in the central portion of the panels to produce a complex, nonuniform stress distribution.

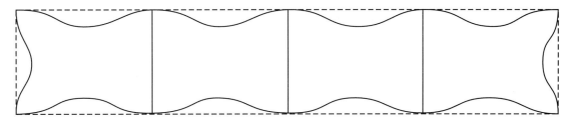

Figure 12.11.8 Flange and stiffener deformation (exaggerated) induced by the diagonal tension field.

To find the diagonal tension angle α, still assuming a uniform stress in the webs, we calculate the complementary virtual work associated with a small virtual change $\delta\alpha$. In the webs, we have uniaxial tension relative to the $\bar{x}\,\bar{y}$ coordinate system of Figure 12.11.2. Thus, from Equation 6.6.4, we have

$$\delta W^*_{\text{int}}\Big)_{\text{panel}} = \iiint\limits_{V_{\text{panel}}} \frac{\bar{\sigma}_x}{E}\delta\bar{\sigma}_x\,dV = \frac{\sigma_p}{E}\delta\sigma_p h l t$$

where σ_p is given by Equation 12.11.4, with the aid of which we find that

$$\delta\sigma_p = 2\tau\frac{d}{d\alpha}\left(\frac{1}{\sin 2\alpha}\right)\delta\alpha = -4\tau\frac{\cos 2\alpha}{\sin^2 2\alpha}\delta\alpha$$

Hence, for each panel, we get

$$\delta W^*_{\text{int}}\Big)_{\text{panel}} = -8\frac{hlt\tau^2}{E}\frac{\cos 2\alpha}{\sin^3 2\alpha}\delta\alpha = -8\frac{P^2l}{Eht}\frac{\cos 2\alpha}{\sin^3 2\alpha} \qquad \textbf{[12.11.15]}$$

For the flanges, we have

$$\delta\sigma_{UF} = \frac{\partial\sigma_{UF}}{\partial\alpha}\delta\alpha \qquad \text{and} \qquad \delta\sigma_{LF} = \frac{\partial\sigma_{LF}}{\partial\alpha}\delta\alpha$$

Substituting Equations 12.11.7 and 8 yields

$$\delta\sigma_{UF} = \delta\sigma_{LF} = \frac{P}{2A_F}\frac{1}{\sin^2\alpha}\delta\alpha \qquad \textbf{[12.11.16]}$$

The complementary virtual work in an upper flange portion of length l is

$$\delta W^*_{\text{int}}\Big)_{UF} = \int_0^l \frac{\sigma_{UF}}{E}\delta\sigma_{UF}A_F\,dx$$

Substituting Equations 12.11.7 and 16 and integrating yields

$$\delta W^*_{\text{int}}\Big)_{UF} = \frac{P^2l}{4A_F E \sin^2\alpha}\left(\frac{l}{h} - \frac{1}{\tan\alpha}\right) \qquad \textbf{[12.11.17]}$$

For a segment of the lower flange, the equation is

$$\delta W_{\text{int}}^*)_{LF} = \frac{P^2 l}{4 A_F E \sin^2 \alpha} \left(-\frac{l}{h} - \frac{1}{\tan \alpha} \right)$$ [12.11.18]

For the interior stiffeners, the virtual stress is obtained from Equation 12.11.12

$$\delta \sigma_{IS} = -\frac{\delta N_{DT_S}}{A_{IS}} \left(1 + \frac{ey}{\rho_{IS}^2} \right)$$

The internal complementary virtual work is therefore

$$\delta W_{\text{int}}^*)_{IS} = \int_0^h \iint_{A_{IS}} \frac{\sigma_{IS}}{E} \delta \sigma_{IS} d A dx = \frac{N_{DT_S} \delta N_{DT_S}}{A_{IS}^2 E} \left[\int_0^h \left(\iint_{A_{IS}} d A + 2 \frac{e}{\rho_{IS}^2} \iint_{A_{IS}} y d A + \frac{e^2}{\rho_{IS}^2} \iint_{A_{IS}} y^2 d A \right) dx \right]$$

Since $\iint_{A_{IS}} y d A = 0$ and $\iint_{A_{IS}} y^2 d A = \rho_{IS}^2 A_{IS}$, this reduces to

$$\delta W_{\text{int}}^*)_{IS} = \frac{N_{DT_S} \delta N_{DT_S}}{A_{IS}^2} \left(A_{IS} + \frac{e^2}{\rho_{IS}^2} A_{IS} \right) = \frac{N_{DT_S} \delta N_{DT_S}}{A_{IS,\text{eff}}}$$ [12.11.19]

where $A_{IS,\text{eff}}$, the effective area of the stiffener, is

$$A_{IS,\text{eff}} = \frac{A_{IS}}{1 + \frac{e^2}{\rho_{IS}^2}}$$ [12.11.20]

From Equation 12.11.10, it follows that

$$\delta N_{DT_S} = \frac{\partial}{\partial \alpha} \left(\frac{Pl}{h} \tan \alpha \right) \delta \alpha = \frac{Pl}{h \cos^2 \alpha} \delta \alpha$$ [12.11.21]

Therefore, the internal virtual work for each of the intermediate stiffeners is

$$\delta W_{\text{int}}^*)_{IS} = \frac{P^2 l^2}{A_{IS,\text{eff}} E h} \frac{\sin \alpha}{\cos^3 \alpha} \delta \alpha$$ [12.11.22]

Similarly, the end stiffener internal virtual work equation is

$$\delta W_{\text{int}}^*)_{ES} = \frac{P^2 l}{4 A_{ES,\text{eff}} E \cos^2 \alpha} \left(\frac{l}{h} \tan \alpha - 1 \right) \delta \alpha$$ [12.11.23]

The internal complementary virtual work for the entire tension field beam is

$$\delta W_{\text{int}}^* = n_{\text{panels}} \times \delta W_{\text{int}}^*)_{\text{panels}} + n_{\text{panels}} \times \delta W_{\text{int}}^*)_{UF} + n_{\text{panels}} \times \delta W_{\text{int}}^*)_{LF} + \left(n_{\text{panels}} - 1 \right) \times \delta W_{\text{int}}^*)_{IS} + \delta W_{\text{int}}^*)_{ES}$$

Since the virtual loading is internal, the external virtual work is zero. Thus, the principle of complementary virtual work requires that

$$n_{\text{panels}} \times \left[\delta W_{\text{int}}^*)_{\text{panels}} + \delta W_{\text{int}}^*)_{UF} + \delta W_{\text{int}}^*)_{LF} \right] + \left(n_{\text{panels}} - 1 \right) \times \delta W_{\text{int}}^*)_{IS} + \delta W_{\text{int}}^*)_{ES} = 0$$ [12.11.24]

Substituting Equations 12.11.15, 17, 18, 22, and 23 on the left, invoking familiar trigonometric identities, combining terms, and simplifying ultimately leads to the following equation for the tension field angle α:

$$\left\{\left[\frac{1}{4n_{\text{panels}}}\left(\frac{A_{IS,\text{eff}}}{A_{ES,\text{eff}}}-4\right)+1\right]\frac{lt}{A_{IS,\text{eff}}}+1\right\}\tan^4\alpha-\frac{1}{4n_{\text{panels}}}\frac{ht}{A_{ES,\text{eff}}}\tan^3\alpha-\frac{1}{2}\frac{ht}{A_F}-1=0 \qquad \text{[12.11.25]}$$

This equation has two complex roots and two real roots, one of them positive and the other negative. We are interested in the positive root.

Example 12.11.1   For the beam in Figure 12.11.9, the aluminum alloy properties are uniform throughout: $E = 10 \times 10^6$ psi, $\nu = 0.25$, and the yield strength in tension and compression is 48,000 psi. Section properties of the flanges and stiffeners are shown. The stiffeners are attached to only one side of the web.

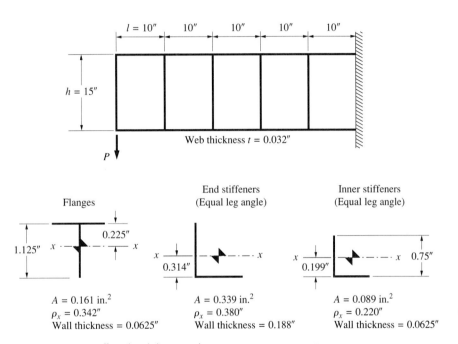

Figure 12.11.9   Stiffened web beam with pertinent section properties.

a. Estimate the value of the load P that will cause the web of the beam to buckle.

b. Assuming that a complete tension field develops in the webs after buckling, calculate the diagonal tension angle α.

c. Estimate the value of the load required to initiate yielding in the flanges.

d. Estimate the value of P required to produce buckling of the stiffeners, assuming no failure of the flanges.

e. Estimate the load required to initiate yielding in the webs, assuming no failures in the stiffeners and flanges.

a. The shear stress in the fully shear resistant panels is

$$\tau = \frac{P}{ht}$$

Substituting this expression into Equation 12.10.1, the buckling formula for a rectangular plate in pure shear, and recalling that b is the short edge of the panel ($b = l$), we get the critical buckling load,

$$P_{cr} = C \frac{\pi^2 E h t^3}{12 \left(1 - \nu^2\right) l^2}$$ [a]

Figure 12.10.1 provides the value of C. For the individual beam panels, $a/b = h/l = 1.5$. If all sides are simply-supported, $k = 7.1$; if all sides are clamped, $C = 11.5$. Let us assume the average of these two extremes, or $C = 9.2$. In this case, the webs will buckle when

$$P = 9.2 \frac{\pi^2 \left(10 \times 10^6\right) (15) (0.032)^3}{12 \left(1 - 0.25^2\right) (10)^2} = 397 \text{ lb}$$ [b]

b. We first calculate the effective areas of the stiffeners, using Equation 12.11.19. The stiffener eccentricity is the distance from the middle of the web to the centroid of the stiffener cross section. Noting that the web thickness is 0.032 in., we use the data in Figure 12.11.9 to obtain the following for the intermediate stiffeners

$$e_{IS} = \frac{0.032}{2} + 0.199 = 0.215 \text{ in.}$$

and for the end stiffener, we get

$$e_{ES} = \frac{0.032}{2} + 0.314 = 0.330 \text{ in.}$$

The effective stiffener areas are therefore

$$A_{IS,\text{eff}} = \frac{0.089}{1 + \left(0.215/0.22\right)^2} = 0.0455 \text{ in.}^2$$ [c]

$$A_{ES,\text{eff}} = \frac{0.339}{1 + \left(0.33/0.38\right)^2} = 0.193 \text{ in.}^2$$ [d]

Substituting these into Equation 12.11.25, together with $h = 15$ in., $l = 10$ in., $t = 0.032$ in., and $n_{\text{panels}} = 5$, yields

$$6.71 \tan^4 \alpha - 0.124 \tan^3 \alpha - 2.49 = 0$$

The positive root of this equation is $\tan \alpha = 0.785$, so the tension field angle is

$$\alpha = 38.1^\circ$$ [e]

c. The maximum stress in the flanges is compression at the wall in the topmost fiber of the lower flange. That stress is the superposition of the uniform stress given by Equation 12.11.8, which is due to the bending moment of the applied load P plus the force of diagonal tension exerted by the web, and the stress due to the bending moment (Equation 12.11.9) produced by the tension field. Thus, if c is the distance from the neutral axis of the flange to the base of the "T" section, the net stress is found to be

$$\sigma_{LF})_{\text{max}} = \left| -\frac{P}{A_F} \left[\frac{L}{h} + \frac{1}{2 \tan \alpha} + \frac{\tan \alpha}{12} \frac{c}{h} \frac{l^2}{\rho_x^2} \right] \right| = \frac{P}{0.161} \left[\frac{50}{15} + \frac{1}{2 \tan 38.1} + \frac{\tan 38.1}{12} \frac{1.125 - 0.225}{15} \frac{10^2}{0.342^2} \right] = 45.4P$$

Setting $\sigma_{LF})_{\text{max}}$ equal to the compressive yield stress of 48,000 psi and solving for P yields

$$P = 1060 \text{ lb}$$ [f]

which is a little more than two and a half times the load at which the webs buckle.

d. The critical stiffeners are the inner ones, since they carry the greater compressive load (Figure 12.11.4) and have a smaller cross section than the one at the left end, where the load is applied. For this beam, $l/h = 0.667$. Therefore, according to Equation 12.11.14, the effective length of a stiffener is

$$L_e = \frac{15}{\sqrt{4 - 2(10/15)}} = 9.18 \text{ in.}$$

Thus,

$$\frac{L_e}{\rho} = \frac{9.18}{0.22} = 41.7 \qquad [\text{g}]$$

Since this slenderness ratio is quite a bit less than 100, it is most likely that the Euler long-column formula is not applicable and we must use the straight-line column strength formula given by Equation 12.6.11

$$\sigma_{cr} = \sigma_{cc}\left[1 - \frac{2\sqrt{3}}{9\pi}\left(\frac{\sigma_{cc}}{E}\right)^{\frac{1}{2}}\left(\frac{L_e}{\rho}\right)\right] \qquad [\text{h}]$$

for which the critical effective length is

$$\left.\frac{L_e}{\rho}\right)_{\text{critical}} = \pi\sqrt{\frac{3}{\sigma_{cc}/E}} \qquad [\text{i}]$$

To find the crippling stress σ_{cc} of an inner stiffener section, we follow the procedure outlined in Section 12.9, segmenting the equal-leg angle section into two rectangles of thickness $t = 0.0625$ in. and length $b = 0.75 - 0.0625 = 0.688$ in., ignoring the corner material. Then, according to Equations 12.9.1, 2, and 3, the critical stress in both of the identical segments is

$$\sigma_{cc}^{(1)} = \sigma_{cc}^{(2)} = \sigma_{cy}\left\{\alpha\left[\frac{1}{12}\frac{\pi^2 C}{1 - \nu^2}\frac{E}{\sigma_{cy}}\left(\frac{t}{b}\right)^2\right]^{1-n}\right\} = 32,200 \text{ psi} \qquad [\text{j}]$$

where $\alpha = 0.8$, $n = 0.6$, $k = 0.425$ (one edge free), and rest of the data has already been provided. From Equation i, we can now determine that

$$\left.\frac{L_e}{\rho}\right)_{\text{critical}} = \pi\sqrt{\frac{3 \times (10 \times 10^7)}{32,200}} = 78.3$$

Comparison with Equation g confirms our expectation that the stiffener is a short column.

Since both segments of the resolved stiffener cross section have the same area, Equation 12.9.6 yields $\sigma_{cc} = 32,100$ psi for the crippling stress of the complete section. Finally, according to Equation h, the critical stress is

$$\sigma_{cr} = 32,200\left[1 - \frac{2\sqrt{3}}{9\pi}\left(\frac{32,200}{10 \times 10^6}\right)^{\frac{1}{2}}\left(\frac{9.18}{0.22}\right)\right] = 22,800 \text{ psi}$$

Multiplying this stress by the stiffener cross-sectional area, we obtain the critical stiffener load:

$$N_{DTS})_{cr} = 22,800 \times 0.089 = 2030 \text{ lb}$$

Substituting this load into Equation 12.11.10 yields the critical value of P for stiffener buckling:

$$P_{cr} = \frac{N_{DTS})_{cr}\, h}{l \tan\alpha} = \frac{2030 \times 15}{10 \times \tan 38.1} = 3880 \text{ lb} \qquad [\text{k}]$$

This is over three and a half times the load at which the lower flange yields.

e. In a complete tension field web, the only nonzero principal stress is the diagonal tension stress σ_p given by Equation 12.11.4, which, solved for P, yields

$$P = \frac{1}{2}\sigma_p ht \sin 2\alpha$$

Using the maximum shear stress theory of failure, we assume the web yields when σ_p reaches the tensile yield stress of the material. Thus,

$$P = \frac{1}{2} \times 48,000 \times 15 \times 0.032 \times \sin 76.2 = 11,200 \text{ lb} \qquad [\text{I}]$$

which is over ten and a half times the load at which the lower flange yields.

In actual stiffened web beams, the panel thickness is such that a compressive stress exists normal to the wrinkling direction after buckling. The web is in a state of *incomplete diagonal tension*, and the structure is referred to as a *semi tension field beam*. For purposes of analysis, it is assumed that the semitension field is the superposition of a state of pure shear and one of complete diagonal tension. Using a *diagonal tension factor k* ($0 \leq k \leq 1$), we apportion the net shear stress to each of the two states of stress, that is,

$$\tau_{\text{pure diagonal tension}} = k\tau \qquad \text{and} \qquad \tau_{\text{pure shear}} = (1-k)\tau \qquad [\text{12.11.26}]$$

so that

$$\tau = \tau_{\text{pure shear}} + \tau_{\text{pure diagonal tension}} \qquad [\text{12.11.27}]$$

For a semitension field beam, the state of stress, depicted in Figure 12.11.10a, is

$$\tau_{xy} = \tau \qquad \sigma_x = \frac{k\tau}{\tan \alpha} \qquad \sigma_y = k\tau \tan \alpha \qquad [\text{12.11.28}]$$

The total shear τ is given in terms of P by Equation 12.11.1, and the expressions for σ_x and σ_y are obtained from Equations 12.11.5 and 6 by using just the diagonal tension portion $k\tau$ of the total shear. Equations 3.6.5 are used to find the components of stress on an element oriented in the diagonal tension direction. These components are illustrated in Figure 12.11.10(b). Observe that the diagonal tension direction is not a principal stress direction, as it is in the case of complete diagonal tension ($k = 1$). Mohr's circle yields the principal stresses and principal stress direction θ_p shown in Figure 12.11.10c. The diagonal tension factor k and the angle of diagonal tension α are found by consulting tables of empirical data.

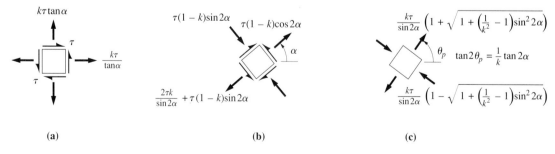

(a)   **(b)**   **(c)**

Figure 12.11.10 (a) State of stress for incomplete diagonal tension. (b) Stress components in the planes of diagonal tension. (c) Principal stresses and directions.

Most of the equations we obtained based on the assumptions of complete tension field beam theory must be modified to reconcile them with actual test data. Bruhn[19] presents a thorough discussion on the analysis and design of diagonal semitension field structures with either flat or curved web systems. See, also, the book by Kuhn[20].

12.12   STATIC WING DIVERGENCE

The lift production by a wing is accompanied by a twisting moment about the elastic axis of the wing box, causing an increase in the angle of attack, which in turn increases the lift and the twisting moment. At a speed called the *divergence velocity*, the effect becomes large enough to cause failure of the wing. Let us investigate this instability phenomena by considering the somewhat oversimplified case of a straight wing with no built-in twist and with a uniform lifting load distribution due to subsonic flow.

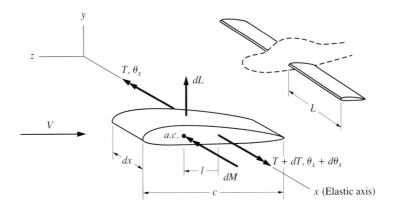

Figure 12.12.1   Differential wing section acted on by external aerodynamic loads (dL and dM) and the internal torsional stress resultants. (a.c. is the aerodynamic center.)

Figure 12.12.1 shows a free-body diagram of a spanwise differential section of a straight wing of constant chord c. The differential aerodynamic lift dL is shown acting at the aerodynamic center of the wing, which is a distance l in front of the elastic axis. The lift is found in terms of the dimensionless section lift coefficient c_l, the differential area cdx, and the dynamic pressure q, by the formula

$$dL = c_l q c dx \qquad \text{[12.12.1]}$$

where $q = \rho V^2/2$ is the dynamic pressure, ρ is the density of the air, and V is the speed of the airplane relative to the atmosphere. Lift is generally accompanied by a pitching moment, so the external differential moment dM is shown acting in the conventionally positive *nose-up* (clockwise) direction. If, as in the figure, dM is measured about the aerodynamic center,[21] then in terms of the dimensionless moment coefficient $c_{m_{a.c.}}$, we have

---

[19] *op. cit.*, Chapter C11.

[20] P. Kuhn, *Stresses in Aircraft and Shell Structures*, New York, McGraw-Hill, 1956.

[21] That point about which the aerodynamic pitching moment is independent of the wing's angle of attack. See, for example, J. D. Anderson, Jr., *Introduction to Flight*, New York, McGraw-Hill, 1989, Chapter 5.

$$dM = c_{m_{a.c.}} qc^2 dx \qquad \text{[12.12.2]}$$

The torque T shown acting on the differential wing section arises, as we know, from the internal stresses required to equilibrate the externally-applied aerodynamic loads. For the free body in Figure 12.12.1 to be in equilibrium, the net moment about the elastic axis must vanish. Thus,

$$(T + dT) - T - ldL - dM = 0$$

Substituting Equations 12.12.1 and 2, cancelling terms, and dividing through by dx, we get

$$\frac{dT}{dx} - c_l qcl - c_{m_{a.c}} qc^2 = 0 \qquad \text{[12.12.3]}$$

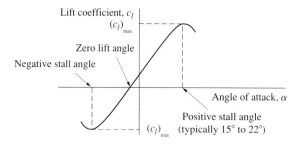

Figure 12.12.2 Variation of wing lift coefficient with angle of attack.

The lift coefficient depends on the wing's angle of attack α, and varies linearly from the zero lift angle to the onset of stall (cf. Figure 12.12.2). We may thus write

$$c_l = \frac{\partial c_l}{\partial \alpha} (\alpha - \theta_x) \qquad \text{[12.12.4]}$$

where $\partial c_l / \partial \alpha$ is the slope of the lift curve, c_l versus α, and α is measured clockwise (here, in radians). If the wing were rigid, then $\theta_x = 0$; that is, there would be no twisting of the structure associated with lift. For the flexible wing under consideration, however, we must subtract the aerodynamically-induced twist θ_x at the section from the angle of attack α which the wing would have if it were rigid. We *subtract* rather than add to remain consistent with our sign convention for positive twist, which, as indicated in Figure 12.12.1, is counterclockwise about the positive x axis, opposite to the direction of increasing α. By definition of the aerodynamic center, the moment coefficient $c_{m_{a.c.}}$ about that point does not depend on the angle of attack: $\partial c_{m_{a.c.}} / \partial \alpha = 0$. Substituting Equation 12.12.4 into Equation 12.12.3 yields

$$\frac{dT}{dx} + qcl \frac{\partial c_l}{\partial \alpha} \theta_x = qcl \frac{\partial c_l}{\partial \alpha} \alpha + c_{m_{a.c.}} qc^2 \qquad \text{[12.12.5]}$$

According to Equations 4.4.1 and 4.4.14, $T = GJ d\theta_x / dx$. If GJ is constant, as in the present case, Equation 12.12.5 becomes a second-order ordinary differential equation in θ_x with constant coefficients:

$$\frac{d^2 \theta_x}{dx^2} + \frac{qcl}{GJ} \frac{\partial c_l}{\partial \alpha} \theta_x = \frac{qcl}{GJ} \frac{\partial c_l}{\partial \alpha} \alpha + \frac{c_{m_{a.c.}} qc^2}{GJ}$$

Introducing the notation

$$\lambda^2 = \frac{qcl}{GJ} \frac{\partial c_l}{\partial \alpha}$$

[12.12.6]

and

$$\theta_a = \alpha + \frac{c_{m_{a.c.}} c}{l(\partial c_l / \partial \alpha)}$$

[12.12.7]

the differential equation can be written in the form

$$\frac{d^2 \theta_x}{dx^2} + \lambda^2 \theta_x = \lambda^2 \theta_a$$

From earlier in this chapter, we recognize the solution of this equation to be

$$\theta_x = \overbrace{A \cos \lambda x + B \sin \lambda x}^{\text{complementary solution}} + \overbrace{\theta_a}^{\text{particular solution}}$$

One boundary condition on the twist angle is that it must vanish at the wing root, where $x = 0$. For that to be true, we must have

$$A = -\theta_a$$

so that

$$\theta_x = B \sin \lambda x + \theta_a (1 - \cos \lambda x)$$

[12.12.8]

The derivative of this expression is

$$\frac{d\theta_x}{d\theta} = \lambda B \cos \lambda x + \theta_a \lambda \sin \lambda x$$

[12.12.9]

At the wing tip the torque T is zero. Since $d\theta_x / dx = T/GL$, it follows that $d\theta_x / dx = 0$ at $x = L$. This, together with Equation 12.12.9, requires that

$$B = -\theta_a \frac{\sin \lambda L}{\cos \lambda L}$$

Substituting this into Equation 12.12.8, we find that

$$\theta_x = \theta_a \left(1 - \cos \lambda x - \frac{\sin \lambda L}{\cos \lambda L} \sin \lambda x \right)$$

Rearranging terms, substituting Equation 12.12.7, and noting the trigonometric identity

$$\cos \lambda L \cos \lambda x + \sin \lambda L \sin \lambda x = \cos \lambda (L - x)$$

we reduce this expression for the twist angle θ_x to

$$\theta_x = \left[\alpha + \frac{c_{m_{a.c.}} c}{l(\partial c_l / \partial \alpha)} \right] \left[1 - \frac{\cos \lambda (L - x)}{\cos \lambda L} \right]$$

[12.12.10]

From this, we can also obtain the torque versus span expression, which is

$$T = GJ\frac{d\theta_x}{dx} = GJ\left[\alpha + \frac{c_{m_{a.c.}}c}{l(\partial c_l/\partial\alpha)}\right]\frac{\lambda \sin \lambda (L-x)}{\cos \lambda L} \tag{12.12.11}$$

The maximum twist occurs at the wing tip, $x = L$, and the greatest torque is at the root, $x = 0$:

$$\theta_x)_{max} = \left[\alpha + \frac{c_{m_{a.c.}}c}{l(\partial c_l/\partial\alpha)}\right]\left(\frac{\cos \lambda L - 1}{\cos \lambda L}\right) \qquad T)_{max} = \left[\alpha + \frac{c_{m_{a.c.}}c}{l(\partial c_l/\partial\alpha)}\right]GJ\lambda \tan \lambda L \tag{12.12.12}$$

Observe that the deflection and torque approach infinity (and the wing fails) when $\cos \lambda L$ approaches zero ($\tan \lambda L$ approaches infinity). This occurs when

$$\lambda L = \frac{\pi}{2}, \frac{3\pi}{2}, \frac{5\pi}{2}, \cdots$$

Obviously, the smallest λ for which the twist and torque become unbounded is $\pi/2L$. At this critical value of λ, Equation 12.12.6 becomes

$$\lambda^2 = \frac{qcl}{GJ}\frac{\partial c_l}{\partial\alpha} = \frac{\pi^2}{4L^2}$$

Setting $q = \rho V^2/2$ and solving for V yields the *divergence speed*, which is

$$V_D = \sqrt{\frac{\pi^2}{(\partial c_l/\partial\alpha)}\frac{J}{clL^2}}\sqrt{\frac{G}{\rho}} \tag{12.12.13}$$

The divergence speed can be raised by increasing the wing torsional rigidity GJ, moving the elastic axis closer to the aerodynamic center (decreasing l), and flying at higher altitudes where the density of the atmosphere is low.

Let us next consider the swept wing shown in Figure 12.12.2. The sweep angle is Ω, and the wing in this case is swept *back*. At any section of the wing, the total angular deflection vector of the flexible wing is

$$\boldsymbol{\theta} = \theta_x\mathbf{i} + \theta_z\mathbf{k}$$

where θ_z is the bending rotation of the elastic axis at the section and θ_x is the twist about the elastic axis. Alternatively, $\boldsymbol{\theta}$ may be resolved into components along the body axes $\bar{x}\bar{z}$ of the airplane and written as

$$\boldsymbol{\theta} = \bar{\theta}_x\bar{\mathbf{i}} + \bar{\theta}_z\bar{\mathbf{k}}$$

Thus,

$$\bar{\theta}_x\bar{\mathbf{i}} + \bar{\theta}_z\bar{\mathbf{k}} = \theta_x\mathbf{i} + \theta_z\mathbf{k}$$

Taking the dot product of both sides of this equation with the unit vector $\bar{\mathbf{i}}$ in the spanwise direction, we get

$$\bar{\theta}_x = \theta_x\left(\mathbf{i}\cdot\bar{\mathbf{i}}\right) + \theta_z\left(\mathbf{k}\cdot\bar{\mathbf{i}}\right) = \theta_x\cos\Omega + \theta_z\sin\Omega \tag{12.12.14}$$

The magnitude of $\bar{\theta}_x$ is the change $\Delta\bar{\alpha}$ in the angle of attack at the section, measured in the plane parallel to the free stream velocity V. Thus,

$$\Delta\bar{\alpha} = -\bar{\theta}_x \tag{12.12.15}$$

The minus sign accounts for the different sign conventions for angle of attack (clockwise positive) and twist (counterclockwise positive). The change $\Delta\alpha$ in angle of attack, as measured in the plane normal to the elastic axis, is

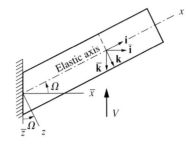

Figure 12.12.3   Plan view of a swept (back) wing.

$$\Delta\alpha = \Delta\bar{\alpha}\cos\Omega \qquad\qquad\qquad \text{[12.12.16]}$$

Substituting Equations 12.12.15 and 12.12.14 into Equation 12.12.16 yields

$$\Delta\alpha = \overbrace{(-\theta_x)}^{\Delta\alpha)_{\text{torsion}}} + \overbrace{(-\theta_z\tan\Omega)}^{\Delta\alpha)_{\text{bending}}} \qquad\qquad \text{[12.12.17]}$$

Since θ_z is generally positive and $\tan\Omega > 0$ in Figure 12.12.3, we see that the wing deflection due to bending *decreases* the angle of attack and therefore acts to prevent static wing divergence. On the other hand, if the wing is swept *forward*, then $\tan\Omega < 0$, and flexure of the wing due to lift acts to *promote* divergence. Additional discussion of static divergence appears in Gatewood,[22] Donaldson,[23] and texts on the subject of aeroelasticity.[24]

EXERCISES

12.1   An unstressed aluminum rod of solid circular cross section is clamped between two walls. What increase in temperature (°C) will cause the rod to buckle?

$$E = 70\text{ GPa} \quad \alpha = 22 \times 10^{-6}\text{ °C}^{-1}$$

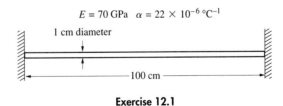

Exercise 12.1

[22]B. E. Gatewood, *Virtual Principles in Aircraft Structures, Volume 1: Analysis,* Chapter 5, Dordrecht, Netherlands, Kluwer Academic Publishers, 1989.

[23]B. K. Donaldson, *Analysis of Aircraft Structures: An Introduction,* New York, McGraw-Hill, p. 812 ff, 1993.

[24]See, for example, R. L. Bisplinghoff, H. Ashley, and R. L. Halfman, *Aeroelasticity,* Reading, MA, Addison-Wesley, 1957; and E. H. Dowell *et al., A Modern Course in Aeroelasticity,* Alphen aan den Rijn, Netherlands, Sijthoff and Noordhoff, 1978.

12.2 By solving the differential equation, determine the buckling load P_{cr} for the pinned-pinned, stepped column. In addition to the boundary conditions,

$$v = 0 \quad \text{at} \quad x = 0 \quad \text{and} \quad x = 2l$$

the following two matching conditions must be satisfied:

$$v^{(1)} = v^{(2)} \quad \text{and} \quad \frac{dv^{(1)}}{dx} = \frac{dv^{(2)}}{dx} \quad \text{at} \quad x = l$$

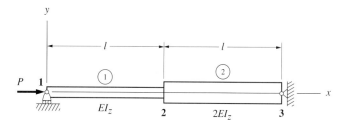

Exercise 12.2

12.3 Repeat Exercise 12.2 with two cubic-displacement beam elements, using the finite element approach.

12.4 An aluminum alloy rod with a solid cross section in the shape of a 1.5 cm by 3 cm rectangle is pinned at each end. If its modulus of elasticity is 72 GPa and its proportional limit is 275 MPa, determine the minimum length for which the Euler buckling formula is valid.

12.5 If the column length found in Exercise 12.4 is increased by 50 percent, what is the Euler buckling load?

12.6 Find an approximate value for the critical buckling load for a slightly tapered fixed-free column of length L in which the minimum cross-sectional area moment of incrtia varies according to the relation $I = I_o \left(1 + \alpha \frac{x}{L}\right)$, where $\alpha << 1$. The term I_o is the moment of inertia at $x = 0$. Use the Rayleigh-Ritz method with a complete cubic polynomial displacement field,

$$v = A + Bx + Cx^2 + Dx^3$$

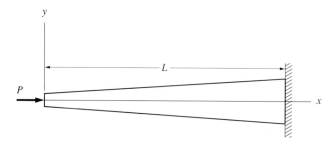

Exercise 12.6

12.7   Repeat Exercise 12.6 using a complete quartic displacement field,

$$v = A + Bx + Cx^2 + Dx^3 + Ex^4$$

12.8   For the beam shown in the figure, determine the shape of the elastic curve if $P_o < P_{cr}$.

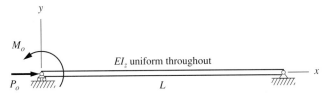

Exercise 12.8

APPENDIX

COMPARATIVE MECHANICAL PROPERTIES OF METAL ALLOYS

Ranges reflect the dependence of properties on specimen condition (e.g., heat treatment), shape and thickness. MILHDBK–5G or other recognized sources should be consulted to select values for use in design.

Alloy (wrought, unless indicated otherwise)		Ultimate Stress (MPa)	Tensile Yield Stress (MPa)	Tensile Elastic Modulus (GPa)	Poisson's Ratio	Density (kg/m3)	Thermal Expansion Coefficient ($10^{-6} C^{-1}$)	Percent Elongation
Steel								
AISI 1025		379	248	200	0.32	7860	10.8	—
AISI 4130		620–1240	690–1120	200	0.32	7860	11.7	5–17
AISI 4140		862–1380	690–1210	200	0.32	7860	11.7	5–17
AISI 4340		862–1790	690–1500	200	0.32	7830	11.7	5–10
5Cr-Mo-V		1650–1930	1380–1650	207	0.36	7780	11.0	4–9
9Ni–4Co–0.2C		1280	1200	198	0.30	7830	11.7	5–10
9Ni–4Co–0.3C		1520	1280–1310	196	—	7750	11.7	10
250 maraged		1690–1760	1640–1720	183	0.31	7920	9.0	5
280 maraged		1900–1930	1800–1860	183	0.31	7920	9.7	5
AF1410		1620	1480	203	—	7830	—	12
AerMet 100		1930–2000	1620–1690	194	—	7890	—	8–10
AM350	(stainless)	1260	1010	200	0.32	7800	11.2	10
AM355	(")	1140–1380	965–1140	200	0.32	7800	11.7	10–12
15–5PH	(")	793–1310	517–1170	194	0.27	7830	11.7	8–18
PH15–7Mo	(")	1280	1140	200	0.28	7670	11.2	2–6
17–4PH	(")	793–1310	527–1170	196	0.27	7830	11.2	10–18
17–7PH	(")	1220–1260	1030–1150	200	0.28	7640	11.3	4–6
AISI 301	(")	503–1280	179–1050	179–200	0.27	7920	15.5	9–40
Aluminum								
2014		393–489	331–434	72.4–74.5	0.33	2800	22.3	1–8
2017		379	220	71.7	0.33	2800	22.0	12
2024		269–510	186–462	72.4–73.8	0.33	2800	22.7	5–15
2090		441–531	386–483	75.8–79.3	0.34	2570	—	3–5
2124		400–469	352–420	71.7	0.33	2768	22.7	1.5–6
2219		345–448	248–365	70.3–72.4	0.33	2850	22.1	4–8

Alloy (wrought, unless indicated otherwise)		Ultimate Stress (MPa)	Tensile Yield Stress (MPa)	Tensile Elastic Modulus (GPa)	Poisson's Ratio	Density (kg/m3)	Thermal Expansion Coefficient (10⁻⁶C⁻¹)	Percent Elongation
Aluminum								
2519		414–455	372–407	72.4	0.33	2820	—	3–6
2618		386–400	248–324	73.8	0.33	2770	20.3	4–7
5052		172–269	65.5–221	69.6	0.33	2680	22.7	2–20
5083		248–345	103–269	70.3	0.33	2660	23.8	6–16
5086		234–345	96.5–255	70.3	0.33	2660	23.8	3–16
5454		214–269	82.7–200	70.3	0.33	2680	23.6	4–18
5456		262–310	103–228	70.3	0.33	2660	24.1	12–16
6013		358	324	69.6	0.33	2710	—	8
6061		179–296	82.7–260	69.6	0.33	2710	22.7	8–16
6151		303	255	69.6	0.33	2710	22.5	6–10
7010		434–524	358–455	70.3	0.33	2820	23.2	2–9
7049		434–510	372–448	70.3	0.33	2820	23.4	2–9
7050		427–552	345–489	70.3–71.0	0.33	2820	23.0	1.5–10
7075		393–586	303–531	65.5–71.7	0.33	2800	22.9	1–10
7150		510–648	462–607	70.3–71.7	0.33	2820	—	1–9
7175		414–593	317–524	69.6–70.3	0.33	2800	23.0	4–9
7475		434–538	345–483	70.0–71.0	0.33	2800	23.0	3–9
A201.0	(cast)	386–414	331–345	71.0	0.33	2800	19.3	1.5–5
354.0	(")	296–345	228–290	73.1	0.33	2710	20.9	2–3
355.0	(")	186	117	71.0	0.33	2710	21.6	2–3
C355.0	(")	241–345	193–276	69.6	0.33	2710	22.3	1–3
356.0	(")	152–172	103–110	71.0	0.33	2680	20.9	0.7–1
A356.0	(")	220–310	152–234	71.7	0.33	2680	20.9	2–5
A357.0	(")	262–345	193–276	71.7	0.33	2680	21.6	3–5
D357.0	(")	310–338	248–283	71.7	0.33	2680	21.6	2–3
359.0	(")	276–324	207–262	72.4	0.33	2680	19.8	3–4
Magnesium								
AZ31B		221–276	103–220	44.8	0.35	1770	25.2	4–12
AZ61A		248–276	110–145	43.4	0.31	1790	25.3	6–9
ZK60A		262–317	138–248	44.8	0.35	1820	25.2	4–7
AM100A	(cast)	117–262	65.5–138	44.8	0.35	1800	—	1–3
AZ91C	(")	117–241	82.7–124	44.8	0.35	1800	25.2	1–4
AZ92A	(")	117–276	89.6–172	44.8	0.35	1820	25.4	1–3

Alloy (wrought, unless indicated otherwise)		Ultimate Stress (MPa)	Tensile Yield Stress (MPa)	Tensile Elastic Modulus (GPa)	Poisson's Ratio	Density (kg/m3)	Thermal Expansion Coefficient ($10^{-6}C^{-1}$)	Percent Elongation
Magnesium								
EZ33A	(")	89.6	75.8	44.8	0.35	1820	25.6	1.5
QE22A	(")	193–276	138–193	44.8	0.35	1800	25.2	1–4
ZE41A	(")	179	121	44.8	0.35	1820	26.1	2
Titanium								
Ti–5Al–2.5Sn		793–944	758–862	107	—	4480	4.1	6–10
Ti–8Al–1Mo–1V		827–1000	758–931	121	0.32	4370	8.8	8–10
Ti–6Al–2Sn–4Zr–2Mo		896–1000	827–931	114	0.32	4540	7.7	8–10
Ti–6Al–4V		862–1100	786–1000	110	0.31	4430	9.0	5–10
Ti–6Al–6V–2Sn		958–1200	848–1100	110	0.31	4540	9.5	6–10
Ti–13V–11Cr–3Al		862–1170	827–1100	100–107	—	4820	8.6	2–10
Ti–15V–3Cr–3Sn–3Al		1000	965	105	—	4760	8.5	7
Ti–10V–2Fe–3Al		1100–1240	1000–1100	110	—	4650	9.7	4–6
Nickel								
Hastelloy X		655–731	276–324	205	0.32	8220	14	29–35
Inconel 600		552–827	207–620	207	0.29	8420	12.6	7–35
Inconel 625		814–883	365–414	205	0.28	8440	13.0	30
Inconel 706		1140–1210	896–1000	210	0.28	8080	13.7	12
Inconel 718		1170–1320	1000–1090	203	0.29	8220	12.4	6–15
Inconel X–750		1030–1140	690–724	211	0.30	8250	13.0	15–20
René 41		1100–1280	827–896	218	0.31	8250	12.6	6–10
Waspaloy		1100–1210	758–827	211	—	8250	12.6	15–20
Beryllium								
Standard grade		324–483	207–345	290	0.1	1850	16.6	1–10
C17200 (copper-beryllium)		1090–1310	855–1140	128	0.27	8250	9.2	1–4

Index

80025 75540